Carlos A. Coello Coello, Gary B. Lamont and David A. Van Veldhuizen

Evolutionary Algorithms for Solving Multi-Objective Problems
Second Edition

Genetic and Evolutionary Computation Series

Series Editors

David E. Goldberg
Consulting Editor
IlliGAL, Dept. of General Engineering
University of Illinois at Urbana-Champaign
Urbana, IL 61801 USA
Email: deg@uiuc.edu

John R. Koza
Consulting Editor
Medical Informatics
Stanford University
Stanford, CA 94305-5479 USA
Email: john@johnkoza.com

Selected titles from this series:

Markus Brameier, Wolfgang Banzhaf
Linear Genetic Programming, 2007
ISBN 978-0-387-31029-9

Nikolay Y. Nikolaev, Hitoshi Iba
Adaptive Learning of Polynomial Networks, 2006
ISBN 978-0-387-31239-2

Tetsuya Higuchi, Yong Liu, Xin Yao
Evolvable Hardware, 2006
ISBN 978-0-387-24386-3

David E. Goldberg
The Design of Innovation: Lessons from and for Competent Genetic Algorithms, 2002
ISBN 978-1-4020-7098-3

John R. Koza, Martin A. Keane, Matthew J. Streeter, William Mydlowec, Jessen Yu, Guido Lanza
Genetic Programming IV: Routine Human-Computer Machine Intelligence
ISBN: 978-1-4020-7446-2 (hardcover), 2003; ISBN: 978-0-387-25067-0 (softcover), 2005

Carlos A. Coello Coello, David A. Van Veldhuizen, Gary B. Lamont
Evolutionary Algorithms for Solving Multi-Objective Problems, 2002
ISBN: 978-0-306-46762-2

Lee Spector
Automatic Quantum Computer Programming: A Genetic Programming Approach
ISBN: 978-1-4020-7894-1 (hardcover), 2004; ISBN 978-0-387-36496-4 (softcover), 2007

William B. Langdon
Genetic Programming and Data Structures: Genetic Programming + Data Structures = Automatic Programming! 1998
ISBN: 978-0-7923-8135-8

For a complete listing of books in this series, go to http://www.springer.com

Carlos A. Coello Coello
Gary B. Lamont
David A. Van Veldhuizen

Evolutionary Algorithms for Solving Multi-Objective Problems

Second Edition

 Springer

Carlos A. Coello Coello
CINVESTAV-IPN
Depto. de Computación
Av. Instituto Politécnico Nacional No. 2508
Col. San Pedro Zacatenco
México, D.F. 07360 MEXICO
ccoello@cs.cinvestav.mx

David A. Van Veldhuizen
HQ AMC/A9
402 Scott Dr., No. 3L3
Scott AFB, IL 62225-5307
dvanveldhuizen@ieee.org

Gary B. Lamont
Department of Electrical and Computer
  Engineering
Graduate School of Engineering
Air Force Institute of Technology
2950 Hobson Way
WPAFB, Dayton, OH 45433-7765
lamont@afit.af.mil

Series Editors:
David E. Goldberg
Consulting Editor
IlliGAL, Dept. of General Engineering
University of Illinois at Urbana-Champaign
Urbana, IL 61801 USA
deg@uiuc.edu

John R. Koza
Consulting Editor
Medical Informatics
Stanford University
Stanford, CA 94305-5479 USA
john@johnkoza.com

ISBN 978-1-4899-9460-8   ISBN 978-0-387-36797-2 (eBook)

Printed on acid-free paper.
© 2007 Springer Science+Business Media, LLC
Softcover re-print of the Hardcover 2nd edition 2007

All rights reserved. This work may not be translated or copied in whole or in part without the written permission of the publisher (Springer Science+Business Media, LLC, 233 Spring Street, New York, NY 10013, USA), except for brief excerpts in connection with reviews or scholarly analysis. Use in connection with any form of information storage and retrieval, electronic adaptation, computer software, or by similar or dissimilar methodology now known or hereafter developed is forbidden.

The use in this publication of trade names, trademarks, service marks, and similar terms, even if they are not identified as such, is not to be taken as an expression of opinion as to whether or not they are subject to proprietary rights.

9 8 7 6 5 4 3 2 1

springer.com

to our wives

Preface to the Second Edition

The response of the multiobjective optimization community to our first edition in 2002 was extremely enthusiastic. Many have indicated their use of our monograph to gain insight to the interdisciplinary nature of multiobjective optimization employing evolutionary algorithms. Others are appreciative for our providing them a foundation for associated contemporary multiobjective evolutionary algorithm (MOEA) research. We appreciate these warm comments along with readers' suggestions for improvements. In that vein, we have significantly extended and modified our previous material using contemporary literature resulting in this new edition, which is extended into a textbook. In addition to new classroom exercises contained in each chapter, the MOEA discussion questions and possible research directions are updated.

The first edition presented an organized variety of MOEA topics based on fundamental principles derived from single-objective evolutionary algorithm (EA) optimization and multiobjective problem (MOP) domains. Yet, many new developments occurred in the intervening years. New MOEA structures were proposed with new operators and therefore better search techniques. The explosion of successful MOEA applications continues to be reported in the literature. Statistical testing methods for evaluating results now offers improved analysis of comparative techniques, innovative metrics, and better visualization tools. The continuing development of MOEA activity in theory, algorithmic innovations, and MOEA practice calls for these new concepts to be integrated into our generic MOEA text. Note that the continuing improvement (speed, memory, etc.) of computer hardware provides computational platforms that permit larger search spaces to be addressed at higher efficiencies using both serial and parallel processing. This phenomenon, in conjunction with user-friendly software interfacing tools, permits an increasing number of scientists and engineers to explore the use of MOEAs in their particular multiobjective problem domains.

With this new edition, we continue to provide an interdisciplinary computer science and computer engineering text that considers other academic fields such as operations research, industrial engineering, and management science. Examples from all these disciplines, as well as all engineering areas in general, are discussed and addressed as to their fundamental unique problem domain characteristics and their solutions using MOEAs. An expanded reference list is included with suggestions of further reading for both the student and practitioner. As in the previous edition, this book addresses MOEA development and applications issues through the following features:

- The text is meant to be both a textbook and a self-contained reference. The book provides all the necessary elements to guide a newcomer in the design, implementation, validation, and application of MOEAs in either the classroom or the field.
- Researchers in the field benefit from the book's comprehensive review of state-of-the-art concepts and discussions of open research topics.
- The book is also written for graduate students in computer science, computer engineering, operations research, management science, and other scientific and engineering disciplines, who are interested in multiobjective optimization using evolutionary algorithms.
- The book is also for professionals interested in developing practical applications of evolutionary algorithms to real-world multiobjective optimization problems.
- Each chapter is complemented by discussion questions and several ideas meant to trigger novel research paths. Supplementary reading is strongly suggested for deepening MOEA understanding.
- Key features include MOEA classifications and explanations, MOEA applications and techniques, MOEA test function suites, and MOEA performance measurements.
- We created a website for this book at:

    http://www.cs.cinvestav.mx/~emoobook

    which contains considerable material supporting this second edition. This site contains all the appendices of the book (which have been removed from the original monograph due to space limitations), as well as public-domain software, tutorial slides, and additional sources of contemporary MOEA information.

This new synergistic text is markedly improved from the first edition. New material is integrated providing more detail, which leads to a realignment of material. Old chapters were modified and a new one was added. As before, the various features of MOEAs continue to be discussed in an innovative and unique fashion, with detailed customized forms suggested for a variety of applications. The flow of material in each chapter is intended to present a natural and comprehensive development of MOEAs from basic concepts to complex applications.

Chapter 1 presents and motivates MOP and MOEA terminology and the nomenclature used in successive chapters including a lengthy discussion on the impact of computational limitations on finding the Pareto front along with insight to MOP/MOEA building block (BB) concepts.

In Chapter 2, MOEA developmental history has proceeded in a number of ways from aggregated forms of single-objective Evolutionary Algorithms (EAs) to true multiobjective approaches such as MOGA, MOMGA, NPGA, NSGA, NSGA-II, PAES, PESA, PESA-II, SPEA, SPEA2 and their extensions. Each MOEA is presented with historical and algorithmic insight. Being aware of the many facets of historical multiobjective problem solving provides a foundational understanding of the discipline. Various MOEA techniques, operators, parameters and constructs are compared. Contemporary MOEA development emphasizes new MOP variable representation, and novel MOEA structures and operators. In addition, constraint-handling techniques used with MOEAs are also discussed. A comprehensive comparison of contemporary MOEAs provides insight to an individual algorithm's advantages and disadvantages.

In Chapter 3, a new chapter, both coevolutionary MOEAs and hybridizations of MOEAs with local search procedures (the so-called memetic MOEAs) are covered. A variety of MOEA implementations within each of these two types of approaches (i.e., coevolution and hybrids with local search mechanisms) are presented, summarized, categorized and analyzed.

Chapter 4 offers a detailed development of contemporary MOP test suites ranging from numerical functions (unconstrained and with side constraints) and generated functions to discrete NP-Complete problems and real-world applications. Our website contains the algebraic description as well as the Pareto fronts (and, if generated by enumeration, the Pareto optimal set as well) of many of the proposed test functions. This knowledge leads to an understanding and ability to select appropriate MOEA test suites based upon a set of desired comparative characteristics.

MOEA performance comparisons are presented in Chapter 5 using many of the test function suites discussed in Chapter 4. Also included is an extensive discussion of possible comparative metrics and presentation techniques. The selection of key algorithmic parameter values (population size, crossover and mutation rates, etc.) is emphasized. A limited set of MOEA results are related to the design and analysis of efficient and effective MOEAs employing these various MOP test suites and appropriate metrics. The chapter has been expanded to include new testing concepts such as attainment functions, elaborated dominance relations, and "quality" Pareto compliant indicator analysis. A wide spectrum of empirical testing and statistical analysis techniques are provided for the MOEA user.

Although MOEA theory is still relatively limited, Chapter 6 presents a contemporary summary of known results. Topics addressed in this chapter include MOEA convergence to the Pareto front, Pareto ranking, fitness sharing, mating restrictions, stability, running time analysis, and algorithmic complexity.

It is of course unrealistic to present every generic MOP application, thus, Chapter 7 attempts to group and classify the multitude of various contemporary MOEA applications via representative examples. This limited compendium with an extensive reference listing provides the reader with a starting point for their own application and research. Specific MOEA operators as well as encodings adopted in many MOEA applications are integrated for algorithmic understanding.

In Chapter 8, research and development of parallel MOEAs is classified and analyzed. The three foundational paradigms (master-slave, island, and diffusion) are defined. Using these three structures, many contemporary MOEA parallel developments are algorithmically compared and analyzed in terms of advantages and disadvantages for different computational architectures. Some general observations about the current state of parallel and distributed MOEAs are also included.

Chapter 9 discusses and compares the two main schools of thought regarding multi-criteria decision making (MCDM): Outranking approaches and Multi-Attribute Utility Theory (MAUT). Aspects such as the operational attitude of the Decision Maker (DM), the different stages at which preferences can be incorporated, scalability, transitivity and group decision making are also discussed. However, the main emphasis is in describing the most representative research regarding preference articulation into MOEAs. This comprehensive review includes brief descriptions of the approaches reported in the literature as well as an analysis of their advantages and disadvantages.

Chapter 10 discusses multiobjective extensions of other search heuristics. The main techniques covered include Tabu search, scatter search, simulated annealing, ant system, distributed reinforcement learning, artificial immune systems, particle swarm optimization and differential evolution.

New examples are integrated throughout the second edition. New algorithms are addressed with special emphasis on the spectrum of MOEA operators and how they are implemented in contemporary and historic MOEAs. Part of the focus is on classifying MOEAs as to implicit or explicit BB types. Other classification features such as probabilistic vs. stochastic are investigated. References are updated to include the current state-of-the-art MOEAs and applications.

Class exercises are integrated into all chapters for pedagogical purposes. Discussion questions within every chapter are updated and expanded. The suggested and focused research ideas from the first edition are brought up-to-date and continue to emphasize the current state-of-the-art horizon.

To profit from the book, one should have at least single-objective EA knowledge and experience. Also, some mathematical knowledge is appropriate in order to understand symbolic functions as well as theoretical MOEA aspects. This knowledge includes basic linear algebra, calculus, probability and statistics. This second edition may be used in the classroom at the senior undergraduate or graduate level depending upon the instructor's purpose. As a class, we suggest that all material could fill a two semester course or with careful selection of topics, a one-semester course. Also, the material in the revised text can be effectively employed by practitioners in many fields.

In support of this text, one can find up-to-date MOEA reference listings of journal papers, conference papers, MOP software, and MOEA software at the Evolutionary Multiobjective Optimization (EMOO) Repository internet web site http://delta.cs.cinvestav.mx/~ccoello/EMOO. This site is continually updated to support the MOEA community and our text. If you have a contribution, please send it to ccoello@cs.cinvestav.mx.

Creating a book such as this requires the efforts of many people. The authors thank Matthew Johnson, Michael Putney, Jesse Zydallis, Tony Kadrovach, Giovani Gómez-Estrada, Dragan Cvetković, José Alfredo López, Nareli Cruz-Cortés, Gregorio Toscano-Pulido, Luis Gerardo de la Fraga, and many others for their assistance in generating computational results and reviewing various aspects of the material. We also thank all those researchers who sent us some of their research papers and theses to enrich the material contained in this edition.

We express our sincere appreciation to Professors David E. Goldberg and John R. Koza for including this book as a volume in their Genetic and Evolutionary Computation book series, published by Springer.

Also, it has been a pleasure working with Springer's professional editorial and production staff. We particularly thank Melissa Fearon and Valerie Schofield for their prompt and kind assistance at all times during the development of this book.

We also want to thank other primary MOEA researchers not only for their innovative papers but for various conversations providing more insight to developing better algorithms. Such individuals include David Corne, Tomoyuki Hiroyasu, Kalyanmoy Deb, Marco Laumanns, Jürgen Branke, Sanaz Mostaghim, Nirupam Chakraborti, Alfredo G. Hernández-Díaz, Julián Molina, Rafael Caballero, Peter Fleming, Carlos Fonseca, Xavier Gandibleux, Yaochu Jin, Kay Chen Tan, Jeffrey Horn, Hisao Ishibuchi, Piero Bonissone, Jonathan Fieldsend, Marco Farina, Arturo Hernández-Aguirre, Lyndon While, Evan J. Hughes, Rajeev Kumar, Shigeru Obayashi, Joshua D. Knowles, J. David Schaffer, Ian Parmee, El-Ghazali Talbi, Hernán Aguirre, Oliver Schütze, Lothar Thiele, and Eckart Zitzler.

The authors also express their gratitude to Antonio Nebro, Enrique Alba, Margarita Reyes-Sierra, Luis V. Santana-Quintero, Ricardo Landa-Becerra, Mario A. Ramírez-Morales, Emanuel Téllez-Enríquez, Richard Day, Charles Haag, and Mark Kleeman for their valuable help at different stages of the development of this second edition. Without their help, this book would had never been finished. Carlos A. Coello Coello also states that his contribution to this book was developed using the computing facilities of the Department of Computer Science of the Centro de Investigación y de Estudios Avanzados from the Instituto Politécnico Nacional (CINVESTAV-IPN) with support provided by CONACyT (the Mexican council of science and technology) to the first author through project no. 45683-Y, which was also greatly appreciated.

Last but not least, we owe a debt of gratitude to our wives for their encouragement, understanding, and exemplary patience.

We hope that the new edition continues to represent not only a comprehensive introduction to MOEAs, but also the contemporary state-of-the-art in MOEA structures, applications, testing and theory.

Carlos A. Coello Coello
Gary B. Lamont
David A. Van Veldhuizen
Spring 2007

Foreword to the Second Edition

Researchers and practitioners alike are increasingly turning to search, optimization, and machine-learning procedures based on natural selection and natural genetics to solve problems across the spectrum of human endeavor. These genetic algorithms and techniques of evolutionary computation are solving problems and inventing new hardware and software that rival human designs. The Springer Series on Genetic and Evolutionary Computation publishes research monographs, edited collections, and graduate-level texts in this rapidly growing field. Primary areas of coverage include the theory, implementation, and application of genetic algorithms (GAs), evolution strategies (ESs), evolutionary programming (EP), learning classifier systems (LCSs) and other variants of genetic and evolutionary computation (GEC). The series also publishes texts in related fields such as artificial life, adaptive behavior, artificial immune systems, agent-based systems, neural computing, fuzzy systems, and quantum computing as long as GEC techniques are part of or inspiration for the system being described.

This is the second (revised and extended) edition of an encyclopedic volume on the use of the algorithms of genetic and evolutionary computation for the solution of multi-objective problems. Multi-objective evolutionary algorithms (MOEAs) are now even more popular than in 2002, when the first edition of this book was published. Researchers and practitioners remain to find an irresistible match between the population available in most genetic and evolutionary algorithms and the need in multi-objective problems to approximate the Pareto trade-off curve or surface.

The authors have kept the remarkable job that distinguished the first edition in collecting, organizing, and interpreting the burgeoning literature of MOEAs in a form that should be welcomed by novices and old hands alike. The volume starts with an extraordinarily thorough introduction, including short vignettes and photographs of many of the pioneers of multi-objective optimization. It continues with as complete a discussion of the many varieties of MOEAs as appears anywhere in the literature. This second edition now adds a new chapter fully devoted to coevolutionary and memetic (i.e., hybrids with local search mechanisms) MOEAs. A discussion of MOEA test suites surveys the important topic of test landscapes and is followed with important chapters on empirical testing and MOEA theory. Such chapters have been considerably extended with respect to the first edition, adding material on state-of-the-art test functions and performance measures (and their limitations), as well as the new developments on the theoretical foundations of MOEAs. Practitioners will especially welcome the thorough survey of real-world MOEA applications, which clearly indicates the growing interest in this field. There is also an ample discussion on parallelization, and a thorough review of mechanisms to incorporate user's preferences in a MOEA (an area called multi-criteria decision making). The final chapter of special topics discusses multi-objective extensions of other methods in soft computation such as simulated annealing, ant colony optimization, and artificial immune systems. These chapters have also been considerably extended and refurbished to reflect the many new developments that have arisen in this field since the publication of the first edition of this book. With about 200 extra pages, a considerable number of new problems and research ideas at the end of each chapter and additional supporting material available through a website, this second edition aims to be adopted as a textbook, while preserving much of its monograph nature.

If you enjoyed the first edition of this book, then you will certainly benefit even more from this second edition. If you still do not know this book, then, I urge you to run—don't walk—to your nearest on-line or off-line book purveyor and click, signal, or otherwise buy this important addition to our literature.

David E. Goldberg
Consulting Editor
University of Illinois at Urbana-Champaign
deg@uiuc.edu
Urbana, Illinois, USA
May 2007

Contents

1 Basic Concepts 1
  1.1 Introduction 1
  1.2 Definitions 3
    1.2.1 Single-Objective Optimization 4
    1.2.2 The Multiobjective Optimization Problem 5
    1.2.3 Multiobjective Optimization Problem 7
    1.2.4 Definition of MOEA Progress 14
    1.2.5 Computational Domain Impact 14
    1.2.6 Pareto Epsilon Model 17
    1.2.7 Decision Maker Impact 18
  1.3 An Example 19
  1.4 General Optimization Algorithm Overview 21
  1.5 EA Basics 24
  1.6 Origins of Multiobjective Optimization 29
    1.6.1 Mathematical Foundations 30
    1.6.2 Early Applications 30
  1.7 Classifying Techniques 31
    1.7.1 *A priori* Preference Articulation 32
    1.7.2 *A Posteriori* Preference Articulation 46
    1.7.3 Progressive Preference Articulation 47
  1.8 Using Evolutionary Algorithms 51
    1.8.1 Pareto Notation 53
    1.8.2 MOEA Classification 54
  1.9 Summary 55

Further Explorations 57

2 MOP Evolutionary Algorithm Approaches 61
  2.1 Introduction 61
  2.2 MOEA Techniques 63
    2.2.1 *A Priori* Techniques 65

              2.2.2 *Progressive* Techniques 70
              2.2.3 *A Posteriori* Techniques 71
              2.2.4 Generic MOEA Goals and Operator Design 77
       2.3 Structures of Various MOEAs 88
              2.3.1 Multi-Objective Genetic Algorithm (MOGA) 88
              2.3.2 Nondominated Sorting Genetic Algorithm (NSGA) 91
              2.3.3 Niched-Pareto Genetic Algorithm (NPGA) 94
              2.3.4 Pareto Archived Evolution Strategy (PAES) 95
              2.3.5 Strength Pareto Evolutionary Algorithm (SPEA) 97
              2.3.6 Multiobjective Messy Genetic Algorithm (MOMGA) ... 99
              2.3.7 Pareto Envelope-based Selection Algorithm (PESA) ...101
              2.3.8 The Micro-Genetic Algorithm for Multiobjective
                     Optimization 102
              2.3.9 Multiobjective Struggle GA (MOSGA) 105
              2.3.10 Orthogonal Multi-Objective Evolutionary Algorithm
                     (OMOEA).. 106
              2.3.11 General Multiobjective Evolutionary Algorithm
                     (GENMOP) 108
              2.3.12 Criticism to Pareto sampling techniques 111
       2.4 Constraint-Handling Techniques 113
       2.5 Critical MOEA Elements 116
              2.5.1 MOEA Comparisons 116
              2.5.2 MOEA Theory 116
              2.5.3 MOEA Fitness Functions 117
              2.5.4 MOEA Chromosomal Representations 117
              2.5.5 MOEA Problem Domains 119
       2.6 MOEA Design Recapitulation 120
       2.7 Summary 121

Further Explorations 123

3 MOEA Local Search and Coevolution 131
       3.1 Introduction 131
       3.2 MOEA Local Search Techniques 131
              3.2.1 Hybrid MOEA Techniques 134
              3.2.2 Comments on Hybrid MOEA Techniques 143
       3.3 MOEA Coevolutionary Techniques 144
       3.4 Coevolution and Symbiosis in EAs 147
              3.4.1 Coevolutionary Algorithms 147
              3.4.2 Cooperative Coevolutionary Genetic Algorithms 149
              3.4.3 Symbiogenetic Coevolution 150
       3.5 Coevolution and Symbiosis in MOEAs 152
              3.5.1 Elitist Recombinative MOGA with Coevolutionary
                     Sharing 152
              3.5.2 Parmee's Co-Evolutionary MOEA 154

		3.5.3 Genetic Symbiosis Algorithm 155
		3.5.4 Interactive GA with Co-evolving Weighting Factors 157
		3.5.5 Multiobjective Co-operative Co-evolutionary GA 158
		3.5.6 Lohn's Coevoluntionary Genetic Algorithm 159
		3.5.7 Distributed Cooperative Coevolutionary Algorithm 161
		3.5.8 Coello's Coevolutionary MOEA 163
		3.5.9 Nondominated Sorting Cooperative Coevolutionary GA 165
	3.6 Applying Coevolutionary MOEAs 165
		3.6.1 Coevolving Multiple MOEAs 166
		3.6.2 Coevolving MOEAs with other Search Algorithms 167
		3.6.3 Coevolving Density Estimators 167
		3.6.4 Coevolving Target Solutions 167
		3.6.5 Coevolving Competing Populations 168
	3.7 Final Comments on Coevoluntionary MOEAs 168

Further Explorations 171

4 MOEA Test Suites 175
	4.1 Introduction 175
	4.2 MOEA Test Function Suite Issues 176
	4.3 MOP Domain Feature Classification 179
		4.3.1 Unconstrained Numeric MOEA Test Functions 182
		4.3.2 Side-Constrained Numeric MOEA Test Functions 187
		4.3.3 MOP Test Function Generators 193
	4.4 Generic Scalable MOP Test Problems 199
		4.4.1 Okabe's Test Functions 207
		4.4.2 Huband's Test Functions 209
	4.5 Combinatorial MOEA Test Functions 220
	4.6 Real-World MOEA Test Functions 222
	4.7 Summary 228

Further Explorations 229

5 MOEA Testing and Analysis 233
	5.1 Introduction 233
	5.2 MOEA Experiments: Motivation and Objectives 235
	5.3 Experimental Methodology 236
		5.3.1 MOP Pareto Front Determination 236
		5.3.2 MOEA Algorithms Testing 238
		5.3.3 Key MOEA Algorithmic Parameters 239
	5.4 MOEA Experimental Measurements 243
		5.4.1 Selection of MOEA Comparison Measures 245
		5.4.2 Generic Attainment Function 245
		5.4.3 Dominance Relations 250
		5.4.4 Primary Quality Indicators 254

| | 5.4.5 Other MOEA Quality Indicators 263
| | 5.4.6 MOEA Experimental Metrics Summary 267
| 5.5 | MOEA Statistical Testing Approaches 268
| | 5.5.1 Statistical Testing Techniques 268
| | 5.5.2 Non-Parametric Statistics (Analysis of Variance) 270
| | 5.5.3 Methods for Presentation of MOEA Results 272
| | 5.5.4 Visualization of Test Results 272
| 5.6 | Software Support of MOEA Testing........................ 273
| 5.7 | Summary.. 276

Further Explorations 277

6 MOEA Theory and Issues 283
6.1 Introduction 283
6.2 Pareto-Related Theoretical Contributions................... 284
6.2.1 Partially Ordered Sets 284
6.2.2 MOEA Convergence 288
6.3 MOEA Theoretical Issues 300
6.3.1 Fitness Landscapes 300
6.3.2 Fitness Functions.................................. 305
6.3.3 Pareto Ranking 307
6.3.4 Pareto Niching and Fitness Sharing 310
6.3.5 Recombination Operators 314
6.3.6 Mating Restrictions................................ 315
6.3.7 Solution Stability and Robustness 317
6.3.8 MOEA Complexity 317
6.3.9 MOEA Scalability 319
6.3.10 Running Time Analysis 320
6.3.11 MOEA Computational "Cost" 326
6.3.12 NFL-Theorem for Multiobjective Optimization Algorithms 326
6.3.13 Alternative Definitions of Optimality 327
6.3.14 Local Search 329
6.4 Summary.. 333

Further Explorations 335

7 Applications 339
7.1 Introduction 339
7.2 Engineering Applications 340
7.2.1 Environmental, Naval and Hydraulic Engineering 340
7.2.2 Electrical and Electronics Engineering 347
7.2.3 Telecommunications and Network Optimization 356
7.2.4 Robotics and Control Engineering 360
7.2.5 Structural and Mechanical Engineering............... 369

	7.2.6	Civil and Construction Engineering 376
	7.2.7	Transport Engineering 377
	7.2.8	Aeronautical Engineering 381
7.3	Scientific Applications.................................. 388	
	7.3.1	Geography 388
	7.3.2	Chemistry 389
	7.3.3	Physics 391
	7.3.4	Medicine 393
	7.3.5	Ecology 396
	7.3.6	Computer Science and Computer Engineering 397
7.4	Industrial Applications 407	
	7.4.1	Design and Manufacture 408
	7.4.2	Scheduling..................................... 416
	7.4.3	Management 424
	7.4.4	Grouping and Packing 426
7.5	Miscellaneous Applications 428	
	7.5.1	Finance 428
	7.5.2	Classification and Prediction 430
7.6	Future Applications.................................... 434	
7.7	Summary... 435	

Further Explorations 437

8 MOEA Parallelization 443
   8.1 Introduction 443
   8.2 pMOEA Fundamental Background 445
        8.2.1 pMOEA Notation 445
        8.2.2 pMOEA Motivation and Issues 446
   8.3 pMOEA Paradigms..................................... 450
        8.3.1 Master-Slave pMOEA Model....................... 452
        8.3.2 Island pMOEA Models........................... 455
        8.3.3 Diffusion pMOEA Model 458
        8.3.4 Hierarchical Hybrid pMOEA Models 459
   8.4 pMOEAs From the Literature............................ 460
        8.4.1 Master-Slave pMOEAs 460
        8.4.2 Island pMOEAs................................ 465
        8.4.3 Diffusion pMOEAs 473
   8.5 pMOEA Analyses and Issues............................. 475
        8.5.1 pMOEA Observations............................ 476
        8.5.2 pMOEA Suitability Issues 476
        8.5.3 pMOEA Hardware and Software Architecture Issues ... 477
        8.5.4 pMOEA Test Function Issues 480
        8.5.5 pMOEA Metric/Parameter Issues 484
   8.6 pMOEA Development Issues 488
        8.6.1 pMOEA Creation Options 490

		8.6.2 Master-Slave Implementation Issues 491

        8.6.2 Master-Slave Implementation Issues 491
        8.6.3 Island Implementation Issues 493
        8.6.4 Diffusion Implementation Issues 499
        8.6.5 Parallel Niching Issues 500
        8.6.6 Parallel Archiving Issues 502
        8.6.7 pMOEA Theory Issues 503
  8.7 A "Generic" pMOEA 503
        8.7.1 Engineering a pMOEA 504
        8.7.2 "Genericizing" a pMOEA 507
  8.8 Conclusions 507

Further Explorations 509

9 Multi-Criteria Decision Making 515
  9.1 Introduction 515
  9.2 Multi-Criteria Decision Making 516
        9.2.1 Operational Attitude of the Decision Maker 517
        9.2.2 When to Get the Preference Information? 518
  9.3 Incorporation of Preferences in MOEAs 520
        9.3.1 Definition of Desired Goals 522
        9.3.2 Utility Functions 526
        9.3.3 Preference Relations 528
        9.3.4 Outranking 531
        9.3.5 Fuzzy Logic 533
        9.3.6 Compromise Programming 535
  9.4 Issues Deserving Attention 536
        9.4.1 Preserving Dominance 537
        9.4.2 Transitivity 537
        9.4.3 Scalability 537
        9.4.4 Group Decision Making 537
        9.4.5 Other important issues 539
  9.5 Summary 540

Further Explorations 541

10 Alternative Metaheuristics 547
  10.1 Introduction 547
  10.2 Simulated Annealing 548
        10.2.1 Basic Concepts 548
        10.2.2 Multiobjective Versions 550
        10.2.3 Advantages and Disadvantages of Simulated Annealing . 556
  10.3 Tabu Search and Scatter Search 557
        10.3.1 Basic Concepts 558
        10.3.2 Multiobjective Versions 559

10.3.3 Advantages and Disadvantages of Tabu Search and
Scatter Search 571
10.4 Ant System.. 572
10.4.1 Basic Concepts..................................... 572
10.4.2 Multiobjective Versions 575
10.4.3 Advantages and Disadvantages of the Ant System 581
10.5 Distributed Reinforcement Learning 582
10.5.1 Basic Concepts..................................... 582
10.5.2 Advantages and Disadvantages of Distributed
Reinforcement Learning 583
10.6 Particle Swarm Optimization 584
10.6.1 Basic Concepts..................................... 584
10.6.2 Multiobjective Versions 585
10.6.3 Advantages and Disadvantages of Particle Swarm
Optimization 593
10.7 Differential Evolution 594
10.7.1 Multiobjective Versions 596
10.7.2 Advantages and Disadvantages of Differential Evolution 604
10.8 Artificial Immune Systems............................... 604
10.8.1 Basic Concepts..................................... 605
10.8.2 Multiobjective Versions 606
10.8.3 Advantages and Disadvantages of Artificial Immune
Systems 611
10.9 Other Heuristics.. 612
10.9.1 Cultural Algorithms 612
10.9.2 Cooperative Search 614
10.10 Summary.. 616

Further Explorations 617

Epilog 623

References 627

Index 761

1
Basic Concepts

Everything has been said before, but since nobody listens we have to keep going back and beginning all over again.

André Gide

1.1 Introduction

Problems with multiple objectives arise in a natural fashion in most disciplines and their solution has been a challenge to researchers for a long time. Despite the considerable variety of techniques developed in Operations Research (OR) and other disciplines to tackle these problems, the complexities of their solution calls for alternative approaches.

The use of evolutionary algorithms (EAs) to solve problems of this nature has been motivated mainly because of the population-based nature of EAs which allows the generation of several elements of the Pareto optimal set in a single run. Additionally, the complexity of some multiobjective optimization problems[1] (MOPs) (e.g., very large search spaces, uncertainty, noise, disjoint Pareto curves, etc.) may prevent use (or application) of traditional OR MOP-solution techniques.

This book is organized in such a way that its contents provides a general overview of the field now called evolutionary multiobjective optimization (EMO), which refers to the use of evolutionary algorithms of any sort (i.e., genetic algorithms [581], evolution strategies [1460], evolutionary programming [499] or genetic programming [905]) to solve multiobjective optimization problems. In fact, we also cover in this book other metaheuristics that have been used to solve multiobjective optimization problems (e.g., particle swarm optimization [840], artificial immune systems [1161], cultural algorithms [1357],

---

[1] Note that the terms "multi-objective" and "multiobjective" are used interchangeably throughout this book.

differential evolution [1525, 1294], ant colony [406], tabu search [572], scatter search [938], and memetic algorithms [661], among others).

Multiobjective optimization problems are attacked today using EAs by engineers, computer scientists, biologists, and operations researchers alike. This book should therefore be of interest to the many disciplines that have to deal with multiobjective optimization problems. At the end of each chapter, we include a section called "Future Explorations", which contains class exercises, class software projects, discussion questions, and possible research directions. Such material aims to provide support for teaching a course, and also delineates some possible topics for developing masters and PhD theses.

This chapter presents the basic terminology and nomenclature for use throughout the rest of the book. Furthermore, a historical overview of multiobjective optimization is also provided, together with a short introduction to evolutionary algorithms. Additionally, we also provide a brief description of the most representative mathematical programming techniques that have been proposed to solve multiobjective optimization problems, including a possible classification of them.

Chapter 2 provides an overview of the different multi-objective evolutionary algorithms (MOEAs) currently available. These techniques go from a simple linear aggregating function to the most popular MOEAs based on Pareto ranking (e.g., MOGA [504], NPGA [709], NSGA [1509], PAES [886], NSGA-II [374], SPEA [1782], SPEA2 [1775] and ϵ-MOEA [372, 373]). Other issues such as chromosomal representations, constraint-handling techniques and the use of secondary populations are also addressed.

Chapter 3 discusses both coevolutionary MOEAs and hybridizations of MOEAs with local search procedures (the so-called memetic MOEAs). A variety of MOEA implementations within each of these two types of approaches (i.e., coevolution and hybrids with local search mechanisms) are presented, summarized, categorized and analyzed.

Chapter 4 presents a detailed development of MOP test suites ranging from numerical functions (both unconstrained and with side constraints) and generated functions to discrete NP-Complete problems and real-word applications. Discussions provide understanding of the MOP domain, and an ability to select appropriate MOEA test suites based upon a set of desired characteristics.

MOEA performance comparisons are presented in Chapter 5. Also, an extensive discussion of possible comparison metrics and presentation techniques are presented. This includes a brief treatment of some recent findings regarding the limitations of unary performance metrics. Results are related to the design and analysis of efficient and effective MOEAs.

Chapter 6 summarizes the (still scarce) MOEA theoretical results found in the literature.

Although it is unrealistic to present every MOP application, Chapter 7 attempts to group and classify the wide variety found in the literature. Problem domain characteristics are presented for each generic application and issues such as genetic operators and encodings are also briefly discussed.

Chapter 8 classifies and analyzes the existing research on parallel MOEAs. The three foundational paradigms (master-slave, island, and diffusion) are discussed, and some general observations about the current state of this area (including its limits and most promising directions) are also presented.

Chapter 9 describes the most representative research regarding the incorporation of preferences articulation into MOEAs. The review is very comprehensive and includes brief descriptions of the approaches reported in the literature as well as an analysis of their advantages and disadvantages.

Chapter 10 discusses multiobjective extensions of other metaheuristics used for optimization. The main techniques covered include tabu search, scatter search, simulated annealing, the ant colony, particle swarm optimization, differential evolution, artificial immune systems, cultural algorithms and distributed reinforcement learning.

The remainder of this chapter is organized as follows. Section 1.2 contains very important concepts such as Pareto optimum, ideal vector, Pareto optimal set, and Pareto front, among others. Section 1.3 aims to put in practice some of the concepts previously covered with an example. Section 1.4 discusses general search and optimization techniques both deterministic and random, and it places evolutionary computation within its appropriate historical context. For those not familiar with evolutionary computation, Section 1.5 offers a short introduction that concludes with a formal definition of an evolutionary algorithm.

Section 1.6 contains a short review of the origins of multiobjective optimization. Then, a taxonomy of the several multiobjective optimization techniques proposed in the OR literature is provided in Section 1.7. Some representative *a priori*, *a posteriori* and *interactive* approaches are also discussed in this section.

Finally, Section 1.8 contains some of the main motivations for using evolutionary algorithms to solve multiobjective optimization problems, as well as some more of the formal notation that is used throughout this book.

Readers who are familiar both with EAs and multiobjective optimization concepts, may want to skip most of this chapter (except for Section 1.2 on nomenclature and the Discussion Questions at the end of the chapter).

1.2 Definitions

In order to develop an understanding of MOPs and the ability to design MOEAs to solve them, a series of formal non-ambiguous definitions are required. These definitions provide a precise set of symbols and formal relationships that permit proper analysis of MOEA structures and associated testing and evaluation. Moreover, they are related to the primary goals for a MOEA:

- Preserve nondominated points in objective space and associated solution points in decision space.
- Continue to make algorithmic progress towards the Pareto Front in objective function space.
- Maintain diversity of points on Pareto front (phenotype space) and/or of Pareto optimal solutions - decision space (genotype space).
- Provide the decision maker (DM) "enough" but limited number of Pareto points for selection resulting in decision variable values.

In order to understand these objectives of multiobjective optimization and their attainment, we start the discussion with single-objective optimization problems.

1.2.1 Single-Objective Optimization

The single-objective optimization problem as presented in Definition 1 continues to be addressed by many search techniques including numerous evolutionary algorithms.

Definition 1 (General Single-Objective Optimization Problem) : *A general* **single-objective optimization problem** *is defined as minimizing (or maximizing) $f(\mathbf{x})$ subject to $g_i(\mathbf{x}) \leq 0$, $i = \{1,\ldots,m\}$, and $h_j(\mathbf{x}) = 0$, $j = \{1,\ldots,p\}$ $\mathbf{x} \in \Omega$. A solution minimizes (or maximizes) the scalar $f(\mathbf{x})$ where $\mathbf{x}$ is a n-dimensional decision variable vector $\mathbf{x} = (x_1,\ldots,x_n)$ from some universe Ω.* □

Observe that $g_i(\mathbf{x}) \leq 0$ and $h_j(\mathbf{x}) = 0$ represent constraints that must be fulfilled while optimizing (minimizing or maximizing) $f(\mathbf{x})$. Ω contains all possible $\mathbf{x}$ that can be used to satisfy an evaluation of $f(\mathbf{x})$ and its constraints. Of course, $\mathbf{x}$ can be a vector of continuous or discrete variables as well as f being continuous or discrete.

The method for finding the global optimum (may not be unique) of any function is referred to as **Global Optimization**. In general, the global minimum of a single objective problem is presented in Definition 2 [72]:

Definition 2 (Single-Objective Global Minimum Optimization) : *Given a function $f : \Omega \subseteq \mathbb{R}^n \to \mathbb{R}$, $\Omega \neq \emptyset$, for $\mathbf{x} \in \Omega$ the value $f^* \triangleq f(\mathbf{x}^*) > -\infty$ is called a* **global minimum** *if and only if*

$$\forall \mathbf{x} \in \Omega : f(\mathbf{x}^*) \leq f(\mathbf{x}) . \tag{1.1}$$

$\mathbf{x}^*$ *is by definition the global minimum solution, f is the objective function, and the set Ω is the feasible region of $\mathbf{x}$. The goal of determining the global minimum solution(s) is called the* **global optimization problem** *for a single-objective problem.* □

Although single-objective optimization problems may have a unique optimal solution, MOPs (as a rule) present a possibly uncountable *set* of solutions, which when evaluated, produce vectors whose components represent trade-offs in objective space. A DM then implicitly chooses an acceptable solution (or solutions) by selecting one or more of these vectors.

1.2.2 The Multiobjective Optimization Problem

The **Multiobjective Optimization Problem** (also called multicriteria optimization, multiperformance or vector optimization problem) can then be defined (in words) as the problem of finding [1218]:

> "a vector of decision variables which satisfies constraints and optimizes a vector function whose elements represent the objective functions. These functions form a mathematical description of performance criteria which are usually in conflict with each other. Hence, the term "optimize" means finding such a solution which would give the values of all the objective functions acceptable to the decision maker."

Decision Variables

The **decision variables** are the numerical quantities for which values are to be chosen in an optimization problem. These quantities are denoted as x_j, $j = 1, 2, \ldots, n$.

The vector $\mathbf{x}$ of n decision variables is represented by:

$$\mathbf{x} = \begin{bmatrix} x_1 \\ x_2 \\ \vdots \\ x_n \end{bmatrix} \qquad (1.2)$$

This can be written more conveniently as:

$$\mathbf{x} = [x_1, x_2, \ldots, x_n]^T, \qquad (1.3)$$

where T indicates the transposition of the column vector to the row vector.

Constraints

In most optimization problems there are always restrictions imposed by the particular characteristics of the environment or available resources (e.g., physical limitations, time restrictions, etc.). These restrictions must be satisfied in order to consider a certain solution acceptable. All these restrictions in general are called **constraints**, and they describe dependences among decision variables and constants (or parameters) involved in the problem. These constraints are expressed in form of mathematical inequalities:

$$g_i(\mathbf{x}) \leq 0 \quad i = 1, \ldots, m \tag{1.4}$$

or equalities:

$$h_j(\mathbf{x}) = 0 \quad j = 1, \ldots, p \tag{1.5}$$

Note that p, the number of **equality constraints**, must be less than n, the number of decision variables, because if $p \geq n$ the problem is said to be **overconstrained**, since there are no degrees of freedom left for optimizing (i.e., in other words, there would be more unknowns than equations). The number of **degrees of freedom** is given by $n - p$. Also, constraints can be **explicit** (i.e., given in algebraic form) or **implicit**, in which case the algorithm to compute $g_i(\mathbf{x})$ for any given vector $\mathbf{x}$ must be known.

Commensurable vs. Non-Commensurable

In order to know how "good" a certain solution is, it is necessary to have some criteria to evaluate it. These criteria are expressed as computable functions of the decision variables,[2] called **objective functions**. In real-world problems, some functions are in conflict with others, and some must be minimized while others are maximized. These objective functions may be **commensurable** (measured in the same units) or **non-commensurable** (measured in different units). The multiple objectives being optimized almost always conflict, placing a partial, rather than total, ordering on the search space. In fact, finding the global optimum of a general MOP is an NP-Complete problem [72].

Attributes, Criteria, Objectives, and Goals

In OR, it is a common practice to differentiate among attributes, criteria, objectives and goals (e.g., [1036]). *Attributes* are often thought of as differentiating aspects, properties or characteristics of alternatives or consequences. *Criteria* generally denote evaluative measures, dimensions or scales against which alternatives may be gauged in a value or worth sense. *Objectives* are sometimes viewed in the same way, but may also denote specific desired levels of attainment or vague ideals. *Goals* usually indicate either of the latter notions. A distinction commonly made in OR is to use the term *goal* to designate potentially attainable levels, and *objective* to designate unattainable ideals.

The convention adopted in this book is the same assumed by several researchers (see for example [706] and [489]) of using the terms *objective*, *criteria*, and *attribute* interchangeably to represent an MOP's goals or objectives (i.e., distinct mathematical functions) to be achieved. The terms *objective space* or *objective function space* are also used to denote the coordinate space within which vectors resulting from evaluating an MOP's solutions are plotted.

---

[2] It is assumed that all functions used in this book are computable.

The objective functions are designated: $f_1(\mathbf{x}), f_2(\mathbf{x}), \ldots, f_k(\mathbf{x})$, where k is the number of objective functions in the MOP being solved. Therefore, the objective functions form a vector function $\mathbf{f}(\mathbf{x})$ which is defined by:

$$\mathbf{f}(\mathbf{x}) = \begin{bmatrix} f_1(\mathbf{x}) \\ f_2(\mathbf{x}) \\ \vdots \\ f_k(\mathbf{x}) \end{bmatrix} \tag{1.6}$$

This can be written more conveniently as:

$$\mathbf{f}(\mathbf{x}) = [f_1(\mathbf{x}), f_2(\mathbf{x}), \ldots, f_k(\mathbf{x})]^T$$

The set of all n-tuples of real numbers denoted by $\mathbb{R}^n$ is called **Euclidean n-space**. Two Euclidean spaces are considered in MOPs:

- The n-dimensional space of the decision variables in which each coordinate axis corresponds to a component of vector $\mathbf{x}$.
- The k-dimensional space of the objective functions in which each coordinate axis corresponds to a component vector $\mathbf{f}_k(\mathbf{x})$.

Every point in the first space represents a solution and gives a certain point in the second space, which determines a quality of this solution in terms of the objective function values.

1.2.3 Multiobjective Optimization Problem

The mathematical definition of a multiobjective problem (MOP) is important in providing a foundation of understanding between the interdisciplinary nature of deriving possible solution techniques (deterministic, stochastic); i.e., search algorithms. The following discussions present generic MOP mathematical and formal symbolic definitions.

The single objective formulation is extended to reflect the nature of multiobjective problems where there is not one objective function to optimize, but many. Thus, there is not one unique solution but a set of solutions. This set of solutions are found through the use of Pareto Optimality Theory [428]. Note that multiobjective problems require a decision maker to make a choice of $\mathbf{x_i}^*$ values. The selection is essentially a tradeoff of one complete solution $\mathbf{x}$ over another in multiobjective space.

More precisely, multiobjective problems (MOPs) are those problems where the goal is to optimize k objective functions simultaneously. This may involve the maximization of all k functions, the minimization of all k functions or a combination of maximization and minimization of these k functions. A MOP global minimum (or maximum) problem is formally defined in Definition 3 [1626]:

Definition 3 (General MOP [1626, 277, 265]) : *A general MOP is defined as minimizing (or maximizing) $F(\mathbf{x}) = (f_1(\mathbf{x}), \ldots, f_k(\mathbf{x}))$ subject to $g_i(\mathbf{x}) \leq 0$, $i = \{1, \ldots, m\}$, and $h_j(\mathbf{x}) = 0$, $j = \{1, \ldots, p\}$ $\mathbf{x} \in \Omega$. An MOP solution minimizes (or maximizes) the components of a vector $F(\mathbf{x})$ where $\mathbf{x}$ is a n-dimensional decision variable vector $\mathbf{x} = (x_1, \ldots, x_n)$ from some universe Ω. It is noted that $g_i(\mathbf{x}) \leq 0$ and $h_j(\mathbf{x}) = 0$ represent constraints that must be fulfilled while minimizing (or maximizing) $F(\mathbf{x})$ and Ω contains all possible $\mathbf{x}$ that can be used to satisfy an evaluation of $F(\mathbf{x})$.* □

Thus, a MOP consists of k objectives reflected in the k objective functions, $m + p$ constraints on the objective functions and n decision variables. The k objective functions may be linear or nonlinear and continuous or discrete in nature. The evaluation function, $F : \Omega \longrightarrow \Lambda$, is a mapping from the vector of decision variables ($\mathbf{x} = x_1, \ldots, x_n$) to output vectors ($\mathbf{y} = a_1, \ldots, a_k$). [1626]. Of course, the vector of decision variables $\mathbf{x_i}$ can also be continuous or discrete.

Ideal Vector

Definition 4 (Ideal Vector) : *Let*

$$\mathbf{x}^{0(i)} = [x_1^{0(i)}, x_2^{0(i)}, \ldots, x_n^{0(i)}]^T$$

be a vector of variables which optimizes (either minimizes or maximizes) the ith objective function $f_i(x)$. In other words, the vector $\mathbf{x}^{0(i)} \in \Omega$ is such that

$$f_i(\mathbf{x}^{0(i)}) = \operatorname*{opt}_{x \in \Omega} f_i(\mathbf{x}) \tag{1.7}$$

Then, the vector

$$\mathbf{f}^0 = [f_1^0, f_2^0, \ldots, f_k^0]^T \tag{1.8}$$

*(where f_i^0 denotes the optimum of the ith function) is ideal for an MOP, and the point in $\mathbb{R}^n$ which determined this vector is the ideal (utopical) solution, and is consequently called the **ideal vector**.* □

In other words, the ideal vector contains the optimum for each separately considered objective achieved at the same point in $\mathbb{R}^n$.

Convexity and Concavity

Definition 5 (Convexity) : *A function $\phi(\mathbf{x})$ is called **convex** over the domain of $\mathbb{R}$ if for any two vectors $\mathbf{x}_1$ and $\mathbf{x}_2 \in \mathbb{R}$,*

$$\phi(\theta \mathbf{x}_1 + (1 - \theta)\mathbf{x}_2) \leq \theta \phi(\mathbf{x}_1) + (1 - \theta)\phi(\mathbf{x}_2) \tag{1.9}$$

where θ is a scalar in the range $0 \leq \theta \leq 1$. □

A convex function cannot have any value larger than the function values obtained by linear interpolation between $\phi(\mathbf{x}_1)$ and $\phi(\mathbf{x}_2)$.

If the reverse inequality of the equation (5) holds, the function is **concave**. Thus $\phi(\mathbf{x})$ is **concave** if $-\phi(\mathbf{x})$ is **convex**. Linear functions are convex and concave at the same time.

A set of points (or region) is defined as a **convex set** in n-dimensional space if, for all pairs of two points $\mathbf{x}_1$ and $\mathbf{x}_2$ in the set, the straight-line segment joining them is also entirely in the set. Thus, every point $\mathbf{x}$, where

$$\mathbf{x} = \theta\mathbf{x}_1 + (1-\theta)\mathbf{x}_2 \quad 0 \le \theta \le 1 \tag{1.10}$$

is also in the set. So, for example, the sets shown in Figure 1.1 are convex, but the sets shown in Figure 1.2 are not.

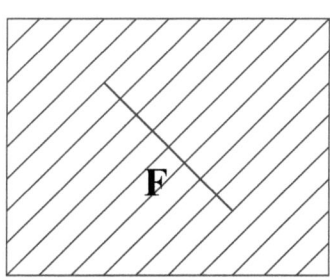
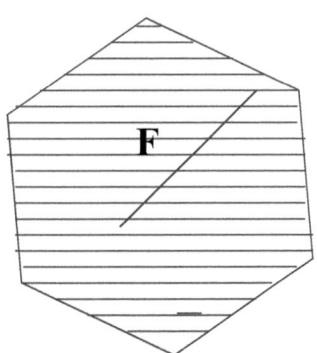

Fig. 1.1. Two examples of convex sets.

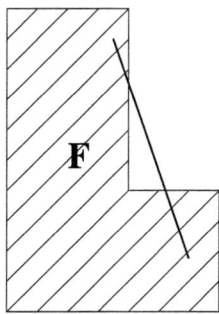
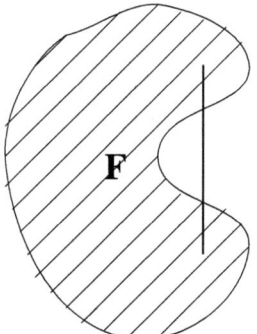

Fig. 1.2. Two examples of non-convex sets.

Pareto Terminology

Having several objective functions, the notion of "optimum" changes, because in MOPs, the aim is to find good compromises (or "trade-offs") rather than a single solution as in global optimization. The notion of "optimum" most commonly adopted is that originally proposed by Francis Ysidro Edgeworth [425] and later generalized by Vilfredo Pareto [1242]. Although some authors call this notion the *Edgeworth-Pareto optimum* (see for example [1517]), the most commonly accepted term is *Pareto optimum*. The formal definition is provided next.

Definition 6 (Pareto Optimality [1626, 277, 265]) : *A solution* $\mathbf{x} \in \Omega$ *is said to be Pareto Optimal with respect to (w.r.t.)* Ω *if and only if (iff) there is no* $\mathbf{x}' \in \Omega$ *for which* $\mathbf{v} = F(\mathbf{x}') = (f_1(\mathbf{x}'), \ldots, f_k(\mathbf{x}'))$ *dominates* $\mathbf{u} = F(\mathbf{x}) = (f_1(\mathbf{x}), \ldots, f_k(\mathbf{x}))$. *The phrase* **Pareto Optimal** *is taken to mean with respect to the entire decision variable space unless otherwise specified.* □

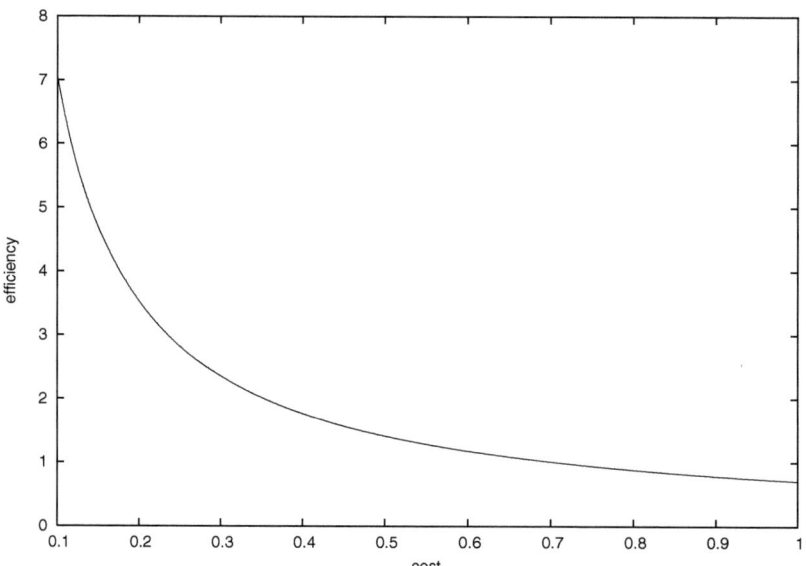

Fig. 1.3. An example of a problem with two objective functions: **cost** and **efficiency**. The Pareto front or trade-off surface is delineated by a curved line.

In words, this definition says that $\mathbf{x}^*$ is Pareto optimal if there exists no feasible vector $\mathbf{x}$ which would decrease some criterion without causing a simultaneous increase in at least one other criterion (assuming minimization).

The concept of Pareto Optimality is integral to the theory and the solving of MOPs. Additionally, there are a few more definitions that are also adopted in multiobjective optimization [1626]:

Definition 7 (Pareto Dominance [1626, 277, 265]): *A vector* $\mathbf{u} = (u_1, \ldots, u_k)$ *is said to* **dominate** *another vector* $\mathbf{v} = (v_1, \ldots, v_k)$ *(denoted by* $\mathbf{u} \preceq \mathbf{v}$*) if and only if* $\mathbf{u}$ *is partially less than* $\mathbf{v}$*, i.e.,* $\forall i \in \{1, \ldots, k\}$, $u_i \leq v_i \land \exists i \in \{1, \ldots, k\} : u_i < v_i$. □

Definition 8 (Pareto Optimal Set [1626, 277, 265]): *For a given MOP, $F(\mathbf{x})$, the* **Pareto Optimal Set**, $\mathcal{P}^*$, *is defined as:*

$$\mathcal{P}^* := \{\mathbf{x} \in \Omega \mid \neg \exists\, \mathbf{x}' \in \Omega\ \ F(\mathbf{x}') \preceq F(\mathbf{x})\}. \tag{1.11}$$

□

Pareto optimal solutions are those solutions within the genotype search space (decision space) whose corresponding phenotype objective vector components cannot be all simultaneously improved. These solutions are also termed *non-inferior*, *admissible*, or *efficient* solutions, with the entire set represented by $\mathcal{P}^*$. Their corresponding vectors are termed *nondominated*; selecting a vector(s) from this vector set (the Pareto front set $\mathcal{PF}^*$) implicitly indicates acceptable Pareto optimal solutions, decision variables or genotypes. These solutions may have no apparent relationship besides their membership in the Pareto optimal set. They form the set of all solutions whose associated vectors are nondominated; Pareto optimal solutions are classified as such based on their evaluated functional values.

Definition 9 (Pareto Front [1626, 277, 265]): *For a given MOP, $F(\mathbf{x})$, and Pareto Optimal Set, $\mathcal{P}^*$, the* **Pareto Front** $\mathcal{PF}^*$ *is defined as:*

$$\mathcal{PF}^* := \{\mathbf{u} = F(\mathbf{x}) \mid \mathbf{x} \in \mathcal{P}^*\}. \tag{1.12}$$

□

When plotted in objective space, the nondominated vectors are collectively known as the **Pareto front**. Again, $\mathcal{P}^*$ is a subset of some solution set. Its evaluated objective vectors form $\mathcal{PF}^*$, of which each is nondominated with respect to all objective vectors produced by evaluating every possible solution in Ω. In general, it is not easy to find an analytical expression of the line or surface that contains these points and in most cases, it turns out to be impossible. The normal procedure to generate the Pareto front is to compute many points in Ω and their corresponding $f(\Omega)$. When there is a sufficient number of these, it is then possible to determine the nondominated points and to produce the Pareto front. A sample Pareto front is shown in Figure 1.3. Note that PF_{true} is used throughout this book interchangeably with $\mathcal{PF}^*$.[3]

Although single-objective optimization problems may have a unique optimal solution, MOPs usually have a possibly uncountable set of solutions on

---

[3] The MOEA literature uses a variety of symbolic notation, but, in contrast, this book has adopted the forms P_{true} and PF_{true}, which are more understandable and precise, and used more in current practice as related to the computational domain.

a Pareto front. Each solution associated with a point on the Pareto front is a vector whose components represent trade-offs in the decision space or Pareto solution space.

The MOP's evaluation function, $\mathbf{f} : \Omega \longrightarrow \Lambda$, maps decision variables ($\mathbf{x} = x_1, \ldots, x_n$) to vectors ($\mathbf{y} = a_1, \ldots, a_k$). This situation is represented in Figure 1.4 for the case $n = 2$, $m = 0$, and $k = 3$. This mapping may or may not be onto some region of objective function space, dependent upon the functions and constraints composing the particular MOP.

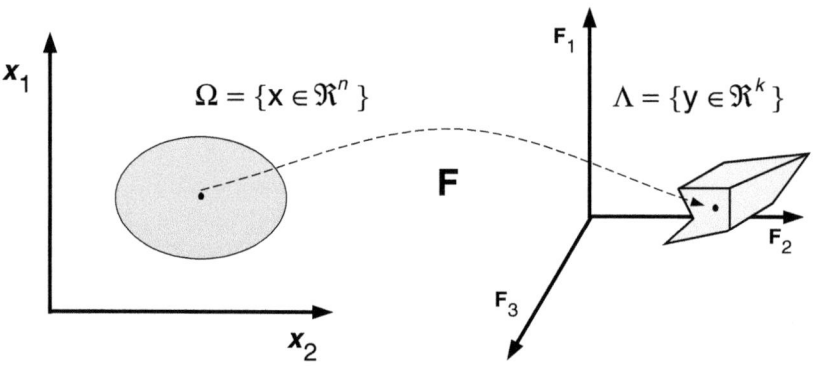

Fig. 1.4. MOP Evaluation Mapping

Note that the DM is often selecting solutions via choice of acceptable objective performance, represented by the Pareto front. Choosing an MOP solution that optimizes only one objective may well ignore solutions, which from an overall standpoint, are "better." The Pareto optimal set contains those better solutions. Identifying a set of Pareto optimal solutions is thus key for a DM's selection of a "compromise" solution satisfying the objectives as "best" possible. Of course, the accuracy of the decision maker's view depends on both the *true* Pareto optimal set and the set presented *as* Pareto optimal.

Weak and Strict Pareto Optimality

Definition 10 (Weak Pareto Optimality) : *A point* $\mathbf{x}^* \in \Omega$ *is a* **weakly Pareto optimal** *if there is no* $\mathbf{x} \in \Omega$ *such that* $f_i(\mathbf{x}) < f_i(\mathbf{x}^*)$, *for* $i = 1, \ldots, k$. □

Definition 11 (Strict Pareto Optimality) : *A point* $\mathbf{x}^* \in \Omega$ *is a* **strictly Pareto optimal** *if there is no* $\mathbf{x} \in \Omega$, $\mathbf{x} \neq \mathbf{x}^*$ *such that* $f_i(\mathbf{x}) \leq f_i(\mathbf{x}^*)$, *for* $i = 1, \ldots, k$. □

Kuhn-Tucker Conditions

Definition 12 (**Kuhn-Tucker Conditions for Noninferiority**) : *If a solution* $\mathbf{x}$ *to the general MOP is noninferior, then there exist* $w_l \geq 0$, $l = 1, 2, \ldots, k$ *(w_r is strictly positive for some $r = 1, 2, \ldots, k$), and* $\lambda_i \geq 0$, $i = 1, 2, \ldots, m$, *such that [910]:*

$$\mathbf{x} \in \Omega \tag{1.13}$$

$$\lambda_i g_i(\mathbf{x}) = 0 \quad i = 1, 2, \ldots, m \tag{1.14}$$

and

$$\sum_{l=1}^{k} w_l \nabla f_l(\mathbf{x}) - \sum_{i=1}^{m} \lambda_i \nabla g_i(\mathbf{x}) = 0 \tag{1.15}$$

□

These conditions are **necessary** for a noninferior solution, and when all of the $\mathbf{f}_l(\mathbf{x})$ are concave and Ω is a convex set, they are **sufficient** as well.

MOP Global Minimum

Defining an MOP's global optimum is not a trivial task as the "best" compromise solution is really dependent on the specific preferences (or biases) of the (human) decision maker. Solutions may also have some temporal dependences (e.g., acceptable resource expenditures may vary from month to month). Thus, there is no universally accepted definition for the MOP global optimization problem. However, an MOP's global optimum is defined to substantiate the material presented in further chapters.

Pareto optimal solutions are those which when evaluated, produce vectors whose performance in one dimension *cannot* be improved without adversely affecting another. The Pareto front $\mathcal{PF}^*$ determined by evaluating $\mathcal{P}^*$ is fixed by the defined MOP and does not change. Thus, $\mathcal{P}^*$ represents the "best" solutions available and allows the definition of an MOP's global optimum.

Definition 13 (**MOP Global Minimum**) : *Given a function* $\mathbf{f} : \Omega \subseteq \mathbb{R}^n \to \mathbb{R}^k$, $\Omega \neq \emptyset$, $k \geq 2$, *for* $\mathbf{x} \in \Omega$ *the set* $\mathcal{PF}^* \triangleq \mathbf{f}(\mathbf{x}_i^*) > (-\infty, \ldots, -\infty)$ *is called the global minimum if and only if*

$$\forall \mathbf{x} \in \Omega : \ \mathbf{f}(\mathbf{x}_i^*) \preceq \mathbf{f}(\mathbf{x}) \ . \tag{1.16}$$

Then, $\mathbf{x}_i^*$, $i = 1, \ldots, n$ is the **global minimum solution set** *(i.e.,* $\mathcal{P}^*$), $\mathbf{f}$ *is the multiple objective function, and the set Ω is the feasible region. The problem of determining the global minimum solution set is called the* **MOP global optimization problem**. □

1.2.4 Definition of MOEA Progress

The generic Pareto definitions presented before can lead to confusion in discussing the algorithmic progress of a MOEA's complex structure. To prevent possible inconsistencies in discussions of MOEAs, Van Veldhuizen [1626] developed Pareto terminology to clarify MOEA computational progress. For example, at any given generation of a MOEA, a "current" set of Pareto solutions (with respect to the *current* MOEA generational population) exists and is termed $P_{current}(t)$, where t represents the current generation number. Because of the manner in which Pareto optimality is defined, $P_{current}(t)$ is generally a non-empty solution set [1626].

Many MOEAs use a secondary population which is referred to as an *archive* or an external archive, in order to store nondominated solutions found through the generational process [1628, 1626]. Since this secondary population contains Pareto solutions generated up to a certain generation, each time another point is considered for addition to the secondary population, the point must be analyzed for nondominance with respect to the points currently in the secondary population. This secondary population is denoted $P_{known}(t)$ by definition. Additionally, $P_{known}(0)$ is defined as the empty set ($\emptyset$) and P_{known} alone as the *final* set of Pareto optimal solutions returned by the MOEA at termination [1626, 1790].

Different secondary population storage strategies exist; the simplest is when $P_{known}(t)$ is updated at each generation (i.e., $P_{current}(t) \bigcup P_{known}(t-1)$). At any given time, $P_{known}(t)$ is thus the set of Pareto solutions *currently found by the MOEA through generation t*. Of course, the *true* Pareto solution set (termed $\mathcal{P}^*$) is defined in the computational domain as P_{true} which is usually a subset of $\mathcal{P}^*$. Both are not explicitly known a priori for problems of any difficulty. P_{true} is defined by the functions composing an MOP and the given computational domain limits.

$P_{current}(t)$, $P_{known}(t)$, and P_{true} are sets of MOEA genotypes where each set's phenotypes form a computational Pareto front set. The associated Pareto front terms for each of these solution sets are defined respectively as $PF_{current}(t)$, $PF_{known}(t)$, and PF_{true}. Thus, when using a computational MOEA to solve MOPs, the implicit assumption is that one of the following holds at termination: $P_{known} = P_{true}$, or $P_{known} \subset P_{true}$, over some norm (Euclidean, RMS, etc.). On the other hand, the limits of the computational domain (finite storage, finite word-length) cause discussion inconsistencies in these set notations as presented in the following section. Observe that various MOEA researchers use a less precise notation referring only to the "approximated" Pareto front or Pareto solution.

1.2.5 Computational Domain Impact

The theory of computation implies that only a countable number of MOP solutions can be computed. Also, the reality of having computers with finite word-length indicates that the accuracy of generated MOP solutions is limited. Note of course that numerical analysis techniques can be employed to define associated error bounds if interested. Now, one could assume that the theoretical $\mathcal{P}^*$ and $\mathcal{PF}^*$ are sets of values represented with infinite word-length. Due to the computation domain (computer) *strictly* using these sets as the *goal* set of Pareto optimal solutions, $\mathcal{P}^*$ and associated optimal vectors $\mathcal{PF}^*$, it is impossible for most MOEAs to converge to the optimal solutions.[4] As developed by Day [344], this phenomenon is due to the computational limitation gap between using an uncountable infinite set (theoretical values) and countable/finite set (computational values). The form of these sets is related respectively to infinite word-length and finite word-length for representing decision variables $\mathbf{x}$, and associated objective vectors, $F(\mathbf{x})$. With this approach, MOEA results and analysis can be better compared to other multiobjective solution techniques.

We use a more precise computational terminology to distinguish between the real-world's computational model, the formal mathematical world's representation of solutions, and the aforementioned P_{true}, PF_{true}, P_{known}, and PF_{known} when solving MOPs. The three cases for each computational set are related to the relationships found in Table 1.1, which lists the three types of theoretical relationships between $\mathcal{P}^*$ and $\mathcal{PF}^*$ set cardinality that must be addressed as related to specific MOP characteristics. Note that countable includes finite.

Table 1.1: Relationships between the $\mathcal{P}^*$ and $\mathcal{PF}^*$ set size.

$|\mathcal{P}^*|$	$|\mathcal{PF}^*|$	Mappings of sets having size 1, $\ddot{n}$[5], and $\ddot{u}$[6]
1. Countable $\rightarrow$	Countable	$\{(1 \rightarrow 1), (\ddot{n} \rightarrow 1), (\ddot{n} \rightarrow \ddot{n})\}$
2. Uncountable $\rightarrow$	Countable	$\{(\ddot{u} \rightarrow 1), (\ddot{u} \rightarrow \ddot{n}), (\ddot{u} \rightarrow \ddot{n})\}$
3. Uncountable $\rightarrow$	Uncountable	$\{(\ddot{u} \rightarrow \ddot{u})\}$

$\mathcal{P}^*$ and $\mathcal{PF}^*$ of course represent the theoretical goal sets for a MOEA search algorithm. However, as indicated before, they may not be computationally achievable in any circumstance. Any goal set having an uncountable $|\mathcal{PF}^*|$ cannot be solved by a Turing Machine because the machine can only generate a countable number of optimal solutions; thus, #3 conditions in Table 1.1 reflect an MOP that cannot be optimally solved by a digital computer. When $|\mathcal{P}^*|$ is uncountable infinite, the MOP reflected in #2 can be computationally solved only under certain problem domain circumstances. Finally, if

---
[4] Examples of finite $|\mathcal{P}^*|$ and $|\mathcal{PF}^*|$ are found in NP-complete problems where the decision variables take on a finite number of values and the associated multi-objective functions likewise only have a finite number of values (usually integer). Further examples are found in deceptive MOPs. This situation is not true for continuous functions of continuous variables.
[5] $\ddot{n}$ represents a countable/infinite or finite set.
[6] $\ddot{u}$ represents an uncountable set.

the set of values contained in $\mathcal{P}^*$ and $\mathcal{PF}^*$ are subsets of computational numbers,[7] then the MOP relationship of #1 can be solved by digital computers with finite word-length and finite storage. Therefore, when defining the goal set, it is important to define a set that the MOEA can converge to; i.e., the goal sets referred to as P_{true} and PF_{true}.

- $P_{\textbf{true}}$: This term is given as the MOP's computational true Pareto Optimal Set (decision variables). Under the following conditions P_{true} is a subset of $\mathcal{P}^*$ assuming that the decision variable values are computational numbers (finite word-length representations).

  1. $\begin{cases} \text{All computational} \\ \text{numbers in } \mathcal{P}^*, & P_{true} \subseteq \mathcal{P}^* \\ \text{and } |\mathcal{P}^*| \text{ finite} \\ \text{At least one non-computational} \\ \text{number in } \mathcal{P}^*, & P_{true} \not\subseteq \mathcal{P}^* \end{cases}$
  2. $P_{true} \not\subseteq \mathcal{P}^*$
  3. $P_{true} \not\subseteq \mathcal{P}^*$

- PF_{true}: This term is given as the MOP's computationally true Pareto Front set (objective values). Under the following conditions PF_{true} is a subset of $\mathcal{PF}^*$ due to the discrete finite word-length decision variables and associated PF_{known} vectors.

  1. $\begin{cases} \text{All computational} \\ \text{numbers in } \mathcal{PF}^*, \text{ with } \mathcal{P}^* \to \mathcal{PF}^* & PF_{true} \subseteq \mathcal{PF}^* \\ \text{and } |\mathcal{PF}^*| \text{ finite} \\ \text{At least one non-computational} \\ \text{number in } \mathcal{PF}^*, & PF_{true} \not\subseteq \mathcal{PF}^* \end{cases}$
  2. $PF_{true} \not\subseteq \mathcal{PF}^*$
  3. $PF_{true} \not\subseteq \mathcal{PF}^*$

- $P_{known}(t)$: This term defines the Pareto solution set best found by the MOEA (finite word-length decision variables) at generation t. P_{known} often does NOT represent the true Pareto Optimal Set; instead, it only represents the best set found by a computational MOEA for a particular MOP.
- $PF_{known}(t)$: This term, PF_{known}, defines a Pareto Front set found by the MOEA at generation t. In general, it may be an intermediate Pareto Front set relative to the MOEA process (*i.e.* objective values at the current generation are probably not as good as objective values in the final generation Pareto Front set PF_{known} found by the MOEA).

Again, a NP-complete problem with finite cardinality of the $\mathcal{P}^*$ Pareto Solution and $\mathcal{PF}^*$ points falls under the set inclusion principle in these descriptions. In this situation, MOEA effectiveness can be explicitly measured as to the obtainment of these points. However, because the MOEA process

---

[7] These values must be a *computational number*; otherwise, a digital computer could not represent the goal sets. A computational number is a number that can be represented or generated within a digital computer with finite word-length.

may never reach PF_{known} for continuous functions, we can define a positive error distance from this computational possibility as ϵ. This provides for a different performance model in the next section for development of Pareto optimal front test metrics.

1.2.6 Pareto Epsilon Model

Considering the fact that in computational domain (finite word length and Turing machine definition), one may never be able to reach the "optimal" Pareto front, the concept of being within a "small" value of the Pareto front is appropriate. Moreover, one can only generate a finite number of points on the Pareto front even though a countable or uncountable number of points exist. The following definitions extend the previous definitions to provide a method of modeling these phenomena. In other words, in the cases where $P_{true} \not\subseteq \mathcal{P}^*$, P_{true} and PF_{true} are in the proximity[8] of $\mathcal{P}^*$ and $\mathcal{PF}^*$ respectively.

Definition 14 (**Pareto epsilon (ϵ) Dominance**) : *A vector* $\mathbf{u} = (u_1, \ldots, u_k)$ *is said to epsilon-dominate another vector* $\mathbf{v} = (v_1, \ldots, v_k)$ *(denoted by* $\mathbf{u} \preceq \mathbf{v}$*) if for some* $\epsilon > 0$ u_i *is partially less than* $v_i + \epsilon$*, i.e.,* $\forall i \in \{1, \ldots, k\}, \ u_i \leq (v_i + \epsilon) \land \exists i \in \{1, \ldots, k\} : u_i < (v_i + \epsilon)$ *where* $\epsilon > 0$. □

Definition 15 (**Pareto epsilon (ϵ) Optimality**) : *A solution* $\mathbf{x} \in \Omega$ *is said to be Pareto epsilon Optimal with respect to* Ω *if and only if there is no* $\mathbf{x}' \in \Omega$ *for which* $\mathbf{v} = F(\mathbf{x}') = (f_1(\mathbf{x}'), \ldots, f_k(\mathbf{x}'))$ *epsilon dominates* $\mathbf{u} = F(\mathbf{x}) = (f_1(\mathbf{x}), \ldots, f_k(\mathbf{x}))$. *The phrase* **Pareto epsilon Optimal** *is taken to mean with respect to the entire decision variable space unless otherwise specified.* □

Definition 16 (**Pareto epsilon (ϵ) Optimal Set**) : *For a given MOP, $F(\mathbf{x})$, the Pareto epsilon Optimal Set, $\mathcal{P}^*_\epsilon$, is defined as:*

$$\mathcal{P}^*_\epsilon := \{\mathbf{x} \in \Omega \mid \neg \exists \ \mathbf{x}' \in \Omega \ F(\mathbf{x}') \preceq F(\mathbf{x})\}. \tag{1.17}$$

□

Definition 17 (**Pareto epsilon (ϵ) Front**) : *For a given MOP, $F(\mathbf{x})$, and Pareto epsilon Optimal Set, $\mathcal{P}^*_\epsilon$, the Pareto epsilon Front ($\mathcal{PF}^*_\epsilon$) is defined as:*

$$\mathcal{PF}^*_\epsilon := \{\mathbf{u} = F(\mathbf{x}) = (f_1(\mathbf{x}), \ldots, f_k(\mathbf{x})) \mid \mathbf{x} \in \mathcal{P}^*_\epsilon\}. \tag{1.18}$$

□

Definition 18 (**Pareto Front width distribution**) : *The width of the Pareto front created by the Pareto epsilon Dominance factor is described by the*

---

[8] Distance of optimal solutions and associated Pareto front vectors to the theoretical true depends upon finite word-length restriction and characteristics of the problem domain. The distance can be defined as ϵ.

Pareto front width distribution. By placing 3D Gaussian distributions (Parzon Windows) on each vector on a Pareto epsilon front, a distribution can be illustrated having a multidimensional Gaussian Distribution Characteristics. See Definition 14 for understanding of Pareto epsilon dominance. □

The following terminology is used to distinguish between the real-world's and mathematical-world's representation of solutions and associated Pareto epsilon front vectors when solving MOPs. In the cases where $P_{true}^{\epsilon} \not\subseteq P_{\epsilon}^{*}$, P_{true}^{ϵ} and $\mathcal{PF}_{true}^{\epsilon}$ are in the proximity of $\mathcal{P}_{\epsilon}^{*}$ and $\mathcal{PF}_{\epsilon}^{*}$ respectively. The three cases listed under each term are related to the relationships found in Table 1.1.

- $P_{\mathbf{true}}^{\epsilon}$: This term is given as the MOP's computational true Optimal epsilon Set (decision variables). Under the following conditions P_{true}^{ϵ} is a subset of $\mathcal{P}_{\epsilon}^{*}$ because the decision variables for the MOP must be discrete.
  1. $\begin{cases} \text{All computational numbers in } \mathcal{P}_{\epsilon}^{*}, \ P_{true}^{\epsilon} \subseteq \mathcal{P}_{\epsilon}^{*} \\ \text{At least one non-computational} \\ \text{number in } \mathcal{P}_{\epsilon}^{*}, \qquad\qquad P_{true}^{\epsilon} \not\subseteq \mathcal{P}_{\epsilon}^{*} \end{cases}$
  2. $P_{true}^{\epsilon} \not\subseteq \mathcal{P}_{\epsilon}^{*}$
  3. $P_{true}^{\epsilon} \not\subseteq \mathcal{P}_{\epsilon}^{*}$

- $PF_{\mathbf{true}}^{\epsilon}$: This term is given as the MOP's computationally true Pareto epsilon Front set (objective values). Under the following conditions PF_{true}^{ϵ} is a subset of $\mathcal{PF}_{\epsilon}^{*}$ due to making the decision variables and objective vectors of the MOP discrete within the computer.
  1. $\begin{cases} \text{All computational numbers in } \mathcal{PF}_{\epsilon}^{*}, \ PF_{true}^{\epsilon} \subseteq \mathcal{PF}_{\epsilon}^{*} \\ \text{At least one non-computational} \\ \text{number in } \mathcal{PF}_{\epsilon}^{*}, \qquad\qquad PF_{true}^{\epsilon} \not\subseteq \mathcal{PF}_{\epsilon}^{*} \end{cases}$
  2. $PF_{true}^{\epsilon} \not\subseteq \mathcal{PF}_{\epsilon}^{*}$
  3. $PF_{true}^{\epsilon} \not\subseteq \mathcal{PF}_{\epsilon}^{*}$

- $P_{\mathbf{known}}^{\epsilon}$: This term defines Pareto epsilon Optimal Set found by the MOEA (decision variables). P_{known}^{ϵ} often times does not represent the true Pareto epsilon Optimal Set; instead, it should represent the best Pareto epsilon Optimal Set found by a MOEA for a particular MOP.

- $PF_{\mathbf{known}}^{\epsilon}$: This term defines the Pareto epsilon Front set found by the MOEA. PF_{known}^{ϵ} may be an intermediate Pareto epsilon Front for the MOEA (*i.e.* a set that is not as good as the final Pareto epsilon Front set found by the MOEA).

1.2.7 Decision Maker Impact

Solutions on the Pareto Optimal Front $\mathcal{PF}^{*}$ represent optimal solutions in the sense that improving the value in one dimension of the objective function vector leads to a degradation in at least one other dimension of the objective function vector. This is also true for PF_{known}. This requires the decision maker to make a tradeoff decision when presented with a large finite number of points on PF_{known}. There exists a difference in terminology between an acceptable compromise solution and a Pareto Optimal Solution [510]. The decision maker typically chooses only a few points in PF_{known} as generated by P_{known}. The associated Pareto Optimal solutions, $\mathbf{u} \in PF_{known}{}^*$, are then the "acceptable" (by the decision maker) compromise solutions. The decision makers base their solution choice on which solutions take into account the non-modelled human's preference. The human preference factor can require engineers and scientists to attempt to find a large number of the points in PF_{known}, since all points may not be weighted equally by the decision maker.

1.3 An Example

To illustrate the application of the previous concepts, one example that has been studied by several researchers [1322, 241] is as follows:

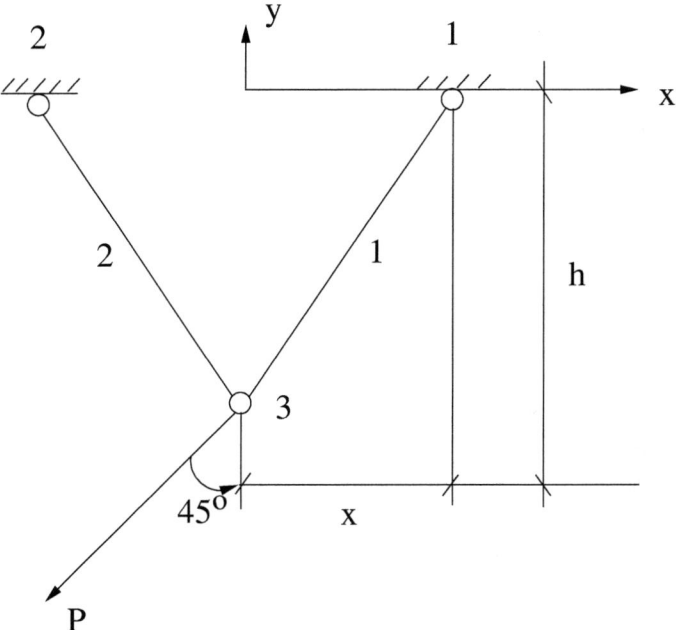

Fig. 1.5. A two-bar plane truss.

The goal is to optimize the two-bar symmetric plane truss shown in Figure 1.5. The design variables are the cross-sectional areas of the two bars. The problem is formulated as follows:

$$\text{Minimize} \begin{cases} f_1(\mathbf{x}) = 2\rho h x_2 \sqrt{1+x_1^2} \\ f_2(\mathbf{x}) = \dfrac{Ph(1+x_1^2)^{1.5}(1+x_1^4)^{0.5}}{2\sqrt{2}Ex_1^2 x_2} \end{cases} \quad (1.19)$$

subject to:

$$g_1(\mathbf{x}) = \frac{P(1+x_1)(1+x_1^2)^{0.5}}{2\sqrt{2}x_1x_2} - \sigma_0 \leq 0 \qquad (1.20)$$

$$g_2(\mathbf{x}) = \frac{P(-x_1+1)(1+x_1^2)^{0.5}}{2\sqrt{2}x_1x_2} - \sigma_0 \leq 0 \qquad (1.21)$$

where $f_1(\mathbf{x})$ is the structural weight of the truss, $f_2(\mathbf{x})$ is the displacement of joint 3 (in Figure 1.5), and $g_1(\mathbf{x})$ and $g_2(\mathbf{x})$ are stress constraints of the members.

In the previous expressions, $x_1 = x/h$, $x_2 = A/A_{min}$, $E=$ Young's modulus, and $\rho=$ density of material. It is assumed that: $\rho = 0.283$ lb/in^3, $h = 100$ in, $P = 10^4$ lb, $E = 3\times 10^7$ lb/in^2, $\sigma_0 = 2\times 10^4$ lb/in^2, $A_{min} = 1$ in^2, and the lower (l) and upper (u) bounds of the design variables are: $x_1^{(l)}=0.1$, $x_2^{(l)}=0.5$, $x_1^{(u)}=2.25$ and $x_2^{(u)}=2.5$.

The true Pareto front of this problem (obtained through an enumerative approach) is shown in Figure 1.6. Once the Pareto front of the problem has been found, the DM is presented the set of Pareto optimal solutions generated and then chooses a point (or points) from this set.

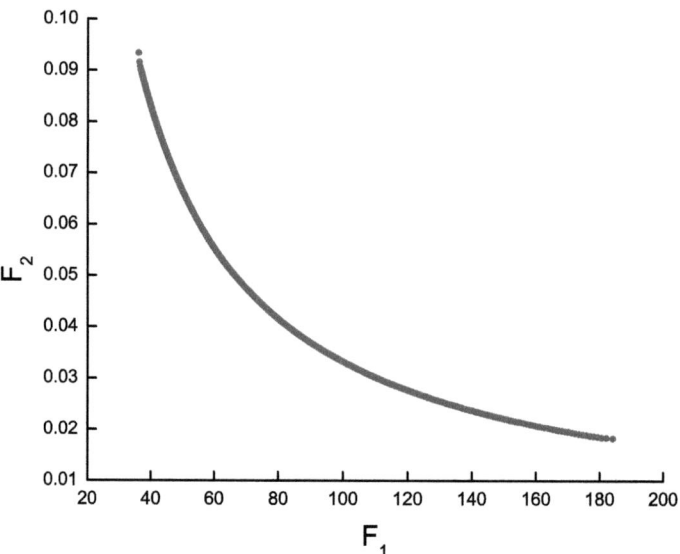

Fig. 1.6. True Pareto front of the two-bar truss problem.

1.4 General Optimization Algorithm Overview

For the purposes of this book, general search and optimization techniques are classified into three categories: enumerative, deterministic, and stochastic (random). Although an enumerative search is deterministic a distinction is made here as it employs no heuristics. Figure 1.7 shows common examples of each type.

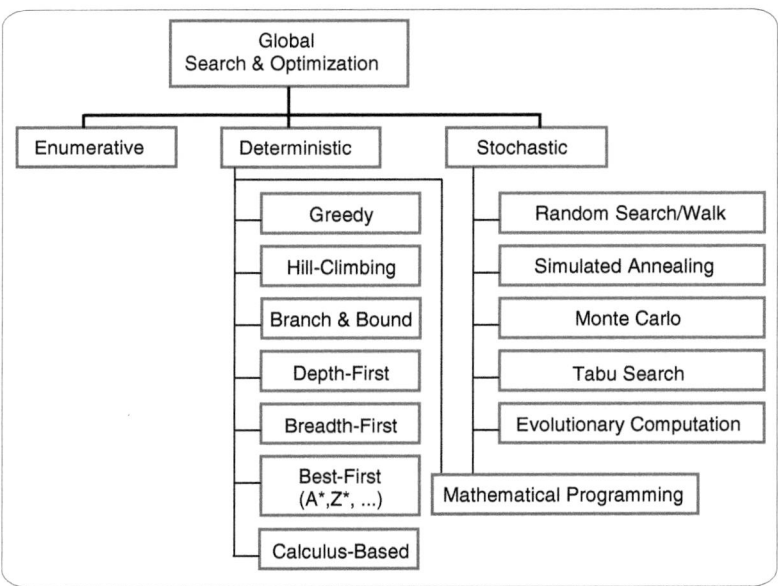

Fig. 1.7. Global Optimization Approaches

Enumerative schemes are perhaps the simplest search strategy. Within some defined finite search space each possible solution is evaluated. However, it is easily seen that this technique is inefficient or even infeasible as search spaces become large. As many real-world problems are computationally intensive, some means of limiting the search space must be implemented to find "acceptable" solutions in "acceptable" time [1101]. Deterministic algorithms attempt this by incorporating problem domain knowledge. Many of these are considered graph/tree search algorithms and are described as such here.

Greedy algorithms make locally optimal choices, assuming optimal subsolutions are *always* part of the globally optimal solution [170, 729]. Thus, these algorithms fail unless that is not the case. Hillclimbing algorithms search in the direction of steepest ascent from the current position. These algorithms work best on unimodal functions, but the presence of local optima, plateaus, or ridges in the fitness (search) landscape reduce algorithm effectiveness [1407]. Greedy and hillclimbing strategies are *irrevocable*. They repeatedly expand a node, examine all possible successors (then expanding the "most promising" node), and keep no record of past expanded nodes [1264].

Branch and bound search techniques need problem specific heuristics/decision algorithms to limit the search space [541, 1264]. They compute some bound at a given node which determines whether the node is "promising;" several nodes' bounds are then compared and the algorithm branches to the "most promising" node [1172]. Basic depth-first search is *blind* or *uninformed* in that the search order is independent of solution location (except for search termination). It expands a node, generates all successors, expands a successor, and so forth. If the search is blocked (e.g., it reaches a tree's bottom level) it resumes from the deepest node left behind [1264]. *Backtracking* is a depth-first search variant which "backtracks" to a node's parent if the node is determined "unpromising" [1172]. Breadth-first search is also uninformed. It differs from depth-first search in its actions after node expansion, where it progressively explores the graph one *layer* at a time [1264]. Best-first search uses heuristic information to place numerical values on a node's "promise"; the node with highest promise is examined first [1264]. A^*, Z^*, and others are popular best-first search variants selecting a node to expand based both on "promise" and the overall cost to arrive at that node.[9] Finally, calculus-based search methods at a minimum require continuity in some variable domain for an optimal value to be found [53].

Greedy and hill-climbing algorithms, branch and bound tree/graph search techniques, depth- and breadth-first search, best-first search, and calculus-based methods are all deterministic methods successfully used in solving a wide variety of problems [170, 581, 1172]. However, many MOPs are high-dimensional, discontinuous, multimodal, and/or NP-Complete. Deterministic methods are often ineffective when applied to NP-Complete or other high-dimensional problems because they are handicapped by their requirement for problem domain knowledge (heuristics) to direct or limit search [500, 541, 581, 1101] in these exceptionally large search spaces. Problems exhibiting one or more of these above characteristics are termed *irregular* [942].

Because many real-world scientific and engineering MOPs are irregular, enumerative and deterministic search techniques are then unsuitable. Stochastic search and optimization approaches such as Simulated Annealing (SA) [861], Monte Carlo methods [1217], Tabu search [572], and Evolutionary Computation (EC) [581, 1100, 72] were developed as alternative approaches for solving these irregular problems. Stochastic methods require a function assigning fitness values to possible (or partial) solutions, and an encode/decode (mapping) mechanism between the problem and algorithm domains. Although some are shown to "eventually" find an optimum most cannot guarantee *the*

---

[9] Note that there has been work regarding the extension of search algorithms such as A^* for multiobjective cases (see for example [1523, 335, 1051]). Such topic, although called "Multiobjective Heuristic Search", will not be covered in this book, since we focus only on stochastic techniques.

optimal solution. They in general provide good solutions to a wide range of optimization problems which traditional deterministic search methods find difficult [581, 729].

A random search is the simplest stochastic search strategy, as it simply evaluates a given number of randomly selected solutions. A random walk is very similar, except that the next solution evaluated is randomly selected using the last evaluated solution as a starting point [1646]. Like enumeration, though, these strategies are not efficient for many MOPs because of their failure to incorporate problem domain knowledge. Random searches can generally expect to do no better than enumerative ones [581, pg. 5].

SA is an algorithm explicitly modeled on an *annealing* analogy, where, for example, a liquid is heated and then gradually cooled until it freezes. Where hill-climbing chooses the *best* move from some node SA picks a *random* one. If the move improves the current optimum it is always executed, else it is made with some probability $p < 1$. This probability exponentially decreases either by time or with the amount by which the current optimum is worsened [1407, 407]. If water's temperature is lowered slowly enough it attains a lowest-energy configuration; the analogy for SA is that if the "move" probability decreases slowly enough the global optimum is found.

In general, *Monte Carlo* methods involve simulations dealing with stochastic events; they employ a pure random search where any selected trial solution is fully independent of any previous choice and its outcome [1460, 1217]. The current "best" solution and associated decision variables are stored as a comparator. *Tabu search* is a meta-strategy developed to avoid getting "stuck" on local optima. It keeps a record of both visited solutions and the "paths" which reached them in different "memories." This information restricts the choice of solutions to evaluate next. Tabu search is often integrated with other optimization methods [572, 1460].

EC is a generic term for several stochastic search methods which computationally simulate the natural evolutionary process. As a recognized research field EC is young, although its associated techniques have existed for about forty five years [497]. EC embodies the techniques of genetic algorithms (GAs), evolution strategies (ESs), and evolutionary programming (EP), collectively known as EAs [496]. These techniques are loosely based on natural evolution and the Darwinian concept of "Survival of the Fittest" [581]. Common between them are the reproduction, random variation, competition, and selection of contending individuals within some population [496]. In general, an EA consists of a *population* of encoded solutions (*individuals*) manipulated by a set of *operators* and evaluated by some *fitness function*.

Each solution's associated *fitness* determines which survive into the next *generation*. Although sometimes considered equivalent, the terms *EA* and *EC* are used separately in this book to preserve the distinction between EAs and other EC techniques (e.g., genetic programming (GP) [905, 89] and learning classifier systems [953, 183]).[10]

MOP complexity and the shortcomings of deterministic search methods also drove creation of several optimization techniques by the Operations Research (OR) community. These methods (whether linear or nonlinear, deterministic or stochastic) can be grouped under the rubric *mathematical programming*. These methods treat constraints as the main problem aspect [1460]. *Linear programming* is designed to solve problems in which the objective function and all constraint relations are linear [681]. Conversely, *nonlinear programming* techniques solve *some* MOPs not meeting those restrictions but require convex constraint functions [1460]. It is noted here that many problem domain assumptions must be satisfied when using linear programming, and that many real-world scientific and engineering problems may only be modeled by nonlinear functions [681, pp. 138,574]. Finally, *stochastic programming* is used when random-valued parameters and objective functions subject to statistical perturbations are part of the problem formulation. Depending on the type of variables used in the problem, several variants of these methods exist (i.e., discrete, integer, binary, and mixed-integer programming) [1460].

1.5 EA Basics

The following presentation defines basic EA structural terms and concepts;[11] the described terms' "meanings" are normally analogous to their genetic counterparts. A *structure* or *individual* is an encoded solution to some problem. Typically, an individual is represented as a string (or string of strings) corresponding to a biological *genotype*. This genotype defines an individual organism when it is expressed (decoded) into a *phenotype*. A genotype is composed of one or more *chromosomes*, where each chromosome is composed of separate *genes* which take on certain values (*alleles*) from some genetic alphabet. A *locus* identifies a gene's position within the chromosome. Thus, each individual decodes into a set of parameters used as input to the function under consideration. Finally, a given set of chromosomes is termed a *population*. These concepts are pictured in Figure 1.8 (for both binary and real-valued chromosomes) and in Figure 1.9.

Just as in nature, Evolutionary Operators (EVOPs) operate on an EA's population attempting to generate solutions with higher and higher fitness. The three major EVOPs associated with EAs are *mutation*, *recombination*, and *selection*. Illustrating this, Figure 1.10 shows *bitwise mutation* on an encoded string where a '1' is changed to a '0', or vice versa. Figure 1.11 shows

---

[10] Although GP and learning classifier systems may be classified as EA techniques, several researchers consider them conceptually different approaches to EC [860].
[11] There is no shortage of introductory EA texts. The general reader is referred to Goldberg [581], Michalewicz [1100], Mitchell [1114], Fogel [496] or Eiben & Smith [435]. A more technical presentation is given by Bäck [72].

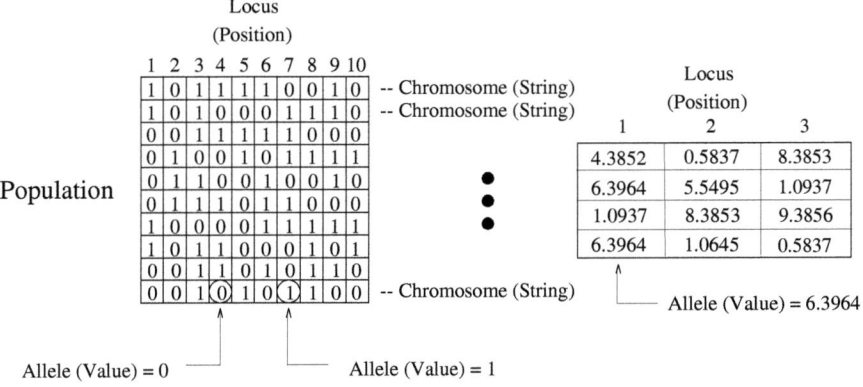

Fig. 1.8. Generalized EA Data Structure and Terminology

single-point crossover (a form of recombination) operating on two parent binary strings; each parent is cut and recombined with a piece of the other. Above-average individuals in the population are *selected* (reproduced) to become members of the next generation more often than below-average individuals. The selection EVOP effectively gives strings with higher fitness a higher probability of contributing one or more children in the succeeding generation. Figure 1.12 shows the operation of the common *roulette-wheel* selection (a *fitness proportional* selection operator) on two different populations of four strings each. Each string in the population is assigned a portion of the wheel proportional to the ratio of its fitness and the population's average fitness.

Real-valued chromosomes also undergo these same EVOPs although implemented differently. All EAs use some subset or variation of these EVOPs. Many variations on the basic operators exist; these are dependent upon problem domain constraints affecting chromosome structure and alleles [72].

An EA requires both an *objective* and *fitness* function, which are fundamentally different. The objective function defines the EA's optimality condition (and is a feature of the problem domain) while the fitness function (in the algorithm domain) measures how "well" a particular solution satisfies that condition and assigns a corresponding real-value to that solution. However, these functions are in principle identical [72, pg. 68] (e.g., in numerical optimization problems).

Many other selection techniques are implemented by EAs, e.g., tournament and ranking [72, 583]. Tournament selection operates by randomly choosing some number q individuals from the generational population and selecting the "best" to survive into the next generation. Binary tournaments ($q = 2$) are probably the most common. Ranking assigns selection probabilities solely on an individual's rank, ignoring absolute fitness values. Two other selection techniques noted in detail are the $(\mu+\lambda)$ and (μ, λ) selection strategies, where μ represents the number of parent solutions and λ the number of children.

26   1 Basic Concepts

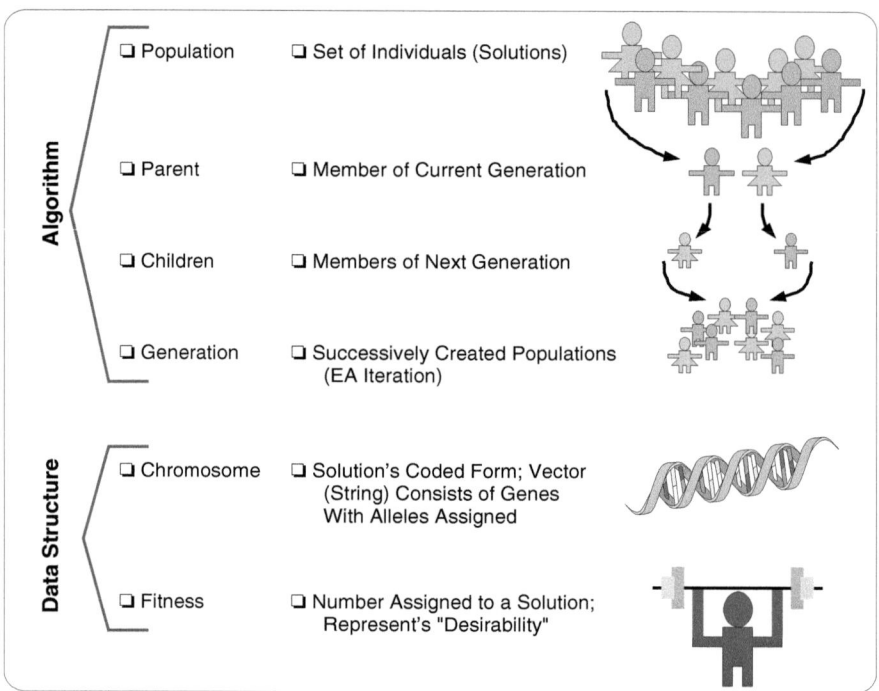

Fig. 1.9. Key EA Components

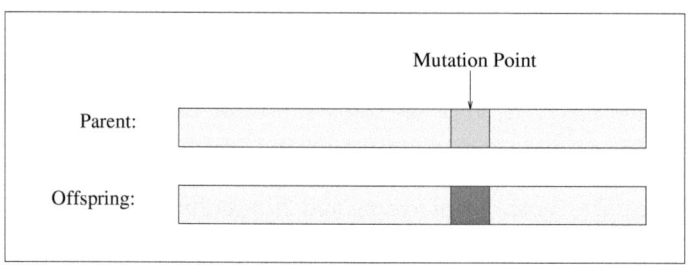

Fig. 1.10. Bitwise Mutation

The former selects the μ best individuals drawing from *both* the parents and children, the latter selects μ individuals from the child population only.

Why is the choice of EA selection technique so important? Two conflicting goals are common to all EA search: *exploration* and *exploitation*. Bäck also offers the analogous terms of convergence reliability and velocity, large and small genotypic diversity, and "soft" and "hard" selection [72, pg. 165]. No matter the terminology, one goal is achieved only at the expense of another. An EA's selective pressure is the control mechanism determining the type of search performed. Bäck's analysis shows a general ordering of selection techniques (listed in order of increasing selective pressure): Proportional, linear

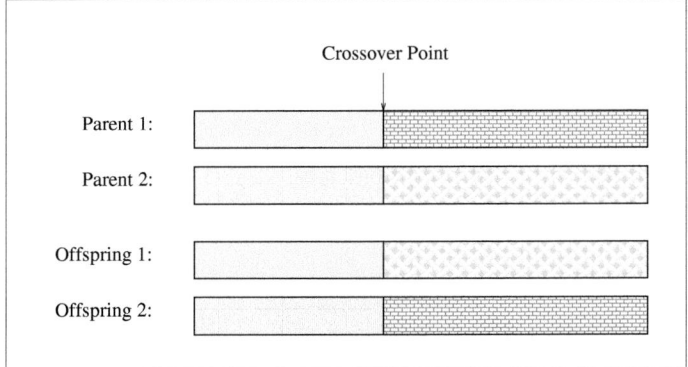

Fig. 1.11. Single-Point Crossover

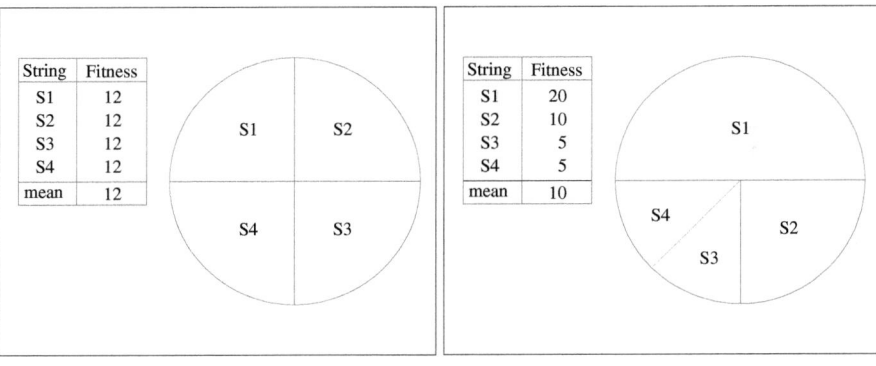

Fig. 1.12. Roulette Wheel Selection

ranking, tournament, and (μ, λ) selection [72, pg. 180]. Finally, an EA's decision function determines when execution stops. Table 1.2 highlights the major differences between the three major EC instantiations.

It is beyond the scope of this book to provide an in-depth analysis of general EVOPs and EA components. Interested readers are directed to the Handbook of Evolutionary Computation [73], which is probably the most comprehensive collection of articles discussing EC, its instantiations, and applications.

Although much room for creativity exists when selecting and defining EA instantiations (e.g., genetic representation and specific EVOPs), careful consideration must be given to the mapping from problem to algorithm domains. "Improper" representations and/or operators may have detrimental effects upon EA performance (e.g., Hamming cliffs [72, pg. 229]). Although there is no unique combination guaranteeing "good" performance [498, 1708], choosing wisely may well result in more effective and efficient implementations.

Table 1.2: Key EA Implementation Differences

EA Type	Representation	EVOPs
EP	Real-values	Mutation and $(\mu + \lambda)$ selection alone
ES	Real-values *and* strategy parameters	Mutation, recombination, and $(\mu + \lambda)$ or (μ, λ) selection
GA	Historically binary; Real-values now common	Mutation, recombination, and selection

To formally define an EA, its general algorithm is described in mathematical terms, allowing for exact specification of various EA instantiations. In this framework, each EA is associated with a non-empty set I called the EA's *individual space*. Each *individual* $\mathbf{a} \in I$ normally represents a candidate solution to the problem being solved by the EA. Individuals are often represented as a vector ($\mathbf{a}$) where the vector's dimensions are analogous to a chromosome's genes. The general framework leaves each individual's dimensions unspecified; an individual ($\mathbf{a}$) is simply that and is modified as necessary for the particular EA instance.

When defining (generational) population transformations Bäck denotes the resulting collection of μ individuals via I^μ, and denotes population transformations by the following relationship: $T : I^\mu \rightarrow I^\mu$, where $\mu \in \mathbb{N}$ [72]. However, some EA variants obtain resulting populations whose size is *not* equal to their predecessors. Thus, this general framework represents a population transformation via the relationship $T : I^\mu \rightarrow I^{\mu'}$, indicating succeeding populations may contain the same *or* different numbers of individuals. This framework also represents all population sizes, evolutionary operators, and parameters as *sequences* [1092]. This is due to the fact that different EAs use these factors in slightly different ways. The general algorithm thus recognizes and explicitly identifies this nuance. Having discussed the relevant background terminology, an EA is then defined as [1092] [72, pg. 66]:

Definition 19 (Evolutionary Algorithm) : *Let I be a non-empty set (the individual space), $\{\mu^{(i)}\}_{i \in \mathbb{N}}$ a sequence in $\mathbb{Z}^+$ (the parent population sizes), $\{\mu'^{(i)}\}_{i \in \mathbb{N}}$ a sequence in $\mathbb{Z}^+$ (the offspring population sizes), $\Phi : I \longrightarrow \mathbb{R}$ a fitness function, $\iota : \bigcup_{i=1}^{\infty} (I^\mu)^{(i)} \longrightarrow \{true, false\}$ (the termination criterion), $\chi \in \{true, false\}$, r a sequence $\{r^{(i)}\}$ of recombination operators $r^{(i)} : \mathbb{X}_r^{(i)} \longrightarrow \mathcal{T}\left(\Omega_r^{(i)}, \mathcal{T}\left(I^{\mu^{(i)}}, I^{\mu'^{(i)}}\right)\right)$, m a sequence $\{m^{(i)}\}$ of mutation operators $m^{(i)} : \mathbb{X}_m^{(i)} \longrightarrow \mathcal{T}\left(\Omega_m^{(i)}, \mathcal{T}\left(I^{\mu'^{(i)}}, I^{\mu'^{(i)}}\right)\right)$, s a sequence $\{s^{(i)}\}$ of selection operators $s^{(i)} : \mathbb{X}_s^{(i)} \times \mathcal{T}(I, \mathbb{R}) \longrightarrow \mathcal{T}\left(\Omega_s^{(i)}, \mathcal{T}\left(\left(I^{\mu'^{(i)} + \chi \mu^{(i)}}\right), I^{\mu^{(i+1)}}\right)\right)$, $\Theta_r^{(i)} \in \mathbb{X}_r^{(i)}$ (the* recombination parameters), $\Theta_m^{(i)} \in \mathbb{X}_m^{(i)}$ (the mutation parameters), and $\theta_s^{(i)} \in \mathbb{X}_s^{(i)}$ (the selection parameters). Then the algorithm shown in Figure 1.13 is called an Evolutionary Algorithm. □

$t := 0;$
initialize $P(0) := \{\mathbf{a}_1(0), \ldots, \mathbf{a}_\mu(0)\} \in I^{\mu^{(0)}};$
while $(\iota(\{P(0), \ldots, P(t)\}) \neq \text{true})$ **do**
    recombine: $P'(t) := r_{\Theta_r^{(t)}}^{(t)}(P(t));$
    mutate: $P''(t) := m_{\Theta_m^{(t)}}^{(t)}(P'(t));$
    select:
        if χ
            then $P(t+1) := s_{(\theta_s^{(t)}, \Phi)}^{(t)}(P''(t));$
            else $P(t+1) := s_{(\theta_s^{(t)}, \Phi)}^{(t)}(P''(t) \cup P(t));$
        fi
    $t := t + 1;$
od

Fig. 1.13. Evolutionary Algorithm Outline

1.6 Origins of Multiobjective Optimization

Multiobjective optimization theory is not as recent as we might think. In fact, some authors (see for example [1516]) indicate that multiobjective optimization is an inherent part of economic equilibrium and, in consequence, it can be traced back to 1776 in which Adam Smith's treatise *The Wealth of Nations* was published.

The general concept of economic equilibrium is often attributed to Léon Walras. However, William Stanley Jevons, Carl Menger, Francis Ysidro Edgeworth and Vilfredo Pareto also did very important work in this regard in the period between 1874 and 1906.

Closely related to multiobjective optimization is also the theory of psychological games and the notion of game strategy (based on analyzing the psychology of the adversary), which is attributed to Félix Édouard Émile Borel.

The so-called *Game Theory* dates back to the work done by Borel in 1921. However, most historians tend to attribute the origins of game theory to a paper from the Hungarian mathematician John von Neumann, which was orally presented in 1926 and published in 1928.

In 1944, John von Neumann and Oskar Morgenstern mentioned that an optimization problem in the context of a social exchange economy was "a peculiar and disconcerting mixture of several conflicting problems" that was "nowhere dealt with in classical mathematics" [1664]. Unfortunately, they did not discuss this problem any further in their book and no real contribution in this regard was made until the 1950s.

In 1951, Tjalling C. Koopmans edited a book called *Activity Analysis of Production and Allocation* [897], where the concept of "efficient" vector was first used in a significant way.[12]

1.6.1 Mathematical Foundations

The origins of the mathematical foundations of multiobjective optimization can be traced back to the period that goes from 1895 to 1906 [1516]. During that period, Georg Cantor [200, 201] and Felix Hausdorff [667] laid the foundations of infinite dimensional ordered spaces. Cantor also introduced equivalence classes and stated the first sufficient conditions for the existence of a utility function. Hausdorff also gave the first example of a complete ordering. However, it was the concept of *vector maximum problem* introduced by Harold W. Kuhn and Albert W. Tucker [910] which made multiobjective optimization a mathematical discipline on its own. The so-called "proper efficiency" in the context of multiobjective optimization was also formulated in this seminal paper that can be considered as the first serious attempt to derive a theory in this area. This same direction was later followed by Kenneth J. Arrow et al. [61] who used the term "admissible" instead of "efficient" points.

However, multiobjective optimization theory remained relatively undeveloped during the 1950s, and the subject was scarcely covered by only a few authors (see for example [895, 896, 689, 696, 864, 821]).

Probably the most important research outcome of the 1950s was *goal programming*, introduced by Abraham Charnes and William Wager Cooper [229] based on an earlier paper [228].

It was until the 1960s that the foundations of multiobjective optimization were consolidated and taken seriously by pure mathematicians when Leonid Hurwicz [728] generalized the results of Kuhn & Tucker to topological vector spaces.

1.6.2 Early Applications

In the 1960s, however, multiobjective public investment problems became more common and "trade-off" became a favorite term used by managers, planners, and decision makers [289]. So, this area arose in a natural fashion in mathematical economics, and many techniques were developed by systems analysts and decision theorists for private and public sector problems, by control theorists for engineering (guidance and design) problems, and by water

---

[12] This monograph played a significant role in bringing the Nobel Prize to Koopmans in 1975 [1519].

resource economists and systems analysts for water resource planning problems. There was also some renewed interest in Kuhn and Tucker's vector maximum theory during the early 1960s, as it is reflected in papers by Zadeh [1746], Klinger [870] and Da Cunha & Polak [324].

The application of multiobjective optimization to domains outside economics began with the work by Koopmans [897] in production theory and with the work of Marglin [1065] in water resources planning. The first engineering application reported in the literature was a paper by Lofti Zadeh in the early 1960s [1746]. However, the use of multiobjective optimization became generalized until the 1970s [1513, 289, 288].

Good reviews of existing mathematical programming techniques for multiobjective optimization can be found in a wide variety of sources [1384, 159, 1036, 1015, 289, 1519, 489, 1704, 732, 1515, 733, 1220, 461, 992, 1111, 428].

1.7 Classifying Techniques

There have been several attempts to classify the many multiobjective optimization techniques currently in use. First of all, it is quite important to distinguish two stages in which the solution of a multiobjective optimization problem can be divided: the optimization of the several objective functions involved and the process of deciding what kind of "trade-offs" are appropriate from the decision maker perspective (the so-called multicriteria decision making process). In this section, some of the many techniques available for these two stages of a multiobjective optimization problem, are discussed, analyzing some of their advantages and disadvantages.

Cohon and Marks [289] proposed one of the most popular classifications of techniques within the Operations Research community:

1. Generating techniques (*a posteriori* articulation of preferences).
2. Techniques which rely on prior articulation of preferences (non-interactive methods).
3. Techniques which rely on progressive articulation of preferences (interaction with the decision maker).

Other classifications are obviously possible (see for example [416]). However, the classification proposed by Cohon & Marks [289] has been adopted for the purposes of this book, because it focuses the classification on the way in which each technique handles the two problems of searching and making (multicriterion) decisions [1631, 706]:

1. ***A priori* Preference Articulation**: make decisions *before* searching (decide $\Rightarrow$ search).
2. ***A posteriori* Preference Articulation**: search *before* making decisions (search $\Rightarrow$ decide).

3. **Progressive Preference Articulation**: integrate search and decision making (decide ⇔ search).

In the following subsections, some of the most representative Operations Research techniques are described, indicating how they fit within these three groups.

1.7.1 *A priori* Preference Articulation

Following Cohon & Marks' classification [289], this group of techniques includes those approaches that assume that either a certain desired achievable goals or a certain pre-ordering of the objectives can be performed by the decision maker prior to the search.

Global Criterion Method

In this method, the aim is to minimize a function which defines a global criterion which is a measure of how close the decision maker can get to the ideal vector $\mathbf{f}^0$. The most common form of this function is [1217]

$$f(\mathbf{x}) = \sum_{i=1}^{k} \left(\frac{f_i^0 - f_i(\mathbf{x})}{f_i^0} \right)^p \tag{1.22}$$

where k is the number of objectives.

For this formula Boychuk and Ovchinnikov [159] have suggested $p = 1$, and Salukvadze [1420] has suggested $p = 2$, but other values of p can also be used. Obviously, the results differ greatly depending on the value of p chosen. Thus, the selection of the best p is an issue in this method, and it could also be the case that any p could produce an unacceptable solution.

Another possible measure of 'closeness to the ideal solution' is a family of L_p-metrics defined as follows

$$L_p(f) = \left[\sum_{i=1}^{k} |f_i^0 - f_i(x)|^p \right]^{1/p}, \quad 1 \leq p \leq \infty \tag{1.23}$$

In general, relative deviations of the form

$$\frac{f_i^0 - f_i(x)}{f_i^0} \tag{1.24}$$

are preferred over absolute deviations, because they have a substantive meaning in any context. The relevant L_p metrics are

$$L_p(f) = \left[\sum_{i=1}^{k} \left| \frac{f_i^0 - f_i(\mathbf{x})}{f_i^0} \right|^p \right]^{1/p}, \quad 1 \leq p \leq \infty \tag{1.25}$$

The value of p indicates the type of distance: for $p = 1$, all deviations from f_i^* are taken into account in direct proportion to their magnitudes, which corresponds to 'group utility' [1742, 419]. For $2 \leq p < \infty$, the larger deviations carry greater weight in L_p; for $p = \infty$, the largest deviation is the only one taken into consideration, which leads to a purely 'individual utility' (min-max criterion), in which all weighted deviations are equal.

Koski [904] has suggested L_p-metrics with a normalized vector objective function of the form

$$f_i(\mathbf{x}) = \frac{f_i(\mathbf{x}) - \min_{x \in F} f_i(\mathbf{x})}{\max_{x \in F} f_i(\mathbf{x}) - \min_{x \in F} f_i(\mathbf{x})} \quad (1.26)$$

In this case, the values of every normalized function are limited to the range [0,1].

Using the global criterion method one non-inferior solution is obtained. If certain parameters w_i are used as weights for the criteria, a required set of non-inferior solutions can be found. Duckstein [416] calls this method **compromise programming**, and his L_p-metric is[13]

$$L_p(\mathbf{x}) = \left[\sum_{i=1}^{k} w_i^p \left| \frac{f_i(\mathbf{x}) - f_i^0}{f_{i\,\max} - f_i^0} \right|^p \right]^{1/p} \quad (1.27)$$

where w_i are the weights, $f_{i\,\max}$ is the worst value obtainable for criterion i; $f_i(\mathbf{x})$ is the result of implementing decision $\mathbf{x}$ with respect to the ith criterion.

The **displaced ideal** technique [1752] which proceeds to define an ideal point, a solution point, another ideal point, etc. is an extension of compromise programming.

Another variation of this technique is the method suggested by Wierzbicki [1703, 1705] in which the global function has a form such that it penalizes the deviations from the so-called reference objective. Any reasonable or desirable point in the space of objectives chosen by the decision maker can be considered as the reference objective.

Let $\mathbf{f}^r = [f_1^r, f_2^r, \ldots, f_k^r]^T$ be a vector which defines this point. Then the function which is minimized has the form

$$P(\mathbf{x}, \mathbf{f}^r) = -\sum_{i=1}^{k}(f_i(\mathbf{x} - f_i^r)^2 + \varrho \sum_{i=1}^{k}(\max(0, f_i(\mathbf{x} - f_i^r)^2) \quad (1.28)$$

where $\varrho > 0$ is a penalty coefficient which in this method can be chosen as constant.

Minimizing (1.28) for the assumed point $\mathbf{f}^r$ a non-inferior solution which is close to this point can be obtained. If for different points $\mathbf{f}^r$ the procedure is carried out, some representation of non-inferior solutions can be found.

---

[13] Metrics for MOEA evaluation are discussed in Chapter 4.

More information on this method can be found in [1217, 1751, 1753].

Goal Programming

Charnes and Cooper [229] and Ijiri [742] are credited with the development of the goal programming method for a linear model, and played a key role in applying it to industrial problems. As mentioned before, this was one of the earliest techniques specifically designed to deal with multiobjective optimization problems.

In this method, the DM has to assign targets or goals that wishes to achieve for each objective. These values are incorporated into the problem as additional constraints. The objective function then tries to minimize the absolute deviations from the targets to the objectives. The simplest form of this method may be formulated as follows [416]:

$$\min \sum_{i=1}^{k} |f_i(\mathbf{x}) - T_i|, \quad \text{subject to } \mathbf{x} \in \Omega \quad (1.29)$$

where T_i denotes the target or goal set by the decision maker for the ith objective function $f_i(\mathbf{x})$, and Ω represents the feasible region. The criterion, then, is to minimize the sum of the absolute values of the differences between target values and actually achieved values. A more general formulation of the goal programming objective function is a weighted sum of the pth power of the deviation $|f_i(\mathbf{x}) - T_i|$ [630]. Such a formulation has been called **generalized goal programming** [738, 739].

Looking again to equation (1.29), the objective function is nonlinear and the simplex method can be applied only after transforming this equation into a linear form, thus reducing goal programming to a special type of linear programming. In this transformation, new variables d_i^+ and d_i^- are defined such that [229]:

$$d_i^+ = \frac{1}{2}\{|f_i(\mathbf{x}) - T_i| + [f_i(\mathbf{x}) - T_i]\}, \quad (1.30)$$

$$d_i^- = \frac{1}{2}\{|f_i(\mathbf{x}) - T_i| - [f_i(\mathbf{x}) - T_i]\}, \quad (1.31)$$

Adding and subtracting these equations, the following equivalent linear formulation may be found:

$$\min \ Z_0 = \sum_{i=1}^{k}(d_i^+ + d_i^-), \quad (1.32)$$

subject to

$$\begin{aligned} \mathbf{x} \in \Omega \\ f_i(\mathbf{x}) - d_i^+ + d_i^- = T_i \\ d_i^+, d_i^- \geq 0, \quad i = 1, \ldots, k \end{aligned} \quad (1.33)$$

Since it is not possible to have both under- and overachievements of the goal simultaneously, then at least one of the deviational variables must be zero. In other words:

$$d_i^+ \cdot d_i^- = 0 \quad (1.34)$$

Fortunately, this constraint is automatically fulfilled by the simplex method because the objective function drives either d_i^+ or d_i^- or both variables simultaneously to zero for all i.

Sometimes it may be desirable to express preference for over- or underachievement of a goal. Thus, it may be more desirable to overachieve a targeted reliability figure than to underachieve it. To express preference for deviations, the DM can assign relative weights w_i^+ and w_i^- to positive and negative deviations, respectively, for each target T_i. If a minimization problem is considered, choosing the w_i^+ to be larger than w_i^- would be expressing preference for underachievement of a goal.

In addition, goal programming provides the flexibility to deal with cases that have conflicting multiple goals. Essentially, the goals may be ranked in order of importance to the problem solver. That is, a priority factor, p_i ($i = 1, \ldots, k$) is assigned to the deviational variables associated with the goals. This is called "lexicographic ordering" by some authors (see for example [1111]). These factors p_i are conceptually different from weights, as it is explained, for example, in [579]. The resulting optimization model becomes

$$\min S_0 = \sum_{i=1}^{k} p_i(w_i^+ d_i^+ + w_i^- d_i^-), \quad (1.35)$$

subject to

$$\begin{aligned} \mathbf{x} \in \Omega \\ f_i(\mathbf{x}) - d_i^+ + d_i^- = T_i \\ d_i^+, d_i^- \geq 0, \quad i = 1, \ldots, k \end{aligned} \quad (1.36)$$

Note that this technique yields a nondominated solution if the goal point is chosen in the feasible domain [416].

More information on this method can be found in [230, 976, 975, 738, 807].

Goal-Attainment Method

In this approach, a vector of weights $w_1, w_2, \ldots, w_k$ relating the relative under- or over-attainment of the desired goals must be elicited from the decision maker in addition to the goal vector $b_1, b_2, \ldots, b_k$ for the objective functions

$f_1, f_2, \ldots, f_k$. To find the best-compromise solution x^*, the following problem is solved [550, 551]:

$$\text{Minimize } \alpha \tag{1.37}$$

subject to:

$$g_j(\mathbf{x}) \leq 0; \quad j = 1, 2, \ldots, m$$
$$b_i + \alpha \cdot w_i \geq f_i(\mathbf{x}); \quad i = 1, 2, \ldots, k \tag{1.38}$$

where α is a scalar variable unrestricted in sign and the weights $w_1, w_2, \ldots, w_k$ are normalized so that

$$\sum_{i=1}^{k} |w_i| = 1 \tag{1.39}$$

If some $w_i = 0$ ($i = 1, 2, \ldots, k$), it means that the maximum limit of objectives $f_i(\mathbf{x})$ is b_i.

It can be easily shown [240] that the set of nondominated solutions can be generated by varying the weights, with $w_i \geq 0$ ($i = 1, 2, \ldots, k$) even for nonconvex problems. The mechanism by which this method operates is illustrated in Figure 1.14. The vector $\mathbf{b}$ is represented by the decision goal of the DM, who also decides the direction of $\mathbf{w}$. Given vectors $\mathbf{w}$ and $\mathbf{b}$, the direction of the vector $\mathbf{b} + \alpha \cdot \mathbf{w}$ can be determined, and the problem stated by equation (1.37) is equivalent to finding a feasible point on this vector in objective space which is closest to the origin. It is obvious that the optimal solution of equation (1.37) is the first point at which $\mathbf{b} + \alpha \cdot \mathbf{w}$ intersects the feasible region in the objective space (denoted by F in Figure 1.14). Should this point of intersection exist, it would clearly be a nondominated solution.

It should be pointed out that the optimum value of α informs the DM of whether the goals are attainable or not. A negative value of α implies that the goal of the decision maker is attainable and an improved solution is then to be obtained. Otherwise, if $\alpha > 0$, then the DM's goal is unattainable.

For more information on this method, refer to [240, 1321].

Lexicographic Method

This is a peculiar method in which the aggregations performed are not scalar. In this method, the objectives are ranked in order of importance by the decision maker (from best to worst). The optimum solution $\mathbf{x}^*$ is then obtained by minimizing the objective functions, starting with the most important one and proceeding according to the order of importance of the objectives.

Let the subscripts of the objectives indicate not only the objective function number, but also the priority of the objective. Thus, $f_1(\mathbf{x})$ and $f_k(\mathbf{x})$ denote

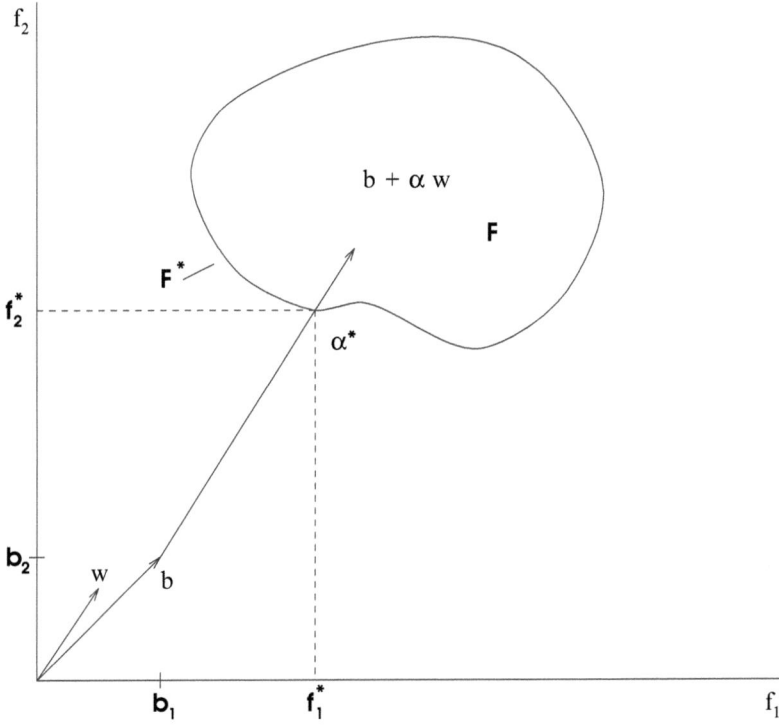

Fig. 1.14. Goal-attainment method with two objective functions.

the most and least important objective functions, respectively. Then the first problem is formulated as

$$\text{Minimize } f_1(\mathbf{x}) \tag{1.40}$$

subject to

$$g_j(\mathbf{x}) \leq 0; \quad j = 1, 2, \ldots, m \tag{1.41}$$

and its solution $\mathbf{x}_1^*$ and $f_1^* = f(\mathbf{x}_1^*)$ is obtained. Then the second problem is formulated as

$$\text{Minimize } f_2(\mathbf{x}) \tag{1.42}$$

subject to

$$g_j(\mathbf{x}) \leq 0; \quad j = 1, 2, \ldots, m \tag{1.43}$$
$$f_1(\mathbf{x}) = f_1^* \tag{1.44}$$

and the solution of this problem is obtained as $\mathbf{x}_2^*$ and $f_2^* = f_2(\mathbf{x}_2^*)$. This procedure is repeated until all k objectives have been considered. The ith problem is given by

$$\text{Minimize} \quad f_i(\mathbf{x}) \tag{1.45}$$

subject to

$$g_j(\mathbf{x}) \leq 0; \quad j = 1, 2, \ldots, m \tag{1.46}$$

$$f_l(\mathbf{x}) = f_l^*, \quad l = 1, 2, \ldots, i-1 \tag{1.47}$$

The solution obtained at the end, i.e., $\mathbf{x}_k^*$ is taken as the desired solution $\mathbf{x}^*$ of the problem.

More information on this method may be found in [1321, 1429].

Min-Max Optimization

The idea of stating the min-max optimum and applying it to multiobjective optimization problems was taken from game theory, which deals with solving conflicting situations. The min-max approach to a linear model was proposed by Jutler [811] and Solich [1503]. It has been further developed by Osyczka [1216], Rao [1320] and Tseng & Lu [1606].

The min-max optimum compares relative deviations from the separately attainable minima. Consider the ith objective function for which the relative deviation can be calculated from

$$z_i'(\mathbf{x}) = \frac{|f_i(\mathbf{x}) - f_i^0|}{|f_i^0|} \tag{1.48}$$

or from

$$z_i''(\mathbf{x}) = \frac{|f_i(\mathbf{x}) - f_i^0)|}{|f_i(\mathbf{x})|} \tag{1.49}$$

It should be clear that for equations (1.48) and (1.49) it is necessary to assume that for every $i \in I$ ($I = 1, 2, \ldots, k$) and for every $\mathbf{x} \in \Omega$, $f_i(\mathbf{x}) \neq 0$.

If all the objective functions are going to be minimized, then equation (1.48) defines function relative increments, whereas if all of them are going to be maximized, it defines relative decrements. Equation (1.49) works conversely.

Let $\mathbf{z}(\mathbf{x}) = [z_1(\mathbf{x}), \ldots, z_i(\mathbf{x}), \ldots, z_k(\mathbf{x})]^T$ be a vector of the relative increments which are defined in $\mathbb{R}^k$. The components of the vector $z(\mathbf{x})$ are evaluated from the formula

$$\forall_{i \in I} z_i(\mathbf{x}) = \max \{z_i'(\mathbf{x}), z_i''(\mathbf{x})\} \tag{1.50}$$

Now the min-max optimum can be defined as follows [1217]:

A point $\mathbf{x}^* \subset \Omega$ is min-max optimal, if for every $\mathbf{x} \in \Omega$ the following recurrence formula is satisfied:
Step 1:

$$v_1(\mathbf{x}^*) = \min_{x \in \Omega}\{z_i(\mathbf{x})\} \quad (1.51)$$

and then $I_i = \{i_1\}$, where i_1 is the index for which the value of $z_1(\mathbf{x})$ is maximal.

If there is a set of solutions $X_1 \subset \Omega$ which satisfies Step 1, then
Step 2:

$$v_2(\mathbf{x}^*) = \min_{x \in X_1}\left(\max_{i \in I, i \notin I_1}\{z_i(\mathbf{x})\}\right) \quad (1.52)$$

and then $I_2 = \{i_1, i_2\}$, where i_2 is the index for which the value of $z_i(x)$ in this step is maximal.

If there is a set of solutions $X_{r-1} \subset \Omega$ which satisfies step $r-1$ then
Step r:

$$v_r(\mathbf{x}^*) = \min_{x \in X_{r-1}}\left(\max_{i \in I, i \notin I_{r-1}}\{z_i(\mathbf{x})\}\right) \quad (1.53)$$

and then $I_r = \{I_{r-1}, i_r\}$, where i_r is the index for which the value of $z_i(\mathbf{x})$ in the rth step is maximal.

If there is a set of solutions $X_{k-1} \subset \Omega$ which satisfies Step $k-1$, then
Step k:

$$v_k(\mathbf{x}^*) = \min_{\mathbf{x} \in X_{k-1}} z_i(\mathbf{x}) \text{ for } i \in I \text{ and } i \notin I_{k-1} \quad (1.54)$$

where $v_1(\mathbf{x}^*), \ldots, v_k(\mathbf{x}^*)$ is the set of optimal values of fractional deviations ordered non-increasingly.

This optimum can be described in words as follows. Knowing the extremes of the objective functions which can be obtained by solving the optimization problems for each criterion separately, the desirable solution is the one which gives the smallest values of the relative increments of all the objective functions.

The point $\mathbf{x}^* \in \Omega$ which satisfies the equations of Steps 1 and 2 may be called the best compromise solution considering all the criteria simultaneously and on equal terms of importance. It should be noticed that even when these equations look quite complicated, in many optimization models, only the first step of this process is necessary to determine the optimum.

Multiattribute Utility Theory

Von Neumann and Morgenstern [1664] developed an axiomatic utility theory to measure individual or group preferences. Utility theory assumes that an individual can choose among the alternatives available in such a manner that the satisfaction derived from the choice made is as large as possible. This, of course, implies the individual is aware of the alternatives available and is capable of evaluating them. Moreover, relative to a vector of objectives it is assumed all information pertaining to the various levels of the objectives can be captured by an individual's **utility function**. In effect, an individual's utility function is a formal, mathematical representation of the individual's preference structure. Multiattribute utility functions, which may be assessed as first proposed in [835, 1315, 122, 836] integrate the objective functions into the preference structure. The highest degree of utility with respect to all the objectives is obtained by maximizing the utility function.

Oppenheimer [1213] distinguishes two approaches to utility maximization: the global and the local approaches.

The **global approach** [834] refers to the above expected utility maximization, and 'may force the decision maker to fit a function not truly representing' the preference function. Nevertheless, the global approach is taken in most multiattribute utility models.

In the **local approach** [556], the above-mentioned problem of locking the decision maker into a given risk attitude is avoided by using a sequence of local linear approximations to the utility function. To each step pertains a trial solution representing an improvement over its predecessor, so that eventually, the sequence reaches its optimum.

The main drawback of this approach is that the DM has to spend a lot of time building single-attribute utility functions. Then, the DM has to make sure that the 'corner utilities' are assessed; the latter makes it possible to combine the single attribute utilities $u_i(x_i)$ into one function $u(\mathbf{x})$.

To illustrate the assessment task, let four attributes, x_1, x_2, x_3, and x_4, be, respectively, the weight W, probabilities of failure $(1 - r) = p_f$, cost k, and deflection Δ of a structure. The first task is to assess the function $u_i(x_i)$, $i = 1, 2, 3, 4$; this can be best done by means of lotteries of the type:

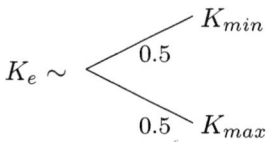

In words, given a lottery in which maximum cost k_{max} and minimum cost k_{min} may be obtained with equal probability 0.5 (for example), which value k_e would the DM accept as a 'certainty equivalent'? Furthermore, if the axioms of von Neumann and Morgenstern are satisfied, it can be proved [1665] that a utility function $u(\cdot)$ exists, leading to the equation:

$$u(k_e) = 0.5 u(k_{max}) + 0.5 u(k_{min}) \tag{1.55}$$

Utility functions are defined within a positive linear transformation and one usually sets $u(k_{max}) = 0$ and $u(k_{min}) = 1$ so that $u(k_e) = 0.5$. This procedure is continued in the intervals (k_{max}, k_e), (k_e, k_{min}), and overlapping intervals, until a satisfactory piecewise-linear approximation of the utility function is obtained.

If the attributes W, $1 - r$, k and Δ are mutually utility independent, the function $u(\mathbf{x})$ is given by [836]:

$$1 + ku(\mathbf{x}) = \prod_{i=1}^{k}[1 + kk_i u_i(x_i)] \tag{1.56}$$

Verification of the utility independence hypothesis, assessment of the k_i ($i = 1, 2, 3, 4$), and consistency check require a further series of lotteries. However, even when a lot of effort is required to construct $u(\mathbf{x})$, it may be worth it in large and costly systems [416].

For more information on this method, refer to [559, 909].

Surrogate Worth Trade-Off

This method, proposed in [631], is a variant of the **trade-off** method in which objective trade-offs are used as the information carrier and the DM responds by expressing a degree of preference over the prescribed trade-offs by assigning numerical values to each surrogate worth function. These functions are used to construct a single objective problem. First, the set of strictly nondominated or efficient solutions is generated, say by any multiobjective optimization technique (normally, the ϵ-constraint technique is used). Then a search along the efficient boundary is performed using a surrogate worth function. Note the difference between this method, which stays on the Pareto optimum boundary, and compromise programming or game theory, in which the Pareto optimum set is approached, respectively, from the infeasible and the feasible regions.

The trade-off function for any two objectives evaluated at a given efficient solution $\mathbf{x}$ is:

$$T_{ij}(\mathbf{x}) = \frac{\partial f_i(\mathbf{x})}{\partial f_j(\mathbf{x})} \tag{1.57}$$

As can be seen from equation (1.57), this method can only be applied when all the objective functions are differentiable.

The surrogate worth function, W_{ij}, $i \neq j$, $i, j = 1, 2, \ldots, n$, is defined as a function of the desirability of the trade-off λ_{ij} on a scale. For example, if a scale ranging from -10 to $+10$ is used, a (-10) would indicate that λ_{ij} marginal units of objective i are worth very much less than one marginal unit of objective j, a $(+10)$ means the opposite, and a zero indicates an even trade-off. The best solution is found when all surrogate worth functions are equal to zero. A complete description of this technique can be found in [630], and an abbreviated version, in [579].

The main advantage of this technique resides in its sound theoretical basis and on its several applications reported in the literature [629, 628, 330, 1209].

On the other hand, computational requirements are non-trivial, and much input is required from the DM.

In general, it can said that the trade-off methods have two main disadvantages [1217]:

- They cannot be used to solve non-convex problems, and
- they allow a satisfactory solution to be found only in a certain region of Pareto optimal solutions, but do not provide a general outlook on the possible range of objectives, and thus the final decision is influenced by the starting point chosen.

ELECTRE

This technique (in its different versions) is applicable to problems that have a discrete predefined set of alternatives in which some of the evaluation criteria are non-quantifiable, i.e., the criteria can only be ranked ordinally or, with additional information, on a ratio or interval scale.

ELECTRE I (elimination and (et) choice translating algorithm) was developed by Benayoun et al. [117]. This technique was improved by Roy [1384] and it has been applied, for example, to water-related problems [336, 419, 417]. The idea is to choose those systems which are preferred for at least a plurality of the criteria and yet do not cause an unacceptable level of discontent for any one criterion. This methodology leans on three concepts: concordance, discordance, and threshold values.

The **concordance** between any two systems i and j is a weighted measure of the number of criteria for which action i is preferred to action j (denoted $i \succ j$) or for which action i is equal to action j (denoted $i \sim j$) and is given as:

$$C(i,j) = \frac{\sum_{k \in A(i,j)} w(k)}{\sum_k w(k)}, \qquad (1.58)$$

where $w(k)$ is the weight of criterion k, $k = 1, \ldots, K$, and $A(i,j) = \{k | i \succeq j\}$. The weights, which are given by the DM, reflect a set of preferences. Concordance may be considered as the weighted percentage of criteria for which one action is preferred to another. Note that, by construction, $0 \leq C(i,j) \leq 1$.

Determination of the **discordance** between i and j requires that an interval scale common to each criterion be defined. The scale is used to compare the discomfort caused between the 'worst' and the 'best' criterion value for each pair of alternatives. A range may be chosen where the 'best' rating would be assigned the highest value of the range and the 'worst' rating would receive the lowest value of the range. Each criterion, however, can have a different range to reflect the 'leeway' available for that criterion [416]. The problem of applying a ratio scale to an ordinal criterion presents theoretical difficulties which are fully addressed in [1753, 1362]. Essentially, evaluations of the type (a, b, c, d) may be assigned in an analogous way in which grades are assigned to students. The **discordance index** is defined as:

$$D(i,j) = \frac{\max\limits_{k=1,K}(Z(j,k) - Z(i,k))}{R^*} \qquad (1.59)$$

where $Z(j,k)$ is the evaluation of alternative j with respect to criterion k, and R^* is the largest of the K criterion scales. Again, by construction, $0 \leq D(i,j) \leq 1$.

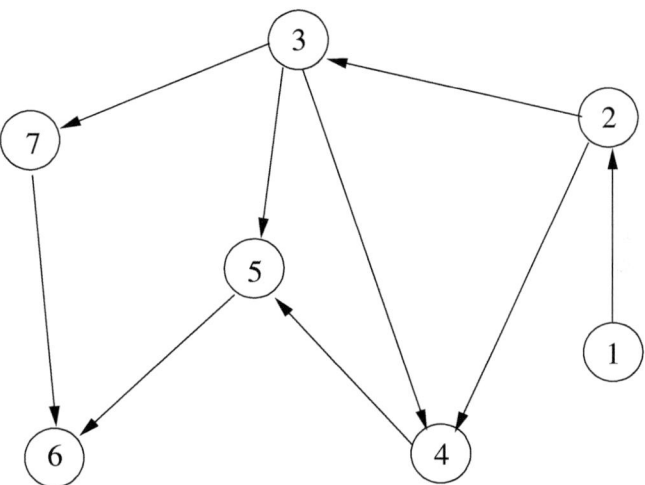

Fig. 1.15. Example of an ELECTRE graph. Each node corresponds to a non-dominated alternative. The arrows indicate preferences. Therefore it can said that alternative 1 is preferred to alternative 2, alternative 4 is preferred to alternative 5, etc.

To synthesize both, the concordance and discordance matrices, **threshold values** (p, q) between zero and one, are defined by the DM. Using a geometric representation, the preference relationships define a transitive and complete graph (G) for each criterion, in which nodes are alternatives and arcs are directed as the preference sign $\succ$. In the case of $i \sim j$, one arc is drawn from i to j and another from j to i. The arc set A of the composite graph (Γ) which synthesizes both concordance and discordance relationships, is given by:

$$a(i,j) \in A \Leftrightarrow (C(i,j) > p) \cap (D(i,j) < q) \qquad (1.60)$$

Figure 1.15 shows an example of the type of graph that ELECTRE I uses. In choosing the value of p, the problem solver specifies how much 'concordance' is wanted: $p = 1$ corresponds to full concordance, which means that i should be preferred or equivalent to j in terms of all criteria. By choosing q, the amount of tolerable 'discordance' is specified: $q = 0$ means no discordance. It is possible that some choices of p and q may eliminate all alternative systems. If this is the case, the values of p and/or q must be restated. It is also possible for cycles to occur in the composite graph (Γ) of ELECTRE I. In such cases the nodes along the cycle are collapsed into one new node, which is equivalent to assigning the same ranking to those systems.

The preference graph (Γ) of ELECTRE I thus yields a partial ordering of the alternative systems. On the other hand, **ELECTRE II** [1384, 417], may be used to obtain a complete ordering, as in [418]. Briefly, ELECTRE II is based on two preference graphs representing the strong preferences (high p and low q) and the weak preferences (lower p and higher q). The weak preferences can be viewed as lower bounds on system performance that the DM is willing to accept. **ELECTRE III** [1390] uses a credibility index, which is modelled by a fuzzy number. This index is used to associate a value to the outranking relation. In **ELECTRE IV** [1390], no weights are assigned to the criteria. In this version, four indices of credibility may be assigned to the values of the outranking relations. There are two other versions of ELECTRE: one called **ELECTRE TRI**, which is customized for a sorting decision problem (it uses conjunctive and disjunctive techniques to assign different alternatives to different categories), and **ELECTRE IS**, which really consists of ELECTRE I plus the use of discriminating thresholds [1390]. The fact that most of the information about some versions of ELECTRE (for example, versions III and IV) is available only in French has certainly limited its use [1386, 1488, 1390]

ELECTRE has been applied to a substantial number of practical problems with a predetermined finite set of alternatives evaluated in terms of ordinal (quantitative or qualitative) criteria, and could be most useful to solve multiobjective optimization problems that have those characteristics, since the technique is robust, simple, requires little input from the DM, and usually leads to plausible results.

The technique, however, has also been criticized. Brans & Vincke [168] criticize the ELECTRE methods precisely because of the parameters that they require. They argue that even though some of these parameters have a real economic meaning and can, therefore, be fixed clearly, some others (such as concordance discrepancies and discrimination thresholds) playing an essential role in the procedures only have a technical character and their influence on the results is not always well understood. Moreover, in some of the ELECTRE methods, the notion of "degree of credibility" is rather difficult for practitioners [168].

More information on this method can be found in [1142, 1143, 1383, 1384, 1389, 579, 1390, 1658, 487].

PROMETHEE

The PROMETHEE methods (Preference Ranking Organization METHod for Enrichment Evaluations) belong to the family of outranking methods (i.e., ELECTRE) introduced by B. Roy. These methods include two phases [169]:

- The construction of an outranking relation on the different criteria or objectives of the problem.
- The exploitation of this relation in order to give an answer to the multi-criteria optimization problem.

In the first phase, a valued outranking relation based on a generalization of the notion of criterion is considered: a preference index is defined and a valued outranking graph, representing the preferences of the DM (six types of functions are used to express these preferences), is obtained.

The exploitation of the outranking relation is realized by considering for each action a leaving and an entering flow in the valued outranking graph: a partial preorder (PROMETHEE I) or a complete preorder (PROMETHEE II) on the set of possible actions can be proposed to the DM in order to solve the decision problem.

In the PROMETHEE methods, Brans proposes an approach that is "very simple and easy to understand by the decision maker" according to him. This method is based on extensions of the notion of criterion. These extended criteria can be easily built by the DM because they represent the natural notion of intensity of preference, and the parameters to be fixed (maximum two) have a real economic meaning.

More information on this method may be found in [168, 337, 420, 1063, 169, 167].

More recently, Huylenbroeck [730] proposed the so-called **Conflict Analysis Model**, that combines the preference function approach of ELECTRE and PROMETHEE with the conflict analysis test of a method called ORESTE (this technique provides an outranking relation by using as its input ordinal evaluations of the alternatives and the ranking of the criteria as a function of their relative importance) [1381, 1261]. Also, Martel et al. [1076] proposed a technique called PAMSSEM (Procédure d'Agrégation Multicritére de type Surclassement de Synthèse pour Évaluations Mixtes), which is based on ELECTRE and PROMETHEE, and can handle non deterministic and fuzzy criteria evaluations [937]. Finally, there is another technique called NAIADE (Novel Approach to Imprecise Assessment and Decision Environments) [1148], which uses a distance operator to obtain pairwise comparisons of the alternatives available to the DM. These comparisons are modelled as fuzzy numbers. The aggregation procedure used by this technique produces incoming and outgoing flows as in PROMETHEE.

1.7.2 *A Posteriori* Preference Articulation

These techniques do not require prior preference information from the DM. Some of the techniques included in this category are among the oldest multiobjective optimization approaches proposed. The reason is that the main idea of these approaches follows directly from the Kuhn-Tucker conditions for noninferior solutions [289].

Linear Combination of Weights

Zadeh [1746] was the first to show that the third of the Kuhn-Tucker conditions for noninferior solutions implies that these noninferior solutions might be found by solving a scalar optimization problem in which the objective function is a weighted sum of the components of the original vector-valued function. That is, the solution to the following problem is, in general, noninferior:

$$\min \sum_{i=1}^{k} w_i f_i(\mathbf{x}) \tag{1.61}$$

subject to:

$$\mathbf{x} \in \Omega \tag{1.62}$$

where $w_i \geq 0$ for all i and is strictly positive for at least one objective. The noninferior set and the set of noninferior solutions can be generated by parametrically varying the weights w_i in the objective function. This was initially demonstrated by Gass and Saaty [546] for a two-objective problem.

Note that the weighting coefficients do not reflect proportionally the relative importance of the objectives, but are only factors which, when varied, locate points in the Pareto optimal set. For the numerical methods that can be used to seek the minimum of equation (1.61), this location depends not only on w_i values, but also on the units in which the functions are expressed.

The ϵ-Constraint Method

This method also follows directly from the Kuhn-Tucker conditions for noninferior solutions. Equation (1.15) can be rewritten as:

$$w_r \nabla f_r(\mathbf{x}) + \sum_{l=1,\ l \neq r}^{k} w_l \nabla f_l(\mathbf{x}) - \sum_{i=1}^{m} \lambda_i \nabla g_i(\mathbf{x}) = 0 \tag{1.63}$$

Since only relative values of the weights are of significance, the rth objective can be selected so that $w_r = 1$. The previous condition defined in equation (1.63) then becomes:

$$\nabla f_r(\mathbf{x}) \sum_{l=1,\ l\neq r}^{k} w_l \nabla f(\mathbf{x}) - \sum_{i=1}^{m} \lambda_i \nabla g_i(\mathbf{x}) = 0 \qquad (1.64)$$

This rewritten condition allows the second term to be interpreted as a weighted sum of the gradients of $k-1$ lower-bound constraints, since there is a plus sign before the summation. This interpretation implies that noninferior solutions can be found by solving:

$$\min f_r(\mathbf{x}) \qquad (1.65)$$

subject to:

$$f_l(\mathbf{x}) \leq \epsilon_l \text{ for } l = 1, 2, \ldots, k \text{ and } l \neq r \qquad (1.66)$$

where ϵ_l are assumed values of the objective functions that must not be exceeded.

The idea of this method is to minimize one (the most preferred or primary) objective function at a time, considering the other objectives as constraints bound by some allowable levels ϵ_l. By varying these levels ϵ_l, the noninferior solutions of the problem can be obtained.

It is important to be aware of the fact that a preliminary analysis is required to identify proper starting values for ϵ_l. To get adequate ϵ_l values, single-objective optimizations are normally carried out for each objective function in turn by using mathematical programming techniques.

This method, also known as **trade-off method**, because of its main concept of trading a value of one objective function for a value of another function, is further explained in [1217, 1019, 211, 733].

1.7.3 Progressive Preference Articulation

These techniques normally operate in three stages [289]: (1) find a nondominated solution, (2) get the reaction of the DM regarding this nondominated solution, and modify the preferences of the objectives accordingly, and (3) repeat the two previous steps until the DM is satisfied or no further improvement is possible.

Probabilistic Trade-Off Development Method

The main motivation of this method (also known as PROTRADE) was to be able to handle risk in the development of the objective trade-offs, and at the same time being able to accommodate the preferences of the DM in a progressive manner [577].

In this case, it is assumed that our multiobjective optimization problem has a probabilistic objective function and probabilistic constraints [576]. According to a 12-step algorithm, an initial solution is found using a surrogate objective function, then a multiattribute utility function is formed leading to a new surrogate objective function and a new solution. The solution is checked to see if it is satisfactory to the decision maker. The process is repeated until a satisfactory solution is reached, as described in [578, 579].

The results of the multiobjective optimization provide not only levels of attainment of the objective function elements (as in the goal attainment method [550]), but also the probabilities of reaching those levels. The technique is interactive, which means that the DM formulates a preference function in a progressive manner, after a trial process [416].

One interesting aspect of this approach is that the DM actually ranks objectives in order of importance (a multi-attribute utility function is used to assist the DM in the articulation of preferences) at the beginning of the process, and later uses pairwise comparisons to reconcile these preferences with the "real" (observed) behavior of the attributes. This allows not only an interactive participation of the DM, but it also allows to gain knowledge about the trade-offs of the problem.

More information on this method may be found in [577, 579].

STEP Method

This method (also known as STEM) is an iterative technique based on the progressive articulation of preferences. The basic idea is to converge toward the 'best' solution in the min-max sense, in no more than k steps, being k the number of objectives. This technique, which is mostly useful for linear problems, starts from an ideal point and proceeds in six steps, as summarized by Cohon [288]:

1. Construct a table of marginal solutions (strictly nondominated if unique), by optimizing each objective function separately.
2. Compute, for each objective:

$$\alpha(i) = \frac{M(i) - m(i)}{M(i)} \left[\sum_{j=1}^{J} c(i,j) \right]^2, \quad (1.67)$$

where
$M(i) = \max f_i(\mathbf{x})$, $m(i) = \min f_i(\mathbf{x})$, and $c(i,j) = $ cost coefficient of ith linear objective.
Let the iteration index $k = 0$
3. Compute $\prod(i) = \alpha(i)/\sum \alpha(i)$ and solve the min-max problem. Call the solution $x(k)$.
4. Show the solution to the DM:
   a) if satisfied, STOP;
   b) if not satisfied and $k < p - 1$, go to Step 5;
   c) if not satisfied and $k > p - 1$, STOP. A different procedure or at least a redefinition of the problem is required.

5. The DM selects an objective satisfied by the solution and determines the amount by which it can be decreased in order to improve the other objectives. If this cannot be done, some other approach is again required.
6. Define a new constraint relaxing the objective selected in Step 5. Set $\alpha(i) = 0$ for that objective, increment k by one, and go to Step 3.

One criticism to this technique is the fact that it assumes that a best-compromise solution does not exist if it is not found after the k steps that the iterative process above described was executed. This does not give any clue to the DM of what to do [289]. Another problem is that it does not explicitly capture the trade-offs between the objectives. The weights in no way reflect a value judgment on the part of the DM. They are artificial quantities, generated by the analyst to reflect deviations from an ideal solution, which is itself an artificial quantity. This definition of the weights serves to obscure rather than capture the normative nature of the multiobjective optimization problems [289].

More details of this technique may be found in [289, 1547, 116].

Sequential Multiobjective Problem Solving Method

This method (also known as SEMOPS) was proposed by Monarchi et al. [1121] and it basically involves the DM in an interactive fashion in the search for a satisfactory course of action.

A surrogate objective function is used based on the goal and aspiration levels of the DM. The goal levels are conditions imposed on the DM by external forces, and the aspiration levels are attainment levels of the objectives which the DM personally desires to achieve. One would say, then, that goals do not change once they are stated, but that the aspiration levels may change during the iteration process. The development of the algorithm is summarized as follows [579]:

The decision problem consists of k goals, n decision variables, and a feasible region Ω. Associated with each of the goals is an objective function which can be used to predict goal attainment or nonattainment. The set of all p objective functions is written as $\mathbf{z} = (z_1, z_2, \ldots, z_k)$, and it is used to judge how well the k goals have been achieved. The range of the ith element of $\mathbf{z}$ is denoted by $\Lambda z_i = [z_{iL}, z_{iu}]$, which is not necessarily defined by the maximum and minimum values of the ith objective function. It is required that Ω be continuous and that all objective and constraint functions be at least first-order differentiable. Thus the constraint or objective functions may be nonlinear. Nondimensionality is achieved by transforming $z_i(\mathbf{x})$ into $y_i(\mathbf{x})$ with a range of values in the interval $[0, 1]$ such that

$$y_i(\mathbf{x}) = \frac{z_i(\mathbf{x}) - z_{iL}}{z_{iu} - z_{iL}} \tag{1.68}$$

Similarly, let $AL = (AL_1, AL_2, \ldots, AL_k)$ denote the vector of aspiration levels. Then, the transformation

1 Basic Concepts

$$A_i = \frac{AL_i - z_{iL}}{z_{iu} - z_{iL}} \qquad (1.69)$$

can be used to define A_i with values in the range $[0, 1]$.

Monarchi [1120] suggests the use of the following transformations:

1. **At most**

$$z_i(\mathbf{x}) \leq AL_i; \ d_i = \frac{z_i(\mathbf{x})}{AL_i} = \frac{y_i(\mathbf{x})}{A_i} \qquad (1.70)$$

2. **At least**

$$z_i(\mathbf{x}) \geq AL_i; \ d_i = \frac{AL_i}{z_i(\mathbf{x})} = \frac{A_i}{y_i(\mathbf{x})} \qquad (1.71)$$

3. **Equals**

$$z_i(\mathbf{x}) = AL_i; \ d_i = \frac{1}{2}\left[\frac{AL_i}{z_i(\mathbf{x})} + \frac{z_i(\mathbf{x})}{AL_i}\right] = \frac{1}{2}\left[\frac{A_i}{y_i(\mathbf{x})} + \frac{y_i(\mathbf{x})}{A_i}\right] \qquad (1.72)$$

4. **Within an interval**

$$AL_{iL} \leq z_i(\mathbf{x}) \leq AL_{iU}; \ d_i = \left[\frac{AL_{iU}}{AL_{iL} + AL_{iU}}\right]\left[\frac{AL_{iL}}{z_i(\mathbf{x})} + \frac{z_i(\mathbf{x})}{AL_{iU}}\right] \qquad (1.73)$$

In each instance, values of $d_i \leq 1$ imply that the ith objective is satisfied. It is also noted that, except for the first type, the d_i are nonlinear functions of the ith objective.

Operationally, SEMOPS is a three-step procedure involving setup, iteration, and termination. Setup involves structuring a principal problem and a set of auxiliary problems with a surrogate objective function. The iteration step involves cycling between an optimization phase (by the analyst), and an evaluation phase (by the DM) until a satisfactory solution is reached, if it exists. The procedure terminates when either a satisfactory solution is found, or the DM concludes that none of the nondominated solutions obtained are satisfactory and gives up in the search.

For the first iteration, then, the principal problem and set of k auxiliary problems shown below are solved:
Principal problem

$$\min s_1 = \sum_{i=1}^{k} d_i \qquad (1.74)$$

subject to

$$\mathbf{x} \in \Omega \qquad (1.75)$$

and the set of auxiliary problems, $l = 1, 2, \ldots, k$.

$$\min s_{1l} = \sum_{i=1, i \neq l}^{k} d_i \qquad (1.76)$$

subject to

$$\mathbf{x} \in \Omega \qquad (1.77)$$

$$z_l(\mathbf{x}) \geq AL_l \qquad (1.78)$$

The resulting solutions are used in the evaluation process to assist the DM to determine the "direction of change" for the next iteration.

More information on this method can be found in [1121, 579].

Other methods that also rely on the progressive articulation of preferences have been proposed in [864, 1433, 1043, 114].

1.8 Using Evolutionary Algorithms

The potential of evolutionary algorithms for solving multiobjective optimization problems was hinted as early as the late 1960s by Rosenberg in his PhD thesis [1375]. Rosenberg's study contained a suggestion that would have led to multiobjective optimization if he had carried it out as presented. His suggestion was to use multiple *properties* (nearness to some specified chemical composition) in his simulation of the genetics and chemistry of a population of single-celled organisms. Since his actual implementation contained only one single property, the multiobjective approach could not be shown in his work.

The first actual implementation of what it is now called a multi-objective evolutionary algorithm (or MOEA, for short) is credited to David Schaffer, who proposed the *Vector Evaluation Genetic Algorithm* (VEGA), in 1984 (in his PhD thesis [1439]). VEGA was mainly aimed for solving problems in machine learning [1439, 1440, 1441]. There is, however, a (rarely mentioned) earlier attempt to use a genetic algorithm to solve a multi-objective optimization problem which dates back from 1983 (see [764]).

Schaffer's work was presented at the *First International Conference on Genetic Algorithms* [1440]. Interestingly, his simple unconstrained two-objective functions became the usual test suite to validate most of the evolutionary multiobjective optimization techniques developed during several of the following years [1509, 709].

Evolutionary algorithms seem particularly suitable to solve multiobjective optimization problems, because they deal simultaneously with a set of possible solutions (the so-called population). This allows to find several members of the Pareto optimal set in a single "run" of the algorithm, instead of having to perform a series of separate runs as in the case of the traditional mathematical programming techniques [261]. Additionally, evolutionary algorithms are less susceptible to the shape or continuity of the Pareto front (e.g., they can easily deal with discontinuous or concave Pareto fronts), whereas these two issues are a real concern for mathematical programming techniques. Also, while many optimization approaches from those described in Section 1.4 were developed for searching intractably large spaces, traditional MOP solution techniques generally assume small, enumerable search spaces [706]. More simply, some MOP solution approaches focus on search and others on multi-criteria decision making (MCDM). MOEAs are then very attractive MOP solution techniques because they address *both* search and multiobjective decision making. Additionally, they have the ability to search partially ordered spaces for several alternative trade-offs. Many researchers have successfully used MOEAs to find good solutions for complex MOPs (see Chapter 7).

A MOEA's defining characteristic is the set of multiple objectives being simultaneously optimized. Otherwise, a task decomposition clearly shows little structural difference between the MOEA and its single-objective EA counterparts. The following definition and figures explain this relationship.

Definition 20 (Multiobjective Evolutionary Algorithm) : *Let $\Phi : I \longrightarrow \mathbb{R}^k$, ($k \geq 2$, a multiobjective fitness function). If this multiobjective fitness function is substituted for the fitness function in Definition 1.13 then the algorithm shown in Figure 1.13 is called a* Multiobjective Evolutionary Algorithm. □

Figures 1.16 and 1.17 respectively show a general EA's and MOEA's task decomposition. The major differences are noted as follows. By definition, Task 2 in the MOEA case computes k (where $k \geq 2$) fitness functions. In addition, because MOEAs expect a *single* fitness value with which to perform selection, additional processing is sometimes required to transform MOEA solutions' fitness *vectors* into a scalar (Task 2a). Although the various transformation techniques vary in their algorithmic impact (see Section 6.3.8 from Chapter 6) the remainder of the MOEA is structurally identical to its single-objective counterpart. However, this does not imply the differences are insignificant.

General EA Tasks

1. Initialize Population
2. Fitness Evaluation
3. Recombination
4. Mutation
5. Selection

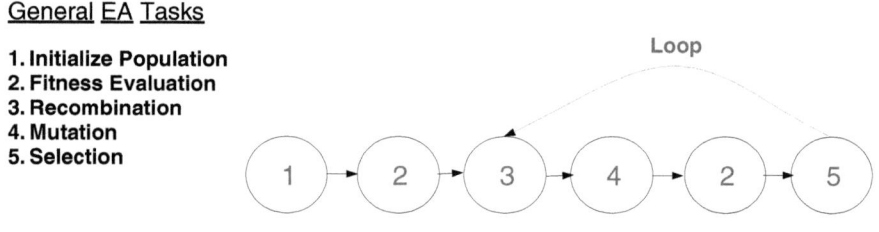

Fig. 1.16. Generalized EA Task Decomposition

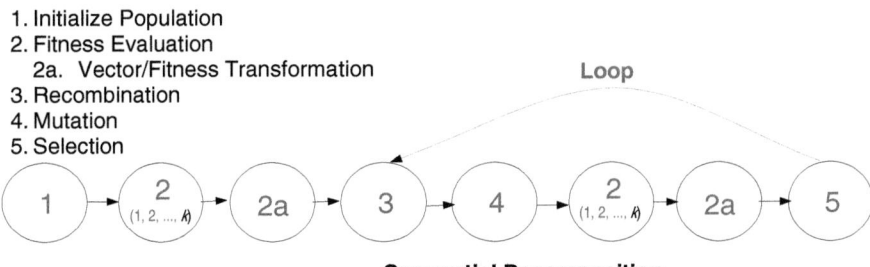

Fig. 1.17. MOEA Task Decomposition

1.8.1 Pareto Notation

A MOEA's algorithmic structure can easily lead to confusion (e.g., multiple, *unique* populations) when identifying or using Pareto concepts. In fact, MOEA researchers have erroneously used Pareto terminology in the literature suggesting a more precise notation is required. During MOEA execution, a "current" set of Pareto optimal solutions (with respect to the *current* MOEA generational population) is determined at each EA generation and termed $P_{current}(t)$, where t represents the generation number. Many MOEA implementations also use a secondary population storing nondominated solutions found through the generations [1628, 1626] (see also Section 2.2.4 from Chapter 2). Because a solution's classification as Pareto optimal depends upon the context within which it is evaluated (i.e., the given set of which it is a member), corresponding vectors of this set must be (periodically) tested and solutions whose associated vectors are dominated removed.

This secondary population is named $P_{known}(t)$. This term is also annotated with t to reflect its possible changes in membership during MOEA execution. $P_{known}(0)$ is defined as the empty set ($\emptyset$) and P_{known} alone as the *final* set of solutions returned by the MOEA at termination. Different secondary population storage strategies exist; the simplest is when $P_{current}(t)$ is added at each generation (i.e., $P_{current}(t) \cup P_{known}(t-1)$). At any given time, $P_{known}(t)$ is thus the set of Pareto optimal solutions *yet found by the MOEA through generation t*. Of course, the *true* Pareto optimal set (termed P_{true}) is not explicitly known for problems of any difficulty. P_{true} is implicitly defined by the functions composing an MOP; it is fixed and does not change. Because of the manner in which Pareto optimality is defined $P_{current}(t)$ is always a non-empty solution set (see Theorem 1 in Chapter 6).

$P_{current}(t)$, P_{known}, and P_{true} are sets of MOEA genotypes;[14] each set's corresponding phenotypes form a Pareto front.[15] The associated Pareto front for each of these solution sets is called $PF_{current}(t)$, PF_{known}, and PF_{true}. Thus, when using a MOEA to solve MOPs, the implicit assumption is that one of the following holds: $P_{known} = P_{true}$, $P_{known} \subset P_{true}$, or $\{\mathbf{u}_i \in PF_{known}, \mathbf{u}_j \in PF_{true} \mid \forall i, \forall j \; \min[distance(\mathbf{u}_i, \mathbf{u}_j)] < \epsilon\}$, where $distance$ is defined over some norm (Euclidean, RMS, etc.).

1.8.2 MOEA Classification

Many successful MOEA approaches are predicated upon previously implemented mathematical MOP solution techniques. As seen in Section 1.6, the OR field proposed several methods well before 1985 [289, 732, 1522]. Their Multiple Objective Decision Making (MODM) problems are closely related to design MOPs. These problems' common characteristics are a set of quantifiable objectives, a set of well-defined constraints, and a process of obtaining trade-off information between the stated objectives (and possibly also between stated or non-stated non-quantifiable objectives) [732].

Various MODM techniques are commonly classified from a DM's point of view (i.e., how the DM performs search and decision making). Cohon & Marks [289] further distinguish methods between two types of DM: a single DM/group or multiple DMs with conflicting decisions. Here, it has been considered that the DM is either a single DM or a group, but a group united in its decisions.

Because the set of solutions a DM is faced with are often "compromises" between the multiple objectives some specific compromise choice(s) must be made from the available alternatives. Thus, the final MOP solution(s) results from both *optimization* (by some method) and *decision* processes. MOEA-based MOP solution techniques are classified here as many OR researchers do, defining three variants of the decision process [289, 732] where the final solution(s) results from a DM's preferences being made known either before, during, or after the optimization process. Thus, the same classification of techniques described in Section 1.7 from this chapter has been adopted for MOEA-based MOP solution techniques.

Basic techniques below this top level of the MODM hierarchy may be common to several algorithmic research fields. However, the discussion is limited to implemented MOEA techniques. A hierarchy of the known MOEA techniques is shown in Figure 1.18 where each is classified by the different ways in which the fitness function and/or selection is treated.

---

[14] Horn [706] uses P_{online}, $P_{offline}$, and P_{actual} instead of $P_{current}(t)$, P_{known}, and P_{true}. The notation presented here is more precise, allowing for each set's generational specification.

[15] Note that when describing MOEAs, genotype refers to decision variable space, whereas phenotype refers to objective function space.

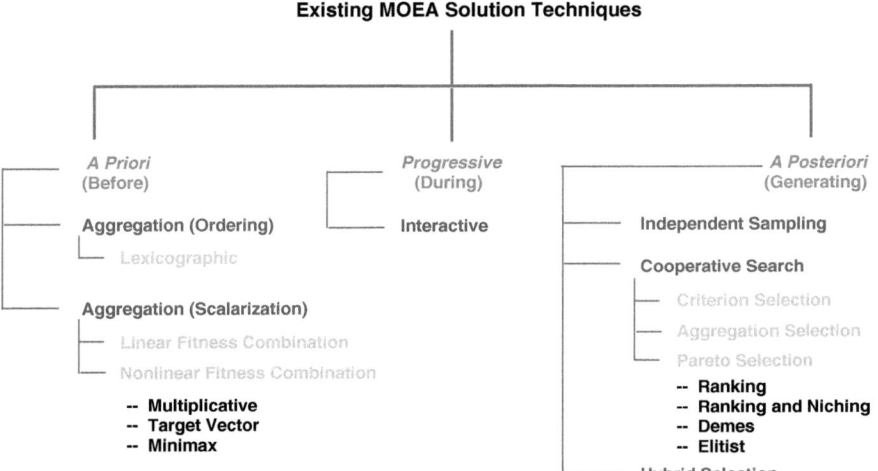

Fig. 1.18. MOEA Solution Technique Classification

1.9 Summary

This chapter contains the basic definitions and formal notation that are adopted throughout this book. Formal definitions of the general multiobjective optimization problem and the concept of Pareto optimum are provided. Other related concepts such as Pareto optimal set, ideal vector, Pareto front, weak and strict Pareto optimality are also introduced. After that, some introductory material on evolutionary computation is discussed, as well as a short historical review of the origins of multiobjective optimization and a taxonomy of approaches suggested by operations researchers to tackle these types of problems. Several representative approaches from this taxonomy are also described and criticized. This chapter ends with a short discussion on the main motivation to use evolutionary algorithms to solve MOPs together with a description of the Pareto notation and MOEA classification adopted throughout this book.

Further Explorations

Class Exercises

1. Enumerate the main advantages and limitations of mathematical programming techniques for multiobjective optimization.
2. Most mathematical programming techniques operate only with a single solution at a time. Do you think that there would be any advantages if a set of solutions was manipulated at a time instead of only one? Would that require any changes in the mathematical programming algorithms that you are familiar with? Discuss.
3. Sketch an algorithm to generate nondominated solutions and execute by hand. Discuss possible ways of improving the computational efficiency of this algorithm (see for example [120, 361]).
4. Discuss the main components of an evolutionary algorithm. Indicate its potential advantages and disadvantages as an optimizer.
5. Indicate possible situations in which an *a priori* preference articulation scheme would be preferred over an *a posteriori* scheme and vice versa.
6. Do you think that interactive approaches (i.e., those using a *progressive* preference articulation scheme) are a better choice (in the general case) than either *a priori* or *a posteriori* approaches? Discuss.

Class Software Projects

1. Implement goal programming using any single-objective optimization technique you wish (see for example [1324]). Test your implementation with the example presented in Section 1.3. How efficient is this technique at finding the true Pareto front of this problem? How much parameter-setting is involved in the process? How many times do you need to run the program to produce a reasonably good Pareto front?
2. Implement a simple genetic algorithm (see [581] for implementation details), and apply it to a single-objective optimization problem. Justify your

choice of encoding (of decision variables), and genetic operators (i.e., type of crossover and mutation). Then, consider a problem with two objective functions, such as the one discussed in Section 1.3. Use a simple linear combination of weights (see Section 1.7.1) to solve this problem with your genetic algorithm. What problems do you see with this approach when attempting to generate the Pareto front of a problem? Suggest a way of keeping "diversity" in the population (i.e., avoid that the entire population converges to a single point). Do you consider this approach appropriate to deal with any number of objectives? One of the main reasons why the use of a simple linear combination of weights is not recommended, is because it can be proved that this approach cannot generate concave portions of the Pareto front regardless of the weights used (see for example [329]). Do you consider this as a major drawback in the sort of multiobjective optimization problems that you wish to solve?
3. Use the same genetic algorithm implemented in the previous problem, and now couple it with compromise programming (see Section 1.7.2). This approach uses a nonlinear combination of weights. Do you see any advantage in doing this with respect to the use of a linear combination of weights? Does compromise programming have the same problems than a simple linear combination of weights? What extra information does the approach require? Is it difficult to obtain it?
4. Repeat the two previous questions, but using a multi-membered evolution strategy. Justify your choice of recombination operator and selection scheme (plus or comma). See [1460] for implementation details.
5. Choose a set of five mathematical programming techniques used for multiobjective optimization (see for example [1111]), and implement them. Then test them using two of the (unconstrained) test functions presented in Chapter 4. Plot the Pareto fronts obtained and compare (graphically) your results with respect to the true Pareto fronts of each test function (obtained by enumeration). What advantages and disadvantages (if any) do you see in these methods? Do they present any limitations? Discuss.

Discussion Questions

1. An obvious problem with multiobjective optimization techniques is that they could generate the same element of the Pareto optimal set several times. Investigate possible ways of dealing with this problem. See for example:
Michael A. Rosenman and John S. Gero, "Reducing the Pareto optimal set in multicriteria optimization", *Engineering Optimization*, Vol. 8, pp. 189–206, 1985.
2. What are the main differences between the multiobjective optimization techniques used for combinatorial optimization problems and those used

for numerical optimization? What is the role of local search in the first class of problems? See for example [429, 430, 431, 532, 879, 1615].

3. Read:
W. Stadler, "Initiators of Multicriteria Optimization", In J. Jahn and W. Krabs, editors, *Recent Advances and Historical Development of Vector Optimization*, pages 3–47. Springer-Verlag, Berlin, 1986.
What areas of research are identified by Stadler in this paper? Describe briefly each of them. Do you think that these research areas are still reasonably active nowadays?

4. Read:
Dylan F. Jones and Mehrdad Tamiz, "Goal Programming in the Period 1990-2000", in Matthias Ehrgott and Xavier Gandibleux (editors), *Multiple Criteria Optimization. State of the Art Annotated Bibliographic Surveys*, pp. 129–170, Kluwer Academic Publishers, 2002.

Explain with your own words the following variants of goal programming:
   a) Weighted Goal Programming
   b) Lexicographic Goal Programming
   c) Tchebycheff Goal Programming

5. Several authors (see for example [416, 841]) have proposed the following criteria to classify multiobjective optimization techniques:
   - marginal vs. non-marginal difference between alternatives
   - quantitative vs. ordinal qualitative criteria
   - prior vs. progressive articulation of preferences
   - interactive vs. non-interactive

   Investigate each of these criteria and classify the approaches discussed in this chapter based on them.

6. Investigate what is the "nadir objective vector" and indicate how can be estimated. Can you find in the specialized literature a multiobjective optimization technique that uses this concept? Can you think of some possible applications for the nadir objective vector?

7. Discuss the GUESS method described by Buchanan [182]. Provide its algorithm and indicate its advantages and disadvantages. Discuss some possible applications of this technique.

8. Look at your local library for some papers on mathematical programming techniques used for multiobjective optimization. Analyze the sort of test functions normally used to validate results and discuss the methodology adopted by operational researchers. Discuss.

9. Investigate what is the chi-square distribution. If it is assumed that there are several elements of the Pareto optimal set of a problem available (obtained, perhaps, by enumeration), how could you use a chi-square distribution to measure the effectiveness of a multiobjective optimization technique?

10. Fliege & Svaiter [492] proposed steepest descent methods for unconstrained multiobjective optimization and a "feasible descent direction" method for constrained problems. Discuss the requirements of the approach and its possible implementation difficulties. What are the main advantages provided by this sort of approach? Discuss.
11. Read the survey on nonlinear multiobjective programming by Tanino & Kuk [1573] and discuss the following issues:
    - Optimality conditions
    - Duality (both Lagrange and Conjugate)
    - Stability and sensitivity analysis
12. Read about interactive nonlinear multiobjective optimization procedures (see for example [1110, 1111]). Discuss at least two approaches not covered in this chapter (for example, light beam search [785, 786] and the reference direction approach [902]).
13. An active application domain in which a considerable amount of work has been done in the last few years is multicriteria scheduling (see for example [1588]). Choose a particular type of multicriteria scheduling problem and discuss its definition, complexity and modeling.

2
MOP Evolutionary Algorithm Approaches

An algorithm must be seen to be believed.

Donald Knuth

2.1 Introduction

Both researchers and practitioners in science, engineering, government, and industry certainly have a strong interest in knowing state-of-the-art multi-objective optimization techniques. For researchers, this is the normal procedure to trigger new and original algorithmic contributions. For practitioners, this knowledge allows them to choose the most appropriate algorithm(s) for their specific multi-objective problem (MOP) domain application. From the decision maker's (DM) perspective, it is desired that only a "few" solutions are available for ease of decision. Thus, as presented in Chapter 1, one is attempting to optimize a vector objective function possibly with constraints resulting in trade-offs between the multiple objectives. This chapter employs the various generic mathematical definitions defined in Chapter 1 for discussing multi-objective evolutionary algorithm (MOEA) design.[1] It is desired that an MOEA generates MOP solutions in P_{true} which provide a trade-off of performance (efficiency, effectiveness) for specific system model objectives (cost/profit, constraints, etc.) that may mutually conflict. For example, the classical multiobjective knapsack problem (profit and weight) and drug development (cost vs. effectiveness) represent vectors of two objectives. Maximizing one objective such as profit usually does not optimize another such as reliability. Many contemporary real-world MOP applications for the practitioner's

---
[1] Note that some MOEA researchers and practitioners use the phrase "Multi-Objective Optimization Problem" (*MOOP*) and "Multi-Objective Optimization" (*MOO*) to associate with the field, instead of MOP and MOEA.

and researcher's critical analysis are discussed in Chapter 7 and [277] with many examples reflected in the current MOEA literature.[2]

Since Evolutionary Algorithms and MOEAs in particular can encode individual solutions in numerous straightforward representations (chromosome data structures) as well as directly compute associated objective values, they have a considerable robust advantage over traditional MOP search techniques (see Chapter 1). That is, traditional techniques may impose restrictions or complex mappings on the problem domain or algorithm domain mathematical model in order to solve the problem. Of course, the No Free Lunch Theorem (NFL) [1708] implies that a MOEA is not a universal robust solution technique for all MOPs. But, MOEAs generally can easily be guided by problem domain information, not having to modify the problem domain model for use with MOEAs. Then, the search process is easier to develop, understand and test in its native form for a given application [1102].

Achieving the exact Pareto front of an arbitrary problem is usually quite difficult. Nevertheless, reasonably good approximations of PF_{true} are generally acceptable within limited computational time (see Chapter 1 for associated notation). MOEAs by definition attempt to find these acceptable but approximate Pareto fronts and Pareto optimal solutions within some implicit or explicit error measure (see Chapter 5).

This chapter addresses the many issues involved in MOP domain and MOEA domain integration from a design perspective. In particular, historic and generally used (MOEA) approaches such as the NSGA [1509, 374], PAES [886], SPEA [1782, 1775], and the MOMGA [1626, 1629, 1790] are detailed and analyzed. In the discussion of various MOEAs, each algorithm is catalogued by recording key elements of its approach, and classified using the structure defined in Chapter 1. The chapter also presents a generic MOEA algorithmic formulation based upon basic evolutionary operators. Related to this generic form, an analysis of currently known MOEA algorithmic design research is given. Many relevant meta-level topics are addressed, highlighting MOEA design concerns which have limited treatment in the literature. For example, discussed are dominance operator differences, diversity operator variations, population structures, impact of MOEA fitness function characteristics, lack of MOEA theory, MOEA chromosomal representations, utility of explicit vs. implicit building block approaches, and other selected topics.

Fundamental MOEA techniques and MOEA design goals along with a generic MOEA structure are presented in Section 2.2. Specific MOEA pseudo code and associated performance is discussed in Section 2.3. Constraint-handling techniques are briefly discussed in Section 2.4. Critical MOEA elements are described in Section 2.5. This leads to Section 2.6 which recapitu-

---

[2] For an up-to-date list of references on evolutionary multi-objective optimization, visit the EMOO repository located at: http://delta.cs.cinvestav.mx/~ccoello/EMOO with a mirror at: http://www.lania.mx/~ccoello/EMOO

lates general MOEA design principles. Section 2.7 presents a summary of the contents of this chapter.

2.2 MOEA Techniques

This section discusses the development of MOEAs and associated techniques and as such is concerned with issues such as the variety of MOEA design efforts, practicality of the various operator techniques, fitness functions and chromosomal representations. EAs, in general are considered as *metaheuristic* problem solvers—top-level general strategies which guide other lower-level heuristics[3] to search for feasible solutions in difficult domains—search landscapes. This treatment of major MOEA research issues provides the interested researcher and practitioner with tools and techniques of the field and their evolution. The following incomplete historical list of EA algorithms for solving MOPs reflects different algorithmic frameworks as well as fitness function and chromosomal representations:

- Vector Evaluated GA (VEGA) [1439, 1440, 1441]
- Lexicographic Ordering GA [518]
- Vector Optimized Evolution Strategy (VOES) [934]
- Weight-Based GA (WBGA) [636]
- Multiple Objective GA (MOGA) [504]
- Niched Pareto GA (NPGA, NPGA 2) [708, 709, 453]
- Nondominated Sorting GA (NSGA, NSGA-II) [1509, 363, 374]
- Distance-based Pareto GA (DPGA) [1225, 1224]
- Thermodynamical GA (TDGA) [863]
- Strength Pareto Evolutionary Algorithm (SPEA, SPEA2) [1782, 1775]
- Multi-Objective Messy GA (MOMGA-I,II,III) [1626, 1629, 1790, 1788, 342, 345, 343]
- Pareto Archived ES (PAES) [885, 886]
- Pareto Envelope-based Selection Algorithm (PESA, PESA II) [301, 299]
- Micro GA-MOEA (μGA, μGA^2) [283, 284, 1597]
- Multi-Objective Bayesian Optimization Algorithm (mBOA) [956, 1265]

It is also noted here that although David Schaffer is credited with the "invention" of the first MOEA in the mid-1980s, other researchers also deserve credit for their contributions during those years. Mainly, it is important to emphasize the early attempt by Ito et al. [764] to use a genetic algorithm to solve a multi-objective optimization problem, which precedes Schaffer's work. Additionally, it is also important to mention the work by Fourman [518], who presented different MOEA implementations at the same conference where Schaffer's work was introduced.

---

[3] *Heuristic* : a problem-solving technique in which the most appropriate local solution or partial solution is selected using comparative rules.

The idea of using Pareto-based fitness assignment was first proposed by Goldberg [581] to solve the problems of Schaffer's approach [1440]. He suggested the use of nondominated ranking and selection to move a population toward the Pareto front in a multiobjective optimization problem. The basic idea is to find the set of strings in the population that are Pareto nondominated by the rest of the population. These strings are then assigned the highest rank and eliminated from further contention. Another set of Pareto nondominated strings are determined from the remaining population and are assigned the next highest rank. This process continues until the population is suitably ranked. Goldberg also suggested the use of some kind of niching technique to keep the GA from converging to a single point on the front [368]. A niching mechanism such as sharing [587] would allow the GA to maintain individuals all along the nondominated frontier. A variety of MOEAs extended these ideas and are discussed next in more detail.

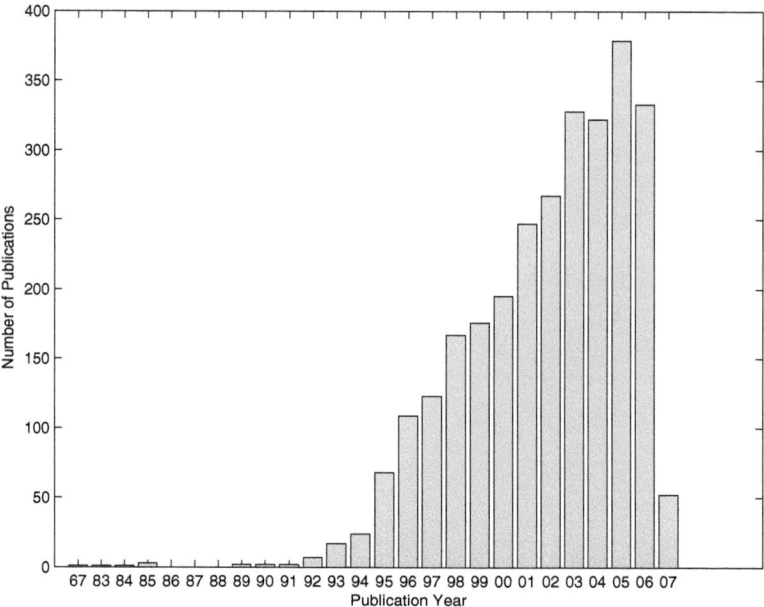

Fig. 2.1. Statistics of the number of publications per year related to evolutionary multiobjective optimization (up to early 2007)

Not until the mid 1990s is there a noticeable increase in published MOEA research. The sheer number of contemporary conference and journal publications and books indicates an active contemporary MOEA research community (see Figure 2.1 and the EMOO repository [266]).

As noted in Section 1.8.2 from Chapter 1, MOEA approaches have been classified into three major categories. These categories and the specific techniques they embody are:

A Priori Techniques: Lexicographic, linear fitness combination, and nonlinear fitness combination.
Progressive Techniques: Progressive techniques or interactive computational steering.
A Posteriori Techniques: Independent sampling, criterion selection, aggregation selection, Pareto-based selection, Pareto rank- and niche-based selection, Pareto deme-based selection, Pareto elitist-based selection, and hybrid selection.

In general, the multiobjective approach to solving MOPs generates "partial orders" of solutions leading to possible multitudes of trade-off solutions in objective space. Note that for single-objective optimization, a "total order" exists. For MOEAs, the concept of dominance is addressed again with reflection on strict partial orders of points in objective space.

Note that with MOPs, an explicit set of objective functions is not required, but only the relative fitness of each solution in a neighborhood and a selection mechanism. Examples of this phenomenon are found in single-objective optimization using simulated annealing [861] and tabu search [572].

There are fundamentally three MOP solution techniques; optimize only the highest priority objective, use an aggregated weight sum of all the objectives, or employ a multiobjective algorithm to find the entire Pareto front (all nondominated points, PF_{true}). Within each of these techniques, there is a multitude of operators that may search from edge to edge of the objective space, move statistically forward towards the Pareto front from an initial set of individuals, or randomly generate and test points. Finding MOP solutions in P_{true} can vary from decision maker priority to attempting to find all solutions. *The essence of multi-criteria optimization is to find the Pareto front.* How does one exploit the objective landscape, different search operators, evaluate with metrics the results, and provide a small number of solutions to the DM? The following subsections expand on generic MOP solution approaches with critical analysis of the various techniques as an attempt to answer these questions.

2.2.1 *A Priori* Techniques

By definition, these a priori techniques require a decision maker (DM) to define the MOP objective relative importance prior to search. This is usually reflected in the weights associated with the aggregated sum of the objectives. In essence, the preferences of the DM are modelled to evaluate and compare solutions in this multicriteria decision making (MCDM) problem. In real-world scientific and engineering problems, it is a non-trivial task to find the one solution of interest to the DM. The ramifications of "bad" objective prioritization choices are easy to understand: the decision maker's "weight" (no matter how defined) could be greater than necessary as more "acceptable" solutions are missed. Optimizing mostly profit could lead to poor quality or reliability, not a good compromise. No matter the optimization algorithm used, this is an inescapable consequence of *a priori* MOEA techniques, which are examined for each sub-technique. These sub-techniques evolve from the general concept of a single scaled function such as represented in a weighted sum of objectives.

A priori techniques can be divided in the following major approaches:

- Lexicographic ordering
- Linear aggregating functions
- Nonlinear aggregating functions

Other approaches include achievement scaling functions [1702, 1707] and the ϵ-constraint method [961, 1318] which also map the MOP to a single-objective optimization problem.

Lexicographic ordering

In this method, the DM is asked to rank the objectives in order of importance. The optimum solution is then obtained by minimizing the objective functions in sequence, starting with the most important one and proceeding according to the assigned order of importance of the objectives. It is also possible to select randomly an objective to be optimized at each generation if the priority is unknown [518].

Criticism of lexicographic ordering - Selecting randomly an objective is equivalent to a weighted combination of objectives, in which each weight is defined in terms of the probability that each objective has of being selected. However, the use of tournament selection with this approach (as Fourman [518] did) makes an important difference with respect to other techniques such as the Vector Evaluated Genetic Algorithm (VEGA) [1440]. This is because the pairwise comparisons of tournament selection make scaling information negligible [505, 507]. This means that this approach may be able to depict concave trade-off surfaces, although that really depends on the distribution of the population and on the problem itself. Its main weakness is that this approach tends to favor more certain objectives when many are present in the problem, because of the randomness involved in the process. This has the undesirable consequence of making the population converge to a particular part of the Pareto front rather than to delineate it completely [260]. The main advantage of this approach is its simplicity and computational efficiency. These two properties make it highly competitive with other non-Pareto approaches such as a weighted sum of objectives or VEGA.

Lexicographic techniques have not found favor with MOEA researchers, as only a few implementations are reported in the specialized literature (see for example [526, 440]). This may be due to the fact that this technique explores objective space unequally, in the sense that priority is given to solutions performing well in one objective over another(s). Or, in other words, one objective is optimized at all costs.

The lexicographic technique appears most suitable *only* when the importance of each objective (in comparison to the others) is clearly known. Of course, trade-offs do exist. On the one hand, any reported solutions are Pareto optimal (by definition and with respect to all solutions evaluated). On the other hand, when is such an "all costs" goal necessary or even appropriate? If one objective is to be optimized regardless of the others' expense, it seems more appropriate to instead use a single objective EA which does not incur the additional overhead of a MOEA.

Linear aggregating functions

The typical form of linear aggregating functions is to compute fitness using:

$$\text{fitness} = \min \sum_{i=1}^{k} w_i f_i(\mathbf{x}) \qquad (2.1)$$

where $w_i \geq 0$ and $i = 1 \ldots k$ are the weighting coefficients representing the DM's relative importance of the k objective functions of the MOP. It is usually assumed for normalization that

$$\sum_{i=1}^{k} w_i = 1 \qquad (2.2)$$

The linear fitness combination technique is a popular scalarizing approach despite its identified shortfalls [329], probably due to its simplicity. In Figure 2.2, observe that one of the equal slope parallel (minimization) lines indicates that the search process finds a single Pareto front point **A** at minimum cost, but only if it is on the *convex hull* of the Pareto front. Although point **B** may be found, it is not retained since a smaller aggregate objective function value is found at point **A**. Observe that different weights reflect different slopes and intersect the points on the convex hull at different points on PF_{true}. Thus, the linear aggregating algorithm usually does not find all Pareto front points of interest. These points are defined as *non-supported* points since they are not on the convex hull of the Pareto front. The variation of weights for a specific MOP can cause large or a very small variation in the number or value of points found on PF_{true}.

Another scalarizing approach is the weighted Tchebycheff model which can find the non-supported points on the Pareto front (i.e., not limited by the convex hull). This search approach uses a reference point, f_i^*. This reference point must be beyond the *ideal point* where each component is less than

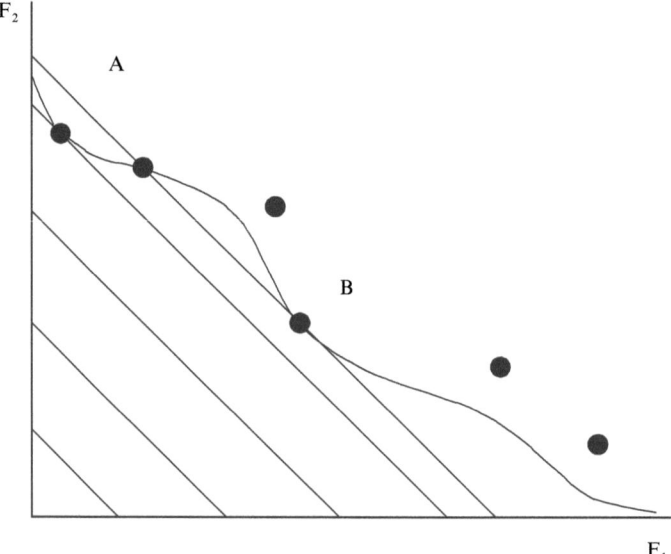

Fig. 2.2. Linear aggregating technique for a bi-objective example with *a priori* selection of weights, $w_1 x_1 + w_2 x_2$.

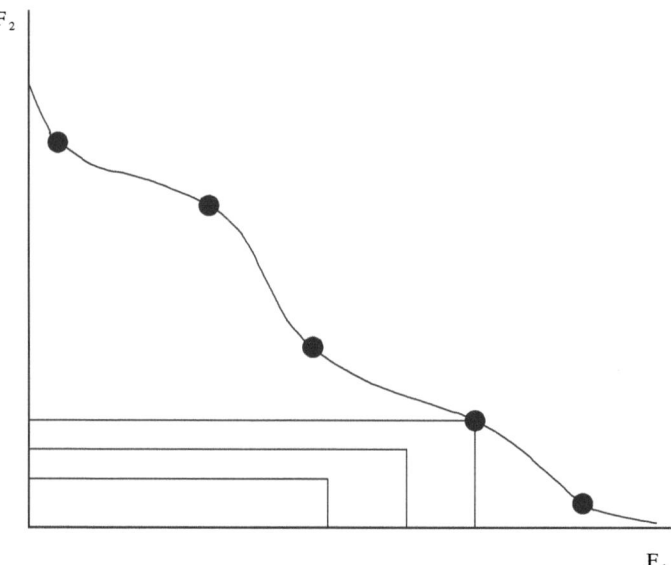

Fig. 2.3. Tchebycheff technique for a bi-objective example with *a priori* selection of weights, $w_1 x_1 + w_2 x_2$.

the minimum value of the ith objective. The weighted Tchebycheff model is reflected in equation (2.3).

$$\text{fitness} = \min \max_i [w_i | f_i(\mathbf{x}) - f_i^* |] \tag{2.3}$$

where $w_i \geq 0$ and $i = 1 \ldots k$. In Figure 2.3, the reference point is the origin. With this technique, assuming that the reference point is properly defined, Pareto points can be found on PF_{true} for appropriate sets of weights in the aggregated objective function. In an attempt to find the appropriate set of weights, a sampling process can be employed.

Criticism of linear aggregating functions - A basic weighted sum MOEA is both easy to understand and implement. The fitness combination technique is also computationally efficient. If the problem domain is "easy" and a sense of each objective's relative worth is known and can be quantified, or even if the time available for search is short, this may be a suitable method to discover an acceptable MOP solution. However, this technique has a major disadvantage due to certain MOP characteristics. Fonseca and Fleming [510] explain that for any positive set of weights and fitness function Φ, the returned global optimum is always a Pareto optimal solution (with regard to all others identified during search). However, if PF_{true} is nonconvex, optima in that portion of the front can *not* be found via this method. This is proved using geometry by Das & Dennis [329]. Thus, blindly using this technique guarantees that some solutions in P_{true} cannot be found when it is applied to certain MOPs. Also note that despite their popularity in the past, linear aggregating functions are nowadays significantly less common than Pareto-based approaches.

Nonlinear aggregating functions

Nonlinear aggregation techniques (e.g., a multiplication of the objective functions) are not very popular in the literature. This may be due to the overhead involved in determining appropriate probability of acceptance or utility functions, and to the various conditions which these objective functions must meet [836]. This additional overhead may not justify resulting solutions' "quality."

Target vector approaches are somewhat more popular than multiplicative approaches, and they may be particularly useful if the DM can specify goals that he/she desires to achieve. The evolutionary algorithm in this case, tries to minimize the difference between the current solution generated and the vector of desirable goals (different metrics can be used for this purpose). Although target vector approaches can be considered as another aggregating approach, they are normally considered separately, because some of target-vector approaches can generate (under certain conditions) concave portions of the Pareto front, whereas approaches based on linear combination of weights cannot.

The most popular target vector approaches are hybrids with: Goal Programming [359, 1702, 1422], Goal Attainment [1707, 1748], and the min-max algorithm [636, 267].

Criticism of nonlinear aggregating functions - Multiplicative approaches are simple and efficient, but they may be troublesome, since the definition of a good nonlinear aggregation function may prove to be more difficult than defining a linear aggregating function.

The same applies to target-vector approaches, which require the definition of goals to be achieved. The computation of these goals normally requires some extra computational effort and can lead to additional problems. Wilson and MacLeod [1707] found that goal attainment could generate, under certain circumstances, a misleading selection pressure. For instance, if there are two candidate solutions which are the same in one objective function value but different in the other, they still have the same goal-attainment value for their two objectives, which means that for an evolutionary algorithm neither of them is better than the other.

An additional problem with target-vector approaches is that they yield a nondominated solution only if the goals are chosen in the feasible domain, and such condition may certainly limit their applicability. Furthermore, just as in all *a priori* techniques, specifying exact goals or weights before search may unnecessarily limit the search space and therefore "miss" desirable solutions.

Note that despite their drawbacks, there are certain problems (particularly, in multiobjective combinatorial optimization problems) in which nonlinear aggregating functions (e.g., based on Tchebycheff weights) can provide very good approximations of the Pareto optimal set (even outperforming Pareto-based approaches) [784, 776, 779, 777, 778, 780, 782, 783, 781].

General Criticism of *a priori* Techniques

It appears that the *a priori* MOEA techniques considered are in general not desirable for general use, except for nonlinear aggregating functions which can be advantageous in certain types of problems (namely in multiobjective combinatorial optimization problems), as discussed in the previous section. If a DM is spending resources to search for MOP solutions, it is reasonable to expect optimal (or "good") solutions. Since these *a priori* techniques arbitrarily limit the search space they may not be able to find all the available solutions in P_{true}. Additionally, implementing "more" effective MOEAs might not be as difficult and involves less overhead than imagined.

2.2.2 *Progressive* Techniques

The fact that there is a relatively small number of cited interactive search efforts in the MOEA literature is surprising (see for example [91]). One would think that no matter what MOP solution technique is implemented, close interaction between the DM and "searchers" can only increase the efficiency (or "desirability") of discovered solutions. It is understandable that a DM's time and effort is at a premium. At least to some level, though, more interaction certainly implies "better" results. Although either *a priori* or *a posteriori* techniques may be used interactively, the latter are more suited to MOPs because they offer a *set* of solutions rather than just one. There is a limit to how much information a DM can process at one time, but surely some greater number of choices beyond one or two is generally more advantageous.

Incorporating DM preferences within and through an interactive search and decision making process may benefit all those involved. Do researchers and/or practitioners feel they don't have the time? Or is it the DM who balks at the additional effort? Real-world applications should surely use this interactive process as the economic implications can be quite significant. In fact, several MOEAs [504, 454, 317, 1346, 719, 598] are able to explicitly incorporate DM preferences within search (see Chapter 9).

General Criticism of Progressive Techniques

The main problem with progressive techniques is that the DM normally has to define goals or a scheme of preferences to bias the search, and this requires an interactive process that may be difficult and inefficient when nothing about the problem is known. Also, under certain circumstances, there might be contradictions in the preferences defined (e.g., when dealing with group preferences). However, when it is desirable to constrain the search within a certain region of interest (something common in complex real-world problems), an interactive process is perhaps the best choice. The main issues here have to do with the way in which the preferences from the DM are incorporated into the MOEA. Any proposal in this direction has to deal with a set of issues such as scalability and intransitivities, among others (see Chapter 9).

2.2.3 *A Posteriori* Techniques

A posteriori techniques are explicitly seeking P_{true} and PF_{true}. Thus, the emphasis is now to perform a search as widespread as possible, as to generate as many different elements of the Pareto optimal set as possible. The decision making process will now take place after completing the search. The following *a posteriori* sub-techniques are examined next:

- Independent sampling techniques
- Criterion selection techniques
- Aggregation techniques (linear, nonlinear)
- ϵ-constraint technique
- Pareto sampling techniques

Independent Sampling Techniques

Various independent sampling approaches generally have reduced effectiveness. This sort of technique uses some fitness combination technique where the weights assigned to each objective are varied over a number of separate MOEA runs. The difference with respect to a priori linear aggregating function is therefore the variability of the weights along the evolutionary process. This variation can allow the generation of larger portions of the Pareto front. But, these points are in most cases not uniformly placed along the Pareto front. Because one does not know a priori a proper selection of weights, the "ideal" uniform distribution of Pareto front points is seldom generated. Again, variations in weights may generate points on PF_{true} close together or at far distances. If these points could be generated using a different approach, then the inverse mapping to the weights would provide the DM with an explicit weighted numerical tradeoff among the objectives, which is a very valuable piece of information. The MOEA techniques discussed generally attempt to generate PF_{true}, but not directly related to the independent sampling weights since finding the inverse mapping is very difficult.

Criticism of independent sampling techniques - The main advantage of this type of technique is its relative simplicity and its efficiency (no Pareto ranking procedure is required). This approach may have limited utility if a low number of objectives is being considered (i.e., two or three). For example, assume a MOEA using a linear fitness combination Tchebycheff approach. If each objective's weight varies from 0 to 1 by 0.05 increments, only 21 MOEA runs are necessary to explore the possible weight combinations and give some picture of PF_{known}. However, even varying the weights at this coarse resolution results in the required number of runs combinatorially increasing with the number of objectives. Thus, its overall usefulness seems quite limited especially as the arbitrary weight combinations may well prevent discovery of some solutions in P_{true}, and also in view of other techniques' strengths. Note however, that this type of approach may be useful to approximate the Pareto front in certain types of problems (e.g., multiobjective combinatorial optimization problems). This is because in certain cases (particularly with convex Pareto fronts) they may produce competitive results with respect to MOEAs based on Pareto ranking at a lower computational cost (see for example [1507]).

Criterion Selection Techniques

The *Vector Evaluated Genetic Algorithm* (VEGA), which was proposed by David Schaffer [1439, 1440, 1441] is normally considered the first implementation of a MOEA. The vector is by definition the vector of k objective functions of the MOP. The VEGA approach is an example of a criterion or objective selection technique where a fraction of each succeeding populations is selected based on *separate* objective performance. The specific objectives for each fraction are randomly selected at each generation. VEGA tends to converge to solutions close to local optima with regard to each individual objective.

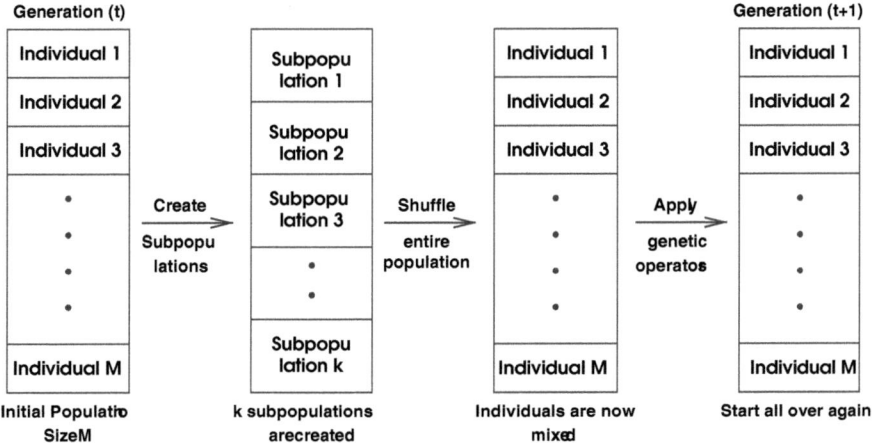

Fig. 2.4. Schematic of VEGA's selection mechanism. It is assumed that the population size is M and that there are k objective functions.

The VEGA concept is that, for a problem with k objectives, k subpopulations of size M/k each would be generated (assuming a total population size of M). Each sub-population uses only one of the k objective functions for fitness assignment. The proportionate selection operator is used to generate the mating pool. These sub-populations are then shuffled together to obtain a new population of size M, on which the GA would apply the crossover and mutation operators in the usual way. Shuffling is done prior to sub-population partitioning in order to reduce positional population bias. This process is illustrated in Figure 2.4. The complexity of VEGA is clearly the same as the single-objective GA.

Schaffer realized that the solutions generated by VEGA were nondominated in a local sense, because their nondominance was limited to the current population. And, while a locally dominated individual is also globally dominated, the converse is not necessarily true [1440]. An individual that is not dominated in one generation may become dominated by an individual who emerges in a later generation. Also, Schaffer noted a problem that in genetics is known as "speciation" (i.e., one could have the evolution of "species" within the population which excel on different aspects of performance). This problem arises because this technique selects individuals that excel in one dimension of performance, without considering other dimensions. The potential danger is that one could have individuals with what Schaffer called "middling" performance[4] in all dimensions, which could be very useful for compromise solutions, but that would not survive under this selection scheme, since they are not in the extreme for any dimension of performance (i.e., they do not

---

[4] By "middling," Schaffer meant an individual with acceptable performance, perhaps above average, but not outstanding for any of the objective functions.

produce the best value for any objective function, but only moderately good values for all of them). Speciation is undesirable because it is opposed to our goal of finding a compromise solution. Schaffer suggested some heuristics to deal with this problem. For example, one could use a heuristic selection preference approach for nondominated individuals in each generation, to protect the "middling" chromosomes. Also, crossbreeding among the "species" could be encouraged by adding some mate selection heuristics instead of using the random mate selection of the traditional GA (i.e., the use of mating restrictions). Per the discussion, VEGA uses a localized criterion for ranking as depicted in Figure 2.5.

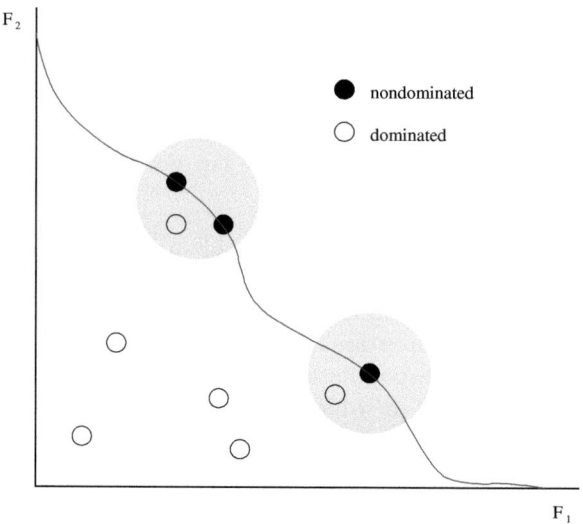

Fig. 2.5. VEGA's criterion-based ranking mechanism.

Norris & Crossley [1190] and Crossley et al. [310] believe this technique reduces the diversity of any given $PF_{current}(t)$. They implemented elitist selection to ensure $PF_{known}(t)$ endpoints (or in other words, $PF_{known}(t)$'s extrema) survive between generations. Otherwise, the MOEA converges to a single design rather than maintaining a number of alternatives. In other attempts to preserve diversity in $PF_{current}(t)$ they also employ a VEGA variant. Here, "k"-branch tournaments (where k is the number of MOP objectives) allow each solution to compete once in each of k tournaments, where each set of tournaments selects $\frac{1}{k}$th of the next population [805].

Criticism of criterion selection techniques - VEGA is very simple and easy to implement, since only the selection mechanism of a traditional GA has to be modified. One of its main advantages is that, despite its simplicity, this sort of approach can generate several solutions in one run of the MOEA. However, note that the shuffling and merging of all the sub-populations that VEGA performs corresponds to averaging the fitness components associated with each of the objectives [587]. Since Schaffer uses proportional fitness assignment [581], these fitness components are in turn proportional to the objectives themselves [507]. Therefore, the resulting expected fitness corresponds to a linear combination of the objectives where the weights depend on the distribution of the population at each generation as shown by Richardson et al. [1360]. This means that VEGA has the same problems than the aggregating approaches previously discussed (i.e., it is not able to generate concave portions of the Pareto front). Nevertheless, VEGA has been found useful in other domains such as constraint-handling, where its biased behavior can be of great help [1543, 275, 274]. Note that these algorithmic developments were in part based upon consideration of the computational hardware performance at the time. Other variations and extensions of the VEGA concept included the Vector Optimized Evolution Strategy (VOES) by Kursawe [934]. His approach of course was based on an evolution strategy along with a fitness evaluation process similar to VEGA. It also employed a diploid chromosome scheme with preservation of nondominated solutions using an elitist approach. The WBGA (weight-based genetic algorithm) proposed by Hajela and Lin [636] is related to VEGA's sampling approach, but it uses a set of weights (each individual is assigned a vector containing such weights). These vectors remain diverse across the population through niching and appropriately selected subpopulations that are evaluated for different objectives in a way analogous to VEGA. Again, this MOEA is simple, but the use of weighted vectors has the same disadvantages as the independent sampling approach.

Aggregation Selection Techniques

Aggregation selection MOEAs incorporate a variety of techniques to solve MOPs such as weighted sums [751], constraint and objective combinations [1017], and hybrid search approaches [358]. However, rather than using static weight combinations for the objectives throughout a MOEA run, the weights are varied between generations and/or each function evaluation. Sometimes the weights are assigned randomly, sometimes they are functions of the particular solution being evaluated, and in other cases are encoded in the chromosome as genes where evolutionary operators (EVOPs) act upon them, too.

Criticism of aggregation selection techniques - As with the criterion selection techniques, aggregation selection approaches can generate a *set* of solutions in a single run of a MOEA. Thus, P_{known} and PF_{known} may be reasonable approximations to P_{true} and PF_{true}, and have required only one MOEA run. These methods are not without their disadvantages, however. When using the weighted sum technique, it is known that certain members of PF_{true} may be missed [329]. Furthermore, both the constraint/objective combination and hybrid search approaches have significant overhead (e.g., solving a linear system of equations to determine an appropriate hyperplane [1764]).

ϵ-Constraint Techniques

The ϵ-constraint technique is based upon selecting a primary objective function and then bounding the others with a separate allowable ϵ-constraint (must be known a priori). The ϵ-constraints are then changed in order to generate another point on the Pareto front (phenotype) and so forth resulting in finding elements in the Pareto optimal set (genotype). Non-uniformity in the distribution of the Pareto front points usually occurs. Examples of this approach can be seen in [1507, 1318, 925, 961].

Criticism of ϵ-constraint technique: Easy to implement, but extensive computation effort is required to generate PF_{known}.

Pareto Sampling Techniques

The disadvantages of aggregation selection techniques make evident that a fitness assignment or selection technique able to "easily" find all members of P_{true} and PF_{true} is desired. Pareto sampling offers this capability, or at least the realistic objective of finding P_{known} and PF_{known}.

Pareto sampling refers to techniques that use the MOEA's population capability to generate several elements of the Pareto optimal set in a single stochastic computational run. Figure 2.6 presents a two objective conceptual understanding of Pareto optimality. Again, one must relate the graphical definition of dominated and nondominated points in objective space and the corresponding solutions in variable space. Because of the strict partial order, various points in the objective space can not be compared to each other with regard to dominance. The intent of many MOEAs of course is to move the nondominated points toward PF_{true} generating a "good" distribution of points on PF_{known}.

Criticism of *A Posteriori* Techniques

These techniques attempt to exploit the population capabilities of evolutionary algorithms to produce a set of elements of the Pareto optimal set in a single run. This can be done either by using a cooperative mechanism (as in VEGA [1440]) or by incorporating directly the concept of Pareto dominance into the selection mechanism of an EA (the most usual way to tackle MOPs with EAs). Scalability is, however, an issue when using Pareto sampling techniques, as indicated before. Also, other types of techniques may be particularly useful within certain specific domains and therefore the importance of knowing about their existence.

Although the *No Free Lunch* (NFL) theorems [1708] indicate that there is no "best" MOEA, certain MOEAs have been experimentally shown to be

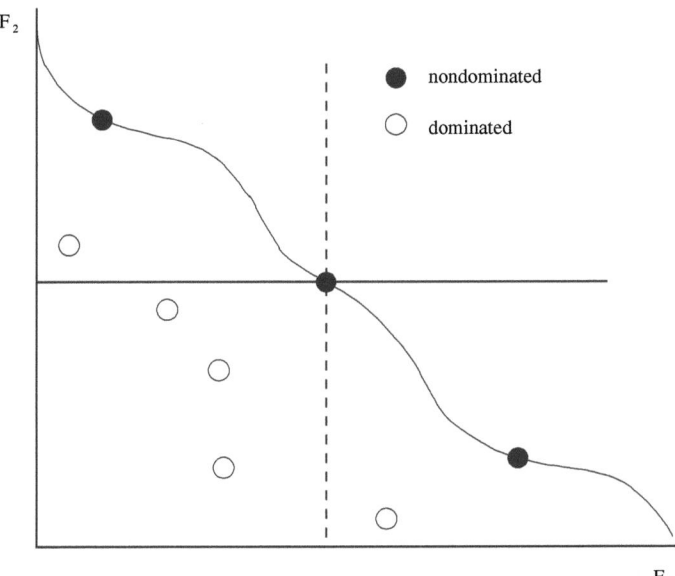

Fig. 2.6. The concept of Pareto optimality as related to nondominance in a maximization MOP

more likely effective (robust) than others for specific MOP benchmarks and certain classes of real-world problems.

2.2.4 Generic MOEA Goals and Operator Design

The basic algorithm design concept is to use Pareto-based fitness assignment to identify nondominated vectors from a MOEA's current population. Regarding this and our previous discussion, the four high-level primary goals of such algorithms for solving MOPs are:

Goal 1. Preserve nondominated points (elitism vs. non-elitism)
with $PF_{current} \to PF_{known}$
Goal 2. Progress or guide PF_{known} towards PF_{true}
Goal 3. Generate and maintain diversity of: points on the Pareto Front, PF_{known} (phenotype) and/or Pareto optimal solutions P_{known} (genotype)
Goal 4. Provide the decision maker (DM) with a limited number of PF_{known} points!

Thus, a MOEA should guide the search towards PF_{true}, generate and maintain a diversity of PF_{known} points, and prevent loss of "good" solutions through archiving. The design of an idealized or generic Pareto-based Multiobjective Evolutionary Algorithm would consist of the following meta-level general procedures:

Step 0: Define the MOP; determine the mathematical form of $F(x) = [f_1(\mathbf{x}), f_2(\mathbf{x}), \ldots f_k(\mathbf{x})]$ and the chromosome representation of $\mathbf{x}$. Define constraints (dynamic, static, linear, nonlinear, etc.). Integrate the "model" into a specific MOEA algorithmic search process.

Step 1: The MOEA generates the Pareto front, PF_{known} (hard part); determine the nondominated sets, generation to generation, via populations. Converge "close" to the true computational Pareto front, PF_{true}; note that what we obtain is an approximation of such a true Pareto front! This is the currently known nondominated population, $PF_{current}$. Execute this same MOEA process for a certain (given) number of generations or until some metric meets some (predefined) threshold.

Step 2: The MOEA attempts to generate a uniform distribution across the known Pareto front, PF_{known}, at the end of each generation.

Step 3: Select several of the "Optimal" points on the Pareto front, PF_{known}, for DM consideration.

Step 4: Determine the associated Pareto Optimal set, P_{known}; implement decision variable values (i.e., our approximation of the Pareto optimal set) as selected by the DM.

Step 5: Visualize algorithm processing and results as appropriate for improving MOEA performance (i.e., efficiency and effectiveness).

Of course, the integration of a specific MOP with selected MOEA software requires insight not only into the problem domain, but into the MOEA operator implementations as well. This *a priori* MOP and MOEA analysis helps support the specific detailed design and implementation; the objective being execution of a MOEA that has a high probability of finding a "good" PF_{known}. A spectrum of MOEAs includes numerous operators which are listed as follows according to their support of the four primary MOEA Goals:

Goal 1. Preserve nondominated points
- Dominance-Based ranking - fitness assignment
- Non-Pareto vs Pareto approaches
- Archiving + elitism of chromosome population

Goal 2. Progress towards points on PF_{true}
- Convergence to true computational Pareto front, PF_{true}
- Generating nondominated phenotype points
- Explicit/Non-Explicit building block manipulation
- Qualitative and Quantitative performance metrics and visual comparisons
- Probabilistic MOEA models; local search incorporation, etc.

Goal 3. Maintain diversity of: points on PF_{known} and/or on P_{known}
- Diversity preservation
- Niching/fitness sharing and crowding on Pareto front (variations)
- Uniform/Diverse nondominated PF_{known}

Goal 4. Provide the DM a limited number of PF_{known} points!

Given the generic MOEA Pareto-based operators, specific variations and aspects of these concepts are presented. Also, MOEAs implementing such detailed operators are referenced.

Dominance-Based Ranking

The dominance relation (or operator), as described in Chapter 1 relates two solutions; therefore, it is a binary operator. The result of this operation for two individual solutions in objective function space has two possibilities: 1) one solution dominates another or 2) the solutions do not dominate each other. There exist various mathematical binary relationships for dominance operators: "reflexive" which the dominance operator is not, "symmetric" which is not, "antisymmetric" which it is not, but it is "transitive." Thus, the dominance operator is ordered since it is not reflexive. It is not a *partial* order but a *strict partial order*. Then, by definition, given a point in objective function space, it could be dominated or not dominated by another point, but it could also be "incomparable" to other points. With this insight, the concept of the Pareto front and Pareto optimal solution are defined in Chapter 1, together with their associated sets. And, a generic algorithm for generating these sets can be formulated as done in Section 2.2.4.

Regarding the generation and selection of the Pareto optimal set, an ordering technique is required. When using an evolutionary algorithm for generating such Pareto optimal set, the fitness values are n tuples (considering n objectives). A scaling technique is required over the tuples via a strict partial order, so that nondominated solutions are generated. Thus, various *ranking* methods have been suggested in the specialized literature. Such methods essentially sort the individuals in objective function space before selection. Each member of the list of possible points (individuals) in objective function space is assigned a rank relative to one of the following dominance definitions:

- **dominance rank**: How many individuals is an individual dominated by (plus 1)?
- **dominance count**: How many individuals does an individual dominate?
- **dominance depth**: At which "front" is an individual located? "Sort."

Computationally implementing one of these ranking approaches in a specific MOEA design is, of course, straightforward. However, given a particular problem domain, performance (efficiency and effectiveness) can have considerable variance. This is due in no small measure to the structure of the fitness landscape being searched! The result of the dominance ranking (see Figure 2.7) is a strict partial ordered list which is used for sorting the points before employing a desired selection operator. As an example of dominance count see

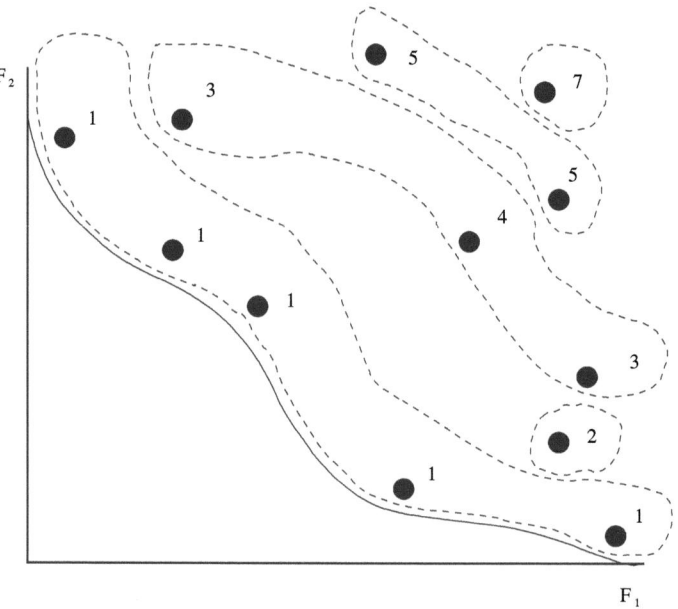

Fig. 2.7. Dominance rank with grouping of equal ranks for sorting.

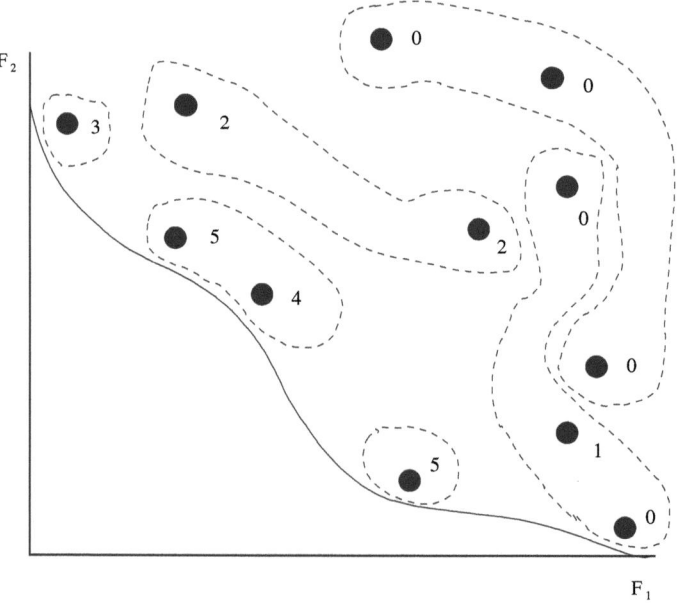

Fig. 2.8. Dominance count with grouping of equal counts for sorting.

Figure 2.8, which imposes a different partial order. In both cases, the different "ranks" are shown within the dotted regions. The "sorting" is based upon "rank depth" and, of course, is different for the two different dominance relationships. The computational order in all cases is usually $O(N^2)$. Specific MOEA examples of dominance ranking use are:

- **MOGA, NPGA** [504, 709]: dominance rank
- **NSGA/NSGA-II** [1509, 374]: dominance depth
- **SPEA/SPEA2** [1782, 1775]: dominance count and dominance rank
- **MOMGA/MOMGA-II** [1626, 1790]: dominance rank

General Diversity Preservation

Another goal of MOEA design is to provide a diversity of PF_{known} or P_{known} points to the DM that have a somewhat uniform distribution across the known Pareto front. Various techniques are available for maintaining diversity in a MOEA, including the Weight Vector Approach, the Fitness Sharing/Niching Approach, Crowding/Clustering, Restricted Mating, and Relaxed Dominance, all of which are discussed next:

- **Weight Vector Approach**: In this case, a vector set in fitness/objective space is used to attempt to diversify points of the Pareto front surface (i.e., the aim is, of course, to generate a uniform distribution of PF_{known}). By changing the weights, different directions are defined, in order to bias the search, and to move solutions away from its neighbors. Weight vector approaches have been found very effective for certain types of applications (for example, multi-objective combinatorial optimization [323, 1617, 1152, 535, 762]).
- **Fitness Sharing/Niching Approach**: In this approach, the size (or radius) of a neighborhood (or niche) is controlled through the σ_{share} value (niche radius). Then, one must count how many solutions are located within the same niche, and the fitness is decreased proportionally to the number of individuals sharing the same neighborhood [587, 368]. This aims to promote the generation of solutions in the least populated regions of the search space (see Figure 2.9). Note the following:
  - The definition of the σ_{share} parameter is critical.
  - In order to apply a fitness sharing function, it is necessary to measure distances [1554, 1769]. Such distances can be measured in genotype or phenotype space.
  - Several MOEAs (e.g., MOGA [504], the NSGA [1509]) adopted this approach, with algorithms $O(N^2)$. However, not all of them applied fitness sharing in the same space (MOGA [504] applied fitness sharing in objective function space, whereas the NSGA [1509] applied it in decision variable space).

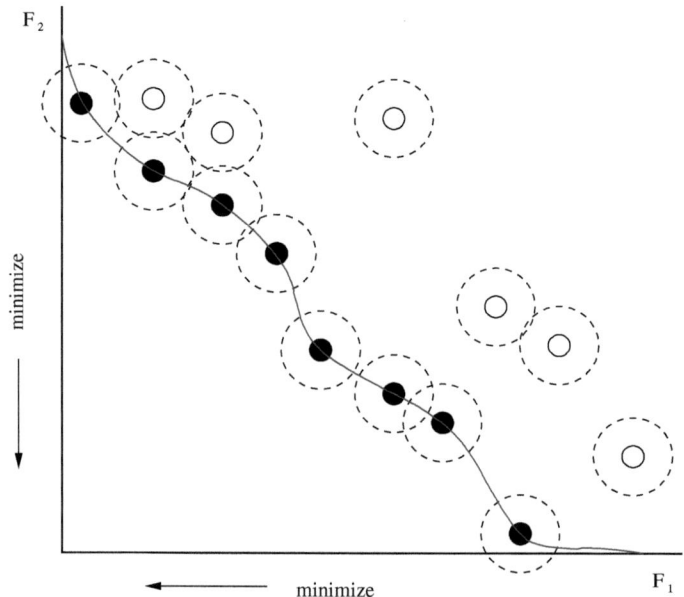

Fig. 2.9. A graphical illustration of fitness sharing

Note that when using niching, it is also possible to adopt different topologies for defining neighborhoods. For example, one could use a grid, and associate the value of σ_{share} to the size of the squares that define such grid (see Figure 2.10). In this example, within each grid square, a desired maximum of one point is kept (the reduction to one point is usually based upon random selection, but other criteria are possible).

The density estimation may be based on several criteria, such as the following:

- **Kernel approach:** The density estimator is based on the sum of f values, where f is a function of the distance (vector) measured either in genotypic or in phenotypic space (e.g., MOGA [504] and the NPGA [709]).
- **Nearest neighbor approach:** The density estimator is based on the volume of the hyper-rectangle defined by the nearest neighbors (e.g., the NSGA-II [374] and SPEA2 [1775]).
- **Histogram approach:** The density estimator is based on the number of solutions that lie within the same hyper-box (e.g., PAES [886] and PESA [301]).

• **Crowding/Clustering:** In this case, we select the surviving solutions according to a region crowdedness metric measured in objective function space (see Figure 2.11). This is an idea similar to fitness sharing, but more

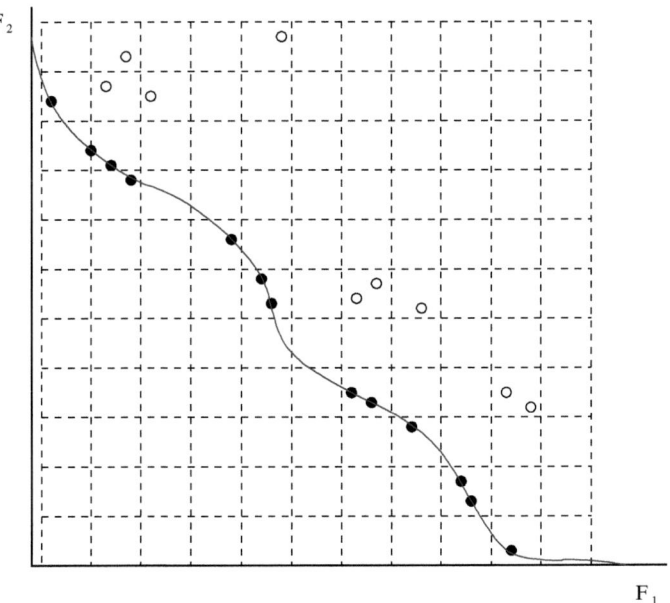

Fig. 2.10. A graphical illustration of a niching scheme based on the use of a grid

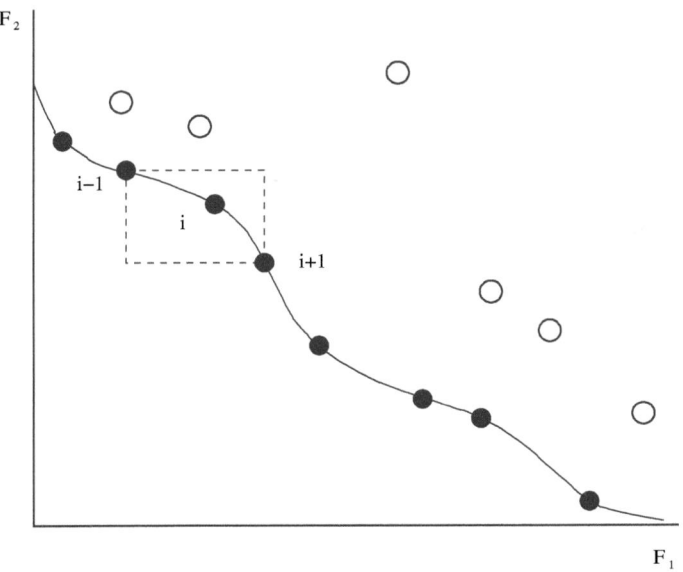

Fig. 2.11. A graphical illustration of crowding

efficient, which algorithms such as the NSGA-II [374] have adopted. It is also possible to use clustering techniques for the same purpose [1775].

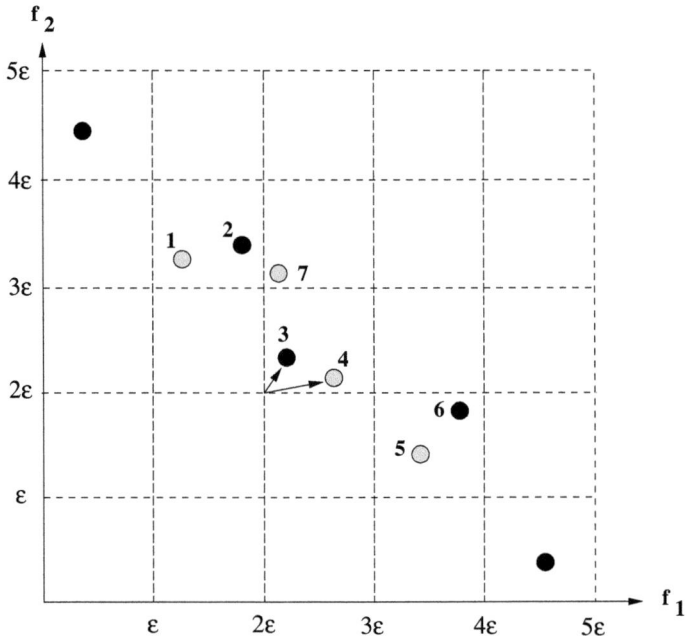

Fig. 2.12. An example of the use of ϵ-dominance in an external archive. Solution 1 dominates solution 2, therefore solution 1 is preferred. Solutions 3 and 4 are incomparable. However, solution 3 is preferred over solution 4, since solution 4 is the closer to the lower left-hand corner represented by point $(2\epsilon, 2\epsilon)$. Solution 5 dominates solution 6, therefore solution 5 is preferred. Solution 7 is not accepted since its box, represented by point $(2\epsilon, 3\epsilon)$ is dominated by the box represented by point $(2\epsilon, 2\epsilon)$.

- **Relaxed forms of dominance**: Use a certain solution x even though it is worse than some solution y in regards to a particular objective (value comparison in objective function space). This relaxation may be compensated by an improvement in other objectives (see for example [470, 471, 798, 1138, 892]).
  Laumanns et al. [959] proposed a relaxed form of Pareto dominance called ϵ-dominance. The main use of this concept in MOEAs has been to filter solutions in an external archive. By using ϵ-dominance, we define a set of boxes of size ϵ and only one nondominated solution is retained for each box (e.g., the one closest to the lower left-hand corner). This is illustrated in Figure 2.12, for a bi-objective case. The use of ϵ-dominance, as proposed in

[959] and illustrated in Figure 2.12, guarantees that the retained solutions are nondominated with respect to all solutions generated during the run.
- **Restricted Mating**: This is quite similar to Crowding/Clustering, but in this case diversity is preserved through the avoidance of certain recombinations. When this approach is adopted, there is normally a parameter (σ_{mate}) which defines the minimum distance that must separate two individuals so that they can mate. It can be used either in serial [1021] or in parallel/distributed implementations [1157].

MOEA Populations

Defining MOEA population structures is directly related to the sets P_{known} and $P_{current}$. Usually, P_{known} is an archival set always updated to retain the best solutions found so far. $P_{current}$, of course, is the set of current generation nondominated solutions. The use is these two populations or sets can be treated differently as to their use as parents at the beginning of each generation. These sets are generally defined as generational (primary or main) and secondary (archival or external) populations.[5]

As Horn [706] indicates, any practical MOEA implementation must include a secondary population composed of all nondominated solutions found so far ($P_{known}(t)$). This is due to the MOEA's stochastic nature which does not guarantee that desirable solutions, once found, remain in the generational population until MOEA termination. This is analogous to elitism but it is emphasized that it is a *separate* population. The question is then how to best utilize this additional population. Is it simply a repository, continually added to and periodically culled of dominated solutions? Or is it an integrated component of the MOEA? Although several researchers indicate their use of secondary populations only a few explain its use in their implementation. As there is no consensus for its "best" use, some of its incarnations are presented next.

A straightforward implementation stores $P_{current}(t)$ at the end of each MOEA generation (i.e., $P_{current}(t) \cup P_{known}(t-1)$). This set must be periodically culled since a solution's designation as Pareto optimal is *always* dependent upon the set within which it is evaluated. How often the population is updated is generally a matter of choice, but as determination of Pareto optimality is an $\mathcal{O}(kM^2)$ algorithm (where k refers to the number of objectives and M to the population size), it should probably not be performed arbitrarily. As this population's size grows comparison time may become significant. This implementation does not feed solutions from $P_{known}(t)$ back into the MOEA's generational population.

---

[5] Note that it is also possible to use a single population in a MOEA (see for example, the NSGA-II [374], in which a plus selection mechanism with an implicit elitist strategy is adopted). However, we will only discuss here the use of two populations because this is the most common practice in the current literature.

Conversely, other published algorithms actively involve P_{known} in MOEA operation. For example, Zitzler and Thiele's [1781, 1780, 1782, 1770] SPEA stores $P_{current}(t)$ in a secondary population and then culls dominated solutions. Solutions from both the MOEA's generational and secondary populations then participate in binary tournaments selecting the next generation. If the number of solutions in $P_{known}(t)$ exceeds a given maximum, the population is reduced by clustering which attempts to generate a representative solution subset while maintaining the original set's ($P_{known}(t)$'s) characteristics. SPEA also uses $P_{known}(t)$ in computing the main population's solutions' fitness; this effectively results in a larger generational population.

Todd and Sen [1591] also insert nondominated solutions from $P_{known}(t)$ into the mating population to maintain diversity, as do Ishibuchi and Murata [750, 752, 751] and Cieniawski et al. [255]. These implementations never reduce the size of $P_{known}(t)$ except when removing dominated solutions. Parks and Miller [1247, 1244, 1245] implement an *archive* of Pareto optimal solutions. However, solutions in $P_{current}(t)$ are not always archived; the process occurs only if a solution is sufficiently "dissimilar" from those already resident. Thus, this also is a form of clustering. If a new solution is added, any archive members no longer Pareto optimal are removed. Like SPEA, the next generation's members are selected from both $P_{known}(t)$ and the current generational population.

Some researchers use secondary populations *not* composed of Pareto optimal solutions. Bhanu and Lee [130] apply a MOEA to adaptive image segmentation; their secondary population is actually a training database from which GA population members are selected. Viennet et al. [1651] use separate GAs to optimize each of the MOP's k functions independently; these "additional" populations are later combined and nondominated solutions removed to provide P_{known}.

A secondary population (of some sort) is a MOEA necessity. Because the MOEA is attempting to build up a (discrete) picture of a (possibly continuous) Pareto front, this is probably a case where at least initially, too many solutions are better than too few. It intuitively seems that a secondary population might also be useful in adding diversity to the current generation and in exploring "holes" in the known front, although how to effectively and efficiently use P_{known} in this way is unknown. Again, it is suggested to experiment directly comparing various secondary population implementations.

Several researchers have studied different aspects of secondary populations in the last few years. See for example [877, 485, 1454, 959, 127, 672].

A Generic MOEA Algorithm

In general, based upon the MOEA Goals, an effective MOEA should incorporate the following generic operations assuming operations on complete individuals:

- An initialization phase generating N individuals in a population P and evaluating fitness. Individual gene encoding from the problem domain could be binary, integer or real.
- *Remove* Pareto dominated individuals from P based upon scalar multiobjective function evaluations such as Pareto ranking; $P \rightarrow P^i$.
- Use a density estimator to limit the number of individuals in P^i that lie on "small" regions of the current PF_{known} or P_{known}. Techniques include *niching, sharing, & crowding* with associated parameter values. The reduced niche count keeps the population at a "reasonable" computational number.
- Perform evolutionary operations (*recombination, mutation*, etc.) to generate new individuals using appropriate parameter values; $P^i \rightarrow P^{ii}$. To *select individuals for recombination* one can use ranking, binary tournament selection, or proportional selection, for example.
- *Select individuals for the next generation* (population P^{iii}) one could operate on $[P^{ii}]$ or $[P^i \bigcup P^{ii}]$ using ranking. P^{iii} is, of course, $P_{current}$. Various selection operators such as binary tournament selection with replacement or elitism can be employed as well for limiting the size of P^{iii}. Elitism in the objective domain seems to generate better results since "good" individuals are retained.
- If a *termination predicate condition* is not met, such as maximum number of generations or convergence criteria, set P^{iii} to P as $P_{current}$.
- *Remove* Pareto dominated and infeasible individuals from P^{iii} or repair infeasible individuals. Set P^{iii} to P as $P_{current}$.
- *Retain* an *archive* of nondominated and feasible individuals by storing P^{iii} in an archive P^{iv}. As the new population P^{iii} is merged with the archive, the nondomination operator is applied to the merged combination. The P^{iv} archive contains P_{known} and associated PF_{known}.
- *Local search* operations in hybrid or memetic MOEAs can also provide good performance by exploring limit regions in objective space [879]; i.e., only moving towards specific regions on the Pareto front.

Examples of MOEAs using archiving are PAES [886], SPEA [1782], SPEA2 [1775], the microGA [284], MOMGA [1632], MOMGA-II [1790], and MOMGA-III [341]. Considering algorithm efficiency, one should evaluate complexity of proposed MOEAs and order them in terms of complexity as a function of population size. An analysis of MOEA complexity is presented in Section 6.3.8.

As implied in the list of possible MOEA operators, most MOEAs follow a pattern of initializing a population of individuals, then executing a generational loop with evolutionary operators, ranking individuals and keeping nondominated solutions in an archive filtered for diversity. Figure 2.13 is a meta-level Generic MOEA pseudo code representation of this concept. Observe that feasibility operations are not included since they could be embedded in the generational loop or outside the loop at the end. The vast majority of

MOEA algorithmic structures adhere to this generic structure, the differences being in the specific operator details as shown in the MOEA pseudo codes of Section 2.3. Applying possible MOEA operator insight then results in a variety of Multi-Objective Evolutionary Algorithm (MOEAs) designs as presented in Section 2.3. In the next section, we address the issues of individual MOEA performance, and we provide an understanding of specific operator utility.

Initialize population P and P^{iv}
Evaluate Objective $F(x)$ values over population
Assign Rank Based on Pareto Dominance
Compute Niche Count
Assign Shared Fitness or Crowding
While not terminal condition (number of generations or other)
    Selection of "good" individuals from $P \rightarrow P^{i}$,
    Recombination, mutation of individuals in $P^i \rightarrow P^{ii}$
    Evaluate Objective Values of Children P^{ii}
    Rank (P^i union P^{ii}) $\rightarrow P^{iii}$ based on Pareto Dominance
    Compute Niche Count
    Assign Shared Fitness or Crowding
    Reduce $P^{iii} \rightarrow P$
    Copy $P^{iii} \rightarrow P^{iv}$ based on Pareto Dominance
End While

Fig. 2.13. Generic MOEA Pseudo code

2.3 Structures of Various MOEAs

Various historical MOEAs are presented noting that many continue to be modified and improved in newer versions. Most generational MOEAs implicitly process building blocks (BBs) while a few others such as the MOMGA [1632] explicitly process BBs. As indicated, BB structures are different on different vectors in the objective space (phenotype space). This situation is reflected in different goal performances for various MOEAs on a given MOP.

2.3.1 Multi-Objective Genetic Algorithm (MOGA)

Carlos M. Fonseca and Peter J. Fleming [504] proposed a variation of Goldberg's technique called "Multi-Objective Genetic Algorithm" (MOGA), in which the rank of a certain individual corresponds to the number of chromosomes in the current population by which it is dominated. Consider, for example, an individual $\mathbf{x}_i$ at generation t, is dominated by $p_i^{(t)}$ individuals in the current generation; thus, an individual is assigned a rank by the following rule: $rank(\mathbf{x}_i, t) = 1 + p_i^{(t)}$ [504]. Figure 2.14 shows a variety of dominated points. Figure 2.15 represents the MOGA dominance based assignment based upon fitness.

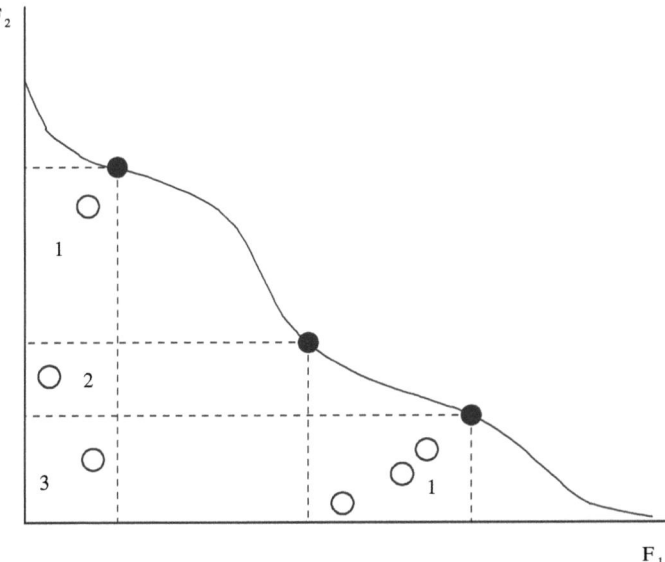

Fig. 2.14. MOGA fitness domination; raw fitness is the number of dominating solutions as shown in the picture

The pseudo code of MOGA is shown in Figure 2.16 with the more formal algorithmic pseudo code in Algorithm 1.[6] Note that the first edition of this book uses the more generic pseudo code form [287]. Observe that $\mathcal{N}'$ refers to the population size, g is the specific generation, $f_j(\mathbf{x}_k)$ is the $j'th$ objective function, $\mathbf{x}_k$ is the $k'th$ individual, $\mathbb{P}'$ the population.

All nondominated MOGA individuals are assigned rank 1, while dominated ones are penalized according to the population density of the corresponding region of the trade-off surface.

Fitness assignment is performed in the following way [504]:

1. Sort population according to rank.

---

[6] Observe that we generally use the algorithmic pseudo code template for additional MOEA descriptions. This is done in order to present a more precise algorithmic description that is useful for understanding and implementation. Also with the two example pseudo code descriptions, one can transform one to the other quickly for pedagogical presentation.

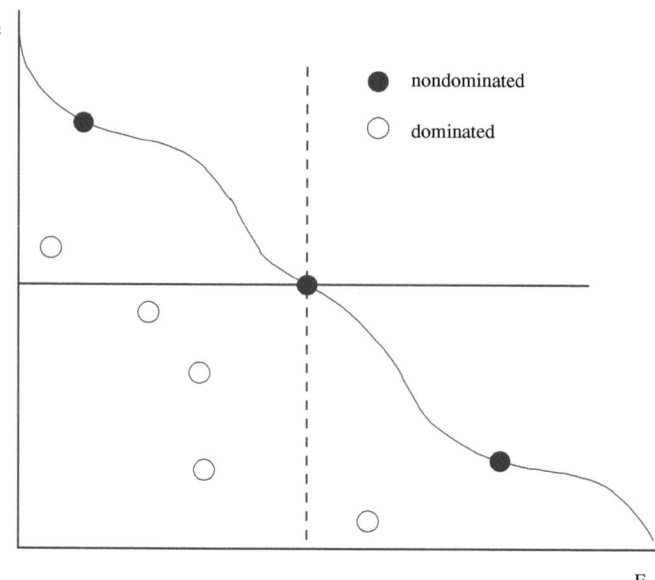

Fig. 2.15. MOGA's Dominance Fitness Assignment

```
Initialize Population
Evaluate Objective Values
Assign Rank Based on Pareto Dominance
Compute Niche Count
Assign Linearly Scaled Fitness
Assign Shared Fitness
For i = 1 to number of Generations
    Selection via Stochastic Universal Sampling
    Single Point Crossover
    Mutation
    Evaluate Objective Values
    Assign Rank Based on Pareto Dominance
    Compute Niche Count
    Assign Linearly Scaled Fitness
    Assign Shared Fitness
End Loop
```

Fig. 2.16. MOGA Pseudo code

2. Assign fitness to individuals by interpolating from the best (rank 1) to the worst (rank $n \leq \mathcal{N}'$) in the way proposed by David E. Goldberg [581] according to some function, usually linear, but not necessarily.

Algorithm 1 MOGA algorithm

1: **procedure** MOGA($\mathcal{N}', g, f_k(\mathbf{x})$) ▷ $\mathcal{N}'$ members evolved g generations to solve $f_k(\mathbf{x})$
2:     Initialize Population $\mathbb{P}'$
3:     Evaluate Objective Values
4:     Assign Rank based on Pareto Dominance
5:     Compute Niche Count
6:     Assign Linearly Scaled Fitness
7:     Shared Fitness
8:     **for** $i=1$ to g **do**
9:         Selection via Stochastic Universal Sampling
10:        Single Point Crossover
11:        Mutation
12:        Evaluate Objective Values
13:        Assign Rank Based on Pareto Dominance
14:        Compute Niche Count
15:        Assign Linearly Scaled Fitness
16:        Assign Shared Fitness
17:     **end for**
18: **end procedure**

3. Average the fitnesses of individuals with the same rank, so that all of them will be sampled at the same rate. This procedure keeps the global population fitness constant while maintaining appropriate selective pressure, as defined by the function used.

As Goldberg and Deb [583] indicate, this type of blocked fitness assignment is likely to produce a large selection pressure that might produce premature convergence. To avoid that, Fonseca and Fleming [504] use a niche-formation method to distribute the population over the Pareto-optimal region, but instead of performing sharing on the parameter values, they use sharing on the objective function values [1509]. Note that MOGA has been also hybridized with neural networks in an attempt to improve its performance [413].

2.3.2 Nondominated Sorting Genetic Algorithm (NSGA)

N. Srinivas and Kalyanmoy Deb [1509] proposed another variation of Goldberg's approach called the "Nondominated Sorting Genetic Algorithm" (NSGA).

The Nondominated Sorting Genetic Algorithm (NSGA) is another modification to the ranking procedure originally proposed by Goldberg [1508]. The pseudo code for this MOEA is given in Figure 2.17 and Algorithm 2.

This NSGA algorithm is based on several layers of classifications of the individuals. Before selection is performed, the population is ranked on the basis of nondomination: all nondominated individuals are classified into one category (with a dummy fitness value, which is proportional to the population size, to provide an equal reproductive potential for these individuals).

```
Initialize Population
Evaluate Objective Values
Assign Rank Based on Pareto Dominance in Each "Wave"
Compute Niche Count
Assign Shared Fitness
For i = 1 to G
    Selection via Stochastic Universal Sampling
    Single Point Crossover
    Mutation
    Evaluate Objective Values
    Assign Rank Based on Pareto Dominance in Each "Wave"
    Compute Niche Count
    Assign Shared Fitness
End Loop
```

Fig. 2.17. NSGA Pseudo code

Algorithm 2 NSGA-I algorithm

1: **procedure** NSGA-I($\mathcal{N}', g, f_j(\mathbf{x}_k)$)   ▷ $\mathcal{N}'$ members evolved g generations to solve $f_k(\mathbf{x})$
2:     Initialize Population $\mathbb{P}'$
3:     Evaluate Objective Values
4:     Assign Rank Based on Pareto dominance in Each *Wave*
5:     Compute Niche Count
6:     Assign Shared Fitness
7:     **for** i=1 to g **do**
8:         Selection via Stochastic Universal Sampling
9:         Single Point Crossover
10:        Mutation
11:        Evaluate Objective Values
12:        Assign Rank Based on Pareto dominance in Each *Wave*
13:        Compute Niche Count
14:        Assign Shared Fitness
15:    **end for**
16: **end procedure**

To maintain the diversity of the population, these classified individuals are shared with their dummy fitness values. Then this group of classified individuals is ignored and another layer of nondominated individuals is considered. The process continues until all individuals in the population are classified. Stochastic remainder proportionate selection is adopted for this technique. Since individuals in the first front have the maximum fitness value, they always get more copies than the rest of the population. This allows for a better search of the PF_{known} regions and results in convergence of the population toward such regions. Sharing, by its part, helps to distribute the population over this region (i.e., the Pareto front of the problem). As a result, one might think that this MOEA converges rather quickly; however, a computational bottleneck occurs with the fitness sharing mechanism. The NSGA was relatively successful during several years (see for example [1692, 145, 1340]), although several comparative studies of the time [260, 1626] indicated that it was outperformed by both MOGA [504] and NPGA [709]. The NSGA was also a highly inefficient algorithm because of the way in which it classified individuals.

Deb et al. [363, 374] have proposed an improved version of the NSGA algorithm, called NSGA-II. The pseudo code of the NSGA-II is shown in Algorithm 3.

Algorithm 3 NSGA-II algorithm

1: **procedure** NSGA-II($\mathcal{N}', g, f_k(\mathbf{x}_k)$)   ▷ $\mathcal{N}'$ members evolved g generations to solve $f_k(\mathbf{x})$
2:     Initialize Population $\mathbb{P}'$
3:     Generate random population - size $\mathcal{N}'$
4:     Evaluate Objective Values
5:     Assign Rank (level) Based on Pareto dominance - *sort*
6:     Generate Child Population
7:         Binary Tournament Selection
8:         Recombination and Mutation
9:     **for** $i = 1$ to g **do**
10:        **for** each Parent and Child in Population **do**
11:            Assign Rank (level) based on Pareto - *sort*
12:            Generate sets of nondominated vectors along PF_{known}
13:            Loop (inside) by adding solutions to next generation starting from the *first* front until $\mathcal{N}'$ individuals found determine crowding distance between points on each front
14:        **end for**
15:        Select points (elitist) on the lower front (with lower rank) and are outside a crowding distance
16:        Create next generation
17:            Binary Tournament Selection
18:            Recombination and Mutation
19:     **end for**
20: **end procedure**

The nondominated sorting algorithm-II (NSGA-II) is a generic non-explicit BB MOEA applied to multiobjective problems (MOPs)–based on the original design of NSGA. As shown in Figure 2.18, it builds a population of competing individuals, ranks and sorts each individual according to nondomination level, applies Evolutionary Operations (EVOPs) to create new pool of offspring, and then combines the parents and offspring before partitioning the new combined pool into fronts. The NSGA-II then conducts niching by adding a crowding distance to each member. It uses this crowding distance in its selection operator to keep a diverse front by making sure each member stays a crowding distance apart. This keeps the population diverse and helps the algorithm to explore the fitness landscape. This MOEA is currently used in most MOEA comparisons. It has also been used as a foundation for other algorithm designs like the multiobjective BOA [845].

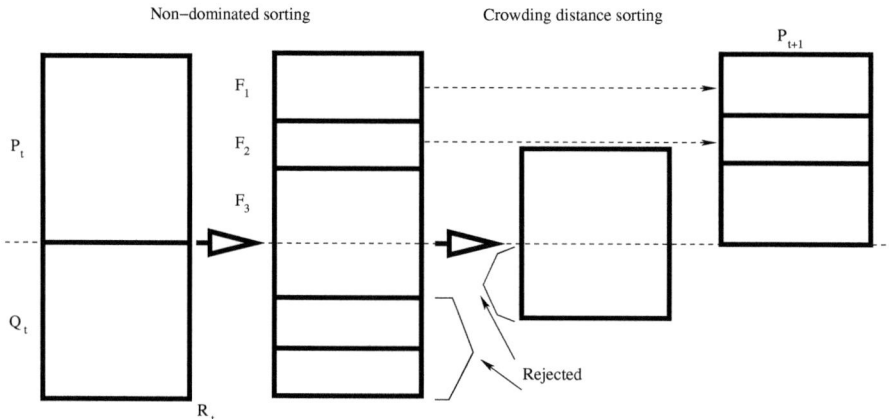

Fig. 2.18. Flow diagram that shows the way in which the NSGA-II works. P_t is the parents population and Q_t is the offspring population at generation t. F_1 are the best solutions from the combined populations (parents and offspring). F_2 are the second best solutions and so on.

2.3.3 Niched-Pareto Genetic Algorithm (NPGA)

Jeffrey Horn and his colleagues [708, 709] proposed a tournament selection MOEA based on Pareto dominance defined as the Niched-Pareto Genetic Algorithm (NPGA) The pseudo code of the NPGA is shown in Algorithm 4 [708]. Two individuals randomly chosen are compared against a subset from the entire population (typically, around 10% of the population). If one of them is dominated (by the individuals randomly chosen from the population) and the other is not, then the nondominated individual wins. When both competitors are either dominated or nondominated (i.e., there is a tie), the result of the tournament is decided through fitness sharing [587]. This is a generational MOEA with implicit BB manipulation.

Horn et al. [708, 709] also suggested a form of fitness sharing in the objective domain, with a metric combining both the objective and the decision variable domains, leading to what the authors called *equivalent class sharing*.

Algorithm 4 NPGA algorithm

1: **procedure** NPGA($\mathcal{N}, g, f_k(\mathbf{x})$)   ▷ $\mathcal{N}'$ members evolved g generations to solve $f_k(\mathbf{x})$
2:     Initialize Population P
3:     Evaluate Objective Value
4:     **for** $i=1$ to g **do**
5:         Specialized Binary Tournament Selection
6:         **Begin**
7:             **if** Only Candidate 1 dominated **then**
8:                 Select Candidate 2
9:             **else if** Only Candidate 2 dominated **then**
10:                 Select Candidate 1
11:             **else if** Both are Dominated or Nondominated **then**
12:                 Perform specialized fitness sharing
13:                 Return Candidate with lower niche count
14:             **end if**
15:         **End**
16:         Single Point Crossover
17:         Mutation
18:         Evaluate Objective Values
19:     **end for**
20: **end procedure**

Erickson et al. [453] proposed the NPGA 2, which uses Pareto ranking but keeps tournament selection (solving ties through fitness sharing as in the original NPGA). The pseudo code of the NPGA 2 is shown in Algorithm 5. Niche counts in the NPGA 2 are calculated using individuals in the partially filled next generation, rather than using the current generation. This is called *continuously updated fitness sharing*, as proposed by Oei et al. [1205].

2.3.4 Pareto Archived Evolution Strategy (PAES)

The Pareto Archived Evolution Strategy (PAES) was designed and implemented by Joshua D. Knowles and David W. Corne [886]. The conceptual approach is quite simple as shown in the pseudo code of Algorithm 6.

PAES consists of a (1+1) evolution strategy (i.e., a single parent that generates a single offspring) in combination with a historical archive that records some of the nondominated solutions previously found. This archive is used as a reference set against which each mutated individual is being compared. This is analogous to the tournament competitions held with the NPGA [709]. PAES also uses a novel approach to keep diversity, which consists of a crowding procedure that divides objective space in a recursive manner. Each solution is placed in a certain grid location based on the values of its objectives (which are used as its "coordinates" or "geographical location"). A map of such grid is maintained, indicating the number of solutions that reside in each grid location. Since the procedure is adaptive, no extra parameters are required (except

Algorithm 5 NPGA 2 algorithm

1: **procedure** NPGA 2($\mathcal{N}', g, f_k(\mathbf{x})$) ▷ $\mathcal{N}'$ members evolved g generations to solve $f_k(\mathbf{x})$
2:    Initialize Population $\mathbb{P}'$
3:    Evaluate Objective Values
4:    **for** i=1 to g **do**
5:      Specialize Binary Tournament Selection **using rank as domination degree**
6:      **Begin**
7:         **if** Only Candidate 1 dominated **then**
8:           Select Candidate 2
9:         **else if** Only Candidate 2 dominated **then**
10:           Select Candidate 1
11:         **else if** Both are dominated or nondominated **then**
12:           Perform specialized fitness sharing
13:           Return Candidate with lower niche count
14:         **end if**
15:      **End**
16:      Single Point Crossover
17:      Mutation
18:      Evaluate Objective Values
19:    **end for**
20: **end procedure**

Algorithm 6 PAES algorithm

1: **procedure** PAES($f_k(\mathbf{x})$)
2:    **repeat**
3:      Initialize Single Population parent, $\mathcal{C}$, and add to archive, $\mathbb{A}$
4:      Mutate $\mathcal{C}$ to produce child $\mathcal{C}'$ and evaluate fitness
5:      **if** $\mathcal{C} \succ \mathcal{C}'$ **then**
6:         discard $\mathcal{C}'$
7:      **else if** $\mathcal{C} \succ \mathcal{C}'$ **then**
8:         replace $\mathcal{C}$ with $\mathcal{C}'$, and add $\mathcal{C}$ to $\mathbb{A}$
9:      **else if** $\exists_{\mathcal{C}'' \in \mathbb{A}}(\mathcal{C}'' \succ \mathcal{C}')$ **then**
10:         discard $\mathcal{C}'$
11:      **else**
12:         apply test $(\mathcal{C}, \mathcal{C}', \mathbb{A})$ to determine which becomes the new current solution and whether to add $\mathcal{C}'$ to $\mathbb{A}$
13:      **end if**
14:    **until** termination criteria is met
15: **end procedure**

for the number of divisions of the objective space). Furthermore, the procedure has a lower computational complexity than traditional niching methods [886]. Figure 2.19 shows a graphical illustration of PAES' adaptive grid. The adaptive grid of PAES and some other issues related to external archives (also

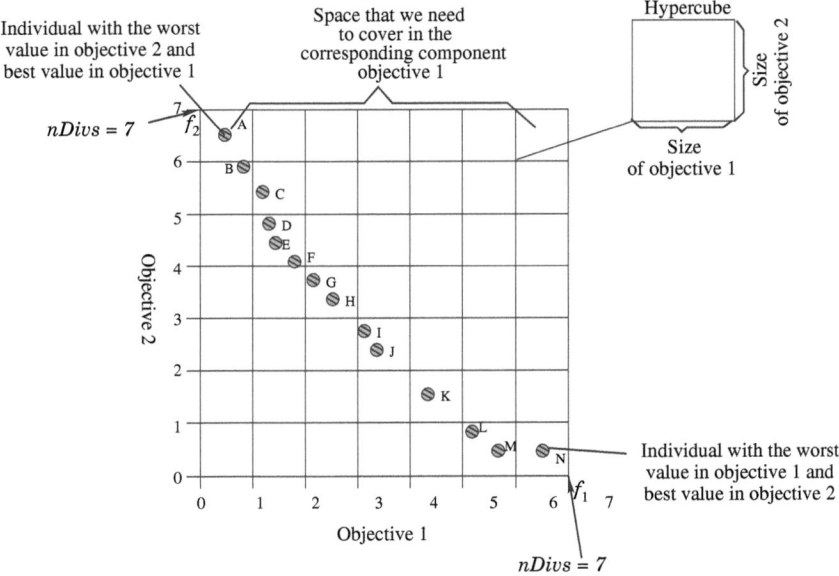

Fig. 2.19. Graphical illustration of the adaptive grid used by PAES.

called "elite" archives) have been studied both from an empirical and from a theoretical perspective (see for example [877, 491]).

Other implementations of PAES were also proposed, namely $(1+\lambda)$-ES and $(\mu + \lambda)$-ES. However, these were deemed to not improve overall performance. A *memetic*[7] version of PAES, called M-PAES was developed as a follow-up to this algorithm [873]. PAES is a convergent *implicit* BB MOEA.

2.3.5 Strength Pareto Evolutionary Algorithm (SPEA)

The *Strength Pareto Evolutionary Algorithm* (SPEA) was introduced by Eckart Zitzler and Lothar Thiele [1782]. This approach was conceived as a way of integrating different MOEAs. The pseudo code of SPEA is shown in Algorithm 7. SPEA uses an external archive containing nondominated solutions previously found (the so-called external nondominated set). At each generation, nondominated individuals are copied to the external nondominated set. For each individual in this external set, a *strength* value is computed. This strength is similar to the ranking value of MOGA [504], since it is proportional to the number of solutions to which a certain individual dominates. In SPEA, the fitness of each member of the current population is computed according to

---

[7] A memetic algorithm donotes the use of local search heuristic with a population-based strategy. The word *memetic* has its roots in the word meme - which was introduced in 1990 by Richard Dawkins in his book "The Selfish Gene" [340]. See Chapter 10 for more details on multi-objective memetic algorithms.

Algorithm 7 SPEA algorithm

1: **procedure** SPEA($\mathcal{N}', g, f_k(\mathbf{x})$)
2:   Initialize Population $\mathbb{P}'$
3:   Create empty external set $\mathbb{E}'$ ($|\mathbb{E}'| < |\mathbb{P}'|$)
4:   **for** $i:=1$ to g **do**
5:     $\mathbb{E}' = \mathbb{E}' \cup \mathcal{ND}(\mathbb{P}')$ ▷ Copy members evaluating to be nondominated of P to E
6:     $\mathbb{E}' = \mathcal{ND}(E)$ ▷ Keep only member evaluating to nondominated vectors in E
7:     Prune $\mathbb{E}'$ (using clustering) if max capacity of $\mathbb{E}'$ is exceeded
8:     $\forall_{i \in \mathbb{P}'}$ Evaluate($\mathbb{P}'_i$) ▷ Evaluate fitness for all member of $\mathbb{E}'$ and $\mathbb{P}'$
9:     $\forall_{i \in \mathbb{E}'}$ Evaluate($\mathbb{E}'_i$)
10:    $\mathcal{MP} \leftarrow \mathcal{T}(\mathbb{P}' \cup \mathbb{E}')$ ▷ Use binary tournament selection with
11:    ▷ replacement to select individuals from $\mathbb{P}' + \mathbb{E}'$
12:    ▷ (multiset union) until the mating pool is full
13:    Apply crossover and mutation on $\mathcal{MP}$
14: **end for**
15: **end procedure**

the strengths of all external nondominated solutions that dominate it. The fitness assignment process of SPEA considers both closeness to the true Pareto front and even distribution of solutions at the same time. Thus, instead of using niches based on distance, Pareto dominance is used to ensure that the solutions are properly distributed along the Pareto front. Although this approach does not require a niche radius, its effectiveness relies on the size of the external nondominated set. In fact, since the external nondominated set participates in the selection process of SPEA, if its size grows too large, it might reduce the selection pressure, thus slowing down the search. Because of this, the authors decided to adopt a technique that prunes the contents of the external nondominated set so that its size remains below a certain threshold. The approach adopted for this sake was a clustering technique called *average linkage method* [1132].

There is also a revised version of SPEA (called SPEA2) whose pseudo code is shown in Algorithm 8 [1775]. SPEA2 has three main differences with respect to its predecessor [1775]: (1) it incorporates a fine-grained fitness assignment strategy which takes into account for each individual the number of individuals that dominate it and the number of individuals to which it dominates; (2) it uses a nearest neighbor density estimation technique which guides the search more efficiently, and (3) it has an enhanced archive truncation method that guarantees the preservation of boundary solutions.

The SPEA2 and NSGA-II are two of the most prominent MOEAs used when comparing a newly designed MOEA. Prevalent in these two MOEAs is the fact that they are *implicit* BB builders and they rely heavily on their density estimator mechanisms.

Algorithm 8 SPEA2 algorithm

1: **procedure** SPEA2($\mathcal{N}', g, f_k(\mathbf{x})$)
2:     Initialize Population $\mathbb{P}'$
3:     Create empty external set $\mathbb{E}'$
4:     **for** $i=1$ to g **do**
5:         Compute fitness of each individual in $\mathbb{P}'$ and $\mathbb{E}'$
6:         Copy all individual evaluating to nondominated vectors $\mathbb{P}'$ and $\mathbb{E}'$ to $\mathbb{E}'$
7:         Use the truncation operator to remove elements from E when the capacity of the file has been extended
8:         If the capacity of $\mathbb{E}'$ has not been exceeded then use dominated individuals in $\mathbb{P}'$ to fill $\mathbb{E}'$
9:         Perform binary tournament selection with replacement to fill the mating pool
10:        Apply crossover and mutation to the mating pool
11:     **end for**
12: **end procedure**

```
For n = 1 to k
    Perform Partially Enumerative Initialization
    Evaluate Each Pop Member's Fitness (w.r.t. k Templates)
    // Primordial Phase
    For i = 1 to Maximum Number of Primordial Generations
        Perform Tournament Thresholding Selection
        If (Appropriate Number of Generations Accomplished)
            Then Reduce Population Size
        Endif
    End Loop
    // Juxtapositional Phase
    For i = 1 to Maximum Number of Juxtapositional Generations
        Cut-and-Splice
        Evaluate Each Pop Member's Fitness (w.r.t. k Templates)
        Perform Tournament Thresholding Selection
            and Fitness Sharing
        P_known(t) = P_current(t) ∪ P_known(t - 1)
    End Loop
    Update k Competitive Templates
        (Using Best Value Known in Each Objective)
End Loop
```

Fig. 2.20. MOMGA Pseudo code

2.3.6 Multiobjective Messy Genetic Algorithm (MOMGA)

The Multiobjective Messy Genetic Algorithm (MOMGA) was proposed by David A. Van Veldhuizen and Gary B. Lamont [1632] as an attempt to extend the messy GA [354] to solve multiobjective optimization problems. The pseudo

Algorithm 9 MOMGA algorithm

1: **procedure** MOMGA($\mathcal{N}, g, f_k(\mathbf{x})$)
2:    **for** $i = 1$ to *epoch* **do**
3:   ▷ PEI Phase
4:       Perform Partially Enumerative Initialization
5:       Evaluate each population member's fitness w.r.t. k templates
6:   ▷ Primordial Phase
7:       **for** $i = 1$ to *Max Primordial Generations* **do**
8:          Perform Tournament Thresholding Selection
9:          **if** Appropriate number of generations accomplished **then**
10:             Reduce Population Size
11:          **end if**
12:       **end for**
13:   ▷ Juxtapositional Phase
14:       **for** $i = 1$ to *Max Juxtapositional Generations* **do**
15:          Cut-and-Slice
16:          Evaluate Each Population member's fitness w.r.t. k templates
17:          Perform Tournament Thresholding Selection and Fitness Sharing
18:          $P_{Known}(t) = P_{current}(t) \cup P_{known}(t-1)$
19:       **end for**
20:       Update k templates   ▷ Using best known value in each objective
21:    **end for**
22: **end procedure**

code of the MOMGA is shown in Figure 2.20 and in Algorithm 9. MOMGA consists of three phases: (1) Initialization Phase, (2) Primordial Phase, and (3) Juxtapositional Phase. In the *Initialization Phase*, MOMGA produces all building blocks of a certain specified size through a deterministic process known as partially enumerative initialization. The *Primordial Phase* performs tournament selection on the population and reduces the population size if necessary. In the *Juxtapositional Phase*, the messy GA proceeds by building up the population through the use of the cut and splice recombination operator.

A revised version of MOMGA (called MOMGA-II) has been proposed by Zydallis et al. [1790]. In this case, the authors extended the fast-messy GA [584]. The pseudo code of the MOMGA-II is shown in Figure 2.21. The fast-messy GA consists also of three phases: (1) Initialization Phase, (2) Building Block Filtering, and (3) Juxtapositional Phase. Its main difference with respect to the original messy GA is in the two first phases. The *Initialization Phase* utilizes probabilistic complete initialization which creates a controlled number of building block clones of a specified size. The *Building Block Filtering Phase* reduces the number of building blocks through a filtering process and stores the best building blocks found. This filtering is accomplished through a random deletion of bits alternated with tournament selection between the building blocks that have been found to yield a population of "good" building blocks. The *Juxtapositional Phase* is the same as in the MOMGA.

This approach is obviously an *explicit BB technique*. Also, the MOMGA-III improves the MOMGA by restructuring the code into an object-oriented form as well as adding better ways of exploring the objective space [341].

For $n = 1$ to k
    Perform Probabilistically Complete Initialization
    Evaluate Each Pop Member's Fitness (w.r.t. k Templates)
// *Building Block Filtering Phase*
    For $i = 1$ to *Maximum Number of BBF Generations*
        If (BBF Required Based Off of Input Schedule)
            Then Perform Building Block Filtering (BBF)
        Else
            Perform Tournament Thresholding Selection
        Endif
    End Loop
// *Juxtapositional Phase*
    For $i = 1$ to *Maximum Number of Juxtapositional Generations*
        Cut-and-Splice
        Evaluate Each Pop Member's Fitness (w.r.t. k Templates)
        Perform Tournament Thresholding Selection
           and Fitness Sharing
        $P_{known}(t) = P_{current}(t) \cup P_{known}(t-1)$
    End Loop
    Update k Competitive Templates
    (Using Best Value Known in Each Objective)
End Loop

Fig. 2.21. MOMGA-II Pseudo code

2.3.7 Pareto Envelope-based Selection Algorithm (PESA)

The Pareto Envelope-based Selection Algorithm (PESA) is suggested by Corne et al. [301]. The pseudo code for the method is given in Algorithm 10. PESA consists of a small internal population and a larger external population. A hyper-grid division of phenotype space is used to maintain selection diversity (application of a crowding measure) as the MOEA runs. Furthermore, this crowding measure is used to allow solutions into the external population via an archive of solutions evaluating to nondominated vectors. A revised version of this MOEA is called PESA-II [299]. The difference between the PESA-I and II is that in the second, selection is region-based and the subject of selection is now a hyperbox, not just an individual (i.e., it first selects a hyperbox, and then it selects an individual within that hyperbox). The motivation behind this approach is to reduce the computational cost associated with Pareto ranking [299]. Finally, these MOEAs are convergent *implicit* BB MOEAs.

Algorithm 10 PESA algorithm

1: **procedure** PESA($\mathcal{N}', f_k(\mathbf{x})$)
2:     Initialize Population $\mathbb{P}'_i$ of size $\mathcal{N}'$ Randomly
3:     Evaluate each member of $\mathbb{P}'_i$
4:     Initialize the external population $\mathbb{P}'_e$ to the empty set
5:     **repeat**
6:         Incorporate individuals evaluating to nondominated vectors from $\mathbb{P}'_i$ into $\mathbb{P}'_e$
7:         Delete the current contents of $\mathbb{P}'_i$
8:         **repeat**
9:             With probability p_c, select two parents from $\mathbb{P}'_e$ ▷ p_c is the probability of crossover
10:            Produce a single child via crossover
11:            Mutate the child created in the previous step
12:            With probability $(1 - p_c)$, select one parent
13:            Mutate the selected parent to produce a child
14:         **until** $\mathbb{P}'_i$ is filled
15:     **until** termination criteria is met
16:     Return($\mathbb{P}'_e$)   ▷ Return the members of $\mathbb{P}'_e$ as the result
17: **end procedure**

2.3.8 The Micro-Genetic Algorithm for Multiobjective Optimization

This approach was introduced by Carlos A. Coello Coello & Gregorio Toscano Pulido [283, 284, 285]. A micro-genetic algorithm is a GA with a small population and a reinitialization process. The way in which the micro-GA works is illustrated in Figure 2.22. First, a random population is generated. This random population feeds the population memory, which is divided in two parts: a replaceable and a non-replaceable portion. The non-replaceable portion of the population memory never changes during the entire run and is meant to provide the required diversity for the algorithm. In contrast, the replaceable portion experiences changes after each cycle of the micro-GA.

The population of the micro-GA at the beginning of each of its cycles is taken (with a certain probability) from both portions of the population memory so that there is a mixture of randomly generated individuals (non-replaceable portion) and evolved individuals (replaceable portion). During each cycle, the micro-GA undergoes conventional genetic operators. After the micro-GA finishes one cycle, two nondominated vectors are chosen[8] from the final population and they are compared with the contents of the external memory (this memory is initially empty). If either of them (or both) remains as nondominated after comparing it against the vectors in this external memory, then they are included there (i.e., in the external memory). This is the

---
[8] This is assuming that there are two or more nondominated vectors. If there is only one, then this vector is the only one selected.

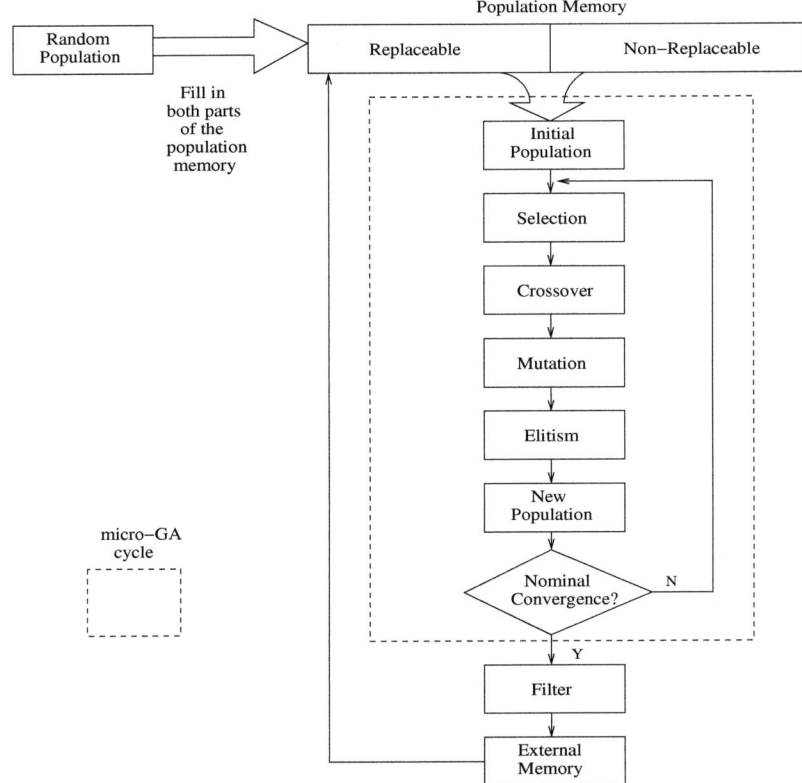

Fig. 2.22. Diagram that illustrates the way in which the micro-GA for multiobjective optimization works [284].

historical archive of nondominated vectors. All dominated vectors contained in the external memory are eliminated.

The micro-GA uses then three forms of elitism: (1) retain nondominated solutions found within the internal cycle of the micro-GA, (2) use a replaceable memory whose contents is partially "refreshed" at certain intervals, and (3) replace the population of the micro-GA by the nominal solutions produced (i.e., the best solutions found after a full internal cycle of the micro-GA).

Although the micro-GA is a very efficient MOEA, its main drawback is that it requires a high number of parameters. This motivated the development of the micro-GA^2 (also called μGA^2) [1597], which uses online adaptation. The way of which the μGA^2 works is illustrated in Figure 2.23. One of the main features of the new approach is the use of a parallel strategy to adapt the crossover operator (i.e., several micro-GAs are executed in parallel). First, the initial crossover operator to be used by each micro-GA is selected. The three crossover operators available are: 1) SBX [362], 2) two-point crossover, and 3) a hybrid crossover operator proposed by the authors of

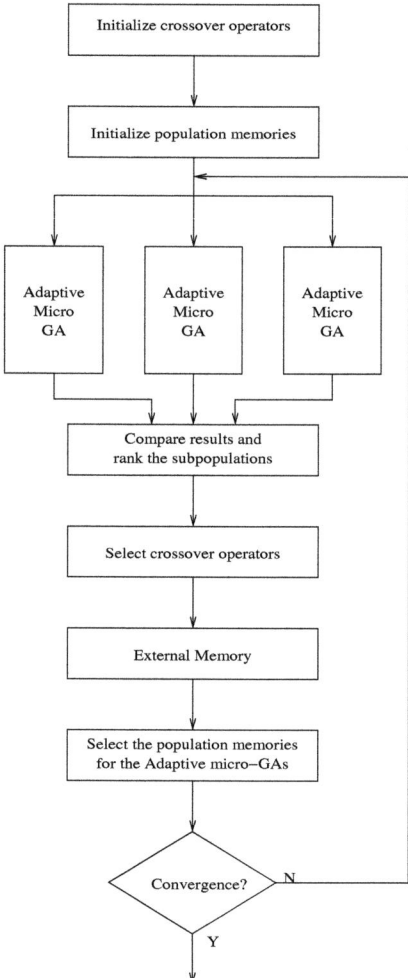

Fig. 2.23. Diagram that illustrates the way in which the μGA^2 works.

this MOEA [1597]. The behavior of this crossover operator depends on the distance between each variable of the corresponding parents: if the variables are closer than the mean variance of each variable, then intermediate crossover is performed; otherwise, a recombination that emphasizes solutions around the parents is applied. These crossover operators were selected because they exhibited the best overall performance in an extensive set of experiments that the authors of this approach conducted. Once the crossover operator has been selected, the population memories of the internal micro-GAs are randomly generated. Then, all the internal micro-GAs are executed, each one using one of the crossover operators available (this is a deterministic process). The nondominated vectors found by each micro-GA are compared against each other and the contribution of each crossover operator is ranked with respect to its effectiveness to produce nondominated vectors. At this point, the crossover operator which exhibits the worst performance is replaced by the one with the best performance. The external memory stores the globally nondominated solutions, and the new population memories (of every internal micro-GA) are filled using this external memory. The new external memories of the micro-GA are identical to this external memory. When all these processes are completed, convergence is checked. For this sake, it is assumed that convergence has been reached when none of the internal micro-GAs can improve the solutions previously reached. The rationale here is that if no new solutions have been found within a certain (reasonably large) amount of time, it is fruitless to continue the search.

The μGA^2 works in two stages: the first one starts with a conventional evolutionary process and it concludes when the external memory of each slave process is full or when at least one slave has reached convergence (as assumed in the previous paragraph). The second stage is finished when global convergence (i.e., when all of the slaves have converged) is reached. An interesting aspect of the μGA^2 is that it attempts to balance between exploration and exploitation by changing the priorities of the genetic operators. This is done during each of the two stages previously described. During the first stage, exploration is emphasized and during the second, exploitation is emphasized. The stages are the following:

- **Exploration stage**: At this stage, mutation has more importance than crossover so that the most promising regions of the search space can be located. At this point, a low crossover rate is adopted and the mutation operator is the main responsible of directing the search. The nominal convergence (i.e., the internal cycle of the micro-GA) is also decreased, since there is no interest in recombining solutions at this point.
- **Exploitation stage**: At this stage, the crossover operator has more importance and therefore nominal convergence is increased to reach better results.

2.3.9 Multiobjective Struggle GA (MOSGA)

The Multiobjective Struggle Genetic Algorithm (MOSGA) [46, 47] combines the struggle crowding genetic algorithm [608] with a Pareto based ranking scheme. The algorithm has the same pattern as the struggle algorithm where two parents are chosen at random from the population, and the normal crossover and mutation is performed to create a child. The child then competes with the most similar individuals in the entire population. The child replaces similar individuals if the child has a better ranking—counteracting genetic drift. The ranking method employed is the same as that adopted in MOGA [504].

Algorithm 11 MOSGA [48]

1: **procedure** MOSGA($\mathcal{N}', g, f_k(\mathbf{x})$)
2:     Initialize Population $\mathbb{P}'$
3:     **repeat**
4:         **for** ($i = 1$ to g) **do**
5:             Randomly Select p parents from $\mathbb{P}'$
6:             Apply EVOPs to create a child
7:             Calculate the rank of the child
8:             Rank the entire population with the new child
9:             Locate the most similar individual
10:            **if** New child's ranking is better than the similar individual **then**
11:                 Replace the similar individual with new child
12:                 Update the ranking of the entire population
13:            **end if**
14:         **end for**
15:     **until** Stopping criterion is met
16: **end procedure**

Although this MOEA has the flavor of being a simple variation of MOGA, the approach is devised to counteract genetic drift which is known to spoil population diversity [46, 47]. An advancement to this algorithm is a technique to assess the robustness of optimal solutions generated by the MOEA. Generational information is extracted from the MOEA to construct a response surface and a good estimate of the robustness of the Pareto front. Again, this algorithm is a generational MOEA and also an *implicit* BB MOEA.

2.3.10 Orthogonal Multi-Objective Evolutionary Algorithm (OMOEA)

The Orthogonal Multi-Objective Evolutionary Algorithm (OMOEA) process begins with a strict definition of the MOP constraints involved for a particular problem to solve. These constraints are considered when Pareto dominance is defined. The algorithm starts by defining a single niche in the decision variable space χ. This niche is recursively split into a group of sub-niches over and over again until a stopping criteria is satisfied. This partitioning forces a uniform search. The pseudo code for OMOEA is given in Algorithm 12 where $\mathbb{P}'$ denotes the global population and Ψ denotes the set of all sub-niches [1755, 1757].

Generally, this MOEA performs well; however, a couple of shortcomings were found by its authors [1756]:

1. Strong interaction (high epistasis) between variables degrades the performance of OMOEA in both precision and distribution of the PF_{known} vectors.
2. As the number of objectives increases, the number of solutions increases exponentially.

Algorithm 12 OMOEA [1755, 1757]

1: **procedure** OMOEA I$(\mathcal{N}', f_k(\mathbf{x}))$ ▷ $\mathcal{N}'$ members evolved until a specified precision is found for $f_k(\mathbf{x})$
2:   Input decision space χ as initial niche.
3:   Evolve niches into $\mathbb{P}'_{\mathcal{N}}(1)$
4:   Split the niche into a group of $\Psi_{\mathcal{N}}$ sub-niches.
5:   Initialize $\mathbb{P}'$ and Ψ
6:   $\mathbb{P}' \leftarrow \mathbb{P}'_{\mathcal{N}}(1); \Psi \leftarrow \Psi_{\mathcal{N}}$
7:   $gen = 1$
8:   **repeat**
9:     **for** (Each $\chi_{\mathcal{N}}^{(s)} \in \chi$) **do**
10:      Evolve $\chi_{\mathcal{N}}^{(s)}$ and yield $P_{\mathcal{N}}^{(s)}(1)$
11:      Split the niche into a group of $\Psi_{\mathcal{N}}$ niches.
12:    **end for**
13:    $\Psi \leftarrow \bigcup \Psi_{\mathcal{N}}; \mathbb{P}' : \mathbb{P}' \leftarrow \bigcup \mathbb{P}_{\mathcal{N}}'^{(s)}(1)$
14:    $gen = gen + 1$
15: **until** (current $\mathbb{P}'$ does not reach the required precision, and the solution number of $\mathbb{P}'$ is not more than a critical value)
16: Output $\mathbb{P}'$ as the satisfying close-to-Pareto-optimal set of MOP
17: **end procedure**

Algorithm 13 OMOEA-II [1756]

1: **procedure** OMOEA-II$(\mathcal{N}, f_k(\mathbf{x}))$
2:   Randomly create population $\mathbb{P}_0$ with size $\mathcal{N}$.
3:   Counter $t \leftarrow 0$
4:   **repeat**
5:    Apply Crossover Operator on $\mathbb{P}_t$ resulting in $\mathbb{P}'_t$ offspring ▷ $|\mathbb{P}_t| = |\mathbb{P}'_t|$
6:    $\mathbb{P}''_t = \mathbb{P}_t \cup \mathbb{P}'_t$
7:    Perform Selection on $\mathbb{P}''_t$ resulting in $\mathbb{P}_{t+1}$
8:    $t = t + 1$
9:   **until** Stopping Criteria Satisfied
10: Output $\mathbb{P}_t$
11: **end procedure**

These shortcomings listed above are not unheard of for MOEAs. As a matter of principle, MOEA designers must recognize both of these problems when developing a new MOEA. To address these limitations, the OMOEA-II was proposed in [1756]. The modification to the OMOEA is to reduce the size of the orthogonal array in order to exploit optimality within a relatively small space. The pseudo code of the OMOEA-II is presented in Algorithm 13. Finally, this is a convergent MOEA that *implicitly* seeks BBs.

Algorithm 14 GENMOP algorithm

1: **procedure** GENMOP($\mathcal{N}, g, f_k(\mathbf{x})$)
2:     Initialize Parent Population $\mathbb{P}_p$ of size $\mathcal{N}$
3:     Evaluate, Rank, Normalize and Save Parent Population
4:     **for** $i=1$ to g **do**
5:         Initialize Children and Mating Pool
6:         Fill Mating Pool with Parents by Rank
7:         **for** $j = 1$ to $size$(children pool) **do**
8:             Statistically select EVOP (weighted section based on previous good/bad children record)
9:             Apply selected EVOP on Children and Mating Pool once
10:            Store EVOP used with new child
11:         **end for**
12:         Mutate new Children
13:         Evaluate new Children
14:         Combine Parents with new Children into a new Parent Pool
15:         Rank, Normalize and Save new Parent Pool
16:     **end for**
17: **end procedure**

2.3.11 General Multiobjective Evolutionary Algorithm (GENMOP)

The General Multiobjective Evolutionary Algorithm (GENMOP) is a general MOEA designed at the US Air Force Institute of Technology (AFIT). GENMOP employs numerous operators to select from when conducting evolutionary operators (EVOPs). As the search progresses, it more often chooses EVOPs that repeatedly produce better solutions. The algorithm works on the supposition that operators that continuously produce better solutions will, in the future, continue to produce good solutions. The pseudo code for GENMOP is given in Algorithm 14. In addition to the pseudo code a program flow/population growth diagram is presented in Figure 2.24 to illustrate the flow population members throughout execution of the MOEA. GENMOP is a generational *implicit* MOEA that can be used on generic problems because it can adapt its operator use to those that provide better solutions.

Other techniques that have been adopted and have not been discussed in this chapter are the following:

- **Pareto deme-based selection**: These are approaches in which Pareto ranking is applied over several subpopulations that are distributed within some sort of geographical structure. The main idea here is to distribute the effort of checking for nondominance by applying Pareto ranking locally within each (presumably small) subpopulation of a (most likely parallelized) MOEA. Then, an additional mechanism has to be used to determine nondominance with respect to the entire population. However, since normally only locally nondominated individuals participate in this

2.3 Structures of Various MOEAs   109

Table 2.1. Summary of EVOPs, fitness, sharing, and representation for discussed implicit BB MOEAs.

MOEA	EVOPS	Fitness	Sharing	$\mathbb{R}$ or $\{0,1\}$	Explicit or Implicit BB
VEGA	c+m	Value of a single objective	Phenotypic Fitness σ_{share}	$\{0,1\}$	Implicit
MOGA	c+m	Linear interpolation using Fonseca and Fleming's Pareto ranking [504]	Phenotypic Fitness σ_{share}	$\mathbb{R}$ $\{0,1\}$	Implicit
NPGA	c+m	Tournament	Phenotypic Fitness σ_{share}	$\{0,1\}$ $\mathbb{R}$	Implicit
NPGA 2	c+m	Rank Dominance	Phenotypic Continuously Update fit. Technique	$\{0,1\}$ $\mathbb{R}$	Implicit
NSGA	c+m	Dummy fitness using Nondominated sorting	Genotypic (σ_{share} - Fitness)	$\{0,1\}$ $\mathbb{R}$	Implicit
NSGA-II	c+m	Nondominated sorting and crowding	Phenotypic	$\{0,1\}$ $\mathbb{R}$	Implicit
SPEA	c+m	Strength value based on dominance and clustering	Phenotypic $\mathbb{R}$	$\{0,1\}$	Implicit
SPEA2	c+m	Strength value based on dominance and clustering	Density function	$\{0,1\}$ $\mathbb{R}$	Implicit
PAES	m	(1+1)single grid	Phenotypic (Hyperbox - sharing)	$\{0,1\}$ $\mathbb{R}$	Implicit
M-PAES	m	(1+1)single grid	Phenotypic	$\{0,1\}$ $\mathbb{R}$	Implicit
PESA	c+m	Pareto ranking	Phenotypic (Hyperbox - sharing)	$\{0,1\}$	Implicit
PESA-II	c+m	Region-based	Phenotypic (Hyperbox - sharing)	$\{0,1\}$	Implicit
μGA	c+m	Pareto ranking	Phenotypic Grid-based	$\{0,1\}$	Implicit
μGA^2	c+m	Pareto ranking	Phenotypic Grid-based	$\{0,1\}$ $\mathbb{R}$	Implicit
MOSGA	c+m	Linear interpolation using Fonseca and Fleming's Pareto ranking [504]	Phenotypic Fitness σ_{share}	$\{0,1\}$	Implicit
OMOEA	c	Based on sub-niche evolution	Genotypic	$\mathbb{R}$	Implicit
OMOEA-II	c	Nondominated sorting	Phenotypic cluster distance	$\mathbb{R}$	Implicit
GENMOP	c+m	Pareto ranking	Phenotypic Fitness σ_{share}	$\mathbb{R}$	Implicit

110   2 MOP Evolutionary Algorithm Approaches

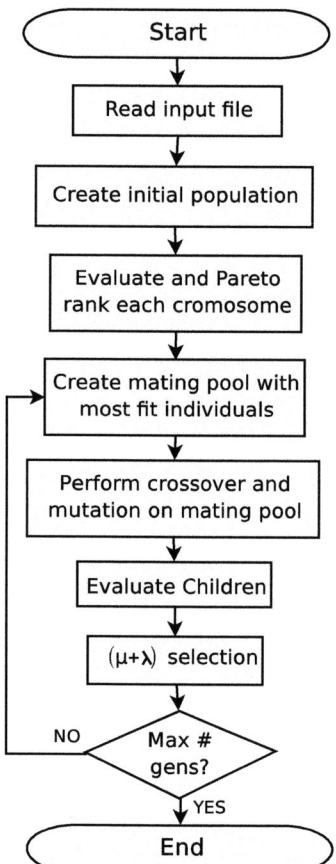

Fig. 2.24. Illustrated is the program flow of the GENMOP. Population of variable length solutions and the evolution process while the algorithm progresses is illustrated. GENMOP pseudo code can be found in Algorithm 14.

mechanism, the procedure is then more efficient than using the entire population of a traditional (sequential) MOEA [1382].
- **Pareto elitist-based selection**: The use of elitism in the context of evolutionary multiobjective optimization has been addressed by several researchers since the mid 1990s. Elitist selection refers to retaining intact the best n individuals ($n \geq 1$) from the current generation to the next one, without applying any operators to them. The use of elitism is known to have great importance when using genetic algorithms to solve single-objective optimization problems [1393]. However, the use of elitism in evolutionary multiobjective optimization is still subject of research [964]. The main idea here is to retain some of the highest ranked individuals in the population (i.e., some nondominated vectors) and then fill the rest of the

population using some other technique (in some cases, dominated vectors are discarded).
- **Hybrid Selection**: These are approaches in which the population capability of a MOEA is exploited, but several selection mechanisms are used along the evolutionary search process. These approaches normally attempt to combine the best of several MOEAs, and combine their selection and/or fitness assignment techniques alternatively at each certain number of generations (the choice of technique to be adopted can be also decided through the use of an uncertainty management technique such as fuzzy logic).

2.3.12 Criticism to Pareto sampling techniques

The main weakness of Pareto ranking in general is that there is no efficient algorithm to check for nondominance in a set of feasible solutions (the conventional process is $O(kM^2)$ for each generation, where k is the number of objectives and M is the population size). Therefore, any traditional algorithm to check for Pareto dominance exhibits a serious degradation in performance as the size of the population and the number of objectives are increased. Also, Pareto ranking becomes inappropriate when dealing with a large number of objectives, because in such cases, all the individuals in the population will soon become nondominated. Additionally, the use of sharing requires to estimate the value of the sharing factor, which is not easy, and the performance of the method normally relies a lot on such value. Nevertheless, and despite its possible disadvantages, Pareto ranking remains as the most popular selection scheme adopted by MOEAs, because of the several advantages that it provides over (linear) aggregating functions.

Besides the general criticism expressed before, there are certain specific comments that have been addressed in the past towards each of the approaches previously discussed:

- **MOGA** : The main criticism towards MOGA has been that it performs sharing on the objective value space, which implies that two different vectors with the same objective function values cannot exist simultaneously in the population under this scheme [1509, 357]. This is apparently undesirable, because these are precisely the kind of solutions that the user normally wants. However, nothing in the algorithm precludes it from performing sharing in parameter value space, and apparently this choice has been taken in some of the applications reported in the literature (see Chapter 7). Also, in its original version, MOGA is a non-elitist MOEA.
  The main advantage of MOGA is that it is efficient and relatively easy to implement [260, 1626]. Its main weakness is that, as all the other Pareto ranking techniques, its performance is highly dependent on an appropriate selection of the sharing factor. However, it is important to note that Fonseca and Fleming [504] have developed a good methodology to compute such value for their approach.

- **NSGA**: Some researchers have reported that NSGA has a lower overall performance than MOGA, and it seems to be also more sensitive to the value of the sharing factor than MOGA [260, 1626]. Other authors [1781] report that the NSGA performed quite well in terms of "coverage" of the Pareto front (i.e., it spreads in a more uniform way the population over the Pareto front) when applied to the 0/1 knapsack problem, but in these experiments no comparisons with MOGA were provided. This is also a non-elitist MOEA.
- **NSGA-II**: The NSGA-II is noticeably more efficient than its previous version, but it also seems to have a questionable exploratory capability. Although the algorithm tends to spread quickly and appropriately when a certain nondominated region is found, it seems to have difficulties to generate nondominated vectors that lie in certain (isolated) regions of the search space [284]. There is also evidence of a notorious search bias of the NSGA-II as the number of objectives increases [1775], although some recent improvements have been introduced in order to deal with this problem (see [912]).
- **PAES**: Despite its efficiency, PAES does not perform well in Pareto fronts that are disconnected. This is due to the fact that PAES is exploratory in nature, and does not keep in the external file the nondominated individuals of the extremes of objective function space. It also stagnates under certain conditions (e.g., in the presence of several disjoint Pareto fronts) [284, 1775].
- **MOMGA, MOMGA-II and MOMGA-III**: Although messy GAs are very powerful, their main drawbacks are related to the exponential growth of their population as the size of the building blocks grows [1114]. Although the fast-messy GA is a good alternative to deal with this problem, it does not solve it completely.
- **Pareto deme-based selection techniques**: To exploit better Pareto deme-based selection techniques, it is desirable to use a parallel MOEA. However, the use of parallelism introduces additional problems to take into account (e.g., the cost of the communication topology adopted). See Chapter 8 for a more detailed discussion of Parallel MOEAs.
- **Pareto elitist-based selection**: The main criticism towards Pareto elitist approaches is that they may not retain diverse enough populations to find and retain a PF_{known} truly representative of PF_{true}, as they retain only $P_{current}(t)$ between generational populations and discard all other solutions. As more and more population members are contained in $P_{current}(t)$ the remaining solutions may not provide enough diversity for effective further exploration. In other words, Pareto elitist approaches, as in single-objective optimization, may introduce a large selection pressure that could cause premature convergence. Therefore, care should be taken of the number of nondominated individuals retained at each generation. Additionally, the use of an efficient approach to maintain diversity is crucial to make effective this sort of technique.

- **Hybrid selection**: Hybrid selection techniques may be advantageous in certain cases (e.g. in multiobjective combinatorial optimization). However, it is by no means obvious how to balance different selection strategies that are applied to the same population. This normally requires an additional mechanism (e.g., fuzzy logic) or additional parameters that may complicate the use of the approach.

2.4 Constraint-Handling Techniques

Handling constraints within a MOEA is an important topic that deserves special attention, particularly when dealing with real-world problems. Most real-world MOPs have constraints that need to be incorporated into our search engine in order to avoid convergence towards infeasible solutions. Constraints can be "hard" (i.e., they must be satisfied) or "soft" (i.e., they can be relaxed) and their proper handling has been a matter of research within single-objective EAs [265].

Normally, the vector $g(\mathbf{x}) \leq 0$ defines the set of MOP constraints (see Section 1.2.2 from Chapter 1). Note that normally, only inequality constraints (i.e., $g(\mathbf{x})$) are considered, because equality constraints can be easily transformed into inequality constraints using, for example:

$$|h(\mathbf{x})| - \epsilon \leq 0 \qquad (2.4)$$

where $h(\mathbf{x}) = 0$ is an equality constraint that we aim to satisfy, and ϵ is the tolerance allowed (a very small value).

The most popular constraint-handling technique both for single-objective and multi-objective EAs are penalty functions [1360]. Exterior penalty functions are the most commonly used in the specialized literature, and their general formulation is the following:

$$\phi(\mathbf{x}) = f(\mathbf{x}) \pm \left[\sum_{i=1}^{n} r_i \times G_i + \sum_{j=1}^{p} c_j \times L_j \right] \qquad (2.5)$$

where $\phi(\mathbf{x})$ is the new (expanded) objective function to be optimized, G_i and L_j are functions of the constraints $g_i(\mathbf{x})$ and $h_j(\mathbf{x})$, respectively, and r_i and c_j are positive constants normally called "penalty factors".

The most common form of G_i and L_j is:

$$G_i = \max[0, g_i(\mathbf{x})]^\beta \qquad (2.6)$$

$$L_j = |h_j(\mathbf{x})|^\gamma \qquad (2.7)$$

where β and γ are normally 1 or 2.

Exterior penalties are usually preferred when using evolutionary algorithms because they do not require an initial feasible solution (as required by interior penalty functions). However, the definition of the penalty factors used in a penalty functions is not straightforward. Ideally, the penalty should be kept as low as possible, just above the limit below which infeasible solutions are optimal (this is called, the *minimum penalty rule* [339, 968, 1490]). This is due to the fact that if the penalty is too high or too low, then the problem might become very difficult for an evolutionary algorithm to solve [339, 968, 1361]. If the penalty is too high and the optimum lies at the boundary of the feasible region, the EA will be pushed inside the feasible region very quickly, and will not be able to move back towards the boundary with the infeasible region. A large penalty discourages the exploration of the infeasible region since the very beginning of the search process. If, for example there are several disjoint feasible regions in the search space, the EA will tend to move to one of them, and will not be able to move to a different feasible region unless such regions are very close from each other. On the other hand, if the penalty is too low, a lot of the search time will be spent exploring the infeasible region because the penalty will be negligible with respect to the objective function [1489]. These issues are very important in EAs, because many of the problems in which they are used have their optimum lying on the boundary of the feasible region [1487, 1490].

The minimum penalty rule is conceptually simple, but it is not necessarily easy to implement. The reason is that the exact location of the boundary between the feasible and infeasible regions is unknown in many of the problems for which EAs are intended (e.g., in many cases the constraints are not given in algebraic form, but are the outcome generated by a simulator [281]).

It is known that the relationship between an infeasible individual and the feasible region of the search space plays a significant role in penalizing such an individual [1360]. However, it is not clear how to exploit this relationship to guide the search in the most desirable direction. Summarizing, the main problem with penalty functions is the definition of good penalty factors that can guide properly the search towards the feasible region. This has triggered a significant amount of research aiming to devise penalty functions that can be easily generalized and that require minimum (or none) parameter tuning (see for example [1404, 475, 1097]).

Note however, that penalty functions are not the only constraint-handling technique available for EAs. Other authors have proposed alternative approaches based, for example, on repairing infeasible solutions in order to make them feasible [1105]. This, however, may be computationally expensive.

Other authors have proposed the use of selection schemes that consider feasible solutions to be superior to infeasible ones (see for example [1291, 360]). This sort of scheme can be easily extended for MOEAs by defining, for example, a binary tournament selection with the three possible cases:

1. If both solutions compared are feasible, use Pareto dominance to define the winner, with the possibility of using a density estimator to break ties (e.g., niche count or crowding distances).
2. If one solution is feasible and the other infeasible, select the feasible one.
3. If both solutions are infeasible, then select the one that is "least" infeasible (e.g., the one with the lowest sum of constraint violation).

Note, however, that when adopting schemes of this sort is important to keep in mind that it is important to preserve at least a few infeasible solutions in the population in order to be able to converge to solutions that lie in the boundary between the feasible and infeasible regions (see for example [1097]).

It is also possible to define a nondominated feasibility ranking technique using a three individual comparison. In this case, a more elaborate set of rules need to be generated to select the tournament winner [1329, 1330].

Despite the important volume of research on constraint-handling techniques for evolutionary algorithms (see for example [1105, 265]), there is a noticeable lack of emphasis on the development of techniques suitable for MOPs. Most researchers assume that techniques used for single-objective EAs are suitable for MOPs as well. This assumption is normally associated with the relative lack of constrained MOPs found in the current literature (see Chapter 4). However, the development of constraint-handling techniques that properly exploit the properties of MOPs is still an open research area.

An interesting research area that has become increasingly popular in the specialized literature is the use of multi-objective optimization concepts to handle constraints in single-objective optimization problems (see [1098] for a survey). The main idea is very intuitive: a constrained single-objective optimization problem is transformed into an unconstrained MOP. From the approaches reported in the specialized literature, we can identify two main ways of performing this transformation:

1. **Bi-Objective Transformation:** In this case, two objectives are considered: the first is the original objective function and the second one is the sum of constraint violation. See for example [1542, 197, 1766, 1736, 1678, 1640].

2. **Multi-Objective Transformation:** In this case, the problem is transformed into an unconstrained MOP, in which we will have $m+1$ objectives, where m is the total number of constraints and the additional objective is the original objective function of our (single-objective) optimization problem. After performing this transformation, any MOEA can be applied to the new problem, and in fact, both population-based approaches (see for example [1250, 264, 274, 990]) and Pareto-based approaches (see for example [511, 262, 279, 50, 1335, 1333, 670, 1229, 797, 793]) have been adopted for this sake. Note, however, that some additional mechanisms are required to guide the search in a proper manner, since constraints and objectives are conceptually different [670]. Thus, when performing

this sort of transformation, it is normally irrelevant to attempt to further optimize a constraint that is already satisfied. Conversely, a solution that represents a good trade-off of the constraints, but remains infeasible, may not be of interest in this case, whereas it would be acceptable in a truly multi-objective problem.

It is important to keep in mind that, when applying these transformations, the aim is to find a single solution (i.e., the constrained global optimum) and not a set of them, as in traditional MOPs.

The main motivations for applying these transformations is to eliminate the need of fine-tuning the penalty factors of a penalty function, and to approach the feasible region in a more efficient way (i.e., requiring a lower number of fitness function evaluations) [1098].

The key for future research in this area is not only to adapt other MOEAs to handle constraints, but to exploit domain knowledge as much as possible (see for example [1329, 947]).

2.5 Critical MOEA Elements

This section contains a brief discussion of the most critical elements associated with MOEAs, including their empirical validation, their theoretical foundations, their fitness function types, their chromosomal representations, and their problem domains.

2.5.1 MOEA Comparisons

MOEA researchers have shown evident concern in developing metrics and performing quantitative comparisons of different techniques. In the origins of evolutionary multiobjective optimization, such comparisons were mainly visual and the test functions had only two or maybe three objectives and a very few decision variables. The Pareto fronts considered were normally convex and had a continuous shape.

Recently researchers have proposed experimental methodologies for *general* MOEA comparative analysis [1781, 1626, 1790]. An extensive discussion on this subject is presented in Chapter 5. Many MOEA publications lacking a more thorough comparative analysis use real-world applications (see Chapter 7). An argument can be made down the lines of "if it works, use it," but in general, using a test problem and/or an application's results to judge comprehensive MOEA usefulness is not conclusive.

2.5.2 MOEA Theory

Less than $1/40^{th}$ of published MOEA papers focus on underlying theoretical analyses of MOEAs. These papers focus mainly on MOEA parameters, behavior, and concepts (see Chapter 6 for a contemporary detailed discussion on MOEA theory). They attempt to further define the nature and limitations of Pareto optimality, the subsequent effects upon MOEA search to determine the necessary conditions to ensure convergence, run-time analysis, fitness landscape analysis, and discuss the characteristics and construction of appropriate MOEA benchmark test function suites. However, this work, although valuable, is evidently insufficient and much more effort in this direction is necessary. As Fonseca and Fleming [506] and Horn [706] have stated: more effort is being spent designing and refining MOEA approaches than on developing accompanying theory.

2.5.3 MOEA Fitness Functions

The catalogued research efforts provide various fitness function types used by MOEAs. Table 2.2 lists several generic fitness function types, their identifying characteristics, and examples of each drawn from the MOEA literature. These listed types are not limited to MOEA applications nor are they the only ones possible. Further examples can be found in [277] and Chapter 7. MOEAs offer the exciting possibility of simultaneously employing different fitness functions to capture desirable characteristics of the problem domain regardless of the implemented MOEA technique.

The fitness functions employed appear limited only by the practitioner's imagination and particular application; several are identified and others must surely exist. However, a fitness function's effectiveness depends on its application in appropriate situations (i.e., it measures some *relevant* feature of the studied problem). The claim by many authors that their particular MOEA implementations are successful imply the associated fitness functions are appropriate for the given problem domains.

Finally, the catalogued efforts clearly show the non-commensurability and independence of many fitness function combinations. For example, optimizing a radio antenna design may involve electromagnetic (energy transmission), geometric (antenna shape), and financial (dollar cost) objectives. The proposed antenna's shape may have no meaningful impact on its cost. Also, these objectives may be measured in megawatts, feet, and euros! These are the factors responsible for the partial ordering of the search space and the subsequent need to develop appropriate MOEA fitness assignment procedures.

2.5.4 MOEA Chromosomal Representations

Theorems exist [498] showing that no intrinsic advantage is provided by any given genetic representation. For any particular encoding and associated cardinality, *equivalent* evolutionary algorithms (in an input/output sense) can be generated *for each individual problem instance*. Although certain gene representations and EVOPs may be more effective and efficient in certain situations, the theorems show that no choice of representation and/or EVOPs operating

Table 2.2. MOEA Fitness Function Types

Category	Characteristic	Examples
Electromagnetic	Energy transfer or reflection	[1117] [1627] [472]
Economic	Production growth	[1470] [607] [84]
Entropy	Information content and (dis)order	[508] [1532] [1382] [924]
Environmental	Environmental benefit or damage	[33] [255] [1607] [414]
Financial	Direct monetary (or other) cost	[66] [1325] [598] [1530] [1443]
Geometrical	Structural relationships	[768] [408] [600]
Physical (Energy)	Energy emission or transfer	[1702] [808] [1247] [346]
Physical (Force)	Exerted force or pressure	[309] [1202] [1646] [669]
Resources	Resource levels or usage	[78] [388] [1470]
Temporal	Timing relationships/Scheduling	[504] [750] [1470] [741]

on one or two parents offers any capability which can not be duplicated by another MOEA instantiation.

The NFL theorems[9] [1708] indicate that if an algorithm performs "well" (on average) for some problem class then it must do worse on average over the remaining problems. In particular, if an algorithm performs better than random search on some problem class then it must perform *worse* than random search on the remaining problems. So, although the NFL theorems imply one MOEA may provide "better" results than another when applied to some problem these other theorems show that that MOEA is not unique. Thus, there appears to be more than one way to skin a cat (or MOP).

Genetic representation is then another MOEA component limited only by the implementor's imagination. The cited efforts indicate the most common representation is a binary string corresponding to some simple mapping from the problem domain. Real-valued chromosomes are also often used in this fashion. And, as in single-objective EAs, combinatorial optimization problems often use a permutation ordering of jobs, tasks, etc. However, some representations are more intricate and therefore notable.

Some MOEAs employ arrays as genome constructs. For example, Baita et al. use a matrix representation to store recessive information [78].[10] Parks and Chow also use matrices as these data structures are more natural representations of their respective problem domains' decision variables [1244, 254]. The Prüfer encoding used by Gen et al. [553] uniquely encodes a graph's spanning tree and allows easy repair of any illegal chromosome. In the known multi-objective Genetic Programming implementations (e.g., [55, 950, 1368, 683, 1408]), a program/program tree representation is used. No matter the representation employed, it is again noted that any claims of "successful" MOEA implementations imply the associated genetic encodings are appropriate for the given problem domain.

2.5.5 MOEA Problem Domains

MOEAs operate on MOPs by definition. A more theoretical discussion of the MOP domain is given in Chapter 4 and elsewhere [1630, 357]. The discussion presented here is in more general terms. When implementing a MOEA it is (implicitly) assumed that the problem domain (fitness landscape) has been examined, and a decision made that a MOEA technique is the most appropriate solution tool for the given MOP. In general, it is accepted that single-objective EAs are useful search algorithms when the problem domain is multidimensional (many decision variables), and/or the search space is very large. Most cited MOEA problem domains appear to exhibit these characteristics.

---

[9] Note that multi-objective extensions of the NFL theorems have been also proposed [300, 298].

[10] As a side note, very few published MOEAs use dominant and recessive genetic information (e.g. [78, 934]).

An overwhelming majority of cited efforts are applied to non-pedagogical problems. This indicates MOEA practitioners are developing and implementing MOEAs as real-world tools. As a quick glance through Chapter 7 shows, these implementations span several disparate scientific and engineering research areas and give credibility to the MOEA's claim as an effective and efficient general purpose search tool.

What differentiates a MOEA from a single-objective EA? What components should be included in a MOEA? When should a MOEA be used? The following section addresses these questions and presents matters of a more philosophical nature raised by the preceding discussion, considering several MOEA design issues.

2.6 MOEA Design Recapitulation

Many MOEAs currently exist (see for example [261, 361, 1219, 290, 1558, 10]). When considering them, those wishing to implement a MOEA may well be asking, "Where do I begin?" An "all purpose" MOEA technique may not be specified (the NFL theorems [1708] do not allow for one). However, certain MOEAs can be suggested as a starting point, and any interested researchers may then select one of these MOEAs to begin their own exploration of the MOP domain.

Definition 13 from Chapter 1 states that an MOP's global optimum is PF_{true}, determined by evaluating each member of P_{true}. Additionally, many *a posteriori* approaches explicitly seek P_{true}. Thus, *a priori* techniques are not *generally* appropriate because they may not be capable of finding each member of P_{true}, and they return only a single solution per MOEA run. The DM's lack of information before search occurs is also a factor.

Although there are several *a posteriori* techniques to consider[11] the focus adopted is on those MOEAs employing elitism and Pareto rank- and niche-based selection. Specifically, it is emphasized to consider the NSGA-II [374], SPEA2 [1775], PAES [886], PESA-II [299], and the microGA for multi-objective optimization [284].

These algorithms stand out because they incorporate known MOEA theory. The Pareto-based selection each employs explicitly seeks P_{true}. All incorporate a density estimator (e.g., niching, crowding, clustering, etc.) in an attempt to uniformly sample PF_{true}. Mating restriction may (or may not) be included in any of the three, as may a secondary population. Finally, their general algorithmic complexity is no higher than other known MOEA techniques, and their source code is (in most cases) available in the Internet (see Appendix H[12]).

---

[11] Progressive approaches incorporate either *a priori* or *a posteriori* techniques; any of the algorithms recommended in this chapter may be used interactively.

[12] All the Appendices of this book are available for download at: http://www.cs.cinvestav.mx/~emoobook

Although each MOEA's authors (and rightly so) point out deficiencies in their own and other MOEAs, any algorithmic approach is bound to have some shortfalls when applied to certain problem classes (c.f., the NFL theorems [1708]). These algorithms' common theme is their respect of known relevant theoretical issues, and their empirical success in both (non-)numeric MOPs and real-world applications. As reported in the specialized literature, these algorithms easily win the title "Most Often Imitated," implying other researchers also see value in them.

Although not straightforward, many existing EA implementations are extendable into the MOEA domain. For example, GENOCOP III [1104] has been readily modified to incorporate both a specialized problem domain code and linear fitness combination technique.

2.7 Summary

This chapter presents an in-depth analysis of MOEA research, discussing in detail several foundational issues such as implemented MOEA techniques and fitness functions, MOEA comparisons and chromosomal representations. More general observations are also made concerning MOEA characteristics and components. Finally, some suggestions regarding how to select a MOEA for a certain application are provided. This analysis identifies appropriate MOEAs recommended for initial use in solving MOPs, and should be used when re-engineering these (or any other) MOEAs to solve particular MOPs.

Further Explorations

Class Exercises

1. Compare (graphically) the ranking schemes adopted by: (1) MOGA [504], (2) NSGA [1509], (3) SPEA [1782], and (4) NSGA-II [374].
2. Apply VEGA [1440] to the following two-objective MOP:

$$\min \quad f_1(x) = x$$
$$\min \quad f_2(x) = (x-5)^2$$
subject to the constraint:
$$-5 \leq x \leq 10$$

   Attempt to do hand-calculations to generate a PF_{known} and P_{known} starting with an initial random population of six individuals.
3. Solve by hand-calculations the MOP of Problem 1 using MOGA [504] and then implement this algorithm and compare its results with respect to VEGA [1440].
4. Consider the following two-objective MOP:

$$\max \quad f_1(x_1, x_2) = x_1 + x_2$$
$$\max \quad f_2(x_1, x_2) = (x_1 + 2)^2 + (x_2)^2$$
subject to the constraints:
$$0 \leq x_1 \leq 1$$
$$0 \leq x_2 \leq 2$$

   Assuming at least four sets of weights, generate the four 3D search landscapes for each weighted fitness function. What are the landscape similarities and differences? Compare smoothness versus roughness.
5. Given the specific two objective Pareto front of exercise 1 and using the linear weighted objective aggregated form, show graphically for selected weights the possible intersects with PF_{true}. How does this relate to possibly generating all the points on the Pareto front using this approach?

6. Adopting the same problem as in the previous exercise, generate points on PF_{true} using the weighted Tcyhebycheff model. Show graphically for selected weights the possible intersect of various equal cost lines (minimization) with the true Pareto front, PF_{true}. How does this relate to possibly generating all the points on the Pareto front using this single objective approach to solving MOPs?
7. Write down at least two different algorithms for ranking a population of individuals in a MOEA, based on Pareto optimality and compare their computational efficiency. Then, analyze the algorithm by Kung et al. [930] and discuss its advantages and disadvantages.
8. Under what circumstances would it be possible to have a Pareto front that consists of a single point? Should a MOEA still work in such case? Discuss.
9. With regard to generating the Pareto optimal set, is a scaling technique required if tournament selection is used? What is the impact of a large range of values as well as fitness values that are very close to each other numerically? How and where would you implement tournament selection in a MOEA?
10. Generate pseudo code for the following MOEAs: 1) Constraint Method-Based Multi-objective EA in [1318] 2) Multi-objective Multi-Criteria EA in [1170], 3) Agent-Based Evolutionary Multi-objective Optimization in [1502], and 4) Simple Multi-Objective Evolutionary Algorithm in [1623].
11. Generate pseudo code for the M-PAES [873] and any other hybrid MOEAs of your choice.
12. Generate flow diagrams for all of the various MOEAs discussed in this chapter and compare flow structures with respect to pseudo code. What are the main similarities and differences?
13. Why does the use of MOEA secondary populations generally permit more exploration of the search space? Relate this to operators also (crossover, mutation, selection).
14. Analyze the time/space complexity (order-of analysis) of MOEAs in the current literature, and compare your analysis with those published (see for example [374, 788]).

Class Software Projects

1. Perform a comparative study of MOEAs whose source code is available in the public domain (e.g., the NSGA-II [374], PAES [886], PESA-II [299], and SPEA2 [1782]).
2. Take from the current literature various pedagogical MOPs that have been solved using various MOEAs. Solve these problems using any of the MOEAs whose source is available in the public domain (see previous project).

3. Implement the OMOEA-II [1756] and compare it with respect to the NSGA-II and SPEA2 using the **DTLZ** test functions [379].
4. Design a graphical interface that facilitates the use of the NSGA-II (e.g., using windows for fine-tuning its parameters).
5. Implement a procedure that reduces the run-time complexity of Pareto ranking (see [788]). Generate time complexity graphs that show the actual gain in CPU time (measured in seconds) with respect to traditional Pareto ranking.

Discussion Questions

1. Select a sample of papers on evolutionary multiobjective optimization from the EMOO repository[13] and perform an analysis that covers, for example, the following aspects:
   - Types of papers (propose a taxonomy for this sake. For example: application-oriented, theoretical, algorithmic design, etc.).
   - Type of algorithm adopted (e.g., the NSGA-II [374], SPEA2 [1775], etc.).
   - Number of objectives of the problem(s) dealt with in each paper.

   See for example the analysis presented in [806, 1626].
2. Some researchers have proposed the use of a micro-genetic algorithm (i.e., a GA with a small population and a reinitialization process) with elitism to solve multiobjective optimization problems [283]. Analyze the computational complexity of this approach. Also, study the role of its (several) parameters on the quality of the solutions produced (consider number of elements of the Pareto optimal set found and closeness to the true Pareto front). Discuss some possible improvements to this algorithm (see for example [1597]).
3. Evolutionary computation researchers have paid relatively little attention to the data structures used to store nondominated vectors. Operation researchers have used (in multi-objective combinatorial optimization), for example, domination-free quad trees where a nondominated vector can be retrieved from the tree very efficiently. Checking if a new vector is dominated by the vectors in one of these trees can also be done very efficiently [626]. Discuss the possible gains (both in terms of algorithm and space complexity) that domination-free quad trees can bring to an evolutionary algorithm that implements Pareto ranking. Also, discuss some of the possible limitations of this data structure.
4. Chen [238] has proposed an algorithm based on a recursive binary division of objective space which is represented by a hyper binary tree structure (the so-called "Pareto Tree Searching Genetic Algorithm" (PTSGA).

---
[13] The EMOO repository is located at:
http://delta.cs.cinvestav.mx/~ccoello/EMOO/

This sort of data structure allows an efficient location (and comparison) of nondominated vectors. What other advantages do you think that this approach may provide? What potential disadvantages does it have? Do you foresee some limitations of this approach? What modifications/improvements do you propose?

5. Knowles et al. [890] have suggested that transforming certain single-objective optimization problems into multiobjective (a process that they call "multi-objectivizing") can remove local optima and therefore, become easier to solve by a hillclimber. Their hypothesis has been validated with certain instances of the traveling salesperson problem. Discuss the main premise of this paper and some of its consequences. Propose another problem (different from the one included in the paper) that you think is appropriate for multi-objectivizing. What are the main drawbacks of the technique? Can you identify some type of problems in which this approach would not work at all? Explain.

6. Analyze the "Incremental Multiple Objective Genetic Algorithm" (IMOGA) presented in [237]. How is this approach different from lexicographic ordering? What do you think that is the source of power of this approach, which presents competitive results with respect to the NSGA-II [374], PAES [886] and SPEA2 [1775], and is able to deal with problems having 4 objectives? Read also about incremental learning [610, 611], which is the idea that inspired the IMOGA.

7. Propose either a variation of an existing MOEA or an entirely new approach. Discuss issues such as inspiration for your algorithm, computational complexity (both time and space complexities), data structures required, performance (using the metrics defined in Chapter 5), main advantages (as compared to traditional MOEAs), and main disadvantages. Was your MOEA designed for a certain type of problems (e.g., combinatorial optimization problems)? Does it use elitism (why or why not?).

8. Perform a comparative study of data structures that have been adopted for secondary (external) populations. Include in your analysis, issues such as computational complexity, memory usage, and ease of implementation.

9. Study the proposal by Alberto and Mateo [30] of using graphs for representing and managing MOEA populations. Implement this scheme and compare it to the use of the adaptive grid of Knowles and Corne [886].

10. What are the advantages and disadvantages of various constraint handling techniques such as additional objectives, static/dynamic penalty functions and time of constraint filtering? (see [265] for a survey on constraint-handling techniques used in evolutionary algorithms). Relate to efficiency (development time, execution time, memory, etc.) and effectiveness (exploration vs. exploitation, "good" Pareto front, etc.)

11. Chen et al. [235] proposed the incorporation of fitness inheritance [1495] to improve the efficiency of a multi-objective evolutionary algorithm (MOEA). Analyze this proposal and criticize it. Compare and contrast elitism with respect to fitness inheritance. Relate fitness inheritance to

the global criterion method discussed in Chapter 1 (Section 1.7.1). Discuss possible ways to extend Chen et al.'s proposal to produce, for example, a more efficient algorithm (see for example [1352, 1354]).
12. Koch and Zell [891] proposed the multi-objective clustering selection evolutionary algorithm. Analyze this proposal and discuss the possible advantages and disadvantages of introducing clustering techniques in a MOEA. Compare this approach to Molyneaux et al.'s proposal [1119]. Discuss computational complexity and parameter fine-tuning of both approaches.
13. Costa and Oliveira [302] proposed an evolution strategy for multiobjective optimization. Analyze this proposal and compare it to other related proposals (see for example [935, 136, 886]).
14. Socha & Kisiel-Dorohinicki [1502] proposed an evolutionary multi-agent system for multiobjective optimization. Compare and contrast this proposal to Menczer et al.'s approach [1088]. Do you see any particular advantages and disadvantages of applying multi-agent systems to multiobjective optimization. Discuss.
15. Valenzuela [1623] proposed a simple evolutionary algorithm for multiobjective optimization. The author of this approach argues that her approach does not require Pareto ranking but only a clever replacement strategy. Analyze this proposal and criticize it. Do you foresee any possible limitations/disadvantages of this algorithm? Compare it to Chakraborti et al.'s [220] algorithm.
16. Mostaghim et al. [1141] discuss three types of quadtrees used to store nondominated vectors and analyze their use in evolutionary multiobjective optimization. Compare this work with the proposal of Everson et al. [462]. Indicate the main motivation to use efficient data structures to store nondominated vectors in the context of evolutionary multiobjective optimization.
17. Current researchers have placed little emphasis in developing approaches in which the number of fitness function evaluations is minimized. This cost reduction is vital in real-world applications. Analyze the strategy proposed by Farina [468] which is based on generalized response surfaces. Compare this strategy to the approach proposed by Duarte et al. [413].
18. Lu and Yen [1021] proposed the Rank-Density based Genetic Algorithm (RDGA). Analyze the ranking strategy adopted by this algorithm as well as the diversity mechanism proposed. Relate the selection and replacement strategies adopted in the RDGA to the cellular genetic algorithm [1699]. Do you see any possible limitations of this algorithm if we consider that it always tries to minimize rank and density values of the population (regardless of the number of objective functions of the problem)? Compare this approach to the non-generational genetic algorithm for multiobjective optimization [1624, 150].
19. Considering gene expression (whether it be real or string valued chromosomes) relate how the biological processes of meiosis and mitosis [74, 75]

are accomplished in MOEAs. Why is it important (or not important) to accurately implement these processes in MOEAs?
20. Analyze the Potential Pareto Regions Evolutionary Algorithm (PPREA) [638, 639], and discuss its main advantages and some of its possible disadvantages.
21. Runarsson and Yao [1405] present two versions of Pareto ranking applied to constraint space, one that considers the objective function value in the ranking process, and another one that does not consider it. These versions are compared to a traditional over-penalized penalty function. The results of this study indicate that Pareto ranking leads to a bias-free search, which led the authors to conclude that this causes to spend most of the time searching in the infeasible region (i.e., the approach was not able to find feasible solutions). Analyze this study and indicate if you agree with the results. If Pareto ranking is not a good way of biasing the search in constrained optimization, why is that there are several (successful) constraint-handling techniques based on Pareto ranking (see for example [670])? Discuss.

Possible Research Ideas

1. Perform a detailed study of the role of crossover in the performance of a certain MOEA (choose anyone you like). Analyze different types of crossover operators (e.g., one-point, two-point, uniform, etc.). Consider a set of benchmark functions such as those discussed in this chapter, and a MOEA that uses elitism and Pareto ranking. See for example [173].
2. Some operation researchers have proposed "unification" procedures that allow to combine several multiobjective optimization techniques under a common framework (see for example [539, 538]). The idea has also been suggested with MOEAs (see for example [963]). Design a common framework where several MOEAs (based, for example, on Pareto ranking) can fit. The framework should also allow to experiment with different genetic operators (i.e., crossover, mutation and elitism variations). Implement this framework and experiment with it. Analyze the potential benefits of such a framework.
3. Propose a new constraint-handling approach for MOEAs. Compare your approach against the use of a penalty function [1360]. Identify advantages and disadvantages of your approach. Then, compare it against techniques that have explicitly been proposed for MOEAs (see for example [1709, 1329]). Use some of the constrained MOPs proposed in the literature [375].
4. Explore the use of methods that allow to reduce the number of evaluations performed by a MOEA. See for example, the use of surrogates [1667, 1334, 818, 819] and the learning of a Gaussian processes model of the search landscape [881, 872].

5. Traditional MOEAs assume that information about the objectives can be obtained with total certainty. However, in real-world applications, is very commonly the case that one is forced to deal with uncertainties [1248]. Investigate the work reported in the literature about uncertainty handling in MOEAs (see for example [1576, 726]). Then, criticize the existing approaches and make your own proposal.
6. Propose a MOEA that uses a target vector approach (see Section 1.7.1 from Chapter 1). Use elitism and an efficient approach to maintain diversity in the population. Conduct an analysis of the complexity of your algorithm and a comparative study of its performance with respect to other MOEAs (e.g., NSGA-II [374], SPEA2 [1775], and NPGA 2 [453]). Can your approach generate Pareto fronts that are concave and discontinuous? Is it competitive with respect to approaches that use Pareto ranking? How efficient is it? What are its main drawbacks?
7. Study the $r(n)$-approximate algorithms proposed by Ehrgott [427] in the context of multiobjective combinatorial optimization. Do you see any relationship of these algorithms with MOEAs? Discuss.
8. Diversity is, with no doubt, a topic that deserves special attention in evolutionary multiobjective optimization. The idea of using diversity as a selection criterion that can guide a MOEA is intriguing, and has been explored by some researchers (see for example the Genetic Diversity Evolutionary Algorithm (GDEA) proposed in [1594]). Devise a new MOEA using this same principle of using diversity as an objective that guides the search.
9. Consider the possible hybridization of evolutionary algorithms with mathematical programming techniques. For example, researchers have proposed hybrids with the ϵ-constraint method (see [1507, 1318, 948]).
10. Analyze the impact of using different fitness assignment schemes in a multi-objective evolutionary algorithm. See for example [188].
11. Study the limitations of MOEAs based on Pareto ranking when dealing with problems that have three or more objective functions (see for example [1304, 1303, 386, 387, 1669, 880]). Compare different schemes that could achieve an objective reduction (see for example [376, 175, 1435]), and propose a new MOEA that can scale properly to a large number of objectives (see for example [1531]).
12. Analyze the possibility of achieving an effective dimensionality reduction that preserves the original nondominance structure of the problem (see for example [175]).
13. Propose a multi-objective evolutionary algorithm that adopts a selection scheme not based on Pareto ranking (see for example [126, 84, 987]).
14. The crowding distance adopted by the NSGA-II [374], despite being very efficient and effective for the bi-objective case, is known to have difficulties when dealing with problems having three or more objectives. Investigate alternative schemes that can overcome this limitation (see for example [912, 901]).

15. One of the current research trends in the design of MOEAs is to base their selection mechanism on a performance measure (see for example the Indicator-based Evolutionary Algorithm (IBEA) [1774, 101, 102] and the S metric selection MOEA (SMS-MOEA) [447, 129]). Discuss this research trend and propose a new selection scheme based on some performance metric that has not been used so far.
16. Analyze several diversity maintenance mechanisms (see for example [1039, 863, 1497, 1619, 984]) and their potential application within a multi-objective evolutionary algorithm.
17. There exist few MOEAs based on a cellular genetic algorithm (see for example [1175, 1152, 1157, 1156]). Propose a new MOEA based on this sort of scheme, and discuss the similarities between the cellular genetic algorithm and the predator-prey scheme [957, 605].
18. The multi-objective extensions of Bayesian Optimization Algorithms are still relatively scarce in the specialized literature (see for example [1613, 1458, 1459, 956, 826, 844, 23]). Propose a new MOEA based on Bayesian Optimization Algorithms and discuss its main advantages with respect to the existing ones.
19. Analyze the possibility of incorporating concepts from thermodynamics to design a new MOEA. See for example the High Performance Multi-Objective Evolutionary Algorithm (HPMOEA) [1784], which adopts Gibbs entropy, and the Thermodynamical Genetic Algorithm (TDGA) [863], which adopts the principles of temperature and entropy as in simulated annealing.
20. The incorporation of concepts from classifier systems into the design of a MOEA has been very scarce (see for example [1624]). Propose a new MOEA based on concepts from this area.
21. In the last few years, there has been an increasing interest in a special type of evolutionary algorithm called "Estimation of Distribution Algorithms" (EDAs), which do not have crossover or mutation, but use instead probability distributions to generate the new population. Such probability distributions are estimated from a set of selected individuals generated at the previous generation [954]. Some researchers have proposed multi-objective versions of EDAs (see for example [1579, 1578, 155]). Propose a new multi-objective EDA and discuss its advantages and possible disadvantages with respect to state-of-the-art MOEAs.
22. Propose a multi-objective extension of a successful single-objective EA. Such an extension should obviously preserve (as much as possible) the advantages of its single-objective counterpart. See for example the MO-CMA-ES [737], that extends the covariance matrix adaptation evolution strategy (CMA-ES) [654], which is a very powerful evolutionary algorithm for real-valued single-objective optimization.
23. Propose a new MOEA inspired on concepts from quantum computing. See for example [855].

3
MOEA Local Search and Coevolution

> When two opposite points of view are expressed with equal intensity, the truth does not necessarily lie exactly halfway between them. It is possible for one side to be simply wrong.
>
> Richard Dawkins

3.1 Introduction

In order to make multiobjective evolutionary algorithms (MOEAs) more beneficial to real-world applications, local search structures have been proposed to drive the search towards the Pareto front more effectively and efficiently. A number of generic local search techniques have been proposed along with problem domain specific methods. These approaches are discussed in this chapter with thoughts on integrating new innovative local search with MOEAs. Another emerging area of MOEA research is applying coevolutionary techniques. Relatively few researchers have explored the idea of combining coevolution with MOEAs. This chapter presents various researchers' algorithmic processes for Coevolutionary MOEAs (CMOEA) with each researcher's efforts summarized, categorized, and analyzed. Some potential concept and future applications of MOEA coevolution are also suggested. Exercises, discussion questions, and possible research directions for MOEA local search and coevolution are presented at the end of the chapter.

3.2 MOEA Local Search Techniques

It turns out that in many multiobjective optimization problems (MOPs), points in PF_{known} are clustered in various regions of objective space. Thus, it may be possible to computationally direct points in such regions as well as isolated points closer to PF_{true} using clever mechanisms that exploit certain properties of the search space. For example, using one or a few of the objectives, it may be possible to adopt a **local search** technique to move a point closer to PF_{true} (i.e., better approximate the Pareto front with PF_{known}). And in addition, based upon MOEA goals, the use of local search (LS) may generate a better distribution of points on PF_{known}. Of course, the LS process starts in decision space or solution space with points in this space mapping to objective space.

Specific local search decision space approaches for consideration would be depth-first search (hill-climbing) [1407], simulated annealing [861], and Tabu search [572]. Since we are combining (hybridizing) global search MOEAs with local search techniques, they are generally defined as *hybrid* or *memetic* MOEAs.[1]

Algorithm 15 represents a generic memetic MOEA with the inclusion of the local search (LS) process noted. The specific position of the LS within a standard MOEA cycle can vary depending upon design (i.e., conducting a LS at every generation or at the end of a certain number of epochs). In general, LS techniques employ decision space neighborhoods whose selected points generate vectors in the objective space (phenotype). Note that doing local search in the phenotype domain is impractical since mapping from nonlinear objective functions back to unique decision variable values is generally impossible.

Balancing global MOEA search with local search for specific MOPs is critical to achieving good results. If fitness function computation in real-world MOPs takes a considerable amount of CPU time, there exist computational tradeoffs between local and global search. Thus, in the design and implementation of a MOEA-LS,[2] specific questions arise relating to LS effectiveness and efficiency:

- How often should the LS be applied based upon a probability, P_{LS}?
- On which k solutions should LS be used given a neighborhood $N(\mathbf{x})$ where $\mathbf{x}$ is a current solution?
- How long should LS be run defined by a time period T?
- How efficient does LS need to be versus effectiveness?

---

[1] Pablo Moscato [1133] introduced the concept of "memetic algorithm" to denote the use of local search heuristics with a population-based strategy. The term "memetic" has its roots in the word "meme", which was first introduced by Richard Dawkins in his classical book "The Selfish Gene" [340]. Dawkins defines a meme as the "unit of imitation" in cultural transmission. Therefore, a **memetic algorithm** can be seen as an approach that tries to mimic cultural evolution rather than biological evolution (like evolutionary algorithms). The main difference has to do with the way in which information is transmitted. Whereas genes are passed intact, memes are typically adapted by the individual who transmits them. For more information on memetic algorithms, see [661].

[2] The terms multiobjective memetic algorithm and MOEA-LS, are used interchangeably in this chapter.

Algorithm 15 Memetic MOEA

1: **procedure** MEMETIC MOEA($\mathcal{N}, g, f_m(\mathbf{x})$)
2:   Randomly initialize population $\mathbb{P}_g$ with $\mathcal{N}$ individuals
3:   Evaluate fitness $f_m(\mathbf{x})$ of each individual $\mathbf{x}$ in $\mathbb{P}_g$
4:   **while** Termination condition false **do**
5:     $g = g + 1$; number of generation
6:     Select $\mathbb{P}'_g$ from $\mathbb{P}_{(g-1)}$ based on fitness $f_k(\mathbf{x})$ ($k = \#$ of objectives)
7:     Apply genetic operators to $\mathbb{P}'_g \to \mathbb{P}''_g$
8:     **Local Search** in $\mathbb{P}''_g$ neighborhood; $\mathbb{P}''_g \to \mathbb{P}'''_g$
9:     Evaluate fitness $f_m(\mathbf{x})$ of each individual in $(\mathbb{P}''_g, \mathbb{P}'''_g)$
10:    Select $\mathbb{P}_g$ from $(\mathbb{P}_{g-1}, \mathbb{P}'_g, \mathbb{P}''_g, \mathbb{P}'''_g)$
11: **end while**
12: **end procedure**

- How could the MOEA recombination and mutation operators relate to the LS operators and parameters?
- Should a Lamarckian or Baldwinian (Baldwin effect) fitness assignment be employed?

Hart [660] analyzed the general aspects of some of the above questions for single objective problems. Others have suggested that the neighborhood structure could be of many forms as well as changing dynamically along with different LS parameter values depending upon MOP insight. For a MOEA-LS, the analysis of local search performance also relates primarily to the values P_{LS}, k, and T. Proper selection of these parameters and other MOEA-LS parameters needs to effectively support the MOEA goals of driving toward PF_{true} with associated diversity of PF_{known} points.

MOEA-LS is employed both to explore and to exploit. In that regard, the last question focuses on two generic fitness assignments to the individuals found by the LS. For a single objective problem, the fitness values attached to an individual (chromosome) in a local search process can be based upon a Lamarckian or a Baldwinian approach. The Baldwinian[3] strategy assigns the best fitness value from the local search in the decision space neighborhood to the neighborhood's starting individual. The Lamarckian[4] strategy, on the

---

[3] James Mark Baldwin proposed in the early XIXth century a mechanism called "organic selection" [79]. This mechanism, which is now known as the "Baldwin effect", refers to a specific selection for general learning ability. Rather than inheriting certain (fixed) abilities, an offspring inherits an increased capacity for learning new skills. This mechanism has been adopted, in different ways, in evolutionary computation.

[4] At around 1801, the French zoologist, Jean Baptiste Pierre Antoine de Monet (knight of Lamarck) started publishing the details of his own evolutionary theory. Lamarck proposed the existence of a mechanism responsible for the changes in the species, which is now known as "Lamarckism". Today, the term "Lamarckism" is used to refer to the theory according to which the characteristics acquired by an

other hand, uses the best fitness value found and the associated individual found in the neighborhood for the next generation. In memetic algorithms, the Lamarckian approach is usually employed since the goal is to use the best individuals to proceed. For multiobjective problems with LS, the Lamarckian strategy is also used in the sense that the nondominated individuals (best) and their set of fitness values are accepted for the next generation. Some MOEA-LS approaches may retain only certain LS nondominated individuals for the next generation based upon additional criteria such as distance or position.

As reflected in some MOEA approaches described in Chapter 2, a simple local search approach uses only one objective at a time and searches in a local neighborhood in decision space with the Lamarckian model. The method searches independently for extreme points in each dimension of the objective space. The more complex local search concept, however, relates to moving objective space points towards PF_{true} as generated from clustered decision space neighborhood regions; thus, possibly providing more population diversity with more nondominated points.

Since the solution of a MOP involves conflicting objectives, then due to computational considerations, the attainment of the hyper-surface Pareto front may suggest a local search focused only on a region of the Pareto surface. Such a region could be motivated by the decision maker as based upon *a priori* knowledge. The following sections discuss various MOEA memetic approaches as related to all of these generic LS goals.

3.2.1 Hybrid MOEA Techniques

Various hybrid MOEAs can be developed by incorporating a specific local search technique within a known MOEA. Examples include Knowles' Memetic PAES (M-PAES) [873], Ishibuchi and Murata's MOGLS [750], Jaszkiewicz's MOGLS [779], Bosman and de Jong's approach [154], Brown and Smith's technique [179], and Kleeman and Lamont's method [867]. All of these are variations of general MOEA global/local search algorithms. For a historical review, Knowles [878] presents a generic evaluation of memetic MOEAs. Note that combining local search with a MOEA has had fair success in solving specific multiobjective combinatorial and continuous optimization problems that exhibit a property called *global convexity* [215]. In these cases, Pareto optimal solutions are clustered (i.e., within a certain neighborhood) in decision variable space. This phenomenon is associated with the concept of "connectedness" in both domains [146]. When connectedness is present, then local search (i.e., neighborhood exploration techniques) should be successful. This phenomenon, however, has been scarcely studied in the context of multiobjective optimization (see for example [803]).

---

individual during its lifetime are directly inherited by its descendants. Although Lamarckism was refuted many years ago, some of the ideas related to this concept have been incorporated, in different ways, in evolutionary computation [1520].

In the following sections, a variety of MOEA local search approaches are discussed along with their advantages and disadvantages for MOPs with various characteristics.

Weighted Vector Methods

In order to select points in objective space based upon local search in decision space, particular MOEA-LS techniques use an aggregating fitness function. A variety of such weighted-sum scalar approaches over the m fitness functions are discussed.

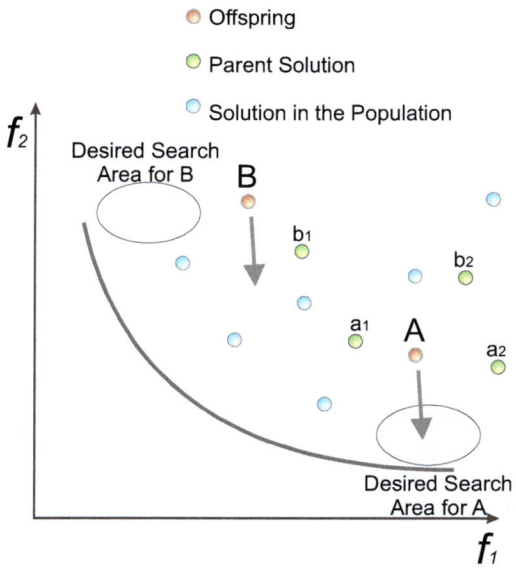

Fig. 3.1. Objective space local search phenomena (adapted from [762])

Ishibuchi and co-researchers have contributed much to the field of memetic MOEAs. Their initial research produced the simple multiobjective genetic local search (S-MOGLS) algorithm [750]. In this early work, a weighted scalar sum of objective functions is adopted by the authors, in order to determine the *best* solution of the local search [1150]. The authors fine tune their algorithm in later work so that local search is only applied to a subset of solutions. An important result from this work is that an appropriate local search direction for each offspring depends on its solution location in the objective space relative to its parents. To illustrate this issue, consider, for example, Figure 3.1, in which the child A of (a_1, a_2) is close to its parents and thus an appropriate local search direction is achieved. However, For B, which is far from its parents

(b_1, b_2), the resulting local search direction is not the desired one. Therefore, local search should be applied to only "good" offspring; that is, only to those solutions in objective function space that are close to parent solutions. This choice is important in achieving the desired goals for a local search engine in objective function space.

In [759], the authors perform a comparison between a weighted local search and Pareto dominance. Then, in further papers [751, 750], a local search approach is suggested with an aggregating fitness approach using randomly generated weights. A local search operator attempts to find a better solution in the single-objective problem of their interest, by using the generated weights. This approach has been extended and improved in a number of further publications [759, 762, 756, 760].

The S-MOGLS algorithm is illustrated by the generic memetic MOEA described in Algorithm 15. The population $\mathbb{P}_g$ is randomly initialized. Pairs of parent solutions are selected from the current population $\mathbb{P}_g$, where $\mathbb{P}'_g$ denotes the set of selected pairs of parent solutions. Genetic operations are applied to each pair in $\mathbb{P}'_g$ to generate an offspring population $\mathbb{P}''_g$. A local search procedure is probabilistically applied to only "good-fitness" offspring in $\mathbb{P}''_g$ in an attempt to form an improved population $\mathbb{P}'''_g$. A new population is constructed from the current population $\mathbb{P}_g$, the offspring population $\mathbb{P}''_g$, and the improved population $\mathbb{P}'''_g$. Each solution is evaluated using Pareto ranking with elitism as well as crowding (an NSGA-II type of structure).

Experimental results on knapsack problems performed by Ishibuchi and Narukawa [756] showed that better results are obtained from the weighted sum-based selection scheme over the Pareto-based scheme. A tournament selection of size 10 based on the weighted sum for parent selection was used for these experiments. The weighted sum-based parent selection does not necessarily work well for flowshop scheduling problems as shown in their work. The superiority of the weighted sum-based selection scheme for knapsack problems may be partially explained by the fact that a pair of similar parents in objective space is likely to be selected when using this selection scheme. Ishibuchi and Narukawa [756] have also demonstrated that better results are obtained for knapsack problems from the three-population model of generation update than from the two-population model. The authors use a specific local search procedure with the randomly weighted sum of the m objectives where λ is defined as the weight vector. The randomly specified weight-sum vector which was used to choose the initial solution for the current local search phase, is also used to calculate the weighted sum of the m objectives for the current solution $\mathbf{x}$ and the candidate solution $\mathbf{y}$. The current solution $\mathbf{x}$ is replaced with the candidate solution $\mathbf{y}$ only when the inequality $f(\mathbf{x}, \lambda) < f(\mathbf{y}, \lambda)$ holds for the weighted sum of the m objectives [760].

In the Memetic Pareto Archived Evolution Strategy (M-PAES) proposed in [873], the next population is constructed from the current population $\mathbb{P}_g$ and the improved population $\mathbb{P}''_g$ as $\mathbb{P}'''_g$. The same mechanism of generation update is used in [749, 813]. Note that one could use not only the current population $\mathbb{P}_g$ and the improved population $\mathbb{P}''_g$ but also the offspring population $\mathbb{P}'_g$.

Ishibuchi et al. [759] also focused on the flowshop scheduling problem which provides considerable insight to MOEA-LS development. Their LS process as designed and implemented can be interfaced not only to the MOGLS MOEA, but it can also be easily integrated into SPEA [1782] and NSGA-II [374] for comparing results. In all their flowshop scheduling examples with three MOEAs the LS approach did quite well. Using a roulette wheel approach with linear scaling for selection of recombination parents and an elitist strategy for population selection, a LS process is applied to the MOEA's population. Termination occurs when a specified number of solutions have been examined. Given a current individual and its neighborhood, a new individual in such a neighborhood is randomly generated. If this neighbor has a better fitness value, then it replaces the current solution in the population. The authors used eight flowshop test problems with the objectives of minimizing makespan and minimizing maximum tardiness. The metric for evaluating the selection of LS parameters was the $D1_R$, a normalized distance measure between points in PF_{known} and *a priori* known points presumed in PF_{true} (see Chapter 5). The ratio of nondominated points is also employed. The extensive testing with LS parameter variation undertaken by the authors provided them general insights and motivation for the importance of understanding the problem domain and the impact of LS search processing. Such insight should help to balance global search and local search in attempting to efficiently achieve effective MOP solutions.

Another weighted sum-based approach was developed by Deb and Goel who have applied local search techniques for engineering shape design [367, 366, 574]. Their work applies a neighborhood search to the NSGA-II [374]. Their initial work applied the local search after the MOEA had completed all generations, and later work compares their earlier results to the same local search being applied after every generation and on every individual. The added LS computational workload evidently impacted efficiency. Again, the selection of the neighborhood in the decision domain is critical to generating and selecting objective space points that move towards PF_{true}. With the vectored method, this is a difficult process at best. While the authors do not explicitly name their algorithm, here it is called memetic NSGA-II (M-NSGA-II) in order to distinguish it from the other algorithms discussed.

Jin and co-authors developed the evolutionary dynamic weighted aggregation (EDWA) algorithm [800, 801]. The main goal of their initial research was to analyze how random dynamic weighting compares to systematic dynamic weighting of the multiple objectives. In [803], Jin also attempts to show empirically that *connectedness* [215, 432] is a key to making EDWA an effective algorithm for continuous MOPs exhibiting global characteristics.

Connectedness relates to the topological structure of P_{true} and PF_{true}. Such solutions are connected in both parameter or decision space and generate connected points in objective space. With these proper MOP characteristics, solutions are distributed regularly in parameter space such that they can be defined with a piecewise linear function (i.e., a neighborhood). By constructing an approximate linear model using the P_{known} points, one searches on PF_{known} close to PF_{true}. That is, by selecting points in the constructed neighborhood of the current solution based upon the piecewise linear function, one may find more points moving along PF_{true}. When PF_{true} is convex, this evolution strategy (ES) approach with dynamic weighting first converges to a point near or on PF_{true}, then moves along or close to the front by changing the dynamic weights per generation. With a concave PF_{true}, the extreme points are initially found and search is conducted in directions near or on PF_{true}. This MOEA-LS is successful in achieving the generic MOEA criteria if the MOP has the required characteristics, even when dealing with a disconnected PF_{true}.[5] If the connectedness property holds both in the objective space and in the decision space, then a LS technique should effectively move PF_{known} towards PF_{true}.

Dominance Methods

As before, in order to select points in objective space based upon local search in decision space, some MOEA-LS techniques use a Pareto dominance-based scheme. Several of such dominance approaches are discussed.

Using Pareto dominance in local search, a candidate solution **y** is generated in the *neighborhood* of the current solution **x**. The current solution **x** is replaced with the candidate solution **y** only when it is better than **x**. Pareto dominance can be used to determine whether **y** is better with respect to the m objectives. In this case, the current solution is replaced with the candidate solution only when **y** dominates (in a Pareto sense) **x**. Note that k candidate solutions can be generated and considered.

Leiva [977] proposed three memetic algorithms based upon the multi-sexual-parents-crossover genetic algorithm (MSPC-GA) [460]. This development, which was originally based on the Multisexual Genetic Algorithm (MSGA) by [1000] applies both a simulated annealing (SA) and a neighborhood search algorithm within their MSPC-GA. The first algorithm, called MSPC-LS1, applies a simulated annealing algorithm to each solution in P_{known} after all the generations have been run. The second algorithm, called MSPC-LS2, applies a simulated annealing algorithm on every individual in every generation. The third algorithm, called MSPC-LS3, applies a neighborhood search on all members of P_{known} after all the generations are run. Results of three standard two-objective test problems are compared for the three memetic algorithms. Pareto fronts of the various algorithms and the number of nondominated points (up to 1200) found at the end of each algorithm's run are compared. The authors report that MSPC-LS3 normally generates the

---

[5] Even if PF_{true} is disconnected, the LS approach is effective because the MOEA population should find at least one point in each segment.

best results with 1200 nondominated points on PF_{true}. However, since the test functions adopted are not too complicated (by today's standards in the specialized literature), the benefit of using local search is not fully evident in this paper, since its use also introduces an extra computational cost. However, more complex MOPs may benefit from the memetic MOEAs introduced by the authors.

Knowles and Corne developed the memetic Pareto archived evolution strategy (M-PAES) algorithm ([873, 883]). This algorithm was based on their previously developed PAES algorithm [885, 886, 884]. Reflecting on the PAES algorithmic structure as discussed in Chapter 2, PAES is a (1+1) multiobjective evolution strategy (ES). PAES uses an external archive, it generates a new child with a Gaussian mutation operator and selects the next generation based upon nondominated comparison of the parent's and child's fitness value. However, M-PAES introduces some changes with respect to its predecessor. M-PAES adopts a $(\mu + \lambda)$-ES (i.e., it has a population), and a crossover operator to recombine local Pareto optimal solutions previously found. Instead of the single external file used by PAES, M-PAES uses two files: one to keep locally nondominated vectors and another one to keep globally nondominated vectors. The algorithm also uses a population of candidate solutions which is the basis to perform local search. Each individual from this initial population is replaced by an improved solution obtained from a local search procedure. To avoid excessive computational costs, M-PAES constrains the search to a certain maximum number of moves; an intermediate population is then generated. This intermediate population is obtained from the recombination of the union of the previous population and the current file of globally nondominated vectors. The same crowding procedure based on geographical location proposed with PAES is adopted with M-PAES. However, in M-PAES, a check against the globally nondominated vectors is also applied for a further filtering of the crowded regions in objective space. Therefore, acceptance of a new individual in the intermediate population requires both global nondominance and residency in a relatively scarcely crowded region of objective space. This algorithm compared well with respect to the Strength Pareto Evolutionary Algorithm (SPEA) [1782] in several instances of the multiobjective 0/1 knapsack problem.

In a further paper, Knowles and Corne [875] proposed to study the details of the fitness landscape of a multiobjective combinatorial problem (the multiobjective quadratic assignment) using different techniques and measures, and then use this information to organize a local search procedure appropriate for that specific problem.

Other MOEA-LS Approaches

There have been other proposals to combine MOEAs with local search procedures. For example, Thomson et al. [1584] proposed a memetic algorithm for the synthesis of nonlinear digital circuits. Sato et al. [1431] proposed a different type of memetic algorithm whose aim is to improve diversity rather than to achieve better convergence (as normally done with multiobjective memetic algorithms).

Kleeman et al. [867] proposed a tiered local search in an attempt to find better laser designs. The tiered local search takes into account problem domain knowledge and focuses the neighborhood search onto different chromosome regions depending upon what generation the search has reached. Through experimentation, it was determined that the electrical field magnitude has a large impact in finding the areas of good solutions. So, earlier searches focus on that neighborhood dimension, in an attempt to reach a "good" search region. But to find the best solutions in a region, the widths of the quantum wells must be adjusted along with finer adjustments of the electrical field. So these neighborhood searches are conducted in the later generations. Thus, a dynamic LS technique could consist of searching certain genes in the beginning, moving towards a specific region of PF_{true}, and then searching a different gene for achieving a "fine" tuning of PF_{known}.

Jaszkiewicz [777] proposed a technique called Random Directions Multiple Objective Genetic Local Search (RD-MOGLS), which uses either a weighted linear utility functions or a weighted Tchebycheff utility function (with weights drawn at random) to combine the objective functions of the problem. Although the technique is based on the use of local search to optimize the aggregating function adopted, it also uses a sample of the best solutions previously found by the algorithm (which is similar to the population of a genetic algorithm) and performs recombination among selected individuals from such a sample (analogous to the use of mating restrictions such as those imposed in MOGA [504]). Also, quad trees are used to efficiently store previously found nondominated solutions [626]. RD-MOGLS is compared against Ishibuchi and Murata's MOGLS [751], and against a multiobjective version of simulated annealing called MOSA [1616] (see Chapter 10) on bi-objective and three-objective instances of the TSP. RD-MOGLS found solutions that have better overall quality than any of the other two approaches, while using a considerably lower number of objective function evaluations. Overall quality is measured in terms of the expected value of a weighted Tchebycheff utility function over a set of normalized weight vectors.

In related work, Jaszkiewicz [779, 774] compares the same algorithm to SPEA [1782] and a greedy heuristic on a set of multiobjective 0/1 knapsack problems. RD-MOGLS is found to be superior with respect to the two metrics adopted: the first computes the volume of the objective space covered by the vectors obtained; the second uses a weighted Tchebycheff scalarizing function to generate "ideal" reference solutions that some multiobjective optimization technique evaluated attempts to achieve. In [775], Jaszkiewicz performs a comprehensive comparative study of local search-based metaheuristics applied to the multiple objective knapsack problem. The author compares five approaches: (1) the Memetic Pareto Archive Evolution Strategy (M-PAES)

[873], (2) Jaszkiewicz's Multiple Objective Genetic Local Search (MOGLS) [779], (3) Serafini's Multiple Objective Simulated Annealing (SMOSA) [1466], (4) Ulungu's Multiple Objective Simulated Annealing (UMOSA) [1616], and (5) Czyzak's Pareto Simulated Annealing (PSA) [323]. For a quantitative comparison of results, the author adopts two performance measures: (1) coverage of two sets [1770] and (2) a weighted Tchebycheff scalarizing function (see Chapter 5). An interesting aspect of this study is that M-PAES is the only algorithm (from the five chosen for the comparison) that relies on Pareto dominance, since all the others use aggregating functions. Results indicated that M-PAES could converge to the true Pareto front, but with a poor spread of solutions. In contrast, the other approaches were able to converge to all the nondominated regions. In [781], Jaszkiewicz performs an even more exhaustive study in which ten metaheuristics are compared in the bi-objective set covering problem. The approaches compared are: (1) Jaszkiewicz's Multiple Objective Genetic Local Search (MOGLS) [779, 777], (2) Ishibuchi and Murata's Multiple Objective Genetic Local Search (IMMOGLS) [751], (3) Serafini's Multiple Objective Simulated Annealing (SMOSA) [1466], (4) Ulungu's Multiple Objective Simulated Annealing (UMOSA) [1616], (5) Czyzak's Pareto Simulated Annealing (PSA) [323], (6) the Nondominated Sorting Genetic Algorithm (NSGA) [1509], (7) the Controlled Elitist Nondominated Sorting Genetic Algorithm (CENSGA) [365], (8) the Strength Pareto Evolutionary Algorithm (SPEA) [1782], (9) the Multiple Objective Multiple Start Local Search (MOMSLS) with random weight vectors [776, 777], and (10) an approach introduced in the paper, which the author calls Pareto Memetic Algorithm (PMA), and is based on Jaszkiewicz's MOGLS [777]. The author adopts a special encoding (a set of columns of a matrix are encoded), a special mechanism to generate the initial set of solutions similar to the one proposed by Eremeev [452], a neighborhood operator which is guided by an aggregating function, a local search mechanism, which is also guided by an aggregating function, and a recombination operator that produces a single offspring from two parents, and is based on the idea of distance preserving crossover [1093]. The performance measure adopted by the author is the average best value of the weighted Tchebycheff scalarizing function over a set of systematically generated normalized weight vectors. Results indicated that the best performers were PMA and MOGLS, and that the recombination operator was mainly responsible for their good performance. The author also reports that both MOSA and PSA are able to generate good results, but are highly sensitive to their parameters settings.

Gandibleux et al. [535] experimented with a GA that keeps all nondominated vectors obtained through Pareto ranking [581]. Then, when selection is performed, it uses a mechanism similar to VEGA [1440] to keep some individuals that excel with respect to each of the objectives. The remainder of the population is filled with individuals not selected by VEGA or Pareto ranking. To select such individuals, the authors use tournament selection based on domination with fitness sharing (a mechanism similar to the NPGA [709]).

Additionally, local search is applied to the best individuals of each generation, in order to improve convergence to PF_{true}. This approach, besides combining three selection mechanisms (Pareto ranking, VEGA and the NPGA), can also be considered a memetic algorithm because of its use of local search. It is important to add that the algorithm does not start with a completely random population, but instead uses a seeding strategy (normally another heuristic or deterministic algorithm) to generate individuals that are good with respect to one objective. This technique has been applied to bi-objective permutation scheduling problems (considering as objectives maximum tardiness and total flow time) and to bi-objective 0-1 knapsack problems.

Gradient knowledge in objective space can also be exploited by a LS algorithm, by generating a gradient that moves a point toward PF_{true}. MOPs consisting of continuous variables are amenable to a gradient approach (or conjugate gradient approach) because derivatives can be obtained (either in an exact algebraic form, or can be numerically approximated) based upon a specific direction from selected nondominating points in PF_{known}. Note however that derivatives in such situations tend to be quite noisy and therefore, a probabilistic noise model can be used to represent the possible stochastic derivative directions (i.e., a direction chosen based upon a random number from the modeled distribution). For example, Bosman and de Jong [153] employ gradient information in numerical MOEAs in an attempt to determine the proper gradient direction for future search. When hybridizing their MOEA, they choose a random direction based upon the gradients of each individual objective. Their results indicate that for five standard numerical test functions, this hybrid technique produces faster and better results. Other hybridizations of MOEAs with gradient information have also been attempted by other researchers (see for example [179, 1456]). Shukla [1485] employs finite difference (FD) and simultaneous perturbation (SP) methods for approximating the gradient. These approximations are used in Schäfflers stochastic method (SSM) [1442] and Timmels population based method (TPM) [1585] for updating the real-valued decision variables (i.e., they are both adopted as mutation operators). These two mutation operators are embedded into the NSGA-II and compared to the original NSGA-II. Four hybrids are produced, since, for each hybrid, both FP and SP are adopted to estimate the gradient. Although the hybrids performed well and were able to outperform the NSGA-II in one case, they failed in multifrontal problems. The author also adopts a stopping criterion based on Kuhn-Tucker conditions.

Marata and Itai [1154] propose a memetic MOEA that enhances the similarity of solutions in different sets of nondominated solutions. The vehicle routing problem (VRP) with three objectives (minimize maximum routing time, minimize the number of vehicles, and maximize solution similarity) is used as a benchmark to evaluate the approach. They use a two-fold MOEA to generate different sets of nondominated solutions via VRP normal and high demand periods. Results indicate that solution similarity is enhanced, but routing time may be increased for various VRP versions.

3.2.2 Comments on Hybrid MOEA Techniques

Of course, many local search variations could be implemented and integrated into a MOEA. For example, its use in an *a priori* MOEA from Chapter 2 is somewhat obvious with a single-objective function and hillclimbing process in the search landscape (objective space).

Considering points in a decision space neighborhood, a "feedback" approach would be of interest. For example, a hyperplane in the decision space is formed by the previous decision point, **x**, and the selected current decision point, **y**. A new decision point is chosen within the hyperplane formed by the points (**x**, **y**). Again, point **x** is the original point with point **y** selected within the neighborhood of point **x**. The selection is based upon feedback from objective space associated with point **y**. The objective space point thus generates the "best" nondominated point in objective space. Of course, there might be a k number of points **y** produced upon this construction. The intent is that this feedback approach uses selected points in a decision space hyperplane that moves objective space points toward PF_{true}. Selection of neighborhood points is critical to the success of this technique. This approach could be applied to the dominated as well as to the nondominated individuals. Also, weighted-vector evaluation could be used instead of a dominance measure.

A summary of the MOEA local search or memetic techniques reviewed in this chapter is presented in Table 3.1. Many of the simple test problems employed in comparing these hybrid MOEA techniques have been convex Pareto fronts of two dimensions (two objectives) making somewhat easy to find PF_{true} if connectedness applies. Other test functions such as the knapsack or scheduling problems provide more of a challenge and have also been adopted in more recent work. As indicated before, most LS approaches are not very successful in general because of the nonlinear mapping from decision space to objective space. However, using specific problem domain characteristics, local search techniques with operators such as restricted mating, the specific Pareto front geometry could be exploited. The size and structure of the search landscape is critical as to the utility of any gradient implementation. Also, the extra LS computational effort may be extensive and may generate few if any better points depending upon the MOP being solved, and it is advisable to take this into consideration before deciding to incorporate LS into a MOEA. Also, the use of gradient local search approaches can be useful for continuous MOPS, but would have limited if any viability to NP-Complete discrete problems for example. The heart of the memetic approach is the local search process in the decision space and the selection of associated objective space points to explore and exploit. The MOEA-LS approaches need to be explored for MOPs with higher dimensionality (i.e., more than two objectives) since local search methods are required to move towards specified regions of the Pareto surface.

Applying MOEA Local Search

It is suggested that MOEA-LS applications should use limited local search in a probabilistic manner, and after a predefined number of generations, instead of doing it at every generation. In general, applying LS at every generation puts too much emphasis on exploitation, possibly leading to premature convergence. In contrast, when adopting a predefined number of generations before applying local search, it is possible to perform more exploration of the search space. Based upon experience, the use of local search may be fruitful, not only regarding convergence, but also diversity, if properly applied.

As to using a dominance technique or a weight-vector approach, such choice depends upon the structure of the search landscape (rugged or smooth) and the Pareto front structure (e.g., concave, convex, connected, disconnected, continuous, discrete). Using the MOEA-LS feedback approach briefly sketched in this chapter, one can incorporate problem domain information via the objective space for doing local search in the decision space. However, in general, and because of the nonlinear mapping between the decision space and the objective space, the use of a simple generic LS approach would probably not result in improved movement toward PF_{true}. Possible computational interactive environments would be of considerable utility in this case.

It is indeed the case that the success of a MOEA-LS relates to connectedness of the MOP objective space and decision space as well as the building block (BB) sizes reflected in different hyperspaces within the objective space. The larger the size of the BB generally the more difficult to select points in the decision space that move points towards PF_{true} (the combinatorial gene issue). The less connectedness, the less effective and efficient the local search process becomes. The use of performance measures to evaluate and compare MOEA-LS results should definitely be part of the design of experiments within this emerging area (also called multiobjective memetic algorithms). Obviously, the MOEA goals may possibly be achieved by executing a MOEA independently of a LS and then combining results at termination with a MOEA-LS, but more sophisticated hybridization schemes are expected to arise within the next few years.

3.3 MOEA Coevolutionary Techniques

In nature there exist organisms that have a symbiotic relationship with other organisms. According to the Merriam-Webster Online Dictionary, **symbiosis** is defined to be the "intimate living together of two dissimilar organisms in a mutually beneficial relationship". **Endosymbiosis** is defined to be "symbiosis in which a symbiont dwells within the body of its symbiotic partner". Some evolutionary algorithms (EAs) and MOEAs have employed symbiotic techniques, although its use is relatively rare. Nevertheless, some of these symbiosis examples reported in the evolutionary computation literature are

3.3 MOEA Coevolutionary Techniques   145

Algorithm	Local Search Used	Where applied	How applied	Method
MSPC-LS1 [977]	Simulated Annealing (SA)	After MOEA	Nondominated individuals	Dominance
MSPC-LS2 [977]	SA	Each Generation	All individuals	Dominance
MSPC-LS3 [977]	Neighborhood Search (NS)	After MOEA	Nondominated individuals	Dominance
M-PAES [873]	NS	Continually	All individuals	Dominance (archive)
Thomson EA [1584]	NS	Mutation operator	Random individuals	Single Objective
Polar Dominance [1431]	Polar dominance	Each Generation	All Subpopulations	Dominance
S-MOGLS [750]	NS	Each Generation	All individuals	Weighted vector
C-MOGLS [1157]	NS	Each Generation	All individuals	Weighted vector
PGS-WLS [756]	NS	Each Generation	All individuals	Weighted vector
WGS-PLS [756]	NS	Each Generation	All individuals	Dominance
PGS-PLS [756]	NS	Each Generation	All individuals	Dominance
RD-MOGLS [777]	NS	Each Generation	Nondominated Random Utility Function individuals	Weighted vector
M-NSGA-II [366]	NS	After MOEA	All individuals	Weighted vector
M-NSGA-II [574]	NS	Each Generation	All individuals	Weighted vector
MIDEA [153]	NS	Each Generation	All individuals	Weighted vector
EDWA [800]	NS	Each Generation After x Generations	Top 10 ranked individuals	Weighted vector
Feedback (this chapter)	NS	Selected Generation	neighborhood	Dominance/vector
GENMOP-MTLS [867]	Tiered NS	Each Generation	Selected chromosomic regions	Dominance

Table 3.1. Memetic MOEAs

reviewed in this chapter, aiming to identify their main advantages and disadvantages.

We call **coevolution** to a change in the genetic composition of a species (or group of species) as a response to a genetic change of another one. In a more general sense, coevolution refers to a reciprocal evolutionary change between species that interact with each other. The term "coevolution" is usually attributed to Ehrlich and Raven who published a paper on their studies performed with butterflies and plants in the mid-1960s [434]. The relationships between the populations of two different species A and B can be described considering all their possible types of interactions. Such interaction can be positive or negative depending on the consequences that such interaction produces on the population. All the possible interactions between two different species are shown in Table 3.2.

	A	B	
Neutralism	0	0	Populations A and B are independent and don't interact
Mutualism	+	+	Both species benefit from the relationship
Commensalism	+	0	One species benefits from the relationship but the other is neither harmed nor benefited
Competition	-	-	Both species have a negative effect on each other since they are competing for the same resources
Predation	+	-	The predator (A) benefits while the prey (B) is negatively affected
Parasitism	+	-	The parasite (A) benefits while the host (B) is negatively affected

Table 3.2. All the possible interactions between two different species.

Evolutionary computation researchers have developed several coevolutionary approaches in which normally two or more species relate to each other using one of the previously indicated schemes. Also, in most cases, such species evolve independently through an evolutionary algorithm (normally a genetic algorithm). The key issue in these coevolutionary algorithms is that the fitness of an individual in a population depends on the individuals of a different population. In fact, we can say that an algorithm is coevolutionary if it has such property.

There are two main classes of coevolutionary algorithms in the evolutionary computation literature:

1. Those based on competition relationships (called **competitive coevolution**): In this case, the fitness of an individual is the result of a series of "encounters" with other individuals [1241, 1377]. This sort of coevolutionary scheme has been normally adopted for games.
2. Those based on cooperation relationships (called **cooperative coevolution**): In this case, the fitness of an individual is the result of a collab-

oration with individuals of other species (or populations) [1288, 1274]. This sort of coevolutionary scheme has been normally adopted for solving optimization problems.

Next, we discuss coevolutionary multiobjective evolutionary algorithms (CMOEAs), including some background knowledge regarding possible instantiations of coevolution in an evolutionary algorithm and some possible future applications of them.

3.4 Coevolution and Symbiosis in EAs

Several papers in the specialized literature investigate the use of coevolution and symbiosis with EAs. In this section, a few (representative) methods are reviewed along with some comments on the novelty of the ideas presented. These reviews provide the reader with better insight into the advantages and disadvantages of using these techniques. They also provide background support for the discussion provided later on, regarding CMOEAs.

3.4.1 Coevolutionary Algorithms

Paredis [1241, 1240] provides a good introduction to coevolution. **Competitive fitness** is the first item discussed. Competitive fitness functions differ from a standard fitness function in that competitive fitness calculations are dependent on the current population to some degree [51]. The dependency can be minimal (only one population member, for example) or comprehensive (where the entire population is aggregated into a single fitness value). Four examples of competitive fitness functions include full competition (all vs. all), bipartite competition (one vs. one or possibly one vs. many), tournament fitness (single elimination binary tournament, not to be confused with tournament selection), and elitist competition (all vs. best). Figure 3.2 shows a graphical representation of these four types of competitive functions. In these competitions, the fitness of an individual is compared to one or more other individuals. The most fit individual "wins" the competition with the others it is compared against.

Competitive fitness is applied extensively in evolutionary games. Some of the games experimented with include Tic Tac Toe [51], Othello [251], Awari [338], Poker [133], Game of Tag [1355], Backgammon [1280], and the iterated prisoner's dilemma [252].

Another form of coevolution is the **predator-prey model**. An example of this model is the work done by Hillis with sorting networks [682]. He used two populations, one consisting of sorting networks and the other was a set of lists with 16 numbers to be sorted. These populations were geographically distributed over a grid of computers. On each computer, a set of networks was applied to a set of lists. The fitness of a network was the percentage of

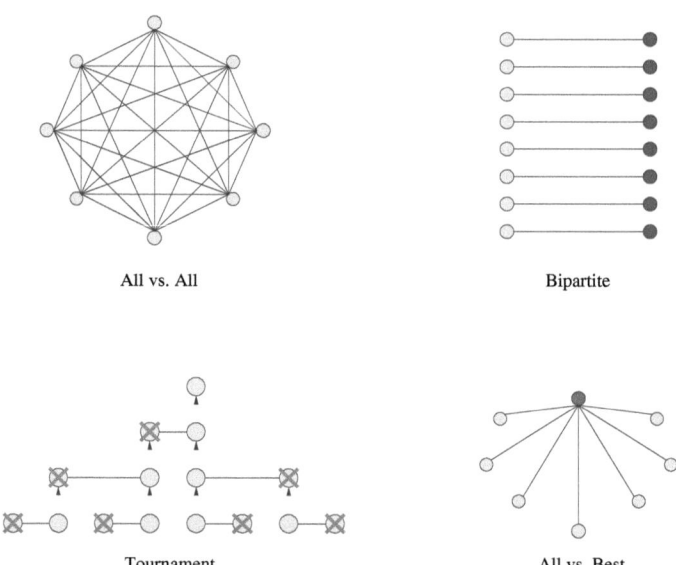

Fig. 3.2. Four types of competitive fitness functions

lists that were sorted while the fitness of a list was the percentage of networks that could not sort it. This type of inverse fitness interaction is typical in a predator-prey model. Hillis found that his coevolved solutions were better than a traditional EA. He attributed this to two reasons. First, since the populations are constantly changing, this encourages more exploration and avoids premature convergence. The other reason is that the fitness testing is more efficient, because the focus is on the lists that it cannot sort correctly.

Another distributed cooperative coevolutionary MOEA is suggested by Tan et al. [1568]. It is essentially an island model with communication through a server. However, each decision vector is decomposed into smaller components. The approach evolves many solutions in the form of cooperative subpopulations (islands). Results over a set of test functions is quite good and it is achieved with a relatively small number of evaluations.

Paredis [1240, 1241] introduced the coevolutionary genetic algorithm (CGA). His algorithm is based on the predator-prey model, in which he has a solution population and a test population. He first initializes the two populations (solutions and tests), and then generates fitness values for each by seeing how well they do compared to 20 random individuals in the opposite population (a 1 or 0 is assigned for each success or failure). These results are stored in a history for each individual and the fitness is the average of these 20 results. In each generation (he calls them cycles), 20 solutions and tests are selected, with the fittest individual being 1.5 times more likely to be chosen over the median fitness. If the solution is successful with the selected test, it receives a 1, or else it gets a 0. The history of both is updated.

The history keeps track of the 20 most recent results of encounters. After 20 encounters, two parents are chosen from the solution population (using the same weighted selection mechanism indicated above). The parents are then recombined and the child is mutated. The fitness of the child is then determined and the child is inserted into the population at the appropriate rank, possibly knocking out one of the parents. This is a $(\mu + 1)$ selection strategy. The parameters the author used are, in his own admission, quite arbitrary. He also does not evolve his tests. He does this on purpose as his tests were specifically designed to shape the search space. Nevertheless, he mentions that there are times when the tests should be evolved as well. Some applications of the CGA include classification, process control, path planning, constraint satisfaction, density classification, and symbiosis. The classification, process control, and path planning examples all evolve neural networks. The classification example uses preclassified, unchanging training examples for the test population, while the other two examples have the test solutions evolving. The constraint satisfaction problem uses constraints as the test population and arrays of variables as the solutions. He finds that the CGA outperforms a traditional GA in these problems. For density classification, he coevolved cellular automata with bit strings in order to classify the density of ones in each bit string. He then moves away from the predator-prey model and applies the CGA in a symbiotic way, where the two populations provide positive fitness feedback instead of negative. He used symbiosis to try to determine the best genetic representation for an individual in a problem. This problem tries to put chromosomes with the tightest linkages next to each other in order to keep good solutions more intact when applying crossover. In his problem, the solutions were evolved together with the representations. This problem actually tries to explicitly link good building blocks in the chromosome and it is successful at creating better solutions.

3.4.2 Cooperative Coevolutionary Genetic Algorithms

An interesting subset of coevolutionary algorithms is a group known as **cooperative coevolutionary genetic algorithms (CCGA)**, which were originally proposed by Potter and de Jong [1288]. This type of algorithm has a symbiotic approach, but instead of solution and test populations, it evolves species populations. To form a solution, an individual from each species is selected and combined with the other selected individuals. The solution is evaluated and the species that made up the solution are scored based on the fitness of the combined solution.

Potter and de Jong proposed two algorithms [1288]: CCGA-1 and CCGA-2. CCGA-1, shown in Algorithm 16, initializes each population of species and assigns the initial fitness by combining each subpopulation member with random individuals from the other subpopulations. Then, each of the species is coevolved in a round robin fashion using a traditional GA. The fitness of each of the evolved subpopulation members is obtained by combining it with the best individual of the other subpopulations and getting the fitness of the individual. This method of credit assignment has several potential problems, such as under sampling and excessive greediness, but it was only created as a starting point to base the effectiveness of further refinements of the algorithm. The results showed that CCGA-1 outperformed the standard GA when the species represented functions that were independent of each other but it performed worse when the function had dependencies.

Algorithm 16 CCGA-1

1: **procedure** CCGA-1$(\mathcal{N}, g, f_k(\mathbf{x}))$
2:    **for** Each species s **do**
3:       $\mathbb{P}'_s(g)$ = Randomly initialize population
4:       Evaluate fitness of each individual in $\mathbb{P}'_s(g)$
5:    **end for**
6:    **while** Termination condition = false **do**
7:       $g = g + 1$
8:       **for** Each species s **do**
9:          Select $\mathbb{P}'_s(g)$ from $\mathbb{P}'_s(g-1)$ based on fitness
10:         Apply genetic operators to $\mathbb{P}'_s(g)$
11:         Evaluate fitness of each individual in $\mathbb{P}'_s(g)$
12:       **end for**
13:    **end while**
14: **end procedure**

In order to overcome this deficiency, a new credit assignment was created. In addition to creating an individual as described before, a second individual is created where a random member of each subpopulation is combined with the individual. The best fitness of the two individuals is the one assigned to the child's fitness. This change in credit assignment produced the Algorithm 17, which did well when the function was independent or when it had dependencies.

Overall, the CCGA design provides a natural mapping onto coarse grained parallel architectures. This type of setup might work well in some applications, such as, for example, the aircraft engine maintenance scheduling problem [865], in which each population member is a total solution.

3.4.3 Symbiogenetic Coevolution

Wallin et al. [1673] introduced a cooperative coevolutionary algorithm that is based on the concept of endosymbiosis. Endosymbiosis deals with a symbiotic relationship between an organism and another that dwells inside the body of the host. In EAs, the use of competitive coevolution can keep populations from stagnating by using another cospecies to maintain evolutionary pressure. Wallin et al. [1673] proposed the **Symbiotic Coevolutionary Algorithm**

Algorithm 17 CCGA-2

1: **procedure** CCGA-2($\mathcal{N}, g, f_k(\mathbf{x})$)
2:    **for** Each species s **do**
3:       $\mathbb{P}'_s(g) = $ Randomly initialize population
4:       Evaluate fitness of each individual in $\mathbb{P}'_s(g)$
5:    **end for**
6:    **while** Termination condition = false **do**
7:       $g = g + 1$
8:       **for** Each species s **do**
9:          Select $\mathbb{P}'_{s1}(g)$ from $\mathbb{P}'_s(g-1)$ based on fitness
10:         Select $\mathbb{P}'_{s2}(g)$ from $\mathbb{P}'_s(g-1)$ at random
11:         **if** $\mathbb{P}'_{s1}(g)$ is most fit **then**
12:            Let $\mathbb{P}'_s(g) = \mathbb{P}'_{s1}(g)$
13:         **else**
14:            Let $\mathbb{P}'_s(g) = \mathbb{P}'_{s2}(g)$
15:         **end if**
16:         Apply genetic operators to $\mathbb{P}'_s(g)$
17:         Evaluate fitness of each individual in $\mathbb{P}'_s(g)$
18:       **end for**
19:    **end while**
20: **end procedure**

(SCA) (see Algorithm 18) which uses two species: hosts and parasites. The hosts are a complete solution while the parasite is a partial solution. For a parasite to be evaluated, it must be paired up with a host. The parasite is made up of two strings. The first string is a gray encoded binary number to direct the parasitic values to a location in the host. The second string is the parasitic value that is used to replace a portion of the host string. Other implementations may include the use of Boolean operators (i.e., AND, OR, XOR) to combine the host and the parasite.

Wallin et al. [1673] generate an equal number of parasites and hosts and perform an all-to-all matching, where each host is paired with each parasite individually and tested. A $(\mu+\lambda)$ selection scheme is employed using a tournament selection operator, where the top $|\mu|$ of the evaluated population forms the mating pool for the next generation of hosts. The k best solutions are stored in an external archive. After the next generation of hosts is formed, the parasites must be evolved. The parasites first reproduce asexually. The algorithm then applies bit mutation to both strings in the parasite. Finally, the program uses roulette wheel selection to pick the next generation of parasites. For their experiments, they compare their SCA with a generational GA on 64-bit and 128-bit decomposable problems, which are deceptive [1673]. All the experiments are averaged over 50 runs and the SCA is tested with parasite sizes of 4, 7, 12, and 17 bits. The authors report that smaller parasite sizes, such as 4 and 7, deliver the best results. While there is a symbiotic link

Algorithm 18 SCA

```
 1: procedure SCA($\mathcal{N}, g, f_k(\mathbf{x})$)
 2:     Initialize host population, $\mathbb{P}'_H$
 3:     Initialize parasite population, $\mathbb{P}'_P$,
 4:     Generate random host population, $\mathbb{P}'_H$, of size $\mathcal{N}$
 5:     Generate random parasite population, $\mathbb{P}'_H$, of size ($\mathcal{N} * 10$)
 6:     Create mating pool, $\mathbb{P}' = \mathbb{P}'_H$
 7:     for $i = 1$ to $g$ do
 8:         Let $\mathbb{P} = \mathbb{P}'$
 9:         for $j = 1$ to $\mathcal{N}$ do
10:             for $k = 1$ to ($\mathcal{N}*10$) do
11:                 Combine $\mathbb{P}_j$ with $(\mathbb{P}'_P)_k$
12:                 Add new child to pool $\mathbb{P}$
13:             end for
14:         end for
15:         Evaluate all members of pool $\mathbb{P}$
16:         Sort pool $\mathbb{P}$
17:         Save top individuals in external archive
18:         Create new mating pool $\mathbb{P}' = \mathbb{P}$ - with top $\mathcal{N}$ individuals
19:         for $j = 1$ to $\mathcal{N}$ do
20:             Randomly select 2 parents from $\mathbb{P}'$ using tournament selection
21:             Create Children $\mathbb{P}'_C$
22:                 Apply Recombination and Mutation
23:         end for
24:         Replace parents with children $\mathbb{P}' = \mathbb{P}'_C$
25:         Evolve Parasite population
26:             Apply Bit Mutation to each parasite using Roulette wheel selection
27:     end for
28: end procedure
```

between the parasites and the hosts, there really is no coevolution taking place. Only the host is being evolved and the parasites are just being mutated.

3.5 Coevolution and Symbiosis in MOEAs

The focus in the previous section was on coevolutionary algorithms within the context of single-objective EAs. In this section, the focus is on how coevolution and symbiosis have been applied with respect to MOEAs. Note that many of the algorithms discussed in this Section are based on concepts originally proposed for single-objective EAs.

3.5.1 Elitist Recombinative MOGA with Coevolutionary Sharing

Neef et al. [1181] developed the **Elitist Recombinative Multiobjective Genetic Algorithm with Coevolutionary Sharing (ERMOCS)**. Their algorithm uses a rank-based selection scheme based on MOGA [504], in which an individual's rank is determined by the number of individuals that dominate it. For selection, they decided to adopt an elitist recombination scheme, where the parents compete with their own children for a spot in the next generation. This type of selection ensures that good solutions are never lost in the search process. Neef et al. [1181] note the importance of niching because it helps to maintain genetic diversity and avoid genetic drift. To this end, the authors mention fitness sharing as prescribed in MOGA [504] and list its two main drawbacks. First, it is difficult to determine the appropriate niche size for a problem, unless one has extensive knowledge of the search space. Secondly, combining tournament selection and a sharing scheme causes chaotic behavior and limits the number of stable niches that are maintained. Since the selection scheme adopted by the authors is very similar to tournament selection, they assume chaotic behavior would ensue if fitness sharing was adopted. Thus, they adopt instead coevolutionary shared niching (CSN), which was originally introduced by Goldberg and Wang [588] as a way to dynamically change the niche size and location of a population. CSN is inspired on the economic model of monopolistic competition. The idea is to create two populations, one of businessmen and another one of customers. The population of customers is in fact the population of solutions to our problem (i.e., members of the Pareto optimal set) that will try to maximize a certain set of criteria, whereas the businessmen will try to locate themselves in such a way that their "profit" can be maximized. Customers will create niches according to their own criteria being optimized. Businessmen will then have to adapt to the current fitness landscape so that they can serve as many customers as possible. By enforcing a competition between these two populations, a uniform spread of the population of customers is expected to emerge. Neef et al. [1181] choose to calculate the fitness of the customers by dividing their rank by the number of customers that use the same business. For the business population, the fitness is the sum of all the ranks of customers that the business serves.

ERMOCS, which is shown in Algorithm 19, has four basic steps:

1. **Population creation**: Both, the customer and the businessmen populations are randomly created.
2. **Fitness calculation**: Customers are ranked using Pareto dominance and businessmen fitness is determined by the sum of the rank-based fitness values of the current customers.
3. **Recombination**: Offspring are created using recombination and mutation and ranked based on the current population. Then, the parents are compared to their offspring and the top best are chosen.
4. **Imprint**: Each businessman is compared to a random customer. If the customer has a better fitness, then it replaces the businessman.

A disadvantage of this algorithm is the fact that a minimum distance parameter needs to be set. Neef et al. [1181] suggest using a combined minimum

Algorithm 19 ERMOCS algorithm

1: **procedure** ERMOCS($\mathcal{N}, g, f_k(\mathbf{x})$)
2:     Initialize Customer population $\mathbb{C}'$
3:     Initialize Businessman population $\mathbb{B}'$
4:     Generate random populations ($\mathbb{C}'$ & $\mathbb{B}'$) - size $\mathcal{N}$
5:     Evaluate Customer Objective Values
6:     Assign Customer rank based on Pareto dominance
7:     Assign Business rank based on Customers served
8:     **for** Each Parent in $\mathbb{C}'$ **do**
9:         Generate $\mathbb{C}'$ Child population
10:           Random Selection (Uniform Probability)
11:           Recombination and Mutation
12:         Assign rank to children based on Pareto dominance with $\mathbb{C}'$
13:         Assign businessmen to children
14:         Calculate shared fitness of children
15:         Compare fitness of children with their parents
16:         **if** Child is fitter than a Parent **then**
17:             Child Replaces Parent in $\mathbb{C}'$
18:             Adjust affected businessman fitness in $\mathbb{B}'$
19:         **end if**
20:     **end for**
21:     Perform imprint operation with $\mathbb{C}'$ and $\mathbb{B}'$
22:     **if** Termination Criteria not met **then**
23:         Go to step 6
24:     **else**
25:         Stop.
26:     **end if**
27: **end procedure**

distance setting and businessman population size measure. So, larger businessman populations would have a smaller distance between them. Another possibility would be to apply a simulated annealing technique [861] to the process, where initially the distance is large between the businessmen. This would promote exploration at the early generations of the algorithm. Then, as the algorithm progresses, the distance would get smaller. This would promote exploration in the later generations.

3.5.2 Parmee's Co-Evolutionary MOEA

Parmee et al. [1252] proposed an algorithm that is referred to as **Parmee's Co-Evolutionary MOEA**. The goal is to develop a preliminary method to identify feasible design regions. It is an effort to initially narrow the search space in the field of airframe design. Instead of implementing a Pareto-based approach, a population for each objective function is evolved simultaneously. This approach, which is shown in Algorithm 20, utilizes a range constraint map. This map must be able to do three things: (1) allow each GA to produce an optimal solution, (2) draw all concurrent searches toward a single design region that best satisfies all objectives, and (3) allow for some flexibility at the end of the run so that better localized solutions can be obtained. The range constraint map uses weights that go from 1.0 (allow for the maximum value of an objective function) to 0.1 (solutions that produce values that are within 10% of the other objective functions). Fitness values of solutions that do not meet the criteria of the range constraint map are penalized adopting an amount defined by the user. The map starts at 1.0 and progresses down to 0.1. Both a linear decrease and a sine function are tested. The authors found that the linear decrease to 0.1 empirically worked the best. It is worth noting that this approach is not meant to be used in traditional multiobjective optimization, but for preliminary design, in which the aim is to quickly locate an interesting design region.

Algorithm 20 Parmee's Co-Evolutionary MOEA

1: **procedure** PARMEE($\mathcal{N}, g, o, f_k(\mathbf{x})$)
2:     Initialize populations $\mathbb{P}'_\kappa$ for each objective o
3:     Generate random populations of size $\mathcal{N}$ for each population $\mathbb{P}'_\kappa$
4:     **for** $i = 1$ to g **do**
5:        **for** $j = 1$ to o **do**
6:            Determine fitness of members in population $\mathbb{P}'_j$
7:            Normalize fitness relative to the min and max values
8:            Apply penalty function (Using range constraint map)
9:            Apply sensitivity analysis
10:              Rank the variables effect on each objective
11:              Adjust fitness values to ensure most influential variables are valid (Using range constraint map)
12:            **for** $k = 1$ to $\mathcal{N} - 1$ **do**
13:                Generate Child population
14:                    Randomly Select Parents (Roulette Wheel Selection)
15:                    Recombination and Mutation
16:            **end for**
17:            $\mathbb{P}'_j = (\mathcal{N} - 1)$ Children + Best individual from current generation
18:        **end for**
19:     **end for**
20: **end procedure**

3.5.3 Genetic Symbiosis Algorithm

Mao et al. [1053, 1054] proposed the **Genetic Symbiosis Algorithm (GSA)** for use with multiobjective optimization problems. The authors basically extend a single objective GSA [685] to handle multiple objective problems. The GSA portion of the algorithm has two features that are different from the standard GA. The first feature is the introduction of a parameter that attempts to represent the symbiotic relationship between individuals. The authors list six different ways that a symbiotic relationship can occur between two individuals. These range from a competitive relationship to a neutral relationship, all the way to a cooperative (benefiting) relationship. The algorithm modifies the fitness of an individual based on its symbiotic relationship to the other individuals. The symbiotic parameter is determined by a fuzzy inference function using the distance between two individuals and the difference in the fitness values of the same two individuals. The modified fitness value is then used as a basis for selection.

For the multiobjective version of the GSA, shown in Algorithm 21, a second symbiotic parameter is added to the algorithm. This parameter describes the interaction of the objective functions. This second parameter functions in the same manner as the initial symbiotic parameter. The sum of all the symbiotic relations of an individual with other individuals, and the sum of the symbiotic relations of the objective function with the other objective functions is combined and multiplied with the initial fitness function [1053]. The initial fitness value depends on the rank of each individual, which is determined in a way similar to the fitness ranking in MOGA [504].

Algorithm 21 GSA for MOP

1: **procedure** GSA FOR MOP($\mathcal{N}, g, f_k(\mathbf{x})$)
2:    Initialize population $\mathbb{P}'$
3:    Generate $\mathcal{N}$ individuals for population $\mathbb{P}'$
4:    **for** $i = 1$ to g **do**
5:        Perform Recombination and Mutation
6:        Calculate Symbiotic Parameters
7:        Modify fitness with symbiotic parameters
8:        Evaluate objective values
9:        Rank individuals in $\mathbb{P}'$ based on Pareto dominance
10:       Generate children
11:   **end for**
12:   **if** Sufficient Learning of Symbiotic Relation **then**
13:       Stop.
14:   **else**
15:       Train Symbiotic Relation with RasID
16:       Go to Step 4
17:   **end if**
18: **end procedure**

To "learn" the symbiotic relationships, inference rules are used that are based on the mean and standard deviation of the distances between individuals and the fitness of all the individuals. This "learning" is actually the training of the parameters so they meet the designer's specifications related to the distribution of individuals. There are five weighting factors that the designer must plug into the algorithm. These values weight the importance of the mean and variance in the phenotype and genotype domains and the weight of the ranks of the solutions. So, the objective of the criterion function is to find solutions that are evenly distributed. But determining these weights means the user might need to have an idea of what the search space looks like. In fact, the simulations that are run by the authors have a known search space. So, one would think that it is difficult to determine if the algorithm would do well when the search space is unknown or if the parameter settings play a big role in the search results. Unfortunately, the authors do not address these issues. The algorithm does its training using a method called **Random Search with Intensification and Diversification (RasID)** [712]. This is a method that was developed by one of the authors as a way to iteratively search both locally and globally. Instead of alternating the searches on a set cycle, the RasID uses an adaptive probability distribution function that changes based on the search history. So, the symbiotic relations are trained using the criterion function and the RasID algorithm. A niche parameter is not needed, but in its place, one needs five other parameters.

3.5.4 Interactive GA with Co-evolving Weighting Factors

Barbosa et al. [91] proposed an **interactive genetic algorithm with co-evolving weighting factors (IGACW)** for the objectives along with the population (see Algorithm 22). This algorithm has two populations, one for the proposed solutions to the problem, and the other for the weight-set population. The problem chosen to solve is a graph layout problem and the goal is to choose the best graph layout based on the chosen weight of importance for each objective function. The populations are evolved in a round robin process. First, the weights are kept constant while the solutions are evolved. Then, after a specified number of generations, a set of solutions are presented to the user for inspection and ranking. The ranking is based on the user's preferences and may be different from the actual fitness assigned to the graphs by the algorithm. Then, the solutions are held constant while the weights are evolved. A graph receives low fitness values when it differs widely from the graphs ranked by the user. After a set number of generations, the best set of weights is chosen based on the user's inputs and the solution sets are evolved again. The user inspects the results and if they are unsatisfactory by his standards, he ranks them again and the process continues. The algorithm finishes once the user is satisfied with the results. To minimize user fatigue, the algorithm is run for several generations with each population before the user is required any input. Also, the user only needs to rank some of the graphs, not all of them. The graphs are ranked from the highest rank down to where the user stopped. The graphs not ranked by the user are ranked behind those ranked by the user, but they maintain their relative order.

To test the algorithm, two experiments are run. The first run is an example of a user who does not care about the number of edge-crossings in the graph. The second run is an example of a user who wants a minimum number of

Algorithm 22 IGACW

1: **procedure** IGACW($\mathcal{N}, g, f_k(\mathbf{x})$)
2:     Initialize layout population, $\mathbb{P}'_L$, randomly
3:     Initialize weight-set population, $\mathbb{P}'_W$, randomly
4:     Compute each criterion for all graph layouts
5:     **while** User not satisfied **do**
6:         Display a sample layout from $\mathbb{P}'_L$
7:         The user ranks the sample
8:         **for** $i = 1$ to g_w (# of gens for weights) **do**
9:             Evaluate weights population $\mathbb{P}'_W$
10:            Generate new weights population $\mathbb{P}''_W$
11:         **end for**
12:         Replace $\mathbb{P}'_W$ with the best set of weights from $\mathbb{P}'_W \bigcup \mathbb{P}''_W$
13:         **for** $i = 1$ to g_l (# of gens for layouts) **do**
14:             Compute each criterion for all layouts in $\mathbb{P}'_L$
15:             Evaluate layout population $\mathbb{P}'_L$
16:             Generate new layout population $\mathbb{P}''_L$
17:         **end for**
18:         Let $\mathbb{P}'_L = \mathbb{P}''_L$
19:     **end while**
20: **end procedure**

edge crossings. The results show that the evolution process for the weights is greatly affected by the user's preferences. The weights associated with the criteria evolve to quite different values in the two runs. The solutions also evolve to greatly different graphs. Since this type of testing is highly subjective on the user's preference, it would be hard, if not impossible, to compare the IGACW algorithm with another MOEA in any way.

3.5.5 Multiobjective Co-operative Co-evolutionary GA

Keerativuttitumrong et al. [837] integrated the multiobjective genetic algorithm (MOGA) designed by Fonseca [504] with the CCGA algorithm (discussed in Section 3.4.2) designed by Potter [1288]. The result is called the **Multiobjective Co-operative Co-evolutionary Genetic Algorithm (MOCCGA)**.

The MOCCGA, shown in Algorithm 23 decomposes the problem into subpopulations based on the decision variables, or a part of the problem that requires optimization. Each individual of every subpopulation is ranked with respect to its subpopulation based on Pareto dominance when it is combined with the best individuals of other subpopulations. After ranking each member of a subpopulation, they are assigned a fitness value based on their rank in the subpopulation. The fitness sharing strategy utilized in this research is applied in the phenotype domain. For selection, the authors adopt stochastic universal sampling. The mutation and crossover rates adopted are 1% and 70%, respectively.

Algorithm 23 MOCCGA

1: **procedure** MOCCGA($\mathcal{N}, g, f_k(\mathbf{x})$)
2:     **for** Each species s **do**
3:         $\mathbb{P}'_s(g)$ = Randomly initialize population
4:         Evaluate fitness of each individual in $\mathbb{P}'_s(g)$
5:         Assign Rank based on Pareto dominance
6:         Compute Niche count
7:         Assign Linearly Scaled Fitness
8:         Assign Shared Fitness
9:     **end for**
10:     **for** $i =$ to g **do**
11:         **for** Each species s **do**
12:             Select $\mathbb{P}'_{s1}(g)$ from $\mathbb{P}'_s(g-1)$ based on fitness
13:             Select $\mathbb{P}'_{s2}(g)$ from $\mathbb{P}'_s(g-1)$ at random
14:             **if** $\mathbb{P}'_{s1}(g)$ is most fit **then**
15:                 Let $\mathbb{P}'_s(g) = \mathbb{P}'_{s1}(g)$
16:             **else**
17:                 Let $\mathbb{P}'_s(g) = \mathbb{P}'_{s2}(g)$
18:             **end if**
19:             Apply genetic operators to $\mathbb{P}'_s(g)$
20:             Evaluate fitness of each individual in $\mathbb{P}'_s(g)$
21:             Assign Rank based on Pareto dominance
22:             Compute Niche count
23:             Assign Linearly Scaled Fitness
24:             Assign Shared Fitness
25:         **end for**
26:     **end for**
27: **end procedure**

MOCCGA is tested against MOGA [504] using the ZDT test functions originally proposed in [1772]. The results indicate that MOCCGA is able to outperform MOGA in all of the test functions adopted, although they both perform poorly in discrete and non-uniform test problems. It is worth noting that a parallel version of the algorithm is also validated (each species is run on its own computer, and the best decision variable is broadcast to the other computers). In this case, small, medium and large test instances are tried, using 1, 2, 4 and 8 processors. Two different broadcast mechanisms are also assessed: (1) MPI broadcast and (2) a customized broadcast. The authors report that the customized broadcast works better and that the use of 4 and 8 nodes provides the most satisfactory speedup.

3.5.6 Lohn's Coevolutionary Genetic Algorithm

Lohn et al. [1007] proposed another approach which is referred to as **Lohn's Coevolutionary Genetic Algorithm** or Lohn's CGA. This algorithm is based on Lohn's earlier work in coevolving fitness schedules in an effort to find good amplifier and antenna solutions [1006, 1008]. For the amplifier problem, there are two populations: a circuits population and a target population. The circuits population consists of individuals that specify the values associated with the components used in the circuit. The target vector is a population of fitness functions. Each of the individuals consists of four elements: gain, bias, power dissipation, and linearity. Both populations are evolved using genetic algorithms. Individuals in the circuit population have their fitness based on how many target vectors they successfully solve as well as the difficulty of those target vectors. The fitness values are normalized between 0 and 1 with 0 being the best fitness. The individuals in the target vector population have their fitness based on how many circuits can achieve their settings. If only one circuit is capable of achieving the specifications in a target vector, then the target vector receives the highest fitness score of 1. If all the circuits or if no circuit meets the specifications, then the target vector receives the worst fitness score of 0. This type of fitness assignment attempts to keep the target vectors from becoming too difficult, but at the same time attempts to eliminate the easy ones. This allows the selection pressure to vary depending on the population of circuits. For a more detailed description of these fitness assignment schemes, see [1006]. The results show that coevolution and a normal GA perform similarly (statistically speaking), so coevolution does not seem to introduce an important gain in this case.

In [1007], Lohn's CGA (shown in Algorithm 24) is evaluated using the ZDT test functions originally described in [1772]. The authors let the target objective vectors (TOV) be the different objectives that the problem is trying to solve. This is an interesting approach since the solutions are guided toward varying fitness goals. Goals that are unattainable or are easy to attain are more likely to be replaced with new ones. This variance of fitness values allows the search to find solutions throughout the whole search space, and not just solutions that are near the optimum value of the various fitness functions. To determine the Pareto front, the author collected all the output TOVs and found the nondominated points from among them. Results are found to be competitive with respect to other MOEAs, although most of them are non-elitist approaches (e.g., NSGA [1509], VEGA [1440], NPGA [709], etc.), except for SPEA [1782]. While this approach can be classified as a cooperative algorithm, it applies a coevolutionary technique that has aspects of both a cooperative model and a competitive (predator-prey) model. The target objective vectors are a lot like the mechanical rabbit that greyhounds chase in a race. It keeps a certain distance from the dogs but never gets too far ahead or too close. This type of relationship is also a symbiotic relationship. The algorithm also has a built-in niching feature. The TOV fitness function is biased toward having fewer solutions. So, TOVs tend to spread out throughout the space in order to distinguish them from the other TOVs. An optimal TOV population would have only one solution for every TOV. So, it seems that the algorithm strives for a uniform spread.

Algorithm 24 Lohn's CGA

1: **procedure** LOHN'S CGA($\mathcal{N}, g, f_k(\mathbf{x})$)
2:     Initialize solution population $\mathbb{P}'_s$
3:     Initialize target objective vector population $\mathbb{P}'_t$
4:     Randomly generate $\mathcal{N}$ individuals for population $\mathbb{P}'_s$
5:     Generate $\mathcal{N}$ individuals for population $\mathbb{P}'_t$ - seed with "easy" target objective vectors
6:     **for** $i = 1$ to g **do**
7:         Determine fitness of individuals in $\mathbb{P}'_s$
8:         Compare each individual in $\mathbb{P}'_s$ to every target objective vector in $\mathbb{P}'_t$
9:         Calculate the fitness of the solutions in $\mathbb{P}'_s$
10:        Calculate the fitness of the target objective vectors in $\mathbb{P}'_t$
11:        Create child solution population $\mathbb{P}''_s$
12:             Randomly Select Parents
13:             Recombination and Mutation
14:        Write $\mathbb{P}'_s$ to an external archive
15:        Set $\mathbb{P}'_s = \mathbb{P}''_s$
16:        Create target objective vector population $\mathbb{P}''_t$
17:             Randomly Select Parents
18:             Recombination and Mutation
19:        Write $\mathbb{P}'_t$ to an external archive
20:        Set $\mathbb{P}'_t = \mathbb{P}''_t$
21:     **end for**
22: **end procedure**

3.5.7 Distributed Cooperative Coevolutionary Algorithm

Tan et al. proposed the **Distributed Cooperative Coevolutionary Algorithm (DCCEA)** [1569]. This algorithm is basically an extension of Keerativuttitumrong et. al.'s work [837] and is shown in Algorithm 25. This approach uses the CCGA cooperative coevolution scheme [1288] and the MOGA ranking scheme [504]. Where this algorithm is particularly different is that it uses a different niching mechanism and an archive for the solutions. The niching scheme is one developed by the authors, so it differs from the one used in [837]. The biggest advantage to this niching scheme is that it is adaptive and the user does not need to set a niche radius *a priori*. As for the archive, it contains all the nondominated solutions up to a user-defined maximum.

Another addition to the algorithm is the extending operator. Since the archive stores the nondominated complete solutions, a niche count can be done on these individuals. The individuals with the smallest niche counts may be in regions that have not be exploited well by the algorithm. To alleviate this, a predetermined number of clones of the individual with the smallest niche count are created and the pieces of the clone are assigned to their various subpopulations. It is hoped that this mechanism can bias the algorithm to areas of the Pareto front that need further exploration.

Algorithm 25 DCCEA

1: **procedure** DCCEA($\mathcal{N}, g, f_k(\mathbf{x})$)
2:     **for** Each species s **do**
3:         $\mathbb{P}'_s(g)$ = Randomly initialize population
4:         Collaborate (Create a complete solution with individuals from each species)
5:         Evaluate fitness of each solution
6:         Assign Rank based on Pareto dominance with external archive
7:         Update individual's rank in $\mathbb{P}'_s(g)$
8:         **if** Solution is nondominated **then**
9:             Update external archive
10:        **end if**
11:        **if** Archive is full **then**
12:            Compute Niche count
13:            Find member with smallest niche count
14:            Clone n copies of archive member to each species
15:        **end if**
16:     **end for**
17:     **for** $i =$ to g **do**
18:         **for** Each species s **do**
19:             Select $\mathbb{P}'_{s1}(g)$ from $\mathbb{P}'_s(g-1)$ based on fitness
20:             Select $\mathbb{P}'_{s2}(g)$ from $\mathbb{P}'_s(g-1)$ at random
21:             **if** $\mathbb{P}'_{s1}(g)$ is most fit **then**
22:                 Let $\mathbb{P}'_s(g) = \mathbb{P}'_{s1}(g)$
23:             **else**
24:                 Let $\mathbb{P}'_s(g) = \mathbb{P}'_{s2}(g)$
25:             **end if**
26:             Evaluate fitness of each solution
27:             Assign Rank based on Pareto dominance with external archive
28:             Update individual's rank in $\mathbb{P}'_s(g)$
29:             **if** Solution is nondominated **then**
30:                 Update external archive
31:             **end if**
32:             **if** Archive is full **then**
33:                 Compute Niche count
34:                 Find member with smallest niche count
35:                 Clone n copies of archive member to each species
36:             **end if**
37:         **end for**
38:     **end for**
39: **end procedure**

The algorithm is also parallelized. The parallelization is a coarse-grained strategy, similar to the one employed by Keerativuttitumrong et al. [837], but more involved. In this case, the subpopulations are combined into peer groups and each peer group is assigned a computer. These peer groups have their own archive and generate complete solutions. After several generations (an exchange interval), each peer group submits its archive and representatives to the central server and downloads the updates from the other peers. If the evolution process in the peers is vastly different with respect to time, the performance deteriorates. To avoid this deterioration, the peers are synchronized according to a user defined synchronization interval. This distributed algorithm is embedded into a distributed computing framework called Paladin-DEC [1561, 1569].

For testing, the authors use five of the six ZDT test functions from [1772] (see Chapter 4). Results are compared with respect to five other MOEAs (PAES [886], PESA [301], NSGA-II [374], SPEA2 [1775] and IMOEA [1563]), based on generational distance (how far the known Pareto front is from the true Pareto front) and spacing (how evenly distributed the members are along the known Pareto front) [287]. The authors find that with respect to generational distance, their results are comparable to the other MOEAs, but the proposed approach does particularly well on the multi-modal and nonuniform test cases. With regard to spacing, the proposed approach consistently obtained the best values.

3.5.8 Coello's Coevolutionary MOEA

Coello Coello and Reyes Sierra [280] proposed a **Coevolutionary MOEA** that the authors labeled the **CO-MOEA**. The CO-MOEA is designed to decompose the problem into competing populations. If a population produces more individuals on the known Pareto front, then it is rewarded by having its population size increased. The goal is to direct the search to the most promising regions of the search space. This approach is shown in Algorithm 26, and it runs in four stages, each consuming the same number of generations. At the first stage, the algorithm explores the entire search space. The algorithm uses MOGA's Pareto ranking scheme [504]. For population diversity, it uses the adaptive grid algorithm proposed by Knowles [877] and found in his PAES algorithm [886]. At the end of the first stage, the algorithm analyzes the current Pareto front values in an effort to determine the critical variables. Each variable is analyzed independently, so there is no effort to determine whether any linkages exist between variables. If a variable plays a role on a small portion of the Pareto front, the algorithm attempts to eliminate intervals that are deemed less fruitful. The algorithm may also decide to subdivide the interval of a variable. The new regions are assigned a different population. This stage basically does an initial global search and then sets the stage for multiple populations to do localized searches in regions which the algorithm deems profitable. By eliminating portions of the search space, the algorithm may have a difficult time finding the optimal solution in deceptive problems, because the selection pressure is too high.

The second stage simultaneously evolves all the populations. Each population focuses on a different region in the search space. At the end of each generation of this stage, each nondominated individual from each population

Algorithm 26 Co-MOEA

1: **procedure** CO-MOEA($\mathcal{N}, g, f_k(\mathbf{x})$)
2:     Set # of gen = 0
3:     Set # of populations = 1
4:     **while** # of gen < g **do**
5:         **if** # of gen = $g/4$ or $g/2$ or $3*g/4$ **then**
6:             Check active populations
7:             Analyze decision variables (compute number of subdivisions)
8:             Construct new subpopulations (update subpopulations)
9:         **end if**
10:        **for** $i = 1$ to # of populations **do**
11:            **if** population i contributes to the current Pareto front **then**
12:                Evolve and compete i
13:            **end if**
14:        **end for**
15:        Apply Elitism
16:        Reassign resources
17:        # of gen = # of gen + 1
18:     **end while**
19: **end procedure**

competes with the other nondominated individuals to create a single Pareto front for that generation. This is done with the adaptive grid algorithm. Each individual has a label, so the algorithm can keep track of which population supplied the individual. The algorithm tracks each population's contribution to the Pareto front. Each population has its population size adjusted according to the number of individuals it placed on the Pareto front. Thus, more productive populations get more individuals, while less productive populations have their size decreased. Populations with no members on the Pareto front are eliminated. So the populations compete with each other in an attempt to increase their membership. This competition is more of an implicit competition, because having more population members is not the goal of each population.

The third and fourth stages are the same. At the beginning of the stage, the algorithm determines how many populations need to be removed. Again, the algorithm analyzes the populations and if they need to be divided any further, the algorithm repeats the process laid out in stage one. A minimum population size determines the size of all populations. The algorithm adjusts the sizes of the populations in the same manner as it did in stage two. This algorithm is compared with three other MOEAs (the microGA [284], PAES [886], and the NSGA-II [374]) using three test functions and four performance measures: Two Set Coverage, Spacing, Generational Distance, and Error Ratio [287]. CO-MOEA is found to be competitive with respect to the other MOEAs, although the authors admit that its main drawbacks include the high selection pressure induced by the algorithm's elitist scheme, and the number of populations that the algorithm may need to use.

3.5.9 Nondominated Sorting Cooperative Coevolutionary GA

Iorio et al. [745] proposed the **Nondominated Sorting Cooperative Coevolutionary Genetic Algorithm (NSCCGA)**. The NSCCGA combines many of the aspects of the CCGA [1288], and the NSGA-II [374]. The algorithm decomposes the problem into subpopulations based on the number of variables (genes) and coevolves each subpopulation in an attempt to find the true Pareto front for a problem.

The NSCCGA is shown in Algorithm 27, and it forms collaborations in a manner slightly different than in the original CCGA [1288]. Instead of choosing the best individual from each subpopulation, a random individual is chosen from a group of "best" individuals. Any individual that was a segment of a nondominated solution is contained in that group. The individuals are ranked based on the NSGA-II Pareto ranking scheme [374]. All generations, except for the first, follow this type of ranking and collaboration scheme. The first generation consists of individuals that have not been ranked yet, so the collaborations are selected at random in this case.

Each subpopulation holds a static number of individuals. So, if a decision must be made between two individuals with the same rank, the algorithm uses the NSGA-II's crowding distance [374] to determine which individual to keep. Each subpopulation uses tournament selection to determine its mating pool. The mating pool has crossover and mutation applied to each individual in order to form a child population. Note that since each subpopulation adopts real-valued variables, the simulated binary crossover (SBX) [370, 362] operator is used. This crossover operator generates the offspring of two parents based on a probability distribution that was derived from a binary encoding.

This algorithm is compared to the NSGA-II on five problems proposed from the ZDT test suite (see Chapter 4) plus a rotated problem proposed by Deb [374]. Zitzler's standard metrics (see Chapter 5) are adopted to compare the average distance of the solution to the Pareto front, the distribution of the points, and the spread of the points. The performance of the NSCCGA is reported to be satisfactory with respect to the NSGA-II in all cases, except for the rotated problem.

3.6 Applying Coevolutionary MOEAs

In this section, we briefly discuss some possible applications of coevolutionary MOEAs which have not been addressed so far in the specialized literature (to the authors' best knowledge).

Algorithm 27 NSCCGA

1: **procedure** NSCCGA($\mathcal{N}, g, f_k(\mathbf{x})$)
2:    **for** Each species s **do**
3:       $\mathbb{P}'_s(g)$ = Randomly initialize population
4:       Collaborate (Create a complete solution with random individuals from each species)
5:       Evaluate collaborations
6:       Assign results to individual undergoing evaluation
7:    **end for**
8:    Assign rank based on Pareto Dominance - *sort*
9:    Calculate crowding distance
10:    Use elitism to keep only the best individuals in each population, $\mathbb{P}'_s(g)$
11:    Create mating pool using Tournament Selection
12:    Create children using mating pool
13:       Randomly Select Parents from mating pool using tournament selection
14:       Recombination and Mutation
15:    **for** $i = 1$ to g **do**
16:       **for** Each species s **do**
17:          Collaborate (Create a complete solution with random individuals from each species)
18:          Evaluate collaborations
19:          Assign results to individual undergoing evaluation
20:       **end for**
21:       Assign rank based on Pareto Dominance - *sort*
22:       Calculate crowding distance
23:       Use elitism to keep only the best individuals in each population, $\mathbb{P}'_s(g)$
24:       Create mating pool using Tournament Selection
25:       Create children using mating pool
26:          Randomly Select Parents from mating pool using tournament selection
27:          Recombination and Mutation
28:    **end for**
29: **end procedure**

3.6.1 Coevolving Multiple MOEAs

A common cliché is "two heads are better than one". This could also be applied to MOEAs. Suppose a MOEA is particularly effective at finding solutions to deceptive problems and another MOEA is effective at solving non-uniform Pareto fronts. If one does not know the phenotype space of the problem, throwing multiple MOEAs with distinctive strengths may be better than using just one. The MOEAs could then be coevolved in such a way that individuals are passed from one algorithm to another in an attempt to seed the algorithms with good solutions.

3.6.2 Coevolving MOEAs with other Search Algorithms

MOEAs have been developed which incorporate local search at some point in the algorithm. But an interesting application may be to coevolve a MOEA with another type of algorithm, such as a local search, Particle Swarm Optimization (PSO) [840], Ant Colony Optimization (ACO) [406], simulated annealing [861], among others. By coevolving, the search can exert a blend of exploitation and exploration that occurs in each generation of the algorithm. There are many ways that this coevolution can take place, but this section limits the discussion to one particular implementation, in which a MOEA is coevolved with a local search algorithm.

Let's assume we have x populations, one is evolved using a MOEA and the other $x-1$ are local search algorithms. At the first generation, the MOEA runs and ranks the top solutions, using a Pareto ranking system and a crowding mechanism. The top solutions are then used to seed the local search algorithms. While the local search algorithms are working to find better solutions, the MOEA is also evolving. After a fixed number of generations, the algorithms cross-pollinate, and they evolve using the new members in their population. A mechanism would have to be devised so that an individual is only submitted to a local search mechanism a certain number of times. This allows for other avenues of exploration in the search. But the populations could be designed so that there are varying degrees in the local search. For example, one local search may flip only one bit or it may focus on only one segment of the chromosome, while another population may flip 4 bits or focus on a separate part of the chromosome.

3.6.3 Coevolving Density Estimators

The use of coevolutionary schemes to handle the diversity estimator of a MOEA is another interesting application that has been only scarcely dealt with in the specialized literature. Coevolutionary niching is a good example of this sort of scheme (see [588, 707]), but it is not the only possibility. The development of coevolutionary crowding and coevolutionary clustering schemes is a topic that certainly deserves attention, too. Additionally, the use of multiple external (elitist) archives that could be coevolved using different density estimators that determine which estimator (or combination of them) is better seems another interesting possibility.

3.6.4 Coevolving Target Solutions

Coevolution could also be applied in a manner similar to the one prescribed in Section 3.5.6. In that research, target objective vectors guide the search toward the Pareto front by increasing the fidelity of the fitness required to be considered a "good" solution. This moving target is an attempt to guide the solutions toward better solutions. This could possibly be done in a manner similar to niching. As a population evolves, the fitness of individuals would be based on their distribution along the Pareto front. Target vectors could then be evolved at the same time and these vectors would be placed in areas throughout the Pareto front. If many points fall into a neighborhood near a target vector, its fitness would be low, but if only one or two vectors fall into a target vector neighborhood, it would achieve a high fitness score. The target vectors would thus evolve to fill less sparse locations along the Pareto front. In turn, the population of solutions would be rewarded with higher fitness values when they are closer to a target vector with only a few points in each neighborhood. This method may prove to be a good alternative to some of the niching methods currently in use.

3.6.5 Coevolving Competing Populations

In Coello's CO-MOEA [280], discussed in Section 3.5.8, populations competed implicitly. A population was rewarded with more individuals when it contributed more to the known Pareto front. But there was no explicit competition among them. A population did not have as a goal to have more individuals than another population; this just happened by chance. But what if there were explicit competition between populations? What if it were the goal of each population to gain more individuals? Next, we briefly discuss a possible scheme for this sort of coevolving competing populations.

One could setup a MOEA that has multiple populations, each with the same initial size. Each population's individuals would be fully instantiated, meaning an individual is a complete solution to the problem. The populations would evolve as any typical GA population does, using selection, mutation, and recombination operators. A global Pareto front would be generated from the best individuals from all the populations. The population that generated the most individuals would be rewarded with more population members and the population with the fewest members would be punished with fewer members. A population that has fewer members would adjust its operator parameters and quite possibly the operators it uses in an effort to improve its search capability. A population that gains members would maintain its *status quo*.

Putting populations to compete against each other and allowing them to vary their operators and parameters could be very beneficial, especially for unknown search landscapes. This self-preservation mechanism would allow the populations to self-adapt and create an algorithm that is better suited to the search space. Plus, if a good population starts to lose individuals, it could adapt in an effort to avoid stagnation.

3.7 Final Comments on Coevolutionary MOEAs

Section 3.5 presents a survey of the many ways researchers use coevolution in MOEAs. Table 3.3 lists MOEAs that implement coevolution, the type of

Algorithm	Symbiotic, Cooperative or Competitive	Subpopulations Used	Where applied in algorithm
ERMOCS [1181]	Cooperative	No	Niching
Parmee's CMGA [1252]	Cooperative	Yes	Population
GSA for MOP [1053]	Symbiotic	No	Niching
IGACW [91]	Cooperative	No	Objective Weighting
MOCCGA [837]	Cooperative	Yes	Population
Lohn's CGA [1007]	Cooperative	No	Fitness Function
DCCEA [1569]	Cooperative	Yes	Population
CO-MOEA [280]	Competitive	No	Population
NSCCGA [745]	Cooperative	Yes	Population

Table 3.3. Coevolutionary Techniques Used in MOEAs

coevolution scheme employed, if the algorithm decomposes the problem into subpopulations, and where the application of coevolution occurs in the algorithm.

Various methods of coevolution have been applied to MOEAs. Some of the advantages and disadvantages of the most representative algorithms of this sort reported in the specialized literature have been discussed. Also, some possible future applications of CMOEAs have been briefly addressed. However, the effectiveness of coevolutionary MOEAs depends upon problem domain knowledge being explicitly used in the search process, and on the design of appropriate schemes that properly exploit the advantages of coevolution. Nevertheless, a variety of innovative directions for coevolutionary MOEAs are possible (see for example [866]).

Further Explorations

Class Exercises

1. Explain the main differences between a memetic MOEA and a Coevolutionary MOEA.
2. Discuss the main advantages and possible disadvantages of a MOEA-LS.
3. Discuss some of the possible metaheuristics that could be used to perform local search in a memetic MOEA. How does the type of encoding (e.g., binary or real-numbers) affect your choice?
4. Why do you think that most MOEA-LS approaches do not move PF_{known} towards PF_{true} by themselves? Discuss.
5. Discuss possible types of coevolutionary MOEAs (based on the types of interaction shown in Table 3.2) that are possible, but were not included in this chapter.
6. How does the connectedness of the solution space P_{true} as well as the connectedness of PF_{true} relate to local search performance?
7. Discuss the main advantages and possible disadvantages of a coevolutionary MOEA.
8. Using the six specific questions from page 132 (in Section 3.2), compare M-PAES [873] and MOGLS [750].

Class Software Projects

1. Implement a memetic multiobjective evolutionary algorithm using Lamarckian fitness assignment. Then, compare this implementation to another one that uses Baldwinian fitness assignment.
2. Implement any of the coevolutionary multiobjective evolutionary algorithms discussed in this chapter and validate it using appropriate test functions and performance measures.

3. Select a two-objective multiobjective optimization problem in which each objective function is highly nonlinear and apply to it one of the MOEA-LS discussed in this chapter. Analyze connectedness, convergence, spread and computational efficiency.
4. Repeat at least two of the MOEA-LS experiments discussed in this chapter, and assess results using performance metrics such as those discussed in Chapter 5. Use different values for k (number of solutions to which the local search is applied) and T (number of iterations during which local search should be applied).
5. Develop an interactive MOEA-LS environment for "steering" local search points to PF_{true}.
6. Implement, execute and evaluate a parallel version of the MOEA-LS approach proposed by Bosman & De Jong [154].
7. Apply Brown and Smith's [179] hyperplane local search approach to one of the test suites described in Chapter 4 (e.g., the ZDT test functions [1772]). Evaluate and compare your results using performance measures and statistical analysis (see Chapter 5).

Discussion Questions

1. Read Knowles and Corne's survey on memetic algorithms used for multiobjective optimization [879]. Discuss some of the future prospects indicated in this paper (e.g., objective correlation, restricted mating schemes, etc.). Propose additional topics for future research in this area.
2. Stochastic local search algorithms are nowadays widely used for combinatorial optimization [704]. Discuss the role of stochastic local search algorithms in multiobjective optimization. How different are these algorithms from traditional MOEAs? See for example [1237, 1238, 1239].
3. Discuss the main issues related to the combination of gradient techniques with MOEAs (e.g., what is the most appropriate stage of the cycle of a MOEA in which such integration should take place?). See for example [179, 1456, 180, 154, 658, 657, 1485]. It is also advisable to read additional references on this topic (see for example [492, 1442]).
4. Analyze the potential of rough sets [1262, 1263] as a possible local search scheme that can be coupled to a MOEA (see for example [1424]).
5. Analyze other predator-prey MOEAs different from those discussed in this chapter (see [957, 986, 1448, 605, 604]).
6. Miconi [1108] presents a study in which he attempts to show the main drawback of coevolutionary algorithms. Analyze this paper and discuss it. Do you agree with the experimental setup provided by Miconi? Do you agree with his conclusions? What type of coevolutionary EA does he consider in his study?

7. What are the basic problems of using a MOEA-LS? Relate to pedagogical test problems and real-world multi-dimensional multiobjective optimization problems.
8. What other metrics besides $D1_R$ would you suggest for evaluating MOEA-LS performance? Justify.
9. Are Coevolutionary MOEAs generally robust, or only suited for problems with certain characteristics? Discuss.
10. When is the tiered MOEA-LS [867] a good approach? Relate to problem characteristics.
11. Analyze other MOEAs that incorporate local search and which were not discussed in this chapter. See for example [949, 663, 1679].
12. Analyze other Coevolutionary MOEAS not discussed in this chapter (see for example [412, 1735]). Incorporate these approaches within the taxonomy proposed in this chapter, and identify strengths and possible weaknesses of each of them.
13. Review and discuss the utility of coevolving MOP solutions using the framework proposed by Lamont et al. [943]. Relate this to the performance of a parallel cooperative coevolutionary approach which uses low-level and high-level teamwork hybrid schemes (see [1174]).
14. Discuss possible interactive visualization techniques for a MOEA-LS. What are the main computational problems that you foresee for such a visualization technique? Does the need of interaction with the user introduce additional problems?
15. What are the main differences and similarities between a local search technique developed for multiobjective continuous problems with respect to another one designed for discrete problems? What are the main features of the search space that a local search approach attempts to exploit in each case?

Possible Research Ideas

1. Using more problem domain information for a many-objective optimization problem, integrate a MOEA-LS with decision maker's preferences to specific objective-space surfaces. Attempt to develop new MOEA operators that improve performance.
2. Attempt to develop a theory of convergence for a generic Memetic MOEA for either continuous or discrete problems.
3. Design an experimental study in which different Coevolutionary MOEAs are applied to similar multi-dimensional multiobjective optimization problems, within a certain class. Based on the results from this study, propose a new Coevolutionary MOEA that combines the main advantages of the approaches compared.
4. Design a "hyperheuristic" for a certain type of multiobjective optimization problem, in which several multiobjective metaheuristics are combined, and

are coordinated by a master control that determines the stage of the search at which any of them should be applied. See for example [186, 187, 189]. Incorporate local search into this scheme.
5. Attempt to develop a theory of convergence for Coevolutionary MOEAs.
6. Propose a Coevolutionary Memetic MOEA, and validate it using standard test functions and performance measures. See for example [1491, 1492].
7. Analyze the role of local search as a mechanism to accelerate the convergence of a MOEA (see for example [12, 11]).
8. Propose an island-based parallel Memetic MOEA that adopts different local search mechanisms within each island (for more information on parallel MOEAs, see Chapter 8).
9. Attempt to relate game theory [1665] with Coevolutionary MOEAs both, from a theoretical and from a practical point of view.
10. Propose a new performance measure (different from those discussed in Chapter 5) specifically designed for assessing performance of coevolutionary MOEAs. How is this performance measure different from the others? Justify your proposal with some theoretical analysis and with benchmark evaluations.
11. Develop a meta-level design for a parallel or distributed coevolutionary MOEA (see Chapter 8 and [1174]). Evaluate your proposed approach using performance measures from both the MOEA and the parallel algorithms literature.

4
MOEA Test Suites

> No amount of experimentation can ever prove me right; a single experiment can prove me wrong.
>
> Albert Einstein

4.1 Introduction

Why test multi-objective evolutionary algorithms (MOEAs)? To evaluate, compare, classify, and improve algorithm performance (effectiveness and efficiency). What is a MOEA test? Should we use a multi-objective optimization problem (MOP) test function, a MOP test suite, pedagogical functions, or a real-world problem? How to find an appropriate MOEA test?

Should we rely on the MOEA literature, on historical use, on test generators, or on well known real-world applications? When to test? Should we adopt and incremental algorithm and test development methodology or should we wait until the final stage of algorithm development to test it?

How should we design a MOEA test? Evidently, several important issues must be taken into consideration. For example: basic assumptions, computational platform selection, statistical tools, performance measures selection, experimental plan, among others. Thus, considerable effort must be spent not only in defining proper MOP tests and in generating the proper design of MOEA experiments, but also in employing the appropriate performance measures and experiment conditions, as well as the proper statistical tools that allow a fair algorithmic comparison. In this chapter, the development of various MOP test suites is discussed in detail.

Many MOEA research efforts select as examples numeric MOP functions to show or judge MOEA performance. In order to appreciate the rational for such selections, a comprehensive discussion of MOP landscape issues and structure is required along with an explanation of why the selected MOPs may be appropriate or inappropriate MOEA test functions. Such MOP characteristics include objective functions structures, constrained vs. unconstrained genotype and phenotype formulations, and the impact of numerical approximation of continuous forms. This chapter precisely addresses all of these issues. Standard suite(s) of test functions exhibiting *relevant* MOP domain characteristics are presented that can provide the necessary common MOEA comparative basis (see [1630, 1628, 1626, 357, 355, 1790, 375, 375, 721, 1207]).

This chapter on MOP development is organized as follows: Section 4.2 discusses general test suite issues. Relevant MOP domain characteristics are presented in Section 4.3 which also proposes appropriate MOPs for MOEA test function suites given the described MOP domain features. Section 4.4 is devoted to scalable multi-objective test problems, describing several test suites found in the current literature. Combinatorial problems are described in Section 4.5, whereas Section 4.6 is devoted to real-world problems.

4.2 MOEA Test Function Suite Issues

The MOEA community has created various test suites as indicated and referenced previously. Specific functions have however been often employed because other researchers did so in their MOEA research, or perhaps because the MOP appears to exercise certain MOEA components. It is not clear that all these particular test functions are appropriate for inclusion into generic MOEA test suites. Explanation is rarely offered as to the specific MOP's origin or *raison d'etre*, yet several appear to be relatively "easy" (see Section 4.3) in the sense of finding the optimal solution. Poloni et al. [1283] also observed the lack of complex mathematical MOEA performance assessment tests. This situation implies that identification of appropriate test function suites to objectively determine MOEA efficiency and effectiveness is required. Other researchers have also noted the need for comprehensive test suites and have presented some ideas and examples [357, 1284, 1646, 1630, 1773, 721, 1207], which are included in this chapter.

Generic test function suites are both condoned and condemned. Any algorithm successfully "passing" all submitted test functions has no guarantee of continued effectiveness and efficiency when applied to real-world problems, i.e., examples prove nothing except as counter examples. When integrating MOP domains and MOEA domains, new and unforeseen situations may arise resulting in undesirable performance for example. A MOEA test suite is then a valuable tool only if relevant issues are properly considered. To motivate the development of MOP test suites, historical single objective EA test functions are addressed first.

Some single objective EA test suites examine an EA's capability to "handle" various problem domain characteristics. These suites incorporate relevant search space features to be addressed by some particular EA instantiation. Some example single-objective EA test suites are:

4.2 MOEA Test Function Suite Issues   177

- De Jong [348], suggests five single-objective GA optimization test functions: sphere-parabolic, Rosenbrock ridge, Rastrigrin steps, Griewank quartic, Schaffer F6 foxholes.
- Michalewicz & Schoenauer [1105] recommend twelve single-objective *constrained* optimization test functions.
- Schwefel [1460], offers a diversity of 62 different landscape functions from the Evolution Strategies literature.
- Whitley et al. [1701], and Goldberg et al. [585], offer other formal GA test suites; informal suites are also used by Yao & Liu [1730, 1731].
- So called deceptive problems include Goldberg's order 3 and 6, bipolar order 6, and Mühlenbein's order 5. Some of these include extensive experimental study of benchmarking functions for GAs.
- Digalakis & Margaritis [392], suggest eight standard functions along with an additional six nonlinear squares problems for single objective EAs.
- Multi-optima examples from the Operations Research (OR) literature include Levy functions [978], Corana functions [296], Freudenstein-Roth and Goldstein-Price functions [1127].
- Others are Ackley's function and Weierstrass function (continuous but not differentiable anywhere).

De Jong's five standard GA test functions reflect the following characteristics [581]: continuous and discontinuous, convex and nonconvex, unimodal and multimodal, quadratic and nonquadratic, low- and high-dimensionality, and deterministic and stochastic. Michalewicz & Schoenauer's test bed addresses the following issues [1105]: type of objective function (e.g., linear, nonlinear, quadratic), number of decision variables and constraints, types of constraints (linear and/or nonlinear), number of active constraints at the function's optimum, and the ratio between the feasible and complete search space size. Particular single objective EAs as well as MOEAs can be subjected to generic test suites based upon these characteristics and then judged on their performance (effectiveness and efficiency). In general, the following characteristics both in the genotype and phenotype domains should be addressed when selecting possible MOPs:

- continuous vs. discontinuous vs. discrete
- differentiable vs. non-differentiable
- convex vs. concave
- modality (unimodal, multi-modal)
- numerical vs. alphanumeric
- quadratic vs. nonquadratic
- type of constraints (equalities, inequalities, linear, nonlinear)
- low vs. high dimensionality (genotype, phenotype)
- deceptive vs. nondeceptive
- biased vs. unbiased portions of PF_{true}

EA and MOEA test suite functions should range in difficulty from "easy" to "hard" as found in pedagogical and generated forms as well as attempt to represent generic real-world situations. Dynamically changing environments can include "moving cones" [1131] with movement ranging from predictable to chaotic to non-stationary and deceptive.

One should consider the following guidelines suggested by Whitley et al. [1701] in developing generic test suites:

- Some test suite problems are *resistant* to simple search strategies.
- Test suites contain nonlinear, unseparable & unsymmetric problems.
- Test suites contain scalable problems.
- Some test suite problems have scalable evaluation cost.
- Test problems have a canonical representation (ease of use).

It should also be noted that Holland states that the use of "sample" pedagogical problems is of little use in understanding the performance of EAs employed in complex real-world engineering and scientific design and analysis problems [700]. That is, sample problems can be used to compare various EA performances, but these results provide little insight to real-world EA applications. The same applies to MOPs for MOEA comparison in general. One must be then very careful in selecting a MOP test suite in regard to defining its purpose. One should contemplate the following: *using a test suite of any kind can be useful from a pedagogical perspective in comparing MOEAs, but in general, may be of little importance when solving "real-world" problems!*

Moreover, the No Free-Lunch (NFL) theorems [1708] imply that if problem domain knowledge is not incorporated into the algorithm domain, no formal assurances of an algorithm's general robust effectiveness exist. NFL theorems in addition imply that incorporating too much problem domain knowledge into a search algorithm reduces its effectiveness on other problems outside and even within a particular class; i.e., non-robust. In selecting a MOP test suite one should consider characteristics from target problem domains. However, as long as a test suite involves only *major* problem domain characteristics, any search algorithm giving effective and efficient results over the test suite might remain broadly applicable to that problem domain. Thus, traits and characteristics common to all or most known MOPs should be completely defined in order to help develop an MOP test suite. A dichotomy may exist between MOEA test suite selection and the possible real-world application!

A MOEA should move towards PF_{true} with considerable diversity in order to converge to a complex Pareto front. Generating diversity only when "near" the PF_{true} generally does not achieve the desired effectiveness. Moreover, if PF_{true} is considerably concave with many regions of the search space not dense in points (due to discretization), then moving through the dense feasible regions requires again a diversity of points in each MOEA generation. This is also true for vacuous regions due, for example, to constraints. The intent of the following discussion is to present possible MOEA test functions and test function suites based upon desired P_{true} and PF_{true} characteristics.

4.3 MOP Domain Feature Classification

Like single-objective EA optimization problems, MOPs may be suitable representatives of real-world multi-objective problems. Most modeled real-world problems are reflected in a mathematically functional structure, but MOPs arguably capture more information about the modeled problem as they allow incorporation of several objective functions. Regardless, modeling a real-world problem may result in a numeric or combinatorial MOP, one that is perhaps simple, perhaps complex. A MOP may contain continuous or discrete or integer-constrained functions or even a mixture. Initially, the discussion is restricted to homogeneously continuous numeric MOPs; other MOP types are discussed in Section 4.5.

Any proposed MOP test suite must offer functions spanning a spectrum of MOP characteristics. Particularly, it must contain "MOEA challenging" functions. In order to then identify appropriate functions for inclusion, relevant MOP domain characteristics must be identified and considered. The variety of known examples from the historical literature are considered as the basis for development; the associated list is found in Tables A.1, A.2, A.3, A.4, A.5 and Section A.5 in Appendix A.[1] These MOPs each incorporate 2-3 functions and 0-12 side constraints. Appendices B and C present a complete set of figures showing P_{true} and PF_{true} for each MOP listed in the tables. These figures are deterministically derived by computing all decision variable combinations possible at a given computational numerical resolution using high performance parallel computers. The purpose is to highlight major structural characteristics of both P_{true} and PF_{true} for use in constructing sound MOEA test function suites.

Some MOP test functions can be built upon commonly used single-objective optimization test functions. For example, Kursawe's MOP incorporates a modified Ackley's function [72, pg. 143] and a modification of one provided by Schwefel [1460, pg. 341]. Poloni's MOP incorporates a modified Fletcher-Powell function [72, pg. 143]. Finally, Quagliarella's MOP uses two versions of Rastrigin's function [231]. Other MOP test generators have been proposed by Deb [357, 355]. The rationale for construction and use of these and many of the other identified MOPs is sometimes unclear.

When implementing a MOEA, it is implicitly assumed that the problem domain has been properly considered. A decision is made that a MOEA is an appropriate search engine for the given MOP. A natural genotype representation defined and efficient exploitation and exploration operators declared. The MOEA's objective is the generation of PF_{known} and P_{known} both of which may be close to PF_{true} and P_{true}, respectively.

Tables 4.1 and 4.2 identify salient MOP domain characteristics viewed from a MOEA perspective and classified under a genotype and phenotype rubric.

---
[1] All the Appendices of this book are available for download at: http://www.cs.cinvestav.mx/~emoobook

Newly identified characteristics can augment the tables as well. Observe that these high-level characteristics were determined from the figures presented in Appendices B, C, D, E and F, whose representation and succeeding interpretation may slightly change based upon underlying computational resolution and graphical presentation of continuous MOPs.

Table 4.1: MOP Numeric Test Function Characteristics

Function	Genotype							Phenotype			
	Connected	Disconnected	Symmetric	Scalable	Solution Type(s)	# Functions	Constraints	Geometry	Connected	Disconnected	Concave	Convex
Binh	x	x			2R	2	2	Curve	x			x
Binh (3)	x				2R	3	2	Point				
Fonseca	x	x			2R	2	0	Curve	x		x	
Fonseca (2)	x		x	x	nR	2	n	Curve	x		x	
Kursawe		x	x	x	nR	2	0	Curve		x	x	
Laumanns	x	x			2R	2	2	Points	x			x
Lis	x	x			2R	2	2	Curve	x			x
Murata	x	x			2R	2	2	Curve	x		x	
Poloni		x			2R	2	2	Curves		x	x	
Quagliarella		x		x	nR	2	n	Points	x			x
Rendon	x	x			2R	2	2	Curve	x			x
Rendon (2)	x	x			2R	2	2	Curve	x			x
Schaffer	x	x			1R	2	0	Curve	x			x
Schaffer (2)			x	x	1R	2	1	Curves	x			x
Viennet	x	x			2R	3	2	Surface	x			x
Viennet (2)	x				2R	3	2	Surface	x	x		
Viennet (3)		x			2R	3	2	Curve	x		x	

The table entries are explained as follows [1630, 1628, 1626]: Each row corresponds to one of the MOPs listed in Appendix A. Each column indicates some genotypic/phenotypic characteristic. P_{true}'s "shape" may be connected, disconnected, symmetric, and/or scalable. PF_{true} may be connected, disconnected, and convex or concave. MOPs exhibiting any of these characteristics are marked with an "x" in the appropriate column. Solution types are notated by the number of decision variables and their type, where "R" indicates real (continuous) decision variables. The number of functions is self-explanatory. Table 4.1 lists MOPs associated with only decision variable constraints, identifying their numbers and types. Table 4.2 lists MOPs which also contain side constraints, identifying both constraint numbers and types. Each MOPs' PF_{true}'s shape is listed, as Pareto fronts may geometrically and/or topologically differ. Also note that only two of these MOPs (Fonseca's second [506] and Schaffer's first [1396]) have known analytical solutions for P_{true}.

Table 4.2: MOP Numeric Test Function (with side constraints) Characteristics

Function	Genotype						Phenotype				
	Connected	Disconnected	Symmetric Scalable	Solution Type(s)	# Functions	Side Constraints	Geometry	Connected	Disconnected	Concave	Convex
Belegundu	x		x	2R	2	2 + 2S	Curve	x			x
Binh (2)	x		x	2R	2	2 + 2S	Curve	x			x
Binh (4)	x			2R	3	2 + 2S	Surface	x		x	
Jimenez	x		x	2R	2	2 + 4S	Curve	x			x
Kita			x x	2R	2	2 + 3S	Curve	x			x
Obayashi	x		x	2R	2	2 + 1S	Curve	x			x
Osyczka		x		2R	2	2 + 2S	Points		x		x
Osyczka (2)		x		6R	2	6 + 6S	Curves		x	x	
Srinivas			x x	2R	2	2 + 2S	Curve	x			x
Tamaki	x		x	3R	3	3 + 1S	Surface	x			x
Tanaka		x		2R	2	2 + 2S	Curves		x	x	
Viennet (4)		x		2R	3	2 + 3S	Surface	x			x

What is P_{true}'s nature? Few MOEA efforts describe an example MOP's underlying multi-dimensional decision variable (genotype) space, i.e., the space where P_{true} resides. Since a MOP is composed of two or more functions, the solution space is obviously restricted by their combined limitations (e.g., decision variable range and side constraints). Within that space, P_{true} may be connected or disconnected, a hyper-area or separate points, symmetric in shape, scalable, and so forth. Solutions may be discrete or continuous, and are composed of one or more decision variables. When solved computationally (and assuming feasible solutions exist), a MOP's P_{true} has only a lower bound (see Theorem 1 in Chapter 6); the upper bound is unknown and varies depending upon the underlying computational resolution.

What is PF_{true}'s nature? PF_{true} lies in objective space and as already noted, may be (dis)connected, convex or concave, and multidimensional. In fact, the structure of *any* Pareto front has theoretical dimensional limitations depending on the number of functions composing the MOP (see Theorem 5 in Chapter 6). PF_{true}'s shape can range from a single vector to a collection of multi-dimensional surfaces [1626].

Test suite functions should encompass combinations of all possible characteristics. Although no guarantor of continued success, any MOEA search algorithm giving effective and efficient results over the test suite should be easily modified to target specific problems. Acceptable results should be evident in the global search space and the local true Pareto front, PF_{true}.

4.3.1 Unconstrained Numeric MOEA Test Functions

Having discussed the MOEA testing requirement and considered the general issues involved in developing a MOEA test function suite, an initial MOP list is proposed for inclusion. As discussed, a complete and sound methodology for constructing MOPs with arbitrary complexity and characteristics still eludes us. Thus, proposed unconstrained numeric test suite MOPs are drawn from the published literature or generated from more basic functions. These MOPs *in toto* address some of the issues discussed in Section 4.2 and reflect the characteristics in Table 4.1. Initially, functions are restricted to those with no side constraints. The mathematical formulations of several of these MOPs, which may be slightly revised from the originals or as elsewhere proposed [1630], are shown in Table 4.3. Figures 4.1 through 4.14 show representations of several of these MOPs' P_{true} and PF_{true}.[2] Note that the graphs' scales for P_{true} may be different than what is stated in Table 4.3 to present P_{true}'s "shape" more clearly.

Schaffer's first (unconstrained) two-objective function is selected for three primary reasons. First is its historical significance; practically all the MOEAs proposed until the mid-1990s were tested using this function. It is also an exemplar of relevant MOP concepts. Second, this MOP allows determination of an analytical expression for PF_{true} [1627]. Third, as noted by Rudolph [1396], this MOP's P_{true} is in closed form so solutions' membership in P_{true} is then easily determined. This MOP's PF_{true} is a single convex Pareto curve (see Figure 4.2) and its P_{true} a line as shown in Figure 4.1. However, its one decision variable implies it may not use the power of a MOEA's search capabilities. This "easy" problem is designated as **MOP1**.

Fonseca's second MOP is also selected because this two-objective function has an advantage of arbitrarily adding decision variables (scalability) without changing PF_{true}'s shape or location in objective space [506]. This MOP's PF_{true} is a single concave Pareto curve (see Figure 4.4) and its P_{true} an area in solution space (see Figure 4.3). Additionally, a closed form for this MOP's P_{true} is claimed [506]. This problem is called **MOP2**.

Next is Poloni's MOP, a maximization problem. This two-objective function's P_{true} consists of two disconnected areas in solution space (see Figure 4.5), while its PF_{true} consists of two disconnected Pareto curves (see Figure 4.6). Its solution mapping into dominated objective space is more convoluted than other MOPs from the literature. This problem is designated as **MOP3**.

---

[2] For a comprehensive list of MOPs and their graphical representations, the reader must consult the Appendices available at: http://www.cs.cinvestav.mx/~emoobook.

4.3 MOP Domain Feature Classification

Kursawe's MOP is included because this two-objective function's P_{true} has several disconnected and unsymmetric areas in solution space. Its PF_{true} consists of three disconnected Pareto curves. Like MOP3, its solution mapping into dominated objective space is also quite convoluted. Like MOP2, its number of decision variables is arbitrary. However, changing the number of decision variables appears to slightly change PF_{true}'s shape and does change its location in objective space. We use it here with three decision variables. Note that both P_{true} (see Figure 4.7) and PF_{true} (see Figure 4.8) are disconnected. This function is renamed **MOP4**.

Viennet's MOP is proposed as the fifth generic test function since this tri-objective function's P_{true} consists of disconnected areas in solution space (see Figure 4.9). And its PF_{true} is a single, convoluted three-dimensional Pareto curve (see Figure 4.10). This function is designated as **MOP5**.

A MOP constructed using Deb's methodology (see Section 4.3.3) is selected. Like MOP4, this two-objective function's P_{true} and PF_{true} are disconnected, although its PF_{true} consists of four Pareto curves (see Figure 4.12). Its solution mapping into dominated objective space is not as convoluted as MOP4's. This problem is used to compare MOEA performance in finding similar phenotypes produced by different MOPs (c.f., MOP4). And this function is now called **MOP6**.

Finally, Viennet's second MOP is also suggested. This tri-objective MOP's P_{true} is a connected region in solution space (see Figure 4.11). Its PF_{true} appears to be a surface and its mapping into objective space appears straightforward (see Figure 4.12). This function is primarily meant to complement MOP5. This function is relabeled as **MOP7**.

Table 4.3: MOEA Test Suite Functions

MOP	Definition	Constraints
MOP1 P_{true} connected, PF_{true} convex	$F = (f_1(x), f_2(x))$, where $f_1(x) = x^2,$ $f_2(x) = (x-2)^2$	$-10^5 \leq x \leq 10^5$
MOP2 P_{true} connected, PF_{true} concave, number of decision variables scalable	$F = (f_1(\mathbf{x}), f_2(\mathbf{x}))$, where $f_1(\mathbf{x}) = 1 - \exp(-\sum_{i=1}^{n}(x_i - \frac{1}{\sqrt{n}})^2),$ $f_2(\mathbf{x}) = 1 - \exp(-\sum_{i+1}^{n}(x_i + \frac{1}{\sqrt{n}})^2)$	$-4 \leq x_i \leq 4;\ i = 1,2,3$

Table 4.3: (continued)

MOP	Definition	Constraints
MOP3 P_{true} disconnected, PF_{true} disconnected (2 Pareto curves)	Maximize $F = (f_1(x,y), f_2(x,y))$, where $$f_1(x,y) = -[1 + (A_1 - B_1)^2 + (A_2 - B_2)^2],$$ $$f_2(x,y) = -[(x+3)^2 + (y+1)^2]$$	$-3.1416 \leq x, y \leq 3.1416$, $A_1 = 0.5 \sin 1 - 2 \cos 1 + \sin 2 - 1.5 \cos 2$, $A_2 = 1.5 \sin 1 - \cos 1 + 2 \sin 2 - 0.5 \cos 2$, $B_1 = 0.5 \sin x - 2 \cos x + \sin y - 1.5 \cos y$, $B_2 = 1.5 \sin x - \cos x + 2 \sin y - 0.5 \cos y$
MOP4 P_{true} disconnected, PF_{true} disconnected (3 Pareto curves), number of decision variables scalable	$F = (f_1(\mathbf{x}), f_2(\mathbf{x}))$, where $$f_1(\mathbf{x}) = \sum_{i=1}^{n-1} (-10 e^{(-0.2) * \sqrt{x_i^2 + x_{i+1}^2}}),$$ $$f_2(\mathbf{x}) = \sum_{i=1}^{n} (|x_i|^a + 5 \sin(x_i)^b)$$	$-5 \leq x_i \leq 5;\ i = 1, 2, 3$ $a = 0.8$, $b = 3$
MOP5 P_{true} disconnected and unsymmetric, PF_{true} connected (a 3-D Pareto curve)	$F = (f_1(x,y), f_2(x,y), f_3(x,y))$, where $$f_1(x,y) = 0.5 * (x^2 + y^2) + \sin(x^2 + y^2),$$ $$f_2(x,y) = \frac{(3x - 2y + 4)^2}{8} + \frac{(x - y + 1)^2}{27} + 15,$$ $$f_3(x,y) = \frac{1}{(x^2 + y^2 + 1)} - 1.1 e^{(-x^2 - y^2)}$$	$-30 \leq x, y \leq 30$
MOP6 P_{true} disconnected, PF_{true} disconnected (4 Pareto curves), number of Pareto curves scalable	$F = (f_1(x,y), f_2(x,y))$, where $$f_1(x,y) = x,$$ $$f_2(x,y) = (1 + 10y) *\left[1 - \left(\frac{x}{1+10y}\right)^\alpha - \frac{x}{1+10y} \sin(2\pi q x)\right]$$	$0 \leq x, y \leq 1$, $q = 4$, $\alpha = 2$

4.3 MOP Domain Feature Classification 185

Table 4.3: (continued)

MOP	Definition	Constraints
MOP7 P_{true} connected, PF_{true} disconnected	$F = (f_1(x,y), f_2(x,y), f_3(x,y))$, where $f_1(x,y) = \frac{(x-2)^2}{2} + \frac{(y+1)^2}{13} + 3,$ $f_2(x,y) = \frac{(x+y-3)^2}{36} + \frac{(-x+y+2)^2}{8} - 17,$ $f_3(x,y) = \frac{(x+2y-1)^2}{175} + \frac{(2y-x)^2}{17} - 13$	$-400 \leq x, y \leq 400$

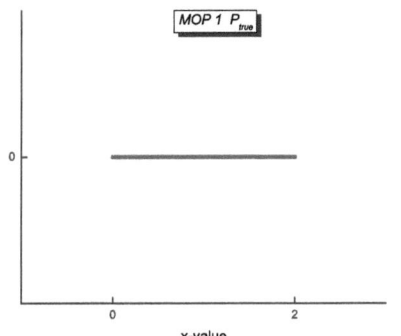

Fig. 4.1. MOP1 P_{true}

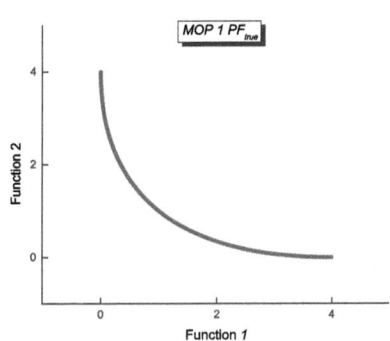

Fig. 4.2. MOP1 PF_{true}

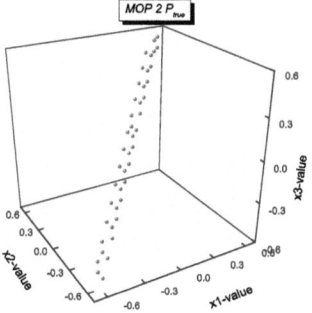

Fig. 4.3. MOP2 P_{true}

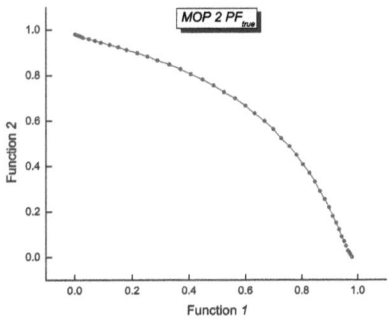

Fig. 4.4. MOP2 PF_{true}

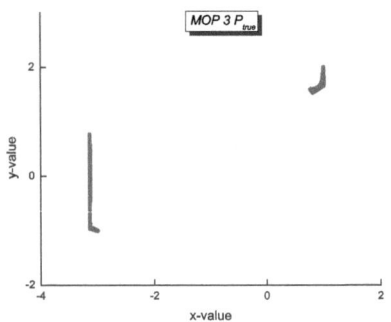

Fig. 4.5. MOP3 P_{true}

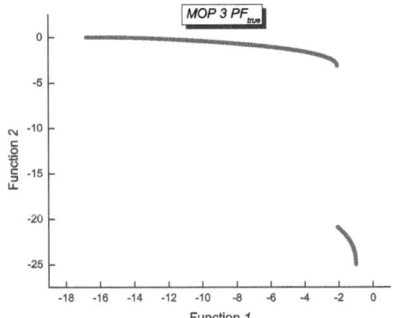

Fig. 4.6. MOP3 PF_{true}

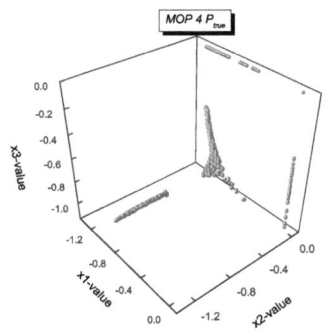

Fig. 4.7. MOP4 P_{true}

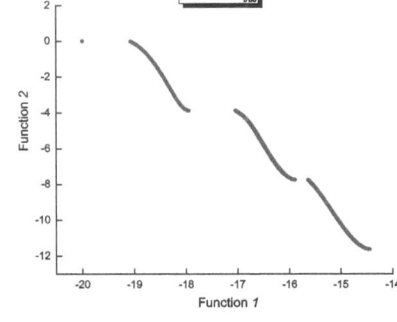

Fig. 4.8. MOP4 PF_{true}

These proposed numeric MOEA test functions in Table 4.3 address the issues mentioned in Section 4.2. MOP1 and MOP2 are arguably "easy" MOPs. MOP2 and MOP4 are scalable as regards decision variable dimensionality. MOP6 is scalable as regarding the number of Pareto curves in PF_{true}. MOP5 and MOP7 are tri-objective MOPs. All are nonlinear, and several show a lack of symmetry in both P_{true} and PF_{true}. Taken together these MOPs begin to form a coherent basis for MOEA comparisons. However, other relevant MOP characteristics (as reflected in Tables 4.1 and 4.2) may also be addressed by other MOPs selected for test suite inclusion. These additional MOPs may need to be constructed in order to exhibit some desired characteristics (see Section 4.3.3). Utilization of a test suite is advantageous to the community in the fact that it presents data that is baselined from a standard test suite [1633]. Another philosophical development of a test suite reflects similar functionality [1772].

Observe that parameters can be added to each function in order to highlight and emphasize PF_{true} characteristics providing more difficulty for a

4.3 MOP Domain Feature Classification   187

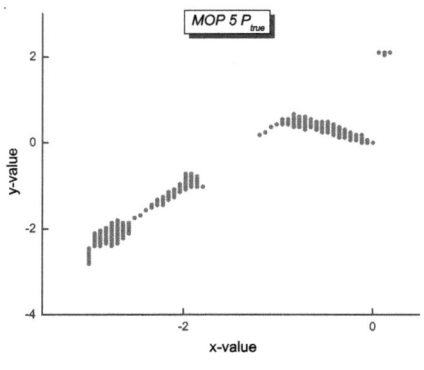

Fig. 4.9. MOP5 P_{true}

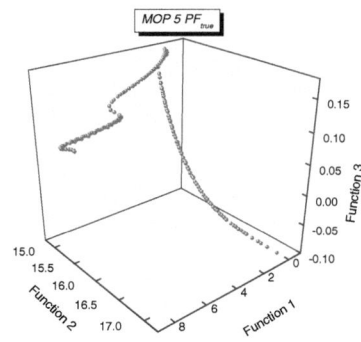

Fig. 4.10. MOP5 PF_{true}

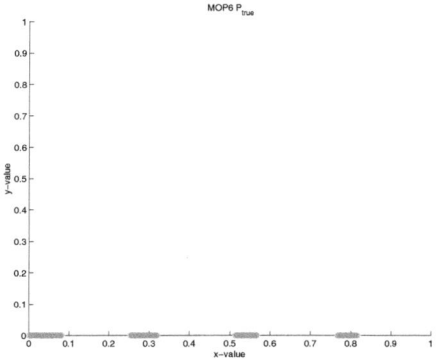

Fig. 4.11. MOP6 P_{true}

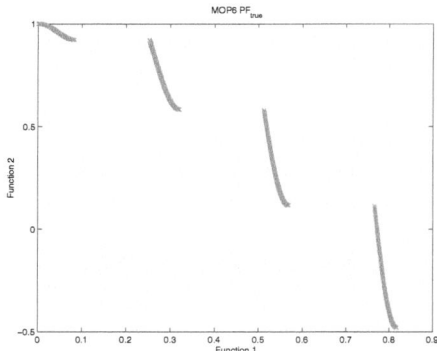

Fig. 4.12. MOP6 PF_{true}

MOEA search. For example, in MOP4, one can vary the parameters a and b. In MOP6, q and α can extend over a range of values.

4.3.2 Side-Constrained Numeric MOEA Test Functions

Side-constrained numeric MOPs should also be in any comprehensive MOEA test function suite. Suitable linear and nonlinear constrained MOPs are proposed as drawn from the published literature as well as generated. Note that solving constrained MOPs with MOEAs brings in other open research issues, most notably how the side constraints are accounted for in the MOEA in order to ensure feasible solutions.

Historically, constraints have been handled by a MOEA through augmenting the objective functions with a penalty function [1360] (i.e., individuals encoding infeasible solutions are less fit than those encoding feasible solutions). This technique generally requires an "unnatural" normalization or a

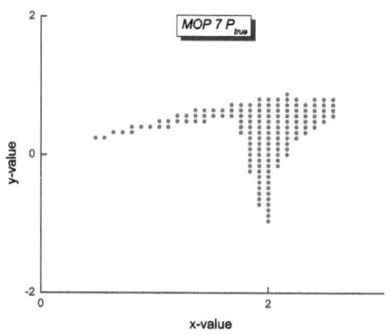
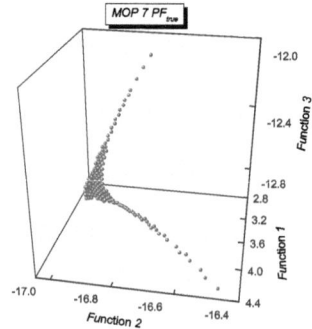

Fig. 4.13. MOP7 P_{true} ****Fig. 4.14.** MOP7 PF_{true}

static/dynamic weighting schedule between these two elements of the new objective form. Another technique [796] uses binary tournament selection comparing two solutions, the feasible solution always being selected. If both are infeasible, the one "closest" to the constraint boundary is used. If both are feasible, then Horn's niched Pareto MOEA is employed. The constraints can be processed separately using the Pareto nondominated concept and then combined with the objective function nondominated set [1329, 1330, 262]. Using these generic approaches, one can classify two constrained MOP solutions using the domination principle if both solutions are feasible (class a); if one is feasible and one is not (class b); or both are infeasible, but one has a smaller constraint violation (class c) over the set of constraints. This classification can then be used in the MOEA processing of constraints when attempting to reach the Pareto front. Of course, the general objective of converging to the true front with a diverse set of points while employing an appropriate constraint handling technique is probably problem dependent. Observe that the MOEA could be applied to the MOP initially without considering the side constraints, but using post-processing to remove infeasible solutions (could take more processing, but reflects simpler algorithmic steps). However, post-processing may delete too many points.

Table 4.4: Side-Constrained MOEA Test Suite Functions

MOP	Definition	Side Constraints
MOP-C1 Binh(2)	$F = (f_1(x,y), f_2(x,y))$, where $$f_1(x,y) = 4x^2 + 4y^2,$$ $$f_2(x,y) = (x-5)^2 + (y-5)^2$$	$0 \leq x \leq 5,\ 0 \leq y \leq 3,$ $0 \geq (x-5)^2 +$ (4.1) $y^2 - 25,$ $0 \geq -(x-8)^2 -$ $(y+3)^2 + 7.7$

4.3 MOP Domain Feature Classification 189

Table 4.4: (continued)

MOP	Definition	Side Constraints
MOP-C2 Osyczka(2)	$F = (f_1(\mathbf{x}), f_2(\mathbf{x}))$, where $f_1(\mathbf{x}) = -(25(x_1-2)^2 + (x_2-2)^2 + (x_3-1)^2 + (x_4-4)^2 + (x_5-1)^2)$, $f_2(\mathbf{x}) = x_1^2 + x_2^2 + x_3^2 + x_4^2 + x_5^2 + x_6^2$	$0 \leq x_1, x_2, x_6 \leq 10$, $1 \leq x_3, x_5 \leq 5$, $0 \leq x_4 \leq 6$, $0 \leq x_1 + x_2 - 2$, $0 \leq 6 - x_1 - x_2$, $0 \leq 2 - x_2 + x_1$, $0 \leq 2 - x_1 + 3x_2$, $0 \leq 4 - (x_3-3)^2 - x_4$, $0 \leq (x_5-3)^2 + x_6 - 4$
MOP-C3 Viennet(4)	$F = (f_1(x,y), f_2(x,y), f_3(x,y))$, where $f_1(x,y) = \frac{(x-2)^2}{2} + \frac{(y+1)^2}{13} + 3$, $f_2(x,y) = \frac{(x+y-3)^2}{175} + \frac{(2y-x)^2}{17} - 13$, $f_3(x,y) = \frac{(3x-2y+4)^2}{8} + \frac{(x-y+1)^2}{27} + 15$	$-4 \leq x, y \leq 4$, $y < -4x + 4$, $x > -1$, $y > x - 2$
MOP-C4 Tanaka	$F = (f_1(x,y), f_2(x,y))$, where $f_1(x,y) = x$, $f_2(x,y) = y$	$0 < x, y \leq \pi$, $0 \geq -(x^2) - (y^2)$ $+1+$ $(a \cos$ $(b \arctan(x/y)))$ $a = 0.1$ $b = 16$

Table 4.4 includes an extensive variety of possible numeric constrained test functions. Based upon the general principles of test function selection indicated previously, various constrained MOPs are suggested. Thus, Binh's second MOP is selected as a side-constrained test function. This two-objective function's P_{true} is an area in solution space and its PF_{true} a single convex Pareto curve. This problem is renamed **MOP-C1**. Next is Osyczka's second MOP, which is a heavily constrained, six decision variable problem. This two-objective function's P_{true}'s shape is currently unknown while its PF_{true} consists of three disconnected Pareto curves. This problem is designated as **MOP-C2**.

Viennet's fourth MOP is also selected for inclusion. This three-objective function's P_{true} is an irregularly shaped area in solution space. Its PF_{true} is a Pareto surface. This problem is designated as **MOP-C3**. These MOPs' mathematical formulations are shown in Table 4.4; figures showing representations of each MOPs' P_{true} and PF_{true} are found in Appendix G. Also suggested is Tanaka's two objective function with two nonlinear constraints which is defined as **MOP-C4** with the indicated genotype and phenotype Pareto curves being the same as shown in Appendix G. Note the disconnected Pareto front regions for **MOP-C4** which can make it difficult for a MOEA to find PF_{true}. As with non-constrained numerical MOPs, parameters can be added to each function in order to highlight and emphasize PF_{true} providing more difficulty for a MOEA search. For example, the parameters a, b of MOP-C4 (Tanaka) from Table 4.4 can be varied over appropriate ranges.

Considering specific variations of these two MOP-C4 parameters along with an *absolute* operator on the last term of the constraint, can result in the following general landscapes:

- standard Tanaka phenotype with $a = .1$ and $b = 16$ (Figure 4.15)
- smaller continuous regions with $a = .1$, $b = 32$ (Figure 4.16)
- increased distance between regions using the absolute value on the last term of the constraint, and $a = .1$, $b = 16$ (Figure 4.17)
- increased distance between regions using the absolute values on the last term of the constraint, and $a = .1$, $b = 32$ (Figure 4.18)
- deeper periodic regions using the absolute value on the last term of the constraint, and $a = .1(x^2 + y^2 + 5xy)$, $b = 32$ (Figure 4.19)
- non-periodic regions on front using the absolute value on the last term of the constraint, and $a = .1(x^2 + y^2 + 5xy)$, $b = 8(x^2 + y^2)$ (Figure 4.20)

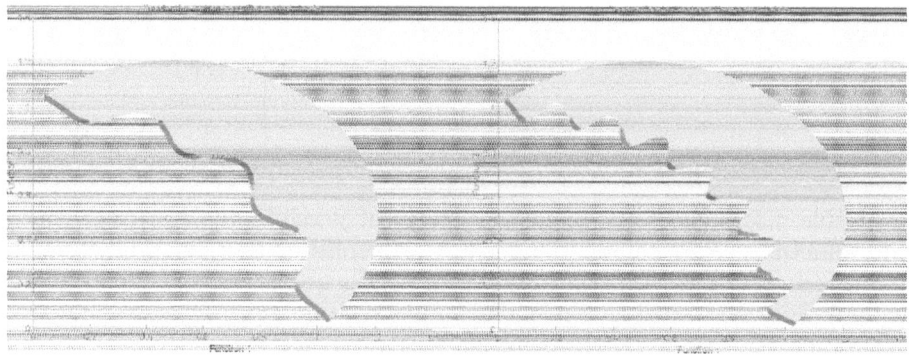

Fig. 4.15. MOP-C4 (Tanaka), $a = .1, b = 16$, Original PF_{true} (P_{true}) regions

Fig. 4.16. MOP-C4 (Tanaka), $a = .1, b = 32$, smaller continuous PF_{true} (P_{true}) regions

By selecting values for the two parameters (a, b) different landscape structures evolve. P_{true} of course is the same by function definition. Although the central Pareto curve in Figure 4.15 appears not to be continuous due to numerical accuracy, it is continuous in reality. The two internal sections of this

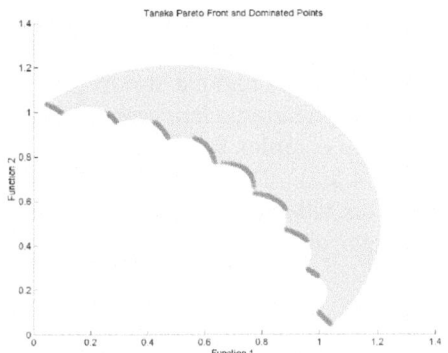
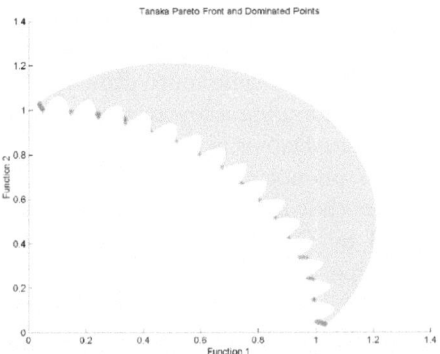

Fig. 4.17. MOP-C4 (Tanaka), using absolute value on the last term of the constraint and $a = .1, b = 16$, increased distance between PF_{true} (P_{true}) regions

Fig. 4.18. MOP-C4 (Tanaka), using absolute value on the last term of the constraint and $a = .1, b = 32$, increased distance between PF_{true} (P_{true}) regions

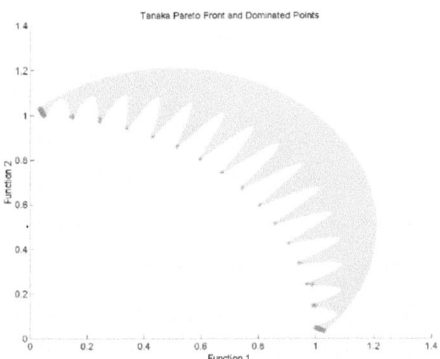
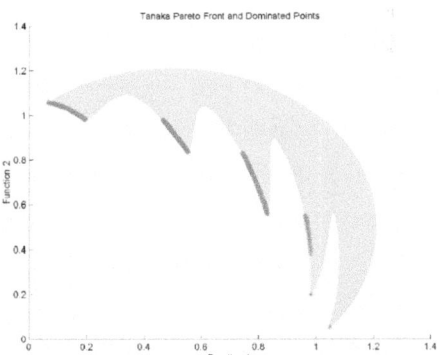

Fig. 4.19. MOP-C4 (Tanaka), using absolute value on the last term of the constraint and $a = .1(x^2 + y^2 + 5xy), b = 32$, deeper PF_{true} (P_{true}) periodic regions

Fig. 4.20. MOP-C4 (Tanaka), using absolute value on the last term of the constraint and $a = .1(x^2 + y^2 + 5xy), b = 8(x^2 + y^2)$, non-periodic regions PF_{true} (P_{true})

curve are very difficult to find numerically because of the near horizontal or vertical slope at the associated points, respectively.

In general, (a, b) controls the length of the continuous region on the Pareto front. As this region is decreased, a MOEA finds fewer points on PF_{true} due to the discretization of **x**; i.e., a more difficult problem. By increasing the value of a, the length of the "cuts" become deeper requiring the search to proceed along a narrower corridor, again more difficult to solve. One can also move away from the periodic nature of the disconnected PF_{true} regions by changing b from the initial value of 16 (more optimal solutions in one direction or the other). Finding all closely packed PF_{true} regions then becomes difficult for the MOEA. As reflected in the figures, such parameter variations can cause MOEA searches to become more difficult as the size of feasible regions are decreased. Just by changing a few parameters very different phenotype landscapes result. Note that Deb et al. [375] have generated a more complex function with quite similar phenotype characteristics (see Section 4.3.3).

Another difficult MOP (**MOP-C5**) with six side constraints is that proposed by Osyczka and Kundu [1225] with two objective functions and six genotype variables as follows:

Minimize $F = (f_1(\mathbf{x}), f_2(\mathbf{x}))$, where

$f_1(\mathbf{x}) = -(25(x_1 - 2)^2 + (x_2 - 2)^2 + (x_3 - 1)^2 + (x_4 - 4)^2 + (x_5 - 1)^2)$
$f_2(\mathbf{x}) = x_1^2 + x_2^2 + x_3^2 + x_4^2 + x_5^2 + x_6^2$

Subject to

$$c_1(\mathbf{x}) = x_1 + x_2 - 2 \geq 0,$$
$$c_2(\mathbf{x}) = 6 - x_1 - x_2 \geq 0,$$
$$c_3(\mathbf{x}) = 2 + x_1 - x_2 \geq 0,$$
$$c_4(\mathbf{x}) = 2 - x_1 + 3x_2 \geq 0,$$
$$c_5(\mathbf{x}) = 4 - (x_3 - 3)^2 - x_4 \geq 0,$$
$$c_6(\mathbf{x}) = (x_5 - 3)^2 + x_6 - 4 \geq 0,$$
$$0 \leq x_1, x_2, x_6 \leq 10, 1 \leq x_3, x_5 \leq 5, 0 \leq x_4 \leq 6 \quad (4.2)$$

The PF_{true} is shown in Figure 4.21 reflecting six regions representing the intersection of specific constraints. Maintaining subpopulations at the different intersections is a difficult MOEA problem. The P_{true} solution values are $x_4 = 0$ and $x_6 = 0$ with the remaining variable values associated with each region presented in Table 4.5.

Table 4.5: MOP-C5 P_{true} solution values [1225]

Region	x₁	x₂	x₃	x₅
AB	5	1	(1,...,5)	5
BC	5	1	(1,...,5)	1
CD	(4.06,...,5)	(0.68,...,1)	1	1
DE	0	2	(1,...,3.73)	1
EF	(0,...,1)	(2,...1)	1	1

In the next section, processes for generating side-constrained numeric MOP functions as well as unconstrained MOP functions are presented.

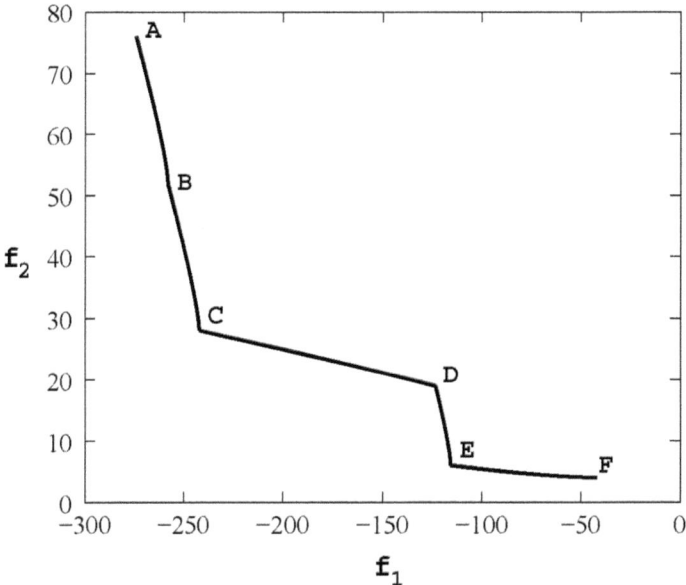

Fig. 4.21. MOP-C5 connected PF_{true} regions [1225]

4.3.3 MOP Test Function Generators

MOP test functions can also be generated by using the single-objective functions. A methodology for constructing MOPs exhibiting desired characteristics has been proposed by Deb [357]. Key issues are addressed in Deb's work. The fundamental generational characteristics include ease of construction and scalability to any number of objective functions or decision variables. Moreover, the associated Pareto fronts have to be known, visualizable in shape and location, and the associated decision variables computationally not difficult to find in PF_{known}. A MOEA should always move toward PF_{true} and P_{true} as well as provide a uniform distribution of points on PF_{known}. Observe that for higher dimensional PF_{known} visualizations, the use of symmetric hyperplanes both at the phenotype and genotype levels, can help in analyzing performance with only 3D plots of limited dimensions.

Deb defines both a *local* and *global* Pareto optimal set. His global Pareto optimal set is what is termed P_{true}; this text's terminology is easily extended to denote a local Pareto optimal set, i.e. P_{local}. However, P_{local} is ill-defined and may be confusing. Consider Deb's definition.

Local Pareto Optimal Set: Given some Pareto optimal set $\mathcal{P}$, if $\forall x \in \mathcal{P}$, $\neg \exists y$ satisfying $\| y - x \|_\infty \leq \epsilon$, where ϵ is a small positive number (in principle, y is obtained by perturbing x in a small neighborhood), and for which $F(y) \preceq F(x)$, then the solutions in $\mathcal{P}$ constitute a local Pareto optimal set.

This definition implies that for some given set of Pareto optimal solutions, each is perturbed in some manner but no new nondominated vectors are found. Deb's purpose here is defining a set of Pareto optimal solutions whose associated front (PF_{local}) is "behind" PF_{true} for the given MOP. Although conceptually possible, any P_{local}'s existence is dependent upon the ϵ selected within which solutions are perturbed. Additionally, too large an ϵ prohibits a P_{local}, too small an ϵ may result in multiple local fronts.

Deb also extends the concepts of multimodality, deception, an isolated optimum, and collateral noise (well known single-objective EA difficulties) to the multiobjective domain. Two of these extensions can be disputed. First, he defines a deceptive MOP as one in which there are at least two optima (PF_{local} and PF_{true}) and where the majority of the search space favors PF_{local}. As stated above this concept depends on P_{local}'s existence. Secondly, Deb defines a multimodal MOP as one with multiple local fronts. This definition mixes terminology. One should use the term multimodal only when referring to a single-objective optimization function containing both local and global minima. As all vectors composing a Pareto front are "equally" optimal there is no Pareto front modality. Perhaps the term "multifrontal" is a better choice to reflect this situation.

Deb also notes some of the same MOP phenotype characteristics as presented in Section 4.3 and in Van Veldhuizen [1626]. He points out that when computationally derived a non-uniform distribution of vectors may exist in some Pareto front. He limits his initial test construction efforts to unconstrained MOPs of only two functions; his construction methodology then places restrictions on the two component functions so that resultant MOPs exhibit desired properties. To accomplish this he defines various generic biobjective optimization problems as reflected in the following two examples:

Minimize $F = (f_1(\mathbf{x}), f_2(\mathbf{x}))$, where

$$f_1(\mathbf{x}) = f(x_1, \ldots, x_m),$$
$$f_2(\mathbf{x}) = g(x_{m+1}, \ldots, x_N) \; h(f(x_1, \ldots, x_m), g(x_{m+1}, \ldots, x_N)) \quad (4.3)$$

where function f_1 is a function of ($m < N$) decision variables and f_2 a function of all N decision variables. The function g is one of ($N-m$) decision variables which are not included in function f. The function h is directly a function of f and g function values. The f and g functions are also restricted to positive values in the search space, i.e., $f > 0$ and $g > 0$.

Deb lists five functions each for possible f and g instantiation, and four for h. These functions may then be "mixed and matched" to create MOPs with desired characteristics.

He states these functions have the following general effect:

f – This function controls vector representation uniformity along the Pareto front.
g – This function controls the resulting MOP's characteristics – whether it is multifrontal or has an isolated optimum.
h – This function controls the resulting Pareto front's characteristics (e.g., convex, disconnected, etc.)

These functions respectively influence search along and towards the Pareto front, and the shape of a Pareto front in $\Re^2$. Deb implies that a MOEA has difficulty finding PF_{true} because it gets "trapped" in the local optimum, namely PF_{local}.

This methodology is not the only way to construct MOPs exhibiting some set of desired characteristics such as curve, surface, convex, non-convex, continuous, discrete, disjoint, scalable, and others. Real-world MOPs may have similar genotype and/or phenotype characteristics but look nothing at all like the examples proposed. Thus the fact a MOEA "passes" all test functions submitted using a generative methodology may have no bearing on its performance in solving real-world MOPs. However, the same can be said of the test suites proposed. Any test functions must be carefully selected to reflect as accurately as possible the problem domain they attempt to represent.

Each of the test functions defined below is structured in the same manner and consists itself of three functions f_1, g, h [357].

$$Minimize: F(\mathbf{x}) = (f_1, f_2),$$
$$subject\ to: f_2(\mathbf{x}) = g(x_2, \ldots, x_m) h(f_1(x_1), g(x_2, \ldots, x_m)),$$
$$where: \mathbf{x} = (x_1, \ldots, x_M). \quad (4.4)$$

The function f_1 is a function of the first decision variable only, g is a function of the remaining $m-1$ variables, and the parameters of h are the function values of f_1 and g. The test functions differ in these three functions as well as in number of variables m and in the values the variables may take.

The six different test functions that follow the scheme given in Equation 4.4 are described next. This benchmark is known as the ZDT (Zitzler-Deb-Thiele) test suite [1772].

Test Problem ZDT1: Has a convex Pareto-optimal front:

$$f_1(\mathbf{x}) = x_1,$$
$$f_2(\mathbf{x}, g) = g(\mathbf{x}) \cdot (1 - \sqrt{f_1/g(\mathbf{x})})$$
$$g(\mathbf{x}) = 1 + \frac{9}{n-1} \cdot \sum_{i=2}^{n} x_i$$

where $n = 30$, and $x_i \in [0, 1]$. The PF_{true} is formed with $g(\mathbf{x}) = 1$. In Figure 4.22 it is shown the PF_{true}, and 10,000 random solutions are plotted in objective space with the PF_{true} in Figure 4.23.

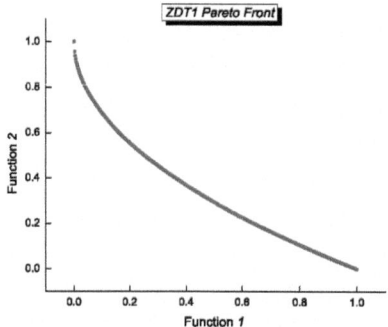

Fig. 4.22. ZDT1 Pareto Front

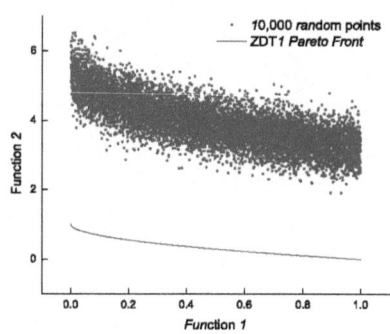

Fig. 4.23. 10,000 random solutions generated are plotted for ZDT1 test problem

Test Problem ZDT2: Has a nonconvex Pareto-optimal front:

$$f_1(\mathbf{x}) = x_1,$$
$$f_2(\mathbf{x}, g) = g(\mathbf{x}) \cdot (1 - (f_1/g(\mathbf{x}))^2)$$
$$g(\mathbf{x}) = 1 + \frac{9}{n-1} \cdot \sum_{i=2}^{n} x_i$$

where $n = 30$, and $x_i \in [0, 1]$. The PF_{true} is formed with $g(\mathbf{x}) = 1$. In Figure 4.24 it is shown the PF_{true}, and 10,000 random solutions are plotted in objective space with the PF_{true} in Figure 4.25.

Test Problem ZDT3: Has a Pareto-optimal front disconnected, consisting of several noncontiguous convex parts:

$$f_1(\mathbf{x}) = x_1$$
$$f_2(\mathbf{x}, g) = g(\mathbf{x}) \cdot \left(1 - \sqrt{\frac{f_1}{g(\mathbf{x})}} - \frac{f_1}{g(\mathbf{x})} \cdot \sin(10\pi f_1)\right)$$
$$g(\mathbf{x}) = 1 + \frac{9}{n-1} \cdot \sum_{i=2}^{n} x_i$$

where $n = 30$, and $x_i \in [0, 1]$. The PF_{true} is formed with $g(\mathbf{x}) = 1$. The introduction of the sin() function causes discontinuity in the Pareto-optimal

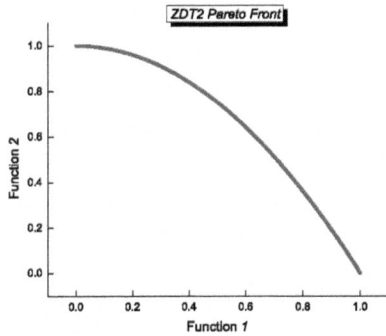

Fig. 4.24. ZDT2 Pareto Front

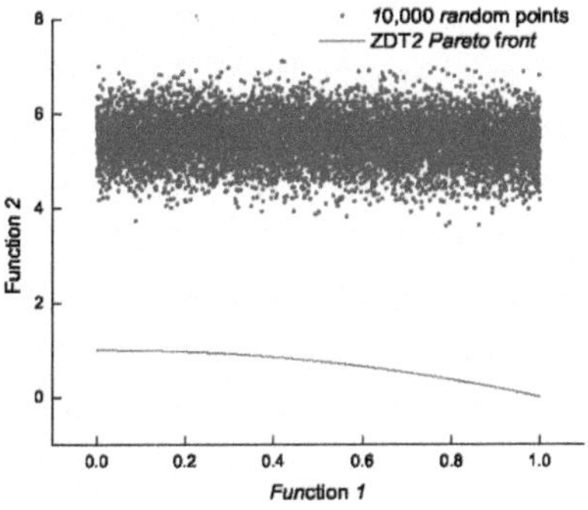

Fig. 4.25. 10,000 random solutions generated are plotted for ZDT2 test problem

front. However, there is no discontinuity in decision variable space. In Figure 4.26 it is shown the PF_{true} , and 10,000 random solutions are plotted in objective space with the PF_{true} in Figure 4.27.

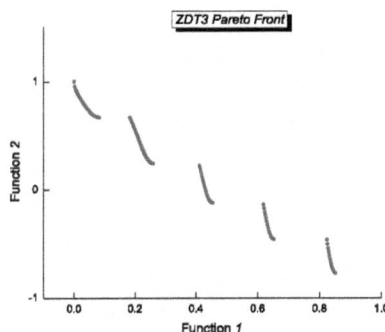
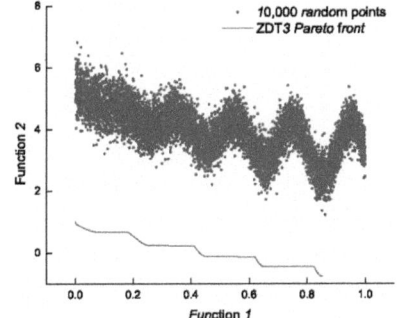

Fig. 4.26. ZDT3 Pareto Front

Fig. 4.27. 10,000 random solutions generated are plotted for ZDT3 test problem

Test Problem ZDT4: Contains 21^9 local Pareto-optimal fronts and, therefore, tests for the EA's ability to deal with multifrontality:

$$f_1(\mathbf{x}) = x_1$$

$$f_2(\mathbf{x}, g) = g(\mathbf{x}) \cdot \left(1 - \sqrt{\frac{f_1}{g(\mathbf{x})}}\right)$$

$$g(\mathbf{x}) = 1 + 10 \cdot (n-1) + \sum_{i=2}^{n}(x_i^2 - 10\cos(4\pi x_i))$$

where $n = 10$, $x_1 \in [0,1]$ and $x_2, \ldots, x_n \in [-5, 5]$. The PF_{true} is formed with $g(\mathbf{x}) = 1$, the best PF_{local} is formed with $g(\mathbf{x}) = 1.25$. In Figure 4.28 it is shown the PF_{true}, and 10,000 random solutions are plotted in objective space with the PF_{true} in Figure 4.29.

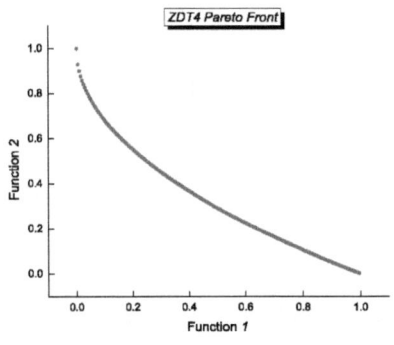

Fig. 4.28. ZDT4 Pareto Front

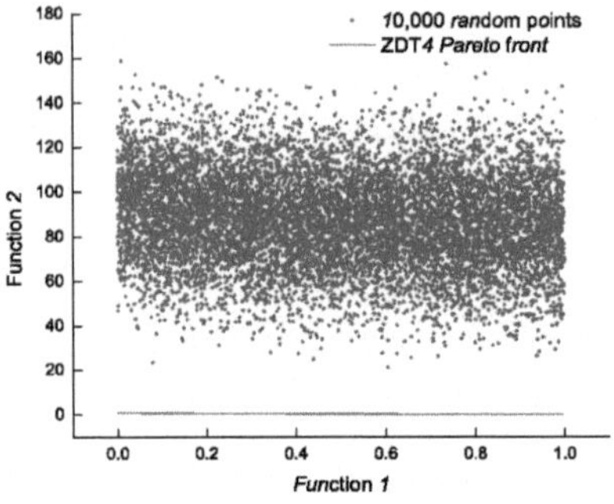

Fig. 4.29. 10,000 random solutions generated are plotted for ZDT4 test problem

Test Problem ZDT5: Describes a deceptive problem and distinguishes itself from the other test functions in that x_i represents a binary string:

$$f_1(\mathbf{x}) = 1 + u(x_1)$$

$$f_2(\mathbf{x}, g) = \frac{g(\mathbf{x})}{f_1(\mathbf{x})}$$

$$g(\mathbf{x}) = \sum_{i=2}^{11} v(u(x_i)),$$

$$v(u(x_i)) = \begin{cases} 2 + u(x_i), & \text{if } u(x_i) < 5; \\ 1, & \text{if } u(x_i) = 5. \end{cases}$$

where $u(x_1)$ gives the number of ones in the bit vector x_i, and $n = 11$, $x_1 \in \{0,1\}^{30}$ and $x_2, \ldots, x_n \in \{0,1\}^5$. The PF_{true} is formed with $g(\mathbf{x}) = 10$, while the best deceptive PF_{true} is represented by the solutions for which

$g(\mathbf{x}) = 11$. The global Pareto-optimal front as well as the local ones are convex. In Figure 4.30 it is shown the PF_{true}, and 10,000 random solutions are plotted in objective space with the PF_{true} in Figure 4.31.

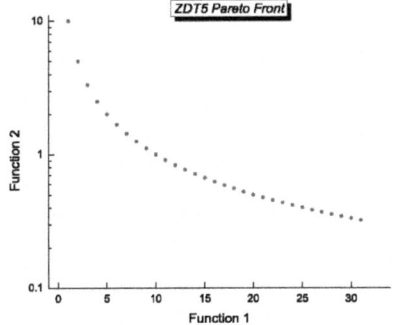

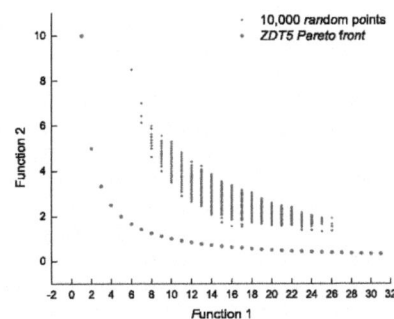

Fig. 4.30. ZDT5 Pareto Front

Fig. 4.31. 10,000 random solutions generated are plotted for ZDT5 test problem

Test Problem ZDT6: Includes two difficulties caused by the nonuniformity of the search space: first, the P_{true} are nonuniformly distributed along the PF_{true} (the front is biased for solutions for which $f_1(\mathbf{x})$ in near one); and second, the density of the solutions is lowest near the PF_{true} and highest away from the front:

$$f_1(\mathbf{x}) = 1 - \exp(-4x_1) \cdot \sin^6(6\pi x_1)$$

$$f_2(\mathbf{x}, g) = g(\mathbf{x}) \cdot (1 - (\frac{f_1}{g(\mathbf{x})})^2)$$

$$g(\mathbf{x}) = 1 + 9 \cdot [\frac{(\sum_{i=2}^{n} x_i)}{9}]^{0.25}$$

where $n = 10$, $x_i \in \{0,1\}$. The PF_{true} is formed with $g(\mathbf{x}) = 1$ and is nonconvex. In Figure 4.32 it is shown the PF_{true}, and 10,000 random solutions are plotted in objective space with the PF_{true} in Figure 4.33.

4.4 Generic Scalable MOP Test Problems

Deb, Thiele, Laumanns and Zitzler [378] have proposed a set of generational MOPs for testing and comparing MOEAs. This suite of benchmarks attempts to define generic MOEA test problems that are scalable to a user defined

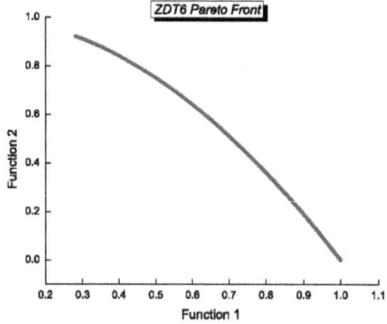
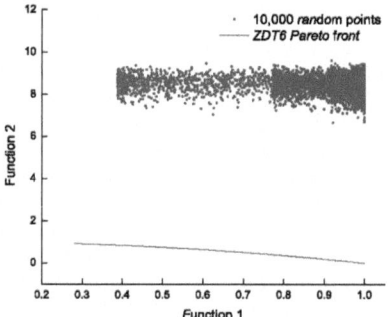

Fig. 4.32. ZDT6 Pareto Front

Fig. 4.33. 10,000 random solutions generated are plotted for ZDT6 test problem

number of objectives. Because of the last names of its creators, this test suite is known as **DTLZ** (Deb-Thiele-Laumanns-Zitzler). They suggest here a representative set of test problems as discussed. Possibly more interesting and useful test problems could be designed using these techniques. An increase in dimensionality of the objective space also causes a random initial population of moderate size to be nondominated to each other, thereby reducing the effect of the selection operator in a MOEA. Thus, this set of test problems is to be used with a large number of objectives and to be used to asses if a MOEA can reach the true Pareto-optimal front. Since the desired front will be known in these test problems, a convergence metric (such as average distance to the front) can be used to track the convergence of an algorithm.

Test Problem DTLZ1: A simple test problem using M objectives; PF_{true} linear Pareto-optimal front, separable, multimodal.

Minimize:
$f_1(x) = \frac{1}{2}x_1 x_2 \ldots x_{M-1}(1 + g(x_M))$,
$f_2(x) = \frac{1}{2}x_1 x_2 \ldots (1 - x_{M-1})(1 + g(x_M))$,

$\vdots$

$f_{M-1}(x) = \frac{1}{2}x_1(1 - x_2)(1 + g(x_M))$,
$f_M(x) = \frac{1}{2}(1 - x_1)(1 + g(x_M))$,
subject to $0 \leq x_i \leq 1$ $\forall$ $i = 1, 2, \ldots, n$
where: $g(x_M) = 100 \left[|x_M| + \sum_{x_i \in x_M}(x_i - 0.5)^2 - \cos(20\pi(x_i - 0.5)) \right]$

For *all* the DTLZ test functions discussed in this chapter, we adopted $M = 3$. The Pareto-optimal solutions correspond to $x_M^* = 0$ and the objective function values on the linear hyper-plane: $\sum_{m=1}^{M} = 0.5$ and it is shown in Figure 4.34. A value of $k = 5$ is suggested here. In the above problem, the total number of variables is $n = M + k - 1$. The difficulty in this problem is to converge to the hyper-plane. The search space contains $(11^k - 1)$ local Pareto-optimal fronts, each of which can attract a MOEA. The problem can be made more difficult by using other multi-modal g functions (using a larger k) and/or replacing x_i by a nonlinear mapping $x_i = N_i(y_i)$ and treating y_i as the decision variables. It is interesting to note that for $M > 3$, all Pareto-optimal solutions on a three-dimensional plot involving f_M and any other two objectives will lie on or below the above hyper-plane.

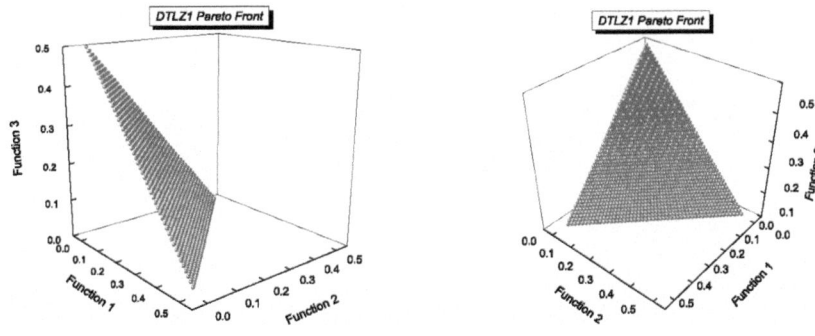

Fig. 4.34. DTLZ1 Pareto Front

Test Problem DTLZ2: This test problem is
Minimize:
$f_1(x) = (1 + g(x_M)) \cos(x_1\pi/2) \cos(x_2\pi/2) \ldots \cos(x_{M-2}\pi/2) \cos(x_{M-1}\pi/2),$
$f_2(x) = (1 + g(x_M)) \cos(x_1\pi/2) \cos(x_2\pi/2) \ldots \cos(x_{M-2}\pi/2) \sin(x_{M-1}\pi/2),$
$f_3(x) = (1 + g(x_M)) \cos(x_1\pi/2) \cos(x_2\pi/2) \ldots \sin(x_{M-2}\pi/2),$
$\vdots \quad \vdots$
$f_{M-1}(x) = (1 + g(x_M)) \cos(x_1\pi/2) \sin(x_2\pi/2),$
$f_M(x) = (1 + g(x_M)) \sin(x_1\pi/2).$
subject to $0 \leq x_i \leq 1 \quad \forall \quad i = 1, 2, ..., n$
where: $g(x_M) = \sum_{x_i \in X_M}(x_i - 0.5)^2$

The Pareto-optimal solutions correspond to $x_i = 0.5$ for all $x_i \in x_M$ and all objective function values must satisfy: $\sum_{i=1}^{M}(f_i)^2 = 1$. PF_{true} is shown in Figure 4.35. It is recommended to use $k = |x_M| = 10$. The total number of variables is $n = M + k - 1$. This function can also be used to investigate a MOEA's ability to scale up its performance in a large number of objectives. Like in DTLZ1, for $M > 3$, the Pareto-optimal solution must lie inside the first quadrant of the unit sphere in a three-objective plot with f_M as one of the axes. To make the problem more difficult, each variable x_i (for $i=1$ to $(M-1)$)

can be replaced by the mean value of p variables $x_i = \frac{1}{p} \sum_{k=(i-1)p+1}^{ip} x_k$.

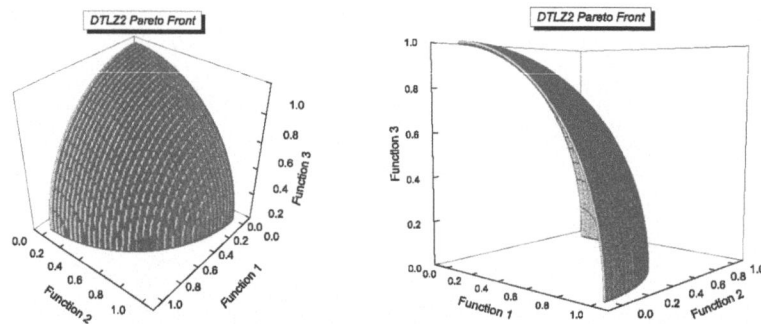

Fig. 4.35. DTLZ2 Pareto Front

Test Problem DTLZ3: This is the same as DTLZ2 except for a new g function: PF_{true} concave, scalable, multimodal. Tests a MOEA's ability to converge to PF_{true}.

Minimize:
$f_1(x) = (1 + g(x_M)) \cos(x_1\pi/2) \cos(x_2\pi/2) \ldots \cos(x_{M-2}\pi/2) \cos(x_{M-1}\pi/2),$
$f_2(x) = (1 + g(x_M)) \cos(x_1\pi/2) \cos(x_2\pi/2) \ldots \cos(x_{M-2}\pi/2) \sin(x_{M-1}\pi/2),$
$f_3(x) = (1 + g(x_M)) \cos(x_1\pi/2) \cos(x_2\pi/2) \ldots \sin(x_{M-2}\pi/2),$
$\vdots \qquad \vdots$
$f_{M-1}(\mathbf{x}) = (1 + g(x_M)) \cos(x_1\pi/2) \sin(x_2\pi/2),$
$f_M(\mathbf{x}) = (1 + g(x_M)) \sin(x_1\pi/2).$
subject to $0 \leq x_i \leq 1 \quad \forall \quad i = 1, 2, \ldots, n$
where: $g(x_M) = 100[|x_M| + \sum_{x_i \in x_M}(x_i - 0.5)^2 - \cos(20\pi(x_i - 0.5))]$

It is suggested that $k = |x_M| = 10$. There are a total of $n = M + k - 1$ decision variables in this problem. The above g function introduces ($3k$ - 1) local Pareto-optimal fronts, and one global Pareto-optimal front. All local Pareto-optimal fronts are parallel to the global Pareto-optimal front and a MOEA can get stuck at any of these local Pareto-optimal fronts, before converging to the global Pareto-optimal front at $g^* = 0$. The global Pareto-optimal front corresponds to $x_M = (0.5, \ldots, 0.5)^T$. The next local Pareto-optimal is at $g^* = 1$. In Figure 4.36, PF_{true} is shown.

Test Problem DTLZ4: This problem uses a modified meta-variable mapping over DTLZ ($y \rightarrow y^\alpha, \alpha > 0$); PF_{true} concave, separable, unimodal. Tests a MOEA's ability to maintain a good distribution of solutions.

4.4 Generic Scalable MOP Test Problems

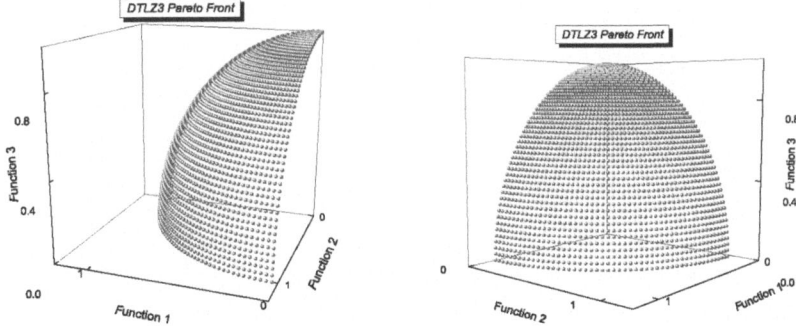

Fig. 4.36. DTLZ3 Pareto Front

Minimize:
$$f_1(x) = (1 + g(x_M))\cos(x_1^\pi\pi/2)\cos(x_2^\pi\pi/2)\ldots\cos(x_{M-2}^\pi\pi/2)\cos(x_{M-1}^\pi\pi/2),$$
$$f_2(x) = (1 + g(x_M))\cos(x_1^\pi\pi/2)\cos(x_2^\pi\pi/2)\ldots\cos(x_{M-2}^\pi\pi/2)\sin(x_{M-1}^\pi\pi/2),$$
$$f_3(x) = (1 + g(x_M))\cos(x_1^\pi\pi/2)\cos(x_2^\pi\pi/2)\ldots\sin(x_{M-2}^\pi\pi/2),$$
$$\vdots \qquad \vdots$$
$$f_{M-1}(\mathbf{x}) = (1 + g(x_M))\cos(x_1^\pi\pi/2)\sin(x_2^\pi\pi/2),$$
$$f_M(\mathbf{x}) = (1 + g(x_M))\sin(x_1^\pi\pi/2).$$
subject to $0 \leq x_i \leq 1 \quad \forall \quad i = 1, 2, \ldots, n$
where: $g(x_M) = \sum_{x_i \in X_M}(x_i - 0.5)^2$

The parameter $\alpha = 100$ is suggested. Here, too, all variables x_1 to x_{M-1} are varied in $(0:1)$. It is also suggested that $k = 10$. There are $n = M + k - 1$ decision variables in the problem. PF_{true} is shown in Figure 4.37. This modification allows a dense set of solutions to exist near the $f_M - f_1$ plane. It is interesting to note that although the search space has a variable density of solutions, the classical weighted-sum approaches or other directional methods may not have any added difficulty in solving these problems compared to DTLZ2.

Test Problem DTLZ5: Again, DTLZ2 is modified by changing all (**y**) with $(1 + 2gf_i)/2(1 + g)$; PF_{true} unimodal, $M < 4$, degenerate fronts occur and PF_{true} is shown in Figure 4.38.
Minimize:
$$f_1(x) = (1 + g(x_M))\cos(\theta_1\pi/2)\cos(\theta_2\pi/2)\ldots\cos(\theta_{M-2}\pi/2)\cos(\theta_{M-1}\pi/2),$$
$$f_2(x) = (1 + g(x_M))\cos(\theta_1\pi/2)\cos(\theta_2\pi/2)\ldots\cos(\theta_{M-2}\pi/2)\sin(\theta_{M-1}\pi/2),$$
$$f_3(x) = (1 + g(x_M))\cos(\theta_1\pi/2)\cos(\theta_2\pi/2)\ldots\sin(\theta_{M-2}\pi/2),$$
$$\vdots \qquad \vdots$$
$$f_{M-1}(\mathbf{x}) = (1 + g(x_M))\cos(\theta_1\pi/2)\sin(\theta_2\pi/2),$$

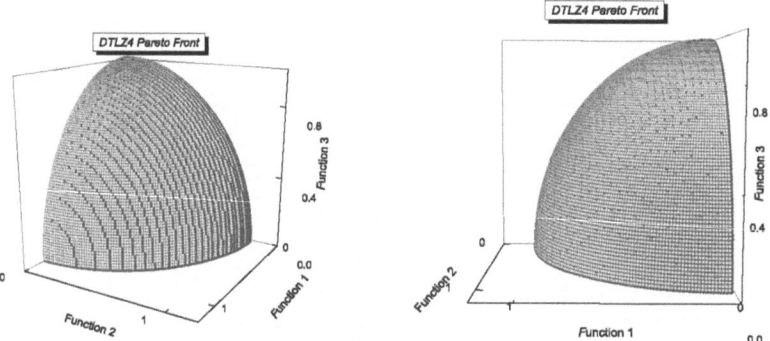

Fig. 4.37. DTLZ4 Pareto Front

$f_M(\mathbf{x}) = (1 + g(x_M))\sin(\theta_1 \pi/2)$.
subject to: $0 \leq x_i \leq 1 \quad \forall \quad i = 1, 2, ..., n$
where: $\theta_i = \frac{\pi}{4(1+g(x_M))}(1 + 2g(x_M)x_i)$, for $i = 2, 3, \ldots, (M-1)$
$g(x_M) = \sum_{x_i \in X_M}(x_i - 0.5)^2$

The g function with $k = |x_M| = 10$ is suggested. As before, there are $n = M + k - 1$ decision variables in this problem and the Pareto-optimal front corresponds to $x_i = 0.5$ for all $x_i \in x_M$ and all objective function values must satisfy: $\sum_{i=1}^{M}(f_i)^2 = 1$. This problem will test a MOEA's ability to converge to a curve and will also allow an easier way to visually demonstrate (just by plotting f_M with any other objective function) the performance of a MOEA. Since there is a natural bias for solutions close to this Pareto-optimal curve, this problem may be easy for an algorithm to solve. Because of its simplicity and ease of representing the Pareto-optimal front, it is recommend that a higher objective ($M \in [5, 10]$) version of this problem is used to study the computational time complexity of a MOEA.

Test Problem DTLZ6: Modifying DTLZ5, a harder problem evolves by changing g with $\sum_{i=1}^{k} z_i^{0.1}$; PF_{true} unimodal, bias, many-to-one-mapping.
  Minimize:
$f_1(x) = (1 + g(x_M))\cos(\theta_1\pi/2)\cos(\theta_2\pi/2)\ldots\cos(\theta_{M-2}\pi/2)\cos(\theta_{M-1}\pi/2)$,
$f_2(x) = (1 + g(x_M))\cos(\theta_1\pi/2)\cos(\theta_2\pi/2)\ldots\cos(\theta_{M-2}\pi/2)\sin(\theta_{M-1}\pi/2)$,
$f_3(x) = (1 + g(x_M))\cos(\theta_1\pi/2)\cos(\theta_2\pi/2)\ldots\sin(\theta_{M-2}\pi/2)$,
$\vdots \qquad \vdots$
$f_{M-1}(\mathbf{x}) = (1 + g(x_M))\cos(\theta_1\pi/2)\sin(\theta_2\pi/2)$,
$f_M(\mathbf{x}) = (1 + g(x_M))\sin(\theta_1\pi/2)$.
subject to: $0 \leq x_i \leq 1 \quad \forall \quad i = 1, 2, ..., n$
where: $\theta_i = \frac{\pi}{4(1+g(x_M))}(1 + 2g(x_M)x_i), \forall i = 2, 3, \ldots, (M-1)$

4.4 Generic Scalable MOP Test Problems

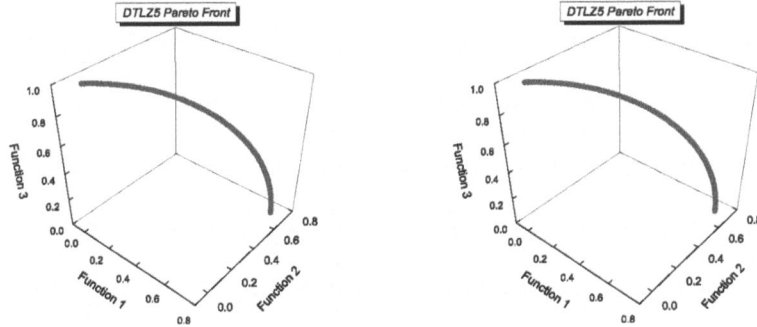

Fig. 4.38. DTLZ5 Pareto Front

$$g(x_M) = \sum_{x_i \in X_M} (x_i)^{0.1}$$

The Pareto-optimal front corresponds to $x_i = 0$ for all $x_i \in x_M$ and is shown in Figure 4.39. The size of the x_M vector is chosen as 10 and the total number of variables is identical as in DTLZ5. The above change in the problem makes it difficult for a MOEA to converge to the Pareto-optimal front as in DTLZ5. The lack of convergence to the true front in this problem causes MOEAs to find a dominated surface as the obtained front, whereas the true Pareto-optimal front is a curve. In real-world problems, this aspect may provide misleading information about the properties of the Pareto-optimal front.

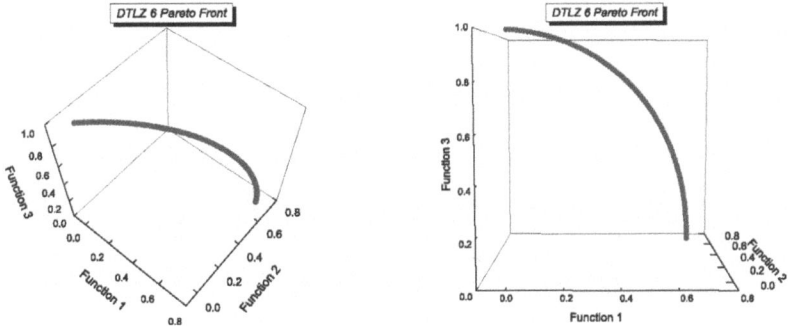

Fig. 4.39. DTLZ6 Pareto Front

Test Problem DTLZ7: PF_{true} disconnected.
Minimize:
$f_1(x) = x_1$,

$f_2(x) = x_2,$

$\vdots \qquad \vdots$

$f_{M-1}(x) = x_{M-1}$
$f_M(x) = (1 + g(x_M)) \cdot h(f_1, f_2, \ldots, f_{M-1}g(x))$
subject to: $0 \leq x_i \leq 1 \quad \forall \quad i = 1, 2, \ldots, n$
where: $g(x) = 1 + \frac{9}{|x_M|} \sum_{x_i \in x_M} x_i,$
$h(f_1, f_2, \ldots, f_{M-1}, g) = M - \sum_{i=1}^{M-1} \left(\frac{f_i}{1+g(x)} (1 + sin(3\pi f_i)) \right)$

This test problem has $2M - 1$ disconnected Pareto-optimal regions in the search space. The functional g requires $k = |x_M j|$ decision variables and the total number of variables is $n = M + k - 1$. It is suggested that $k = 20$. The Pareto-optimal solutions corresponds to $x_M = 0$ and the Figure 4.40 shows the PF_{true}. This problem will test an algorithm's ability to maintain individuals in different Pareto-optimal regions. The problem can be made harder by using a higher frequency sine function or using a multi-modal g function.

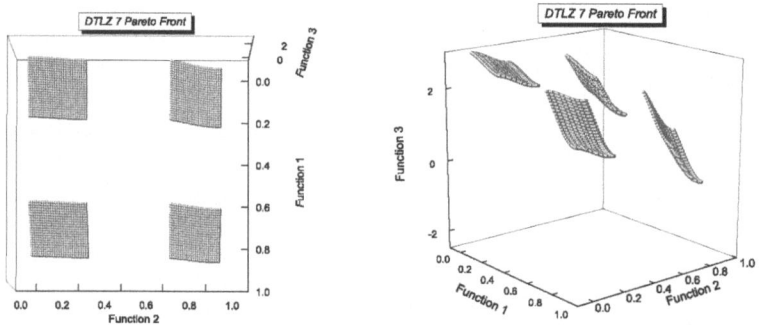

Fig. 4.40. DTLZ7 Pareto Front

Test Problem DTLZ8: In this problem, a constraint surface is defined.
Minimize:
$f_j(x) = \frac{1}{\lfloor n/M \rfloor} \sum_{\lfloor i=(j-1)\frac{n}{M} \rfloor}^{\lfloor j\frac{n}{M} \rfloor} (x_i), \forall j = 1, 2, \ldots, M,$
subject to: $0 \leq x_i \leq 1 \quad \forall \quad i = 1, 2, \ldots, n$
where: $g_j(x) = f_M(x) + 4f_j(x) - 1 \geq 0, \forall j = 1, 2, \ldots, (M-1)$
$g_M(x) = 2f_M(x) + min_{i,j=1, i \neq j}^{M-1} [f_i(x) + f_j(x)] - 1 \geq 0,$

The number of variables is considered to be larger than the number of objectives $n > M$. It is suggested $n = 10M$. In this problem, there are a total of M constraints. The Pareto-optimal front is shown in Figure 4.41 and is a combination of a straight line and a hyper-plane. The straight line is the intersection of the first $(M-1)$ constraints (with $f_1 = f_2 = \ldots = f_M - 1$ and the hyper-plane is represented by the constraint g_M). MOEAs may have difficulties in finding solutions in both regions of this problem and also in maintaining a good distribution of solutions on the hyper-plane.

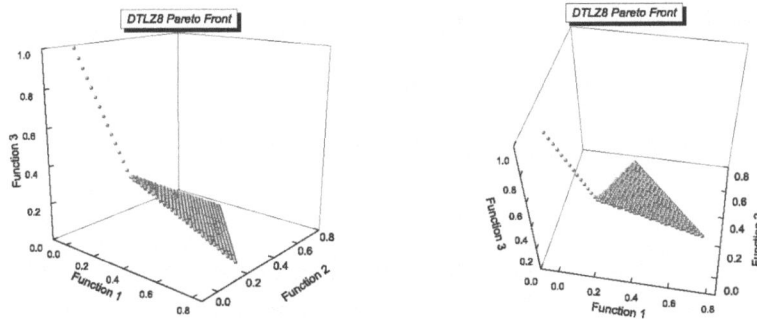

Fig. 4.41. DTLZ8 Pareto Front

Test Problem DTLZ9: This test problem is also created using the constraint surface approach.

Minimize:
$$f_j(x) = \frac{1}{\lfloor n/M \rfloor} \sum_{\lfloor i=(j-1)\frac{n}{M} \rfloor}^{\lfloor j\frac{n}{M} \rfloor} \left(x_i^{0.1}\right), \forall j = 1, 2, \ldots, M,$$
subject to: $0 \leq x_i \leq 1 \quad \forall \quad i = 1, 2, \ldots, n$
where: $g_j(x) = f_M^2(x) + f_j^2(x) - 1 \geq 0, \forall j = 1, 2, \ldots, (M-1)$

The number of variables is considered to be larger than the number of objectives. For this problem, it is suggested $n = 10M$. The Pareto-optimal front is a curve with $f_1 = f_2 = \ldots = f_M - 1$, similar to that in DTLZ5. However, the density of solutions gets thinner towards the Pareto optimal region. The Pareto-optimal curve lies on the intersection of all $(M-1)$ constraints. This feature of this problem may cause difficulties to a MOEA. However, the symmetry of the Pareto-optimal curve in terms of $(M-1)$ objectives allows an easier way to illustrate the obtained solutions. In Figure 4.42, we show PF_{true} with $M = 3$.

4.4.1 Okabe's Test Functions

Tatsuya Okabe et. al [1207] propose a new methodology to generate multi-objective test functions based on the mapping of probability density functions

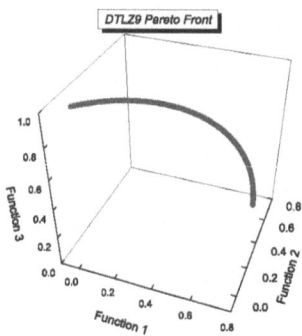 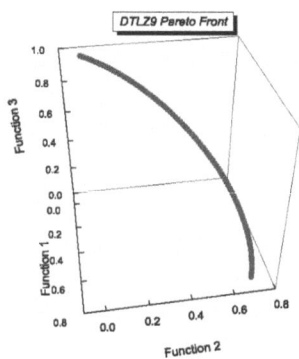

Fig. 4.42. DTLZ9 Pareto Front

from decision to objective space, and give two examples to illustrate its use. The basic idea to construct the problems is to depart from a starting space (called S^2) between decision space and objective space, and from there, one has to construct both the decision space and the objective space by applying appropriate functions to S^2. Namely, the authors propose to use the inverse of generation operation, i.e. deformation, rotation and shift. The procedure is described in detail in [1207].

With the framework presented in [1207], a variety of test functions can be generated. However, as indicated before, only two examples are described in [1207]. Both of them are described next.

Test Problem OKA1:

Minimize:
$f_1 = x'_1$,
$f_2 = \sqrt{2\pi} - \sqrt{|x'_1|} + 2|x'_2 - 3\cos(x'_1) - 3|^{\frac{1}{2}}$,
where:
$x'_1 = \cos(\pi/12)x_1 - \sin(\pi/12)x_2$,
$x'_2 = \sin(\pi/12)x_1 + \cos(\pi/12)x_2$,
subject to:
$x_1 \in [6\sin(\pi/12), 6\sin(\pi/12) + 2\pi\cos(\pi/12)]$,
$x_2 \in [-2\pi\sin(\pi/12), 6\cos(\pi/12)]$,

The P_{true} is at: $x'_2 = 3\cos(x'_1 + 3)$ and $x'_1 \in [0, 2\pi]$. P_{true} is shown in Figure 4.43. PF_{true} is at: $f_2 = \sqrt{(2\pi)} - \sqrt{f_1}$ and $f_1 \in [-\pi, \pi]$ and is shown in Figure 4.44. The Distribution indicator is:

$$D_{x \to f} = \frac{3}{2}|x'_2 - 3\cos(x'_1) - 3|^{\frac{2}{3}} \qquad (4.5)$$

4.4 Generic Scalable MOP Test Problems   209

The distribution indicator ($D_{x\to f}$) measures the amount of distortion the probability density in the decision space suffers under the mapping from decision space to objective space.

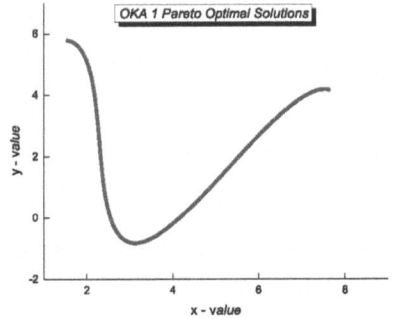

Fig. 4.43. Oka1 Pareto Optimal Set

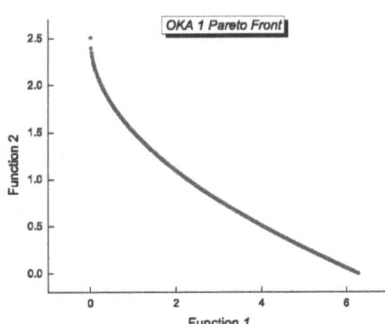

Fig. 4.44. Oka1 Pareto Front

Test Problem OKA2:
  Minimize:
$f_1 = x_1$,
$f_2 = 1 - \frac{1}{4\pi^2}(x_1 + \pi)^2 + |x_2 - 5\cos(x_1)|^{\frac{1}{3}} + |\frac{x}{3} - 5\sin(x_1)|^{\frac{1}{3}}$,
subject to:
$x_1 \in [-\pi, \pi]$,
$x_2, x_3 \in [-5, 5]$,

The P_{true} is at: $(x_1, x_2, x_3) = (x_1, 5\cos(x_1), 5\sin(x_1))$ and $x_1 \in [-\pi, \pi]$. P_{true} is shown in Figure 4.45. PF_{true} is at: $f_2 = 1 - \frac{1}{4\pi^2}(f_1 + \pi)^2$ and $f_1 \in [-\pi, \pi]$ and is shown in Figure 4.46. The Distribution indicator is: $D_{x\to f} = 9|x_2 - 5\cos(x_1)|^{\frac{2}{3}}|x_3 - 5\sin(x_1)|^{\frac{2}{3}}$.

These test functions are very difficult for any MOEA, because the closer the population gets to the Pareto front, the more sparse the probability density becomes.

4.4.2 Huband's Test Functions

Another scheme to generate scalable test functions is the one proposed by Huband et al. [721]. The authors propose a set of transformations that are sequentially applied to the decision variables, where each transformation adds a desired characteristic to the problem. All the problems generated with this methodology follow this format:

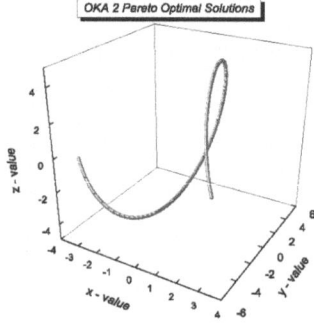

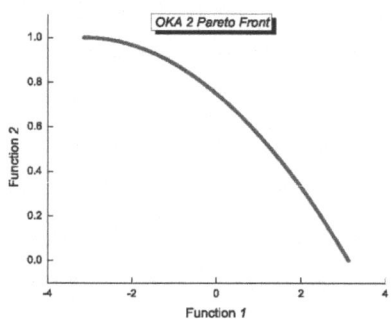

Fig. 4.45. Oka2 Pareto Optimal Set

Fig. 4.46. Oka2 Pareto Front

$$\text{Given } \mathbf{z} = [z_1, \ldots, z_k, z_{k+1}, \ldots, z_n]$$
$$\text{Minimize } f_{m=1:M}(\mathbf{x}) = Dx_M + S_m h_m(x_1, \ldots, x_{M-1})$$
$$\text{where } \mathbf{x} = [x_1, \ldots, x_M]$$
$$= [\max(t_M^p, A_1)(t_1^p - 0.5) + 0.5, \ldots,$$
$$\max(t_M^p, A_{M-1})(t_{M-1}^p - 0.5) + 0.5, t_M^p]$$
$$\mathbf{t}^p = [t_1^p, \ldots, t_M^p] \hookleftarrow \mathbf{t}^{p-1} \hookleftarrow \ldots \hookleftarrow \mathbf{t}^1 \hookleftarrow \mathbf{z}_{[0,1]}$$
$$\mathbf{z}_{[0,1]} = [z_{1,[0,1]}, \ldots, z_{n,[0,1]}]$$
$$= [z_1/z_{1,\max}, \ldots, z_n/z_{n,\max}]$$

where $\mathbf{z}$ is the vector of decision variables with $0 \leq z_i \leq z_{i,\max}$, D, $A_{1:M-1}$ and $S_{1:M}$ are constants to modify position and scale of the Pareto front. The flexibility of this methodology lies on the $h_{1:M}$ functions, and the transformations to obtain the transition vectors $\mathbf{t}^{1:p}$, which keep desired characteristics of the problem as separate design decisions.

First, the $h_{1:M}$ functions define the shape of the Pareto front, which can be linear, convex, concave, mixed (convex and concave), and disconnected. Such characteristics are obtained with the following set of functions:

$$\text{linear}_1(x_1, \ldots, x_{M-1}) = \prod_{i=1}^{M-1} x_i$$

$$\text{linear}_{m=2:M-1}(x_1, \ldots, x_{M-1}) = \left(\prod_{i=1}^{M-m} x_i\right)(1 - x_{M-m+1})$$

$$\text{linear}_M(x_1, \ldots, x_{M-1}) = 1 - x_1$$

$$\text{convex}_1(x_1,\ldots,x_{M-1}) = \prod_{i=1}^{M-1}(1-\cos(x_i\pi/2))$$

$$\text{convex}_{m=2:M-1}(x_1,\ldots,x_{M-1}) = \left(\prod_{i=1}^{M-m}(1-\cos(x_i\pi/2))\right) \cdot$$
$$(1-\sin(x_{M-m+1}\pi/2))$$

$$\text{convex}_M(x_1,\ldots,x_{M-1}) = 1-\sin(x_1\pi/2)$$

$$\text{concave}_1(x_1,\ldots,x_{M-1}) = \prod_{i=1}^{M-1}\sin(x_i\pi/2)$$

$$\text{concave}_{m=2:M-1}(x_1,\ldots,x_{M-1}) = \left(\prod_{i=1}^{M-m}\sin(x_i\pi/2)\right) \cdot$$
$$\cos(x_{M-m+1}\pi/2)$$

$$\text{concave}_M(x_1,\ldots,x_{M-1}) = \cos(x_1\pi/2)$$

$$\text{mixed}_M(x_1,\ldots,x_{M-1}) = \left(1-x_1-\frac{\cos(2A\pi x_1+\pi/2)}{2A\pi}\right)^\alpha$$

$$\text{disc}_M(x_1,\ldots,x_{M-1}) = 1-x_1^\alpha\cos^2(Ax_1^\beta\pi)$$

For a mixed Pareto front, when $\alpha > 1$ the overall shape is concave, when $\alpha < 1$ it is concave, and when $\alpha = 1$ it is linear; the number of segments concave-convex is defined by A. For a disconnected Pareto front, α controls the overall shape in the same way it does it for a mixed front; β controls the location of the disconnected segments, and the number of such disconnected segments is defined by A.

With this set of functions is easy to design the shape of the Pareto front, but what about the fitness landscape? The rest of the characteristics are added through a set of transformations. Huband et al. distinguish between three types of transformations, based on the characteristics they emphasize as important when designing multiobjective problems. Bias transformations produce a bias in the fitness landscape, and are used to produce polynomial bias, flat regions, or other type of bias depending of the values of another variables; shift transformations move the location of optimal values, and are used to apply a linear shift, or to produce deceptive and multimodal problems; and reduction transformations, which combine the values of several variables into a single one. They are used to produce unseparability of the problem (dependency between variables). The set of transformations is the following:

$$\text{b\_poly}(y,\alpha) = y^\alpha$$

$$\text{b\_flat}(y, A, B, C) = A + \min(0, \lfloor y - B \rfloor) \frac{A(B-y)}{B} - \min(0, \lfloor C - y \rfloor) \frac{(1-A)(y-C)}{1-C}$$

$$\text{b\_param}(y, u(\mathbf{y}'), A, B, C) = y^{B+(C-B)\left(A-(1-2u(\mathbf{y}'))\left|\lfloor 0.5 - u(\mathbf{y}')\rfloor + A\right|\right)}$$

$$\text{s\_linear}(y, A) = \frac{|y - A|}{||A - y| + A|}$$

$$\text{s\_decept}(y, A, B, C) = 1 + (|y - A| - B)\left(\frac{\lfloor y - A + B\rfloor\left(1 - C + \frac{A-B}{B}\right)}{A - B} + \frac{\lfloor A + B - y\rfloor\left(1 - C + \frac{1-A-B}{B}\right)}{1 - A - B} + \frac{1}{B}\right)$$

$$\text{s\_multi}(y, A, B, C) = \left(1 + \cos\left((4A + 2)\pi\left(0.5 - \frac{|y - C|}{2(\lfloor C - y\rfloor + C)}\right)\right) + 4B\left(\frac{|y - C|}{2(\lfloor C - y\rfloor + C)}\right)^2\right) \bigg/ (b + 2)$$

$$\text{r\_sum}(\mathbf{y}, \mathbf{w}) = \frac{\sum_{i=1}^{|\mathbf{y}|} w_1 y_i}{\sum_{i=1}^{|\mathbf{y}|} w_i}$$

$$\text{r\_nonsep}(\mathbf{y}, A) = \frac{\sum_{j=1}^{|\mathbf{y}|}\left(y_j + \sum_{k=0}^{A-2} |y_j - y_{1+(j+k)\bmod|\mathbf{y}|}|\right)}{\frac{|\mathbf{y}|}{A}\lceil\frac{A}{2}\rceil\left(1 + 2A - 2\lceil\frac{A}{2}\rceil\right)}$$

With this set of transformations and shape functions, Huband et al. [721] propose a set of scalable problems for algorithm performance evaluation and analysis. It is easy to obtain the characteristics of each, based on the transformation and shape functions used, but they are mentioned explicitly after the problems.

WFG1:
Minimize

$$f_{m=1:M-1}(\mathbf{x}) = x_M + S_m \text{convex}_m(x_1, \ldots, x_{M-1})$$
$$f_M(\mathbf{x}) = x_M + S_M \text{mixed}_M(x_1, \ldots, x_{M-1})$$

where

$$y_{i=1:M-1} = \text{r\_sum}([y'_{(i-1)k/(M-1)+1}, \ldots, y'_{ik/(M-1)}],$$
$$[2((i-1)k/(M-1)+1), \ldots, 2ik/(M-1)])$$
$$y_M = \text{r\_sum}([y'_{k+1}, \ldots, y'_n], [2(k+1), \ldots, 2n])$$
$$y'_{i=1:n} = \text{b\_poly}(y''_i, 0.02)$$
$$y''_{i=1:k} = y'''_i$$
$$y''_{i=k+1:n} = \text{b\_flat}(y'''_i, 0.8, 0.75, 0.85)$$
$$y'''_{i=1:k} = z_{i,[0,1]}$$
$$y'''_{i=k+1:n} = \text{s\_linear}(z_{i,[0,1]}, 0.35)$$

This problem is separable and unimodal, but it has a flat region and is strongly biased toward small values of the variables, which makes it very difficult for some MOEAs. Two different views of PF_{true} for this problem are shown in Figures 4.47 and 4.48. From these figures, we can see that this problem has a mixed PF_{true}.

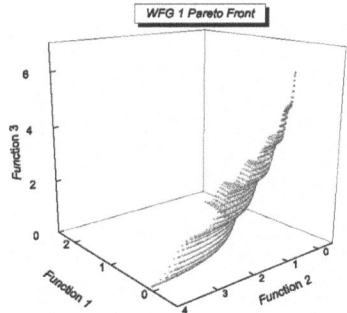

Fig. 4.47. WFG1 Pareto Front

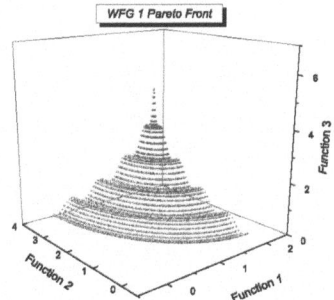

Fig. 4.48. WFG1 Pareto Front

WFG2:
Minimize

$$f_{m=1:M-1}(\mathbf{x}) = x_M + S_m \text{convex}_m(x_1, \ldots, x_{M-1})$$
$$f_M(\mathbf{x}) = x_M + S_M \text{disc}_M(x_1, \ldots, x_{M-1})$$

where

$$y_{i=1:M-1} = \text{r\_sum}([y'_{(i-1)k/(M-1)+1}, \ldots, y'_{ik/(M-1)}], [1, \ldots, 1])$$
$$y_M = \text{r\_sum}([y'_{k+1}, \ldots, y'_{k+l/2}], [1, \ldots, 1])$$
$$y'_{i=1:k} = y''_i$$
$$y'_{i=k+1:k+l/2} = \text{r\_nonsep}([y''_{k+2(i-k)-1}, y''_{k+2(i-k)}], 2)$$
$$y''_{i=1:k} = z_{i,[0,1]}$$
$$y''_{i=k+1:n} = \text{s\_linear}(z_{i,[0,1]}, 0.35)$$

This problem is unseparable and multimodal. Two different views of PF_{true} for this problem are shown in Figures 4.49 and 4.50. Note that PF_{true} is disconnected.

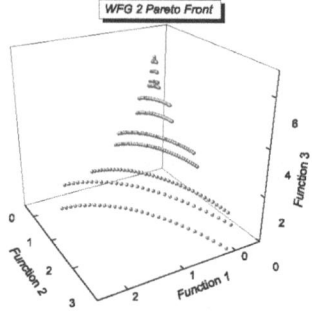

Fig. 4.49. WFG2 Pareto Front   **Fig. 4.50.** WFG2 Pareto Front

WFG3:
Minimize

$$f_{m=1:M}(\mathbf{x}) = x_M + S_m \text{linear}_m(x_1, \ldots, x_{M-1})$$

where

$$y_{i=1:M-1} = \text{r\_sum}([y'_{(i-1)k/(M-1)+1}, \ldots, y'_{ik/(M-1)}], [1, \ldots, 1])$$
$$y_M = \text{r\_sum}([y'_{k+1}, \ldots, y'_{k+l/2}], [1, \ldots, 1])$$
$$y'_{i=1:k} = y''_i$$
$$y'_{i=k+1:k+l/2} = \text{r\_nonsep}([y''_{k+2(i-k)-1}, y''_{k+2(i-k)}], 2)$$
$$y''_{i=1:k} = z_{i,[0,1]}$$
$$y''_{i=k+1:n} = \text{s\_linear}(z_{i,[0,1]}, 0.35)$$

Again, this problem is unseparable but unimodal. It has a degenerated Pareto front (the dimensionality of the Pareto front is $M-2$). In Figures 4.51

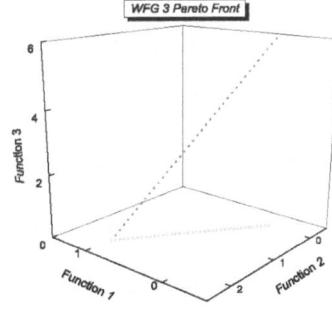 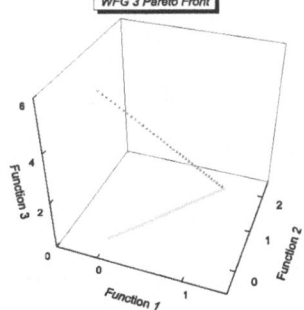

Fig. 4.51. WFG3 Pareto Front   **Fig. 4.52.** WFG3 Pareto Front

and 4.52, the Pareto front is unidimensional, even when it was generated for a problem with three objectives.

WFG4:

Minimize

$$f_{m=1:M}(\mathbf{x}) = x_M + S_m \text{concave}_m(x_1, \ldots, x_{M-1})$$

where

$$y_{i=1:M-1} = \text{r\_sum}([y'_{(i-1)k/(M-1)+1}, \ldots, y'_{ik/(M-1)}], [1, \ldots, 1])$$
$$y_M = \text{r\_sum}([y'_{k+1}, \ldots, y'_n], [1, \ldots, 1])$$
$$y'_{i=1:n} = \text{s\_multi}(z_{i,[0,1]}, 30, 10, 0.35)$$

In this case, the problem is separable, but highly multimodal. This, and the rest of the problems from this benchmark have concave Pareto fronts, as the one shown in Figures 4.53 and 4.54.

WFG5:

Minimize

$$f_{m=1:M}(\mathbf{x}) = x_M + S_m \text{concave}_m(x_1, \ldots, x_{M-1})$$

where

$$y_{i=1:M-1} = \text{r\_sum}([y'_{(i-1)k/(M-1)+1}, \ldots, y'_{ik/(M-1)}], [1, \ldots, 1])$$
$$y_M = \text{r\_sum}([y'_{k+1}, \ldots, y'_n], [1, \ldots, 1])$$
$$y'_{i=1:n} = \text{s\_decept}(z_{i,[0,1]}, 0.35, 0.001, 0.05)$$

A deceptive problem, separable. PF_{true} for this problem is shown in Figures 4.55 and 4.56.

216   4 MOEA Test Suites

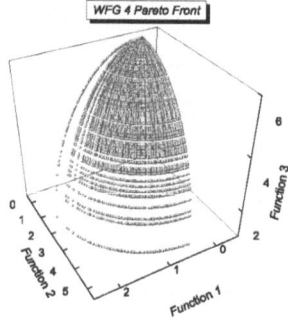

Fig. 4.53. WFG4 Pareto Front

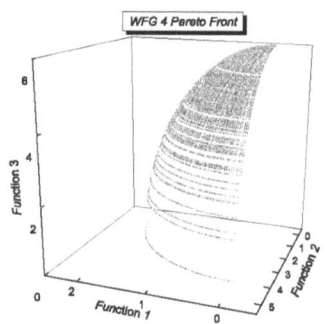

Fig. 4.54. WFG4 Pareto Front

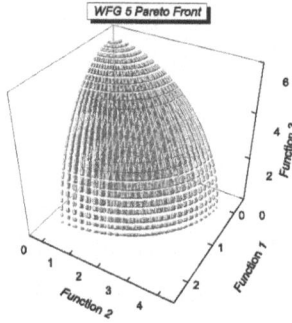

Fig. 4.55. WFG5 Pareto Front

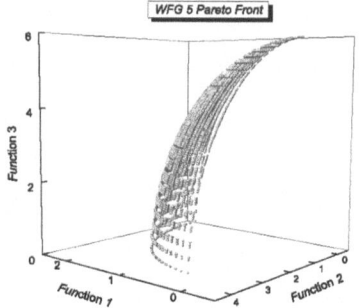

Fig. 4.56. WFG5 Pareto Front

WFG6:
Minimize

$$f_{m=1:M}(\mathbf{x}) = x_M + S_m \text{concave}_m(x_1, \ldots, x_{M-1})$$

where

$$y_{i=1:M-1} = \text{r\_nonsep}([y'_{(i-1)k/(M-1)+1}, \ldots, y'_{ik/(M-1)}], k/(M-1))$$
$$y_M = \text{r\_nonsep}([y'_{k+1}, \ldots, y'_n], l)$$
$$y'_{i=1:k} = z_{i,[0,1]}$$
$$y'_{i=k+1:n} = \text{s\_linear}(z_{i,[0,1]}, 0.35)$$

This problem is unseparable. Two views of PF_{true} of this problem are shown in Figures 4.57 and 4.58.

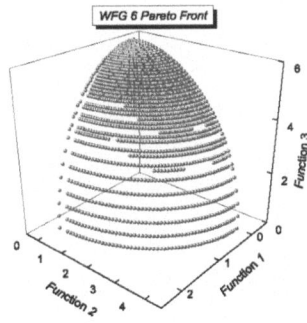

Fig. 4.57. WFG6 Pareto Front

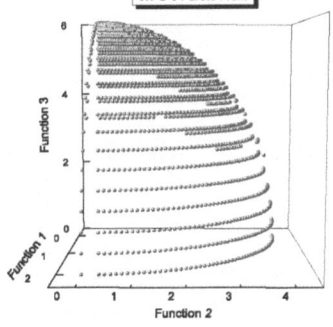

Fig. 4.58. WFG6 Pareto Front

WFG7:
Minimize
$$f_{m=1:M}(\mathbf{x}) = x_M + S_m \text{concave}_m(x_1, \ldots, x_{M-1})$$

where

$$y_{i=1:M-1} = \text{r\_sum}([y'_{(i-1)k/(M-1)+1}, \ldots, y'_{ik/(M-1)}], [1, \ldots, 1])$$
$$y_M = \text{r\_sum}([y'_{k+1}, \ldots, y'_n], [1, \ldots, 1])$$
$$y'_{i=1:k} = y''_i$$
$$y'_{i=k+1:n} = \text{s\_linear}(y''_i, 0.35)$$
$$y''_{i=1:k} = \text{b\_param}(z_{i,[0,1]}, \text{r\_sum}([z_{i+1,[0,1]}, \ldots, z_{n,[0,1]}],$$
$$[1, \ldots, 1]), 0.98/49.98, 0.02, 50)$$
$$y''_{i=k+1:n} = z_{i,[0,1]}$$

Having a parameter dependent bias, this problem is also separable and unimodal. Two views of PF_{true} for this problem are shown in Figures 4.59 and 4.60.

WFG8:
Minimize
$$f_{m=1:M}(\mathbf{x}) = x_M + S_m \text{concave}_m(x_1, \ldots, x_{M-1})$$

where

218   4 MOEA Test Suites

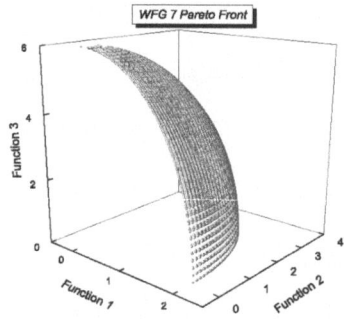
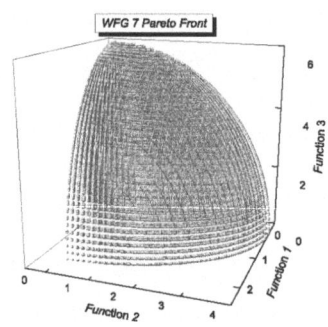

Fig. 4.59. WFG7 Pareto Front **Fig. 4.60.** WFG7 Pareto Front

$$y_{i=1:M-1} = \text{r\_sum}([y'_{(i-1)k/(M-1)+1}, \ldots, y'_{ik/(M-1)}], [1, \ldots, 1])$$
$$y_M = \text{r\_sum}([y'_{k+1}, \ldots, y'_n], [1, \ldots, 1])$$
$$y'_{i=1:k} = y''_i$$
$$y'_{i=k+1:n} = \text{s\_linear}(y''_i, 0.35)$$
$$y''_{i=1:k} = z_{i,[0,1]}$$
$$y''_{i=k+1:n} = \text{b\_param}(z_{i,[0,1]}, \text{r\_sum}([z_{1,[0,1]}, \ldots, z_{i-1,[0,1]}],$$
$$[1, \ldots, 1]), 0.98/49.98, 0.02, 50)$$

This problem also has a parameter dependent bias, but is also unseparable. Two views of PF_{true} for this problem are shown in Figures 4.61 and 4.62.

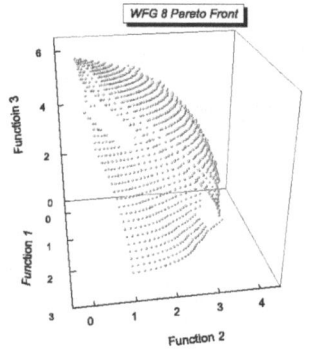
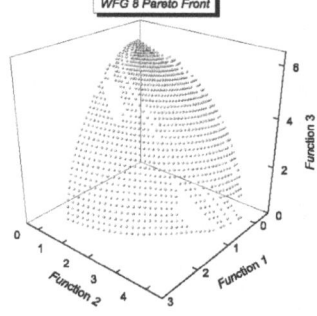

Fig. 4.61. WFG8 Pareto Front **Fig. 4.62.** WFG8 Pareto Front

WFG9:
Minimize
$$f_{m=1:M}(\mathbf{x}) = x_M + S_m \text{concave}_m(x_1, \ldots, x_{M-1})$$
where
$$y_{i=1:M-1} = \text{r\_nonsep}([y'_{(i-1)k/(M-1)+1}, \ldots, y'_{ik/(M-1)}], k/(M-1))$$
$$y_M = \text{r\_nonsep}([y'_{k+1}, \ldots, y'_n], l)$$
$$y'_{i=1:k} = \text{s\_decept}(y''_i, 0.35, 0.001, 0.05)$$
$$y'_{i=k+1:n} = \text{s\_multi}(y''_i, 30, 95, 0.35)$$
$$y''_{i=1:n-1} = \text{b\_param}(z_{i,[0,1]}, \text{r\_sum}([z_{i+1,[0,1]}, \ldots, z_{n,[0,1]}],$$
$$[1,\ldots,1]), 0.98/49.98, 0.02, 50)$$
$$y''_n = z_{n,[0,1]}$$

The last problem of the suite is unseparable, multimodal, deceptive, and has a parameter dependent bias. All these features make it a very difficult problem. Two views of PF_{true} for this problem are shown in Figures 4.63 and 4.64.

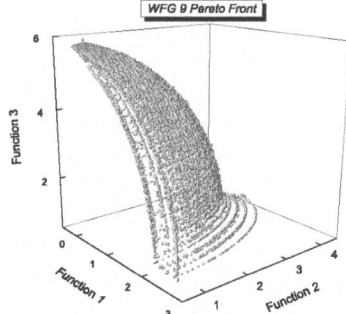

Fig. 4.63. WFG9 Pareto Front

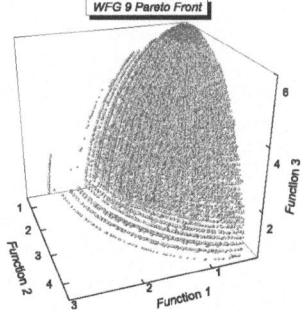

Fig. 4.64. WFG9 Pareto Front

The authors suggest to use these problems with 24 variables ($k = 4$ and $n = 24$). The following constants hold for all the problems:

$$z_{i=1:n,\max} = 2i$$
$$S_{m=1:M} = 2m$$
$$A_1 = 1$$
$$A_{2:M-1} = \begin{cases} 0, \text{ for WFG3} \\ 1, \text{ otherwise} \end{cases}$$
$$D = 1$$

The whole approach proposed by Huband et al. [721] is very versatile, because it is very easy to design new test problems with desired properties. The resulting problems, are very difficult to solve for most current MOEAs. Thus, they are a good choice for testing the performance of new algorithms.

4.5 Combinatorial MOEA Test Functions

Although most MOP test functions found in the MOEA literature are numeric, some combinatorial problems are used that provide differing algorithmic challenges. A combinatorial optimization problem is mathematically defined as follows [541]:

Combinatorial Optimization Problem: A combinatorial optimization problem π is either a minimization or maximization problem consisting of three parts:
1. A domain D_π of instantiations;
2. For each instance $I \in D_\pi$ a finite set $S_\pi(I)$ of candidate solutions for I; and
3. A function m_π that assigns a positive rational number $m_\pi(I, \sigma)$ to each candidate solution $\sigma \in S_\pi(I)$ for each instance $I \in D_\pi$. $m_\pi(I, \sigma)$ is called the solution value for σ.

A MOEA is able to search these finite (discrete) solution spaces but may require specialized EVOPs ensuring only feasible solutions (i.e., $S_\pi(I)$) are generated for evaluation (i.e., a repair function). However, the phenotype domain of combinatorial MOPs is slightly different than that of its numeric counterparts. These MOPs' mapping into objective space is discrete and offers only isolated points (vectors) in objective space. As only a finite number of solutions exist, only a finite number of corresponding vectors may result. Although these vectors may appear to form a continuous front when plotted, the genotype domain's discrete nature implies no solutions exist that map to vectors between those composing PF_{true}.

Various combinatorial MOPs are reflected in the MOEA literature. Horn & Nafpliotis [708, 709] and Deb [357] present combinatorial (unitation[3]) MOPs. Louis & Rawlins [1018] convert a deceptive GA problem into a MOP. NP-Complete problems are combinatorial optimization problems and many can be formulated as NP-Complete MOP test functions. For example:

- Some researchers focus on the use of fuzzy logic and MOEAs in solving Multiobjective 0-1 Programming problems (e.g., [825, 1415, 1480, 1413]).
- Several efforts investigate Multiobjective Solid Transportation Problems (e.g., [192, 792, 795, 736, 989, 988]).

---

[3] A *unitation function* is a function whose value depends only upon the number of ones and zeroes in the string on which it acts.

4.5 Combinatorial MOEA Test Functions

- A number of researchers have focused their efforts on solving Multiobjective Flowshop Scheduling Problems (e.g., [751, 1551, 173, 1554, 174, 57, 62]).
- Analogously, Multiobjective Job Shop Scheduling Problems are also relatively popular in the literature (e.g., [991, 76, 77, 107, 1557, 459, 458, 540]). Some researchers have reported that the use of linear aggregating functions with adaptive weights are highly competitive (and computationally efficient) when dealing with certain types of Multi-Objective Combinatorial Optimization (MOCO) problems such as the traveling salesman problem and job shop scheduling problems [777, 206, 1727, 1238, 1237]. The importance of using mating restrictions and local search in MOCO problems has also been stressed by some researchers [776, 1554].
- Finally, in recent years, Multiobjective Knapsack Problems have become popular, particularly among operations researchers (e.g., [1415, 850, 1782, 779, 1781, 292, 918, 593]).

For the multi-objective 0/1 knapsack problem with n knapsacks and m items, the objective is to maximize

$$f(\mathbf{x}) = (f_1(\mathbf{x}), ... f_n(\mathbf{x})) \qquad (4.6)$$

where

$$f_i(\mathbf{x}) = \sum_{j=1}^{m} p_{i,j} x_j \qquad (4.7)$$

and where $p_{i,j}$ is the profit of the jth item in knapsack i and x_j is 1, if selected. The constraint is

$$\sum_{j=1}^{m} w_{i,j} x_j \leq c_i \text{for all } i \qquad (4.8)$$

where $w_{i,j}$ is the weight of item j in knapsack i and c_j is the capacity of knapsack j.

Another NP-complete problem for MOEA benchmarking is the multiobjective minimum spanning tree problem [887], defined as mo-MST or mc-MST (multi-criteria). In this problem multiple weights are assigned to each edge of the graph. Each edge has K associated non-negative real numbers representing K attributes; a K tuple. The generic multiobjective problem is to find a single spanning tree solution that is minimal over all K edge tuple components. Since a single solution is generally not possible, the requirement is to find a set of spanning trees called the Pareto optimal set where each tree in the set is nondominated by another via the weighted edge sequence. Such problems can be parameterized to provide benchmark problems for MOEA comparative studies. Generators can be employed for generating various non-Euclidean mc-MST problems based upon particular selection of edge weights: random uncorrelated, correlated, anti-correlated, and m-degree vertex correlated. Also concave Pareto fronts or other geometries can be generated from specific sparse or complete graphs. If there is a constraint on the maximum vertex degree in each spanning tree, then the problem is defined as the multiobjective degree-constrained minimum spanning tree problem.

In essence, these problems are constrained minimization problems with the additional constraint on **x** such that it is only able to take on discrete values (e.g., integers). The use of these combinatorial MOPs in any proposed MOEA test suite should also be considered. On the one hand, EAs often employ specialized representations and operators when solving these NP problems which usually prevents a general comparison between various MOEA implementations. On the other hand, NPC problems' inherent difficulty should present desired algorithmic challenges and complement other test suite MOPs. Table 4.6 outlines possible NP-Complete MOPs for inclusion. Databases such as *TSPLIB* [1345], *MP-Testdata* [1786], and the *OR Library* [109], exist for these NP-Complete problems. On another note, the landscapes for various NP-Complete problems vary over a wide range with the knapsack problem usually reflecting a somewhat smooth landscape and the travel salesperson problem exhibiting a many-faceted landscape. The latter then being more difficult to search for an "optimal" Pareto front. Other NP-Complete problem databases are also available.

Table 4.6: Possible Multiobjective NP-Complete Functions

NP-**Complete Problem**	**Example**
Traveling Salesperson	Min energy, time, and/or distance; Max expansion
Coloring	Min number of colors, number of each color
Set/Vertex Covering	Min total cost, over-covering
Maximum Independent Set (Clique)	Max set size; Min geometry
Vehicle Routing	Min time, energy, and/or geometry
Scheduling	Min time, deadlines, wait time, resource use
Layout	Min space, overlap, costs
NPC-Problem Combinations	Vehicle scheduling and routing
0/1 Knapsacks - Bin Packing	Max profit; Min weight
Minimum Spanning Trees	tuple weighted edges; minimum weighting

4.6 Real-World MOEA Test Functions

Finally, real-world applications should be considered for inclusion in any comprehensive MOEA suite. Many of the applications presented in Chapter 7 could be selected. This type of MOP may be numeric, non-numeric, or both, and usually has more constraints (in terms of resources) than the numeric test problems considered in previous sections. Note that many real-world applications employ extensive computational fitness function software (e.g., computational fluid dynamics or computational electromagnetic or bioinformatics) requiring data interchange and data structure mapping (c.f., [1045, 805, 152, 1245, 1580, 1200, 1307])

Possible application MOP test suite examples include (see Chapter 7):

- **Aeronautical Engineering**: wing and airfoil design [1198, 1046, 42, 701, 118, 1171, 1301, 52, 448].
- **Chemical Engineering**: modelling of chemical processes, polymer extrusion [683, 544, 131, 823, 45, 70, 936, 14, 132].
- **Electrical Engineering**: planning of electrical power distribution system, circuit design [1317, 1166, 1707, 275, 8, 1374, 65, 9].
- **Hydraulic and Environmental Engineering**: water quality control, placement of wells and pumps [232, 1325, 708, 1339, 1338, 1349, 967, 854, 833, 1074].
- **Mechanical and Structural Engineering**: plane trusses, gear train, spring, welded beam [366, 269, 636, 242, 1409, 1716, 1254, 1314, 853, 1001, 1002, 1034, 1035, 619].
- **Computer Science**: distributed database management, coordination of agents, machine learning [886, 206, 678, 799].
- **Finance**: investment portfolio, ranking stocks [1012, 226, 1638, 1206, 1444, 1446, 1445, 1443, 1212].
- **Scheduling**: flowshop scheduling, production scheduling, job shop scheduling, time-tabling [1551, 1425, 1593, 1231, 258, 761, 62, 99, 1478, 245].

A possible test suite example application MOP involves logistics research in resource allocation. The effort is directed at developing a mission-resource value assessment for rationally assigning relative value to resources and identifying alternative resource mixes to logistics and operational planners [1670]. In general, it is desirable to develop a distributed computing architecture that links current and planned logistics information systems to the deliberate and priority action planning processes, databases, and policies–an end-to-end system linking operations and logistics. Note that a mission ready resource (MRR) is a combination of an asset type and its resources, for example, aircraft, pilot, fuel, support equipment and personnel, etc., that is designed to have a certain suitability for a single task. A combination of MRR types is defined to be an MRR set or resource mix. *The objective is that given a choice among time-phased asset sets, simultaneously minimize resource consumption (cost) and maximize asset set suitability over time.*

The symbolic MOP Formulation is given m tasks and n MRR types, the solution set is an $m \times n$ matrix. A matrix element is a decision variable, $x_{i,j}$, that represents the number of MRRs of type j allocated to tasks of type i. Assuming that each task is satisfied by exactly one MRR, and that no interactions exist between differing MRR types, then the suitability, S, for all MRRs is defined by:

$$S = \sum_{j=1}^{n} \sum_{i=1}^{m} a_{i,j}\, x_{i,j} \qquad (4.9)$$

where $a_{i,j}$ is the suitability of MRR j for Task i and $x_{i,j}$ is the number of MRRs j allocated to task type i.

The requirement that all tasks $i = 1, \ldots, n$ must be satisfied at a particular resource level (RL) k is:

$$RLtask_{k,i} = \sum_{j=1}^{n} x_{i,j} \qquad (4.10)$$

Since the desired capability for a task is set by the decision maker and defined to be static, the left-hand side of equation (4.10) is an equality constraint.

$$RLmrr_{j,k} \geq \sum_{i=1}^{m} x_{i,j,k} \qquad (4.11)$$

In this application, the decision variables are allowed to take on any nonnegative integer value as long as they do not exceed the specified resource level. Therefore, the left-hand side of equation (4.11) is an inequality constraint.

The maximum number of efforts per day for a particular asset, A, is given by its *turn rate*, t. For a quantity d of asset A, the total turn rate is

$$TTR_A = (d_A)(turn\ rate_A) \qquad (4.12)$$

Given that A has P configurations corresponding to P MRR types, the upper bound for any combination of the P MRR types is

$$TTR_A \geq \sum_{r=1}^{P} \sum_{i=1}^{m} x_{i,P_r} \qquad (4.13)$$

It is difficult to determine what the actual logistical footprint is for a given asset set. At the very least, it is clear that for each additional asset deployed, there is a corresponding increase in cost for additional resources, e.g. fuel, supplies, etc. Assuming that consumption is linear and without interaction, the weight consumption, W, and volume consumption, V, for all MRRs are

$$W = \sum_{j=1}^{n} \sum_{i=1}^{m} \beta_j\ x_{i,j} \qquad (4.14)$$

and

$$V = \sum_{j=1}^{n} \sum_{i=1}^{m} \lambda_j\ x_{i,j} \qquad (4.15)$$

β_j and λ_j are the weight and volume consumed by a single MRR j.

The form of the *suitability maximizing / lift minimizing MOP* with A asset types, m tasks, n MRR types, at a resource level k, and decision variables $(x_{1,1}, x_{i,j}, \ldots, x_{m,n})$ is to maximize:

minimize:
$$S = \sum_{j=1}^{n} \sum_{i=1}^{m} a_{i,j}\, x_{i,j} \qquad (4.16)$$

$$W = \sum_{j=1}^{n} \sum_{i=1}^{m} \beta_j\, x_{i,j} \qquad (4.17)$$

and minimize:
$$V = \sum_{j=1}^{n} \sum_{i=1}^{m} \lambda_j\, x_{i,j} \qquad (4.18)$$

subject to:
$$\sum_{j=1}^{n} x_{i,j} = RLtask_{k,i} \text{ for } i = 1\ldots m \qquad (4.19)$$

$$\sum_{i=1}^{m} x_{i,j} \leq RLmrr_{k,j} \qquad (4.20)$$

for $A = 1\ldots a$ and $P_a = $ number MRR types for a

The number of constraints resulting from equation (4.19) is equal to the number of tasks. These constraints ensure that the total number of efforts for Task i is exactly the desired capability at that resource level. The maximum value for any decision variable is found by using equation (4.19) and allocating all task capability to one MRR type. The number of constraints resulting from equation (4.20) is equal to the number of MRR types. These constraints ensure that no MRR type can be allocated a number of efforts that exceeds the given resource level. These constraints are also used when there are restrictions on the available number of any MRR type, e.g. attrition or changes in asset turn rate. It is important to note that each constraint refers to a single MRR type.

The specific number of tasks in Table 4.7 and MRR types in Table 4.11, along with their task suitabilities, are defined. The suitabilities reflect notional but reasonable values that clearly differentiate the MRR types. The same can be said for the consumption values in Table 4.11. To keep the number of task capability decisions by the decision maker at a reasonable level, five resource levels in Table 4.8 are specified, equating to 15 separate task preference decisions. These preferences are reflected in Table 4.9. When the ratios are applied to their respective resource level values, the result is the capability matrix in Table 4.9. The values are rounded to a whole number so that the sum across tasks is equal to the resource level. The values of the resource levels were chosen to create solution spaces of increasing size. Given three tasks and five MRR types and a resource level of 300 efforts per day, the worst case number of possible resource mixes is approximately 9.72 x 10^{19}. For the target MOP, it is assumed that there is no restriction on the available number of any MRR type; no attrition; and that each asset has one associated MRR type, i.e. one effort per day. These simplifying assumptions are made to provide an opportunity to explore the basic problem solution and complexity.

Table 4.7: Tasks

Index	Nomenclature
1	Air-to-Air (AA)
2	Air-to-Ground (AG)
3	Precision Locating (PL)

Table 4.8: Resource Levels

Index	RL (efforts per day)
1	16
2	32
3	75
4	150
5	300

Table 4.9: Desired Task Capability Ratios

	Percent to Task		
Index	AA	AG	PL
1	60	30	10
2	30	60	10
3	25	60	15
4	20	50	30
5	20	30	50

Table 4.10: Desired Capability Matrix

	TASK (efforts per day)			
Index	AA	AG	PB	Decision Space Cardinality
1	10	5	1	630,630
2	10	20	2	159,549,390
3	19	45	11	$\approx 2.56 x 10^{12}$
4	30	75	45	$\approx 1.48 x 10^{16}$
5	60	90	150	$\approx 4.37 x 10^{19}$

4.6 Real-World MOEA Test Functions

Table 4.11: Task Suitability / Lift Consumption Matrix

Index	MRR Type	Task Suitability			Lift Consumption	
		AA	AG	PL	Weight (Short Tons)	Volume (Cubic feet)
1	F_A	0.800	0.400	0.001	20.2	1650.0
2	F_B	0.300	0.800	0	001 28.5	2475.0
3	F_C	0.600	0.600	0.100	35.7	2887.5
4	B_1	0.001	0.001	0.800	19.9	1705.0
5	B_2	0.001	0.001	0.400	22.5	2200.0

The complete symbolic MOP formulation is as follows:
Decision variables- Number of MRR j assigned to Task $i = (x_{1,1}, \ldots, x_{i,j})$
Maximize:

$$S = 0.8x_{1,1} + 0.3x_{1,2} + 0.6x_{1,3} + 0.001x_{1,4} + 0.001x_{1,5} \quad (4.21)$$
$$+ 0.4x_{2,1} + 0.8x_{2,2} + 0.6x_{2,3} + 0.001x_{2,4} + 0.001x_{2,5}$$
$$+ 0.001x_{3,1} + 0.001x_{3,2} + 0.1x_{3,3} + 0.8x_{3,4} + 0.4x_{3,5}$$

Minimize:

$$W = 20.2(x_{1,1} + x_{2,1} + x_{3,1}) + 28.5(x_{1,2} + x_{2,2} + x_{3,2}) \quad (4.22)$$
$$+ 35.7(x_{1,3} + x_{2,3} + x_{3,3}) + 19.9(x_{1,4} + x_{2,4} + x_{3,4})$$
$$+ 22.5(x_{1,5} + x_{2,5} + x_{3,5})$$

and minimize:

$$V = 1650(x_{1,1} + x_{2,1} + x_{3,1}) + 2475(x_{1,2} + x_{2,2} + x_{3,2}) \quad (4.23)$$
$$+ 2887.5(x_{1,3} + x_{2,3} + x_{3,3}) + 1705(x_{1,4} + x_{2,4} + x_{3,4})$$
$$+ 2200(x_{1,5} + x_{2,5} + x_{3,5})$$

subject to:

$$(x_{1,1}, \ldots x_{3,5}) \geq 0 \quad (4.24)$$
$$(x_{1,1}, \ldots x_{3,5}) \in I \quad (4.25)$$
$$x_{1,1} + x_{1,2} + x_{1,3} + x_{1,4} + x_{1,5} = RLtask_{m,1} \quad (4.26)$$
$$x_{2,1} + x_{2,2} + x_{2,3} + x_{2,4} + x_{2,5} = RLtask_{m,2} \quad (4.27)$$
$$x_{3,1} + x_{3,2} + x_{3,3} + x_{3,4} + x_{3,5} = RLtask_{m,3} \quad (4.28)$$
$$x_{1,1} + x_{2,1} + x_{3,1} \leq RLmrr_{m,1} \quad (4.29)$$
$$x_{1,2} + x_{2,2} + x_{3,2} \leq RLmrr_{m,2} \quad (4.30)$$
$$x_{1,3} + x_{2,3} + x_{3,3} \leq RLmrr_{m,3} \quad (4.31)$$
$$x_{1,4} + x_{2,4} + x_{3,4} \leq RLmrr_{m,4} \quad (4.32)$$
$$x_{1,5} + x_{2,5} + x_{3,5} \leq RLmrr_{m,5} \quad (4.33)$$

where m is the Resource Level index for the current problem.

This specific resource allocation MOP application has three objective functions that produce a PF_{known} set of discrete points. Of course the objective function domain is not represented by a continuous line or surface as shown in Figures 4.65 and 4.66. In fact, the entire genotype space consist of integers and the phenotype space is also discrete but somewhat dense as reflected in Figure 4.65. Note that the integer aspect of this problem along with the equality constraints provide an interesting discrete MOEA test problem that is similar to the continuous Osyczka and Kundu function in Section 4.3.2.

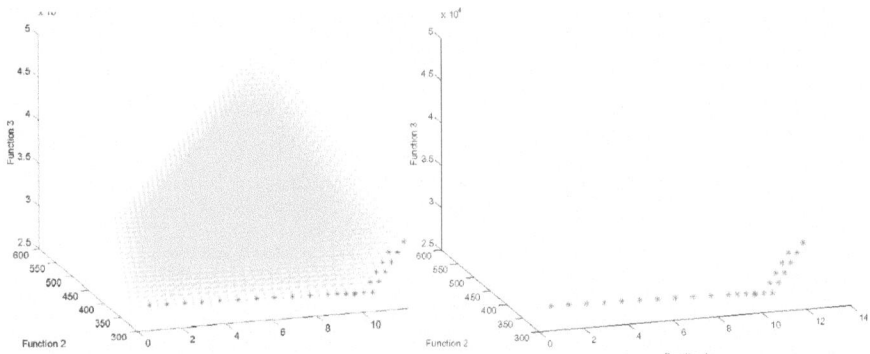

Fig. 4.65. MOP-ALP Discrete 3D Integer Search Space, PF_{true} on left and along bottom

Fig. 4.66. MOP-ALP PF_{true} for the three functions as directly related to Figure 4.65

4.7 Summary

In the tradition of providing test suites for evolutionary algorithms, an extensive list of specific MOEA test functions is proposed. The development of this list is based upon accepted and historic EA test suite guidelines. Specific MOEA test suites can evolve from this proposed list based upon individual research objectives and problem domain characteristic classifications. With generic MOEA test suites, researchers can compare their multiobjective numeric and combinatorial optimization problem results (regarding effectiveness and efficiency) with others, over a spectrum of MOEA instantiations. Example MOEA comparisons in the next chapter include a variety of presentation techniques. Using the test suite functions, MOEA comparisons can be made more precise and their results more informative via metrics and statistical analysis. But again some say that *using a test suite of any kind can be useful from a pedagogical perspective in comparing MOEAs, but in general, may be of little importance when solving "real-world" problems!*

Further Explorations

Class Exercises

1. Discuss the importance of scalability (both in decision variable space and objective function space) in the context of MOEA validation.
2. When scaling a test function in its number of decision variables, does it become more difficult for a MOEA? Why?
3. When scaling a test function in its number of objective functions, does it become more difficult for a MOEA? Why?
4. Discuss possible limitations of current MOEA test suites.
5. How would you improve the current methodology adopted to validate MOEAs?
6. Do you consider important to have benchmarks with dynamic MOPs? Discuss.
7. What would be the main issues that you would consider to be the main justification to propose a new test suite for MOEAs?
8. Can you think of a possible source of difficulty for modern MOEAs that is not covered by any of the test suites provided in this chapter?

Class Software Projects

1. Write a program that generates the true Pareto front of a problem using an enumerative approach. Test your program with some of the most simple test functions discussed in this chapter (2 decision variables, 2 objectives).
2. Write an interface that allows your program to invoke GNUplot or a similar public-domain library to generate graphical representations of both P_{true} and PF_{true} (when dimensionality allows it).
3. Implement a constraint-handling technique to a MOEA that does not include such type of mechanism (e.g., PAES [886]) and test it with some of the constrained MOPs presented in this chapter.

4. Write a program that uses spatial data structures (e.g., quadtrees) to store the nondominated solutions generated. Analyze the efficiency of this program, when generating PF_{true} by enumeration, with respect to traditional implementations (i.e., using a flat file).
5. Write a program using parallel processing (implemented in MPI [1230]) for generating the PF_{true} of a problem by enumeration. Test your program with the test functions included in this chapter.
6. Implement the real-world test problem proposed by Gaspar-Cunha and Covas in [543], and use any of the MOEAs discussed in Chapter 2 to solve it.
7. Modify one of the sets of test functions discussed in this chapter, so that they become epistatic.[4]
8. Implement two MOEA test problems adopting superspheres, as proposed by Emmerich and Deutz [449]. Use the NSGA-II [374] to attempt to solve these test problems. Write a report detailing all your work and your statistical analysis of results (see Chapter 5).

Discussion Questions

1. What is a "good" MOEA test suite? Relate to desired characteristics and classes of problems.
2. Can a "standard" set of test functions embody a complete set of MOP problem characteristics? Discuss.
3. What is the advantage (if any) of adding more decision variables to a continuous multi-objective optimization test problem? Discuss.
4. Discuss the impact of test function characteristics as an advantage for various MOEA search techniques.
5. Do you consider appropriate to adopt benchmarks with pedagogical test functions to assess performance of a MOEA? Do you think that such sort of validation is of any practical use when using a MOEA for solving a real-world problem? Discuss.
6. Do you think that scalable test functions are really useful to assess performance of a MOEA? Can scalability provide any insights regarding the performance (and possible limitations) of a MOEA? Discuss.
7. For side-constrained MOPs, is it better to embed in the MOEA the feasibility processing or do post processing to remove the infeasible solutions?
8. How could the genotype variables be discretized in a manner that would provide better convergence to PF_{true}?
9. Knowles and Corne [876] made publicly available two problem instance generators for a multi-objective version of the quadratic assignment problem. Develop an experimental study using these problem instance generators. It is of particular interest to analyze how difficult is to move towards

---

[4] Epistasis in evolutionary computation refers to a strong interaction among the genes (or decision variables) in a chromosome [1100].

the true Pareto front. Present a report that discusses your findings in detail.
10. Consider the so called *multiple multi objective problem* (M-MOP) as defined by Ponweiser and Vincze [1285]. Discuss how is this problem different from a traditional MOP.

Possible Research Ideas

1. Continue to develop a detailed classification of the application problem domain for generic MOEA test purposes. One should include a wealth of dimensions and variation. The objective is to present a multi-dimensional table where one can attempt to map their specific problem characteristics into a particular region of the table. This table would therefore also include appropriate MOEAs and associated parameter values for that region. Thus, such table would be able to provide expert advice to the engineer or scientist regarding a "good" algorithmic approach for their MOPs.
2. Develop a formal mathematical model of simple and difficult Pareto front problems due to quantization (discretization) of variables and their associated fitness landscape.
3. When considering NP-Complete type problems with multiple objectives, define or classify the search space landscape characteristics regarding the discrete Pareto front.
4. Develop a set of test suite functions for dynamic (changing) fitness environments and evaluate specific MOEAs. Attempt to develop insight to environmental dynamic parameter variables as affecting MOEA performance (effectiveness and efficiency). See [473, 474].
5. Address the issue of "stretching" the MOP landscape to improve effectiveness for a minimization problem. This technique for a single objective function first identifies a local optimum, functionally makes all regional points above such local optimum disappear in a revised fitness function, and then stretches upward the modified landscape in the region of the local minimum. Then, the search continues on the new landscape [1255].
6. Deb et al. [377] proposed the use of explicit linkages among variables as a way of generating difficult multi-objective optimization problems. To exemplify their methodology, the authors generated more difficult versions of some of the ZDT [1772] and DTLZ [379] test problems discussed in this chapter. Propose other multi-objective test functions that exploit this concept.
7. Emmerich and Deutz [449] proposed test problems whose Pareto optimal sets and Pareto fronts are closed-form expressions based on Lamé superspheres. These test problems are scalable in the number of objectives and decision variables, and also in what the authors call *resolvability of conflict*

(one can move from low conflict resolvability to high conflict resolvability). Use the test problems proposed in this paper as building blocks for generating a more complex test suite. Additionally, explore the use of hyper-ellipsoids or combinations of other geometric structures different from the superspheres adopted by Emmerich and Deutz.
8. Develop an experimental design that can help to identify sources of difficulty (for state-of-the-art MOEAs) in a multi-objective optimization problem. Rely on performance measures and statistical analysis to perform such a study.
9. Rudolph et al. [1402] point out the need (that sometimes arises in real-world problems) of being able to preserve Pareto subsets of equivalent quality. Build test problems that present this property and propose a novel way of dealing with them.

5
MOEA Testing and Analysis

> It doesn't matter how beautiful your theory is, it doesn't matter how smart you are. If it doesn't agree with experiment, it's wrong.
>
> Richard P. Feynman

5.1 Introduction

Regarding the scientific method of experimentation, it is desirable to construct an accurate, reliable, consistent and non-arbitrary representation of multi-objective evolutionary algorithm (MOEA) architectures and performance over a variety of multi-objective optimization problems (MOPs). In particular, through the use of standard procedures and criteria, one should attempt to minimize the influence of bias or prejudice of the experimenter when testing a MOEA hypothesis. The design of each experiment must conform then to an accepted "standard" approach as reflected in any generic scientific method. When employing the scientific method, the detailed design of MOEA experiments can draw heavily from outlines presented by Barr et al. [93] and Jackson et al. [765]. These generic articles discuss computational experiment design for heuristic methods, providing guidelines for reporting results and ensuring their reproducibility. Specifically, they suggest that a well-designed experiment follows the following steps:

1. Define experimental goals;
2. Choose measures of performance - metrics;
3. Design and execute the experiment;
4. Analyze data and draw conclusions;
5. Report experimental results.

The scientific method as a more generic approach has four steps: observation[1], hypothesis, predict using hypothesis, and testing [1706, 911, 94]. Another very important experimental goal is determining how well the test problems and proposed metrics capture essential MOP and MOEA characteristics and performance. This chapter follows all these generic guideline concepts in developing experimental MOEA testing procedures. Such comparative experiments use appropriate MOP benchmarks or test suites as developed in Chapter 4.

The main goal of testing is usually to compare MOEA effectiveness over various chosen MOPs by measuring solution quality[2]. Once a meta-level testing process has been designed using the guidelines, specific MOPs and metrics must be selected. Observe that metrics usually fall into two performance categories: (1) **Efficiency** (measuring computational effort to obtain solutions, e.g., CPU time, number of evaluations/iterations - use of spatial and temporal resources), and (2) **Effectiveness** (measuring the *accuracy* and *convergence* of obtained solutions and the data interface to the environment). Effectiveness includes *Robustness* (measuring how well the code recovers from improper input), *Scalability* (measuring how large a class of problems the code can solve as related to the increasing problem dimension) and *Ease of use* (measuring the amount of effort required to use the software - user friendliness). Of course, there is always the trade-off between effectiveness (solution quality) and efficiency (execution time).

In order to study and analyze the dynamics of MOEA execution, generational population measurement of PF_{known} and P_{known} is necessary. In all cases, the associate measures provide qualitative date that is usually reduced to qualitative statements through metrics. Such metrics are usually based upon the concept of Pareto dominance as discussed in Chapter 1. Also, empirical stochastic distributions for effectiveness and efficiency can be evolved from these measurements using first-order (mean) and second-order (variance) and higher-order statistics. Using just the mean and variance values of course is explicitly assuming an underlying normal distribution. Note that the distribution form is characterized by estimating the cumulative distribution function.

Many MOEA researchers' *modus operandi* is an algorithm's comparison (possibly the researcher's own new and improved variant) against some other MOEA by analyzing results for specific MOP(s). Results are often "clearly" shown in visual graphical form indicating the new algorithm is more effective for selected MOPs. These empirical, relative experiments are of course incomplete as regarding robustness and general MOEA comparisons. The litera-

---

[1] Observation in this case is used as a foundation to model and approximate the real-world problem.

[2] Although MOEAs are classified as "stochastic multi-objective optimizers," the attainment of a MOP's true Pareto optimal front can be difficult and may even be impossible. Thus, the quantitative measurement of performance relies on PF_{known}, an attainment set by definition.

ture's history of visually comparing MOEA performance on non-standard and possibly unjustified numeric MOPs does little to determine a given MOEA's actual generic efficiency and effectiveness. Extensive experimentation, analysis, and metrics concerning MOEA parameters, components, and approaches are required as a minimum. For example, statistical analysis techniques beyond mean and variance include Kruskal-Wallis hypothesis testing, analysis of covariance, and F-ratio testing as detailed in Section 5.5.1.

It is not the intent to indicate that one MOEA is better or more robust than another, but to describe general experimental methodology, appropriate metrics, statistical analysis and presentation techniques for a wide range of MOEA testing. In fact, different MOEA performances are directly associated with the specific operators each MOEA employs as discussed in the previous chapter. The philosophy, rational and goals of generic MOEA experiments are presented in Section 5.2 of this chapter. The MOEA experimental methodology is proposed in Section 5.3. Section 5.4 formally develops and analyzes attainment functions as a graphical effectiveness evaluation technique. More explicit Pareto dominance relations are presented in order to develop Pareto compliant generic quality indicators. The section concludes with an extensive listing and discussion of such indicators. Statistical testing approaches coupled with presentation methods appropriate for comparing MOEAs is reflected in Section 5.5. Availability of MOEA testing and metric software environments is discussed in Section 5.6. Following this discussion is a listing of class exercises, MOEA software experiments, discussion questions, and research ideas regarding MOEA testing and analysis.

5.2 MOEA Experiments: Motivation and Objectives

The major goal of MOEA experiments is to compare well-engineered algorithms in terms of effectiveness and efficiency as regards *carefully selected MOP test problems* through the use of appropriate metrics. Such metrics include quality indicators which attempt to summarize comparative experimental outcomes. Also, generation of the attainment function which approximates effectiveness regarding the probability density of PF_{known}. These measurement approaches should suffice to validate MOEA feasibility and promise.

One should not claim that MOEAs are the only algorithms able to solve these test problems efficiently and effectively. Consider, for example, Tabu search, dynamic programming, simulated annealing, depth-first search, breath-first search, and heuristic search as comparative algorithms. The No Free Lunch Theorem indicates that these search algorithms as well as MOEAs are not individually robust over all problems by definition. Regarding MOEAs, one desires to see if a given MOEA performs "better" than another over a specific problem domain class or classes, and if so determine why. If all MOEAs perform equally well, one may wish to determine the reason, as that situation implies MOEA implementation choice may not be crucial. Other interesting performance observations may also arise during experiment execution and analysis.

MOEAs (described in Chapter 2 and briefly in Section 5.3.2) are all based on similar evolutionary mechanisms. These MOEAs should be tested on various numeric problems, constrained MOPs, experimental benchmarks, and selected scientific and engineering applications. Examples of course prove nothing, but generally most MOEAs have "good" performance for various MOPs based upon a finite number of appropriately defined metrics. Each MOEA's performance should be statistically compared in solving these carefully selected MOPs. The intent of this chapter is to present a methodology spectrum addressing the elements of MOEA test and analysis.

Relevant *quantitative* MOEA performance based on appropriate MOP experiments is discussed in the next chapter. Many comparisons cited in the contemporary literature visually compare such algorithmic results through quality indicators or metrics as defined in this chapter. As experimental MOPs' P_{true} and PF_{true} are often not known, these conclusions are only relative to PF_{known} (see Chapter 1). In particular, the definition of dominance (weak, strict, etc.) and the concept of PF_{known} also defined as the *Pareto front approximation set*, are presented in order to discuss comparison techniques. This statistical methodology provides a basis for *absolute* and relative conclusions regarding MOEA performance based upon various metrics.

5.3 Experimental Methodology

Having discussed MOEA domains previously in Chapter 2 with MOPs test suites presented in Chapter 4, meaningful MOEA experiments can be conducted. Although test suite functions do provide a common basis for MOEA comparisons, results are empirical unless the global MOP Pareto optima are known. However, there are ways to determine P_{true} and PF_{true} for certain problems, either theoretically or with exhaustive search! See Chapter 1 for a theoretical discussion of modelling known and unknown Pareto fronts. Teaming this data with appropriate metrics then allows desired quantitative and qualitative MOEA comparisons related to P_{true} and PF_{true} and respectively P_{known} and PF_{known}. In general, these metrics are *unary quality indicators* in that they map the PF_{known} set of points in objective space to a real value; e.g., a mapping of a random set to a random value. In Section 5.4 a formal development of appropriate metrics is given the stage.

5.3.1 MOP Pareto Front Determination

When the real and continuous world is modelled (e.g., via objective functions) on a computer (a discrete machine with finite word length), there is a fidelity loss between the (possibly) continuous mathematical model and its discrete representation. Any formalized continuous MOP being computationally solved suffers this fate as the results are *approximation sets* such as PF_{known}. At a "standardized" computational resolution and representation, MOEA results can be quantitatively compared not only against each other but against certain benchmark MOPs' PF_{true}. Thus, whether or not the selected MOP's PF_{true} is actually continuous or discrete is not of experimental concern, as the representable P_{true} and PF_{true} are fixed based on certain assumptions (see Chapter 1 for a more detailed discussion) such as an underlying computational grid. Of course, PF_{known} is in reality a random set of points in the objective space.

Computational Grid Generation of PF_{true}: A *computational grid* is defined by placing an equidistantly spaced grid over decision variable space, allowing a uniform sampling of possible solutions. Each grid intersection point (computable solution) is then assigned successive numbers using a binary representation. Of course, Grey scale encoding could also be applied. Given a fixed length binary string, decision variable values are determined by mapping the binary (sub)string to an integer *int* and then solving the following for each x_i:

$$x_i = l + \frac{int * (u - l)}{2^n - 1}, \qquad (5.1)$$

where l and u correspond to the lower and upper decision variable bounds and n is the length of the binary string (for each x_i). For example, given the binary string 1011100001, x_1 represented by the first three bits and x_2 by the last seven, and upper and lower bounds for both variables set at 4.0 and -4.0 respectively, *int* for $x_1 = 5$ and $x_1 = 1.714$, while *int* for $x_2 = 97$ and $x_2 = 2.110$.

Binary encodings have been identified with shortfalls; e.g., Hamming cliffs [72, pg. 229], so other encodings can be used in MOEAs. Although restricting MOEA genetic representation to binary strings may result in less effective results it does allow for desired standard comparisons between MOEAs. If one algorithm uses real-valued genes, its computational grid's "fidelity" is much finer, giving it a search advantage because it is able to "reach" more discrete points in the solution space, but possibly taking more time. Additionally, different computational platforms may allow different resolutions (i.e., different ϵ values – the smallest computable difference between 1 and the next smallest value) and different numbers of distinct values (i.e., how many distinct numbers can be computed). As discussed in Chapter 1, PF_{known} lies somewhere on this computational grid.

Thus, even though a binary representation restricts a search space's size it allows for a quantitative MOEA comparison, determination of an MOP's P_{true} (at some resolution; P_{known}), and an enumeration method for deterministically searching a solution space (see the next section). The underlying resolution may be increased/decreased as desired, at least up to some point where computation becomes impractical or intractable. The methodologies presented are designed for experimentation and can be used to make judgments about proposed MOEA architectures and their implementations.

Search Space Enumeration of PF_{true} : Enumerative exhaustive deterministic search may be the only viable approach to solving irregular or chaotic problems [1186]. Harnessing ever-expanding computational capability to obtain the desired solutions should provide better solutions. Such software has been constructed executing on parallel high-performance computers whose purpose is to find P_{true} and PF_{true} for several numeric MOPs [1626, 287]. The resulting sets are still only a discrete representation of their continuous counterparts, but are the "best possible" at a given computational resolution for a given computational platform. Various high performance computational platforms can be employed to deterministically enumerate all possible solutions for a given MOP at a given computational resolution. References [1626, 287] present data for a considerable number of selected MOPs. The program is written in "C" and uses the Message Passing Interface (MPI) to distribute function evaluations among many processors. Using this P_{true} database, various MOEA results can be compared not only against each other, but also against the *true* MOP optimum. However, these MOEAs should use a binary encoding and mapping. At least for selected MOPs, a quantitative comparison is then possible. This methodology allows more absolute performance observations. Although, the resulting data could just be P_{known} and PF_{known} at a very high resolution resulting from a Monte-Carlo approach.

5.3.2 MOEA Algorithms Testing

Various MOEAs need to be selected for MOEA testing. Such algorithms and their original *raison d'etre* are discussed in detail in Chapter 2 as well as the referenced literature. In general, one should select a set of MOEA algorithms because they specifically incorporate what appear to be key theoretical problem/algorithm domain aspects such as Pareto ranking, niching, and fitness sharing (see Chapters 2 and 6). Other researchers appear to share these thoughts as the MOGA [504], the NPGA [709], the NSGA-II [374], the SPEA [1782], the SPEA2 [1775] and PAES [886] are some of the literature's most cited and imitated. As all these MOEA architectures move towards each other temporally in terms of data structures, parameters, and operators, the affirmation that one MOEA is "better" than another is probably ill-conceived even when considering the NFL theorems [1708]. In reality, it is the incorporation of specific operators, which we measure in these performance comparisons.

Most current MOEAs (e.g., MOGA [504], NPGA [709], NPGA 2 [453], NSGA [1509], NSGA-II [374], PAES [886], M-PAES [873], SPEA [1782], and SPEA2 [1775]) are based on "traditional" GAs with chromosome individuals; the MOMGA [1626], MOMGA-II [1788], and MOMGA-III [341] are based on messy GA (mGA) building blocks and can be considered somewhat nontraditional. along with the IMOEA [234]. However, the conceptual evolutionary process modeled by each algorithm is basically the same and gives the basis for their direct performance comparison. Table 5.1 lists various MOEAs' key characteristics which are addressed as to testing in the next section. For more algorithmic insight, discussion of specific MOEAs along with pseudocode are presented in Chapter 2.

Other MOEA algorithms could be considered for inclusion in comparative experiments. These include random search, VEGA [1440], the microGA [284], the microGA^2 [1597], the OMOEA [1757], and MOSGA [48], among many others. Note that several MOEA comparisons have shown random search performs *much* worse than other tested algorithms over complex MOPs due to "rough" fitness or search landscapes.

Observe that VEGA could be excluded because it is biased towards solutions performing "well" in only one dimension [706], and because several efforts indicate VEGA performs "worse" than other proposed MOEAs. SPEA is included, observing that because of its explicit incorporation of a secondary population in the fitness assignment process [1781, 1782] it may unfairly impact comparative performance (See Chapter 2). Of course, these and other alternative MOEAs can be considered in other experiments. Observe that we are not considering aggregating functions (e.g., [636]), since they are not considered true MOEAs by many researchers in this area. Nevertheless, it is worth indicating that there exist some aggregating schemes that can overcome the main limitation of linear aggregating functions (i.e., its incapability for generating non-convex portions of the Pareto front). For example, Jin et al. [800], uses evolutionary strategies and changes aggregated weights dynamically through uniformly distributed random weight combinations. Probabilistic modelling MOEAs have also been developed including the mBOA [844] and the IMOEA [234] which also manipulate explicitly building blocks.

5.3.3 Key MOEA Algorithmic Parameters

Many evolutionary algorithm experimenters vary key algorithmic parameters and associated characteristics (see Table 5.1) in an attempt to determine the most effective and efficient implementation for a particular problem instantiation or class. Although suggested in general, a complete parameter analysis investigating effects of differing parameter values is generally beyond the scope of experiments. It could be to some extent included in comparative studies. The purpose of such experiments is to determine general MOEA performance and to explore the specific algorithm domain, not to "tune" MOEAs for good performance on some MOP. Many algorithms execute with default parameter values as reported in the literature, implementing each MOEA as "out of the box."

The MOEA literature typically reports using default single-objective EA parameter values, except perhaps for population size. Because MOEAs track a set of solutions, and because more objectives imply the possibility of more Pareto optimal solutions (by definition when using a discrete representation),

Table 5.1. Key Experimental MOEA Characteristics

	EVOPs	Fitness Assignment	Sharing Niching Crowding	Population
IMOEA [234]	Crossover and Mutation (p_c, p_m)	Tournament Elitism Random	Phenotypic (σ_{share} - Fitness)	Deterministically initialized; $N_{max}1$
MOGA [504]	Crossover and Mutation ($p_c = 1$, $p_m = \frac{1}{k}$)	Linear interpolation using [504], Pareto ranking	Phenotypic (σ_{share} - Fitness)	Randomly initialized; $N = 50$
MOMGA [1626]	"Cut and splice" ($p_{cut} = 0.02$, $p_{splice} = 1$)	Tournament ($t_{dom} = 3$)	Phenotypic (σ_{share} - Domination)	Deterministically initialized; $N = 100$
MOMGA-II [1788] and MOMGA-III [341]	"Cut and splice" ($p_{cut} = 0.02$, $p_{splice} = 1$)	Tournament ($t_{dom} = 3$)	Phenotypic (σ_{share} - Domination)	Deterministically initialized; $N = 100$
NPGA [709] and NPGA 2 [453]	Crossover and Mutation ($p_c = 1$, $p_m = \frac{1}{k}$)	Tournament ($t_{dom} = 5$)	Phenotypic (σ_{share} - Domination)	Randomly initialized; $N = 50, 100$
NSGA [1509]	Crossover and Mutation ($p_c = 1$, $p_m = \frac{1}{k}$)	"Dummy" fitness using [581], Pareto ranking	Phenotypic (σ_{share} - Fitness)	Randomly initialized; $N = 50, 100$
NSGA-II [374]	Crossover and Mutation ($p_c = 1$, $p_m = \frac{1}{k}$ v	Nondominated ranking and crowding	Phenotypic crowding on various fronts	Randomly initialized; $N = 50, 100$;
PAES [886] and M-PAES [873]	Crossover and Mutation ($p_c = 1$, $p_m = \frac{1}{k}$)	(1+1)- single grid	Phenotypic (σ_{share} - Fitness)	Randomly initialized; Update archive $N = 50, 100$
PESA [301] and PESA-II [299]	Crossover and Mutation ($p_c = 1$, $p_m = \frac{1}{k}$)	(1+1)- single grid	Phenotypic (σ_{share} - Fitness)	Randomly initialized; Update archive $N = 50, 100$
VEGA [1440]	Crossover and Mutation ($p_c = 1$, $p_m = \frac{1}{k}$)	Based on a single objective	Phenotypic (σ_{share} - Fitness)	Randomly initialized; $N = 50, 100$
SPEA [1782] and SPEA2 [1775]	Crossover and Mutation ($p_c = 1$, $p_m = \frac{1}{k}$)	"Dummy" fitness using [581], Pareto ranking	Phenotypic (σ_{share} - Fitness)	Randomly initialized; $N = 50, 100$

researchers sometimes enlarge the MOEA's generational population to an upper bound. Note that the purpose of most experiments is MOEA performance comparison and not determination of ideal parameter settings for some (class of) MOPs. If possible, key MOEA parameter values should be kept identical. A discussion of these key parameters follows:

Population Size: The MOGA [504], NPGA [709], NSGA-II [374], PAES [886], SPEA2 [1775] and MOMGA-II [1788] generally use a random population initialization scheme. That is, given some genetic representation, all solutions in the initial generational population are uniformly selected from the solution space. The MOMGA uses a deterministic scheme. For each era (signified by k) the MOMGA generates all possible building blocks (BBs) of size k. Thus, its initial population composition is always known. However, the initial competitive templates are randomly generated. The NSGA-II, for example, distributes a constant population across a sequence of ranked fronts. On the other hand, the IMOEA [234] uses a dynamic population sizing technique based upon the current number of individuals on PF_{known} and a desired population density. Additionally, micro-MOEAs limit their population size to a few individuals given a large objective function computation cost [284].

Mating Restrictions: Mating restrictions have both its proponents and opponents. Existing empirical experimental results sometimes indicate it is necessary for good performance, and at other times various MOEA implementations seem to operate well without it. These empirical results indicate the NFL theorems are alive and well [1708]. Observe that incorporating mating restrictions in some experimental MOEA software usually requires major code modifications, and because of its uncertain usefulness in the MOP domain, mating restriction are *not* incorporated in most experimental MOEAs.

Fitness Assignment: The MOMGA [1626] and NPGA [709], for example, employ tournament selection and so require no specific solution fitness manipulation besides those values returned by the MOP fitness function. The MOGA first evaluates all solutions, then assigns fitness by sorting the population on rank ('0' being the best and 'N' the worst – see Chapter 2). Fitness is assigned linearly to each ordered solution; final fitness is determined by averaging the fitness values for identically ranked solutions and then performing fitness sharing. The NSGA [1509], for example, evaluates and sorts the population by rank. However, it assigns some large "dummy" fitness to all solutions of the best rank. Then, a lower "dummy" fitness value is assigned to the solutions of the next best rank, and so on. Note here that all experimental MOEAs employ fitness scaling as each objective dimension's magnitude may be vastly different.

Fitness Sharing: Most experimental MOEAs incorporate phenotypic-based sharing using the "distance" between objective vectors for consistency.

For the MOGA and NSGA, σ_{share} is computed and a sharing matrix formed via the standard sharing equation [581]. Fitness sharing occurs only between solutions with the same rank [511, 1509]. Other approaches use, instead, crowding (e.g., the NSGA-II [374]).

Various MOEAs such as the NPGA and MOMGA use a slightly different sharing scheme. Two solutions undergoing tournament selection are actually compared against those in a small comparison set. Sharing occurs only if both solutions are dominated or nondominated with respect to the comparison set. A σ_{share} value is used, however, the associated niche count is simply the number of vectors within σ_{share} in phenotypic space rather than a degradation value applied against unshared fitness. The solution with the smaller niche count is selected for inclusion in the next generation. Horn et al. [709] label this *equivalence class sharing*. An identical scheme is implemented in the MOMGA as it also uses tournament selection. Per Horn's recommendation, continuously updated sharing is used by both the NPGA and the MOMGA due to the observation that chaotic niching behavior may result when combining fitness sharing and tournament selection [708].

σ_{share} represents how "close" two individuals must be in order to decrease each other's fitness. This value commonly depends on the number of optima in the search space. As this number is generally unknown, and because PF_{true}'s shape within objective space is also unknown, σ_{share}'s value is assigned using Fonseca's suggested method [511]:

$$N = \frac{\prod_{i=1}^{k}(\Delta_i + \sigma_{share}) - \prod_{i=1}^{k}\Delta_i}{\sigma_{share}^k}, \qquad (5.2)$$

where N is the number of individuals in the population, Δ_i is the difference between the maximum and minimum objective values in dimension i, and k is the number of distinct MOP objectives. As all variables but one are known σ_{share} can be easily computed. For example, if $k = 2$, $\Delta_1 = \Delta_2 = 1$, and $N = 50$, the above equation simplifies to:

$$\sigma_{share} = \frac{\Delta_1 + \Delta_2}{N - 1} = 0.041. \qquad (5.3)$$

This appears a reasonable way to obtain σ_{share} values, although Horn also presents equations bounding PF_{true}'s possible size [708] but leaves the user to choose specific σ_{share} values. Finally, as each MOP's objective values may span widely disparate ranges all objective values are scaled before σ_{share} is computed. This action is meant to prevent unintentional niching bias.

Representation and Evolutionary Operators (EVOPs): As described in Section 5.3.1, experimental methodology generally assumes that each MOEA uses a binary representation. Thus, if all MOEAs are assumed to use an l-bit ($l = 24$) string for each solution and identical minimum/maximum values in each decision variable dimension, we ensure identical "reachability" of the test algorithms for a given MOP. The bit length could be increased in later experiments to examine larger search spaces. However, some MOEAs employ different binary-value to real-value mappings. The MOMGA [1626], MOMGA-II [1788], NPGA [709], and deterministic enumeration programs use the mapping shown in equation (5.1); the MOGA [504], NSGA [1509], and NSGA-II [374] execute as part of a larger program (see Section 5.5) that uses a different mapping. This may result in differing mapped values due to truncation or round-off errors as the schemes are implemented. There is not yet a "default" MOEA crossover rate but various experiments used crossover probabilities in the range $p_c \in [0.7, 1.0]$ [708, 512, 1509]. Thus, other experimental MOEAs use single-point crossover with $p_c = 1.0$. All but the MOMGA used a mutation rate of $p_m = \frac{1}{l}$ where l is the number of binary digits.

Termination, Solution Evaluations, and Population Size: When should a MOEA stop executing? The easy answer is after convergence occurs – but when is that? Some "best guess" is normally made and appropriate termination flags set. This technique is used in this experimental series with terminate search based on the number of solution evaluations.

For various MOPs, a MOEA can be executed first and the number of executed solution evaluations per run determined. Other MOEAs (each with population size of, for example, $N = 50$) can be then set to execute the same number of evaluations (N multiplied by the number of generations), ensuring a very nearly equivalent computational effort for each tested MOEA.

The literature sometimes indicates that more objectives imply a larger generational population size is necessary. However, as experiments generally involve only bi- and tri-objective MOPs, population size is left at the suggested single-objective GA default size of 50 [72, pg. 123] or some small factor. Again note that the purpose of these experiments is to explore MOEA performance and not to determine ideal parameter settings over the selected test functions.

5.4 MOEA Experimental Measurements

When comparing MOEAs numerous "standardized" metrics have been identified and employed by various researchers. Initial developers of MOEA benchmarks include [1626, 355, 1770]. Early use of individual metrics is found for example in [1452, 1509, 503, 323, 261]. Others have extended such metrics, added other metrics, developed more formal metric descriptions, and provided analytical insight [1631, 874, 287, 882]. Summary of "good" metrics that can be used to statistically compare MOEAs is found in [1631, 287, 1788, 882, 1783]. In this section, a more formal development of MOEA comparison measures is presented as explicitly related to the concepts of *attainment functions, dominance definitions* and characterization of *quality indicators/metrics*. A MOEA experimenter then can select an appropriate set of approaches to evaluate MOEA performance based upon a sound foundation.

Table 5.2: Dominance relations on objective vectors and approximation sets when working with compatibility and completeness. For a graphic illustration of how these relations are used, please refer to Figure 5.3 on page 255.

Relation	Objective Vectors	
Strictly Dominates	$f_\tau(\mathbf{x}^1) \succ\succ f_\tau(\mathbf{x}^2)$	$\forall_{i\in\mathcal{F}}$, ($f_\tau(\mathbf{x}^1)$ is better than $f_\tau(\mathbf{x}^2)$)
Dominates	$f_\tau(\mathbf{x}^1) \succ f_\tau(\mathbf{x}^2)$	$f_\tau(\mathbf{x}^1)$ is not worse than $f_\tau(\mathbf{x}^2)$ in all objs and better in at least one objective
Weakly Dominates	$f_\tau(\mathbf{x}^1) \succeq f_\tau(\mathbf{x}^2)$	$f_\tau(\mathbf{x}^1)$ is not worse than $f_\tau(\mathbf{x}^2)$ in all objs
Incomparable	$f_\tau(\mathbf{x}^1) \parallel f_\tau(\mathbf{x}^2)$	neither $f_\tau(\mathbf{x}^1)$ weakly dominates $f_\tau(\mathbf{x}^2)$ nor $f_\tau(\mathbf{x}^2)$ weakly dominates $f_\tau(\mathbf{x}^1)$
Indifferent	$f_\tau(\mathbf{x}^1) \sim f_\tau(\mathbf{x}^2)$	$f_\tau(\mathbf{x}^1)$ has the same value $f_\tau(\mathbf{x}^2)$ in each objective
	Approximation Sets	
Strictly Dominates	$A \succ\succ B$	every $a^2 \in B$ is strictly dominated by at least one $x^1 \in A$
Dominates	$A \succ B$	every $x^2 \in B$ is dominated by at least one $x^1 \in A$
Better[3]	$A \triangleright B$	every $a^2 \in B$ is weakly dominated by at least one $x^1 \in A$ and $A \neq B$
Weakly Dominates	$A \succeq B$	every $x^2 \in B$ is weakly dominated by at least one $x^1 \in A$
Incomparable	$A \parallel B$	neither A weakly dominates B nor B weakly dominates A
Indifferent	$A \sim B$	A weakly dominates B and B weakly dominates A

---

[3] Indicates that the indicator is dominance compliant and the left side results with respect to the indicator is better than the right side.

5.4.1 Selection of MOEA Comparison Measures

In general, theoretical MOEA analysis can be achieved, but extensive simplifying assumptions reduce the utility of the results. Evaluating realistic MOEA effectiveness requires experimental assessment by executing numerous runs and applying statistical analysis of the results. Measures of this nonlinear phenomena is qualified through the use of metrics. Standardized metrics have been proposed by [1626, 1631, 1788] for MOEA comparison. These include seven quality indicators based upon mappings from PF_{known} data, an experimental attainment set. Knowles and others on the other hand recommend two complementary approaches for comparing MOEA results [882]. They suggest that an *empirical attainment* function and set of Pareto *dominance-compliant quality indicators* be used to evaluate and compare the approximation sets, PF_{known}, from multiple runs. Ziztler also supports this suggestion of having a reduced set of quality indicators and indicates that it is advantageous to have metrics that are both *complete* and *compatible* to make good comparisons between MOEA results [882]. He suggests the following three metrics: ϵ indicator and the R_2 and R_3 utility indicators [1777]. Thus, in addition to the seven standardized metrics mentioned by Van Veldhuizen and Zydallis, three more metrics can be added to drive toward a compatible comparison for stochastic multiobjective algorithms [341]. All these various metrics are detailed in this section.

Observing in the literature that there are other metrics for evaluating MOEA performance, the ones suggested are adequate in that they encompass different aspects of MOEA performance. The later part of this section however addresses for historical completeness these other metrics that are similar to those suggested.

Understanding and selecting "quality" MOEA metrics requires the development of precise definitions. In order to properly select a specific set of MOEA measurements, the definition of a generic attainment function is formally discussed and rigorous dominance-compliant (Pareto-compliant) quality indicators suggested. There are three generic aspects for evaluating MOEA effectiveness performance. They include the Generic Attainment Functions, Dominance Relationships, and Quality Indicators over Approximate Pareto front sets. The definitions in the following sections are provided in order to prepare the reader for the multitude of terminology, insight, and application for MOEA testing. Observe that the suggested measurements can be used to evaluate different types of *multi-objective optimizers* besides MOEAs.

5.4.2 Generic Attainment Function

Generic attainment function development requires statistical data from a MOEA generational process in order to present a graphical effectiveness measure. The attainment function is by definition a first order moment measure usually for evaluating MOEA results. The experimental data could provide information about MOEA performance differences given any specific metric as well as providing statistical multivariate distributions and possibility higher moments [882, 513] for analysis. The attainment function is usually not evaluated for a specific metric in part due to the large amount of memory required for each generational state storage.

However, its use in generating statistical distributions is reflected in a graphical evaluation of performance based upon empirical PF_{known} data. An empirical attainment function defines a surface that divides objective space into goals (PF_{true} for example) that have been obtained and those that have not with a frequency bounded below of some a priori defined percent. Thus, this function provides for each vector in the objective space, the probability that it is weakly dominated (see Table 5.2) by an approximation set, PF_{known}. The empirical attainment function summarizes the outcomes of multiple runs of the MOEA. Thus in essence it is an implicit form of visualization. However, the experimental focus is usually on MOEA performance in terms of efficiency and effectiveness and not where the process differences lie.

The PF_{known} outcome of any multiobjective optimizer is a set of nondominated objective vectors evaluated during one or more executions. If the optimizer is stochastic, a MOEA for example, the associated Pareto-set approximations are random, and their distribution is of interest as a performance metric. In order to develop a more detailed understanding of an attainment function and its utility, consider the following definitions [513]:

Definition 21 (Random nondominated point set) : *A random point-set*

$$X \equiv \{(X_1,,..,X_M)\epsilon\mathbb{R}^d : P(X_i \leq X_j) = 0, i \neq j\} \quad (5.4)$$

where both the number of elements M and the elements X, themselves are random and the probability the $P(0 \leq M < \infty) = 1$, is called a random nondominated point set ($RNP-set$) □

Observe that the minimization of all objective functions is assumed without loss of generality. Random Pareto-set approximations produced by stochastic multiobjective optimizers such as MOEAs on d-objective problems are thus RNP-sets in $\mathbb{R}^d$. This definition is used to define an "attained set."

Definition 22 (Attained set) : *The random set*

$$\begin{aligned} Y &\equiv \{y\epsilon\mathbb{R}^d|X_1 \leq y \vee \ldots \vee X_M \leq y\} \\ &= \{y\epsilon\mathbb{R}^d|X \triangleleft y\} \end{aligned} \quad (5.5)$$

is the set of all goals $y\epsilon\mathbb{R}^d$ attained by the RNP-set X. □

For a two-dimensional MOP objective space, the nondominated points are defined by $RNP-set\{X\}$. By construction then, Y is defined as the area in which all the dominated points lie between the nondominated points and a preselected reference point (boundary point) in objective space.[4] The attained set is then the set of goals (nondominated PF_{true} points) possibly attained in some manner from the set of points in X. Using this definition, an "attainment indicator" focuses on the nondominated points through a mapping of a selected quality indicator, I, related to obtaining the goal PF_{true}.

Definition 23 (Attainment indicator) : . *Let $I\{\cdot\} \equiv I\{.\}(z)$ denote an indicator function. Then, the random variable $b_X(z) = I(X \triangleleft z)$ is called the attainment indicator of X at goal $z \epsilon \mathbb{R}^d$.*
□

The set of all attainment indicators indexed by $z \epsilon \mathbb{R}^d$ is the binary random field $b_X(z), z \epsilon \mathbb{R}^d$. For the deterministic case, this binary field fully characterizes a single Pareto-set approximation, as one can always be obtained from the other. As an infinite-dimensional quality indicator, it could be used to construct a comparison method which is *complete and compatible* with respect to weak-dominance [1783] as depicted in Table 5.2. Section 5.4.3 presents a precise meanings of complete and compatible. This symbolism supports a MOEA quality indicator formalism as an assessment tool for stochastic algorithmic evaluation.

Definition 24 (Attainment function) : . *The function $\alpha_X : \mathbb{R}^d \mapsto [0,1]$ with*

$$\alpha_X(z) = P(b_X(z) = 1) \qquad (5.6)$$

is called the attainment function of X.
□

As identified in Grunert da Fonseca et al. [609], the attainment function is the first-order moment measure of the binary random field $b_X(z), z \epsilon \mathbb{R}^d$ derived from Y (the set attained from X) and, as such, it offers a useful description of the location of the distribution of Y (and also of X). Note that for $M = 1$, the optimizer produces a single random objective vector X per run, and the attainment function reduces to the usual multivariate distribution function $F_X(z) = P(X \leq z)$. A natural empirical counterpart of the (theoretical) attainment function $a(.)$ may be defined as follows:

Definition 25 (Empirical attainment function)) : . *Let $b_i(z)...b(z)$ be n realizations of the attainment indicator $b_X(z), z \epsilon \mathbb{R}^d$. Then, the function defined as $\alpha_n : \mathbb{R}^d \mapsto [0,1]$ with*

$$\alpha_n(z) = \frac{1}{n} \sum_{i=1}^{n} b_i(z) \qquad (5.7)$$

is called the empirical attainment function of X (EAF).
□

The realizations $b_i(z), ..., b_n(z)$ correspond to n runs of the optimizer. In Figure 5.1, contour plots of the $EAFs$ obtained from various independent

---

[4] The distributions of both random sets, X and Y are equivalent, i.e. knowledge of the distribution of X automatically provides a characterization of the distribution of Y, and vice versa given the boundary point.

runs each of two simple multiobjective genetic algorithm variants (MOGA-A and MOGA-B) on a bi-objective optimal control problem are depicted. While the theoretical attainment function is continuous, the EAF is a discontinuous function. Thus, it exhibits transitions not only at the data points but also at other points, the coordinates of which are combinations of the coordinates of the data points. This is much like in the case of the multivariate empirical distribution function [790]. In the figure, the "grand" empirical attainment surfaces (the same in both plots) indicate the borders beyond which the goals are never attained or always attained. They are computed from the combined collection of approximation sets. Differences in the frequency with which certain goals are met by the respective algorithms A and B are represented in the region between these two surfaces. In the left part of the figure, darker regions indicate goals that are attained more frequently by MOEA A than by MOEA B. In the right side, the reverse is shown. The intensity of the shading can correspond to either the magnitude of a difference in the sample probabilities, or to the level of statistical significance of a difference in these probabilities (which is better?).

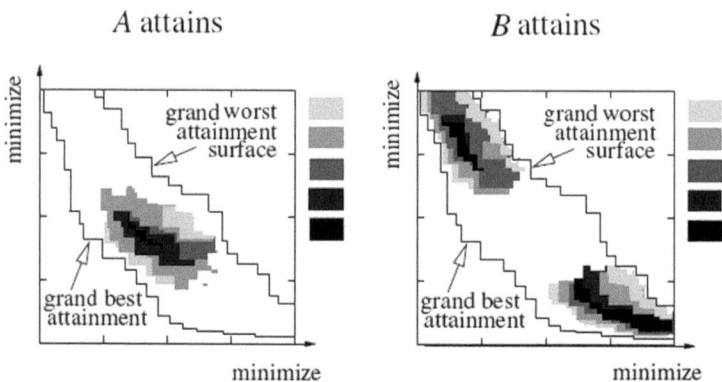

Fig. 5.1. Individual empirical attainment surfaces differences between the probabilities of attaining different goals on a two-objective minimization problem with optimizer A and optimizer B [882].

The EAF thus serves as an estimator for the theoretical attainment function $\alpha_X(z)$, in the same way as the multivariate empirical distribution function estimates the (theoretical) multivariate distribution function $F_X(z)$, for all $z \epsilon \mathbb{R}^d$.

The optimizer performance on MOP, related to the corresponding $RNP-$ set distribution, is evaluated via EAF estimates. The farther lower left the weight of the graphical attainment function, the greater is the probability of attaining tighter goals, and the better is the algorithm's performance. The performance of two (or more) optimizers operating on the same optimization MOP problem can be compared by comparing the corresponding attainment functions. A suitable, Smirnov-like, statistical testing procedure based on (two) *EAFs* has been applied by [1474]. Rejecting the statistical *null-hypothesis* of equal attainment functions indicates that the optimizers (MOEAs) exhibit different performance. However, if such a null hypothesis cannot be rejected, optimizers may still exhibit different performance, not only because of the statistical error involved, but also because of the $RNP - set$ distribution. Thus, the performance of the specific MOEA is not completely characterized by the attainment function.

Whereas the attainment function (first-order moment) describes the distribution of the $RNP - set X$ in terms of location, it does not address the dependence structure within the nondominated elements of X. Thus, a second order moment attainment function is required for that purpose [513]. A second-order moment type allows pairwise relationships to be analyzed between elements of an approximation set X. This extended model can thus provide more definitive statistical analysis of MOEA performance. Thus, assuming a minimization MOP, the second-order attainment function is defined as:

Definition 26 (Second-order attainment function) : . *The function defined as* $\alpha_X^{(2)} : \mathbb{R}^d x \mathbb{R}^d \mapsto [0,1]$, *with*

$$\alpha_X^{(2)}(z_1, z_2) = P(b_X(z_1) = 1 \wedge b_X(z2) = 1) \quad (5.8)$$

is called the second-order attainment function of X. □

The second-order attainment function is the second, non-centered, moment measure of the binary random field $\{b_X(z), z \epsilon \mathbb{R}^d\}$ derived from the attained set.

In random set theory terminology, the second-order attainment function would be called the covariance of the attained set (see, for example, Stoyan et al. [1527]). This function expresses the probability of the elements of the same Pareto-set approximation X simultaneously attaining two different goals, z_1, $z_2 \epsilon R^d$. The second-order attainment function is symmetric in its arguments, and includes all the information of the (first-order) attainment function, as $a(z, z) = ox(z)$ for all $z \epsilon \mathbb{R}^d$, and $o(z_l, z_2) = o(z_2, z_l) = ox(z_l)$ for all $z_1 < z_2 \epsilon IP$. A natural empirical counterpart of the (theoretical) second-order attainment function can be defined as follows:

Definition 27 (Second-order empirical attainment function) : . *Let* $b_1(z), \ldots b_n(z)$ *be realizations of the attainment indicator* $b_X(z), z \epsilon \mathbb{R}^d$. *Then, the function* $\alpha_X^{(2)} : \mathbb{R}^d v \mathbb{R}^d \mapsto [0,1]$ *with*

$$\alpha_X^{(2)}(z_1, z_2) = \frac{1}{n} \sum_{n=1}^{n} b(z_1).b_2(z) \quad (5.9)$$

is called the second-order empirical attainment function of X (second − order EAF). ☐

The second-order EAF is a discontinuous function, with the values $\alpha_x^{(2)}(z_1, z_2)$ representing the proportion of optimization runs (Pareto-set approximations) which attained goals z_1 and z_2 simultaneously.

The visualization of the second-order EAF is more difficult than that of the first-order EAF, since it is defined in $\mathbb{R}^{2d}$. Even with only two objectives, this results in four dimensions, and direct visualization is impossible. A useful approach is to fix one goal.

Another method of evaluating an optimizer's second-order behavior is to define a covariance function [513] and generate an empirical covariance function formulation. This provides an alternative method of computation for analyzing second-order behavior.

5.4.3 Dominance Relations

The general MOEA assessment technique is based on pairwise comparisons of approximation sets, similarly to Pareto dominance-based fitness assignment comparisons. Approximation sets of PF_{known} generated by different MOEA optimizers are collected, pairwisely compared, and ranked according to the number of approximation sets by which a specific set is dominated. As a result, each MOEA is associated with a sample of ranks where the ranks are to be minimized. Statistically, the various rank samples are compared. The advantage of this approach is that it is based only on Pareto dominance relations between sets and a ranking procedure, and so is not biased with respect to preferences. But, it also represents the least informative among itself, attainment functions, and quality indicators. these approaches as differences in quality cannot be localized. Also, associated first-order and second-order statistical measures are the Mann-Whitney rank sum for two MOEAs and the Kruskal-Wallis rank test for more than two MOEAs (see Section 5.5.1).

Since "nondominance" is a critical measure in MOEA operator execution, one is interested in precisely defining dominance relations. Table 5.2 presents an expanded definition of new dominance symbols from Chapter 1, including strict dominance, dominates, weakly dominates, incomparable, and indifferent. Figure 5.2 indicates that there exists a overlapping of these dominance relationships that permit the ordering of approximation set pairs, (A, B), for which $A \triangleleft B$.

Let there exist a multi-objective problem having k objective functions: $\mathbf{F}^1 = \{f_1(\mathbf{x}), \cdots, f_k(\mathbf{x})\}$. $\mathbf{F}^1$ is to be minimized. Assume also that each objective function assigns every solution $\mathbf{x}$ in the search space Ω a real value $z_i = f_i(x)$ reflecting the merit according to the ith criteria for a particular solution $\mathbf{x}$. Thus, every $\mathbf{x} \in \Omega$ is mapped to a vector $\mathbf{z} = \{z_1, \cdots, z_k\} \in \Omega$. Note that the sets of mutually incomparable objective vectors are the Pareto front approximations, and for mutually incomparable solutions, the Pareto set approximation or PF_{known} is the result of a MOEA search. Accordingly,

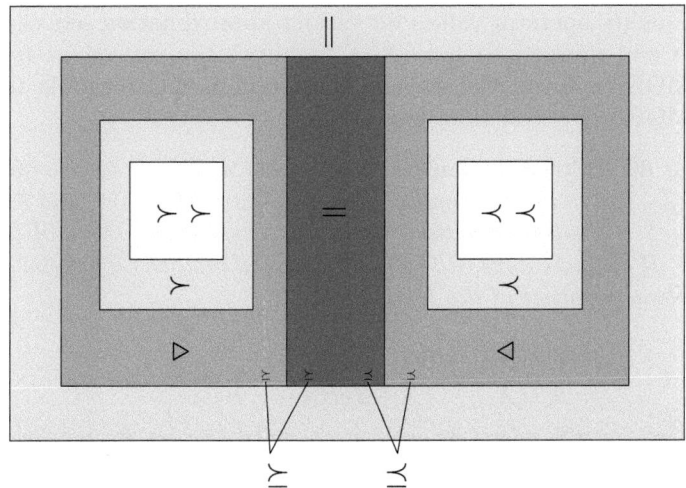

Fig. 5.2. The set of overlapping ordered pair relationships between approximation sets is partitioned based upon the different dominance relationships

the approximation set or generally PF_{known}, is defined in the following Definition 28.

Definition 28 (Approximate Set) : Let $\hat{A} \subseteq \Omega$ be a set of all objective vectors. $\hat{A}$ is called an approximation set if any evaluated individual of $\hat{A}$ evaluates to a vector that does not weakly dominate (see Table 5.2 on page 244 for the definition of weak domination) any other objective vector produced by evaluating individuals in $\hat{A}$. The set of all approximation sets is denoted as Ω. □

Given PF_{known} approximation sets, then an approximation set A dominates an approximation set B if and only if for each element of B there exists at least one element of A that dominates the element of B as presented in Chapter 1. Of course this definition could be applied to a specific MOEA process as well as for comparing the results of different MOEA results. This definition of approximation sets could also be applied to P_{known}, the Pareto solution set (genotype level), as well.

Since dominance evaluations are independent allowing objective vectors or approximation sets, they can be compared even though the objectives may be non-commensurable (not measurable by the same measure).

Statistical evaluation of mapping approximation sets to quality indicator values can result for example in associated means and standard deviations. A more formal definition of a quality metric follows:

Definition 29 (Quality Indicator (metric)) : A h-ary quality indicator is a function $\mathcal{I}_i : \Omega^h \mapsto \mathbb{R}$, which assigns each vector $(\hat{A}_1, \hat{A}_2, \cdots, \hat{A}_{\dot{h}})$ of $\dot{h}$ approximation sets a real value $\mathcal{I}_i(\hat{A}_1, \cdots, \hat{A}_{\dot{h}})$. □

These quality operator values for various approximation sets can formally be defined and ordered to provide a "quality" comparison or relationship between MOEAs. Such effectiveness relationships may be able to indicate that one MOEA is better than another.

Definition 30 (Comparison method) : Let $A, B \in \Omega$ be two approximate sets, $\mathcal{I} = (\mathcal{I}_1, \mathcal{I}_2, \cdots, \mathcal{I}_j)$ a combination of quality indicators, and $E{:}\mathbb{R}^j \mathrm{x} \mathbb{R}^j \mapsto \{false, true\}$ a Boolean function which takes two real vectors of length j as arguments. If all indicators in $\mathcal{I}$, the comparison method $\ddot{\mathcal{C}}_{\mathcal{I},E}$ defined by $\mathcal{I}$ and E is a Boolean function of the form

$$\ddot{\mathcal{C}}_{\mathcal{I},E} = E(\mathcal{I}(A), \mathcal{I}(B)) \qquad (5.10)$$
$$= E(\mathcal{I}(A_1, \cdots, A_{\hat{i}}), \mathcal{I}(B_1, \cdots, B_{\hat{j}}))$$
$$= E(\ \{\mathcal{I}_1(A_1, \cdots, A_{\hat{i}}), \cdots, \mathcal{I}_j(A_1, \cdots, A_{\hat{i}})\},$$
$$\quad \{\mathcal{I}_1(B_1, \cdots, B_{\hat{j}}), \cdots, \mathcal{I}_j(B_1, \cdots, B_{\hat{j}})\}\ )$$

□

Compatible and Complete: Ideally, selection of quality indicators that are both compatible and complete is something that MOEA statisticians strive for when selecting metrics for comparing MOP optimization heuristics; nevertheless, one can define *unary and binary quality indicators* as a mapping from a set (approximation, vectors, etc.) to the set of real numbers for MOEA evaluation. In order to develop and understand these quality indicators, a more precise definition is needed.

Definition 31 (Compatibility and Completeness) : Let ▶ be a binary relation on approximation sets. The comparison method $\ddot{\mathcal{C}}_{\mathcal{I},E}$ is denoted as ▶-compatible if either for $A, B \in \Omega$

$$\ddot{\mathcal{C}}_{\mathcal{I},E} \Rightarrow A \blacktriangleright B$$

or for any $A, B \in \Omega$

$$\ddot{\mathcal{C}}_{\mathcal{I},E} \Rightarrow B \blacktriangleright A$$

The comparison method $\ddot{\mathcal{C}}_{\mathcal{I},E}$ is denoted as ▶-complete if either for any $A, B \in \hat{\Omega}$

$$A \blacktriangleright B \Rightarrow \ddot{\mathcal{C}}_{\mathcal{I},E}$$

or for any $A, B \in \Omega$

$$B \blacktriangleright A \Rightarrow \ddot{\mathcal{C}}_{\mathcal{I},E}$$

Thus, if a set of indicators, $\mathcal{I}$, operating on two approximation sets, A and B, are said to have compatibility and completeness then the following is true:

$$\ddot{\mathcal{C}}_{\mathcal{I},E} \Rightarrow A \blacktriangleright B \Leftrightarrow A \blacktriangleright B \Rightarrow \ddot{\mathcal{C}}_{\mathcal{I},E} \ \text{ and } \ \ddot{\mathcal{C}}_{\mathcal{I},E} \Rightarrow B \blacktriangleright A \Leftrightarrow B \blacktriangleright A \Rightarrow \ddot{\mathcal{C}}_{\mathcal{I},E}$$

Zitzler showed that one metric cannot possibly have this quality [1777]. He showed the best situation is compatibility without any completeness (*i.e.*, $\succ\succ$-compatibility without any completeness, or $\rhd$-compatibility in combination with $\rhd$-completeness.). That means either a strong statement can be made (the evaluated individuals of A strongly dominate the evaluated individuals of B) for only a few pairs of evaluated members of $A \rhd B$; or weaker statements can be made (the evaluated individuals of A are not worse than the evaluated individuals of B (*i.e.*, $A \succeq B$ or $A \parallel B$) for all pairs $A \succ B$ [1777]). See Table 5.2 on page 244 for definitions of symbols described.

Knowles presented a similar idea on the performance assessment of stochastic multiobjective optimizers [882] in which he recommends having quality indicators that are only *Pareto dominance compliant* or *Pareto compliant*, $\rhd$, to guarantee that one algorithm's results are at least better than another before calling that algorithm itself better. This test is presented within Table 5.2 and called the *Better* relation. Indicators should be chosen to reduce the dimension of approximation sets while respecting the dominance compliance. Various quality indicators are defined in the next section.

Thus, any metric that yields a preference for an approximation set A over another approximation set B, when $A \rhd B$, is unreliable. But when the indicator cannot yield a preference for an approximation set A over B, when $A \rhd B$, then it is reliable [882]. Nevertheless, a set of MOEA quality metrics or indicators is suggested. As indicated in the beginning of this chapter, the term performance measure usually refers to both effectiveness (quality) and efficiency (time), while the measures proposed in the literature usually capture only the former aspect. Therefore, some authors use the term "quality indicator" instead of "metric" [874].

It is generally desired to maintain consistency with the inherent structure of the optimization problem under consideration. Thus, the total order of Ω imposed by the choice of the quality indicator function should not contradict the partial order of Ω that is imposed by the weak Pareto dominance relation. That is, whenever an approximation set A is preferable to an approximation set B with respect to weak Pareto dominance, the indicator value for A should be at least as good as the indicator value for B; such indicators are *Pareto compliant*.

Definition 32 (Pareto Compliant) : *An indicator $I : \Omega \to \mathbb{R}$ is Pareto compliant if for all $A, B \in \Omega : A \preceq B \Rightarrow I(A) \geq I(B)$, assuming that greater indicator values correspond to higher quality (otherwise $A \preceq B \Rightarrow I(A) \leq I(B)$). In the context of order theory, a Pareto compliant indicator I is an order-preserving function from $(\Omega, \preceq)$ to $(\mathbb{R}, \geq)$ (respectively $(\mathbb{R}, \leq)$).*

Pareto compliant indicators define refinements of the partial order induced by weak Pareto dominance. Observe that many of the indicators that are employed in the MOEA literature are not Pareto compliant. Several popular indicators are designed to assess just one isolated aspect of an approximation set's quality, such as its proximity to the Pareto optimal front, or its spread in objective space, or the evenness with which the points in it are distributed. These quality indicators, sometimes referred to as 'functionally-independent' indicators are by definition Pareto noncompliant.

Definition 33 (Pareto Noncompliant) : *Any indicator that can yield for any approximation sets $A, B \in \Omega$ a preference for A over B, when B is preferable to A with respect to weak Pareto dominance ($B \preceq A \wedge \neg A \preceq B$, or $B \triangleleft A$ for short), is Pareto noncompliant.* □

Various indicators in the following sections are classified as Pareto compliant and Pareto noncompliant. Of course, Pareto noncompliant indicators are still useful for optimization scenarios based on (weak) Pareto dominance; for instance, they may be used to refine the preference structure of a Pareto compliant indicator for approximation sets having identical indicator values. Furthermore, there may be other multiobjective optimization problems that are not based on weak Pareto dominance and for which such an indicator is appropriate - provided it does not contradict the partial order definition. Finally, note that this discussion is restricted to unary quality indicators only, although an indicator can take an arbitrary number of approximation sets as arguments. Several quality indicators have been proposed that assign real numbers to pairs of approximation sets [1783]. For instance, the unary hypervolume indicator can be extended to a binary quality indicator by defining $I_H(A, B)$ as the hypervolume of the subspace of the objective space that is dominated by A but not by B.

Observe that although the concept and definition of Pareto dominance is always scale independent and normalization independent, scaling and normalization of quality indicator functions is normally necessary to allow different objectives to contribute equally to comparative indicator values.

5.4.4 Primary Quality Indicators

Suggested MOEA primary quality indicators are discussed in this section. Some are Pareto compliant, others are Pareto noncompliant. Other possible quality indicators are presented in Section 5.4.5 in that they have measurement overlap with the primary indicators. In general, MOEA indicators reflect a cardinality value related to the number of solution points, a real value for proximity to the Pareto front or other set of points, or real values for measuring PF_{known} characteristics (diversity of points, uniformity of points, etc.). Also, observe that some metrics are easy to understand and compute, others need to be normalized and scaled (linear, nonlinear) in order to generate comparative values. Still others have issues associated with measuring distances for problem non-commensurable objectives. For some NP-complete problems such as the knapsack MOP, one can use the absolute values of each objective.

In order to select an appropriate set of *quality indicators* with associated statistical values requires insight to formal quality metric relationships as discussed. Again, observe that many indicators or metrics are not reliable in that they can violate dominance ordering. Combining unreliable metrics does not of course make the overall measurement better. Nevertheless, various quality indicators provide performance insight. Of course, dominance can be measured by ranking or nondominated sorting as discussed in the previous chapter resulting in different statistical measured values. Ranking is generally preferred since it is a finer grain approach. A rank test can be applied to determine if there is a statistical significant distribution difference between MOEA generations or comparative effectiveness of various MOEAs.

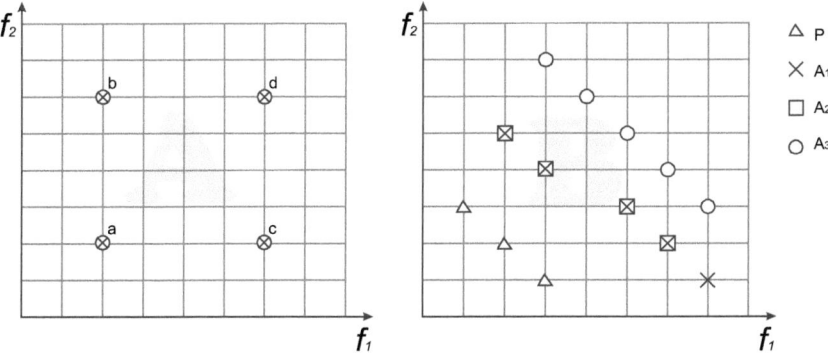

Fig. 5.3. Graphic examples of minimization dominance relations on objective vectors and approximation sets.

From the symbols defined in Table 5.2 and Figure 5.3, the following objective vector relationships hold: $a \succ b$, $a \succ c$, $a \succ d$, $b \succ d$, $c \succ d$, $a \succ\succ d$, $a \succeq a$, $a \succeq b$, $a \succeq c$, $a \succeq d$, $b \succeq b$, $b \succeq d$, $c \succeq c$, $c \succeq d$, $d \succeq d$, and $b \parallel c$.

Also from the symbols defined in Table 5.2 and Figure 5.3 the following approximation set dominance relationships hold for algorithm results A_1, A_2, and A_3 having a PF_{true} of P: $A_1 \succ A_3$, $A_2 \succ A_3$, $A_1 \succ\succ A_3$, $A_1 \succeq A_1$, $A_1 \succeq A_2$, $A_1 \succeq A_3$, $A_2 \succeq A_2$, $A_2 \succeq A_2$, $A_3 \succeq A_3$, $A_1 \triangleright A_2$, $A_1 \triangleright A_3$, and $A_2 \triangleright A_3$.

Error Ratio (ER): The Error Ratio (ER) metric reports the number of vectors in PF_{known} that are not members of PF_{true} [1626, 1630]. This metric which is Pareto compliant, requires that PF_{true} is known and that the MOEA approaches the Pareto front. Mathematically, this metric is represented in equation (5.11):

$$\mathbf{ER} \triangleq \frac{\sum_{i=1}^{|PF_{known}|} e_i}{|PF_{known}|} \quad (5.11)$$

where e_i is zero when the ith vector of PF_{known} is an element of PF_{true} or e_i is one if the ith vector of PF_{known} is not an element of PF_{true} [287].

If $ER = 0$, the PF_{known} is the same as PF_{true}; but when $ER = 1$, this indicates that none of the points in PF_{known} are in PF_{true}. A lower ER is better. The example in Figure 5.4 has $E = \frac{2}{3}$. A similar metric [1781, 1782] measures the "percentage of solutions" in some set (e.g., P_{known} or PF_{known}) dominated by another solution set's members (e.g., P_{true} or PF_{true})

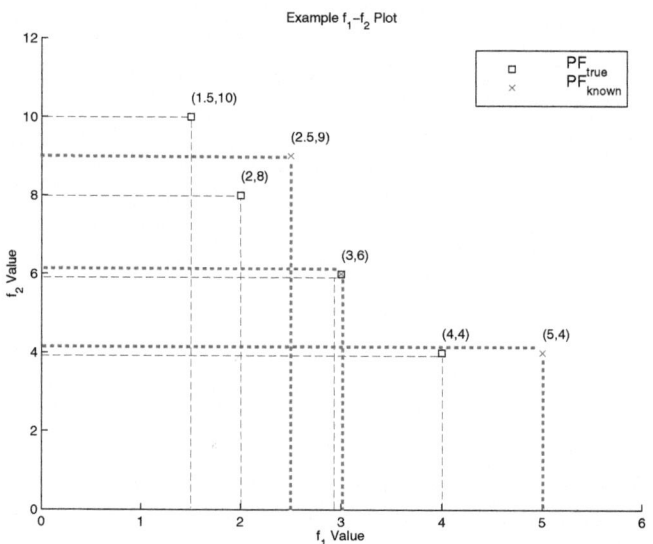

Fig. 5.4. PF_{known}/PF_{true} Example

Generational Distance (GD): The Generational Distance (GD) reports how far, on average, PF_{known} is from PF_{true} [287, 1627, 1630]. This metric which is Pareto noncompliant requires that the PF_{true} be known. It is mathematically defined in equation (5.12).

$$\mathbf{GD} \triangleq \frac{(\sum_{i=1}^{n} d_i^p)^{1/p}}{|\text{PF}_{known}|} \quad (5.12)$$

where $|\text{PF}_{known}|$ is the number of vectors in PF_{known}, $p = 2$, and d_i is the Euclidean phenotypic distance between each member, i, of PF_{known} and the closest member in PF_{true} to that member, i. When $GD = 0$, $\text{PF}_{known} = \text{PF}_{true}$. The example in Figure 5.4 has $d_1 = \sqrt{(2.5-2)^2 + (9-8)^2}$, $d_2 = \sqrt{(3-3)^2 + (6-6)^2}$, $d_3 = \sqrt{(5-4)^2 + (4-4)^2}$, and $G = \sqrt{1.118^2 + 0^2 + 1^2}/3 = 0.5$.

Observe that with $p = 1$ the absolute error is employed in the metric which has a linear relationship. For any p value, regions of the Pareto front can use the same equation and then a weighted sum can be obtained across the entire Pareto front. Also, the kernel can be modified as $(drel_i - d_{ave})$ for a relative comparison where $drel_i$ is the relative distance between two consecutive PF_{known} fronts for the last two generations, and d_{ave} is the average of the distances $drel_i$ across a region. This is similar to an empirical convergence metric. Some authors have propose the use of an Inverted Generational Distance metric [273], in which distances are measured from the true Pareto front to the Pareto front obtained by a MOEA. This aims to reduce some of the main problems of this metric in cases in which, for example, PF_{known} has very few points, but they all are clustered together.

Schott proposes a "7-Point" average distance measure that is similar to generational distance [1452]. In his experiments neither P_{true} or PF_{true} are known, so he generates seven points (vectors) in objective space for comparison. Assuming a bi-objective minimization MOP and an (f_1, f_2) coordinate system with origin at $(0,0)$. We first have to determine the maximum value in each objective dimension. Two equidistantly spaced points are then computed between the origin and each objective's maximum value (on the objective axis). The "full" measure is then created by averaging the Euclidean distances from each of the seven axis points to the member of PF_{known} closest to each point. Given a general bi-objective minimization MOP $F(\mathbf{x}) = (f_1(\mathbf{x}), f_2(\mathbf{x}))$, the seven points are:

$$\{(0, (\max f_2(\mathbf{x}))/3), (0, 2*(\max f_2(\mathbf{x}))/3), (0, (\max f_2(\mathbf{x}))), (0,0),$$
$$((\max f_1(\mathbf{x}))/3, 0), (2*(\max f_1(\mathbf{x}))/3, 0), ((\max f_1(\mathbf{x})), 0)\} \quad (5.13)$$

Hyperarea and Ratio (HA,HR): The hyperarea (hypervolume) and hyperarea ratio metric which are Pareto compliant relate to the area of coverage of PF_{known} with respect to the objective space [287, 1781] for a two-objective MOP. This equates to the summation of all the rectangular areas, bounded by some reference point and $(f_1(\mathbf{x}), f_2(\mathbf{x}))$. Mathematically, this is described in equation (5.14):

$$\mathbf{HA} \triangleq \left\{ \bigcup_i area_i | vec_i \in \mathrm{PF}_{known} \right\} \quad (5.14)$$

where vec_i is a nondominated vector in PF_{known} and $area_i$ is the area between the origin and vector vec_i. The example in Figure 5.4 has an $HR = \frac{33.5}{29} = 1.155$. It is important to note that if PF_{known} is not convex, the results may be misleading [1627]. It is assumed that the reference point for the hyperarea is the minimum value for each objective. Note, the Hypervolume (HV) and hyperarea measurements are similar, except the HV can be used with dimension above two.

Also proposed is a *hyperarea ratio* metric defined as:

$$\mathrm{HR} \triangleq \frac{H_1}{H_2}, \qquad (5.15)$$

where HA_1 is the PF_{known} hyperarea and HA_2 is the hyperarea of PF_{true}. Using the Pareto fronts in Figure 5.4 as an example, the rectangle bounded by $(0,0)$ and $(4,4)$ has an area of 16 units. The rectangle bounded by $(0,0)$ and $(3,6)$ then contributes $(3*(6-4)) = 6$ units to the measure, and so on. Thus, PF_{true}'s $H = 16+6+4+3 = 29$ units^2, and PF_{known}'s $H = 20+6+7.5 = 33.5$ units^2. Zitzler and Thiele do note that this metric may be misleading if PF_{known} is non-convex [1781]. They also implicitly assume the MOP's objective space origin coordinates are $(0, \ldots, 0)$, but this is not always the case. The vectors in PF_{known} can be translated to reflect a zero-centered origin, but as each objective's ranges may be radically different between MOPs, optimal H values may vary widely.

Implementation of the hyperarea metric is considered only for maximization MOPs. Thus, **HR** values less than one indicate a found Pareto front that is not as good as the true Pareto front. When HR equals one, then $\mathrm{PF}_{known} = \mathrm{PF}_{true}$. Of course, this metric generally requires that PF_{true} be known.

Since Pareto fronts are generally unknown, two approximation reference set techniques are suggested [882]. First, all approximation sets generated by the algorithms under consideration are combined, and then the dominated objective vectors are removed from this union. The remaining points, which are not dominated by any of the approximation sets, form the reference set. Second, it is proposed to use a reference set that dominates 50% of solutions in the search space as a kind of median reference set. To this end, a certain number of points (e.g., 1000) are randomly created, each one representing the outcome of one run of a random search strategy, and then the 50% attainment surface of these 1000 artificial runs is taken as the reference set.

The advantage of the first approach is that the reference set weakly dominates all approximation sets under consideration; however, whenever additional approximation sets are included in the comparison, the reference set needs to be re-computed. The second approach avoids this problem, but also gives a slightly different picture: it measures the quality of an approximation set with respect to a reference set independent of any algorithm; it is also 'nearly' independent of how many points are sampled, so that 1000 points should be enough for its location to have converged. When communicating the results of an experiment, it is strongly advised to communicate the reference point and reference sets used, in addition to the indicator values. This means that others can compare their indicator values, computed using the same reference point and reference set, directly with the ones reported in the study, without access to the approximation sets.

5.4 MOEA Experimental Measurements

Spacing (S): The spacing (S) metric numerically describes the spread of the vectors in PF_{known} [287, 1452]. This Pareto noncompliant metric measures the distance variance of neighboring vectors in PF_{known}. Equations (5.16) and (5.17) define this metric.

$$\mathbf{S} \triangleq \sqrt{\frac{1}{|PF_{known}| - 1} \sum_{i=1}^{|PF_{known}|} (\bar{\mathbf{d}} - \mathbf{d_i})^2} \qquad (5.16)$$

and

$$\mathbf{d_i} = min_j(|f_1^i(\mathbf{x}) - f_1^j(\mathbf{x})| + |f_2^i(\mathbf{x}) - f_2^j(\mathbf{x})|) \qquad (5.17)$$

where $d_i = \min_j(|f_1^i(\mathbf{x}) - f_1^j(\mathbf{x})| + |f_2^i(\mathbf{x}) - f_2^j(\mathbf{x})|)$, $i, j = 1, \ldots, n$, $\bar{d}$ is the mean of all d_i, and n is the number of vectors in PF_{known}. When $S = 0$, all members are spaced evenly apart. Note that this becomes important in the deception problems where all Pareto front vectors are equally spaced. This metric does *not* require the researcher to know PF_{true}, although it is normally assumed that a MOEA has already converged prior to applying this metric. Most experimental MOEAs perform fitness sharing (niching or crowding) in an attempt to spread each generational population ($PF_{current}(t)$) evenly along the known front. Because PF_{known}'s "beginning" and "end" are known, a suitably defined metric judges how well PF_{known} is distributed.

Overall Nondominated Vector Generation (ONVG): The Overall Nondominated Vector Generation (ONVG) measures the total number of nondominated vectors found during MOEA execution [1630, 1626]. This Pareto noncompliant metric is defined as:

$$\mathbf{ONVG} \triangleq |PF_{known}| \qquad (5.18)$$

Overall Nondominated Vector Generation Ratio (ONVGR): Overall Nondominated Vector Generation Ratio (ONVGR) measures the ratio of the total number of nondominated vectors found PF_{known} during MOEA execution to the number of vectors found in PF_{true}. Van Veldhuizen [1626] defines this Pareto noncompliant metric as shown in equation (5.19):

$$\mathbf{ONVGR} \triangleq \frac{|PF_{known}|}{|PF_{true}|} \qquad (5.19)$$

When $ONVGR = 1$, this states only that the same number of points have been found in both PF_{true} and PF_{known}. It does not infer that $PF_{true} = PF_{known}$. This metric requires that the researcher knows PF_{true}. Schott [1452], uses this unreliable metric (although defined over the Pareto optimal set, i.e., $|P_{known}|$). Genotypically or phenotypically defining this metric is probably a matter of preference, but again note multiple solutions may map to an identical vector, or put another way, $\mid P_{known}\mid\geq\mid PF_{known}\mid$. Although counting the number of nondominated solutions gives some feeling for how effective the MOEA is in generating desired solutions, it does not reflect on how "far" from PF_{true} the vectors in PF_{known} are. Additionally, too few vectors and PF_{known}'s representation may be poor; too many vectors may overwhelm the distance measure.

It is difficult to determine what good values for $ONVG$ might be. PF_{known}'s cardinality may change at various computational resolutions as well as differing (perhaps radically) between MOPs. Reporting the ratio of PF_{known}'s cardinality to the discretized P_{true}'s gives some feeling for the number of nondominated vectors found versus how many exist to be found.

The example in Figure 5.4 has an $ONVG = 3$ and an $ONVGR = 0.75$.

Maximum Pareto Front error (ME): It is difficult to measure how well a set of prototype vectors compares to another. For example, in comparing PF_{known} to PF_{true}, one wishes to determine how far "apart" the two sets are and how well they conform in shape. The Maximum Pareto Front error (ME) measures how well a set of vectors compares to another [1630, 1626]. This particular Pareto noncompliant metric determines a maximum error band which when considered with respect to PF_{known}, encompasses every vector in PF_{true}. More specifically, it measures the largest minimum distance between each vector in PF$_{known}$ and the corresponding closest vector in PF$_{true}$. Equation (5.20) presents this metric mathematically.

$$\mathbf{ME} \triangleq max_j \left\{ \left\{ min_i \left(\sum_{k=1}^{m} |f_k^i(\mathbf{x}) - f_k^j(\mathbf{x})|^p \right)^{\frac{1}{p}} \right\} \right\} \quad (5.20)$$

where $i = \{1,\cdots,|\text{PF}_{known_1}|\}$ and $j = \{1,\cdots,|\text{PF}_{known_2}|\}$ index vectors in PF$_{known}$ and PF$_{true}$, respectively. A resultant of 0 indicates PF$_{known} \subseteq$ PF$_{true}$. Any other resultant value indicates that at least one vector of PF$_{known}$ is not in PF$_{true}$. The vectors in Figure 5.4's P_{known} are $1.118, 0$, and 1 units away from the closest vector in P_{true}. Thus, $ME = 1.118$.

The *coverage error* of Sayin [1437] is identical to ME but is generalized to a continuous Pareto front. It is also unreliable.

Hypervolume (HV): The hypervolume Pareto compliant indicator is defined as the area of coverage of PF$_{known}$ with respect to the objective space [287, 1781] for a two-objective MOP. This equates to the summation of all the rectangular areas, bounded by some reference point and $(f_1(\mathbf{x}), f_2(\mathbf{x}))$. Mathematically, this is described in equation (5.21):

$$\mathbf{HV} \triangleq \left\{ \bigcup_i vol_i | vec_i \in \text{PF}_{known} \right\} \quad (5.21)$$

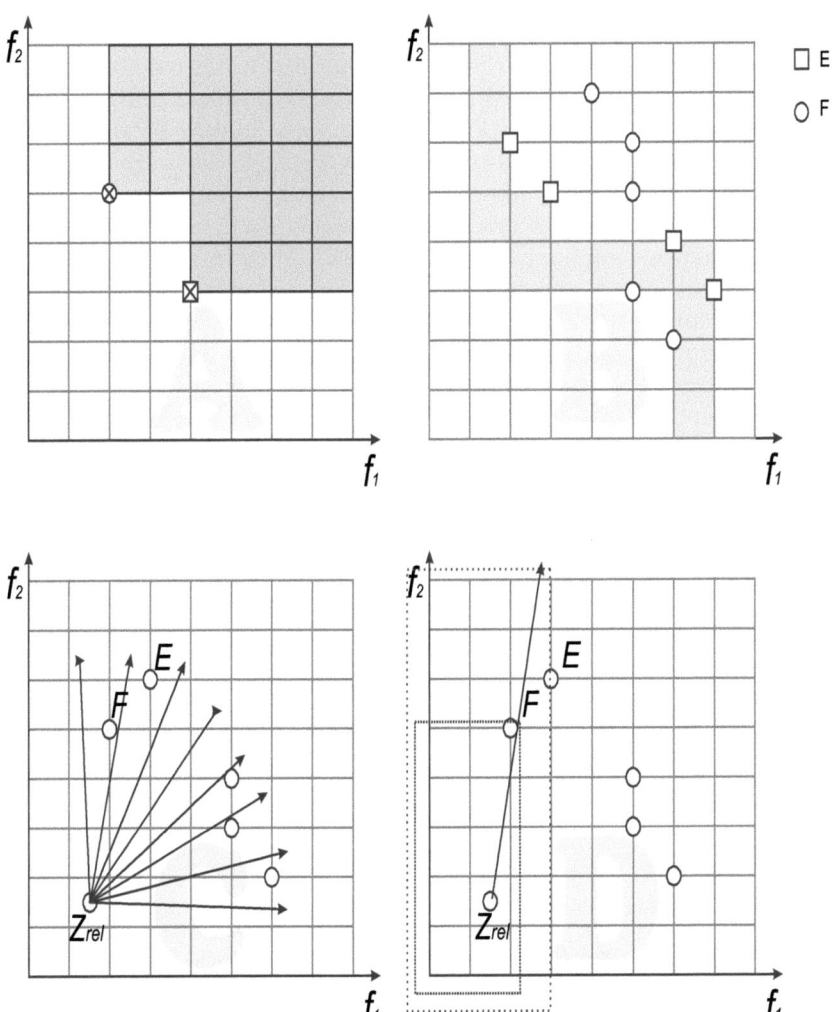

Fig. 5.5. Illustrated in this figure are the hypervolume, R_2, R_3 and ϵ indicators as described in a presentation held at EMO 2005 [882]. Graphic "A" indicates a minimization MOP and an example of how the hypervolume is calculated. Graphic "B" shaded area indicates how the ϵ indicator calculates how far Pareto front A must move in each objective to cover Pareto front B (*i.e.*, How far must the vectors resulting from evaluating individuals of A be moved to dominate the vectors resulting from evaluating individuals of B in all objectives). Finally, graphics "C" and "D" illustrate how the utility functions of R_2 and R_3 are rendered. Graphic "C" illustrates how the vectors are evenly spread out from the worst reference point to the best reference point. Graphic "D" illustrates the difference is calculated with respect to each vector.

This indicator is the same as the hyperarea metric discussed above, but this indicator does go beyond two dimensions and substitutes vol_i for the $area_i$ in equation (5.21). Graphic **A** in Figure 5.5 illustrates how the hyperarea is calculated for a minimization MOP from two approximation sets, A and B. All MOPs are translated to maximization MOPs if not already defined as such; therefore, the areas are summed up from the bottom left.

The hypervolume indicator, described in Section 5.4.4, can be used as the basis of a dominance compliant comparison [1777, 882]. In fact, given the results of two algorithms, E and F, it is shown that all cases where E is better than F are detected by this indicator while respecting the dominance compliance. Also, suggested indicators are the epsilon indicator, R_2, and R_3 indicators, which are described next. Each indicator is based on different preference information; therefore, using them all provides a range comparisons rather than just one.

ϵ-indicator: Given two approximate sets, A and B, this ϵ-indicator measures the smallest amount, ϵ, that must be used to translate the set, A, so that every point in B is covered. The **B** in Figure 5.5 illustrates how far A must move to cover B. This is a Pareto compliant quality indicator. Formally, this measure is defined as a:

Definition 34 (ϵ-indicator) : *Let $A, B \subseteq X$. Then, the ϵ-indicator $I_\epsilon(A, B)$ is defined as the minimum $\epsilon \in \mathbb{R}$ such that any solution $b \in B$ is ϵ-dominated by at least one solution $a \in A$:*

$$I_\epsilon(A, B) = \min\{\epsilon \in \mathbb{R} | \forall b \in B \exists a \in A : a \succ_\epsilon b\} \tag{5.22}$$

□

So, when $I_\epsilon(A, B) < 1$, all solutions in B are dominated by a solution in A. If $I_\epsilon(A, B) = 1$ and $I_\epsilon(B, A) = 1$, then A and B represent the same Pareto front approximation. If $I_\epsilon(A, B) > 1$ and $I_\epsilon(B, A) > 1$, then A and B are incomparable (i.e., they both contain solutions not dominated by the other set).

R_R Indicators: The final two suggested Pareto compliant indicators are R_2 and R_3 utility indicators [653]. There is a third indicator, R_1, in this same class; however it is not a member of the basic test criteria (see next section). The utility, $u(A, \tilde{\lambda})$, of the approximation set A, on scalarizing vector, $\tilde{\lambda}$, is the minimum distance of a point in the set, A, from the reference point. Equations (5.23) and (5.24) mathematically define these two indicators.

$$\mathcal{I}_{R2} = \frac{\sum_{\tilde{\lambda} \in A} u(\tilde{\lambda}, B) - u(\tilde{\lambda}, A)}{|\tilde{\lambda}|} \tag{5.23}$$

$$\mathcal{I}_{R3} = \frac{\sum_{\tilde{\lambda} \in A} [u(\tilde{\lambda}, B) - u(\tilde{\lambda}, A)]/u(\tilde{\lambda}, B)}{|\tilde{\lambda}|} \tag{5.24}$$

5.4 MOEA Experimental Measurements

Graphically, R_2 and R_3 are illustrated in Figure 5.5.C and 5.5.D. These utility functions, u, require a reference point and a user-specified number of scalarizing vectors, $\tilde{\lambda}$. Vectors are uniformly distributed across the objective space. The distance of the point (in each set) that is closest to the reference point is measured and the differences in these distances are added up. In order to obtain an indicator from these two indicators, the set, B, is replaced with a reference set containing the true Pareto front points, R. These indicator functions then effectively measure the difference in the mean distance of the attainment surfaces A and R from a user-defined reference point.

Table 5.3 lists the suggested seven MOEA metrics and the three indicators to use when comparing MOEAs. The table indicates whether each metric/indicator requires PF_{true} and explicitly compares results from one generation to another.

Table 5.3. Summary of the ten Suggested MOEA Metrics/Indicators

	Metric Name	PF_{true} required?	Generational Metric?
1	Error Ratio (ER)*	Yes	No
2	Generational Distance (GD)**	Yes	Yes
3	Hyperarea Ratio (HR)*	Yes	No
4	Spacing (S)**	No	No
5	Overall Nondominated Vector Generation (ONVG)**	No	Yes
6	ONVG Ratio (ONVGR)**	Yes	Yes
7	Max PF error **	Yes	No
$\mathcal{I}1$	ϵ indicator *	No	No
$\mathcal{I}2$	Utility R_2 indicator *	Yes/No	No
$\mathcal{I}3$	Utility R_3 indicator *	Yes/No	No

* - Pareto Compliant, ** - Pareto Noncompliant

With these ten MOEA quality measures, a generic MOEA metric set is provided that in general permits extensive comparison of MOEAs across a multitude of MOPs. Although other measures in the next section have also been suggested that reflect individual views.

5.4.5 Other MOEA Quality Indicators

What other metrics might adequately measure a MOEA's results or allow meaningful comparisons of specific MOEA implementations? Additional metrics can be selected upon which to base MOEA performance claims, and as the literature offers few quantitative MOEA metrics, proposed metrics must be carefully defined to be useful. Additionally, no single metric can entirely capture total MOEA performance, as some measure algorithm effectiveness and others efficiency. Temporal effectiveness and efficiency may also be judged, e.g., measuring a MOEA's progress at each generation. All may be considered when judging a MOEA against others. Following are possible metrics developed for use in analyzing these experiments, but they should not be considered a complete list although they represent in part those found in the literature as indicated.

The metrics identified in this section generally measure performance in the phenotype domain. Whereas Benson and Sayin indicate many OR researchers attempt to generate P_{true} and thus implicitly measure performance in genotype space [119], MOEA researchers have mainly focused on generating PF_{true} (and thus measure performance in phenotype space). As there is a direct correspondence between solutions in P_{true} and vectors in PF_{true} one method may not be "better" than another. However, note that multiple solutions may map to an identical vector.

Although described in terms of measuring final MOEA performance, many of these metrics may also be used to track performance of generational populations. This utilization indicates performance during execution (e.g., rate of convergence to the MOEA optimum) in addition to an overall performance metric. Most examples presented use two-objective examples and associated metrics for ease of manipulation. Of course, the metrics may be extended to MOPs with an arbitrary number of objective dimensions. However, in so doing, the reliability of the metric becomes questionable just due to the accuracy of the computations.

Two Set Coverage (CS): Zitzler et al. [1772] propose a MOEA comparative metric which can be termed *relative coverage comparison of two sets*. Consider $X', X'' \subseteq X'$ as two sets of phenotype decision vectors. CS is defined as the mapping of the order pair (X', X'') to the interval $[0, 1]$ per equation (5.25).

$$CS(X', X'') \triangleq \frac{|\{a'' \epsilon X''; \exists a' \epsilon X' : a' \succeq a''\}|}{|X''|} \quad (5.25)$$

If all points in X' dominate or are equal to all points in X'', then by definition $CS = 1$. $CS = 0$ implies the opposite. In general, $CS(X', X'')$ and $CS(X'', X')$ both have to be considered due to set intersections not being empty. Of course, this metric can be used for X = P_{known} or PF_{known}. The advantage of this Pareto compliant metric is that it is easy to calculate and provides a relative comparison based upon dominance numbers between generations or MOEAs. Observe that it is not a distance measure of how close these sets are as that is a different metric.

Average Pareto Front Error: This Pareto noncompliant metric also attempts to measure the convergence property of a MOEA by using distance to PF_{true}. From each solution in PF_{known}, its perpendicular distance to PF_{true} is determined by approximating PF_{true} as a combination of piecewise linear segments with the average of these distances defining the metric value. For example, Deb et al. [363, 374] use 500 segments. One could also use the medium.

Distributed Spacing (ι): Srinivas & Deb [1509], define a similar measure expressing how well a MOEA has distributed Pareto optimal solutions over a nondominated region (the Pareto optimal set). This Pareto noncompliant metric is defined as:

$$\iota \triangleq (\sum_{i=1}^{q+1} (\frac{n_i - \overline{n}_i}{\sigma_i})^p)^{1/p} , \qquad (5.26)$$

where q is the number of desired optimal points and the $(q+1)$-th subregion is the dominated region, n_i is the actual number of individuals in the ith subregion (niche) of the nondominated region, $\overline{n}_i$ is the expected number of individuals in the ith subregion of the nondominated region, $p = 2$, and σ_i^2 is the variance of individuals serving the ith subregion of the nondominated region. They show that if the distribution of points is ideal with $\overline{n}_i$ number of points in the ith subregion, the performance measure $\iota = 0$. Thus, a low performance measure characterizes an algorithm with a good distribution capacity. This metric may be modified to measure the distribution of vectors within the Pareto front. In that case both metrics (S and ι) then measure only uniformity of vector distribution and thus complement the generational distance and maximum Pareto front error metrics.

Progress Measure (P, RP): Bäck defines a parameter used in assessing single-objective EA convergence velocity called a *Progress Measure* [72], which quantifies *relative* rather than *absolute* convergence improvement by:

$$P \triangleq \ln \sqrt{\frac{f_{max}(0)}{f_{max}(T)}}, \qquad (5.27)$$

where $f_{max}(i)$ is the best objective function value in the parent population at generation i.

To account for the (possible) multiple solutions in P_{known}, this definition is modified as follows:

$$RP \triangleq \ln \sqrt{\frac{G_1}{G_T}}, \qquad (5.28)$$

where G_1 is the generational distance at generation 1, and G_T the distance at generation T.

Generational Nondominated Vector Generation (GNVG): This Pareto noncompliant metric tracks how many nondominated vectors are produced at each MOEA generation and is defined as:

$$GNVG \triangleq | PF_{current}(t) | . \qquad (5.29)$$

Nondominated Vector Addition (NVA): As *globally* nondominated vectors are sought, one hopes to add new nondominated vectors (that may or may not dominate existing vectors) to PF_{known} each generation. This Pareto noncompliant metric is then defined as:

$$NVA \triangleq |\,PF_{known}(t)\,| - |\,PF_{known}(t-1)\,|\,. \tag{5.30}$$

However, this metric may be misleading. A single vector added to PF_{known} (t) may dominate and thus remove several others. PF_{known} (t)'s size may also remain constant for several successive generations even if $GNVG \neq 0$.

R1 Indicator: The $R1$ metric [651] calculates the probability that an approximation set A is better than a set B over a set of utility functions, $\mathbf{U}$, and $R1_R$ is identical to $R1$ when it is used with a reference set.

$$R1(A, B, \mathbf{U}, p) = \left\{ \int_{u \in \mathbf{U}} C(A, B, u) p(u) du, \right\}$$

subject to:

$$\left[C(A, B, u) = \begin{cases} 1 & : \ if \ u(A) > u(B) \\ 1/2 & : \ if \ u(A) = u(B) \\ 0 & : \ if \ u(A) < u(B) \end{cases} \right] \tag{5.31}$$

With this measure, A is better than B if $R1 > \frac{1}{2}$, A is not worse than B if $R1 \geq \frac{1}{2}$. Notice that $R1(A, B, U, p) = 1 - R1(B, A, U, p)$. With A and B the two approximation sets, U is a set of utility functions, $u : \mathbb{R}^k \mapsto \mathbb{R}$. This function maps each point in objective space into a measure of utility, $p(u)$ is the probability density of the utility $u \in \mathbf{U}$, and $u(A) = max_{z \in A}\{u(z)\}$ and also for $u(B)$. Joshua Knowles suggests that the $R1$ indicator requires a set of utility functions which must be defined. Recently, Fonseca et. al. [513] presented a method for defining the utility functions for $R1$. Furthermore, Zydallis [1788] stated that an indicator such as $R1_R$ uses low computational resources and can differentiate between different levels of complete outperformance if given a reference set. In addition, he stated that these indicators are somewhat complex to understand and require the use and determination of utility functions, reducing the attractiveness of the metric [1788]

Although not mathematically modelled or discussed in detail, other unreliable metrics that may be considered for MOEA performance evaluation are, for example: *average best weight combination, distance from reference set, fraction of Pareto front covered, chi-square deviation indicator, maximum spread, maximum distance of two solutions, deviation from uniform distribution, number of distinct choices* [1783, 882], among others.

5.4.6 MOEA Experimental Metrics Summary

Attainment functions can be generated for graphical analysis and quality indicators mapped from PF_{known}. Although implemented in the phenotype domain these experimental quality indicators or metrics may also be defined in a genotypic fashion. For example, the error ratio, generational distance, spacing, and overall nondominated vector generation metrics are valid when modified to reflect a genotypic basis. Here, PF_{true} is replaced with P_{true} values and dimensionality. However, note that decision variable dimensionality may easily exceed the number of objective dimensions, which may require further metric refinement. In addition, Schott uses three other metrics [1452]: cost function evaluations, clone proportion, and total clones identified. These measures although not relevant to the effectiveness experiments discussed here, they relate to algorithm efficiency.

When experimenting, the number of function (solution) evaluations should be a constant between MOEAs ensuring "equal" computational effort by each; Schott [1452] appears interested only in measuring the results of a single MOEA. No effort to identify clones (previously evaluated solutions) is done during execution. Generally, MOEAs execute quickly for "simple" computational MOPs. However, when compared to many real-world MOPs, where each fitness evaluation may take from minutes to hours, it makes no sense to incorporate the overhead of clone identification within these experiments. Thomas' use of MOEAs in submarine stern design, where each individual's fitness evaluation took about 10 minutes, is a case where clone identification is more useful [1580]. As no clones are identified in these experiments clone proportion is not considered. Additional experiments can easily include these and other metrics as appropriate.

Observe that the metrics (E,G,ME, ONVGR, and RP) require that PF_{true} be known, whereas, the metrics (S, ONVG, P, GNVG, and NVA) are relative to PF_{known} or $PF_{current}$. Thus, the later set of metrics is generally more useful for real-world problems where PF_{true} is not known.

Note that combining quality indicators is reflected in much of MOEA comparative literature. Generally this technique provides excellent quality assessments provided that the indicators are Pareto compliant. In particular, if two Pareto compliant indicators contradict one another on the preference ordering of two approximations sets, then this implies that the two sets are incomparable. However if the indicators used are Pareto noncompliant, a common approach is to assess isolated aspects of a decision maker's preferences with respect to approximation sets, e.g., their proximity to the Pareto front, diversity, evenness, and cardinality, in terms of distinct quality indicators. However, it is possible that all of the indicators judge that the approximation set A is preferable to B, when in fact B is better than A according to Pareto dominance. Thus, the combination of several indicators does not nullify or minimize the impact of their Pareto non-compliance: on the contrary, it can give an unjustified sense of 'security' to the interpretations made.

5.5 MOEA Statistical Testing Approaches

The mean and standard deviation is collected on each metric for each set of results found by both MOEAs. *The central limit theorem allows for the assumption that, after a large number of measurements are taken, a normal distribution for a particular measurement can be assumed.* Once a normal distribution is assumed either a student-t or z test may be performed to compare results. Note that these tests can be superseded because the non-parametric Kruskal-Wallis test (KWtest) errs on the side of caution and can be employed in their place.

If possible, all MOEAs to be statistically compared should be executed on the same computational platform for consistency (compiler, finite word length, speed, etc.). This platform should thus be described. The spectrum of possible performance metrics and statistical testing techniques are presented along with performance visualization approaches.

Timing results are usually not of specific experimental concern except for fitness functions that take considerable computation and fitness function search landscapes that are very rough. For most MOP benchmarks, each MOEA usually executes in a matter of minutes. Empirical observations indicate that MOEAs run in a *MATLAB* environment are slower than those that are compiled codes. Other results are found in referenced papers or reports. Note that MOEAs exhibit roughly the same polynomial computational complexity (see Chapter 6).

5.5.1 Statistical Testing Techniques

MOEAs generate complex phenomena and as such are difficult to predict their performance (effectiveness and efficiency). As indicated, the gap between MOEA theory and practice is wide (see Chapter 6). Of course, one desires to make performance measurements of MOEA process execution and associated results. This is suggested for purposes of dynamic process control via algorithm parameters, preprocessing of data input for testability, and statistical valid interpretation of results. Associated tools could for example be online during MOEA execution or offline (post processing). The section on MOEA software tools provides friendly computational environments for statistical processing of MOEA experimental quantitative data. The statistical analysis then permits qualitative inference as to comparative performance.

Statistical testing of MOEAs begins with understanding of basic statistical concepts such as statistical population, statistical sample, and statistical significance. Basic techniques include order statistics (mean, average, variance, median, max, min, nth quartile deviation,...) confidence intervals, correlation measures (correlation coefficient, student's t test, multiple regression) and testing propositions (F-test and others) as found in copious textbooks. Figure 5.6 presents various methods of displaying individual experimental results per variable value. Such displays give the viewer an understanding of the data variation using the above statistics. One can also add outliers to each vertical element. Non-overlapping variances indicate experimental significance. Also, measurements of genotype and phenotype diversity, fitness variance, fitness histograms, and entropy can be employed. Many examples exist in the literature reflecting the use of these icons to display the MOEA distributions at termination [1773]. Although comparison of distributions is desired, such data may not be available from empirical testing. Normal distributions can not be assumed i.e., use of means and standard deviations first and second order statistics) is not appropriate in many cases. Thus, the use of selected metrics from the proposed list above are appropriate in attempting to represent the effectiveness/efficiency of a MOEA for a given MOP. In addition, the sensitivity of each selected metric should be addressed given the specific MOP regarding PF_{true} approximations.

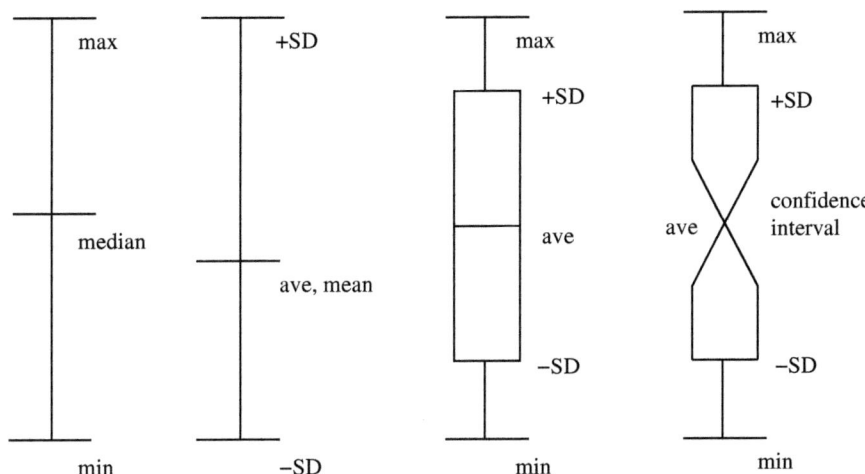

Fig. 5.6. Various techniques of presenting data and variations

As indicated in the Introduction to this chapter, appropriate design of MOEA experiments should draw heavily from guidelines presented by Barr et al. [93] and Jackson et al. [765]. These articles discuss computational experiment design for heuristic methods using the indicated set of steps. In this process, one should determine the use of average and standard deviation vs medium and max and min for a particular test. It is also important to address the number of runs in order to acquire enough statistical data; usually 30 or greater based upon the central limit theorem, but can vary from 10 to 30+ depending upon the MOP domain knowledge.

Also, the experimental results should address the use of analysis-of-variance testing, the t test and others for determining the logic of a given hypothesis. Hypothesis testing techniques indicate that one MOEA is more effective than another over a problem set/class as reflected in Veldhuizen &

Lamont [1630] and the Mann-Whitney rank-sum test [32]. *Hypothesis testing* as to the consistency of statistical data to support the desired *null hypothesis* is computed using a statistical inference test. This test generates the probability, p-value, of supporting the null hypothesis. The user defined significance level, α is the largest acceptable p-value. If the p-value generated by the statistical test is larger than α, the null hypothesis is rejected and an alternative hypothesis is accepted. Regarding MOEA testing, it is desired to evaluate one MOEA's approximation set distribution over another MOEA's distribution using this hypothesis evaluation approach.

5.5.2 Non-Parametric Statistics (Analysis of Variance)

Required for comparisons of MOEA performance is a non-parametric hypothesis inference test because the distribution of the population for metric results is unknown. That is, one assumes that the distribution does not represent a normal distribution for example where only first-order (mean) and second-order statistics (variance) are required. Furthermore, if the population for metric results does turn out to be a normal distribution, other methods may be more accurate at deciding if there is a difference; however, the non-parametric statistics are inaccurate only in errors on the side of caution. Thus, there is no error made if the nonparametric statistic concludes that there is a difference between the two algorithms.

Nonparametric tests consist of two meta-level approaches: rank tests and permutation tests. Rank tests pool the values from several samples and convert them into ranks by sorting them, and then employ tables describing the limited number of ways in which ranks can be distributed (between two or more algorithms) to determine the probability that the samples come from the same source. Permutation tests use the original values without converting them to ranks but estimate the likelihood that samples come from the same source explicitly by Monte Carlo simulation. Rank tests are the less powerful but are also less sensitive to outliers and computationally inexpensive. Permutation tests are more powerful because information is not discarded, and they are also better when there are many tied values in the samples, however they can be expensive to compute for large samples.

There are many statistical tests for MOEA quality indicators that can be used when comparing if two or more algorithms are different (better or worse) from one another. One of the most common non-parametric tests is the Wilcoxon Rank-Sum (Mann-Whitney) test for two independent samples. If matched samples have been collected, then the Wilcoxon signed rank test [1477] or Fisher's matched samples test [1477] can be used instead of the Mann-Whitney rank sum test or respectively Fisher's permutation test. The more general form of the Mann-Whitney test is called the Kruskal-Wallis statistic where h independent samples can be compared. These statistical tests can be performed for each comparable quality indicator using gathered experimental data.

The Kruskal-Wallis H test (KWtest) is the main statistical method used in the determination if two samples are from the same population. An alternative to the one-way independent-samples Analysis of Variance (ANOVA) is the Kruskal Wallis Test. This test is primarily used when no knowledge of the type of distribution is known; however, it can be shown that the sampling distribution of H is nearly a chi-squared[5] distribution with $h-1$ degrees of freedom, given that $\mathcal{N}_1, \mathcal{N}_2, \ldots, \mathcal{N}_h$ sum to at least 5. In all KWtests accomplished, both the Chi-squared-statistic and the F-statistic[6] are evaluated. The definition of the Kruskal-Wallis H Test reflected in equation (5.32).

$$H = \frac{12}{\mathcal{N}(\mathcal{N}-1)} \sum_{i=1}^{h} \frac{\tilde{R}_i^2}{\mathcal{N}_j} - 3(\mathcal{N}+1) \qquad (5.32)$$

- Given
  h sample sizes $\mathcal{N}_1, \mathcal{N}_2, \ldots \mathcal{N}_h \therefore \mathcal{N} = \sum_{i=1}^{h} \mathcal{N}_i$
  h samples are ranked together according to size, therefore the ranks are $\tilde{R}_1, \tilde{R}_2, \ldots, \tilde{R}_h$

Upon calculation of H using equation (5.32), this value, H, is treated as though it were a value of chi-square sampling distribution with the degrees of freedom (df) $= h - 1$. This nonparametric method for analysis of variance is used for a one-way classification, or one-factor experiments, and generalizations can be made.

The Fisher permutation test is a non-parametric test for evaluating differences between two independent (or two matched pair) samples with the result being p-value.

Other more extensive statistical techniques related to analysis of variance include analysis of covariance (ANCOVA); one-way, two-way, and n-way ANOVA; F-ratio related to significance of a null hypothesis; Bonferroni-adjusted multiple t-tests; Dunn test; Sidak test; q range statistic;HSD test (honestly significant difference); Newman-Kuels test; Ryan-test; Least significant difference test (LSD); Scheffe test; Tukey-Kramer test, Miller-Winer test; etc., as well as the use of Latin hypersquares to reduce the number of observations among others necessary to compute ANOVA. One needs to consult with a statistician as to the utility of such detailed statistics for comparing MOEA performance.

The confidence levels resulting from a statistical testing procedure for measuring the differences between distributions only has a meaning if certain assumptions are true. One of these assumptions, which is easy to overlook, is that the data on which the test has been carried out is not being used to make more than one inference. The situation is made even worse if only the

---

[5] Under the null hypothesis that the positive and negative values are equally likely, the test statistic follows the chi-square distribution with $h-1$ degree of freedom.

[6] The F test assumes known population variances of approximately normal distribution and the population variances are homogeneous.

cases reported are where the null hypothesis was rejected, and not reported are other tests: in this case, results can look convincing when, in fact, they are not significant.

Multiple testing issues in the case of assessing stochastic multiobjective optimizers can arise for at least two different reasons: 1) there are more than two algorithms requiring inferences about performance differences between all or a subset of them. 2) there are multiple hypotheses test that are required with the same data, e.g., differences in the distributions of more than one indicator. In order to address this issue six possible approaches are considered: (1) execute all tests as normal (with uncorrected p-values) but report all tests done openly and indicate that the significance levels are not, therefore, Pareto noncompliant; (2) for the special case with multiple algorithms but just one statistic (e.g. one indicator), use a statistical test that is designed explicitly for assessing several independent samples; (3) for multiple statistics (e.g. multiple different indicators) for just two algorithms, relate to an inference derived per-sample from all statistics, (e.g. test the significance of a difference in hypervolume between those pairs A and B where the diversity difference between them is positive), then the permutation test can be used to derive the null distribution, as usual; (4) minimize the number of different tests carried out on the same data by carefully choosing which tests to apply before collecting the data; (5) for each test, generate new independent data which is not to be used for any other test; (6) apply the tests on the same data but use methods for correcting the p-values for the reduction in confidence associated with data re-use [882]. When comparing MOEAs for worst case or best case performance, statistically inference can be found using Fisher's permutation test.

5.5.3 Methods for Presentation of MOEA Results

Pareto front point data and Pareto solution set data can be presented with MOEA comparison tables for any of the selected metrics (see Figure 5.6) as well as histograms comparing MOEAs across metrics, and scatter plots for MOEA comparison across a set of MOPS. Also useful for insight are static and dynamic representations of P_{true} and PF_{true} curves or surfaces for one or more MOEAs per generation or at termination. Use of icons, color and shading for specific MOEA response are appropriate. Since many MOPs are multi-dimensional, slicing across three phenotype or genotype variables is appropriate in 3D. Such visualization can be accomplished with MATLAB, or using other similar tools. Packages such as EXCEL provide spread-sheet statistical data analysis with associated graphical constructs. Many of the graphs in this text were generated using these tools.

5.5.4 Visualization of Test Results

Visualization is considered to be one of the elementary ways to distinguish the difference between two approximation sets. When using this technique the researcher visually looks at the graphical representation of the PF_{known} and PF_{true} set and determines if the results are good/bad or indifferent. For example, graphical analysis can help the researcher determine the cardinality of the set, total number of disjoint fronts, and structure of the front. Advantages of this method are many and can be concluded faster than statistically analyzing the approximation sets. Note that the concept of the *Attainment Function* as presented in Definition 24 is a specific technique for MOEA approximate set visualization.

Many MOEA researchers recognize that visualization of MOEA results provides an easy mechanism to evaluate the general MOEA performance when compared to another reference set. A more detailed analysis using other metrics, like those suggested within this chapter is necessary to statistically compare the performance of multiple MOEAs. One disadvantage to using visualization techniques is that as the dimensionality increases the ability to visually see difference between approximation sets decreases. Since three dimensions is typically the maximum that one can easily visualize (3D is also difficult to recognize differences), it is generally suggested that visual analysis be limited to three objectives.

5.6 Software Support of MOEA Testing

Various archives on the internet provide MOEA software including most of the MOEAs measurements in this and the previous Chapter. Also, the integration of MOEA software with various MOPs and statistical routines is an option available in integrated environments.

For example, The MOMGA and NPGA are extensions of existing algorithms and specific software (the mGA and SGA-C) from the Illinois Genetic Algorithms Laboratory [744]. The NPGA is the original code used by Horn in his MOEA research [708, 709]. Both the MOMGA and NPGA are written in "C" and are compiled using the Sun WorkShop Compiler version C 4.2. Much of the associated research and related experimentation employs the GEATbx v2.0 for use with *MATLAB* [1279]. This toolbox offers the user several "default" EA instantiations (e.g., real- or binary-valued GA, ES, EP) and excellent visualization output to aid in analysis. GEATbx requires only a limited amount of user effort to implement a specific EA. Thus, the MOGA and NSGA are written as self-contained "m-files" using other pre-defined toolbox routines. They were constructed using definitions given in the literature [511, 1509]. These MOEAs can also be executed within the *MATLAB* 5.2 environment.

With the variety of MOEA optimization strategies and evaluation tools, it is increasingly difficult for an application engineer to choose, implement, and apply state-of-the art algorithms without in-depth programming knowledge and expertise in the optimization domain. Also, for a developer of MOEA optimization, methods to test and compare algorithms on different benchmark test problems take considerable software development. In both cases, the main problems are the implementation overhead and potential implementation errors. Contemporary optimization methods usually involve complex operations and require a considerable programming effort; the same holds for the application and test problem side. Thus various MOEA and general optimization software environments have been developed. They include KEA, PISA, Guimoo, and iSIGHT. In most cases, parts of the software environment can be downloaded for integration into the user's unique optimization package.

KEA: A software package for development, analysis and application of multiobjective evolutionary algorithms called KEA (*Kit for Evolutionary Algorithms*) provides an object-oriented design that offers a good suitable environment for various kinds of optimization tasks [96]. It provides an interface to evaluate multi-objective fitness functions written in Java or C/C++ using a variety of multi-objective and single-objective evolutionary algorithms; SPEA2, NSGA-II, MOPSO/DOPS, and Simplex. In addition KEA contains several state-of-the-art comparison methods for performance measure of algorithms. Furthermore KEA is able to display the progress of optimization in a dynamic display or just to display the results of optimization in a static visualization mode. KEA includes pre-defined state-of-the art (e.g., NSGA-II or SPEA2) that can be easily applied to the problem at hand. On the other hand, the researchers who design their own algorithms are interested in comparing and tuning their algorithms. Due to these two scenarios, the following targets have been addressed in the design of KEA:

1. Provide a library of:
   - some state-of-the-art algorithms, which are able to solve MOP,
   - test problems and
   - visualization modes
2. Support of useful analysis methods
3. Extensibility

KEA uses a structure based on classes. The main part of the internal management of the system is done by the KEA - class. This class has the main functionality to coordinate algorithms as well as problems and to control the preparation and formatting of data. The menu driven user interface KEA GUI takes care of the presentation of the functionality that KEA provides. The Parser is used to analyze the input string. In general KEA is a command line based tool. It includes several MOEA test benchmarks as presented in Chapter 4 along with code for the various quality indicators discussed in this chapter. KEA is available at: http://ls11-www.cs.uni-dortmund.de/people/schmitt/Daten/Kea/kea.jsp.

PISA: This *Platform and Programming Language Independent Interface for Search Algorithms* (PISA) includes a text-based interface that allows one to separate the algorithm-specific part of an optimizer from the application specific part [141, 931]. These parts are implemented as independent programs forming freely combinable modules. It is therefore possible to provide these modules as ready-to-use packages. As a result, an application engineer can easily exchange the optimization method and try different variants, while an algorithm designer has the opportunity to test a search algorithm on various problems without additional programming effort. PISA uses a control flow process and a data flow process that the user must integrate for their specific MOP. Again, the package includes such MOEAs as the NSGA-II [374], SPEA2 [1776] and the Indicator Based Evolutionary Algorithm [1774]. Various benchmark test problems are available as libraries. Many of the quality indicators discussed in this paper are integrated along with visualization selections. PISA is available at: http://www.tik.ee.ethz.ch/pisa/.

Guimoo: The *Graphical User Interface for Multi-objective Optimization* (Guimoo) is free software dedicated to the analysis of results in multi-objective optimization. Its main features enable on-line visualization of approximate Pareto frontiers. Such information could be used by the expert to build more efficient metaheuristics. A Pareto frontier may be characterized by its (dis)continuity, (dis)convexity, modality and others. Some metrics for quantitative and qualitative performance evaluation include S-metric, R-metrics, contribution, entropy, generational distance, spacing, size of the dominated space, coverage of two sets and coverage difference. Guimoo aims to be generic and its architecture permits one to easily customize it in order to provide the user more functionalities, as a specific problem is tackled. Beyond generic benchmark problems, problems with related files are supplied for demonstration. They deal with the 'Vehicle Routing Problem', 'Flow Shop Scheduling' and 'Radio Network Optimization.' Guimoo is available at: http://guimoo.gforge.inria.fr/.

iSIGHT: iSIGHT is a desktop productivity tool for engineers and scientists that allows them to integrate key steps in the design process and then automate the execution of those steps through diverse optimization techniques. Various metaheuristics (e.g., the multi-island genetic algorithm and Adaptive Simulated Annealing) and MOEAs (i.e., the NSGA-II [374] and the Neighborhood Cultivation Genetic Algorithm [1684]) are integrated in the package along with other stochastic and deterministic optimization approaches. One can also easily integrate other MOEAs. A considerable number of output graphical techniques are available. iSIGHT is available at: http://www.engineous.com/product_iSIGHT.htm.

PARADISEO: The *PARAllel and DIStributed Evolving Objects* (ParadisEO) is a free C++ white-box object-oriented framework dedicated to the reusable design of parallel metaheuristics for (multi-objective) optimization [193]. It is basically an extension of the EO (Evolving Objects) evolutionary computation framework. It provides a broad range of new features including local searches (Hill Climbing, Simulated Annealing and Tabu Search), the most common parallel models (based on the walk, the solution and the objective function) and some hybridization mechanisms. ParadisEO is based on a clear conceptual separation of the solution methods from the problems they are intended to solve. This separation confers to the user a maximum code and design reuse. A first implementation relies on a multi-programmed layer (Posix threads) and some communication libraries (LAM-MPI or PVM) for execution on dedicated parallel and/or distributed computational resources. Another implementation relies on Athapascan and Inuktitut for the dynamic scheduling on a dedicated grid environment. A recent release is now available. It is based on Condor and the Master/Worker API for High Throughput Computing and Grid Computing on volatile non dedicated resources. There is also a new library called MOEO (*Multi-Objective Evolving Objects*), which provides a multi-objective package on top of EO.[7] PARADISEO is available at: http://www2.lifl.fr/~cahon/paradisEO/index.html.

5.7 Summary

This chapter presents an experimental methodology for quantitatively and qualitatively comparing MOEA performance following the presented generic testing guidelines and techniques. After motivating generic experiments, key methodology components are discussed. Appropriate quality indicators or metrics are proposed, classified, and analyzed, and selected based upon certain criteria. Considering attainment functions, quality indicators, and dominance ranking, there is of course no best MOEA performance assessment approach. It is recommended therefore that the unique and complementary characteristics of all be employed in MOEA testing. Note that a collection of these approaches are found in various software MOEA computational environments discussed at the end of this chapter.

Because of the continuing improvements in the variety of MOEAs, their structures and operators are generally moving towards generating similar results over a large set of indicator values. Consequentially there is no one best "robust" MOEA, but any contemporary sophisticated and evolving MOEA should be able to perform well on real-world multi-objective problems based upon similar operators. Employing appropriate testing techniques as presented should reflect this situation.

---

[7] EO is available at: http://eodev.sourceforge.net/.

Further Explorations

Class Exercises

1. Choose a set of (non-Pareto compliant) quality indicators from those discussed in this chapter, and describe the process required to compare the performance of two MOEAs (e.g., NSGA-II [374] and SPEA2 [1775]) using the following test problem:
   MOP: P_{true} is connected and PF_{true} is concave. The problem will be considered with 3 decision variables, although it is scalable. $F = (f_1(\mathbf{x}), f_2(\mathbf{x}))$, where

   $$f_1(\mathbf{x}) = 1 - \exp(-\sum_{i=1}^{n}(x_i - \frac{1}{\sqrt{n}})^2),$$

   $$f_2(\mathbf{x}) = 1 - \exp(-\sum_{i+1}^{n}(x_i + \frac{1}{\sqrt{n}})^2)$$

   $$-4 \leq x_i \leq 4; \ i = 1, 2, 3$$

   Justify your choice of quality indicators. If PF_{true} is known, does that affect your choice of quality indicators? Discuss.

2. Using the framework of the previous exercise, analyze performance for different values of crossover rate, mutation rate and population size. Show statistical results using mean, median, variance, and quartiles. Do the selected indicators portray the actual performance of the MOEA?

3. Add to Table 5.1 other MOEAs and their associated operators and parameters (see Chapter 2).

4. In Figure 5.2, define those dominance relationships that are contained within others.

5. Repeat problems 1 and 2, but using only Pareto compliant quality indicators.

6. What statistics should be used to test and compare MOEAs? Does the choice depend of the problem domain? Why is it important to use Pareto compliant performance measures?
7. When assessing performance of a MOEA, what are the main issues involved? What do we aim to measure?
8. Analyze performance of a MOEA in a NP-Complete problem (e.g., TSP or knapsack), using a set of selected Pareto compliant indicators such as R_2, R_3, and hypervolume.
9. Generate a table of all possible statistical analysis techniques for evaluating MOEAs and indicate pros and cons of each using appropriate criteria.
10. Review the paper by Knowles and Corne [874] on the utility of MOEA performance metrics as employed in problem 2. Is the human eye a better evaluator in general? Why? or Why not?
11. Regarding the previous problem, describe in as much detail as possible the informal process used by a human eye to evaluate MOEA performance. Hint: Consider what metrics are used by an evaluator when performing a 'sight' evaluation.
12. Setup a MOEA test plan for a specific MOP domain using appropriate guidelines and analysis techniques. What are the various phases of your approach? How much is it going to cost (time and space) to undertake such test plan?
13. How does the "No Free Lunch" (NFL) theorem [1708] relate to performance assessment of MOEAs?

Class Software Projects

1. Download the KEA environment and use it with a set of test functions from those described in Chapter 4. Study the process of problem domain and algorithm domain integration. Also, select appropriate MOEA quality indicators/metrics and evaluate results. Present a written report that details all your implementations and your results with its corresponding analysis.
2. Experiment with PISA [141] using some of the benchmark problems included therein. Add a MOEA not included, and compare its performance to algorithms that are already included (e.g., NSGA-II [374] and SPEA2 [1775]). Write a report in which you comment about ease of use, portability and features of PISA.
3. Use Guimoo[8] to solve the DTLZ test problems [379]. Discuss its ease of use and its flexibility to incorporate new test functions.
4. Evaluate the visualization facilities of iSIGHT[9] using several of the test problems included therein.

---

[8] Guimoo is available at: http://guimoo.gforge.inria.fr/
[9] iSIGHT is available at: http://www.engineous.com/product_iSIGHT.htm

5. Evaluate ParadisEO-MOEO[10] [993], which is a white-box object-oriented generic framework whose aim is to facilitate the design of MOEAs. Discuss its ease of use, its flexibility and the main highlights of its overall architecture.
6. Implement a software tool that allows users to apply any of the performance measures studied in this chapter to the results produced by a MOEA. Design a user interface that facilitates the use of this tool.
7. Analyze the MO-JGA (multi-objective Java Genetic Algorithm) framework proposed in [1086] for rapid development of MOEAs. Adopt this framework to implement a MOEA not currently available, and write a report where you indicate the difficulties (if any) that you had to face when developing your implementation.

Discussion Questions

1. Why might one MOEA be better than another via testing (consider different techniques)? Discuss advantages and disadvantages in achieving the desired inference result of the different techniques considered.
2. How are quality metrics, dominance ranking, and attainment functions affected primary by MOEA architectures and operator parameter values? Discuss.
3. Why are the relatively "simple" MOEA structures adopted in the mid-1990s being replaced by more complex variants nowadays? Identify "generic" MOEA structures in modern approaches. How do they relate to our choice of quality indicators? Discuss.
4. What is the utility of an attainment function over a set of quality indicators? Relate to real-world problems.
5. Try to devise a "new" (Pareto compliant) MOEA performance measure and argue its usefulness and structure. Consider variants of the performance measures discussed in this chapter, as well as innovative ones.
6. What performance measures (from those discussed in this chapter) would you consider appropriate to assess performance of a parallel MOEA (see Chapter 8)? What other performance measures would you propose?
7. How might one test and evaluate a MOEA designed for a specific real-world application in which PF_{true} is unknown? How would you determine if the MOEA is performing well? What sort of performance measures would you adopt in this case?
8. How does weak dominance play a role in the selection of quality indicators for comparing MOEAs?
9. Consider a problem that has only been solved considering it as single-objective and is now modeled as a multi-objective problem. Would you consider appropriate to compare the results obtained by a MOEA for the

---

[10] ParadisEO-MOEO is available at: http://paradiseo.gforge.inria.fr

multi-objective case with those obtained by an EA for the single-objective case? Why?
10. Some authors have shown that, under certain circumstances, an evolutionary algorithm using a linear aggregating function can provide better results than a MOEA (see for example [757]). Discuss possible situations in which one could observe this type of behavior. Why do you think that this happens? How would you compare the performance of a single-objective evolutionary algorithm with respect to that of a MOEA?
11. Consider a set of MOEAs to be compared over a set of test problems, and a set of performance measures to be adopted to assess performance. How would you use averages, medians, max/min, standard and quartile deviations, confidence intervals, correlation measures (i.e., correlation coefficient, student's t test and multiple regression) and testing propositions (e.g., F-test) in assessing performance?
12. Discuss the inclusion of measurements of genotype and phenotype diversity, fitness variance, fitness histograms, and entropy to compare MOEAs and metrics. Relate to analysis of variance (ANOVA).
13. How would one employ the concepts of mean-squared error or some other norm to measure the bias of a curve estimator for a nondominated set of vectors? How do these results relate to the generational distance metric?
14. What test suite characteristics would you suggest for comparative analysis of a variety of MOEAs to make quantitative or qualitative judgments?
15. What sort of quality indicators would you consider appropriate to measure the impact of local search in a MOEA? Discuss.
16. How would you measure the dynamics of a MOEA population as it moves towards PF_{true}?
17. Why is dominance ranking an independent MOEA assessment method for comparing and ranking pairwise PF_{known} approximations? How are the various rank samples compared?
18. Discuss and compare the usefulness of dominance ranking, quality indicators and attainment functions for evaluating MOEA performance?
19. Select MOEA papers from the literature and analyze and criticize their statistical approaches and performance assessment analysis.
20. Despite being a unary performance measure, the hypervolume indicator [1781] has several properties that make it a relatively popular quality indicator [888]. However, the computational time required to compute this quality indicator may limit its use in practice. That is the reason why several researchers have proposed a variety of algorithms for computing this indicator [1713, 491, 1695, 1696, 1697, 514]. Analyze these proposals and discuss their main differences and their computational efficiency. Discuss.
21. Analyze and criticize the Pareto-rank histograms proposed by Kumar & Rockett [923] as a performance measure for MOEAs.
22. Address the measurement problems associated with decision variable space and objective function space scalability using the approaches sug-

gested in this chapter. What performance approximations may permit efficient computation for MOEA comparison?
23. Use MOEAs and other appropriate algorithms to solve a very difficult, complex, and high-dimensional engineering or scientific problem. Assess and compare results with respect to other approaches. Perform a statistical analysis of results.
24. Which performance measure would you adopt to determine if a MOEA has converged if the exact location of PF_{true} is unknown? Discuss.

Possible Research Ideas

1. Classify the existing metrics for evaluating MOEAs based on their main features. Propose new metrics based on aspects of performance not covered by existing metrics.
2. What is the relationship between performance measures and incorporation of user's preferences? Propose a framework that integrates both. See for example [1771].
3. Extend the attainment function concept to higher-order statistical measurements. See for example [513].
4. Develop an integrated inferential methodology using attainment functions, dominance relations, and quality indicators for MOEA comparison.
5. Develop and evaluate Pareto compliant measurement operators for different yet specific genotype MOEA chromosome representations. Consider binary, real and integer encodings.
6. Propose efficiency measures (related to time and space complexity) and analyze their performance using a set of modern MOEAs and test functions. Show in graphical form the behavior of these efficiency measures during MOEA execution. Do these measures assess efficiency as originally aimed?
7. Create a set of new quality indicators and discuss their Pareto compliance as well as utility. Modifying known quality indicators is possible.
8. Analyze the performance measure based on entropy proposed by Ali Farhang-Mehr [466, 467]. Propose a different performance measure based on entropy.
9. Chiam et al. [244] argue on the usefulness of a more general view of the empirical evaluation of MOEAs, which includes a structural algorithmic development plan and a general theory of adequacy. Analyze the framework proposed by the authors and develop a case study to illustrate its use. Then, identify possible weaknesses of this framework and propose appropriate modifications to overcome them.
10. Analyze the running performance measures proposed in [369]. Propose a new running performance measure.

11. Analyze visualization techniques for displaying results produced by MOEAs (particularly in higher dimensions). See for example [1199, 1014, 1299, 1726].

6
MOEA Theory and Issues

> He who loves practice without theory is like the sailor who boards ship without a rudder and compass and never knows where he may cast.
>
> Leonardo da Vinci

6.1 Introduction

Many MOEA development efforts acknowledge various facets of underlying MOEA theory, but make limited contributions when simply citing relevant issues raised by others. Some authors, however, exhibit significant theoretical detail. Their work provides basic MOEA models and associated theories. Table 6.1 lists contemporary efforts reflecting MOEA theory development. In essence, a MOEA is searching for optimal elements in a partially ordered set or in the Pareto optimal set. Thus, the concept of convergence to P_{true} and PF_{true} is integral to the MOEA search process.

As observed, MOEA theory noticeably lags behind applications, at least in terms of published papers. This is even clearer when noting few of these categorized papers (see Chapter 2) *concentrate* on MOEA theoretical concerns. Others discuss some MOEA theory but do so only as regarding various parameters of their respective approaches. This quantitative lack of theory is not necessarily bad but indicates further theoretical development is necessary to (possibly) increase the effectiveness and efficiency of existing MOEAs. The rest of this chapter is organized as follows. Section 6.2 presents various MOEA theoretical definitions, theorems, and corollaries. Appropriate mathematical definitions are provided to support understanding of theorems and corollaries. Also, note that certain theorems require specific MOEA structures in order to prove convergence. Section 6.3 discusses MOEA issues as related to contemporary theoretical results including fitness functions, fitness landscapes, Pareto ranking, niching, mating restriction, running time analysis and stability. The chapter concludes with some research ideas and discussion questions.

6 MOEA Theory and Issues

Table 6.1: MOEA Theory

Researcher(s)	Paper Focus
Fonseca and Fleming [510]	MOEA mathematical formulations
Rudolph [1396], Rudolph & Agapie [1401]	MOEA convergence
Veldhuizen and Lamont [1627]	MOEA convergence and Pareto terminology
Veldhuizen and Lamont [1630]	MOEA benchmark test problems
Hanne [643]	MOEA convergence and Pareto terminology
Laumanns et al. [959], Knowles and Corne [877]	Archiving techniques
Deb & Meyarivan [371], Deb et al. [375]	Constrained test problems
Rudolph [1395, 1397]	MOEA search under partially ordered sets
Ehrgott [427]	Analysis of the computational complexity of multiobjective combinatorial optimization problems
Rudolph [1399]	Limit theory for EAs under partially ordered fitness sets
Laumanns et al. [962, 960]	Running time analysis
Laumanns et al. [958]	Mutation control
Hanne [644]	Convergence to the Pareto optimal set
Aguirre and Tanaka [16], Knowles and Corne [875]	Fitness landscapes

6.2 Pareto-Related Theoretical Contributions

Pareto-based theorems and definitions have been developed to support research objectives and other theoretical results as reflected in Table 6.1. Many MOEAs assume each generational population contains Pareto optimal solutions (with respect to that population). For example, Theorem 1 substantiates this assumption. As the MOEA literature offers little guidance concerning possible Pareto front cardinality and dimensionality, Theorems 4 and 5 provide an upper bound. Thus, these and other theoretical Pareto contributions as referenced further bounding both problem and algorithm domains. Appropriate ones are presented here for coherence. Some theorems are presented without proof, but such proofs can be found in the associated references. Others have been reworded in order to reflect the nomenclature used in this book. A limited number of symbols have also been employed for ease of initial understanding. But, first the associated definitions of partially ordered sets are discussed as related to individual fitness function values.

6.2.1 Partially Ordered Sets

Partially ordered sets are an integral aspect of moving a population towards the Pareto front using dominance. The underlying concept of a mathematical relation provides the basic foundation for studying partially ordered sets over MOEA fitness functions. Note that optimality is assumed to reflect a minimization MOP in the continuing discussion.

Definition 35 (Relation) : *Let x, y and z be in X, some set satisfying a binary relation R such that xRy satisfies one or more of the following properties:*

reflexive (xRx is true), antireflexive (xRy implies that $x \neq y$), symmetric (xRy implies yRx), antisymmetric (xRy and yRx imply x=y), asymmetric (xRy implies negation of yRx is true) and transitive (xRy and yRz implies that xRz), for all x, y and z in X. □

Definition 36 (Equivalence) : *If R is reflexive, symmetric, and transitive, then R is an equivalence relation.* □

Definition 37 (Partial Order) : *If R is reflexive, antisymmetric, and transitive, then R is a partial order relation.* □

Definition 38 (Dominance) : *Per the discussion in Chapter 1 using Pareto partial order set relations.* □

Definition 39 (Poset) : *(X,$\leq$), where $\leq$ is a partial order in X, is called a poset or a partial ordered set.* □

Definition 40 (Minimal Element) : *An element x^* of X is said to be minimal of the poset (X,$\leq$) if there is no x in X such that x is less than x^* in terms of the partial order relation $\leq$ ($x \leq x^*$). The set of all minimal elements is defined as M(X,$\leq$) and is complete if for each x in X there is at least one x^* in M(X,$\leq$) such that x^* is less than x ($x^* \leq x$).* □

If X is a finite set, then the completeness of M(X,$\leq$) is guaranteed. If the poset (X,$\leq$) is infinite then the set of minimal elements may be incomplete. Sufficient conditions for completeness in case of an infinite set exist as discussed at the end of this section. However, here it is assumed that the cardinality of the minimal set is finite. Note that a decision vector is denoted by x or **x** in the discussion and usually X = $\mathbb{R}^n$. Likewise F defines the fitness vector space, $\mathbb{R}^q$.

Pareto Optimal Set Minimal Cardinality

Because of the manner in which Pareto optimality is defined, any non-empty finite solution set contains at least one Pareto optimal solution (with respect to that set); i.e., a minimal element. As this may be non-intuitive, and because it is assumed in many MOEA implementations, the various theorems are presented based on a finite set W.

Theorem 1: Given an MOP with feasible region Ω in X = $\mathbb{R}^n$ and any non-empty finite solution set $W \subseteq \Omega$, there exists at least one solution $\mathbf{x} \in W$ that is Pareto optimal with respect to W [1627, 1626]. □

Proof. Label the k-dimensional objective vectors resulting from evaluating each $\mathbf{x}_i \in W$ in non-decreasing, lexicographic order as $v_1, v_2, \ldots, v_n$ with $v_i = (v_{i,1}, v_{i,2}, \ldots, v_{i,k})$. If all v_i are equal then v_1 is nondominated because of the partial order. Otherwise, there exists a smallest $j \in \{1, \ldots, k\}$ such that for some $i \in \{1, \ldots, n-1\}$, $v_{1,j} = v_{2,j} = \ldots = v_{i,j} < v_{i+1,j} \leq v_{i+2,j} \leq \ldots \leq v_{n,j}$. This shows that $v_{i+1}, v_{i+2}, \ldots, v_n$ do not dominate v_1.

If $i = 1$, then v_1 is nondominated. On the other hand, if $i \neq 1$ and $j = k$, $v_1 = v_2 = \ldots = v_i$ and v_1 is again nondominated. Otherwise, there exists a smallest $j' \in \{j+1, \ldots, k\}$ such that for some $i' \in \{1, \ldots, i-1\}$, $v_{1,j'} = v_{2,j'} = \ldots = v_{i',j'} < v_{i'+1,j'} \leq v_{i'+2,j'} \leq \ldots \leq v_{i,j'}$. If either $i' = 1$, or $i' \neq 1$ and $j' = k$, v_1 is nondominated. Otherwise continue this process. Because k is finite, eventually v_1 is nondominated and therefore there is at least one solution that is Pareto optimal with respect to W.

Another approach to showing that $P_{known}(t)$ is non-empty is through the use of the *positive variation kernel*. This kernel relates to the sequential generation of MOEA populations consisting of the $P_{known}(t)$ set[1] (which consist of minimal elements). The associated theorems are presented without proof and are basically constructive in nature. The reader is referred to the references for further details.

Definition 41 (Positive Variation Kernel) : *Given the number of and specific parents participating in producing a single child, the positive variation kernel (> 0) of a general EA is a function mapping the transition probability of the parents to the possible child over the search space. The joint transition probabilities for mating and mutation operators is guaranteed if the operators' probabilities are bounded.* □

Rudolph's variation kernel (i.e., transition probability function) is equivalent to a reachability condition (appropriate mutation and recombination operators allowing every point in the search space to be visited). Rudolph also refers to at least one sequence leading to an associated point on P_{true}, as compared to this work which indicates that through Pareto ranking *all* decision variable sequences lead towards P_{true}; likewise, these variables' phenotypical expressions lead towards PF_{true}.

Theorem 2: With a finite search set, a MOEA with a "positive variation kernel" and an elitist selection strategy generates a sequence of populations ($P_{known}(t)$) such that at least one population individual enters into the set of minimal elements of the poset in finite time with probability one [1395, 1397]. □

Assuming a MOEA has a positive variation kernel and a strong elitist selection strategy over a finite search set, a set of minimal elements is generated.

Theorem 3: (sufficient conditions) If the variation kernel of the MOEA is positive, then the final population, P_{known}, completely consists of minimal

---

[1] In this case, the population of the MOEA at time t consists exclusively of the elements of $P_{known}(t)$.

elements after a finite number of generations with probability one [1395, 1397].
□

Theoretical PF_{true} and P_{true} bounds are useful in defining a given problem domain. Theoretical statements exist defining the structural bounds any Pareto front may attain. For example, Corollary 6.1 provides a lower bound for the cardinality of the Pareto front.

Corollary 6.1. *Given a MOP with feasible region Ω in $X = \mathbb{R}^n$ and any nonempty finite solution set $W \subseteq \Omega$, its Pareto front PF_{true} is a set containing at least one vector. This result follows directly from Theorem 1 [1627, 1626].*

Theorem 4: The Pareto front PF_{true} of any MOP is composed of at most an uncountably infinite number of vectors [1627, 1626]. □

Proof. The Pareto front's cardinality is bounded above by the cardinality of the objective space.

Theorem 4 provides an upper bound on the cardinality of the Pareto front for MOPs with Euclidean objective spaces (spaces containing all n-tuples of real numbers, $(x_1, x_2, \ldots, x_n)$, denoted by $\mathbb{R}^n$). This includes many MOPs of interest. One can use the following definition in bounding the Pareto front's dimensionality [34, pg. 174]:

Definition 42 (Box-Counting Dimension) : *Let $\mathbb{R}^k$ be partitioned by a grid of k-dimensional boxes of side-length ϵ, where the boxes' sides are parallel to the objective axes. A bounded set S in $\mathbb{R}^k$ has box-counting dimension*

$$boxdim(S) = \lim_{\epsilon \to 0} \frac{\ln N(\epsilon)}{\ln(\frac{1}{\epsilon})}, \qquad (6.1)$$

where the limit exists and where $N(\epsilon)$ is the number of boxes that intersect S.
□

Theorem 5: For a given MOP with X, F, and a Pareto optimal set P_{true}, if the Pareto front PF_{true} is bounded, then it is a set with box-counting dimension no greater than $(q-1)$ [1627, 1626]. □

Proof. Without loss of generality assume PF_{true} is a bounded set in $[0,1]^q$. Take S to be the closure of PF_{true}. Because $[0,1]^q$ is closed, S is a bounded set in $[0,1]^q$. By hypothesis, $[0,1]^q$ is partitioned by a grid of k-dimensional boxes of side-length ϵ, where the boxes' sides are parallel to the objective axes. For each $r \in R \triangleq \{0, \epsilon, 2\epsilon, \ldots, \lfloor \frac{1}{\epsilon} \rfloor \epsilon\}^{q-1}$ define $R_r = [r_1, r_1+\epsilon] \times [r_2, r_2+\epsilon] \times \cdots \times [r_{k-1}, r_{k-1}+\epsilon] \times [0,1]$. If $S \cap R_r \neq \emptyset$, define p_r to be the point that minimizes

f_q over R_r and B_r to be any box that includes p_r. Also define $S_\epsilon = \{p_r\}$ and $B_\epsilon = \cup_r B_r$. Then B_ϵ covers S_ϵ. Because S is closed $\lim_{\epsilon \to 0} S_\epsilon = S$, and $B \triangleq \lim_{\epsilon \to 0} B_\epsilon$ covers S. Because $PF_{true} \subseteq S$, B also covers PF_{true}. Hence, $N(\epsilon) = |R| = \lceil \frac{1}{\epsilon} \rceil^{q-1}$, and the box-counting dimension of PF_{true} is

$$\lim_{\epsilon \to 0} \frac{\ln(\lceil \frac{1}{\epsilon} \rceil^{q-1})}{\ln(\frac{1}{\epsilon})} \leq \lim_{\epsilon \to 0} \frac{\ln((\frac{2}{\epsilon})^{q-1})}{\ln(\frac{1}{\epsilon})}$$
$$= \lim_{\epsilon \to 0} \frac{(q-1)[\ln 2 + \ln(\frac{1}{\epsilon})]}{\ln(\frac{1}{\epsilon})}$$
$$= \lim_{\epsilon \to 0} [\frac{(q-1)\ln 2}{\ln(\frac{1}{\epsilon})} + (q-1)]$$
$$= q - 1 \qquad (6.2)$$

In practice, the Pareto front PF_{true} is a collection of $(q-1)$ or lower dimensional surfaces termed Pareto surfaces. The special case where $q = 2$ results in surfaces termed Pareto curves. Horn & Nafpliotis [708], and Thomas [1580] state that a q-objective MOP's Pareto front *is* a $q - 1$ dimensional surface. But this, by Theorem 5, has been shown to be incorrect; the front is *at most* a $(q-1)$ dimensional surface [1626]. Although asymptotic bounds are useful, researchers must also account for the Pareto front's possible shape within those bounds. Theorem 5 then implies that any proposed MOEA benchmark test function suite should contain MOPs with Pareto fronts composed of Pareto curve(s), Pareto surface(s), or some combination of the two (see Chapter 1).

6.2.2 MOEA Convergence

Using the theorems of the previous section, the convergence of PF_{known} to PF_{true} can be addressed. To provide insight to the development, the supporting background of single objective EAs is initially presented using the concepts of total order and reachability.

Given that $\mathbf{x}$ is a single-objective optimization decision variable, I the space of all feasible decision variables, F a fitness function, and t the generation number, Bäck proves [72, pg. 129] that an EA converges with probability one if it fulfills the following conditions:

$$\forall \mathbf{x}, \mathbf{x}' \in I, \mathbf{x}' \text{ is reachable from } \mathbf{x}$$
$$\text{by means of mutation and recombination;} \qquad (6.3)$$

and the population sequence $P(0), P(1), \ldots$ is monotone, i.e.,

$$\forall t : \min\{F(\mathbf{x}(t+1)) \mid (\mathbf{x}(t+1) \in P(t+1))\}$$
$$\leq \min\{F(\mathbf{x}(t)) \mid (\mathbf{x}(t) \in P(t))\} \qquad (6.4)$$

Bäck's definition of monotonicity, appropriate in the context of single objective EAs, is fitness-based and assumes that the objective space is totally ordered. Neither of these restrictions is appropriate in the context of MOEAs. A solution's Pareto-based fitness depends on the set within which it is evaluated, and consequently may vary from one generation to the next. Also, the objective space for a MOEA is partially and not necessarily totally ordered. As previously discussed, a convergence theorem for MOEAs requires a more general definition of monotonicity that is both fitness independent and appropriate for objective spaces that are not totally ordered. The search for the Pareto optimal set is a search in partially ordered sets for optimal multiobjective problem (multi-criteria) solutions. Using a population-based MOEA, an increasing movement towards the true Pareto front PF_{true} from PF_{known} (t) can usually be achieved. One such definition is given by the condition

$$PF_{known}(\text{t+1}) = \text{M}(PF_{current}(\text{t}) \cup PF_{known}(\text{t}), \preceq) \qquad (6.5)$$

where $\preceq$ represents the Pareto dominance partial order and $PF_{known}(0) = \emptyset$. It can be shown by induction on t that under this condition, $PF_{known}(t)$ consists of the set of solutions evaluated through generation t that are Pareto optimal with respect to the set of all such solutions. Thus, $PF_{known}(t+1)$ either retains or improves upon solutions in $PF_{known}(t)$. In this sense, Condition (6.5) ensures that $PF_{known}(t)$ monotonically moves towards PF_{true}. Employing these conditions results in global convergence to the Pareto optimal solution set.

Theorem 6: A MOEA satisfying (6.3) and (6.5) converges to the global optimum (P_{true}) of a MOP with probability one, i.e.,

$$Prob\left[\lim_{t \to \infty}\{P_{true} = P_{known}(\text{t})\}\right] = 1.$$

□

Proof. A MOEA may be viewed abstractly as a Markov chain consisting of two states. In the first state, $P_{true} = P_{known}(t)$, and in the second state this is not the case. By Condition (6.5), there is zero probability of transitioning from the first state to the second state. Thus, the first state is absorbing. By Condition (6.3), there is a non-zero probability of transitioning from the second state to the first state. Thus, the second state is transient. The theorem follows immediately from Markov chains theory [32].

Another form of P_{true} convergence is associated with the variation kernel hypothesis of Theorem 2, although the population can grow to a large number.

Theorem 7: If Theorem 2 holds, the EA population completely converges to the set of optimal elements P_{true} if the set of minimal elements at $P_{known}(t)$ is the union of parents and generated children [1395, 1397]. □

Rudolph's other EA convergence theorems employ the concept of the homogeneous stochastic matrix G instead of the probabilistic variation kernel. The use of the matrix G can define a chain of population movements. Using this concept, convergence can then be extended to MOEAs. Initially, the generalization of finding optimal elements in a partially ordered set is addressed since multiobjective optimization is a specialized class of such a set. Note that in general Rudolph's 2000 theorems are clarifications of his 1998 theorems as referenced.

Definition 43 (Stochastic Matrix): *$G_{m \times n}$ is a stochastic matrix if the sum of each of its rows is equal to 1, that is, if $\sum_{j=1}^{n} g_{ij} = 1$, $i = 1, ..., m$.* □
The transition probabilities of the population of a MOEA, from the current to the next generation, can be described by means of a stochastic matrix G.

Definition 44 (MOEA convergence with probability 1 to minimal set): *A MOEA is said to converge to the entire set of minimal elements P_{true} with probability one if*

$$d_1(PF_{true}, PF_{known}(t)) \to 0 \text{ with probability one as } t \to \infty,$$

where $d_1(PF_{true}, PF_{known}(t))$ is a distance function between PF_{true} and $PF_{known}(t)$, and it is defined as:

$$d_1(PF_{true}, PF_{known}(t)) = |PF_{true} \cup PF_{known}(t)| - |PF_{true} \cap PF_{known}(t)|$$

Or an EA is said to converge to the set of minimal elements if the distance function

$$d_2(PF_{true}, PF_{known}(t)) = |PF_{known}(t)| - |PF_{true} \cap PF_{known}(t)|$$

satisfies:

$$d_2(PF_{true}, PF_{known}(t)) \to 0 \text{ with probability one as } t \to \infty.$$

[1399]. □

Similar to Theorem 6, another convergence theorem can be proved using the stochastic transition matrix G.

Theorem 8: Let G be the homogeneous stochastic matrix describing the transition behavior of a specific MOEA whose population at time t is $P_{known}(t)$ provided that $PF_{known}(t) = M(PF_{known}(t-1) \cup PF_{current}(t-1), \preceq)$ ($\preceq$ represents the Pareto dominance partial order). If matrix G is positive then the distance function $d_1(PF_{true}, PF_{known}(t))$ goes to zero with probability one as $t \to \infty$ [1399]. □

Note that this approach usually generates larger and larger populations. Regarding the computational desire for a smaller population, consider limiting the number of minimal child individuals passed to the next generation.

Theorem 9: Let G be the homogeneous stochastic matrix describing the transition behavior of a specific MOEA, as in Theorem 8, but such that it only draws a limited number of new unique minimal individuals from the children, for the next generation. If matrix G is positive then the distance $d_2(PF_{true}, PF_{known}(t))$ goes to zero with probability one as $t \to \infty$ [1399]. □

Constraining the population $P_{known}(t)$ size to be constant from generation to generation, convergence hypotheses can still be stated.

Theorem 10: Let G be the homogeneous stochastic matrix describing the transition behavior in a specific MOEA, as in Theorem 8, but such that it limits the population size (population size remains constant by control). If the matrix G is positive then the distance $d_2(PF_{true}, PF_{known}(t))$ goes to zero with probability one as $t \to \infty$ [1399]. □

Regarding explicit multiobjective problem formulations, Rudolph has shown the following:

Corollary 6.2. *If F is a vectored valued objective MOP function associated with Theorem 9, then the population converges with probability one to the Pareto set, P_{true}. Moreover, the population size converges to the minimum of the preset number or the cardinality of P_{true}.*

Corollary 6.3. *If F is a vectored valued objective MOP function associated with Theorem 10, then the constant population size converges with probability one to elements in P_{true}.*

Rudolph's Corollary 3 [1395] guarantees that given a countable infinite MOEA population and a MOP, at least one decision variable (x_k) sequence exists such that $f(x_k)$ converges in the mean to PF_{true}. Note Rudolph's nomenclature is different than that which is presented here.

Rudolph [1396], also proved that a specific multiobjective ES with $(\mu + \lambda = 1+1)$ converges with probability one to a member of P_{true} of the MOP. His distance metric is in the genotype domain, as compared to Van Veldhuizen's work, which is phenotypically based. The EVOPs in his model are not able to search the entire space (in a probabilistic sense) since a step size restriction is placed upon the probabilistic mutation operator. Thus, convergence only occurs when the ES's step size is proportional to the distance to the Pareto set as shown in the elaborate proof. However, this distance is obviously unknown in problems of high complexity which is typical of most real-world problems.

Some theorems are for a specific EA and MOP instantiation with constrained EVOPs while others requires a less-specific EA structure. The theorems show that what one seeks is possible – given MOEAs do converge to an optimal *set*, although some theorems define a genotypic optimum and others a phenotypic one. Using phenotypical information may often be more appropriate as a decision maker's costs and profits are more accurately reflected in attribute space.

Another MOEA convergence theorem variation by Hanne considers continuous objective functions. His approach employs underlying geometric structures based primarily on convex sets. Such representations can be found in the mathematical and operations research literature. For proving theorems Hanne uses concepts of cones, efficient sets, efficiency preservation, and other sophisticated mathematics. Hanne has embedded these concepts into a MOEA implementation called LOOPS (Learning Object-Oriented Problem Solver) [642, 644, 645]. Readers are directed to the associated references for a detailed description and associated theorem proof details. Some of these constructs are presented here in support of the major convergence theorem. Note that the previous theorems assumed a finite cardinality or countable search space, here, Hanne assumes a continuous F and X.

Definition 45 (Cone) : *K (a vector space) is a cone if and only if αx is in K for all x in K, $\alpha >$ zero. If K is convex and a cone, then K is a convex cone. A cone K is pointed if and only if $K \cap -K = \{0\}$. Cones can be used to induce ordering relations similar to the previous partial order definitions. For example, if K is pointed, then $\leq_K$ is a partial order operator defined as $x \leq_K y$ if and only if y-x is in K (not including zero).* □

Thus, K as a cone by definition has the properties of a partial order or a Pareto order. X is of course the decision space and F is the vector space of fitness functions, and by construction the vector space F(X) is in K. Note that F(X) is assumed to be a compact and convex set as defined by the MOP constraints imposed.

Definition 46 (Efficient/Pareto set) : *A set (X,F) is efficient in X ($x \epsilon X$) with respect to the fitness function F if and only if $(F(\hat{x}) + K) \cap F(X) = \{F(\hat{x})\}$. The efficient set is encoded as E(X,F). The efficient set is equivalent to the Pareto optimal set, P_{true} [643].* □

That is, the intersection of $F(\hat{x}) + K$ with $F(X)$ is the set of some points on the Pareto optimal front. The fact that $F(X)$ is contained in K guarantees that the resulting set is nonempty and is $F(\hat{x})$, a point in P_{true}. Note that $F(\hat{x})$ is a boundary point of the compact and convex set $F(X)$.

Definition 47 (Regular MOP) : *A multiobjective problem is regular if and only if the fitness function is continuous, the decision space X is closed and bounded (compact), the topological interior of X is not zero, and the cone K is a nontrivial closed convex cone with interior of K not zero, $F(X) \subset K$.* □

Hanne defines the dominating set symbolically as $DOM(M^t)$ where the argument M^t is the current population at t which can be compared to $P_{known}(t)$. And $DOM(M^t)$ refers to those points in X that of course dominate M^t, but therefore specifically reflects those points in P_{true} that can be reached from M^t.

Definition 48 (Efficiency preserving MOEA) : *A MOEA is efficiency preserving if and only if for all $t >$ zero and $\emptyset \neq M^0$, $Dom(M^{t+1})$ is included in $Dom(M^t)$. That is, as t increases the set cardinality of reachable P_{true} points from M^t decreases or stays the same.* □

From these definitions, Hanne has developed the following decision variable convergence theorem for multiobjective EAs based upon an evolution strategies MOEA approach:

Theorem 11: If (X,F) is regular and the MOEA is efficiency preserving, then $Prob\,[\emptyset \neq (lim_{t \to \infty} M^t) \subseteq \text{E(X,F)}] = 1$ [643]. □

This theorem relates to a "convergence in probability" to within the Pareto optimal set P_{true} or E(X,F) from P_{known} or equivalently M^t. Again, P_{known} (t) converges to a nonempty subset of P_{true} based upon the assumptions of an efficiency preserving MOEA and the regularity of the MOP. In fact, the convergence may be to only one point. Note that this theorem may be extensible to other MOEA types with other MOEA selection functions and heuristics.

Convergence of Alternative Heuristics

In [1653, 1655], Villalobos et al. present the asymptotic convergence analysis of Simulated Annealing (SA), an Artificial Immune System (AIS) and a General Evolutionary Algorithm (GEA), for multiobjective optimization problems. SA is a heuristic search technique based on some analogies with an annealing process in which a crystal is produced. On the other hand, the AIS algorithm is a technique that simulates in a computer certain aspects of an immune system. For the mathematical model, Villalobos et al. consider the SA proposed in [1467], and the AIS for multiobjective optimization proposed in [311]. For these metaheuristics, that use a uniform mutation rule, they show that the associated Markov chain converges geometrically to its stationary distribution, but not necessarily to the optimal solution set. Convergence to the optimal solution set is ensured if elitism is used (by means of an external archive or *elite set* of limited size). In fact, Villalobos et al. assume that the size of the external archive is smaller than the size of the population.

The metaheuristics are modeled as Markov chains with transition probabilities that use uniform mutation and possibly other operations. For the case of SA, Villalobos et al. define $G_{ij}(c_k)$ as the generation probability of state j from state i, where c_k is a parameter that simulates the temperature, and develop a convergence proof of SA, for multiobjective optimization problems. They prove that the SA algorithm converges with probability one (as defined in Theorem 6), if the transition matrix $G(c)$ associated with the generation probabilities $G_{ij}(c_k)$ is irreducible. For the MISA and GEA algorithms,

Villalobos et al. prove the convergence with probability one, when using elitism, by providing demonstrations similar, though more detailed, to that presented for Theorem 6.

In [1723], Xue et al. perform a mathematical modeling and convergence analysis of a continuous Multi-Objective Differential Evolution (C-MODE) algorithm. The MODE algorithm is an extension of the Differential Evolution algorithm, to the multiobjective context. Differential Evolution (DE) is a type of evolutionary algorithm proposed by Storn and Price [1526] for optimization problems over a continuous domain. DE is similar to a (μ, λ)-ES in which mutation plays the key role.

The basic idea of DE is to adapt the search step inherently along the evolutionary process in a manner that trades off exploitation against exploration. In [1723], the convergence properties of C-MODE are studied in a similar manner to the work presented by Hanne in [643]. In this way, Xue et al. prove the convergence of the population of the C-MODE to the Pareto optimal solutions with probability one by means of a theorem analogous to Theorem 11. On the other hand, Xue et al. study the C-MODE operators and their effects on the convergence properties of the algorithm, under the Gaussian initial population assumption.

They first examine the population evolution of the C-MODE with only reproduction operators, and introduce later the selection factor. They show that the limiting properties of C-MODE depend on the factor $(2KF^2 + (1-\gamma)^2)$, where K, F and γ are the parameters associated to the approach. If this factor is greater than 1, the population variance matrix explodes, and C-MODE with selection successfully identifies the optimal solution set; otherwise, the population variance matrix vanishes.

Xue et al. confirm the mathematical results developed by simulation results obtained by applying C-MODE to numerical examples with different parameter settings. Also, they conduct simulation results on complicated continuous benchmark functions and show that the C-MODE performs better when the parameters are set to meet the obtained conditions. In this way, the results obtained by Xue et al. can also be used to guide the parameter setting of the C-MODE when applied in real world applications.

In [1724, 1720], Xue et al. extend their theoretical work by modeling a discrete version of MODE, D-MODE, in the framework of Markov processes and develop the corresponding convergence properties. They study the Markov model for the D-MODE with finite population size. Two situations are considered: one with a population large enough to contain all the Pareto optimal solutions while the other is the opposite. In the second situation, an external archive is needed to store all visited Pareto optimal solutions. In both cases, Xue et al. prove the convergence with probability one of D-MODE to the set of Pareto optimal solutions by providing demonstrations similar to those presented for Theorems 6 and 10.

Archiving Strategies

As we can see, the algorithms (and theorems) proposed by Rudolph [1395], Rudolph and Agapie [1401] and Hanne [643] guarantee convergence to the true Pareto optimal front, however, they do not guarantee a good distribution of Pareto optimal solutions.

In [959], Laumanns et al. discuss this behavior and also provide some analysis to show that other algorithms proposed after those which ensure convergence, like PAES [886], SPEA [1782] or NSGA-II [374] (where selection is based only on diversity or density measures), suffer from possible deterioration and thus, convergence can no longer be guaranteed. Deterioration occurs when elements of a solution set at a given time are dominated by a solution set the algorithm maintained some time before. According to Laumanns, this can happen using the standard Pareto-based selection schemes, even under elitism.

Based on a new concept named ϵ-dominance, Laumanns at al. propose new archiving strategies that overcome this fundamental problem and provably lead to MOEAs which have both the desired convergence and distribution properties. The work of Laumanns solely deals with the updating process of an archive A of limited size, maintained by a *generic* search algorithm that generates just one new sample at each iteration (by means of a black box optimization process). Similar to the definition of Pareto dominance (see Definition 7), they define the concept of ϵ-dominance in the following way:

Definition 49 (ϵ-Dominance) : *A vector* $\mathbf{u} = (u_1, \ldots, u_k)$ *is said to ϵ-dominate another vector* $\mathbf{v} = (v_1, \ldots, v_k)$ *for some* $\epsilon > 0$, *denoted by* $\mathbf{u} \preceq_\epsilon \mathbf{v}$, *if and only if* $\forall i \in \{1, \ldots, k\}$, $(1 - \epsilon)u_i \leq v_i$. □

On the limit $\epsilon \to 0$, and the ϵ-dominance becomes the normal dominance. The concepts of ϵ-approximate Pareto set and ϵ-Pareto set are also defined in the following.

Definition 50 (ϵ-approximate Pareto set) : *Let* $F \subset \mathbb{R}^k$ *and* $\epsilon > 0$, *a set* $\mathcal{P}_\epsilon$ *is called ϵ-approximate Pareto set of* F, *if any vector* $x' \in F$ *is ϵ-dominated by at least one vector* $x \in \mathcal{P}_\epsilon$. □

Definition 51 (ϵ-Pareto set) : *Let* $F \subset \mathbb{R}^k$ *and* $\epsilon > 0$, *a set* $\mathcal{P}_\epsilon^*$ *is called ϵ-Pareto set of* F, *if* $\mathcal{P}_\epsilon^*$ *is an ϵ-approximate Pareto set of* F *and* $\mathcal{P}_\epsilon^*$ *contains only Pareto optimal solutions ($\mathcal{P}_\epsilon^* \subset \mathcal{P}^*$, see Definition 8).* □

In this way, Laumanns et al. provide the following theorem related to an updating strategy which converges and preserves diversity of the solution vectors.

Theorem 12: Let $A^{(t)}$ the archive obtained by a generic algorithm that updates $A^{(t)}$ at each generation in such a way that new points are only accepted if they are not ϵ-dominated by any other point of the current archive (of course, dominated points are removed). Then $A^{(t)}$ is an ϵ-approximate

Pareto set and its size is bounded by:

$$|A^{(t)}| \leq \left(\frac{\log K}{\log(1+\epsilon)}\right)^m \qquad (6.6)$$

□

To guarantee that the points in $A^{(t)}$ are Pareto optimal solutions, Laumanns et al. propose to discretize the search space by a division into boxes, where each vector uniquely belongs to one box.

Definition 52 (ϵ-box) : *The ϵ-box of a point $\mathbf{f} = (f_1, ... f_k)$ is defined by a box index vector $b := (b_1, ..., b_k)$ where*

$$b_i := \left\lfloor \frac{\log f_i}{\log(1+\epsilon)} \right\rfloor.$$

□

Theorem 13: Let $A^{(t)}$ be the archive obtained by a generic algorithm that updates $A^{(t)}$ at each generation in such a way that it maintains a set of nondominated ϵ-boxes (represented by vectors) and at most one element is kept in each ϵ-box, which is only replaced by a dominating one. Then $A^{(t)}$ is an ϵ-Pareto set and its size is bounded according to Equation 6.6. □

In this way, it is guaranteed that $A^{(t)}$ contains only elements which are nondominated by any of the generated vectors. Also, the archiving algorithms suggested by Laumanns are generic and enable convergence with a guaranteed spread of solutions. The use of ϵ-dominance also makes the algorithms practical by allowing a decision maker to control the resolution of the Pareto set approximation by choosing an appropriate ϵ value. In [959], Laumanns et al. present some additional simulation results to demonstrate the behavior of the algorithms proposed, and perform comparisons against the SPEA [1782] and NSGA-II [374] algorithms, where it is visible that the updating selection process of the SPEA and NSGA-II approaches suffer from the problem of partial deterioration. Finally, they provide some discussion about other definitions of ϵ-dominance and several mechanisms to dynamically adapt the value of ϵ.

Although according to Definitions 50 and 51, the sets obtained by the algorithms proposed by Laumanns et al. represent optimal approximations, it may be possible that not all regions of the Pareto set are equally represented, specially the extremes of the Pareto front.

In [877], Knowles and Corne analyze several archiving algorithms including those proposed by Rudolph and Laumanns et al. and, as previously noted by Laumanns, they conclude that the strategies proposed by Rudolph do not encourage the storage of a good distribution of vectors, although they guarantee convergence. On the other hand, they discuss that the methods proposed by Laumanns et al. have certain problems concerning the initial setting of important parameters (ϵ).

According to Knowles and Corne, with the ϵ-methods one must either preset the value of ϵ, or bound the size of the archive and use an adaptive setting of ϵ (which are also provided by Laumanns et al. in [959]). In the former case, the number of points in the archive is bounded only by some function of the (unknown) objective space ranges. In the latter case, ϵ may become arbitrarily large and so the final set achieved may be a poor representation of the sequence of points presented to the archiving algorithm. Specifically, the number of points in the final archive may be far fewer than desired. However, in either case, convergence is guaranteed.

In this way, Knowles and Corne propose an Adaptive Grid Archiving (AGA) algorithm which addresses some of the problems with the archiving strategies considered previously. The algorithm is based on the archiving method used in PAES [886], which maintains an archive of bounded size, encourages an even distribution of points across the Pareto front, is computationally efficient, and provably converges under certain conditions. The basic principle of AGA is that as points in the objective space are generated and archived, the location and size of a grid in the k-dimensional objective space is adapted so that it just envelops the points. The grid is used to aid in selecting which points to remove from the archive, should the latter reach its capacity bound. In this case, a point from the maximally crowded region(s) is selected, so long as it is not an extremal point. According to Knowles and Corne, this strategy ensures archived points cover a wide extent in objective space and are "well distributed".

The rules of AGA do not constitute an algorithm that converges in the sense of reaching a state in which the members of the archive do not change for all future iterations (this is because nondominated vectors can be removed and cycles of entry and removal of these vectors can ensue). Nevertheless, Knowles and Corne provide a convergence result for AGA[2]: when (if) the grid boundaries converge (i.e. stop moving) a set of regions in the grid become constantly occupied over time by points in the archive and these grid regions contain the true Pareto front. In this way, the archive contains points that are at most a distance l from the true Pareto front, where l is the large diagonal of a grid region.

Knowles and Corne prove that the lower boundaries of the grid converge in all cases (assuming a minimization case) and, since such fact is not true for the upper boundaries in general, they show that the special cases when the upper boundaries can be proved to converge are when there are only two objectives, and when the true Pareto front is as broad as the entire search space in all objective dimensions, that is, it spans the feasible objective space in all

---

[2] All proofs of convergence presented by Knowles and Corne rely on the fact that, at every time step, the generating process gives every point in the search space a nonzero probability of being generated.

objectives. Thus, the convergence of the AGA algorithm is not guaranteed in the general case because the upper boundaries of the grid may fluctuate.

This occurs when the Pareto front has a smaller extent than the whole objective space [877]. This causes a problem because points not in the true Pareto front may cause the grid ranges to be extended, then at some later time these must be reduced again when the points are removed because they become dominated. In this way, AGA suffers from a problem related to those of the adaptive methods proposed by Laumanns et al.

In fact, according to Knowles and Corne, if the ranges of the grid were set in advance, that is, no adaptive scheme were used, and the archive bound and number of grid regions were set appropriately, then a result similar to that of Laumanns et al. algorithm is obtained.

More recently, Knowles and Corne provide a discussion about the properties of an imaginary ideal archiving algorithm and investigate the extend to which these properties can be, in principle, attained [878]. In [878], the main thesis of Knowles and Corne is that those methods that do not assure a formally guaranteed ϵ-approximation level may perform much better in practice. This is especially true under a "blind, one-shot scenario", where we are completely ignorant of Pareto front ranges and have only a single algorithm run from which to collect our solutions. According to Knowles and Corne, the properties of an ideal archiving algorithm, in terms of the archive A^t that it produces at generation t, are the following:

P1 A is itself a nondominated set (A is shorthand for A^t).
P2 $|A| \leq N$, i.e., the size of A is bounded.
P3 The archive A converges to a stable set in the limit of t.
P4 The archive A contains only Pareto optimal points from the sequence generated by the optimization process.
P5 All extrema Pareto optimal solutions, from the generated sequence, are in A.
P6 The size of the archive is "as close as possible" to N or the number of Pareto optimal points from the generated sequence.
P7 For every Pareto optimal point of the generated sequence, there is a point in A "nearby".

In [878], Knowles and Corne inquire if any archiving algorithm exists that can guarantee to produce archives satisfying all of the properties. They discuss that although P1, P2, P4 and P5 are straightforward and well-defined, P3, P6 and P7 are not. P3 expresses a feature of the archive in the limit of t and it is essential because it expresses the requirement that the archive should eventually converge. On the other hand, P6 and P7 capture the ideas, respectively, of the final archive not becoming too small, and of being well-distributed. Regarding property P6, Knowles and Corne provide the following theorem.

Theorem 14: No archiving algorithm can maintain the size of the archive equal to the minimum between the bound N and the number of Pareto optimal solutions from the generated sequence. □

In this way, Knowles and Corne showed that trying to keep the minimum size of the archive high (as large as possible) is impossible in a strict sense. On the other hand, considering that property P7 means that A should be an ϵ-approximate set of size up to N that minimizes ϵ, they provide the following theorems.

Theorem 15: No archiving algorithm can maintain an ideal archive, that is, an ϵ-approximate set of size up to N that minimizes ϵ. □

Theorem 16: No archiving algorithm can maintain an ϵ-approximate set of size up to N with an ϵ that is less than a constant κ of the ideal value of ϵ, without additional knowledge of the extent of the Pareto front of the generated sequence. □

Thus, Knowles and Corne demonstrated the existence of some theoretical limitations on algorithms for maintaining bounded archives. In [878], they also provide a case of study where they empirically test the archiving performance of AGA and of the ϵ-based archiving algorithms proposed by Laumanns et al. (Theorems 12 and 13). For their experiments, they use several different sequences of points designed to be difficult for archiving algorithms. The sequences include a small Pareto front in a large objective space, a discontinuous Pareto front, and a highly non-uniformly space Pareto front. They conclude that, in these situations, AGA performs better than the ϵ-approximate methods proposed by Laumanns et al. [959].

In this way, Knowles and Corne showed that for general sequences of points, no algorithm that respects one bound can maintain an ideal *minimum* number of points in the archive. That is, no archiving algorithm can, in general, be expected to maintain an "optimal" representation of the Pareto front when the size of that set is larger than the archive bound.

As shown, MOEA convergence theorems can use different but similar hypotheses based upon underlying decision space and assumptions on the MOP fitness landscapes and MOEA EVOPs. Minor algorithmic variations in population generation operators are assumed in each theorem and proof. However, convergence results are stated in very similar generic terms. Such generic convergence is guaranteed to either PF_{true} or P_{true} in infinite time per the previous theorems.

For finite time, the major characteristic of convergence rate is probably and uniquely associated with the specific MOP. Note that the metrics of Chapter 5 relate to this characteristic. A generic theoretical rate of convergence formulation is still unknown. The more important issue is the rate at which an MOEA converges to PF_{true}, and whether $PF_{known}(t)$ uniformly represents PF_{true} as $t \to \infty$. The MOEA literature is largely silent on these issues, although Rudolph shows the convergence rate for the specific (1+1) EA structure is sub-exponential [1396]. Also, as we will see in Section 6.3.10, some researchers have studied simple algorithms when applied to simple problems and have obtained the corresponding running time. However, further theoretical developments are still needed on this issue.

6.3 MOEA Theoretical Issues

MOEA researchers indicate that MOEA theory is lagging behind MOEA implementations and applications. For example, until recently no proof was offered showing a MOEA is capable of converging to P_{true} or PF_{true} (see Section 6.2.2). Chapter 2 implies that although the number of MOEA variations and implementations is significant, this fact alone does not indicate a corresponding *depth* of associated theory (as reflected by Table 6.1).

Why is there a lack of underlying MOEA theory? Although some mathematical foundations exist the current situation seems akin to Goldberg's comparisons of engineer and algorithmist [582]. He likens algorithms to "conceptual machines" and implies scientists are hesitant to move forward without exact models precisely describing their situation. On the other hand, he claims a design engineer often accepts less accurate models in order to build the design. MOEA researchers certainly seem to have taken this approach!

Realizing that simple assumptions are sometimes made in order to develop limited theoretical results, the foundations of single-objective EA theory are well-established. The *Handbook of Evolutionary Computation* [73] devotes entire chapters to theoretical EC results established up to 1997. Sample topics include EA types, selection, representation, crossover, mutation, fitness landscapes, and so on. There are also other (more recent) books that are entirely devoted to EA theory (see for example [814, 1343]). Several foundational textbooks are also available, such as those by Goldberg [581], Michalewicz [1100], Mitchell [1114], Bäck [72], and by Eiben & Smith [435]. Although much of this theory is (may be?) valid when regarding MOEAs, some is not. The evolution of the above Handbook into two volumes (*Evolutionary Computation 1 & 2*) [74, 75] although serving an excellent practitioners perspective, offers little in the way of theory. This section presents contemporary knowledge concerning selected MOEA theoretical issues.

6.3.1 Fitness Landscapes

The concept of *fitness landscape* was introduced by Sewall Wright in 1932 [1712] as a metaphor to describe multiple domains of attraction in evolutionary dynamics. The idea of Wright was to see the search space as having multiple peaks towards which the population would evolve by climbing up. Wright was mainly interested in looking at the way in which populations could escape from local optima through stochastic fluctuations. Thus, this is certainly one of the earliest proposals to use stochastic processes for optimizing multimodal functions [364].

Wright's concept was conceived as an aid to visualize the behavior of the selection and variation operators during the evolutionary process. Thus, fitness landscapes are normally used to study the efficacy of evolutionary algorithms [1500]. Particularly, researchers have adopted the so-called *NK fitness landscape* proposed by Stuart Kauffman [828, 827, 829] to explore the way in which epistasis[3] controls the ruggedness of an adaptive landscape. Kauffman's idea was to specify a family of fitness functions having a ruggedness which could by tunable through the manipulation of a single parameter.

Although several researchers have used landscape analysis in a single-objective optimization context (see for example [1500, 364, 1637]), little work currently exists for multiobjective problems. Next, we will review the most relevant work in this regard.

Knowles and Corne [875] introduce the multiobjective quadratic assignment problem (mQAP) and investigate some methods for measuring the corresponding multiobjective combinatorial landscape. Since local search moves can be quickly evaluated on the QAP, hybrids of global optimization schemes and local search techniques are favored for solving the QAP problem. In this way, they study the correlation between nearby optima in order to design the best overall search strategy (the best way to move about in the multiobjective landscape): approach first PF_{true} and then spread around from there, start repeatedly from new random points, or using a gradual approach towards PF_{true} from all directions in parallel.

According to Knowles and Corne, the best approach will depend on whether or not points nearby in objective space are also nearby in the permutation space. With this aim, they provide a number of landscape measurement methods based on measures previously applied to fitness landscapes of QAP instances: diameter of a population, entropy, fitness distance correlation and flow dominance. All such methods are mainly based on distances in the parameter (permutation) space and are applied on the information obtained after performing several local search runs using a set of evenly distributed scalarizing vectors. The information recorded consists of the starting points and the local optimum reached, along with their corresponding multiobjective evaluations and the number of local search moves applied. For the problems used, they conclude that it seems to be easy to move along P_{known} than to get to P_{true} in the first place, and to find solutions on P_{true} once one has been found.

---

[3] A gene is *epistatic* if its presence suppresses the effect of another one at another location in the chromosome [1100].

The methods proposed by Knowles and Corne represent an important contribution since they seem to be generally appropriate for multiobjective combinatorial optimization.

Another approach to study the fitness landscape of multiobjective combinatorial optimization problems is the one proposed by Stadler and Flamm [1512]. They show that geometrical properties of ordinary landscapes like saddle points, barriers and basins can be extended to the poset-valued case of multiobjective optimization in a meaningful way, and describe an algorithm that efficiently extracts these features from an exhaustive enumeration of a given generalized landscape. In this way, they assume that they can enumerate the decision space X exhaustively.

According to Stadler and Flamm, a landscape is formally a triple $(X, \mathcal{X}, f)$ consisting of a set of configurations X, a topological structure $\mathcal{X}$ that determines the mutual accessibility of configurations and a cost (or fitness) function $f : X \to \mathbb{R}$. The neighborhood relation $\mathcal{X}$ is typically defined by the move set of a search heuristic. In this way, their studies are restricted to the simplest case in which the configuration space $(X, \mathcal{X})$ is a finite undirected graph $G = (X, E)$ with vertex set X and edge set E (edges connect configurations that can be inter-converted by a single move).

The algorithm described by Stadler and Flamm, called *flooding algorithm*, arranges the local minima and the saddle points in a unique hierarchical structure which is conveniently represented as a tree, termed *barrier tree* (which in the case of partially ordered sets is in fact a forest). As a consequence, the Pareto solutions can be identified from the tips of the barrier forest. Stadler and Flamm provide a few examples to illustrate the type of information that can be gained, by applying their algorithm to a knapsack problem, a web access problem and a RNA secondary structures problem.

In [17], Aguirre et al. present an extension of Kauffman's NK-Landscapes to multiobjective MNK-Landscapes in order to use them as a benchmark tool and as a mean to understand better the working principles of MOEAs. Kauffman's NK-Landscapes model of epistatic interactions, particularly, have been the center of several theoretical and empirical studies both for the statistical properties of the generated landscapes and for their EA-hardness [829]. However, according to Aguirre et al., the effects of epistasis and NK-Landscapes in the context of MOEAs are almost unexplored topics.

Aguirre et al. define a MNK-Landscape as a vector function

$$f(\mathbf{x}) = (f_1(x), f_2(x), \cdots, f_M(x)) : \mathcal{B}^N \to \mathbb{R}^M$$

where M is the number of objectives, $f_i(\mathbf{x})$ is the i-th objective function, $\mathcal{B} = \{0, 1\}$ and N is the bit string length. Also, they define a set of integers $K = \{K_1, K_2, \cdots, K_M\}$ where K_i is the number of bits in the string $\mathbf{x}$ that epistatically interact with each bit in the ith-landscape. Each f_i is a non-linear function of $\mathbf{x}$ expressed by a Kauffman's NK-Landscape model of epistatic interactions.

6.3 MOEA Theoretical Issues

Besides defining N and K_i for each f_i, it is also possible to arrange the epistatic pattern between bit x_j and the other K_i interacting bits. That is, the distribution D_i of K_i bits among N. Thus, M, N, K and $D = \{D_1, \cdots, D_M\}$, completely specify a multiobjective MNK-Landscape and by varying them we can analyze the properties of the multiobjective landscapes and study the effects of the number of objectives, size of the search space, intensity of epistatic interactions, and epistatic pattern on the performance of multiobjective combinatorial optimization algorithms.

Aguirre et al. introduce an elitist multiobjective Random Bit Climber moRBC($\delta : 1+1$) that at all times keeps one parent individual from which it creates one offspring. The child is created by cloning the parent and flipping one bit, then the child is evaluated and replaces the parent if it dominates the parent. Child creation, evaluation, and (possibly) parent replacement are repeated N times. If no replacements are detected, a dominance local optimum has been found and the moRBC($\delta : 1+1$) opts for restart.

The restarting process replaces the parent with one individual chosen from a population which contains up to δ solutions, which are nondominated by the parent and amongst themselves. If the population is empty, the parent is replaced with a newly created random string. This process continues until a given number of evaluations has been performed. Additionally, the nondominated solutions found throughout the search are kept in an archive of limited size, and the ones with better crowding distance are preferred (crowding is also used to maintain fixed the size of the population).

The performance of the moRBC($\delta : 1+1$) algorithm is compared using the hypervolume measure against the NSGA-II and SPEA2 algorithms, on scalable random epistatic problems. Aguirre et al. conduct their study on MNK-Landscapes with 2 and 3 objectives and 100 bits, varying the number of epistatic interactions K_i from 0% to 50% in all objectives, and random epistatic patterns among bits for all objectives (D_i random). They study the effect of population size on the moRBC($\delta : 1+1$) and conclude that when K_i increases the search performance is worse, but the use of the population improves it. Also, they conclude that, as the number of objectives increase, the size of the population also needs to be increased in order to improve the performance. Finally, for the problems used, they conclude that the overall performance of the moRBC($\delta : 1+1$) is better than that of the NSGA-II and SPEA2 algorithms.

In [16], Aguirre and Tanaka extend the work presented in [17] by studying the working principles, behavior and performance of the NSGA-II, in order to clarify its poor performance when tested on MNK-Landscapes, as observed in [17].

First, Aguirre and Tanaka extend the study performed in [17] using NSGA-II and SPEA2, by varying the value of $N = 20, 50, 100$. As in [17], they conclude that the performance of NSGA-II and SPEA2 is worse when K_i increases. Then, they focus on NSGA-II and look into the effects of selection, drift, recombination and mutation.

Aguirre and Tanaka study the effects of selection by comparing the original NSGA-II($\mu + \lambda$) against a version of NSGA-II(μ, λ), in order to analyze the impact of elitism. They observe that if elitism is not included there is a severe deterioration in performance for all values of K and M, except for $K = M = 5$. Thus, Aguirre and Tanaka conclude that elitism is a very important feature for multiobjective combinatorial optimization.

However, they indicate that elitism can also bring some undesired side effects that could severely affect the efficacy and efficiency of the algorithms. For example, the presence of elitism increases selection pressure making elitist algorithms more prone to the effects of *genetic drift*. Genetic drift is a phenomenon that emerges from the stochastic operators of selection, recombination and mutation. It refers to the change on bit frequencies due to chance alone especially in small populations. It is well-known that genetic drift affects negatively the performance of EAs especially if a strong selection pressure is used. Aguirre and Tanaka study the effects of genetic drift by proposing an enhanced version of NSGA-II that eliminates fitness duplicates from the population.

According to Aguirre and Tanaka, duplicates hinder exploration and selection as well, since they accumulate rapidly, decreasing the likelihood that the algorithm will explore a larger number of different candidate solutions during a run. In this way, after comparing both versions of NSGA-II, they conclude that elimination of duplicates improves the performance of NSGA-II in two and five objectives.

On the other hand, Aguirre and Tanaka study the effects of recombination by comparing the performance of NSGA-II with and without recombination. They conclude that NSGA-II including recombination and mutation performs better than using only mutation for small values of K, but, for other values of K, recombination does not contribute to performance. Also, Aguirre and Tanaka study the effects of elitism and mutation by assigning an age limit to elite solutions and bias selection accordingly. Aguirre and Tanaka test two different versions of NSGA-II. The first version does not include recombination and increases by one the age of an elite solution each time it is selected for mutation. Age is also incremented by one at each iteration and individuals with age greater than N are eliminated from the population. The second version uses age to guide mutation: the mutation segment is chosen at random and the bit within the segment is given by the age of the individual. After the experiments, Aguirre and Tanaka conclude that the new versions increase substantially the performance of NSGA-II. Finally, Aguirre and Tanaka perform comparisons among the original NSGA-II, NSGA-II without duplicates and with elitist-mutation mechanisms and moRBC($\delta : 1 + 1$). They observe that the enhanced NSGA-II has a very similar performance to that of moRBC($\delta : 1+1$) in two and three objectives. However, in problems with five objectives moRBC($\delta : 1 + 1$) still performs better. The work of Aguirre and Tanaka constitutes a guide for practitioners on how to set up their algorithms and gives useful insights on how to design more robust and efficient MOEAs.

6.3.2 Fitness Functions

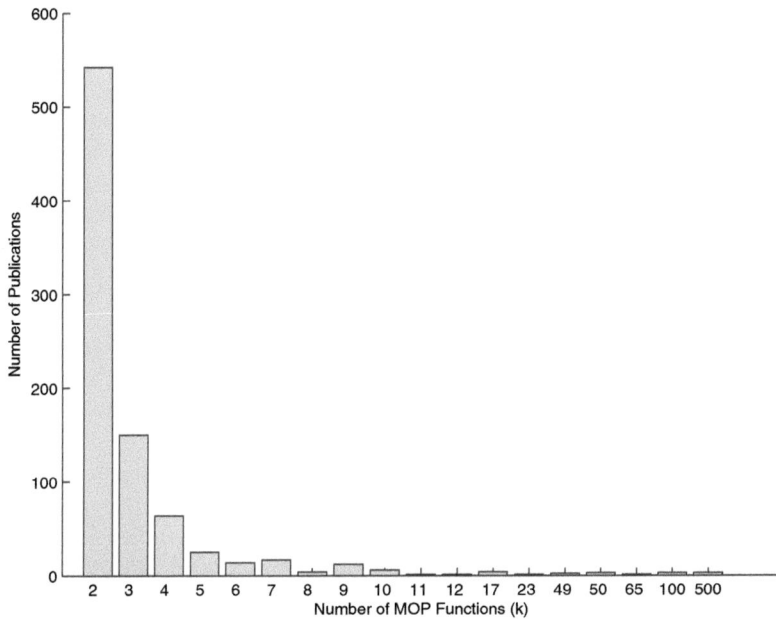

Fig. 6.1. MOEA Citations by Fitness Function using a sample of papers taken from the EMOO repository (http://delta.cs.cinvestav.mx/~ccoello/EMOO) (up to early 2007)

The general manner of fitness function implementation is two-fold. This is reflected by the work of Wienke et al. [1702] and Fonseca & Fleming [508], who each solved MOPs with seven fitness functions. Wienke et al. [1702] essentially used seven copies of an identical objective function, which was to meet atomic emission intensity goals for seven different elements. Although the elements and associated goals are each different, the fitness functions are conceptually identical. This does not make the MOP "easier" but perhaps makes the objective space somewhat easier to understand. On the other hand, Fonseca and Fleming's MOP's seven objectives appear both incommensurable and independent. Both P_{known} and PF_{known} are hard to visualize, as are their interrelationships. For example, when considering the mathematical polynomial model constructed by their MOEA, it is unclear how the number of terms affects the long-term prediction error and how that error may affect variance and model lag.

With that said, Figure 6.1 shows the number of citations employing a given number of fitness functions (as of the beginning of year 2007) and using a sample of references from the EMOO repository. The overwhelming majority use only two fitness functions, most probably for ease and understanding.

Several use three to nine, and a few go beyond ten. The currently known maximum is 500 objective functions within a single MOEA. In this approach, both a linear combination of weights and the ϵ-constraint method were used to coordinate agents [206]. In a distant second place is the approach by Coello Coello and Hernández Aguirre [274] in which up to 65 objective functions are handled, using a population-based approach similar to VEGA [1440]. The latter authors also report other applications in constraint-handling and combinational circuit design with 49, 23, and 17 objectives [264, 275, 274]. Of the only two other works that report using 17 objectives, one does not indicate the specific objectives [1305] and the other implements conceptually identical objectives [1336]. The highest number of conceptually different implemented fitness functions is found in a linkage design problem [1422] where nine objectives are used.

How many fitness functions are enough? How many objectives are generally required to adequately capture an MOP's essential characteristics? *Can all characteristics be captured?* The cataloged efforts imply most real-world MOPs are effectively solved using only two or three. There is a practical limit to the maximum number of possible objective functions, as the time to compute several complex MOEA fitness functions quickly becomes unmanageable. A theoretical limit exists as far as Pareto optimality is concerned. As additional objectives are added to an MOP more and more MOEA solutions meet the definition of Pareto optimality. Thus, as Fonseca and Fleming indicate for most Pareto MOEAs [507], the size of $P_{current}(t)$, $PF_{current}(t)$, $P_{known}(t)$, and $PF_{known}(t)$ grows, and Pareto selective pressure decreases. However, some confusion results from both theirs and Horn's [706] statements implying that the size of PF_{true} grows with additional objectives. The Pareto front is composed of Pareto curve(s), Pareto surface(s), or some combination of the two (see Chapter 1). And, as Cantor proved [612], the infinity of points on a line, surface, cube, and so on are the same (represented by $\aleph_1$). Thus, the cardinality of PF_{true} does *not* grow with the number of objectives, only (possibly) its topological dimension and associated MOEA operators. However, since MOEAs deal with *discretized* numerical representations the number of possible solutions (and therefore the number of computable vectors composing PF_{known}) may increase as more objectives are added.

Finally, some limit to human understanding and comprehension exists. The human mind appears to have a limited capacity for simultaneously distinguishing between multiple pieces of information or concepts. Perhaps this is best noted by Miller's seminal paper proposing a human one-dimensional span of judgment and immediate memory of 7 ± 2 [1113]. He notes that adding objective dimensions increases this capacity but at a decreasing rate. This seems to argue a "more the merrier" viewpoint for the number of MOP objectives, but visualizing and understanding objective interrelationships becomes more difficult as their numbers grow. Thus, certain techniques are designed to map high-dimensional information to two or three dimensions for better understanding (e.g., Sammon mapping [1421] and profiles [352]). Fonseca and

Fleming [504, 508, 512], often use profiles (or tradeoff graphs) to show MOEA solution values and their interrelationships. Figure 6.2 is an example profile for an MOP with seven objectives; the lines simply connect each solution's objective values.

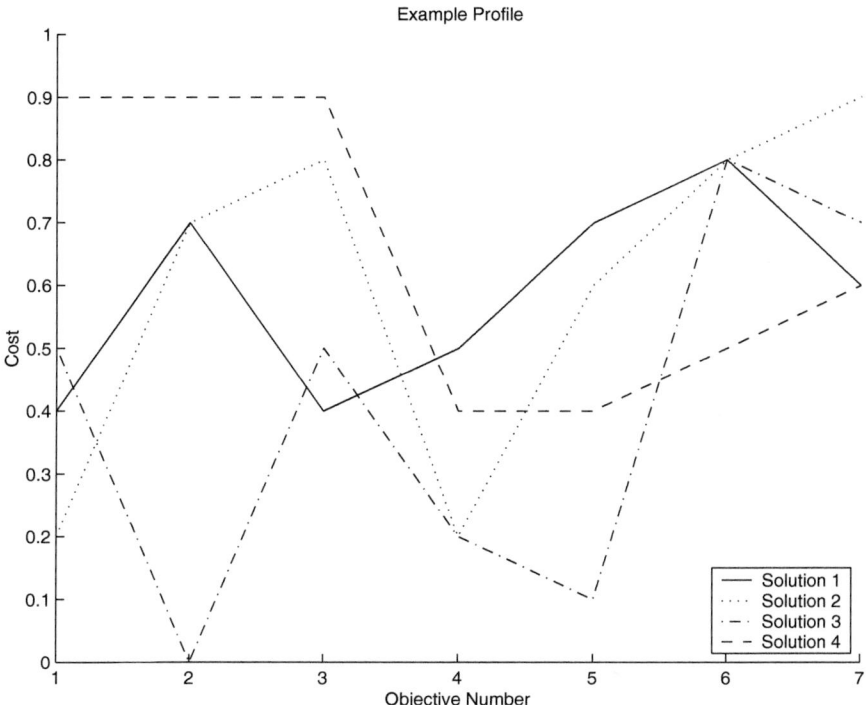

Fig. 6.2. Example MOP Profile

Many contemporary MOEA implementation results imply that two or three objectives are "satisfactory" for most problem domains. Thus, MOEA application to a given MOP should begin with two or three primary objectives in an effort to gain problem domain understanding. One may be able to ascertain how the different objectives affect each other and an idea of the fitness landscape's topology. Other fitness functions may then be added in order to capture other relevant problem characteristics.

6.3.3 Pareto Ranking

Two Pareto fitness assignment methods are primarily used in MOEAs although variations do exist (see Chapter 2). Although the elitist approach has limited theoretical insight via Theorems 2 and 7, it has found application favor with good results (see Chapters 2 and 4). This approach as well as the other fitness assignment operators should also be addressed from a more general theoretical perspective for providing convergence to the Pareto front. In general, all preferred (P_{true}) solutions are assigned the same rank based upon fitness value and other solutions assigned some higher (less desirable) rank. With the scheme proposed by Goldberg [581], where a solution x at generation t has a corresponding objective vector x_u, and N is the population size, the solution's rank is defined by the algorithm in Figure 6.3.

```
curr_rank = 1
m = N
while N ≠ 0 do
    For i = 1 : m do
        If x_u is nondominated
            rank(x, t) = curr_rank
    End For
    For i = 1 : m do
        If rank(x, t) = curr_rank
            Store x in temporary population
            N = N − 1
    End For
    curr_rank = curr_rank + 1
    m = N
End While
```

Fig. 6.3. Rank Assignment Algorithm

The second technique, proposed by Fonseca & Fleming [507] operates somewhat differently. As before, a solution x at generation t has a corresponding objective vector x_u. Let $r_u^{(t)}$ signify the number of vectors associated with the current population dominating x_u; x's rank is then defined by:

$$\text{rank}(x, t) = r_u^{(t)} . \tag{6.7}$$

This ensures all solutions with nondominated vectors receive rank zero.

Some approaches simply split the population in two, e.g., assigning solutions with nondominated vectors rank 1 and all others rank 2 [112]. Using the same notation, this ranking scheme is defined by:

$$\text{rank}(x, t) = \begin{cases} 1 & \text{if } r_u^{(t)} = 0, \\ 2 & \text{otherwise.} \end{cases} \tag{6.8}$$

When considering Goldberg's and Fonseca and Fleming's ranking schemes, it initially appears that neither is "better" than the other, although it is mentioned in the literature that Fonseca and Fleming's method, which effectively assigns a cost value to each solution, might be easier to mathematically analyze [510]. Horn [706] also notes this ranking can determine more ranks (is finer-grained) than Goldberg's (assuming a fixed population size).

Another ranking method using Pareto optimality as its basis is proposed by Zitzler & Thiele [1781, 1782]. Their rank assignment algorithm is lengthy. Thus, the reader is instead referred to the original sources for implementation details. Their MOEA implementation uses a secondary population whose solutions are directly incorporated into the generational population's fitness assignment procedure. Effectively, each Pareto optimal solution (at each generation) is assigned a fitness equal to the proportion of evaluated vectors its associated vector dominates. Because of the secondary population's inclusion in the fitness assignment process this method's complexity may be significantly higher than the other methods. Additionally, this method has a known shortfall. Deb [357] presents a geometric argument that this fitness assignment method has inherent bias. Pareto optimal solutions whose associated vectors dominate more vectors (or dominate a larger portion of objective space) receive higher fitness than other Pareto optimal solutions. However, each Pareto optimal solution should receive equal fitness. This method is then biased, as it may result in some Pareto optimal solutions receiving preference over others in the selection process.

There is currently no clear evidence as to the benefit(s) of any of these ranking schemes over another. Only one experiment whose purpose is directly comparing any of these schemes is reported in the literature. Thomas compared Fonseca and Fleming's and Goldberg's Pareto ranking schemes in a MOEA applied to submarine stern design [1580]. He concludes both outperformed tournament selection, and that Fonseca and Fleming's ranking appears to provide a fuller, smoother PF_{known}. However, he cautions that this is a singular data point. On a similar note, only one paper in the MOEA literature presents data on the number of population "fronts" using Goldberg's ranking. Vedarajan et al. [1638] present a graph showing the number of fronts found in each generation. With a population size of 300 individuals the first generation has over 40 fronts. This quickly drops from generations 10 to 100, and it oscillates between 20 and 25.

Analyzing these schemes' mathematical complexity is revealing. Table 6.2 (showing each scheme's best and worst case) and the following analysis only consider population size in computing complexity, where N is the size of the generational population and N_1 is the size of P_{known}. Assume that as comparisons are performed appropriate counter or fitness value assignments are made or updated. Thus, the binary, Fonseca and Fleming's, and Zitzler's ranking schemes require only one "pass" through the population(s) regardless of the number of nondominated solutions. Their worst and best case complexities are identical. Goldberg's scheme, however, requires at most $N-1$ "passes" through the population if there is only one Pareto optimal solution per (reduced) population. In addition, Zitzler's scheme's complexity increases if P_{known}'s size is much larger than the generational population's. Thus, Goldberg's and Zitzler and Thiele's ranking schemes (potentially) involve significantly more overhead than do the others.

Table 6.2: MOEA Fitness Ranking Complexities

Technique	Best Case	Worst Case
Binary	$N^2 - N$	$N^2 - N$
Fonseca	$N^2 - N$	$N^2 - N$
Goldberg	$N^2 - N$	$\frac{1}{3}(N^3 - N)$
Zitzler	$(N + N_1)^2 - N - N_1$	$(N + N_1)^2 - N - N_1$

It is also instructional to look at the possible value ranges for each ranking scheme. The binary scheme (equation (6.8)) offers only two values, $\Phi \in [0, 1]$. Both Fonseca and Fleming's (equation (6.7)) and Goldberg's scheme (Figure 6.3) offer N possible values, $\Phi \in [0, 1, \ldots, N-1]$. However, in practice Goldberg's scheme uses some subset of these values (resulting in a "coarser" ranking). Zitzler's scheme offers (possibly non-integer) values $\Phi \in [1, N)$. Using Fonseca's second function as an example (see Appendix A[4]), Figure 6.4 shows the resultant solution rankings of three Pareto ranking schemes.

Further clouding the issue is the fact that rank itself is often *not* directly used as a solution's fitness. For example, Fonseca and Fleming first used their ranking scheme in a MOEA implementation named the MOGA [504]; Srinivas and Deb implement Goldberg's scheme in the NSGA [1509]. Both transform assigned rank before selection occurs. The MOGA sorts solutions by rank and assigns fitness via linear or exponential interpolation, while the NSGA uses "dummy" fitness assignment, ensuring only that each "wave" of Pareto optimal solutions has a maximum fitness smaller than the preceding wave's minimum value.

6.3.4 Pareto Niching and Fitness Sharing

Due to stochastic errors associated with its genetic operators, evolutionary algorithms tend to converge to a single solution when used with a finite population [368]. As long as the goal is to find the global optimum (or at least a very good approximation of it), this behavior is acceptable. However, there are certain applications in which the aim is to find not one, but several solutions. Multiobjective optimization is certainly one of those applications, because the goal is to find the entire (or at least a considerable portion of the) Pareto front of a problem, and not only a single nondominated solution. The question is then how to keep the EA from converging to a single solution.

Early evolutionary computation researchers identified this convergence phenomenon of EAs, called *genetic drift* [348], and found that it happens

---

[4] All the Appendices of this book are available for download at: http://www.cs.cinvestav.mx/~emoobook

6.3 MOEA Theoretical Issues 311

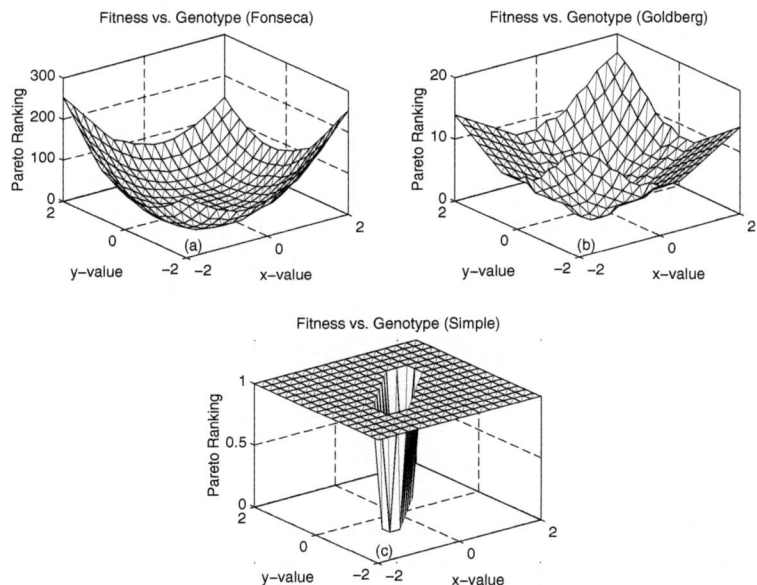

Fig. 6.4. Pareto Ranking Schemes

in nature as well. They correctly stated that the key to solve this problem is to find a way of preserving diversity in the population, and several proposals, modeled after natural systems were made. Holland [699] suggested the use of a "crowding" operator, which was intended to identify situations in which more and more individuals dominate an environmental niche, since in those cases the competition for limited resources increases rapidly, which results in lower life expectancies and birth rate. De Jong [348] experimented with such a *crowding* operator, which was implemented by having a newly formed offspring to replace the existing individual more similar to itself. The similarity between two individuals was measured in the genotype, by counting the number of bits along each chromosome that were equal in the two individuals being compared. De Jong used two parameters in his model: generation gap (G) and crowding factor (CF) [368]. The first parameter indicates the percentage of the population that is allowed to reproduce. The second parameter specifies the number of individuals initially selected as candidates to be replaced by a particular offspring [348]. Therefore, CF=1 means that no crowding takes place, and as the value of CF is increased, it becomes more likely that similar individuals replace one another [348].

Goldberg and Richardson [587] used a different approach in which the population was divided in different subpopulations according to the similarity of the individuals in two possible solution spaces: the decoded parameter space (phenotype) and the gene space (genotype). They defined a sharing function $\phi(d_{ij})$ as follows [587]:

$$\phi(d_{ij}) = \begin{cases} 1 - \left(\frac{d_{ij}}{\sigma_{sh}}\right)^\alpha, & d_{ij} < \sigma_{share} \\ 0, & \text{otherwise} \end{cases} \quad (6.9)$$

where normally $\alpha = 1$, d_{ij} is a metric indicative of the distance between individuals i and j, and σ_{share} is the sharing parameter which controls the extent of sharing allowed. The fitness of an individual i is then modified as:

$$f_{s_i} = \frac{f_i}{\sum_{j=1}^{M} \phi(d_{ij})} \quad (6.10)$$

where M is the number of individuals located in the vicinity of the i-th individual.

Deb and Goldberg [368] proposed a way of estimating the parameter σ_{share} in both phenotypical and genotypical space. In phenotypical sharing, the distance between two individuals is measured in decoded parameter space, and can be calculated with a simple Euclidean distance in an p-dimensional space, where p refers to the number of variables encoded in the EA; the value of d_{ij} can then be calculated as:

$$d_{ij} = \sqrt{\sum_{k=1}^{p} (x_{k,i} - x_{k,j})^2} \quad (6.11)$$

where $x_{1,i}, x_{2,i}, \ldots, x_{p,i}$ and $x_{1,j}, x_{2,j}, \ldots, x_{p,j}$ are the variables decoded from the EA.

To estimate the value of σ_{share}, Deb and Goldberg [368] proposed the expression:

$$\sigma_{share} = \frac{r}{\sqrt[p]{q}} = \frac{\sqrt{\sum_{k=1}^{p}(x_{k,max} - x_{k,min})^2}}{\sqrt[p]{2q}} \quad (6.12)$$

where r is the volume of a p-dimensional hypersphere of radius σ_{share} and q is the number of peaks that the EA aims to find.

In genotypical sharing, d_{ij} is defined as the Hamming distance between the strings and σ_{share} is the maximum number of different bits allowed between the strings to form separate niches in the population. The experiments performed by Deb and Goldberg [368] showed sharing as a better way of keeping diversity than crowding, and indicated that phenotypic sharing was better than genotypic sharing. Several other proposals exist (see [1039] for a more detailed review of approaches to keep diversity).

Several MOEA Pareto niching and fitness sharing variants have been proposed with the same goal as in traditional single-objective optimization – finding and maintaining *multiple* optima. However, MOEAs use sharing in an attempt to find a uniform (equidistant) distribution of vectors *representing* PF_{true}, i.e., one in which PF_{known}'s shape is a "good" approximation of PF_{true}. One can compare selected implementations of this concept.

Fonseca and Fleming's MOGA [511] uses restricted sharing, in the sense that fitness sharing occurs only between solutions with identical Pareto rank. They measure niching distance in phenotypic space, i.e., the distance (over some norm) between two solutions' evaluated fitness vectors is computed and compared to σ_{share} (the key sharing parameter). If the distance is less than σ_{share} the solution's associated niche count is then adjusted. Srinivas and Deb's NSGA [1509] implements a slightly different scheme, where distance is measured (over some norm) in genotypic space, i.e., the distance between two solutions is compared to σ_{share}.

Horn and Nafpliotis define niching differently in their MOEA named the NPGA [708, 709], which performs selection via binary Pareto domination tournaments. Solutions are selected if they dominate both the other and some small group (t_{dom}) of randomly selected solutions. However, fitness sharing occurs only in the cases where both solutions are (non)dominated. Each of the two solution's niche counts is computed not by summing computed sharing values, but by simply counting the number of objective vectors within σ_{share} of their evaluated vectors in phenotype space. The solution with a smaller niche count (i.e., fewer phenotypical neighbors) is then selected. Horn et al. [709] term this *equivalence class sharing*.

Another fitness sharing variant uses the NSGA's rank assignment scheme (i.e., Goldberg's Pareto ranking [581]) but adopts phenotypic-based sharing [1107]; another combines both genotypic and phenotypic distances in determining niche counts [1382]. Fitness sharing may also be applied to solutions regardless of rank instead of restricting sharing between equally ranked solutions. The revised version of the NSGA, called NSGA-II (see Chapter 2), implements a crowding mechanism that does not require a σ parameter [363].

Most of these methods require setting explicit values for the key sharing parameter σ_{share}, which can affect both MOEA efficiency and effectiveness. Fitness sharing's performance is also sensitive to the population size N. Assigning appropriate values to σ_{share} is difficult as it usually requires some *a priori* knowledge about the shape and separation of a given problem's niches. However, as phenotypic-based niching attempts to obtain equidistantly spaced vectors along PF_{known}, both Fonseca and Fleming [511] and Horn et al. [708, 709] are able to give guidelines for determining appropriate MOEA σ_{share} values. These values are based on known phenotypical extremes (minimum and maximum) in each objective dimension. Horn & Nafpliotis [708] also suggest appropriate values for the NPGA's tournament size parameter (t_{dom}).

To determine σ_{share}'s value using Fonseca and Fleming's method, one uses the number of individuals in the population (which implicitly determines the number of niches), scales the known attribute values, and determines the extreme attribute values in each objective dimension. These parameters are then used to derive σ_{share}. Horn et al.'s guidelines use the above parameters to define bounds for σ_{share}'s value.

How does one find each objective dimension's extreme values? One suggested approach is by computing objective values using each decision variables' minimum and maximum value. This is not feasible because decision variable extrema may not correspond to attribute extrema; the combinatorics and unknown relationships between different decision variable values is an additional factor. Thus, the minimum and maximum values of either the generational or a secondary population may be used. Fonseca and Fleming [511], indicate that recomputing σ_{share} at each generation (using current generational extrema) yields good results. Note that the MOEA's stochastic nature may not preserve these values between generations, i.e., the associated solutions may not survive. Thus, it is better to select objective extremes from the secondary population if one is incorporated in the MOEA. By definition, this population contains each objective dimension's extrema *so far*, ensuring the "ends" of PF_{known} are not lost.

As with the proposed Pareto ranking schemes, there is then no clear evidence as to the benefit(s) of one Pareto niching and sharing variant over another. Nor are experiments reported in the literature comparing key components of these different approaches (e.g., σ_{share} value assignment).

Note the following in regard to the appropriate sharing domain. Horn et al. indicate sharing should be performed in a space that one "cares more about" [708, 709]. Phenotypic-based sharing does make sense if one is attempting to obtain a "uniform" representation of PF_{true}. On the other hand, Benson and Sayin indicate many operations researchers "care more about" obtaining a "uniform" representation of P_{true} [119], in which case genotypic-based sharing seems appropriate. The end representation goal should drive the sharing domain.

6.3.5 Recombination Operators

Most of the current literature on MOEAs focuses mainly on studying the effect of different selection operators. However, few studies that focus on the effect of the variation operators (e.g., recombination) currently exist, and they tend to be normally empirical (see for example [174]). Next, we will review one of these studies which has a more theoretical orientation.

In [605], Grimme and Schmitt propose new variation operators for MOEAs. They start from the predator-prey model of Laumanns et al. [957], and examine the employment of the different search operators in a much better mode. In the predator-prey model of Laumanns et al. [957], the predators move across the spatial structure according to a random walk and chase the prey according to one of the optimization criteria. The worst prey within the neighborhood of the predator is "eaten" and its position is refilled by a prey created by recombination (discrete or intermediate) of those preys within its neighborhood.

In this way, the selection neighborhood and the recombination neighborhood are not the same. The selection neighborhood is constructed from the predator and the recombination neighborhood is constructed from the prey. According to Grimme and Schmitt, in the original study of the predator-prey model of Laumanns et al. two major problems were observed: the loss of diversity and stagnation of the process of convergence to the true Pareto front.

To handle the above problems, Grimme and Schmitt perform two modifications to the model: (1) They consider the selection neighborhood as being equal to the reproduction neighborhood and (2) they introduce the use of a mutation operator and incorporate a second group of predators in order to achieve the independent execution of mutation and recombination. Furthermore, in addition to discrete and intermediate recombination, the weighted intermediate recombination is also used. After some experiments on a very simple test problem, Grimme and Schmitt conclude that the proposed mechanisms along with the weighted intermediate recombination lead to both good convergence and a diverse set of solutions. However, the proposed mechanisms fail on a problem that does not possess a Pareto set parallel to the axes of the coordinate system. For this reason, Grimme and Schmitt develop a new recombination mechanism based on the geometric form of a n-simplex, which is insensitive to the rotation of the Pareto set.

By using this new recombination operator, Grimme and Schmitt improve the approximation of the set of efficient solutions significantly. In this way, the changes proposed by Grimme and Schmitt yield significantly better results than the original model proposed by Laumanns et al. Finally, Grimme and Schmitt analyze the influence of variation operators and they conclude that the recombination operators favor convergence towards the Pareto set, though they tend to collapse the population for a balance compromise. On the other hand, they also conclude that the mutation operator is essential for introducing new information since without mutation, the convergence is limited to the bounds of the area of search space covered by the initial population.

6.3.6 Mating Restrictions

The idea of restricted mating is not new. Goldberg [581], first mentions its use in single-objective optimization problems to prevent or minimize "low-performance offspring (lethals)." In other words, restricted mating biases how solutions are paired for recombination *in the hopes of* increasing algorithm effectiveness and efficiency. Goldberg presented an example using genotypic-based similarity as the mating criteria.

Deb and Goldberg [368] suggested the use of restrictive mating with respect to the phenotypic distance. The idea is to allow two individuals to reproduce only if they are very similar (i.e., if their phenotypic distance is less than a factor called σ_{mate} measured over some metric). This is intended to produce distinct "species" (mating groups) in the population [1114]. *Island model* GAs also implement restricted mating but in a geographic sense where solutions mate only with neighbors residing within some restricted topology [202]. It is also noted [261] that other researchers believe restricted mating should allow recombination of *dissimilar* (over some metric) individuals to prevent lethals. However defined, restricted mating is also incorporated within many MOEAs in an attempt to reduce unfit (e.g., non-Pareto optimal) offspring.

For example, Baita et al. [78], and Loughlin & Ranjithan [1017], place solutions on a grid and restrict the area within which each solution may mate. Lis & Eiben [1000], allow mating only between solutions of different "sexes." Jakob et al. [768], restrict mating to solutions within a particular deme. Hajela & Lin [636], implement a unique form of mating restriction. In their linear fitness combination (weighted-sum) MOEA formulation, they apply restricted mating based on a solution's associated weighting variables to prevent crossover between designs with radically different weight combinations. When considering general MOEAs phenotypic-based restricted mating between similar solutions is of more interest to us. Several MOEA researchers state in their published reports [504, 506, 1781]: "Following the common practice of setting $\sigma_{mate} = \sigma_{share}$ ..."

This may be a common practice, but no background is cited in the literature. As σ_{share} attempts to define a region within which all vectors are "related," setting σ_{mate} equal to σ_{share} is intuitive. The same rationale holds in genotypic sharing and mating restriction. Currently only empirical explanations are offered for the implementation (or lack) of restricted mating in various MOEA approaches. In fact, Fonseca and Fleming [507] noted that "... the use of mating restriction in multiobjective EAs does not appear to be widespread." Obviously, some researchers believe restricted mating is necessary or they would not have implemented it, but others indicate it is of no value.

Zitzler & Thiele [1781], state that for several different values of σ_{mate}, no improvements were noted in their test problem results (a MOP with two - four objectives) when compared against those with no mating restriction. Shaw & Fleming [1470], report the same qualitative results for their application (a MOP with three objectives) whether or not mating restriction was incorporated. Horn et al. [709], offer empirical evidence directly contradicting the basis for mating restriction. They note that recombining solutions whose associated vectors are on different portions of $PF_{known}(t)$ *can* produce offspring whose vectors are on $PF_{known}(t+1)$ but between their parents. They also claim that for a specific MOP, a constant (re)generation of vectors through recombination of "dissimilar" parents maintains PF_{known}. They believe most recombinations of solutions in P_{known} also yield solutions in P_{known}.

Thus, as in single-objective optimization, no clear quantitative evidence regarding restricted mating's benefits exists. The empirical evidence presented in the literature can be interpreted as an argument either for or against this type of recombination and leaves the MOEA field in an unsatisfactory predicament. This issue clearly benefits from experiments directly comparing its algorithmic inclusion/exclusion. One must also consider the NFL theorems [1708], realizing that mating restriction may not always be effective (or needed) for every problem (class).

6.3.7 Solution Stability and Robustness

Both EAs and MOEAs search for some problem's optima. At least for MOPs it has been noted [740] that P_{true} may not, and often *is not*, the most desirable solution set because its members are "unstable" (e.g., due to engineering tolerances, nonlinear response). It is also suggested that these solutions are often on the "edge" of optimality and/or feasibility. Thus, just as in single-objective optimization, any solutions returned as optimal must be evaluated with respect to any constraints not explicitly considered in the objective function(s). Or, perhaps a suitably defined sensitivity objective (e.g., engineering tolerances) may be incorporated into the MOEA.

6.3.8 MOEA Complexity

It is well known that fitness function evaluation (for many real-world problems) dominates EA execution time. Thus, when discussing various MOEAs' algorithmic complexity one is concerned mainly about the number of fitness evaluations. Solution comparisons and additional calculations are considered, as this overhead is not found in simple GA (SGA) implementations. The complexity of the evolutionary operators is ignored for the current purpose.

Table 6.3: MOEA Solution Technique Complexity

MOEA Technique	Computational Complexity
SGA	$T_f G n$
Lexicographic	$T_f G n k + G n^2 k - G n k$
Linear Combination	$T_f G n k + G n k - G n$
Multiplicative	$T_f G n k + G n k - G n$
Target Vector	$T_f G n k + G k^2 + 2 G k$
Minimax	$T_f G n k + 3 G n k$
Independent Sampling	$c[T_f G n k + G n k - G n]$
Criterion Selection	$T_f G n k + G n$
Aggregation Selection	$T_f G n k + G n k - n$
Pareto Rank	$T_f G n k + G n^2 k - G n k$
Pareto Niche and Share	$T_f G n k + G n^2 k - G n k + n^2$
Pareto Demes	$T_f G n k + G \frac{m^2 k}{n^2} - G \frac{mk}{n} + \frac{m}{n} T_{comm}$
Pareto Elitist	$T_f G n k + G n^2 k - G n k$

MOEA complexity is generally greater than that of SGAs. After fitness evaluation in an SGA, resultant values are stored in memory and no further computation is (normally) required as far as fitness is concerned. However, a

MOEA sometimes combines and/or compares these stored values which adds algorithmic complexity. As a reference, the complexity of the various MOEA techniques is presented in Table 6.3; SGA complexity is included for comparison. Each technique's "worst-case" was used to generate these figures.

The table's notation is as follows. Population size is denoted by n and the number of generations by G. T_f represents fitness computation time (assumed here to be equal for each objective). The number of fitness functions is designated by k and the number of solutions per processor (the Pareto demes case) by m. All table entries are based upon a single generational population, i.e., no secondary populations are used. All techniques are assumed to store a solution's evaluated fitness making selection's computational cost inconsequential. All listed techniques have the identical basic cost of $T_f G n k$ fitness computations. Finally, independent sampling's complexity was computed using several runs of a linear fitness combination technique. Randomly assigned weights (in the fitness functions) were used for the aggregation technique's complexity determination. Table 6.3 shows MOEA techniques explicitly incorporating Pareto concepts are the most computationally expensive; this is due primarily to the $\mathcal{O}(n^2)$ cost of determining which solutions in some set are Pareto optimal.

MOEA storage requirements are problem dependent. Like other EAs these requirements are mandated by the specific data structures used. Required storage increases linearly with the number of fitness functions used, and when a secondary population is brought into play.

It is noted here that MOEA complexity may be a moot issue in real-world applications. As fitness function evaluation (for many real-world problems) dominates EA execution time, the overhead involved in any of the presented techniques may be miniscule in comparison. If that is the case the complexity issue "goes away" as long as the technique appears effective and efficient.

MOEA population storage requirements

Although several researchers have explored the use of alternative data structures to store the external population of a MOEA (see for example [1454, 1141, 485]), very little work has dealt with the use of alternative data structures for the main population of a MOEA (arrays are normally adopted for that sake).

In [30], Alberto and Mateo present a new tool for representing and managing populations of MOEAs, sorted by using Pareto ranking orders, by means of the use of directed graphs that keep the information on the relations among the individuals of the populations. The main idea is to build a domination graph (DG) based on the dominance relations among the individuals of the population, and to work by inserting and deleting nodes and arcs in this graph. A DG has its nodes associated with the individuals of the population of a MOEA, and by means of the arcs, the dominance relations between individuals are represented in such a way that if there exits an arc from the node associated with individual i to the node associated with individual j, then individual i dominates individual j. Since there is not only one DG associated with one population of a MOEA, Alberto and Mateo select the DG which has the fewest number of arcs, in order to make the management of the DG as efficient as possible. This DG is called irreducible domination graph (IDG) and, according to Alberto and Mateo, is unique given a population of a MOEA and can be constructed using the layers classification proposed by Goldberg [581].

Alberto and Mateo develop algorithms for the construction and updating (addition and removal of individuals in the population) of the IDG and analyze their theoretical complexities. They show that the storage requirements of the entire process is $O(m)$, where m is the maximum between the number of arcs at the beginning and at the end of the process. Also, they show that the complexity of the algorithms for constructing the initial IDG, removal and addition of nodes is $O(n^3 + n^2q + C(q,n))$, $O(mn + n2)$ and $O(m + nq)$, respectively, where $C(q,n)$ is the complexity process for calculating the layers, q is the number of objectives, and n and m are the number of nodes and arcs, respectively.

On other hand, Alberto and Mateo present a computational analysis of the algorithms proposed and they verify that, in practice, the arc number of the graph is (in almost all cases) $O(n^b)$ ($b \approx 1$), the IDG construction algorithm requires a time $O(n^b)$ ($b \approx 2$) and the removal and addition times have a practical behavior that can be considered linear.

The methodology proposed by Alberto and Mateo has the important characteristic that it can be used to re-implement existing operators or algorithms. With this regard, they perform a simplification of the selection process of the NSGA-II algorithm and show how the dynamic updating of the proposed graph for making the ranking and selection behaves more efficiently. According to Alberto and Mateo, this is because when, for example, a new population is constructed, if some elements are retained from the old one, it is only necessary to make the calculations corresponding to the new individuals because the information of the individuals that are retained is already stored in the graph. Furthermore, when adding one individual to the population, with their structure, in general, it will not be generally necessary to compare it with all the existing ones.

6.3.9 MOEA Scalability

Another interesting topic of research that has been only scarcely studied is scalability of MOEAs. Some researchers have proposed multi-objective estimation of distribution algorithms[5] (MOEDAs) with the aim of carrying over the

---

[5] Estimation of Distribution Algorithms (EDAs) do not use crossover or mutation, but adopt instead probability distributions to generate the new population. Such probability distributions are estimated from a set of selected individuals generated at the previous generation [954].

polynomial scalability of single-objective EDAs to boundedly-difficult multi-objective search problems [1430]. As Sastry et al. [1430] indicate, MOEDAs have been shown to outperform traditional MOEAs in efficiently searching and maintaining Pareto optimal solutions on such problems. However, the usual scalability approach used for single-objective EDAs does not work for multiobjective problems, and it is easy to get into combinatorial difficulty.

In [1430], Sastry et al. analyze the scalability of MOEDAs on a class of boundedly-difficult additively decomposable problems ($m - k$ deceptive *trap* problems). They measure the scalability of MOEDAs as the minimum number of function evaluations required to maintain at least one copy of all the Pareto-optimal solutions for problems of different sizes.

In addition, Sastry et al. investigate the population size required to maintain at least one copy of all the Pareto solutions. They demonstrate that even if the building blocks are correctly identified, the combinatorial explosion of the number of Pareto optimal solutions can overwhelm the niching capability and, as expected, this leads to exponential scalability. Also, since the number of Pareto optimal solutions grows exponentially, in order to maintain at least one copy of all the global solutions, exponentially large population sizes would be required.

Finally, Sastry et al. show that MOEDAs scale polynomially with problem size only if the multiple objectives share common building blocks and have a limited number of building blocks that are different. That is, there is an imposed limit on the type of additively decomposable problems MOEDAs can solve in polynomial time.

6.3.10 Running Time Analysis

It is also of theoretical interest to perform a quantitative analysis of a multi-objective evolutionary algorithm for a given class of problems. Of particular interest is to be able to determine the expected running time, and the success probability of a MOEA for a given optimization time. Other researchers have performed this sort of study for single-objective evolutionary algorithms (see for example [1394]).

In [962], Laumanns et al. presented a running time analysis of multiobjective evolutionary algorithms for a discrete optimization problem. They define a simple pseudo-Boolean problem called LOTZ (Leading Ones-Trailing Zeroes), which maps n binary decision variables into 2 objective functions:

$$\text{LOTZ}(x_1, x_2, ..., x_n) = \left(\sum_{i=1}^{n} \prod_{j=1}^{i} x_j, \sum_{i=1}^{n} \prod_{j=i}^{n} (1 - x_j) \right) \quad (6.13)$$

Based on this definition of the LOTZ problem, they investigate the time required to find the entire set of Pareto optimal solutions.

They shown that different multiobjective generalizations of a (1+1) EA (like the multi-start option or the random walk performed by PAES [886]) as well as the Simple population-based Evolutionary Multiobjective Optimizer (SEMO) (proposed by them) need on average at least $\Theta(n^3)$ steps to optimize this function. Also, they propose the Fair population-based Evolutionary Multiobjective Optimizer (FEMO) (which improves the selection mechanism of SEMO) and prove that this algorithm is able to find the whole Pareto set in $\Theta(n^2 \log n)$ steps (function evaluations). In this way, Laumanns et al. showed for the first time that the concept of population leads to a provable advantage on a multiobjective optimization problem compared to standard approaches based on scalarizing functions. The SEMO algorithm, proposed by Laumanns et al., contains a population of variable size that stores all individuals that are not dominated by any other individual found so far. At the beginning, the population is initialized with a single element, which is drawn at random from the decision space. In each iteration, one parent individual x is drawn from this population uniformly at random and mutated. The mutation operator consists of flipping a bit randomly chosen from the individual x. The child x' is added to the population if it is not already contained. All individuals that are dominated by the child are in turn deleted from the population. On the other hand, the *fair* sampling strategy implemented by FEMO guarantees that at the end all individuals receive about the same number of samples. The selection strategy of FEMO counts the number of offspring each individual produces and deterministically chooses the individual which has produced the least number of offspring so far; ties are broken randomly.

According to Giel [560], the mutation operator used by SEMO implies that SEMO searches locally in the manner of a hill climber, and, since FEMO differs from SEMO only in the selection operator, both SEMO and FEMO can only be applied if we have some intuition of the optimization problem that suggests that such strategies are not very likely to get trapped in local optima. In this way, in [560], Giel proposes a variant of SEMO, called global SEMO, which modifies the mutation operator by flipping each bit of an element of the population (also uniformly selected) independently with probability $1/n$, where n is the dimension of the search space. Thus, global SEMO searches globally and its population will not get stuck in local optima forever.

Giel analyzes the running time of global SEMO for the Boolean decision space and proves that the expected running time is $\mathcal{O}(n^n)$ for all objective functions $\{0,1\}^n \to \mathbb{R}^m$. Also, Giel shows that the expected running time of global SEMO for the LOTZ function (equation (6.13)) is $O(n^3)$. Finally, Giel adapts the function $x \to (x^2, (x-2)^2)$, commonly used to test algorithms in the continuous decision space, to the Boolean space in two different variants, and shows that the expected running time of global SEMO, under certain conditions, is $O(n \log n)$ and $\Theta(n^{(k+1)})$ (where $0 \leq k \leq n-1$), respectively.

More recently, Laumanns et al. extended their work presented in [962], by introducing two pseudo-Boolean model problems (which are scalable in the number of decision variables and number of objectives), defining a new population-based MOEA, proposing methods to analyze these algorithms and presenting complexity results regarding the expected running time of the different algorithms for the different problems analyzed [960, 955]. According to Laumanns et al., the LOTZ problem has a particular feature: all non-Pareto-optimal decision vectors only have 1-bit Hamming neighbors that are either better or worse, but never incomparable to it. This fact facilitates the analysis of the population-based algorithms, which certainly cannot be expected from other multiobjective optimization problems. Therefore, in [960, 955], they present another simple multiobjective problem, called COCZ (Count Ones Count Zeroes), where this condition does not hold:

$$\text{COCZ}(x_1, x_2, ..., x_n) = \left(\sum_{i=1}^{n} x_i, \sum_{i=1}^{n/2} x_i + \sum_{i=n/2+1}^{n} (1 - x_j) \right) \quad (6.14)$$

where $n = 2k$ and $k \in \mathbb{N}$.

Based on this definition of the COCZ problem, and on the hypothesis that given a dominated solution x, the mutation operator applied to x produces a dominating decision vector x' with probability bounded by $p(x) > 0$, Laumanns et al. prove that the expected running time of SEMO applied to COCZ is $O(n^2 \log n)$. Also, they prove that the expected running time of FEMO applied to COCZ is bounded above by $O(n^2 \log n)$ and below by $\Omega(n^2)$. On the other hand, Laumanns et al. prove that the expected running time of a multistart $(1+1)$-EA based on the ϵ-constraint method is $(1/2)(n^3 + n^2)$ for LOTZ and $\Theta(n^2 \log n)$ for COCZ. In this way, the running time of SEMO for both the LOTZ and the COCZ problems is of the same order as the multistart $(1+1)$-EA.

In [960, 955], Laumanns et al. extended the FEMO algorithm in order to achieve maximum progress toward the Pareto front. In this way, they proposed the GEMO (Greedy Evolutionary Multiobjective Optimizer) algorithm, which uses a greedy selection mechanism whose main idea is to allocate all search effort to offspring of the most recently successful mutant. As long as only mutually nondominating individuals are found, GEMO acts like FEMO, in order to spread out the population and, hence, the search effort, fairly and equally. However, when further progress toward the Pareto front is achieved (a new individual is found that dominates elements of the current population), all other remaining population members are disabled (they cannot produce any offspring). When GEMO finally reaches the Pareto front and no further progress is possible, it will again behave like FEMO.

Laumanns et al. mention that the behavior of GEMO in the LOTZ problem is identical to that of FEMO. However, they prove that the expected running time of GEMO applied to COCZ is bounded by $\Theta(n^2)$. Furthermore, in [960, 955], Laumanns et al. generalize both the LOTZ and the COCZ problems to arbitrary objective space dimensions. According to Laumanns et al., the LOTZ problem can be generalized to an arbitrary even number of objectives m by concatenating $m/2$ biobjective LOTZ problems of $2n/m$ bits each:

$$m\text{LOTZ}(x_1, x_2, ..., x_n) = (f_1, f_2, ...f_m) \qquad (6.15)$$

with:

$$f_k = \begin{cases} \sum_{i=1}^{n'} \prod_{j=1}^{i} x_{j+n'(k-1)/2} & \text{if } k \text{ is odd} \\ \sum_{i=1}^{n'} \prod_{j=i}^{n'} (1 - x_{j+n'(k-2)/2}) & \text{else} \end{cases}$$

where $m = 2m'$, $m' \in \mathbb{N}$, and $n = m'n'$, $n' \in \mathbb{N}$.

Based on this definition of the mLOTZ problem, Laumanns et al. prove that the expected running time of the multistart $(1+1)$-EA, the SEMO and FEMO, and the GEMO algorithms applied to mLOTZ is bounded by $\Theta(n^{m/2}n^2)$, $O(n^{m+1})$ and $O(n^{m/2}n \log n)$, respectively.

On the other hand, Laumanns et al. define the mCOCZ problem in the following way:

$$m\text{COCZ}(x_1, x_2, ..., x_n) = (f_1, f_2, ...f_m) \qquad (6.16)$$

with:

$$f_j = \sum_{i=1}^{n/2} x_i + \begin{cases} \sum_{i=1}^{n'} x_{i+n/2+(j-1)n'/2} & \text{if } j \text{ is odd} \\ \sum_{i=1}^{n'} (1 - x_{i+n/2+(j-2)n'/2}) & \text{else} \end{cases}$$

where $m = 2m'$, $m' \in \mathbb{N}$, and $n = mn'$, $n' \in \mathbb{N}$.

Based on this definition of the mCOCZ problem, Laumanns et al. prove that the expected running time of the multistart $(1+1)$-EA, the SEMO and FEMO, and the GEMO algorithms applied to mCOCZ is bounded by $\Theta(n^{m/2}n \log n)$, $O(n^{m+1})$ and $O(n^{m/2}n \log n)$, respectively.

Finally, in [960, 955], Laumanns et al. provide a short discussion about the disadvantage of the SEMO, FEMO and GEMO algorithms, derived from the use of the one-bit mutation operator, as it was already noted in the work of Giel [560]. Thus, they discuss the possible effects of the independent-bit mutation operator, that is, a mutation operator in which each bit is flipped independently with probability $1/n$.

In [917], Kumar and Banerjee propose a simple multi-objective evolutionary algorithm based on an archiving strategy, which is adapted to work efficiently for problems where the Pareto optimal solutions are Hamming neighbors uniformly distributed over the Pareto front. They call their algorithm Restricted Evolutionary Multiobjective Optimizer (REMO).

The REMO algorithm uses a restricted mating pool or population of only two individuals and a separate archive for storing all other points that are likely to be produced during a run of the algorithm. The two individuals of the population are selected from the union of the current population and the archive. With a probability of 0.5, the individuals are selected at random. Otherwise, the selection is based on a special function that assigns to each individual a fit value equal to the number of its Hamming neighbors, and selects the two individuals with the smallest values of fit. The rest of the individuals are transferred to the archive. Such a mechanism assures that the individuals selected for mutation are more likely to produce new Hamming neighbor individuals. In this way, the expected waiting time till the desired individual is selected for mutation is considerably reduced. The algorithm in its main loop selects an individual from the population at random, and creates a new individual by mutating a single bit randomly chosen. The new individual is added to the population if it is not dominated by the population and the archive, and the dominated individuals are removed.

Kumar and Banerjee present a rigorous complexity analysis and prove that the expected running time of the REMO algorithm when applied to the LOTZ function is $O(n^2)$ with a probability of $1 - e^{-\Omega(n)}$. However, since they define the running time of an algorithm in terms of the number of iterations, the corresponding expected running time of REMO in terms of the number of evaluations performed when applied to the LOTZ function is $O(n^4)$. Also, Kumar and Banerjee prove that the expected running time of the REMO algorithm when applied to the boolean Quadratic function (QF) defined as:

$$QF : ((\|x\| - a)^2, (\|x\| - b)^2) \quad \|x\| = \sum_{i=1}^{n} x_i$$

is $O(n \log n)$ ($O(n^2)$, in terms of the evaluations performed).

Finally, Kumar and Banerjee analyze REMO when applied to the well-known bi-objective 0-1 knapsack problem. They perform a partition of the decision space into fitness layers (by partitioning the items of the knapsack problem into blocks) and prove that the expected running time of REMO is $O(\frac{n^{2m+1} P_{knap}}{m^{2m+1}})$, where n is the number of items, m is the number of blocks into which the items can be divided and P_{knap} refers to the sum of the profits of the items in the knapsack problem ($n \neq m$).

Since an important open problem is to understand the role of populations in MOEAs, in [561], Giel and Lehre present a simple biobjective problem which emphasizes the case in which populations are needed. Rigorous runtime analysis point out an exponential runtime gap between the population-based algorithm SEMO and several single individual-based algorithms on this problem.

All the single individual multi-objective evolutionary algorithms considered by Giel and Lehre are instantiations of a scheme in which initially one individual x is randomly chosen from $\{0,1\}^n$ and, at each iteration, a new individual x' is obtained by applying a mutation operator to x; x' replaces x if x' is better than x. The algorithms differ in the choice of the mutation and selection operators. Giel and Lehre use two different mutation operators: local mutation (flips a randomly chosen bit) and global mutation (flips each bit independently with probability $1/n$). On the other hand, four selection operators are used:

- *Weakest selection operator*: favors x' over x if x' weakly dominates x, or x' and x are incomparable.
- *Weak selection operator*: favors x' over x if x' weakly dominates x.

- *Strong selection operator:* favors x' over x if x' dominates x.
- *ϵ-constraint selection operator:* if $f_1(x) < \epsilon$, then the operator favors x' over x if $f_1(x') \geq f_1(x)$. If $f_1(x) \geq \epsilon$, then the operator favors x' over x if $f_1(x') \geq \epsilon$ and $f_2(x') \geq f_2(x)$.

In this way, Giel and Lehre obtain eight different single individual-based MOEAs. For the case of SEMO, Giel and Lehre consider two versions: local SEMO (with local mutation) and global SEMO (with global mutation). The objective function used by Giel and Lehre is defined in terms of the *active block* of a string. Let $n = mk$ and let x be a bit string of length n. We say that bit string x is divided into k blocks, where each block has length $m \geq 2$.

Definition 53 (**Block Value, Active Value**) : *Given a search point $x \in \{0,1\}^{km}$ and an integer i, $0 \leq i \leq k-1$. Then the i-th **block value** of x, denoted by $|x|_i$, is defined as*

$$|x|_i := \sum_{j=mi}^{m(i+1)-1} x_j$$

*The **active block** of a search point x is the leftmost block with lowest block value. The number*

$$j := \min \, argmin_{0 \leq i \leq k-1}\{|x|_i\}$$

denotes the active block index. □

The multi-objective function $f : \{0,1\}^n \to \mathbb{N} \times \mathbb{N}$ used by Giel and Lehre is defined as:

$$f(x) := (2^{jm}(|x|_j + 1), 2^{(k-j-1)m}(|x|_j + 1))$$

where j is the active block index of x. The aim is to maximize f. As the entire Pareto front of f is non-convex, all methods that require a convex Pareto front are not applicable to f.

In [561], Giel and Lehre prove that search points with the same active block index are comparable whereas search points with distinct block indices are incomparable. From this result, they also prove that all search points selected by either the weak or the strong selection operators have the same active block index as the initial search point. In this way, Giel and Lehre are able to prove that there is a large fraction of the Pareto front such that, with an overwhelming probability, the algorithms that use either the weak or the strong selection operator have to be started $e^{\Omega(n)}$ times before finding any Pareto optimal point from this fraction.

In this way, both weak and strong selection turn out to be inadequate. The weakest selection operator alleviates this problem by allowing to change the active block index. However, Giel and Lehre show that this is not sufficient and prove that the algorithms that use the weakest selection operator need with overwhelming probability an exponential time to find any Pareto optimal solution.

For the case of the algorithms that use the ϵ-constraint selection operator, Giel and Lehre prove a result similar to that proved for the weak and strong selection operator's algorithms. Finally, Giel and Lehre prove that, within polynomial time, the SEMO population covers the entire Pareto front. More precisely, Giel and Lehre prove that the expected running time until the SEMO population covers the Pareto front is $O(nk^2 \log m)$.

In this way, among the algorithms considered, only the population-based MOEA is successful and all the other algorithms fail. That is, only the population-based algorithm SEMO finds the Pareto front in expected polynomial time. This result demonstrates the importance of populations for certain types of multi-objective problems. Also, this result of Giel and Lehre improves the result provided by Laumanns et al. in [962], since that result yields only a small polynomial runtime gap, whereas Giel and Lehre provide an exponential gap.

6.3.11 MOEA Computational "Cost"

When practically considered, MOP evaluation cost limits MOEA search. The most "expensive" EA component in many real-world MOPs is the fitness function evaluation. Since all algorithms must eventually terminate the number of fitness evaluations is then often selected as the finite resource expended in search, i.e., the choice is made *a priori* for an EA to execute n fitness evaluations. The "best" solution found is then returned. Assuming solutions are not evaluated more than once (no clones) a total of n points (possible solutions) in the search space are explored.

Now consider a k-objective function. Here, k fitness evaluations are performed for each possible solution (one for each objective). Assuming resources are still limited to n fitness evaluations and that each objective evaluation is equally "expensive", only $\lfloor \frac{n}{k} \rfloor$ points in the search space are now explored. All else held equal, a k-objective optimization problem may then result in a k-fold decrease in search space exploration. Note also that in the context of MOEAs, this implies using the term "fitness function evaluations" to measure computational effort may be somewhat misleading. The term "solution evaluations" is clearer.

This result implies a MOEA may require longer (than a single-objective EA) "wall clock" execution times for good performance. Further search is never guaranteed to return the optimal answer but one wishes as much exploration as possible in the time allowed. This increases the sense of confidence one has found the true, and not a local, optimum.

6.3.12 NFL-Theorem for Multiobjective Optimization Algorithms

As we mentioned before, the No Free Lunch Theorems (NFL) [1708] imply that, if problem domain knowledge is not incorporated into the algorithm domain, no formal assurances of an algorithm's general robust effectiveness exist.

In [898], Köppen extends the NFL theorems to the case of multiobjective optimization through of the following theorem:

Theorem 17: For any two deterministic algorithms a and b, any performance value $k \in \mathbb{R}$, and any performance measure c:

$$\sum_f \delta(k, c(m, a, f)) = \sum_f \delta(k, c(m, b, f))$$

where δ is the Kronecker delta and (m, a, f) represents a sequence of m successive applications of the algorithm a to the problem f. □

In other words, the NFL theorem states that, on average, each algorithm has the same performance when applied to all possible problems f, provided that no *a priori* knowledge of the problem is assumed. Nevertheless, Köppen also shows that even in cases of *a priori* knowledge, when the performance measure is related to the set of extrema point sample so far, the NFL theorems still hold.

According to Köppen, the NFL theorem can also be seen as stating the impossibility to obtain a concise mathematical definition of algorithm performance. However, he shows that a procedure for obtaining function-dependent algorithm performance can be constructed, the so-called tournament performance, which is able to gain different performance measures for different multiobjective algorithms. Finally, Köppen proposes a heuristic procedure to measure algorithm performance:

1. Let algorithm a run for k evaluations of cost function f and take the set M_1 of nondominated points obtained by the algorithm.
2. Select k random domain points and compute the Pareto set M_2 of the corresponding f values.
3. Compute the set of M_3 of elements of M_2 that are not dominated by any element of M_1.

The relation of $|M_1|$ to $|M_3|$ gives a measure of how algorithm a performs with respect to random search.

6.3.13 Alternative Definitions of Optimality

According to Farina and Amato [471], when dealing with multi-objective optimization problems, the concepts of Pareto dominance and Pareto optimality may be inefficient in modeling and simulating human decision making. This occurs when problems with a number of objectives relatively large (more than two or three) are considered, since the set of Pareto optimal solutions becomes large and unmanageable. In [471], Farina and Amato discuss that, when comparing two solutions according to Pareto definitions, the following three aspects are not taken into consideration:

- the number of improved (or decreased) objectives,
- the size of such improvements (or decreases) and
- the decision maker preferences between objectives (if any).

Also, they mention that these three issues are crucial in the human decision making process and may lead to several degrees of dominance, when two solutions are compared and, consequently, to several degrees of optimality among Pareto optimal solutions. For these reasons, with the aim of generalizing the definition of Pareto optimality,

Farina and Amato introduce different fuzzy-based definitions of dominance and optimality [471]. They provide three definitions: k-optimality, k_F-optimality (with fuzzy numbers) and fuzzy optimality. Each one is a sound extension both of the previous one and of Pareto optimality. Moreover, such definitions can be extended to the case of fuzzy objectives and fuzzy constraints. The hypothesis underlying the given definitions is that the satisfaction of the human decision maker linearly increases with the increase of the improved objectives.

Definition 54 (($1-k$)-**Dominance**) : v_1 is said to $(1-k)$-dominate v_2 if and only if
$$\begin{cases} n_e < M \\ n_b \geq \frac{M-n_e}{k+1} \end{cases}$$
where n_e is the number of objectives, v_1 and v_2 are equal, n_b is the number of objectives, v_1 is better than v_2, and $0 \leq k \leq 1$. □

If we consider n_w as the number of objectives where v_1 is worst than v_2, we have $n_b + n_w + n_e =$total number of objectives.

Definition 55 (k-**Optimality**) : $v*$ is k-optimum if and only if there is no $v \in \Omega$ such that v k-dominates $v*$. □

It can be seen that the former is a loose version of Pareto dominance (1-dominance) and the latter is a strong version of Pareto optimality (0-optimality). An extension of the previous definitions is to substitute crisp relations with fuzzy ones.

Definition 56 (($1-k_F$)-**Dominance**) : v_1 is said to $(1-k_F)$-dominate v_2 if and only if
$$\begin{cases} n_e^F < M \\ n_b^F \geq \frac{M-n_e^F}{k_F+1} \end{cases}$$
where n_b^F, n_w^F and n_e^F are defined in terms of the membership functions μ_b, μ_w and μ_b, respectively, and $0 \leq k_F \leq 1$. □

Definition 57 (k_F-**Optimality**) : $v*$ is k_F-optimum if and only if there is no $v \in \Omega$ such that v k_F-dominates $v*$. □

Farina and Amato consider two possible membership shapes: linear and Gaussian. Both membership shapes require some parameters that can be provided by the human decision-maker knowledge of the system, defining the practical meaning of equality and improvement. In addition, Farina and Amato obtain a more general procedure through the introduction of a fuzzy definition for the dominance relation itself, and not only for the quantities n_b^F, n_w^F and n_e^F.

Definition 58 (Fuzzy-Dominance) : *Let*

$$\mu_D(v_1, v_2) \triangleq f_{\mu_D}(n_b^F(v_1, v_2), n_w^F(v_1, v_2), n_e^F(v_1, v_2))$$

a membership function defined in terms of a membership function (or a fuzzy system) f_{μ_D} that depends on n_b^F, n_w^F and n_e^F. Then μ_D is a fuzzy dominance relation if for any $\alpha \in [0, 1]$, $\mu_D(v_1, v_2) > \alpha$ implies that v_1 k_F-dominates v_2. □

A membership function for optimality μ_O can be implicitly defined through its α-cuts:

Definition 59 (Fuzzy optimality) : *A membership function μ_O represents the fuzzy optimality relation if for any $k_F \in [0, 1]$ $v*$ belongs to the k_F-cut of μ_O if and only if there is no $v \in \Omega$ such that $\mu_D(v, v*) > k_F$.* □

It is very important to note that Farina and Amato exclude the preferences of the decision maker from the optimization process and consider fuzzy memberships as a tool for the numerical formalization and treatment of the size of improvements in the dominance definition. That is, in the work developed by Farina and Amato, fuzzy set theory is not related to treatment of preferences among objectives, but it is related to the size of improvements, with the underlying hypothesis that all objectives have equal importance.

Based on this definitions, different subsets of Pareto optimal solutions can be computed using simple and clear information provided by the decision maker and using a parameter value ranging from zero to one (k). When the value of the parameter is zero, the introduced definitions coincide with classical Pareto dominance and optimality. When the parameter value is increased, different subsets of Pareto optimal solutions can be obtained corresponding to higher degrees of optimality. In [471], Farina and Amato test their definitions on analytical cases, in order to show their validity and nearness to human decision making.

6.3.14 Local Search

The use of local search within a MOEA is another interesting topic that has been only scarcely studied.[6] Although memetic MOEAs[7] have existed for some

---

[6] Local search in the context of MOEAs is discussed in more detail in Chapter 3.
[7] Memetic MOEAs are discussed in Chapter 10.

time (see for example [879]), most of the hybridizations between a MOEA and a local search mechanism are relatively straightforward algorithmic designs, lacking a careful analysis of the trade-offs involved (i.e., evidently, the use of local search is expected to improve quality of the solutions, but also introduces extra computational costs). Additionally, the use of more elaborate local search mechanisms is still a subject of ongoing research.

In [658], Harada et al. propose a local search method, called Pareto Descent Method (PDM), which finds Pareto descent directions and moves solutions in such directions thereby improving all objective functions simultaneously.

According to Harada et al. a descent direction is defined in the following way:

Definition 60 (Descent Direction) : *Denote by $\nabla f_i(\mathbf{x})$ ($i = 1, 2, ..., m$) the gradients of the objective functions at a solution $\mathbf{x} = (x_1, x_2, ..., x_n)$. A direction $\mathbf{d} = (d_1, d_2, ..., d_n)$ is said to be a **descent direction** if it satisfies:*

$$\mathbf{d} \cdot (-\nabla f_i(\mathbf{x})) \geq 0 \quad (i = 1, 2, ..., m). \tag{6.17}$$

□

There are often multiple descent directions and not all descent directions are similarly capable of improving all objective functions.

Definition 61 (Pareto Descent Direction) : *A descent direction $\mathbf{d}$ is a **Pareto Descent Direction** iff $\mathbf{d}$ can be expressed as a convex combination of the steepest descent directions of objective function, i.e. there exist $\alpha_i \geq 0$ ($i = 1, 2, ..., m$) such that:*

$$\mathbf{d} = \sum_{i=1}^{m} \alpha_i (-\nabla f_i(\mathbf{x})) \tag{6.18}$$

□

In this way, a Pareto descent direction is a descent direction to which no other descent direction is superior in improving all objective functions. There are often multiple Pareto descent directions, and none of the Pareto descent directions is better than any of the others. PDM assumes that the objective functions are differentiable, and that local Pareto optimal solutions that are not Pareto optimal solutions do not exist. Since Equation 6.17 is a simultaneous linear inequality, the complete set of descent directions and Pareto descent directions forms a convex cone pointed at the origin in the vector space. PDM imposes a linear constraint on the convex cone ($\sum_{i=1}^{m} \alpha_i \leq 1$) and obtains a convex polyhedron which has the origin as one of its vertices. Then, PDM calculates directions in the convex cone, by finding the vectors from the origin to the vertices of the convex polyhedron. With this aim, PDM obtains the vertices of the convex polyhedron by solving the corresponding linear programming problems (by means of the Simplex method). Thus, PDM finds a set of feasible Pareto descent directions or feasible descent directions,

6.3 MOEA Theoretical Issues

as appropriate, for solutions inside feasible regions or on feasible region boundaries.

Having found a feasible Pareto descent or a feasible descent direction at a solution, PDM moves the solution in that direction until just before any of the objective functions deteriorate or any of the constraints are violated. If no feasible descent directions can be found, the corresponding solution is Pareto optimal. Harada et al. compare the performance of PDM against three different local search algorithms taken from the literature: Random Direction Search (RDS, which is an ES-like simple local search method), Weighted Steepest Descent Method (WSDM, which forms a scalar function and optimizes it with steepest descent method) and Combined Objectives Repeated Line-Search (CORL, which utilizes the gradients of objective functions to calculate the convex cone of Pareto descent directions) [658]. The experiments using two different test functions with two and three objectives indicate that PDM has good speed convergence and does not bias the solutions to a limited portion of the Pareto optimal set.

In [154], Bosman and de Jong present an adaptive resource-allocation scheme that uses three gradient techniques in addition to the variation operator in a MOEA. They investigate the combined use of three gradient techniques by an adaptive means of choosing how often to use each gradient technique, in order to have the benefits of all of them. The three gradient techniques used are: Random-Objective Conjugate Gradients (ROCG) which applies the conjugate gradients algorithm to a randomly chosen objective; Alternating-Objective Repeated Line-Search (AORL) in which the objective that is searched locally can be altered during search; and Combined-Objectives Repeated Line-Search (CORL) (also used by Harada et al. in [658]). Bosman and de Jong use a generational scheme in which all gradient techniques are applied to one or more solutions in the population at the end of each generation. Each gradient technique is applied only as long as the ratio of the number of evaluations required by that specific gradient technique and the total number of evaluations required so far is smaller than a ratio parameter ρ_e ($0 \leq \rho_e < 1/3$).

According to Bosman and de Jong, a fixed ratio approach is not optimal in general because if one gradient technique is clearly superior to another gradient technique, it is more efficient to allow the more superior technique to spend more search effort. In this way, Bosman and de Jong propose an adaptive allocation of resources such that during the search the most effective gradient technique is assigned the largest probability; or, if no technique is efficient compared to the baseline MOEA, it reduces the use of the gradient techniques to a minimum. At each generation, the effectivity of each gradient technique and the variation operator of the MOEA is assessed in terms of the number of improvements per evaluation. The number of improvements obtained by the variation operator of the baseline MOEA is equal to the number of offspring solutions that are nondominated but also dominate at least one solution in the population. On the other hand, the number of improvements obtained by a gradient technique is equal to the number of offspring solutions that are nondominated but also dominate the solution that the gradient technique started from. Since this notion of improvement is strict, Bosman and de Jong also consider (as a different version of their approach) counting improvements when a new nondominated solution is created that does not necessarily dominates the solution(s) it was created from.

Regarding the number of evaluations used by each operator, for the gradient techniques this number is equal to the number of evaluations used in the most recent generation. However, in order to make a fair comparison and to allow an increase in the number of calls to the gradient techniques, the number of evaluations of the variation operators of the baseline MOEA is obtained by calculating the sum of all evaluations backwards over previous generations until the number of evaluations is at least as large as the largest number of evaluations used by any gradient technique (in the most recent generation). In this way, after calculating the effectivity of each operator, the total number of evaluations used in the most recent generation is then redistributed proportionally to each operator for the next generation.

Bosman and de Jong perform tests on a few well-known benchmark problems with specific gradient properties. They compare the results of a MOEA with the adaptive resource-allocation scheme against the results of the same MOEA without the use of gradient techniques and with a scheme in which resource allocation is constant. The baseline MOEA used by Bosman and de Jong is the so-called naive MIDEA [155], which is an EDA designed for multi-objective optimization. The experiments show that the proposed scheme makes proper use of the gradient techniques only when required and thereby leads to results that are close to the best results that can be obtained by fine-tuning the resource allocation for a specific problem. Also, it is observed that the scheme for counting improvements that provides the best results is the less strict one, which improves diversity along the entire Pareto front.

In [1455], Schütze proposes numerical methods for the approximation of the entire set of solutions of multi-objective optimization problems. More precisely, he proposes algorithms for the computation of tight coverings of such sets. He proposes algorithms for different assumptions of smoothness of the underlying models. Schütze proposes three basic algorithms which can be used on general MOPs followed by extensions both for non-smooth and for smooth objectives leading to particular continuation methods. Continuation methods can be used to compute solution sets efficiently. However, these techniques are of local nature. Given an initial set S_0 of (local) Pareto solutions, all further solutions computed by these methods are restricted to the connected components of the set of (local) Pareto points contained in S_0.

The three basic algorithms developed by Schütze are the Subdivision Algorithm, the Sampling Algorithm and the Recovering Algorithm. The Subdivision Algorithm generalizes a known subdivision algorithm for the computation of invariant sets of single dynamical systems. This algorithm has the advantage of being very robust with respect to errors by the use of the descent direction. However, all the gradients of the objectives have to be available and the algorithm is unable to distinguish between a local and a global Pareto point. Also, Schütze discusses that the subdivision techniques are restricted in use to moderate dimensionality of the parameter space. The Recovering Algorithm uses a kind of "healing" process which allows to recover those substationary points which have been previously lost. The recovering algorithms are local in nature, but on the other hand are not restricted to moderate dimensions like the algorithms based on subdivision techniques.

Since the main drawback of the two previous algorithms is that the gradients of the objectives are needed, Schütze proposes the Sampling Algorithm which takes only the function values of the objective functions into account. The results of this algorithm show it works quite well, in particular when the dimensionality of the MOP is moderate. However, this algorithm is not as robust to errors as the first two. In order to obtain a best overall performance, Schütze proposes a combination of the three algorithms (assuming that the gradients of all objectives are available): start with the subdivision algorithm, apply the recovering algorithm in order to fill the gaps which have possibly been generated before, and use the sampling algorithm to tighten the extended covering.

In cases where the MOP is not continuous (non-differentiable) and/or the dimension of the parameter space is large, Schütze proposes combinations of MOEAs with both the subdivision and the recovering techniques. Since MOEAs typically generate very quickly good approximations of Pareto points, they only have to run for a short time and the recovering techniques can merely be applied to improve the obtained results. Schütze proposes the EA-subdivision algorithm, which uses the sampling algorithm combined with a short MOEA. The only task of the MOEA is to find as fast as possible one good approximation of the Pareto set. Also, Schütze proposes two versions of the recovering algorithm (static and dynamic) for the corresponding process of "healing".

6.4 Summary

This chapter presents various MOEA theorems and corollaries providing insight to the Pareto front convergence possibilities in a given MOP application. Also, general observations concerning MOEA theoretical issues as related to operator selection and insight are discussed. Finally, other aspects such as computational complexity, running time analysis, landscape analysis, and computational cost of a MOEA are briefly discussed.

Further Explorations

Class Exercises

1. Provide the basic elements required to prove convergence of a MOEA, as indicated in [1401]. How is this different from the proof provided in [1653]?
2. Describe the basic concepts associated to Markov chains that are required to prove convergence of a simple (single-objective) genetic algorithm (see [1393]). What extra elements are required for a multi-objective genetic algorithm?
3. Why is it a difficult problem to define bounds on the convergence of a MOEA? Discuss possible ways to tackle this problem.
4. Can you sketch a possible way of incorporating self-adaptation in a theoretical model of convergence of a MOEA? Discuss.
5. Discuss scalability (in the number of objective functions) and its implications for Pareto dominance. Can Pareto ranking be affected by scalability? How? See for example [880].
6. Does the use of an external archive affect in any way the requirements to prove convergence of a MOEA? Discuss.
7. Indicate the difference between the traditional Pareto dominance and ϵ-dominance [959]. Can we prove convergence of a MOEA if ϵ-dominance is adopted instead of Pareto dominance? Discuss.
8. Why do you think that is difficult to derive accurate convergence rates and convergence bounds for a MOEA?
9. What is the usual mapping of "multimodality" into multi-objective optimization problems? Do you think that this feature increases the degree of difficulty of a problem?

Class Software Projects

1. If convergence is measured in terms of obtaining elements of the Pareto optimal set (assuming that the true Pareto optimal set is known), write

a computer program that shows in a graphical form the convergence of a MOEA in a discrete problem. Analyze the behavior of the MOEA when different Pareto ranking schemes and variation operators (i.e., crossover and mutation) are used.
2. Choose a set of test functions that are considered as "very difficult" to solve for state-of-the-art MOEAs (see for example [721]). Design an empirical study in which the behavior of a MOEA is statistically analyzed when attempting to solve such difficult test functions. Can you derive some "sources of difficulty" from this study? Discuss.
3. Introduce variable linkages in a problem (see for example [377]) and analyze (in an empirical way), how does this variable linkage affect the behavior of a MOEA. Write a report that details your experimental setup, your results and your analysis of such results.
4. Using a program that you have written, trace the convergence of a MOEA to the true Pareto front of a discrete multi-objective optimization problem.
5. Write a program to generate, in graphical form, the fitness landscape of a multi-objective optimization problem. Use your program to analyze the fitness landscapes of simple problems (i.e., with two objectives and two decision variables).
6. Implement 3 different diversity preservation mechanisms (e.g., fitness sharing [368], clusters [1782] and the adaptive grid [886]) within any MOEA of your choice (see Chapter 2) and compare their performance using the DTLZ test functions [379] (see Chapter 4).

Discussion Questions

1. Discuss two possible ways in which the development of a specific MOEA may provide insights regarding the generation of new theoretical results.
2. Develop a series of MOEA experiments to evaluate various population selection operators on (μ, λ) and $(\mu+\lambda)$ multi-objective evolution strategies and relate to anticipated convergence results based on theoretical analysis.
3. Regarding positive variation kernels and strong elitist preservation properties, evaluate five MOEAs with respect to their relationship to this model.
4. Discuss some other possible applications of concepts related to MOEAs in evolutionary computation. For example, Rudolph [1398] proposed the use of partially ordered sets (and MOEAs) as an alternative way to deal with noisy fitness functions.
5. Analyze the convergence proofs for multi-modal multi-objective optimization presented by Bonnemay et al. [149]. How is this work related to the proofs described in this chapter? Discuss.
6. Analyze the runtime analysis of a MOEA in the multi-objective minimum spanning tree problem presented by Neumann [1183]. What sort of MOEA is adopted for this study? What is the expected runtime derived by the author? What types of Pareto fronts are considered in this analysis?

Possible Research Ideas

1. Consider the use of recombination and mutation adaptive operators as well as fitness sharing in the development of an extended MOEA convergence theory.
2. Relax the concept of positive variation kernel or the strong elitist preservation and attempt to generate new lemmas and theorems related to convergence.
3. Relax the concept of partially order sets for MOEAs and generate new lemmas and theorems related to convergence.
4. Generate necessary conditions for MOEA convergence for finite search spaces and infinite search spaces.
5. Attempt to generate necessary and sufficient MOEA convergence conditions for finite search spaces.
6. Extend Hanne's theoretical convergence results for other selection functions.
7. Formalize selected MOEA algorithms in a generalized Back's or Merkle's notation and develop associated theorems of reachability and convergence [72, 1092].
8. Consider the extension of a generalized theory of MOEA convergence for fitness functions with bounded additive noise [1399].
9. Consider the use of the concept of entropy to monitor and control the diversity of the population of a MOEA. Propose a theoretical framework that properly models this sort of approach. See some related proposals such as [312, 465].
10. Attempt to generate the necessary and sufficient conditions to ensure convergence of a multi-objective particle swarm optimizer (MOPSO). Is this similar to proving convergence for a multi-objective genetic algorithm? Why?
11. Provide a runtime analysis of a (binary) multi-objective particle swarm optimizer in a discrete multi-objective optimization problem.
12. Use statistical mechanics to study the population dynamics in a MOEA. See for example [1469].

7
Applications

> In computing, turning the obvious into the useful is a living definition of the word "frustration".
>
> Alan Perlis

7.1 Introduction

Although the application of classical multiobjective optimization techniques to solve problems in different areas (e.g., management, engineering and science) started as early as 1951 (see Section 1.6.2 from Chapter 1), Multi-Objective Evolutionary Algorithms (MOEAs) were applied for the first time until the mid-1980s [764, 1440, 518]. However, since the late 1990s, there has been a considerable increase in the number of applications of MOEAs. This has been mainly originated by the success of MOEAs in solving real-world problems.[1] MOEAs have generated either competitive or better results than those produced using other search techniques. This has made the task of classifying MOEA applications difficult and subjective. Trying to deal with this problem, it was decided to use a rather simple and general classification in this chapter, trying to fit each paper reviewed within the closest category according to the focus of the work. For example, a paper that is related to scheduling and naval engineering but is more focused on the second subject, is classified under "environmental, naval and hydraulic engineering". This avoids overlapping to a certain extent, but can be confusing for some people. Therefore, it was decided to add as many entries as possible to the analytical index provided at the end of this book to facilitate the search. Additionally, *italics* characters are used throughout this chapter to indicate the specific name of an application, in an attempt to facilitate the search of specific information.

---

[1] In fact, there are several recent surveys on applications of MOEAs in specific areas (see for example [393, 1445]), and there is even a recent book entirely devoted to real-world applications of MOEAs [277].

Besides indicating the specific application studied, brief discussions about the type of MOEA used (including encoding, genetic operators and type of evolutionary algorithm) are provided (if such information is available). Additionally, if the approach adopted is compared to any other technique, a brief discussion of the results obtained is also provided.[2]

To facilitate location of information, each section contains a table with a summary of the applications reviewed in that section. The table describes briefly the type of specific application reviewed, the references related to it, and the type of MOEA adopted in each case.[3]

For practical purposes, applications of MOEAs have been divided in four main groups: engineering, scientific, industrial and miscellaneous. Each of these has been further divided into subgroups, as we will see later on.

7.2 Engineering Applications

Engineering is, by far, the most popular application domain area within the MOEAs literature. This is mainly because engineering applications normally have "good" mathematical models (equations/inequalities) that can directly be associated with a MOEA search. To understand better the particular areas of interest within this domain, engineering applications have been further divided into eight subgroups: (1) environmental, naval & hydraulic, (2) telecommunications and network optimization, (3) structural & mechanical, (4) aeronautical, (5) electrical and electronics, (6) robotics and control, (7) civil and construction, and (8) transport.

7.2.1 Environmental, Naval and Hydraulic Engineering

Table 7.1: Summary of environmental, naval and hydraulic engineering applications

Specific Applications	Reference(s)	Type of MOEA
Groundwater pollution remediation	[1364]	VEGA, GA with Pareto ranking
	[708]	NPGA
	[255]	VEGA, GA with Pareto ranking, GA with Tchebycheff weighting method
	[1325]	Multi-Niche Crowding GA
	[453]	NPGA 2
	[542]	GA with a linear aggregating function
Water quality control	[232]	GA with a nonlinear aggregating function
	[1340, 1341, 1342]	NSGA
	[1507]	Noninferior Surface Tracing Evolutionary Algorithm

---

[2] When no mention to any comparisons is made, it means that the authors did not report any validation of their approach with respect to other techniques.
[3] See Chapter 2 for information on each specific type of MOEA.

7.2 Engineering Applications

Table 7.1: (continued)

Specific Applications	Reference(s)	Type of MOEA
Pumping scheduling	[1457, 1432]	GA with Pareto ranking
Water distribution network	[637]	Structured messy GA with Pareto ranking
	[515]	NSGA-II
	[833]	CAMOGA
Gas supply network	[1543, 1542]	VEGA hybridized with Pareto ranking
Air quality management	[1017, 1016]	GA with the Neighborhood Constraint Method
Calibration of hydrologic models	[1733, 1734, 622]	MOCOM-UA
	[849]	Accelerated Convergence Genetic Algorithm with a nonlinear aggregating function
Design of marine vehicles	[1580, 1581]	GA with Pareto ranking, MOGA and NPGA
	[178]	GA with Pareto ranking
	[969]	GA coupled with the ϵ-constraint method
Planning of containership layouts	[1591]	MOGA variant
Location of site retail and service facilities	[616, 615]	GA with a linear aggregating function

Ritzel et al. [1364] use VEGA [1440] and Pareto ranking (also called Pareto GA) to solve a *groundwater pollution containment problem* in which two objectives are considered: reliability and cost of the hydraulic containment system. The decision variables considered are the number of wells to install, their locations and the amount of pumping required from each of them. They employ binary representation and deterministic binary tournament selection with both VEGA and the Pareto GA. Results of both techniques are compared with respect to each other and against those produced with the Mixed Integer Chance Constrained Programming (MICCP) method [1128]. The Pareto GA was found to be superior to VEGA in terms of generation of the trade-off curve. The Pareto GA also compared well against MICCP, but several problems were found to fine-tune the parameters of the GA and to decrease its computational cost (initially higher than that of MICCP).

Horn et al. [708, 709] use the NPGA to solve a similar problem: *optimal well placement for groundwater containment monitoring*. The problem has two objectives: find the placement of a set of wells so that the number of detected leak plumes from a landfill into the surrounding groundwater is maximized, while the volume of cleanup involved is minimized.

Cieniawski et al. [255] work on the same problem as Horn et al. [708] assuming uncertainty. In their study, they consider two objectives: maximize reliability and minimize contaminated area at the time of first detection. They compare the results produced by VEGA, a Pareto GA (with and without sharing), a GA using a Tchebycheff weighting method as proposed by Steuer [1522] and simulated annealing using the same Tchebycheff weighting method adopted with the GA. They also experiment with a combined technique in which VEGA is run for a certain number of generations after which Pareto ranking is applied. The combined approach between VEGA and Pareto ranking provides the best results (in terms of coverage of the Pareto front) together with the weighted GA approach (except for some non-convex portions of the Pareto front that are not found by the algorithm). Apparently, the use of pure Pareto ranking could not find the extremes of the Pareto curve, and the authors indicate how their weighted simulated annealing approach could outperform the GA in terms of CPU time required (while finding the same points of the Pareto front as the GA). Their preliminary experiments with sharing led them to advise the use of sharing in objective space rather than in decision variable space. For these experiments, an integer representation and binary tournament selection are adopted.

Vemuri and Cedeño [1325, 1326] use a Multi-Niche Crowding GA (MNC GA) to determine the *optimum placement of pumping (and recharge) wells* and *optimum pumping schedules during the remediation of contaminated groundwater aquifers*. Three objectives are considered: minimize remediation cost, maximize the amount of contaminant removed, and minimize the concentration of contaminant leaving the site. The objective is to find a set of wells whose cost stays within certain budget restrictions. Instead of using Pareto ranking, they compute ranks of each solution for each particular objective and determine fitness by adding these rankings (e.g., considering three objectives, an individual whose encoded solution ranks second for the first and second objectives and fourth for the third objective has a fitness of $2 + 2 + 4 = 8$). The representation used is a variable-length integer string.

Erickson et al. [453] use the NPGA 2 to *design groundwater remediation systems*. Two objectives are considered: minimize cost and maximize cleanup performance. The NPGA 2 uses Pareto ranking in a similar way as MOGA [504], but it keeps tournament selection from its original version [709]. It also continues using fitness sharing (in objective space), calculating niche counts using the partially filled next generation population instead of the current one. Continuously updated fitness sharing is adopted to decrease the high selection pressure of tournament selection [1205]. Results are compared against an enumerated random search approach and against the use of a simple GA in which the objective function is to minimize cost and cleanup performance is handled as a constraint. The quality of results is assessed in terms of closeness to PF_{true} and spread of solutions along PF_{true}. The conclusions indicated that as the problems increased in complexity (i.e., as the number of decision variables increased), the NPGA 2 became more effective and efficient than the two other methods considered, in terms of finding more nondominated vectors that were closer to PF_{true} and better distributed along it.

Garrett et al. [542] use a GA with a linear combination of weights for *bioremediation optimization of trichloroethylene-contaminated groundwater*. Four objectives are considered: aquifer pumping flow, oxygen injection rate and time period, toluene (bioremediation enzyme) pulsing rate, and well separation. The complex bioremediation mathematical model incorporates multi-dimensional flow, advection and dispersion, equilibrium or rate-limited sorption, and biodegradation. Using a real-coded GA and this very high-dimensional varying-weight model on a parallel computation platform, parameter results were found to be very similar to those proposed by consulting experts.

Chen and Chang [232] use a GA with an aggregating approach (a multiplication of the aspiration levels of each objective) for a *water quality control problem*. Three objectives are considered: maximize the assimilative capacity of the river, minimize the treatment cost for water pollution control and maximize the economic value of the river flow corresponding to recreational aspects. A nonlinear fuzzy membership function is used to model the uncertainty involved in the computation of the objectives and a penalty function is used to incorporate the constraints of the problem into the fitness function.

Reed et al. [1342, 1340, 1341] use the NSGA to perform *cost effective long-term groundwater monitoring*. Two objectives are minimized: sampling costs and local concentration estimation errors. The approach employs binary representation, phenotypic sharing, and elitism (defined in this case in such a way that only one individual per niche is retained). Historical data at a single snapshot in time are also used to identify potential spatial redundancies within the monitoring network under study. The original NSGA is compared against an elitist version, concluding that the second version performs considerably better than the first. Additionally, some guidelines are derived to: (1) identify the most appropriate population size, (2) choose proper niche sizes, (3) setting up the elitist selection pressure, and (4) avoid genetic drift.

Srigiriraju [1507] uses the Noninferior Surface Tracing Evolutionary Algorithm (NSTEA) to solve an *estuary water quality management problem*. Two objectives are considered: minimize cost of BOD control, and maximize equity with respect to levels of treatment among the different dischargers. NSTEA consists of a genetic algorithm that uses a technique similar to the ϵ-constraint method (see Section 1.7.2 from Chapter 1). The idea is to use one of the objectives of the problem to cause convergence of a GA to a certain nondominated vector and handle the other objectives as constraints. Then, each objective is used sequentially, but to reduce computational costs, the same population of a previous run of the GA is used as the starting point for generating the following nondominated vector. This is then similar to the use of an aggregating function in which the weights are adapted after convergence has been achieved to a certain nondominated vector. The authors use real-numbers representation, uniform crossover and non-uniform mutation. Results compared well with respect to those produced by the ϵ-constraint method (using linear programming) in terms of coverage and spread across the Pareto front. The NSTEA compared well in terms of accuracy, coverage and spread with respect to the use of linear programming. The NSTEA is also compared to VEGA [1440], NPGA [709], NSGA [1509] and SPEA [1782] on several test functions and some instances of the 0/1 knapsack problem. The NSTEA outperformed almost all the other methods in terms of the same three metrics previously mentioned. The only exception was SPEA that outperformed the NSTEA in some instances of the 0/1 knapsack problem.

Schwab et al. [1457] and Savic et al. [1432] use a GA with Pareto ranking in a *pump scheduling problem* in which the objective is to minimize marginal costs of supplying water while staying within certain physical and operational constraints. Two costs are considered: energy consumption and pump switching (i.e., the cost caused by the switching of the pumps). They use binary representation and probabilistic tournament selection.

Halhal et al. [637] use Pareto ranking with a structured messy GA to choose the best possible improvements to make to a *water distribution network* with a limited budget. Two objectives are considered: minimize capital cost and maximize benefits. The approach is tested with two problems: a small looped network consisting of 15 pipes, 9 nodes and 7 loops, and a real water distribution network for a town in Morocco, which consists of 115 nodes and 167 pipes. The proposed technique outperforms a simple GA in terms of quality of the solutions produced.

Formiga et al. [515] use the NSGA-II [374] hybridized with a hydraulic simulator based on the method of Nielsen [1185] for the *optimal design of a water distribution system*. Three objectives are considered: minimize investment costs, maximize the entropy of the system and maximize the system demand supply ratio. The authors adopt BLX-α crossover [457] and Random Mutation [1100]. Two test networks are adopted to validate the proposed approach: (1) a two-loop fictitious network and (2) the Hanoi-Vietnam water distribution system. In both cases, the authors adopt two performance measures to validate their approach: two set coverage [1770] and Schott's spacing[1452].

Keedwell and Khu [833] use the cellular automaton and genetic approach to multi-objective optimization (CAMOGA) to *optimize water distribution networks*. Two objectives are minimized: cost and total head deficit. CAMOGA is a multiobjective extension of an approach previously proposed by the same authors, which is called Cellular Automaton for Network Design Algorithm (CANDA) [832]. CANDA is used for single-objective optimization of water distribution systems, and it provided good results. However, the deterministic nature of CANDA makes it unsuitable for large distribution networks. However, the authors realized that a CANDA run could be used as a seed for a standard genetic algorithm, which was able to improve such results [831]. So, CAMOGA is basically this same approach, but instead of using a simple genetic algorithm, the NSGA-II [374] is adopted as the search engine. The authors compare the results obtained by CAMOGA with respect to those produced by the NSGA-II alone. For that sake, the S-metric [1781] is adopted. CAMOGA is found to outperform the NSGA-II both in terms of Pareto dominance and in terms of spread in two industrial problems. More important is the fact that CAMOGA only requires a very low number of network simulations to achieve its results because of the cellular automaton adopted.

Surry et al. [1543, 1542] use VEGA with Pareto ranking based on constraint violation to *optimize a gas supply network*. In this case, the use of a MOEA is focused on the constraint-handling problem, and two objectives are minimized: cost of the network and constraint violation. The representation used is a variable cardinality integer string with parameterized uniform crossover and non-cyclic creep mutation. Results compared fairly with those produced by a simple GA with a penalty function.

Loughlin and Ranjithan [1017, 1016] use the Neighborhood Constraint Method (NCM) in an *air quality management application*. Two objectives are considered: minimize the cost of controlling air pollutant emissions and maximize the amount of emissions reduction. The application is a combinatorial optimization problem with a considerably large search space (5^{300}). Real-numbers representation and tournament selection are used. The approach is compared to Pareto ranking and a hybrid between MOGA [504] and NPGA [709]. The NCM was able to produce a better spread of solutions along the Pareto front than the other methods.

Yapo [1733], Yapo et al. [1734] and Gupta et al. [622] use the MOCOM-UA method to *calibrate hydrologic models*. In the first case, two objectives are minimized: the unbiased, minimum variance estimator, and the maximum likelihood estimator. In the second case, three objectives are minimized: residual standard deviation, residual bias, and residual whiteness. MOCOM-UA uses Pareto ranking starting with a feasible population. Using a downhill simplex search strategy, a sample of points is improved. The simplex strategy acts as crossover, but reproduction is not sexual because more than two parents intervene to generate offspring (this is called panmictic reproduction). A triangular probability distribution is used to select the candidates for reproduction.

Khu [849] uses a nonlinear aggregating approach called *Accelerated Convergence Genetic Algorithm* (ACGA) [999] to *calibrate the NAM rainfall-runoff model*. Two objectives are considered (out of a total of five possible criteria to be considered): minimize peak flow root mean square error and minimize the overall root mean square error. Two approaches are studied: one where objectives are chosen beforehand, and another where objectives are chosen interactively. The author uses binary representation, one-point crossover, uniform mutation and a special selection procedure that enforces that only the fittest individuals can mate.

Thomas [1580, 1581] uses Pareto ranking, MOGA and NPGA to investigate the *feasibility of full stern submarines*. Three objectives are considered: maximize internal volume, minimize power coefficient for ducted propulsor submarines, and minimize cavitation index. Binary representation and different selection techniques are used. The author also proposes the use of multiple, non-interbreeding species[4] to allow simultaneous optimization of independent and mutually exclusive options (i.e., rotor-only, rotor-stator, stator-rotor or stator-rotor-stator). He also uses phenotypic sharing. To compare the different algorithms under study, several population-based factors are analyzed: mean volume, mean power coefficient, number of nondominated solutions, and cost

---

[4] Each species refers in this case to a propulsor configuration.

function evaluations. Conclusions indicate that MOGA outperforms the other methods in all of the aspects considered.

Brown and Thomas [178] use a GA with Pareto ranking for *naval ship concept design*. Two objectives are considered: maximize overall measure of effectiveness (this factor represents customer requirements and relates ship measures of performance to mission effectiveness) and minimize life cycle cost. Binary representation and roulette wheel selection with stochastic universal sampling are used.

Lee [969] uses the ϵ-constraint method coupled with a genetic algorithm in the *preliminary design of a marine vehicle*. Two objectives are minimized: building cost and operating cost. Constraints of three types are considered: legal (i.e., related to regulations), environmental and technical. The GA is used to find the separate optima required by the ϵ-constraint technique. In fact, the GA is used only for coarse-grained optimization, and it is coupled with the direct search method to find the global optima of the problem under study. Binary representation and roulette wheel selection are used. An interesting aspect of this work is that the author extracts knowledge from the historical runs performed by the hybrid optimization technique used in order to fine tune the parameters required by both the GA and the direct search method.

Todd and Sen [1591] use a variant of MOGA for the *preplanning of containership layouts* (a large scale combinatorial problem). Four objectives are considered: maximize proximity of containers, minimize transverse center of gravity, minimize vertical center of gravity, and minimize unloads. Binary representation and roulette wheel selection with elitism based on non-dominance are used. A multi-attribute decision making tool is used to select solutions that most closely meet the stevedores' requirements. They use the same algorithm in the *shipyard plane cutting shop problem* [1592, 1590]. Two objectives are considered: minimize makespan and minimize total penalty costs (every manufactured part is given a due date obtained from the assembly schedule, and they receive a certain cost penalty for missing such due date). In a further paper, Todd and Sen [1593] apply the same schedule builder to a complex *job shop problem* that considers 4 machines, and 20 jobs with a maximum of 5 stages and 6 types of operations. Two criteria are considered in this case: minimize makespan and minimize average job time (i.e., reduce the number of jobs that are in progress at one time). A special integer representation scheme is used together with proportional selection.

Guimarães Pereira et al. [616, 615] use a GA with a linear aggregating function combined with fuzzy logic to *generate alternatives for site retail and service facilities*. The location of these stores or service facilities has a direct impact on their accesibility for consumers, and their environmental impact, among other things. Five criteria are considered: height, geology, aspect, land use and the distance from two urban centers. Binary representation, two-point crossover, proportional selection and elitism are used. The authors only propose a methodology, but do not present any results. The characteristics of the territory under study are stored in a Geographic Information System (GIS) coupled to a GA.

7.2.2 Electrical and Electronics Engineering

Table 7.2: Summary of electrical and electronics engineering applications

Specific Applications	Reference(s)	Type of MOEA
Symbolic layout compaction	[518]	GA with lexicographic ordering
VLSI cell placement	[1412]	Simulated evolution with fuzzy rules
	[63, 64]	GA with a linear aggregating function
Design of DSP systems	[171, 172]	GA with Pareto ranking
Optimal planning of an electrical power distribution system	[1317, 1316]	GA and evolutionary programming with Pareto ranking
Design of a voltage reference circuit	[1166]	Evolutionary programming with Pareto ranking
Power dispatch	[1607]	Hybrid of a GA and simulated annealing with a linear aggregating function
Economic load dispatch	[1763]	Multi-objective particle swarm optimizer
System-level synthesis	[142]	GA with a linear aggregating function
	[390]	MOGA
	[389]	Parallel recombinative simulated annealing and MOGA
	[391]	MOGA
Design of electromagnetic devices	[36]	Evolution strategy with a linear aggregating function
	[1117]	GA with a linear aggregating function
	[1419]	GA with a linear aggregating function
	[152]	Evolution strategy with a linear aggregating function
	[1692]	NSGA
	[1691]	GA with Pareto ranking, NPGA, NSGA
Design of antennas	[1689, 1690]	NSGA
	[1582]	MOGA, simulated annealing
Design of a three-phase induction motor	[857]	Evolution strategy with a linear aggregating function
Fault tolerant system design	[1452]	NPGA
Synthesis of CMOS operational amplifiers	[1604]	Variation of the NSGA-II (MO-Turtle GA)
	[1749]	GA with a target vector approach
Design of filters	[659]	GA with Pareto ranking
	[1750]	GA with a target vector approach
	[1707]	VEGA, a GA with goal attainment, a GA with a linear aggregating function, a GA with Pareto ranking
	[1450]	Evolution strategy and a dominance-based tournament selection scheme

Table 7.2: (continued)

Specific Applications	Reference(s)	Type of MOEA
Design of lamps	[437, 438, 439]	GA with an aggregating function
Microprocessor design	[1518]	GA with Pareto ranking
Shape design of a single-phase reactor	[385]	Nondominated sorting evolution strategy
Design of combinational circuits	[1369]	Multi-objective genetic programming
	[275, 1030, 1029]	VEGA
Design of an electro-mechanical system	[1344]	NSGA-II
Coordinated design of power system stabilizers and static var compensators	[1785]	MOGA
Design of land grid array solder joints	[641]	multi-objective differential evolution

Fourman [518] uses a genetic algorithm with lexicographic ordering for *symbolic layout compaction*. The two main objectives considered are: minimize cost and satisfy certain design rules. The problem is approached in different ways. First, the objectives are solved sequentially: remove design-rule violations first, and then reduce the area of the layout. To avoid the search bias produced by nearly feasible individuals, Fourman adds an extra criterion: shorter chromosomes have a larger fitness value. Then, he experiments with trade-offs between the objectives using a scoring function (i.e., a linear combination of weights). Finally, he decides to select randomly the objective to be optimized at each generation, and finds this approach to work surprisingly well. Symbolic variable length representation is used in the GA, which is implemented in ML.

Sait et al. [1412] use simulated evolution [869] and fuzzy rules for *VLSI standard cell placement*. Three objectives are minimized: wire-length, power dissipation and circuit delay. Layout width is considered an additional constraint. The proposed approach is compared against a GA with a fuzzy aggregating function. Results indicate that the GA is able to produce circuits with a better performance for small instances. However, as the number of cells increases, simulated evolution has a better performance. Additionally, the GA requires a considerably larger execution time than simulated evolution.

Arslan et al. [63, 64] use a GA with a linear combination of objectives for *structural synthesis of cell-based VLSI circuits*. Three objectives are considered: maximize functionality, minimize delay and minimize physical size of the circuit. In further work, Bright [171] and Bright and Arslan [172] use Pareto ranking for *high-level low power design of DSP systems*. Two objectives are minimized: power consumption and area. Binary representation and roulette wheel selection are used. Results are compared against some DSP benchmark designs reported in the literature. The GA proposed was able to produce feasible circuits that were considered competitive (with respect to those of the benchmark used), but not necessarily better. The GA, however, was able to converge to these solutions relatively fast as compared to traditional approaches.

Ramírez Rosado et al. [1317, 1316] use a GA and evolutionary programming (EP) with Pareto ranking for the *optimal planning of an electrical power distribution system*. Two objectives are considered: minimize global economic costs and minimize the amount of expected energy not supplied. The GA uses an integer representation and a "filter" operator that allows to determine a maximum allowed limit of the global economic cost of the distribution system solutions. This is used as preference information to avoid the generation of solutions that are too expensive. When comparing the GA to EP, none of the two algorithms is found to be superior to the other.

Nam et al. [1166] use Evolutionary Programming with Pareto ranking to optimize the *design of a voltage reference circuit*. Two objectives are considered: minimize the reference voltage at room temperature and minimize the temperature variation effect. Results are compared to four other (single-objective optimization) techniques. The approach generates competitive solutions with respect to the other algorithms (solutions are sub-optimal, but represent better trade-offs between the two objectives).

Tsoi et al. [1607] use a hybrid of a GA and simulated annealing with a linear combination of weights in a *power dispatch problem*. Two objectives are considered: total fuel cost and total emissions (environmental impact). The authors use an incremental GA [494] with floating point representation, two-point crossover, roulette wheel selection and uniform mutation. Simulated annealing is used to maintain diversity in the population through a replacement policy.

Zhao & Cao [1763] use a multi-objective particle swarm optimizer (MOPSO) for *economic load dispatch*. Three objectives are minimized: fuel cost, emission and total real power loss. The problem has three types of constraints: power balance, generation capacity and security. The authors adopt a linear membership function to incorporate the preferences from the user in the ranking process as to select a single solution from the Pareto optimal set. The authors perform two experiments. In the first, only two objectives are considered (fuel cost and emission) and results are compared with respect to a multi-objective evolutionary algorithm.[5] Based on a visual comparison, the authors determine that their MOPSO provides better diversity and better nondominated solutions than the MOEA. The (single) best compromise solution found by each of the two approaches (MOPSO and MOEA) are also presented by the authors. In a second experiment, the three objectives are considered. In this case, however, no comparisons of any type are performed.

Blickle et al. [142] use a GA with a linear combination of weights to solve a *system-level synthesis problem* (i.e., map a task-level specification onto a heterogeneous hardware/software architecture) in an optimal way. Two objectives are minimized: cost and latency of the implementation. The approach

---

[5] The authors do not indicate exactly what sort of MOEA is adopted.

is applied to a video codec for image compression using the H.261 standard and to a PDE-integrator. A special encoding (consisting of lists of integers representing allocations) with a repair algorithm and restricted tournament selection is used. Constraints are handled through the use of a penalty function. In a further paper, Blickle et al. [143] use Pareto ranking to perform design space exploration of the same problem.

Dick and Jha [390] use MOGA for the *co-synthesis of hardware-software embedded systems*. Two objectives are minimized: price and power consumption. The authors use integer representation, a technique to create clusters of solutions (similar to niches), cluster-level operators (mutation) and solution-level operators (crossover). Reproduction is restricted to individuals within the same cluster. Partial domination is considered to rank clusters. Individuals that violate hard constraints are removed and those which violate soft constraints are handled through a penalty function. In further work, Dick and Jha [389] use again a system which incorporates parallel recombinative simulated annealing [1039] and MOGA for *co-synthesis of hardware-software of embedded systems* using dynamically reconfigured FPGAs. In this approach, clusters are used as well, and each of them is assigned a rank (the sum of ranks of all the architectures contained within it). Boltzmann trials are used to select architectures within the same cluster. A global temperature-dependent criterion is used to keep diversity in the population. Two objectives are considered: price and deadline violation. In another paper, Dick and Jha [391] use MOGA for a specific application of *co-synthesis of hardware-software of embedded systems*: the *core-based single-chip system synthesis*. Three objectives are considered in this case: price, IC area and power consumption.

Alotto et al. [36] use an evolution strategy with a linear combination of weights to *optimize electromagnetic devices*. Two objectives are considered: make the value of energy as close as possible to the design goal, and minimize the flux density RMS error. Results are compared to the use of single-objective simulated annealing and a global search algorithm. All the algorithms tested were able to generate optimal configurations close to the design specifications. However, the evolution strategy required the lowest number of fitness function evaluations.

Mohammed and Üler [1117] use a GA with a linear combination of weights to *design electromagnetic devices*. The approach is applied to a pot core problem with two objectives: reduce the size of the device while maintaining a certain magnetic flux density. The authors use binary representation with Gray coding, linear fitness scaling, stochastic remainder selection (without replacement), uniform crossover and uniform mutation. Results are compared against the use of dynamic search. The GA produced competitive (and in some cases better) results but did not consider eddy currents and losses as dynamic programming did.

Saludjian et al. [1419] use a GA with a linear weighted sum to *optimize an electromagnetic superconducting device*. Two objectives are considered: the stored energy in the device has to comply with a previously defined value, and the magnetic induction along two lines of the device has to be as small as possible. The authors use floating point representation, proportional selection with linear ranking, four crossover operators (one-point, two-point, uniform and arithmetic) and two mutation operators (uniform and non-uniform). They also use a local search operator (a hillclimber).

Borghi et al. [152] use an evolution strategy and a linear combination of weights to *reduce the torque ripple of permanent magnet actuators*. Three objectives are considered: minimize e.m.f. harmonic content, maximize the e.m.f. fundamental component, and minimize the cogging torque. The authors use a (1+1) evolution strategy coupled with a filled function acceleration technique [548].

Weile et al. [1692] use the NSGA [1509] to *design electromagnetic devices* (namely, microwave absorbers). Two objectives are considered: reflectance and thickness. Results are compared to the use of a linear combination of weights. The NSGA was able to converge to a set of points whereas the aggregating function converged to a single solution. From their experiments, the authors conclude that the use of a linear combination of weights is not appropriate for the application of interest to them. In related work, Weile et al. [1691] use Pareto ranking, NPGA and NSGA to design *multilayer microwave absorbers*. Two objectives are considered: minimize thickness and minimize reflection. Pareto ranking is implemented using tournament selection and crowding. Binary representation is adopted in all cases, and phenotypic sharing is used with both NPGA [709] and NSGA (Pareto ranking is implemented without niching or fitness sharing). Sharing is done on objective function space in all cases. NSGA was found (by graphical inspection) to provide the highest quality Pareto fronts and to preserve diversity for a longer number of generations, but it was also found to be the most expensive algorithm in terms of CPU time. The NSGA is also compared to the use of simulated annealing and a Simple Genetic Algorithm adopting a weighted Tchebycheff procedure [1522]. The NSGA was found to produce better Pareto fronts than these two other techniques. The approach is applied to the design of microwave absorbers with five layers of materials selected from representative databases of available materials in the 0.2-2GHz, 2-8 GHz and 9-11 GHz bands. In further work, Weile and Michielssen [1689, 1690] use the NSGA to design *thinned antenna arrays with digital phase shifters*. Two objectives are minimized: bandwidth and maximum reduced sidelobe level. Integer representation and triangular sharing [581] are used. The approach is applied to the design of a 200 element symmetric linear array of 80 isotropic elements with three bit phase shifters separated by one-half wavelength.

Thompson [1582] uses MOGA [504] and multiobjective simulated annealing to *design an antenna tuning unit*. Two objectives are considered: minimize the mismatches between the source and load impedances and minimize the power delivered to the load at the harmonic frequency. The authors use binary representation, multi-point crossover and uniform mutation. Results produced by MOGA, a multiobjective version of simulated annealing and MOGA with elitism (called EMOGA) are compared using the two metrics proposed by Zitzler and Thiele [1782]. Multiobjective simulated annealing produced better solutions (in terms of the metric adopted) and at a lower computational cost than any of the other approaches. Also, the use of elitism is shown to be beneficial to MOGA. Simulated annealing was found to be superior to both MOGA and EMOGA.

Kim et al. [857] use an evolution strategy and a linear combination of weights to *design a three-phase induction motor*. Two objectives are considered: efficiency and power density. The authors use a (1+1) evolutionary strategy where simulated annealing is adopted to perform mutation, and a "shaking" process is employed to maintain diversity. The process consists of changing the mutation step length whenever there is diversity loss in the population.

Schott [1452] uses the NPGA [709] for *fault tolerant system design*. Two objectives are minimized: unavailability and purchase cost. The NPGA is compared to the ϵ-constraint method on two examples. Some of the performance criteria considered in the comparison are the number of dominated points, the number of fitness function evaluations, the number of nondominated points, etc. Two-point crossover, binary representation, uniform mutation, steady-state selection, equivalence class sharing (on phenotypic space), variable population size and clone replication are adopted in the GA implemented. The NPGA was found to be superior for three main reasons: (1) it is difficult to handle three or more objectives with the ϵ-constraint method, (2) it is more expensive (computationally speaking) to perform several runs of the ϵ-constraint method to match the results of a single run of the NPGA, and (3) the uncertainty involved in choosing a proper efficient set for the ϵ-constraint method is considered an important disadvantage of this technique.

Trefzer et al. [1604] use a variation of the NSGA-II (called *MO-Turtle GA*) for the *synthesis of operational amplifiers*. This approach adopts the variation operators of the *Turtle GA* [1605] (namely, the *Random Wires* mutation and the *Implanting Block of Cells* crossover). The implementation of these two operators is done in hardware, in a field programmable transistor array (FPTA). The authors experiment with different combinations of objectives, going up to 11 objectives. In all cases, the manually made operational amplifiers are better than the solutions generated by the MO-Turtle GA in terms of both *distortion* (noise) and *resource consumption*. However, the evolved solutions present similar performance in terms of other objectives such as *offset*, *slew-rate*, and *settling-time* and provide better results in terms of *phase-margin*, which is the main objective considered in this research. The authors also analyze certain pairwise combinations of objectives and realize that some of these combinations provide better fitness values over time (e.g., *magnitude* versus *offset*). The authors conclude that the MO-Turtle GA is able to provide designs competitive with the solutions produced by humans, but fails at synthesizing additional gain-stages.

Zebulum et al. [1749] use a GA with a target vector approach (with adaptive weights) for the *synthesis of low-power operational amplifiers*. Seven objectives are considered: GBW, gain, linearity, power consumption, area, phase margin, and slew-rate. GBW, gain and phase margin are maximized, and the others are minimized. A target vector is defined for all objectives, and weights are assigned such that large values are used for objectives for which the average fitness is far from the desired (target) value and small values are adopted for those objectives whose average fitness is close to the desired value. The approach is similar to the goal-attainment method. Integer representation, uniform mutation and proportional selection are used. Results are compared against other approaches used to optimize operational amplifier designs. The GA-based approach produced competitive circuits that, at least in some cases, improved the solutions generated by the other techniques compared. In further work, Zebulum et al. [1750] use the same approach for the *synthesis of analog active filters*. Four objectives are considered in this case: maximize frequency response, minimize power dissipation, maximize the Maximal Symmetric Excursion (MSE), and minimize integrated output noise.

Harris and Ifeachor [659] use a GA with Pareto ranking to *design non-linear Finite Impulse Response (FIR) filters*. Two objectives are considered: minimize the maximum error between the actual response and the desired response template, and minimize the error from linearity of the phase response in a specified region of the passband. Binary representation is used, but the paper focuses on explaining why the GA has a poor performance in this problem and the use of a higher cardinality representation is suggested.

Wilson and Macleod [1707] use goal attainment, VEGA, a weighted sum of objectives and Pareto ranking to *design multiplierless IIR filters*. Two objectives are considered: minimize response error and implementation cost. Binary and gray coding representations with stochastic universal sampling and linear ranking are used. The different techniques used are compared against each other. The comparison is both graphical and based on statistical measures of performance. Goal-attainment produced the best results, but the weighted sum of objectives was found to provide consistently good results as well. Interestingly, Pareto ranking has a poor performance in this application.

Schnier et al. [1450] use an evolution strategy and a dominance-based tournament selection scheme (similar to the NPGA [709]) to *design digital filters*. Three objectives are minimized: passband maximum amplitude deviation, passband maximum delay deviation, and the inverse of maximum amplitude in stopband. A mechanism that adapts a constraint vector dynamically to the current fitness values in the population is employed. The authors use real-numbers representation, mating restrictions, Cauchy mutation, and a form of implicit fitness sharing applied over the combined (i.e., aggregated) fitness values of each individual. An interesting aspect of this work is that the authors adopt different schemes for selecting parents than for selecting individuals for survival into the next generation. In the first case, dominance-based tournaments are used (ties are solved through niche counts). For the second case, all nondominated individuals generally survive, as long as they satisfy the constraints, and depending on their number in the population (i.e., if there are too many nondominated individuals in the population, then some of them are removed based on the aggregated value of their different objective functions). This is obviously done to maintain diversity. Results are found to be competitive with those generated by a human designer.

Eklund and Embrechts [437, 438, 439] use a GA with an aggregating function to *design optical filters for lamps*. Three objectives are considered: minimize the color difference between the desired and the actual light, maximize relative efficiency of the lamp, and have filters that are relatively "smooth", with only a few notches. In practice, only two objectives are considered, since the third one (smoothness) can be automatically achieved when the two other objectives are properly met. The authors use floating point representation, three types of crossover (one-point, arithmetic and heuristic) and three types of mutation (uniform, non-uniform and boundary). Heuristic knowledge obtained from previous runs is employed to accelerate convergence (something that the authors call "smoothing the chromosomes").

Stanley and Mudge [1518] use a GA with Pareto ranking to solve a *microprocessor design problem*. The approach considers two design constraints: chip area and chip power dissipation, and one design objective: maximize performance (number of clock cycles required to execute a certain number of instructions). An interesting aspect of this application is the fact that the authors incorporate the constraints of the problem as additional objectives that have to be satisfied. The implementation is done using a parallel asynchronous scheme (this is necessary since workstations of different types and speeds are used to distribute the task of evaluating the population fitness of the GA). The authors use two-point crossover, uniform mutation, steady state generational replacement, and a mechanism to eliminate duplicates.

Di Barba et al. [385] use a multiobjective evolution strategy (called Nondominated Sorting Evolution Strategy) to *design the shape of a single-phase series reactor for power applications*. Two objectives are considered: minimize the material cost of the reactor and minimize the mean radial component of magnetic induction in the cross-section of the winding. The problem has also constraints related to the induction in the core, the current density in the winding and the insulation gaps between winding and core. Emphasis is placed on efficiency. The authors use a $(1+1)$ evolution strategy with a parallel implementation of three processes: mutation, an annealing procedure that is applied right after mutation, and the generation of new individuals. Nondominated sorting is used to classify individuals giving higher (dummy) fitness values to nondominated solutions (as in the NSGA [1509]). Fitness sharing on objective function space is used.

Rodríguez Vázquez and Fleming [1369] use genetic programming with MOGA [504] to *design combinational logic circuits* (namely, a 6-multiplexer). Two objectives are considered: correctness (functionality of the circuit) and optimality (reduce nodes of the parse tree encoding of a circuit). The number of input variables is incorporated as a constraint. Results are compared to the use of a single-objective GP implementation. The multiobjective GP implementation found better compromises than the single-objective version. Furthermore, the frequency of generating functional circuits was higher when using the multiobjective approach.

Coello Coello et al. [275] & Coello Coello and Hernández Aguirre [274] use VEGA [1440] to *design combinational circuits*. Since each value from the truth table to be matched by the circuit generated by a genetic algorithm is considered as an objective, up to 65 objectives are used. The authors use integer representation to encode a matrix representing a circuit. Each gene represents either one of the inputs of the circuit (only two inputs per gate are allowed), or a gate (from a set previously defined by the user). They also use two-point crossover, elitism and tournament selection. Since results are really combinational circuits in which the number of gates is minimized, direct comparisons with other circuit-design techniques are possible. The approach generates circuits that are either equivalent or more compact (in terms of the number of gates) than those generated by human designers using Karnaugh maps [822] and the Quine-McCluskey method [1309, 1084]. In further work, Luna et al. [1030] use VEGA in the same problem. However, this time particle swarm optimization is adopted as the search engine. An interesting aspect of this work is the use of two integer-based encodings, one of which is proposed by the authors. The authors compare PSO with several types of encoding against a version of VEGA that uses a genetic algorithm [274]. The results indicate that the PSO-based version of VEGA that adopts the proposed integer encoding outperforms all the other techniques. An interesting aside from this research is that binary PSO was found to be a poor performer for combinational circuit design.

Régnier et al. [1344] use the NSGA-II [374] for the *optimal design of an electromechanical system*. The case study consists of the design of an inverter-permanent magnet motor-reducer-load association. Two objectives are minimized: global losses and the mass of the system. The authors also perform parametric sensitivity analysis in order to determine the effect of the constraints on the Pareto optimality of the solutions produced. Finally, the authors also illustrate the *a posteriori* nature of MOEAs, by providing a simple example in which cogging torque is adopted as an additional criterion so that a single nondominated solution may be chosen from the Pareto optimal set. Although this application is a simple academic problem, the authors indicate that their goal is to illustrate how MOEAs can be used, in general, in the design of electromechanical systems.

Zou et al. [1785] used a MOEA similar to MOGA [504] for the *coordinated design of power system stabilizers (PSS) and static var compensators (SVC)*. The aim of the proposed approach was to improve the power angle stability while maintaining the voltage quality of the power system. The authors adopt fitness sharing and a criterion based on the use of the progress ratio to stop the MOEA. The results are not compared with respect to any other approach.

Han et al. [641] use a multi-objective differential evolution approach based on Lampinen's work [946] for *reliability-based design optimization of land grid array solder joints under thermodinamical load*. The idea is to present a methodology for second level electronic package solder joints. Two objectives are considered: maximize the system performance and minimize the performance variance. These two objectives aim to produce robust designs. A probability analysis of the system output is performed on the response surface using a Quasi-Monte Carlo simulation method. A very robust design was produced using this approach (with a reliability of 99.73%), but no direct comparisons with respect to other techniques are provided.

7.2.3 Telecommunications and Network Optimization

Table 7.3: Summary of telecommunications and network optimization applications

Specific Applications	Reference(s)	Type of MOEA
Network Design	[1572]	GA with Pareto ranking
	[1096]	GA with Pareto ranking
	[1740, 1741]	Simulated evolution and fuzzy rules
	[919]	Pareto converging genetic algorithm
	[1124]	GA with an aggregating function
	[868]	Modified NSGA-II
Multicast flows	[401]	SPEA
Improve wire-antenna geometries	[1634]	GA with an aggregating function
Adaptive distributed database management	[886, 889]	PAES
Offline routing	[885, 886, 889]	PAES
Production process planning	[1764]	Evolution strategy with an aggregating function
Minimum spanning tree problem	[1765]	Evolution strategy with an aggregating function
Broadcasting in mobile networks	[1031]	Archive-based Scatter Search

Tang et al. [1572] use a GA with Pareto ranking to *design a Wireless Local Area Network (WLAN)*. Four objectives are minimized: number of terminals with their path loss higher than a certain threshold, number of base-stations required, the mean of the path loss predictions of the terminals in the design space, and the mean of the maximum path loss predictions of the terminals. A hierarchical GA (HGA) with two types of genes (control and parameter) is used. The HGA has a lot of resemblance with the Structured Genetic Algorithm (stGA) proposed by Dasgupta and McGregor [334].

Meunier et al. [1096] use a GA with Pareto ranking to *design a mobile telecommunication network*. Three objectives are considered: minimize the number of sites used, maximize the amount of traffic held by the network, and minimize the interferences. Additionally, constraints related to coverage of the area and handover of every cell of the network are considered as well. A steady state GA with fitness sharing (applied in the objective space) is used by the authors. A parallel implementation is adopted to speed up the search in this complex multiobjective combinatorial optimization problem. The authors also use a multilevel encoding (similar to the Structured GA [333]), *ad-hoc* genetic operators and a penalty function to handle the constraints of the problem.

Youssef et al. [1740, 1741] use simulated evolution [869] and fuzzy rules to *design the topology of a campus network*. Three objectives are minimized: monetary cost, average network delay and the maximum number of hops between any source-destination pair in the network. Additionally, three constraints are also considered: the traffic flow of any link is not allowed to exceed a certain threshold, the number of clusters attached to a network device is not allowed to exceed its port capacity, and certain hierarchies on the network devices can be enforced by the designer. The topology of the network is treated as a spanning tree. Therefore, this becomes a multiobjective combinatorial optimization problem. Simulated evolution operates in three steps: evaluation, selection and allocation. For evaluation, a fuzzy rule is employed. Selection is based on the goodness of the links of each candidate topology. Allocation consists of removing links and trying new ones in such a way that they contribute to the best possible overall solution. A second fuzzy rule that combines the three objectives of the problem is used for the allocation process. In [1740], simulated evolution is found to produce better results (in terms of monetary cost) than simulated annealing, although the second algorithm has slightly shorter execution times. In [1741], results between two versions of simulated evolution are proposed: the first is the traditional version (previously used by the same authors), and the second uses Tabu search for the allocation phase. From this comparison, it was concluded that the second version of the algorithm provided better performance because it exploits the good exploratory capabilities of Tabu search.

Kumar et al. [919] use the Pareto Converging Genetic Algorithm (PCGA) [924] to *design the topology of mesh communication networks*. Two objectives are minimized: cost and average packet delay. The authors use selection based on Pareto ranking and rank-histograms for assessing convergence to PF_{true}. The chromosomes adopted encode the topology of the network (including link-capacities, details of the link types and the interface components) and a routing vector that provides the path between every pair of nodes for the topology encoded.

Montana and Redi [1124] use a genetic algorithm with a linear aggregating function to *optimize the parameters of a mobile ad hoc network protocol*. The authors adopt real-numbers encoding, uniform crossover, flip mutation, exponential selection and a steady-state replacement policy (the new individual produced after each generation replaces to the worst one in the population). Two objectives are minimized: dropped packets and transmission delay. Results are compared with respect to an approach in which the parameters of the network are set by hand. The authors found that the GA adopted produced results significantly better (for the two objectives considered) than those found by the manual tuning procedure.

In Kleeman et al. [868], the NSGA-II [374] is is modified and extended to solve a variation of the *multicommodity capacitated network design problem* (MCNDP). This architectural variation represents a hybrid communication network with multiple objectives including costs, delays, robustness, vulnerability, and reliability. Nodes in such systems can have multiple and varying link capacities and rates as well as information (commodity) quantities to be delivered and received. Each commodity has an independent prioritized bandwidth requirement. Monte Carlo techniques provide insight to the complex fitness landscape and motivated the use of a novel MOEA initialization procedure and a mutation method which efficiently generated effective PF_{known} solutions.

Donoso Meisel [401] apply an algorithm based on SPEA [1782] to *optimize static multicast flows*. Eleven objectives are considered: maximal link utilization, total hop count, hop count average, maximal hop count, maximal hop count variation for a flow, total delay, average delay, maximal delay, maximal delay variation for a flow, total bandwidth consumption, and number of subflows. An interesting aspect of this work is the encoding adopted, which the author claims to be the first to allow representing several flows (unicast and/or multicast) with as many splitting subflows as necessary. This encoding automatically satisfies the constraints of the problem, such that an additional procedure to handle such constraints becomes unnecessary. Two *ad-hoc* crossover operators are adopted: flow crossover and tree crossover. An *ad-hoc* mutation operator was also adopted. In order to validate the approach, several network scenarios were considered and two metrics were adopted: overall nondominated vector generation and overall nondominated vector generation ratio [1626]. The multi-objective evolutionary algorithm adopted produced the best overall results, improving on the results obtained with analytical models and other heuristics.

Van Veldhuizen et al. [1634] use a GA with a weighted sum to *improve wire-antenna geometries*. Four objectives are considered: radiated power gain, azimuthal symmetry of radiated power, input resistance and input reactance. The authors use a steady state GA with real-numbers representation, tournament selection and a variation of arithmetic crossover. Death penalty is applied to infeasible designs (i.e., they are given a fitness value of zero).

Knowles et al. [886, 889] use PAES to solve the *adaptive distributed database management problem* [1195]. The problem consists of finding an optimal choice of client/server connections given the current client access rates, basic server speeds, and other general details of the communications matrix [886]. Two objectives are minimized: the worst response time (measured in milliseconds) seen by any client, and the mean response time of the remaining (non-worst) clients. Integer representation and uniform crossover are used. PAES is compared against several variations of the NSGA [1509] and the NPGA [709]. A variation of the statistical technique proposed by Fonseca and Fleming [509] is used to allow a quantitative comparison of the MOEAs.

Results indicate that PAES is able to outperform the other MOEAs implemented in this problem, both in terms of the metric adopted and in terms of the computational time required. In related work, Knowles and co-workers [885, 886, 889] use PAES to solve the *offline routing problem*. Two objectives are minimized: communication costs and congestion. A straightforward integer representation is adopted in this case. Results are compared against a steady state version of the NPGA using Fonseca and Fleming's statistical technique previously mentioned [509]. The authors indicate that PAES was again able to outperform the other MOEAs implemented. In this case, diploid chromosomes and specially designed genetic operators are used.

Zhou and Gen [1764] use an evolution strategy with an adaptive evaluation function to solve a *production process planning problem* (stated as a network flow problem). The approach consists of a weighted sum of the objectives in which the weights are defined (and modified during the evolutionary process) in such a way that the evolution strategy approaches the ideal vector (computed beforehand). This procedure reduces the spread of the population, but provides a relatively easy (and relatively efficient) way to solve the two-objective problem used by the authors. The authors use a $(\mu + \lambda)$-ES with a special state permutation encoding, together with a mutation procedure based on a neighborhood search technique (no crossover operator is used in this case). In related work, Zhou and Gen [1765] use a genetic algorithm with the same adaptive evaluation function to solve a *minimum spanning tree problem*. However, in this case, the NSGA [1509] is also used in the same problem. They incorporate fitness sharing on the objective function space to keep diversity with the NSGA. The authors use a special permutation representation (using Prüfer numbers [1298]), roulette wheel selection, uniform crossover and uniform mutation. Both approaches are compared against each other and against an enumeration procedure. The comparison criterion adopted is the percentage of Pareto optimal solutions found by each method. Both GAs outperformed the enumeration procedure. However, regarding the two approaches proposed, the authors suggest to use the adaptive function approach in cases where a single solution is desired (as tends to happen when using mathematical programming approaches) and the NSGA when the whole Pareto front is needed. If possible, a combination of both GA-based approaches is also recommended.

Luna et al. [1031] used the Archive-based Scatter Search (AbSS) approach for *optimizing the broadcasting strategy of a metropolitan mobile ad-hoc network (MANET)*. Three objectives were considered: maximizing the network coverage, minimizing the network usage, and minimizing the makespan. Results were compared with respect to those produced by a cellular multiobjective genetic algorithm (cMOGA) using three performance measures: (1) the number of Pareto optimal solutions found, (2) set coverage [1770] and (3) hypervolume [1781]. AbSS was found to outperform cMOGA with respect to set coverage and hypervolume, while slightly improving the number of Pareto optimal solutions found.

7.2.4 Robotics and Control Engineering

Table 7.4: Summary of robotics and control engineering applications

Specific Applications	Reference(s)	Type of MOEA
Robot path planning	[408]	Fuzzy tournament selection
	[526]	Evolution strategy with lexicographic ordering
	[679]	MOGA
Fault diagnosis	[1058]	MOGA
	[1060]	Parallel version of MOGA
	[1059]	MOGA
Nonlinear system identification	[1370, 1367, 56]	Multiobjective genetic programming
Elevator car routing problem	[1612]	GA with an aggregating function
Controller design	[1453]	MOGA
	[400]	GA with an aggregating function
	[673, 675]	Multiobjective robust control design GA
	[830]	GA with the Pareto partitioning method
	[763]	MOGA
	[137, 138]	Multiobjective evolution strategy
	[1040]	GA with a fuzzy population ranking method
Design of control systems	[248, 249]	MOGA
	[1694]	MOGA
	[1567]	Hybrid of MOGA and the NPGA
	[1566, 1565]	Incremented multiobjective evolutionary algorithm
	[413]	MOGA combined with a neural network
	[939]	NSGA-II
	[1602, 1603]	GA with fuzzy logic and an aggregating function
	[327, 326, 247]	MOGA
	[145]	NSGA
Robotic manipulator problems	[851]	GA with a predator fitness approach
	[1223]	GA with a variation of the Pareto set distribution method
	[768]	Parallel GA with an aggregating function
	[1214]	Evolution strategy with an aggregating function
	[270]	GA with a weighted min-max approach
Generation of fuzzy rule systems	[804]	Evolution strategy with an aggregating function
	[794]	GA with Pareto ranking

Dozier et al. [408] use Fuzzy Tournament Selection (FTS) for *evolutionary path planning*. Three objectives are minimized: distances from start to destination, sums of the changes in slope, and average changes in slope. The membership functions developed by the authors allow to adapt the focus of the algorithm on each of the objectives in a similar way as lexicographic ordering. The authors use a GA with steady state selection, real-numbers representation, flat crossover with Gaussian mutation, and uniform mutation. Results are compared against the same system without the use of FTS. The paths generated by the approach proposed are not only optimal (or at least sub-optimal), but also diverse. This is considered very important by the authors, since such diversity can be used as an error-recovery mechanism for situations in which replanning is necessary.

Gacôgne [525] uses a GA with a subpopulation scheme based on lexicographic ordering to *tune a fuzzy controller for the guidance of an autonomous vehicle in an elliptic road*. Between two and four objectives are considered from the following: maximize the number of steps, minimize the "road-holding", maximize the ratio between the performed distance and the maximum distance for a vehicle with maximum speed, and maximize the number of meaningful symbols in the set of rules. The author uses variable-length chromosomic strings that encode different coefficients and rules using mixed variable types (real numbers and integers). He also employs specialized crossover and mutation operators to deal with this mixed encoding. In related work, Gacôgne [526] uses an evolution strategy with lexicographic ordering in a similar problem: *optimize rules of a fuzzy controller used by a robot for obstacle avoidance*. The approach, in this case, uses self-adaptation of the parameters of the GA.

Higashihara and Atsumi[6] [679] use MOGA to *acquire sensory-action network of a mobile robot*. The approach is applied to simulations using a corridor following task and a garbage collecting task. Both problems are considered with two objectives related to the movements of the robots and their ability to perform the task desired. The authors actually use a GA to balance the weights of a neural network. They employ integer representation, special crossover and mutation operators, and ranking selection. During the search, they keep two subpopulations based on the ranking of each individual: one has the highest ranked individuals and the other the lowest ranked. Both subpopulations are allowed to mix during crossover. Mutation is only applied to the lowest ranked individuals. MOGA [504] is slightly modified in such a way that a certain ranking area is established according to a threshold defined by the user. The use of this threshold allows to decide which individuals fall into each subpopulation. Results compared well with respect to the use of a single-objective optimization approach in terms of trade-offs generated.

Marcu [1058] uses MOGA for *fault diagnosis* thought of as a problem of pattern recognition. Goal attainment is used to accomodate goal information that allows to alter the way in which individuals are compared against each other. The number of objectives is dependent upon the number of classes in a problem, and are classified in two groups: volumes of decision components and misclassification errors. Both are minimized. Real-coded representation with stochastic universal sampling is adopted. The approach is used for feature

---

[6] Thanks to Dr. Tomoyuki Hiroyasu for his help interpreting the contents of this paper, which was published in Japanese.

selection and classifier design in a three-tank system. To identify the number of components required to approximate a decision region, the author uses a clustering technique based on fuzzy logic. Results are compared against the Unknown Input Fault Detection Observer (UIFDO) technique. The author indicates that MOGA produces solutions with a better performance (i.e., more faults are detected in the system) than the UIFDO technique.

In a further paper, Marcu and Frank [1060] use a parallel version of MOGA in the same three-tank problem previously mentioned (see Chapter 8 for more information on parallel MOEAs). Several recombination operators are considered in this case (discrete, intermediate and linear) together with non-uniform mutation [1100]. Three migration topologies are investigated as well: ring topology, neighborhood migration and unrestricted migration.

Finally, in another paper, Marcu et al. [1059] use MOGA to *design a dynamic artificial neural network employed for fault diagnosis*. Six objectives are minimized: the sum of squared errors that characterize a certain neural network architecture and its training set, the number of coefficients of active synaptic filters in the hidden layer, the number of coefficients of active internal filters in the hidden layer, the number of active back-connections in the hidden layer, the number of coefficients of active synaptic filters in the output layer, and the number of coefficients of active internal filters in the output layer. A Structured GA [333, 334] with binary representation is used in this case. The same three-tank problem mentioned before is used to test the approach.

Rodríguez Vázquez et al. [1370] use genetic programming with MOGA (the so-called Multi-Objective Genetic Programming approach) for *nonlinear system identification*. Several objectives are considered: number of terms in the model, model degree (of nonlinearity), model lag, residual variance, long-term prediction error, the autocorrelation function of the residuals and the crosscorrelation function between the input and the residuals. In some experiments, tree size is considered as an additional objective. A hierarchical tree representation is used to incorporate preference information (a vector of goals specified by the decision maker is used for that sake). Results are compared to traditional nonlinear system identification techniques (e.g., stepwise regression and orthogonal regression). The proposed approach produced model structures that are either equivalent or (in some cases) better than those originated with traditional techniques. The same approach has been applied to a simple Wiener process, to a Surge Tank System and to an aircraft gas turbine engine [1367, 56].

Tyni and Ylinen [1612] use a genetic algorithm with an aggregating function to *control a group of elevators* (the so-called *elevator car routing problem*). The proposed approach is called Evolutionary Standardized-Objective Weighted Aggregation Method. Two objectives are minimized: average passenger waiting time and energy consumption. The proposed approach is really a linear aggregating function in which the objectives are normalized. The decision maker expresses his/her preferences through the use of weights (from zero to one). However, in order to be able to generate non-convex portions of the Pareto front, the authors adopt a controller which adjusts the weights using a target vector approach (i.e., a nonlinear aggregating function) in which the passenger waiting time is used as the target value to be attained. An interesting aspect of this application is that it is really a real-time optimization problem, since the optimization has to be performed twice a second. Thus, the proposed approach is able to come up with a solution within the allowable 500 microseconds time frame. The proposed approach is validated using a simulator. The results indicate that the proposed approach is able to reduce by 20% the daily energy consumption with respect to a traditional approach, while also reducing the average passenger waiting time.

Schroder et al. [1453] use MOGA to *design active magnetic bearing controllers*. The approach is applied to a Rolls-Royce marine turbo machine's rotor suspended by active magnetic bearings. Nine objectives are minimized: steady state error, compliance at 4 Hz, maximum current, and noise susceptibility for two bearings (this makes a total of 8 objectives), plus the minimization of controller complexity. Two extra objectives are treated as constraints: minimize the maximum real part of the eigenvalues of the closed loop system, and minimize the reciprocal of the length of the simulation time. The authors use a Gray coded logarithmic representation and fitness sharing. Besides the large number of objectives of this problem, it has a fairly large search space (2^{402}).

Donha et al. [400] use a GA with a linear combination of weights for H_∞ *controller design*. Four objectives are minimized: overshoot, controller roll-off frequency, rise time and settling time. Binary representation, tournament selection, and elitism are used. Results are compared to a conventional design methodology. The authors indicate that the proposed approach is more efficient (computationally speaking) than designing by trial-and-error. The designs generated are used as suboptimal solutions that an experienced designer can further improve.

Herreros López and co-workers [673, 675, 674] use a hybrid (called MRCD GA, where MRCD stands for "Multiobjective Robust Control Design") between the tournament selection used by the NPGA and the ranking procedure used by MOGA to *design robust controllers*. Two algorithms are proposed, each designed for a specific controller design problem (one of them considers uncertainties and the other one does not). Problems with two, three and four objective functions are considered, particularly related to mixed $\mathcal{H}/\mathcal{H}_\infty$ controller design. The implementation has the following features: parallelism (different subpopulations with different search limits each), migration, dynamical search space boundaries, one-point crossover, uniform mutation, elitism, and incorporation of constraints as additional objectives. The two algorithms are compared against LMI (linear matrix inequalities) methods and against several other MOEAs (MOGA, MOMGA, NPGA and NSGA). Results indicate that the MRCD GA is able to generate highly competitive results and can even outperform the other methods (from a control perspective) in some cases.

Kawabe and Tagami [830] use a GA with the Pareto partitioning method to *design a robust PID* (Proportional, Integral and Derivative actions) *controller with two degrees of freedom*. Two objectives are minimized: the maximum disturbance response and the reference response. They use binary representation with Gray coding, roulette wheel selection with linear scaling, uniform bit mutation and uniform crossover.

Istepanian and Whidborne [763] use MOGA to *design finite word-length feedback controllers*. Two objectives are minimized: the difference between the closed loop system and the original closed loop system, and the cost of the implementation. The authors use binary representation, one-point crossover, uniform mutation and stochastic universal sampling selection.

Binh and Korn [137, 138] use a multiobjective evolution strategy to *design a multivariable controller*. The approach enforces satisfaction of the soft constraints of the problem before checking for non-dominance. A multimodal function with eight local minima, and a bi-objective problem are used to test the approach. The authors use two-point crossover, ranking selection, and an extended representation that includes a *life environment* (i.e., the information about constraint violation), and the *personal experience* (i.e., reproduction capabilities) of an individual. Results are compared against other techniques on a benchmark problem. The proposed approach was able to produce solutions that were either equivalent or better than those generated by traditional techniques used to solve this problem.

Mahfouf et al. [1040] use a GA with a fuzzy population ranking method to *optimize the performance index table of a self-organizing fuzzy logic controller*. Two objectives are optimized: the integral of absolute error plus the integral of absolute value of control effort. Each individual is ranked according to its performance on each objective, and then is labeled and placed on a fuzzy decision table. This decision table gives the overall rank of each rule. The authors use binary representation, selective breeding, one-point crossover and uniform mutation. The approach is compared to the use of Pareto ranking [581]. The results indicate that the proposed approach performed better than Pareto ranking in terms of output rise-time and final produced rule-base surface around the center area of the controller designed.

Chipperfield and Fleming [248, 249] use MOGA to *design a multivariable control system for a gas turbine engine*. The goal is to find a set of pre-compensators that satisfy a number of time-response design specifications while minimizing the interactions between the loops of the system. Nine objectives related to the foreaft differential thrust and the total engine thrust are considered. A Structured GA [333, 334] with real-numbers representation is used in combination with intermediate recombination, mating restrictions, and fitness sharing on the objective domain.

Whidborne et al. [1694] use MOGA for *control system design*. Three examples with three, nine, and seven objectives are considered (e.g., rise-time, overshoot, bandwidth, and other functionals of the system step response). MOGA is compared against the moving boundaries process (a technique based on

Rosenbrock's hill climbing algorithm [1376]), and the Nelder-Mead Minimax method [1182]. Although the results are inconclusive, the authors indicate that MOGA required the largest number of fitness function evaluations of all the methods compared. Also, MOGA was unable to distribute solutions properly along PF_{true} (this happened because the authors do not implement any niche formation technique).

Tan and Li [1567] use a hybrid of MOGA and the NPGA for *time and frequency domain design unification of linear control systems*. Nine objectives are considered: stability, closed-loop sensitivity, disturbance rejection, plant uncertainty, actuator saturation, rise time, overshoot, settling time, and steady state error. The authors compute costs for each individual using MOGA's ranking, but use a tournament selection scheme (like in the NPGA) to improve the efficiency of the algorithm. Fitness sharing and mating restrictions are used as well. The GA uses decimal encoding, and two-point crossover. The approach is used to design an ULTIC controller that satisfies a number of time domain and frequency domain specifications. In further work, Tan et al. [1566, 1565] use an incremented multiobjective evolutionary algorithm [1564] to *design control systems*. The following objectives are considered: close-loop stability, tracking thumbprint specification, robust margin, high frequency gain, and minimum controller order. The authors use an evolutionary algorithm designed by them which employs a dynamic population size (i.e., the population size is adapted over time) together with a fuzzy boundary local perturbation scheme that performs the "incrementing" of previously found nondominated individuals as to fill up the gap among them (i.e., it aims to get a better coverage of the Pareto front). The authors also use a dynamic sharing that does not require to compute niche sizes [1562]. Finally, the approach also allows the incorporation of preferences (or goals) from the decision maker. Results are compared against those produced by another evolutionary multiobjective optimization technique previously proposed by the same authors [1562]. Results indicated that the solutions produced by the incremented multiobjective evolutionary algorithm dominated the solutions generated by the previous approach of the same authors.

Duarte et al. [413] use MOGA combined with a neural network that approximates the values of the objectives (in order to reduce the computational cost) to *design a simple control system (a cascade compensator)*. Six objectives are considered, related to the design specifications for the compensator. Five Radial Basis Function (RBFs) neural networks are used to estimate five of these objectives (one could be obtained directly at a low computational cost). The use of neural networks significantly reduces the computational cost, producing also more nondominated solutions. However, the computational cost associated with the use of the neural networks themselves is not considered by the authors.

Lagunas Jiménez [939] uses the NSGA-II [374] for the *fine-tuning of a robust PID controller*. Three objectives are minimized: the speed of the response when facing a reference signal, the sensitivity when facing changes in the structure of the plant, and the effect of the noise in the output signal. The author uses 13 plants to validate his approach (some of these plants were taken from the PhD thesis of Alberto Herreros López [675]). In his experiments, the author adopts both binary and real-numbers encoding, and a penalty function to incorporate the constraints of the problem. An interesting aspect of this work is that the author implements a filtering approach in order to retain only the best four solutions based on two criteria: best response to changes in the reference signal (step) and best attenuation to perturbations in the control signal. The results are found to be competitive with respect to those reported by Herreros López [675].

Trebi-Ollennu and White [1602, 1603] use a GA with fuzzy logic and a linear combination of weights to *design nonlinear control systems*. The objectives considered are: depth, pitch angle, stern and bow thruster input, and hydroplane input. However, only one of them is considered at each iteration, with respect to time. Fuzzy logic is used to deal with the uncertainties involved in the design. A weighting strategy is adopted for the membership function. This membership function is used to deal with the multiple objectives of the problem and its constraints. The authors use proportional selection with linear ranking, one-point crossover, uniform mutation, and binary representation with Gray coding.

Dakev et al. [327] and Chipperfield et al. [247] use MOGA to *design an electromagnetic suspension system for a maglev vehicle*. Seven objectives corresponding to the performance parameters of the system are considered. The authors use real-valued representation, intermediate recombination, breeder mutation, sharing, and mating restrictions (these last two are applied in the objective domain). In further work, Dakev et al. [326] use MOGA to solve *optimal control problems*. Three examples are considered: (1) determine time-optimal trajectories for a two-link articulated manipulator, (2) determine time-optimal following path control of a two-link manipulator, and (3) perform the optimal control for a lifting reentry space vehicle. The first problem consists of a planar two-link articulated manipulator for which the authors want to determine the time-optimal trajectories for given initial and final angle configurations. Two feasible controls have to be found such that time is minimized and torque constraints are satisfied. In the second example, a model of a robot with one revolute and one prismatic joint is considered. The goal is to minimize the final time along the trajectories of two controls satisfying a set of inequality and equality constraints. The last example is based on a simplified model of a space vehicle flight through the Earth's atmosphere. The goal is to find a vehicle control angle minimizing the total stagnation point convective heating per unit area for given initial and terminal values of the velocity of the vehicle, its flight path angle, and its altitude. Real-coded GAs are used in all cases. Results are compared to single-objective optimal control approaches. The approach proposed is able to find an optimal solution to the problem using four independent randomly chosen control parameters. In contrast, the single-objective optimal control approach adopted requires a close approximation which has to be found empirically. Without such an approximation, the method does not work properly.

Blumel et al. [145, 144] use the NSGA in a *robust trajectory tracking problem*. The problem is addressed for a highly nonlinear missile, taking into account the design of an autopilot. Four objectives are considered: rising time, steady state error, overshoot and settling time. The NSGA is used to determine the membership function distribution within the outer loop control system of the airframe aerodynamics. A hybrid (binary and real) representation is used. Intermediate crossover is used for the real part of the chromosomes and multi-point crossover is used for the binary part. Fitness sharing is also applied. A normalization process is applied to each objective function, based on a target vector function (i.e., a function that uses the ideal values for each criterion considered separately).

Khwaja et al. [851] use a GA with a predator fitness approach to solve the *inverse kinematics problem of an arbitrary robotic manipulator*. Three objectives are considered: minimize the Euclidean distance of the tool center point to its desired position, minimize the rotation angle necessary to achieve the desired orientation, and minimize the discrete joint velocities. This approach uses a *predator function*[7] to keep the population size at a certain level by trying to terminate individuals at certain times (i.e., delete a fraction of the population at certain intervals to keep diversity). An objective and a chromosome are randomly chosen in the current population and a random value between zero and one is generated. If this random value is higher than the survival probability of the selected chromosome (for the objective selected), the chromosome is deleted. This scheme operates within binary tournament selection and is applied until the desired population size is achieved. Survival probabilities of a chromosome are computed using an utility function that combines the two first objectives (maintain position and orientation accuracy of the end-effector). Binary representation, one-point crossover and uniform mutation are used by the authors.

Osyczka et al. [1223] use a GA with a variation of the Pareto set distribution method [1226] to *optimize the design of robot grippers*. Two objectives are minimized: the difference between maximum and minimum gripper forces and assumed range of the gripper ends displacement, and the force transmission ratio between the gripper actuator and the gripper ends. A tournament selection based on feasibility is used to handle constraints. In related work, Osyczka and Krenich [1222] use the same approach together with the indiscernibility interval method to reduce the number of solutions within the Pareto optimal set. This allows a considerable reduction in the computational cost of the multiobjective optimization algorithm. The approach is applied to the same problem of designing robot grippers.

---

[7] This "predator function" is a nonlinear aggregating function similar to compromise programming (see Section 1.7.1 from Chapter 1).

Jakob et al. [768] use a parallel GA with a weighted sum for two *task planning and learning applications*. The first application is to move the tool center point of a simulated industrial robot to a given location on a "good" path avoiding obstacles. In this case, five criteria are used: avoid failure checks (overstep and collisions), get a certain accuracy in reaching the target position, obtain a smooth path more or less on a straight line from start to end, get a short travel time, and get a short action chain or a minimum of energy consumption. The second application is to control the behavior of an autonomous vehicle. Five criteria are also considered in this case: the distance between the final and target positions, the time needed for the travel, the deviation from the direct path between the starting and final points, the number of collision monitor activations while performing a Collision Avoidance action Sequence (CAS), and the number of emergency monitor activations. The authors experiment with a wide variety of different parameters for the GA (e.g., survival rules, population sizes, and ranking parameters).

Ortmann and Weber [1214] use a (μ, λ) evolution strategy with a linear combination of weights to *optimize the trajectory of a robot arm*. Six objectives are considered, each of them corresponding to one joint of a robot arm. Three types of weighting criteria are compared: fixed weighting, adaptive weighting depending on one gene and adaptive weighting depending on all genes. Since the problem is decomposable, all the joints are optimized in parallel. Aiming to produce a smooth trajectory, three types of transition functions are considered: linear, logarithmic, and squared. The best results were obtained using adaptive weighting depending on one gene, combined with a logarithmic transition function.

Coello Coello et al. [270] use a GA with a weighted min-max approach to optimize the *counterweight balancing of a Puma-560 robot arm*. Four objectives are considered, related to the minimization of torques and joint forces of the robot arm. Both binary and integer representation are used, together with deterministic tournament selection. Results are compared against MOGA, VEGA, NSGA, Hajela and Lin's weighted min-max technique, two Monte Carlo methods, the global criterion method, a conventional weighted min-max technique, a pure weighting method, a normalized weighting method, and a GA with a linear combination of objectives. Using as a metric the closeness to the ideal vector, the proposed approach produced better results than any of the other techniques.

Jin et al. [804] use an evolution strategy with an aggregating function to *generate fuzzy rule systems*. Three objectives are considered: completeness, consistency and compactness. The approach is tested on the design of a distance controller for cars. Similar work is reported by Jiménez et al. [794]. In this case, however, an evolutionary algorithm with Pareto ranking is used. Three objectives are also considered: compactness, transparency and accuracy of the fuzzy model. Compactness refers to the number of rules of the system, the number of fuzzy sets and the number of inputs for each rule. Transparency refers to linguistic interpretability and to locality of the rules. Accuracy refers to how appropriate are the rules generated to model the problem. This is easy to determine since there is normally an output that it is expected to be matched. The authors use an evolutionary algorithm with variable-length chromosomes, real-numbers representation, and an initialization process that ensures that all the individuals from the initial generation satisfy the constraints of the problem. They also experiment with several crossover and mutation operators and use elitism and a niche formation technique. The approach is tested on the fuzzy modeling of a second order nonlinear plant.

7.2.5 Structural and Mechanical Engineering

Table 7.5: Summary of mechanical and structural engineering applications

Specific Applications	Reference(s)	Type of MOEA
Truss design	[242]	GA with Pareto ranking
	[384, 1323]	GA with cooperative game theory
	[1190, 310]	Two-branch tournament GA
	[1422]	GA coupled with goal programming
	[1003, 1671]	GA with an aggregating function
	[260, 269, 636]	GA with a weighted min-max approach
	[1167, 68]	MOGA
Beam design	[112, 113]	GA with Pareto ranking
	[558, 557]	GA with Pareto elitist-based selection
	[1221, 1222]	GA with the Pareto set distribution method
	[260, 268]	GA with a weighted min-max approach
	[1714]	GA with Pareto ranking
Plate design	[1504]	GA with an aggregating function
Motorcycle's frame design	[1366]	GA with an aggregating function
Structural control systems	[857, 858]	GA with an aggregating function
	[927, 928, 1225, 1224]	GA with compromise programming
Packing problems	[602, 601]	Iterative GA
Gear-box design	[933]	MOGA
Leg mechanism design	[380]	NSGA-II
Micromechanical densification modeling parameters	[1337, 1336]	Fuzzy logic based multiobjective genetic algorithm
Blade design	[640]	Differential evolution with Pareto-based selection
Operational cost optimization of a steel plant	[697]	Nash GA [1463] and the predator-prey GA [986]
Training a feedforward neural network using noisy data from an industrial blast furnace	[1272]	predator-prey GA [986]

Cheng and Li [242] use a GA with Pareto ranking and a fuzzy penalty function to optimize a *four-bar pyramid truss* (the objectives considered are: minimize structural weight and control effort), a *72-bar space truss* (the objectives are:

minimize structural weight and strain energy), and a *four-bar plane truss* (the objectives are: minimize structural weight and vertical displacement). Results are compared against traditional single-objective optimization techniques. The authors show how their GA is able to produce not only a variety of nondominated solutions, but also that these solutions dominate those generated by traditional single-objective optimization methods. Binary representation, stochastic remainder selection, uniform crossover and a penalty function are adopted.

Rao [1323] uses a GA in combination with cooperative game theory for the optimization of a *two-bay truss* and a *six-bay truss*. In both cases three objectives are considered: minimize weight, minimize deflection and maximize fundamental frequency of vibration. He also solves the problem of *selection of actuator locations* in an actively controlled structure. In this second problem, two objectives are minimized: energy dissipation and weight. Binary representation and proportional selection are used. Results are compared against the global optima produced by a mathematical programming technique (used to optimize each of the objectives separately). The GA was able to approximate the global optima in one case (actuator locations) but not in the other. An inappropriate discretization of the decision variables is considered to be the cause of this last problem.

A similar approach is used by Dhingra and Lee [384] to optimize a *25 bar truss*, and a *two-bay truss*. The same three objective functions as before are used in this case (i.e., minimize weight and deflection and maximize fundamental frequency of vibration). A bargaining model based on an utility function is adopted for the cooperative game theoretic approach implemented. The model also allows the incorporation of degrees of importance to each objective function. The authors use binary representation, proportional selection, one-point crossover, linear scaling, a penalty function, and uniform mutation. Results are compared against branch-and-bound (used to optimize each of the objectives separately). Results indicate that the GA was able to approximate the ideal vector at a considerably lower computational cost than branch-and-bound.

Crossley and co-workers [1190, 310] use the two-branch tournament GA to optimize a *ten-bar plane truss* (two objectives are minimized: weight and vertical displacement) made both of a single and of several materials. Results are compared against mathematical programming techniques considering each objective separately. The GA compared fairly against classical optimization techniques, but it was evident that at least in some cases a portion of the Pareto front could not be generated by the proposed approach. However, the lower computational cost of using a MOEA is argued by the authors as the main advantage of their approach over classical techniques. Binary representation with Gray coding is used.

Sandgren [1422] uses goal programming coupled with a GA to optimize *plane trusses* and to design a *planar mechanism*. Several examples (each with different amount of objectives and constraints) are considered: a three-bar plane truss with two objectives (minimize volume and peak stress); the design of several planar mechanisms with nine objectives (minimize the distance between the actual motion generated by the mechanism and the desired linear path, minimize the difference between the desired velocity and the actual coupler point velocity, minimize the closeness of the coupler point to two points previously defined at the beginning and at the end of the required motion range, minimize size of the linkage, minimize weight of the linkage, maximize distance of the mechanism from a position where the mechanism does not assemble over the required motion range, the change in velocity of the coupler point over the required range of motion has to be as insensitive as possible to small fluctuations in the angular velocity of the input link, and the coupler point path traced has to be as insensitive as possible to: a) a small change in the length of the input link, and, b) a small change in the location of the coupler point position). Finally, the author also presents the design of a *ten-bar plane truss* with 7 objectives (minimize weight of the structure, minimize the maximum stress of any member, minimize the maximum displacement of any node, the maximum stress has to be as insensitive as possible to a change in loading magnitude or direction, the level of stress in each member has to be as uniform as possible, minimize the number of different beam sections). Binary representation and proportional selection are adopted.

Liu et al. [1003] use a GA with an aggregating function to minimize the linear regulator quadratic control cost, the robustness, and the modal controllability of the integrated *topology/control system of a 45-bar plane truss* with four actuators, subject to total weight, asymptotical stability and eigenvalue constraints. They use binary representation and roulette wheel selection.

Wallace et al. [1671] use a GA with a nonlinear aggregating function to solve *design problems*. The approach is applied in a *truss design problem* where the following criteria are considered: two normal stress safety factors, a design safety factor for buckling, cost, and two diameter ratios. The same approach is also applied to a *specification optimization problem* with three criteria: desired energy levels to be encouraged by an environmentally-friendly label, cost, and percentage of existing market within the energy criterion. Entropy (i.e., information contents) is used to determine the contribution of each specification to the fitness of an individual. The authors use binary representation, one-point crossover, uniform mutation, stochastic remainder selection, a penalty function, and elitism.

Coello Coello and co-workers [260, 269] use a GA with a weighted min-max approach to optimize a *25-bar space truss* and a *200-bar plane truss*. In both cases the objectives are to minimize the structural weight, the displacement of each free node and the stress that each member of each of the trusses has to support. Both binary and integer representation are used, together with deterministic tournament selection. Results are compared against MOGA, VEGA, NSGA, Hajela and Lin's weighted min-max technique, two Monte Carlo methods, the global criterion method, a conventional weighted min-max technique, a pure weighting method, a normalized weighting method, and a

GA with a linear combination of objectives. The min-max approach proposed performs better than the other MOEAs with respect to the generation of solutions that are closest to the ideal vector.

Hajela and Lin [636] use a GA with a weighted min-max approach to optimize a *10-bar plane truss* with two loading cases (two objectives are considered: minimize structural weight and vertical displacement), and a *wing-box structure* (two objectives are considered: minimize structural weight and maximize the sum of its first two natural frequencies). Binary representation and proportional selection with mating restrictions are used in both examples.

Azarm and co-workers [68, 1167] use a variation of MOGA to *optimize a two-bar plane truss*, and a *vibrating platform design*. The four main differences of this approach with respect to MOGA are the following: (1) the approach is interactive, (2) an L_2-norm is used to decide when to stop the evolutionary process (when no improvement is detected), (3) the same norm is used to detect spread uniformity of the Pareto set, and (4) only nondominated solutions have their constraints evaluated (a penalty function is used in case of constraint violation). The approach also uses a filtering procedure during each generation. This procedure deletes some individuals from each niche to encourage diversity. Mating restrictions based on phenotypic distance are also used. The two problems used to test the approach have two objectives each. In the case of the truss, the objectives are: minimize volume and stress. In the case of the vibrating platform, the objectives are: minimize cost and fundamental frequency of the platform. The authors use binary representation, stochastic universal selection, one-point crossover, uniform mutation and fitness sharing. Results are compared against MOGA (without the modifications proposed by the authors). The authors indicate that the modified version of MOGA has a faster convergence to PF_{true} (with a lower computational cost) and produces nondominated solutions that have a more uniform distribution.

Belegundu et al. [112, 113] use a GA with a variation of Pareto ranking in which dominated and infeasible solutions are removed from the population and replaced by newly (randomly) generated individuals. They apply their approach to optimize the design of *turbomachinery airfoils* considering two objectives: minimize the torsional resonant amplitude and maximize the torsional flutter margin. They also use their approach to minimize thermal residual stresses and cost of a *5-ply symmetric laminated composite*. Binary representation with roulette wheel selection are used.

Gero et al. [558, 557] use a GA with a Pareto Elitist-based selection to optimize a *beam section* in which two criteria are considered: maximize moment of inertia and minimize perimeter. The authors use exhaustive generate and test to combine the two objectives considered, and from that set of solutions, they extract those that are Pareto optimal. The authors also optimize the inverse problem to obtain the so-called attainable criteria set. Binary representation with proportional selection are used.

Osyczka and Krenich [1221, 1222] use a GA with the Pareto set distribution method to *design a beam of variable cross-sectional size*. Two objectives are minimized: the volume and the displacement of the beam under a certain force. A penalty function is used to incorporate the stress constraints of the problem and the indiscernibility interval method is used to reduce the number of elements of the Pareto optimal set.

Coello Coello and co-workers [260, 268] use a GA with a weighted min-max approach to optimize an *I-beam* (two objectives are minimized: its cross-sectional area and its static deflection). Both binary and integer representation are used, together with deterministic tournament selection. Results are compared against MOGA, VEGA, NSGA, Hajela and Lin's weighted min-max technique, two Monte Carlo methods, the global criterion method, a conventional weighted min-max technique, a pure weighting method, a normalized weighting method, and a GA with a linear combination of objectives. The min-max approach proposed performs better than the other MOEAs with respect to the generation of solutions that are closest to the ideal vector.

Wu and Azarm [1714] use a GA with Pareto ranking to *design a vibrating platform*. Two objectives are considered: maximize the fundamental frequency of the beam and minimize the material cost. A primary-secondary fitness approach is used to handle constraints. Primary fitness is used to measure individuals' performance, and secondary fitness is adopted to interpret the "matching" of two individuals (this is a mating restriction mechanism whose purpose is to generate offspring better than their parents). The constraint-handling technique is compared to a traditional penalty function. The proposed approach performs better in terms of closeness to PF_{true}, spread and a metric proposed by the authors, called "inferiority index" (the ratio between dominated and nondominated solutions generated by a MOEA).

Soremekun [1504] uses a GA with a linear combination of weights to *optimize simply supported composite plates comprised of two materials*. The two objectives minimized are: laminate weight and laminate cost. The author uses integer representation, one-point crossover, roulette wheel selection with linear ranking and uniform mutation with *ad-hoc* operators to add, delete or swap ply stacks. A penalty function is used to incorporate constraints into the fitness function. The user of the program decides what trade-offs are the most appropriate based on experience. However, because of the use of a linear combination of weights, certain (desirable) portions of the Pareto front can not be obtained.

Rodriguez et al. [1366] use a GA with a linear aggregating function to design a *motorcycle frame*. Two objectives are considered: minimize the mass and the maximum structural stress of the frame. An interesting aspect of this application is that the chromosomes contain a hybrid encoding, since some of the decision variables are discrete (e.g., the tube diameters commercially available) and others are continuous (e.g., angles and fillets). Thus, the chromosome consists of both real-numbers and integers and a suitable mutation operator is adopted for each of these two encodings (the same crossover operator is adopted for both of them). In order to evaluate the fitness of each design produced, the authors use a finite element analysis commercial package. One of the limitations of the approach is that a single Pareto optimal solution is obtained for each run of the algorithm (which is computationally expensive). Nevertheless, some of the solutions obtained were found to be better than the solutions obtained by a traditional design approach (i.e., based on manual intervention to fine tune the design parameters and run the simulation iteratively).

Kim and Ghaboussi [857, 858] use a GA with an aggregation approach (a multiplication of the objective functions) to *optimize a controller of a civil structure*. Three objectives are considered, related to the peak accelerations, peak displacements, RMS accelerations and RMS displacements of an active mass driver. Two constraints (maximum displacement and acceleration of AMD) are incorporated into the fitness function using a penalty function. The authors use binary representation, roulette wheel selection, two-point crossover and uniform mutation. Results are compared to the use of sample optimal control. The proposed method's performance in the response reduction was found to be superior to that of the sample optimal control.

Kundu and co-workers [927, 928, 1224, 1225] use a GA with compromise programming to *design a structural control system* for seismic vibration isolation. Two objectives are considered, related to the performance of the system. The authors maintain a set of nondominated solutions at each generation, all of which have the same fitness value (i.e., the Pareto front found so far).

Grignon and Fadel [601] use the Iterative GA [602] to *solve free-form packing problems*. The aim is to optimize multiple system level assembly characteristics of complex mechanical assemblies by placement of their components. Three objectives are maximized: compactness, static and dynamical balanced loading and maintainability (this is actually the sum of two objectives: accessibility of an object and its ease of removal from the system). Mechanical functional constraints and interference constraints are incorporated into the ranking process through the use of a penalty function. The approach is tested with two engineering configuration problems: the design of a satellite and the design of a car engine. The authors use a steady state GA with binary representation and roulette wheel selection with sigma truncation.

Kurapati and Azarm [933] use MOGA to solve a *gear-box design problem*. Two objectives are minimized: the speed reducer volume (or weight) and the stress in one of the shafts. This approach is aimed for multidisciplinary design optimization, and it therefore considers the exploration of hierarchical systems. The approach uses MOGA [504] to solve each of the multiobjective optimization problems associated with each hierarchically decomposed system to be solved. Then, the artificial immune system is used to coordinate the various subsystems present, and to handle the interactions among them. The best individuals obtained from each subsystem are selected and are considered *antigens*. A population of randomly created *antibodies* is created and is compared against the antigens using Hamming distances to measure "similarity" among them. The antibodies are evolved using a simple genetic algorithm, and copies of the resulting antibodies are inserted back into each subsystem.

Deb & Tiwari [380] use the NSGA-II [374] to *design a two degree-of-freedom leg mechanism consisting of a four-bar crank rocker and a pantograph*. Additionally, the compound leg mechanism was considered with spring elements. Three objectives are minimized in this case: the peak crank torque required to operate the mechanism for one complete cycle, the vertical actuating force required and the normalized leg size, defined as the overall frontal area required for the leg mechanism compared with the rectangular space within which the foot point is allowed to move. The problem has also 17 inequality constraints, related to the geometry of the mechanism (to ensure a smooth operation within a minimal fluctuation in the operating parameters), and to the foot path curvature to be followed by the leg mechanism. In fact, the authors indicate that the constraints define a very small feasible region, which makes this problem still more challenging. An interesting aspect of this work is that the authors adopt a hybrid approach in which the solutions obtained from an evolutionary algorithm are further processed by using a local search method, as suggested in [361]. This is done by applying a k-means clustering algorithm to identify the most representative regions of nondominated solutions found by the NSGA-II. After that, a simple GA with a combined weighted objective is adopted to try to improve the reference solutions identified by the clustering algorithm. The authors argue that this is a recommended procedure when applying a multi-objective evolutionary algorithm to a high-complexity problem. Results are compared with respect to a previous study in which only a few nondominated solutions had been found. The NSGA-II clearly outperforms these previous results, and is able to handle a more complex instance of the problem (adding spring elements to the leg mechanism) which the original study could not properly solve.

Reardon [1336] uses a fuzzy logic based multiobjective genetic algorithm to *optimize micromechanical densification modeling parameters for warm isopressed beryllium powder*. Seventeen objective functions corresponding to experimental data points are considered. The approach consists of an aggregating function that incorporates all the objectives into a single value, using fuzzy rule sets. This aggregating function determines fitness for each individual in the population of the GA. The author uses binary representation, one-point crossover, uniform mutation, tournament selection and phenotypic fitness sharing. In a related paper, Reardon [1337] uses the same approach to *optimize the micromechanical densification modeling parameters for copper powder*. In this case, six objectives related to the differences between the calculated densification values and six data points reported in the literature, are used as objectives.

Hampsey [640] uses differential evolution with a Pareto-based selection operator to *design wind turbine blades*. Three objectives are considered: starting performance, peak power production and induced stresses. The problem is also subject to different constraints (many of them geometric). The author proposes a mechanism to encode bi-cubic B-spline blade surfaces as real-valued vectors of geometry parameters. The approach is able to find *new* blades whose simulated performance indicates that they may outperform current state-of-the-art small wind turbine blade designs.

Hodge et al. [697] use the Nash GA [1463] and the predator-prey GA [986] for *optimizing the operational cost of the primary end of an integrated steel plant containing two blast furnaces utilizing both pellets and sinters, an electric furnace and a basic oxygen furnace*. Two constraints are minimized: the total cost of steel production and the total constraint violation. The authors use the predator-prey GA [986] to obtain the Pareto front of the problem, and the Nash GA to obtain the Nash equilibrium point. This Nash equilibrium point is determined considering a game between two players with individual search spaces, and it is used by the authors as a bound (not reachable in their example) similar to the ideal vector. The authors argue that the Nash equilibrium point can be used to assess how good is the Pareto front obtained. However, they also indicate its possible disadvantages, since they were unable to compute the Nash equilibrium point for another instance of the problem in which three objectives were considered. In a related paper, Pettersson et al. [1272] use the predator-prey GA [986] for *training a feedforward neural network using noisy data from an industrial iron blast furnace*. Two objectives are minimized: the training error of the neural network and the required number of active connections in the lower part of it. The decision variables are the architecture of the lower part of the network and their corresponding weights. The authors adopt an *ad-hoc* crossover operator that swaps similarly positioned hidden nodes of two parents. They also adopt a mutation operator taken from the differential evolution algorithm [1294].

7.2.6 Civil and Construction Engineering

Table 7.6: Summary of civil and construction engineering applications

Specific Applications	Reference(s)	Type of MOEA
Building construction planning	[476]	GA with Pareto ranking
	[842, 843]	GA with a target vector approach
Design of a thermal system for a building	[1710, 1711]	MOGA
City planning	[86, 87, 83, 85]	GA with a target vector approach

Feng et al. [476] use a GA with Pareto ranking to *solve construction time-cost trade-off problems*. Two objectives are minimized: project duration and cost. Integer representation and proportional selection are used. Nondominated individuals at each generation are retained as a way to maintain diversity in the population. The approach is tested on an 18-activity CPM (Critical Path Method) network and compared to the results produced by exhaustive enumeration. The GA is considered very efficient in this application because it finds 95% of the nondominated solutions generated by enumeration, but exploring only 0.00042% of the search space.

Balling et al. [86, 87] use a GA with a target vector approach (similar to min-max) for *optimal future land use and transportation plans for a city*. Three objectives are minimized: travel time, cost and change. Proportional selection, integer representation, one-point crossover and uniform mutation are used in this work. Death penalty is adopted to incorporate several constraints to the problem (i.e., solutions that violate constraints are discarded). In related work, Balling [83, 85] use a GA with a target vector approach to *find the optimum land use and transportation plans for two adjacent cities*. Two objectives are minimized: the cost minus revenues and the change from the status quo. The approach proposed is called "maximin fitness function" and is designed to favor nondominated vectors that are properly distributed along the Pareto front. The maximin fitness function produced better spread along the Pareto front than Pareto ranking in the application chosen. The authors use binary representation, tournament selection (with a tournament size of six), one-point crossover, non-uniform mutation, and elitism.

Khajehpour and co-authors [842, 843] use a GA with a target vector approach for *optimal cost revenue conceptual design of rectangular office buildings*. Three objectives are considered: minimize capital cost, minimize operating cost and maximize income revenue. Death penalty is used to handle the constraints of the problem. The GA implemented uses binary representation, roulette wheel selection, two-point crossover, uniform mutation, and elitism.

Wright and co-authors [1710, 1711] use MOGA for *designing thermal systems for a building*. In a first study, two objectives are minimized: daily energy cost and occupant thermal discomfort for a summer design day. Each trial solution is evaluated by running a simulation of the performance of the building. Such a simulation is based on a finite difference model of the building and a steady model of the heating, ventilating and air conditioning (HVAC) system. In a second study, the authors minimize two other objectives: daily energy cost and capital cost. MOGA is able to produce good trade-offs, but the authors report some difficulties to handle the constraints of the problem, since they can not find feasible solutions when performing a low number of evaluations.

7.2.7 Transport Engineering

Table 7.7: Summary of transport engineering applications

Specific Applications	Reference(s)	Type of MOEA
Train systems	[225]	GA with an aggregating function
	[965]	Unified model for multiobjective evolutionary algorithms
Road systems	[625]	NPGA with an outranking approach

Table 7.7: (continued)

Specific Applications	Reference(s)	Type of MOEA
	[613, 614, 617]	GA with an outranking approach
	[78]	GA with local geographic selection
	[40]	MOGA
	[1305]	GA with a target vector approach
Transportation problems	[1728, 736, 243]	GA with an aggregating function
	[554, 555, 552]	GA with fuzzy logic and an aggregating function
Vehicle routing	[809]	parallel NSGA hybridized with Tabu search
	[945]	GA with Pareto ranking

Chang et al. [225] use a GA with a linear combination of weights for the *operating optimization of electrified railway systems*. Two objectives are considered: minimize high power recovery and minimize load sharing. Binary representation and roulette wheel selection are used. The weights used are varied in the multiple runs performed. Results are compared to an approach that considers only one objective. The authors indicate that their proposed approach improves the power recovery and load sharing by controlling the firing angles of the traction system under study.

Laumanns et al. [965] use the Unified Model for Multi-objective Evolutionary Algorithms to *design road trains*. Ten objectives are considered, related to overall weight, gear box, engine and driving strategy minimizing fuel consumption, optimizing the driving performance and increasing driving convenience. The authors use an evolution strategy that applies only mutation, since the high interdependence of the design variables in every part of the objective space make recombination too disruptive.

Haastrup and Guimarães Pereira [625] use the NPGA combined with an outranking decision making method based on PROMETHEE (Preference Ranking Organization METHod for Enrichment Evaluations [168]) to *optimize the planning of a traffic route*. Two objectives are minimized: the number of people affected by traffic through minimization of driven distance, and the amount of noise generated by traffic. Binary representation, tournament selection and phenotypic fitness sharing are used.

In further work, Guimarães Pereira [613, 614, 617] uses a GA with an outranking approach [1391] combined with preference indexes (inspired by ELECTRE I [1383]) to compute weights that are linearly combined in a fitness function. This approach is used to *generate alternative motorway routes*. Five criteria are considered: height, geology, aspect, land use and distance from urban centers. Binary representation and tournament selection are used. Phenotypic sharing is used to keep diversity. The author experiments both with one-point and uniform crossover, and with or without one-bit mutation. Since a single fitness value is adopted for each individual, the author is able to measure on-line and off-line performances of the GA [348].

Baita et al. [78] use a GA with Local Geographic Selection (LGS) to solve a *vehicle scheduling problem*. Two objectives are minimized: the number of vehicles and the number of deadheading trips. LGS is based on the idea that the population has certain spatial structure (a toroidal grid with one individual per grid location, in this case). Selection takes place locally on this grid, and each individual competes with its nearby neighbors. Basically, an individual finds its mate during a random walk starting from its location and the individual with the highest fitness found during the walk is selected. This approach makes unnecessary the use of fitness sharing, since it creates niches in a natural way. The authors reproduce 85% of the population using crossover based on LGS and the remaining by means of tournament selection (a small percentage of the population is evolved using mutation only, to preserve diversity). The technique is applied to the urban public transportation of the city of Mestre, in Venice. An integer, multiploid chromosomic representation is used. Results are compared to those produced using a custom-made program developed by the Mass Transportation Company of Venice (this program performs single-objective optimization). The proposed MOEA is able to produce solutions with the same number of vehicles than single-objective optimization, but with a much reduced number of deadheading trips.

Anderson et al. [40] use MOGA to design a *fuzzy logic traffic signal controller*. Several objectives are considered, but they really deal with only two objectives at a time: maximize average travel time and minimize emissions of pollutants. In a further paper, Sayers et al. [1436] concentrate on two conflicting objectives: minimize delay of both vehicles and pedestrians. As a secondary criterion to solve ties, the authors use variability of delay across vehicle streams. The aim is to identify a range of robust (i.e., that can perform well in different simulated traffic scenarios) parameter sets corresponding to different policy objectives. MOGA is used again, with integer representation, tournament selection and niching.

Qiu [1305] uses a GA with a weighted sum of goal deviations to *prioritize and schedule road projects*. Seventeen objectives related to maximize the investment effectiveness subject to the current budget constraints are adopted. The author uses permutation-based encoding, binary tournament selection, elitism, partially mapped crossover (PMX), and uniform mutation. Results are compared to the use of goal programming, and the GA is able to find solutions for a type of problem in which goal programming could not be applied.

Cheng et al. [243] use a GA with a linear combination of weights to solve a *linear transportation problem*, a *minimum spanning tree problem*, and an *interval programming problem*. Two objectives are considered in all cases. Weights are computed in terms of the maximum and minimum values for each objective at each generation. An adaptive penalty approach is used to incorporate the constraints of each problem. For the linear transportation problem, the authors use a matrix encoding and the special crossover and mutation operators proposed by Vignaux and Michalewicz [1652]. For the minimum spanning tree problem, the authors use a Prüfer number encoding, uniform crossover and perturbation mutation. For the interval programming problem, the authors use integer representation, uniform crossover and perturbation mutation. Only the results of the linear transportation problem are compared to the method of Aneja and Nair [49]. The authors indicate that their approach is much more effective than the method of Aneja and Nair because it finds more nondominated vectors and they are more uniformly spread.

Yang and Gen [1728] use a GA with a linear combination of weights to solve a *bicriteria linear transportation problem*. Two objectives are minimized: total transportation cost and total deterioration. The approach uses problem-specific knowledge in the form of appropriate data structures and specialized genetic operators (an "evolution program", as defined by Michalewicz [1100]). The authors use the matrix encoding of Vignaux and Michalewicz [1652], a specialized initialization procedure, special crossover and mutation operators, and mating restrictions (called extinction and immigration by the authors). Results are compared to two other techniques in terms of which method produces the best compromise solution. The GA proposed was found to perform better than the other techniques. In related work, Ida et al. [736, 988, 989] use the same type of GA with a linear combination of weights to solve a *multiobjective chance-constrained solid transportation problem*. In this case, the cost coefficients are exactly known and the production and demands (or transportation capacities) cannot be determined exactly. Although several objectives are considered, only results with two or three objectives are reported. The authors use a 3D integer representation, elitism, and specialized genetic operators to direct the search towards a compromise solution (the Technique for Order Preference by Similarity to Ideal Solution [941] is used to find such a "compromise solution"). The authors have also used fuzzy logic and ranking for this problem [988, 989]. Gen et al. [554, 555, 552] extend this approach to allow more than two objectives (three objectives are considered in this case), and add fuzzy logic to handle the uncertainty involved in the decision making process. A weighted sum is still used in this approach, but it is combined with a fuzzy ranking technique that helps to identify Pareto solutions. The coefficients of the objectives are represented with fuzzy numbers reflecting the existing uncertainty regarding their relative importance.

Jozefowiez et al. [809] use a parallel version of the NSGA [1509] hybridized with Tabu search for *solving bi-objective vehicle routing problems*. Two objectives are minimized: the total distance and the difference between the length of the longest tour and the length of the shortest tour. The authors adopt an elitist version of the NSGA in which a new diversification mechanism is introduced. This mechanism consists in maintaining several external archives. For their bi-objective problem, the authors maintain 3 archives: one contains the Pareto optimal solutions considering the minimization of f_1 and f_2, the second considers that f_1 is minimized and f_2 is maximized, and the third considers that f_1 is maximized and f_2 is minimized. The solutions contained in these archives are moved to the main population of the MOEA. However, since parallelism is used, the authors only attach one of the 3 types of archive to each island and when performing migration, the islands exchange their archives, aiming to explore new regions of the search space during this exchange of information. The authors experiment with route-based crossover (RBX) [1289], split crossover [1296, 1297] and sequence-based crossover [1289], and use Or-opt mutation (this operator moves from one to three consecutive customers from a tour to another position in the same tour or to another tour). They also use a 2-opt local search mechanism, whose aim is to improve the total length of the tours produced. The authors also adopt clustering techniques (the average linkage method [1132], as in SPEA [1782]) for pruning their external archives in order to keep their size small. The authors also evaluate a hybridization of their parallel MOEA with Tabu search. This approach is called Π^2-TS (Parallel Pareto Tabu Search), and it takes the Pareto optimal solutions found by the MOEA as its starting point. It uses Or-opt as its neighborhood operator, and it acts as a local search mechanism that aims to improve the solutions obtained by the MOEA. As expected, the use of this local search mechanism improves the convergence of the approach. Results are compared with respect to the best known total lengths for a set of instances of the bi-objective vehicle routing problem.

Lamont et al. [945] propose a parallel genetic algorithm based on Pareto ranking to *support the design of a comprehensive mission planning system for swarms of autonomous aerial vehicles* (UAV). This is a three dimensional vehicle routing problem. Four objectives are considered: cost, encompassing distance traveled, amount of climbing a vehicle does, and risk resulting from flying through areas of threat. A one-point midpoint crossover and three distinct mutation operators are applied with equal probability. Infeasible solutions are repaired. The approach also adopts an external archive to store the nondominated solutions found during the search, and an elitist selection scheme. The developed system consists of a terrain-following path planner (based on a parallel MOEA) and a vehicle router (based on an evolutionary algorithm). Visualized results indicate effective performance of the proposed MOEA.

7.2.8 Aeronautical Engineering

Table 7.8: Summary of aeronautical engineering applications

Specific Applications	Reference(s)	Type of MOEA
Constellation design	[443]	Two-branch tournament GA
	[1079, 1080]	Variation of the NSGA
	[664, 663, 662]	Variation of the NSGA
Helicopter design	[309]	GA with a Kreisselmeir-Steinhauser function
	[308]	Two-branch tournament GA and Pareto ranking
	[493]	GA with a modified version of Pareto ranking
Aerodynamic optimization	[1676]	GA coupled with game theory
	[1281, 1282, 1283, 1284]	NPGA

Table 7.8: (continued)

Specific Applications	Reference(s)	Type of MOEA
	[1200, 1196, 1202, 1201, 1197, 1198, 1203, 1228, 1227]	MOGA
	[564]	GA with Pareto ranking
	[1057, 1045, 1046]	NSGA
	[1371, 1372]	VEGA
	[42, 44, 43, 1645, 1646, 1647]	GA with Pareto ranking
	[1252]	Co-evolutionary multiobjective GA
	[41]	GA with an aggregating function
	[1306, 1307]	Virtual Subpopulation Genetic Algorithm
	[1308]	Parallel genetic algorithm
	[666, 665]	evolution strategy with an aggregating function
	[1168]	NSGA-II [374]

Ely et al. [443] use the two-branch tournament GA for *constellation design for zonal coverage using eccentric orbits*. Two objectives are minimized: the maximum altitude of the constellation, and the total number of satellites in the constellation. Binary representation and uniform crossover are used.

Mason et al. [1079, 1080] use a variation of NSGA [1509] called MINSGA (Modified Illinois NSGA) to *design optimal earth orbiting satellite constellations*. Two objectives are considered: minimize number of satellites and maximize continuous coverage. The modifications introduced are the use of stochastic universal selection (instead of stochastic remainder selection), and the use of a normalized parameter set in the sharing scheme. Binary representation and fitness sharing are used. Results are compared against NSGA (without the changes proposed by the authors). The MINSGA was found to exhibit a performance similar to the NSGA in unconstrained problems, and a better performance in constrained problems. In further work, Hartmann et al. [663, 664, 662] use another variation of the NSGA to *optimize low-thrust interplanetary spacecraft trajectories*. Three objectives are considered: maximize mass delivered to target, minimize time of flight and maximize heliocentric revolutions. The genetic algorithm is coupled with a calculus-of-variations based trajectory optimizer, called SEPTOP (*Solar Electric Propulsion Trajectory Optimization Program*). The hybrid adopted uses a "memetic algorithm" (i.e., a localized search algorithm used to improve the characteristics of a previously obtained solution) [1134] (the resulting algorithm is called Nondominated Sorting Memetic Algorithm, or NSMA for short). Baldwinian learning was also used (i.e., only the improved fitness values obtained are recorded) [1700]. The authors use binary representation, one-point crossover, and fitness sharing (mutation is not applied). The approach found novel Earth-Mars trajectories for a Delta II 7925 launch vehicle with a single 30-cm xenon engine for spacecraft propulsion. Results are not compared to any other approach, but the GA was able to produce novel trajectories with very high performance and non-intuitive structures.

Crossley [309] uses a GA with a Kreisselmeir-Steinhauser function that combines all the objectives and constraints into a single envelope function. This approach is used to *design a helicopter rotor system*. Two objectives are minimized: the hover power required and the weight of the rotor system. Binary representation and proportional selection are used. Results are compared to a GA with a linear combination of weights that the author uses in a previous paper [307]. In another paper, Crossley [308] applies a two-branch tournament GA and Pareto ranking to the same problem.

Flynn and Sherman [493] use a GA with a modified version of Pareto ranking to *optimize the design of helicopter flat panels*. Four objectives are considered: the buckling of a panel and the buckling of a bay have to remain within a specified range; minimize the panel weight, and minimize the number of frames and stiffeners. Dominance is checked independently for each objective function, and zero is assigned to a counter when the individual is nondominated with respect to that objective; otherwise, one is added to the counter for each individual that dominates the current solution under evaluation. Then, the ranking of an individual is given by the addition of its counters for each objective, so that nondominated individuals with respect to the entire population get a range of zero. The authors use one-point crossover, binary representation with Gray coding, elitism, and uniform mutation. Results are compared to rule-based optimization and to constrained, partially optimized designs (the proposed approach performed better in both cases).

Wang and Periaux [1676] use a GA with concepts of game theory for *multipoint aerodynamic optimization*. Two objectives are considered, related to the slat and flap position of a three-element airfoil system. The concepts of Nash equilibrium [1169] and Stackelberg equilibrium [1013] are used by two GAs to model a two-person competitive game. Results between the so-called Nash GA and the Stackelberg GA indicate that the second approach produces better results (in terms of quality), but at a higher computational cost.

Poloni et al. [1281, 1282, 1283, 1284] use the NPGA for the *design of a multipoint airfoil*. Two objectives are considered: to have a high lift at low speed and a low drag at transonic speed. A parallel GA is used in this case, placing individuals in a toroidal grid with one individual per grid location. A local selection mechanism is used, so that each individual has to compete against its neighbors. A random walk is used to determine the mate with which an individual reproduces. Two-point crossover and uniform mutation are adopted.

Marco et al. [1057] use a parallel implementation of the NSGA in a problem of *computational fluid dynamics* in which the goal is to exhibit a family of profiles that have a smooth transition in shape from a certain initial profile to another. Two objectives are considered: minimize the cost of the low-drag and the high-lift profiles of a 2D airfoil. Binary representation and binary tournament selection are used.

Obayashi et al. [1200] use MOGA coupled with an inverse optimization technique to *design a transonic wing for mid-size regional aircraft*. Two objectives are minimized: the difference between the spanwise lift distribution and an elliptic distribution for a specified total lift, and the induced drag. Apparently, they use the same sort of representation reported in related work (i.e., real numbers). In a related paper, Oyama et al. [1228] use MOGA for *aerodynamic design of a transonic wing*. Two objectives are considered: maximize lift and minimize drag of a wing. The Taguchi method [1380] is used to analyze the epistasis of the design variables of the problem. A scheme known as best-N selection [1610] is adopted for elitism. Under this scheme, the best N individuals from a pool of N parents and N children are selected for the next generation. The authors use fitness sharing, real numbers representation, tree-structure encoding of the design variables, one-point crossover, and uniform mutation. In further work, Obayashi et al. [1204, 1201, 1197, 1198, 1196, 1202] use MOGA for the multidisciplinary optimization of *wing planform design*. Three objectives are considered: minimize aerodynamic drag, minimize wing weight and maximize fuel weight stored in the wing. Real numbers representation and stochastic universal selection are used. They also experiment with coevolutionary shared niching [588] which they find to be superior to fitness sharing. In related work, Obayashi et al. [1203] use the same approach to *optimize a wing for supersonic transport*. Three objectives are considered as well: minimize the drag for supersonic cruise, the drag for transonic cruise and the bending moment at the wing root for supersonic cruise. The problem has high dimensionality (66 variables are considered in this case) and the search space is constrained (e.g., there are wing area and wing thickness constraints). In this case, fitness sharing on phenotypic space is adopted and niche sizes are computed as suggested by Fonseca and Fleming [504]. Average crossover and non-uniform mutation are used with the real-coded GA adopted. Results are compared against an existing aerodynamic design of the supersonic wing for the National Aerospace Laboratory's Scaled Supersonic Experimental Airplane. The proposed approach was able to produce a novel wing design.

Oyama and Liou [1227] use MOGA to *optimize turbopumps of cryogenic rocket engines*. Two objectives are considered: maximize total head and minimize input power. The authors use floating point representation, fitness sharing, best-N selection [1610], and blend crossover. The approach is applied to the redesign of a single-stage centrifugal pump and the redesign of a multistage liquid oxygen pump.

Giotis and Giannakoglou [564] use a GA with Pareto ranking to solve *transonic airfoil design problems*. Two objectives are minimized: drag and the difference between a given lift and a target value. Fitness sharing in objective function space is adopted, together with binary representation. A multilayer perceptron is used to reduce the CPU cost of the GA. The approach is tested with the optimization of the RAE-2822 isolated profile under inviscid flow conditions. Results are compared to those produced optimizing the two objectives separately. The authors indicate that Pareto ranking can produce better trade-offs and the additional use of neural networks considerably reduces the computational cost of the approach.

Mäkinen et al. [1045, 1046] use the NSGA in an *airfoil shape optimization problem*. Two objectives are considered: minimize the drag coefficient and amplitude of the backscattered wave while the lift coefficient is not less than a certain given value. A quadratic penalty function is used to incorporate the lift constraint into the fitness function used for each objective. The authors use tournament selection, phenotypic sharing on parameter value space and floating point representation. The approach is used to optimize an NACA64A410 airfoil, using 15 design variables.

Rogers [1372, 1371] uses a GA with a population-based approach (similar to VEGA [1440]) to find the *optimum placement of aerodynamic actuators for aircraft control*. Three objectives are considered: provide uncoupled pitch, roll and yaw moments. Each of these objectives is considered as a separate subproblem and a separate population is allocated for each of them. A composite fitness function is formed from the members of the three subpopulations which satisfy their corresponding constraints. Arrays are built to store any individuals that are feasible in any dimension (i.e., in any of the objectives), so that they can be combined in different ways. The fitness function consists of an OR function which is applied to the binary strings of the three members selected (i.e., the three objectives) for an individual. The result obtained (i.e., a binary sum without carry) is adopted as the fitness value of an individual. The author uses one-point crossover, incest prevention [456], and uniform mutation. Both a sequential and a parallel version of the algorithm are tested. The author only refers to the computational time saved by his approach with respect to exhaustive search.

Anderson and Lawrence [43] and Anderson et al. [44] use a GA with Pareto ranking to *extract munition aerodynamic characteristics and initial (launch) conditions* from high quality position and altitude data. Two objectives are minimized: the root-mean-square (RMS) position error and the RMS Euler angle error. Binary representation and tournament selection are used.

Anderson and Gebert [42] use a GA with Pareto ranking for *preliminary wing subsonic design*. Three objectives are maximized: the lift-to-drag ratio, area ratio and lift ratio. Binary representation and tournament selection are used.

Vicini and Quagliarella [1647, 1646, 1645] use a GA with Pareto ranking hybridized with a conjugate gradient technique whose fitness function is a linear combination of objectives to *optimize the shape of a wing for transonic flow conditions*. Two objectives are considered: minimize aerodynamic drag and minimize structural weight. The same approach is also used for *airfoil design*, considering two objectives: minimize wave drag and maximize thickness. Binary representation and a "simple random walk" selection scheme (applied to nondominated parents) are used. Results are compared to those produced by a conjugate gradient technique using a linear combination of objectives. Pareto ranking was considered better in terms of robustness and quality of the solutions produced.

Parmee and Watson [1252] use a co-evolutionary multiobjective genetic algorithm for *preliminary design of airframes*. The idea of this approach is to use separate GAs for each of the objective functions of the problem. Such GAs run concurrently. Then, the fitness values for the individuals within each of these GAs is adjusted through a comparison with respect to the other GAs. This is, in fact, similar to a cooperative game theoretical approach such as the one described by Barbosa [92]. The method is really created to converge to a single (ideal) trade-off solution. However, through the use of penalties the algorithm can maintain diversity in the population. These penalties relate to variability in the decision variables' values. The authors also store solutions produced during the evolutionary process so that the user can analyze the historical paths traversed by the algorithm. Three objectives are considered: subsonic specific excess power, ferry range, and attained turn rate. The authors use roulette wheel selection and perform some sensitivity analysis of the parameters used.

Anderson [41] uses a GA with a linear combination of weights to *design subsonic wings*. Four objectives are considered: maximize the lift/drag ratio, maximize the lift/weight ratio, meet design lift goals, and maintain structural integrity. The most important objective is the first (lift/draft ratio). The author uses binary representation, one-point crossover and uniform mutation.

Quagliarella and Vicini [1306, 1307] use the Virtual Subpopulation Genetic Algorithm (VSGA) for *wing design*. Two objectives related to aerodynamic and structural requirements are initially defined. The resulting wing is then modified to further reduce its aerodynamic drag. The authors use a parallel genetic algorithm with several subpopulations distributed on a single toroidal topology. The selection scheme is based on random walk: for each subpopulation, random walk is used to select the parents to mate. The Pareto front is generated by extracting nondominated vectors with respect to the whole population. The authors use binary representation, one-point crossover, elitism, and a migration policy based on boundaries defined through a certain probability function.

In a further paper, Quagliarella and Vicini [1308] use a parallel GA with the same approach to *design high-lift airfoils*. Two objectives are considered: maximize the lift force and minimize the pitching moment coefficient. Results are compared to the use of an "enhanced random walk" selection scheme in which nondominated parents are selected in two phases: in the first, locally nondominated individuals are randomly marked as possible parents; in the second phase, two ranking criteria are used: (a) minimum Euclidean distance of a solution to the current Pareto front, and (b) minimum value of a given objective between the solutions belonging to the set of nondominated vectors. One of these two criteria is used (with a certain probability) to select parents for the following generation. The parallel GA has a better performance (in terms of CPU time required) than the scalar GA, but no improvement in the quality of the results is achieved.

Hasenjäger et al. [666, 665] use an evolution strategy with covariance matrix adaptation (the so-called CMA-ES) and a linear aggregating function to *optimize gas turbine stator blades*. Two objectives are minimized: mass-averaged total pressure loss and variation of the circumferential static pressure distribution. The authors indicate that the main difficulty when optimizing 3D aerodynamic designs such as this one is the high computational cost associated with 3D computational flow analysis. So, the problem to be solved is high-dimensional and has a very high computational cost associated with the evaluation of each fitness function. The authors use a parallelized 3D Navier-Stokes flow solver (which adopts MPI [1230]), and a master-slave model of the optimization loop is implemented using the Parallel Virtual Machine (PVM) library [549]). Two versions of the problem are solved: a single-objective one (in which mass-averaged total pressure loss is minimized), and a multi-objective version in which the second objective previously indicated is also considered. For the multi-objective version of the problem, the dynamic weight aggregation approach of Jin et al. [800, 801] is adopted. The main goal of this work was to identify new design concepts from the optimization results, rather than improving on a certain (previously known) design. According to the authors, this goal is achieved in their work.

Nariman-Zadeh et al. [1168] use the NSGA-II [374] for the *multiobjective optimization of the thermodynamic cycle of ideal turbojet engines*. Four objectives are considered: minimize specific thrust, minimize specific fuel consumption, maximize propulsive efficiency, and maximize thermal efficiency. The decision variables are the input Mach number and the pressure ratio of the compressor of the turbojet engines being optimized. The authors adopt the so-called ϵ-elimination diversity approach, which consists of a procedure that removes duplicates and individuals that are identical (both in decision variable space and objective function space) within a certain threshold ϵ. An interesting aspect of this work is that the multiobjective optimization task is not the final goal. The authors use another evolutionary algorithm together with singular value decomposition for the optimal design of both connectivity configuration and the values of coefficients involved in group method of data handling (GMDH)-type neural networks. These GMDH-type neural networks are used for the inverse modeling of the input-output data table obtained as the best Pareto front. A special encoding (consisting of symbolic strings) and appropriate operators are adopted to encode the GMDH-type networks. The polynomial neural network that is produced by the approach is validated using 760 data samples which are randomly chosen for training purposes, while 250 more data samples are used for testing purposes. Results indicated a match between the predicted and the computed values with an accuracy of two significant digits.

7.3 Scientific Applications

Scientific applications occupy the third place in terms of popularity (after industrial applications). Scientific applications have been subdivided in six groups: geography, chemistry, physics, medicine, ecology and computer science & computer engineering. Not surprisingly, computer science & computer engineering is, by far, the most popular subdiscipline. Medicine and chemistry are in (distant) second and third places, respectively. The less usual scientific application area that resulted from the bibliography survey presented in this chapter is ecology, from which only two applications are reviewed.

7.3.1 Geography

Table 7.9: Summary of geography applications

Specific Applications	Reference(s)	Type of MOEA
Environmental modeling	[772]	GA with an aggregating function
Site-search problems	[1716]	MOEA based on nondominated sorting
Land use planning	[1083]	MOGA
	[1524]	GA with goal programming

Jarvis et al. [772] use a GA with a linear combination of weights to *choose the most characteristic pre-located site data from a wider set* in the context of environmental modeling. This is a combinatorial optimization problem in which the aim is to move the initial data available closer to a form useful in meeting two main sampling criteria: 1) "representativeness" of sample data relative to full data set (e.g., to have a distribution of heights representative of total range within the domain used) and 2) sampling requirements related to interpolation tasks (e.g., nearby sites are required for successful production of variogram). These two criteria involve the use of eight objective functions. A real-coded GA with one-point crossover, uniform mutation and inversion is used. Results are compared to those found using a deterministic methodology in which sites are 'weeded' in a sequential fashion according to each criterion in turn, ranked according to the importance assigned by the user. The approach is applied to meteorological data for England and Wales, aiming to choose 200 sites out of a possible 985. The authors indicate that the GA produced competitive trade-off solutions with respect to those generated by an enumerative approach.

Xiao et al. [1716] use a multi-objective evolutionary algorithm based on nondominated sorting to *generate alternatives for multi-objective site-search problems*. Two objectives are minimized: the total cost for the site, the mean distance between the site and the facility. A special encoding (and associated crossover and mutation operators) based on undirected graphs is adopted to represent feasible solutions. Results are only compared with respect to a random search, but the authors argue about the superiority of the MOEA adopted.

Matthews et al. [1083] use MOGA for *rural land use planning*. Two objectives are maximized: net present value of the property over 60 years and the Shannon-Wiener index, which measures the diversity and evenness of land use. A non-generational genetic algorithm with a mechanism to eliminate duplicates and restrict mating is implemented. Two representations are used: one has a fixed-length genotype and the other uses a variable-length order-dependent representation. Fitness sharing on phenotypic space is adopted. The authors use uniform crossover and uniform mutation for the fixed-length representation, and several *ad-hoc* operators are defined for the order-dependent representation. The initial population benefits from the application of other heuristics and specific domain knowledge. Both evenness and extent of coverage (of the Pareto front) are used as comparison criteria to determine the quality of solutions produced with the two representations adopted. Both representations produced a good number of nondominated vectors. However, the variable-length representation was found to be more susceptible to the niche and population sizes.

Stewart et al. [1524] use a GA with goal programming for *land use planning*. Three objectives were considered: maximize the natural value of the area, maximize the recreational value of the area and minimize the cost of changing land use. The authors use a linear aggregating function and a goal programming formulation to solve the multiobjective (nonlinear) combinatorial optimization problem arising from the model adopted for land use planning. They adopt a genetic algorithm with *ad-hoc* crossover and mutation operators. Payoff tables are constructed for the objectives, so that they can compute the ideal vector and from them, they approximate the nadir values. Both of them (ideal vector and nadir values) are used as the lower and upper level goals for each objective. An interesting aspect of this work is the use of a factor to encourage aggregations into clusters, and another one to encourage achievement of the target numbers for each land use. These factors also act as specific mechanisms to satisfy certain requirements of the problem that otherwise the GA could not properly handle. After solving some randomly generated instances, the authors solve a real-world problem related to the management of the Jisperveld region in the Netherlands. The GA was able to produce a meaningful range of alternative plans responsive to the changing preferences of the users, which was very satisfactory for its designers.

7.3.2 Chemistry

Table 7.10: Summary of chemistry applications

Specific Applications	Reference(s)	Type of MOEA
Intensities of emission lines of trace elements	[1702]	GA coupled with goal programming

Table 7.10: (continued)

Specific Applications	Reference(s)	Type of MOEA
Modeling of a chemical process	[683]	Multiobjective genetic programming
Search of molecular structures	[808]	GA with an aggregating function
Polymer extrusion optimization	[544]	GA with the reduced Pareto set approach
	[545]	NSGA and a GA with the reduced Pareto set approach
Design of a chemical plant	[221]	Pareto Converging Genetic Algorithm [923]

Wienke et al. [1702] use a GA with goal programming to *optimize the intensities of six emission lines of trace elements* in alumina powder as a function of spectroscopic excitation conditions. Seven objectives are considered (the goal is to find the best compromise combination of seven relative emission intensities). Binary representation and proportional selection are used. Results are verified by an overlapping resolution map and by a control experiment. The authors indicate that the GA is able to match the results of the overlapping resolution map and also suggest the use of sharing to maintain diversity.

Hinchliffe et al. [683] use Multi-Objective Genetic Programming (MOGP) to *model a steady state chemical process system*. Their approach is based on MOGA and they also use fitness sharing in the objective domain and the concept of preferability based on a given goal vector as proposed by Fonseca and Fleming [504]. Four objectives are minimized: root mean square (RMS) error on training data set, residual variance, correlation between residuals and the process output and the model string length. Two industrial case studies are analyzed: the development of an inferential estimator for bottom product composition in a vacuum distillation column and the development of a model for the degree of starch gelatinization in an industrial cooking extruder. Linear ranking selection and a maximum tree size of 500 characters are used. Results are compared to a GP approach that considers only one objective, based on the RMS error. MOGP did not produce significantly better predictions than the single-objective GP implementation against which it was compared. However, the capability of generating good trade-offs under several objectives was considered as the main advantage of MOGP.

Gaspar Cunha et al. [544] use a GA with a ranking procedure called the Reduced Pareto Set (RPS) algorithm to solve a *polymer extrusion optimization problem*. Four objectives are considered: minimize melt temperature, minimize length of screw required for melting, minimize power consumption and maximize mass output. Binary representation, fitness sharing and roulette wheel selection are used. Results are compared to the NPGA [709] and a GA with a linear combination of weights. The fact that the RPS algorithm produces less elements of PF_{true} than the NPGA is considered by its authors as an advantage, since they argue that this facilitates the decision-making process. The GA with a linear aggregating function was outperformed by the two other approaches.

In a further paper, Gaspar Cunha et al. [545] also compare the RPS algorithm to the NSGA [1509] in the same problem. The authors indicate that the RPS algorithm produces slightly better results than the NSGA, but the difference is marginal.

Jones et al. [808] use a GA with a linear combination of weights for *conformational search of 3D molecular structures*. Two objectives are considered: the distance range between each pair of atoms in the pharmacophore and the energy of conformation which has to be within certain limits. The authors use binary representation, one-point crossover, uniform mutation and roulette wheel selection. Results are compared against a deterministic procedure called SYBYL CSEARCH [328]. The GA was found to be more effective and more efficient than CSEARCH since it retrieved more hits and in less time.

Chakraborti et al. [221] use the Pareto Converging Genetic Algorithm (PCGA) [923] to design the Williams and Otto Chemical Plant. Two objectives are maximized: annual return and low constraint violation. The approach adopts binary encoding, binary tournament selection, one- and two-point crossover, and both Creep and Jump mutations. Results are compared with respect to several mathematical programming techniques and against both a single-objective (sequential) genetic algorithm and a single-objective parallel (island model) genetic algorithm.

7.3.3 Physics

Table 7.11: Summary of physics applications

Specific Applications	Reference(s)	Type of MOEA
Reflector backscattering	[1266, 1267]	Nash-GA
Optimization of multilayered anti-reflection coatings	[1028]	GA with Pareto ranking
Analysis of experimental spectra	[591, 590]	NPGA
Design of a water reactor	[1244, 1245, 1246]	GA with Pareto ranking
Electrical impedance tomography problem	[711]	GA with Pareto ranking
Design of quantum cascade lasers	[867]	GENMOP-LS

Périaux et al. [1266, 1267] use the Nash-GA (based on non-cooperative game theory) to find an optimal distribution of active control elements in order to *minimize the backscattering of a reflector*. Two objectives are minimized: the radar cross section (RCS) considering a +45° incidence wave and the RCS considering a -45° incidence wave. Binary representation and genotypic sharing are used. Results are compared to the use of the NSGA. The authors conclude that the NSGA [1509] produces better solutions, but they argue that the solutions of the Nash-GA are more stable.

Lum et al. [1028] use a GA with Pareto ranking for *constrained optimization of multilayered anti-reflection coatings*. Two objectives are minimized:

reflection and thickness. The authors adopt fitness sharing, a penalty function to handle the constraints of the problem, and binary encoding (with Gray codes). In their experiments, the authors consider a two-layer structure of magnetic materials, and consider three cases: (1) single objective (reflectivity is the only objective), (2) linear aggregating function to combine the two objectives, and (3) a multi-objective approach. Interestingly, the Pareto front produced by the GA with Pareto ranking from the third case did not include nor covered the solutions generated by the GAs adopted for the two previous cases. This indicated a lack of spread of the MOEA adopted, which is an issue that the authors recognized and that they aimed to address in their future work.

Golovkin et al. [591, 590] use the NPGA to *analyze experimental spectra and monochromatic images* (i.e., spectroscopic analysis). Two objectives are minimized: the difference between the data and the fits for both emissivities and spectra. The main goal of the work is to estimate plasma temperature and density gradients by performing simultaneous analysis of experimental X-ray spectra and monochromatic images. Binary representation, uniform crossover, uniform mutation and equivalence class sharing are used. The authors also allow competition between parents and offspring for selection purposes.

Parks [1245, 1244, 1246] uses a GA with Pareto ranking to *design a pessurized water reactor*. Three objectives are considered: minimize the enrichment of the fresh fuel, maximize the burn-up of the fuel to be discharged, and minimize the ratio of the peak to average assembly power throughout a cycle. A matrix representation (i.e., two-dimensional arrays) is used together with the so-called heuristic tie-breaking crossover (HTBX) operator [1286]. An archive of nondominated solutions is kept and updated during the evolutionary process, using reactivity distributions as a criterion to determine similarity between two solutions. Also, nondominated individuals which are sufficiently "dissimilar" to current solutions are retained to maintain diversity.

Hsiao et al. [711] use a GA with Pareto ranking for solving *an electrical impedance tomography problem*. This is an inverse problem with nonlinear equations that are normally solved using iterative optimization techniques. The problem is normally handled as a single-objective optimization problem, but in this case, two objectives are considered, related to the errors between the predicted solution and the measured data. The authors compare the Davidon-Fletcher-Powell method [1292] a simple genetic algorithm, and a Pareto genetic algorithm. The preliminary results obtained led the authors to propose a hybrid approach in which the genetic algorithm is used as a global optimizer, and the Davidon-Fletcher-Powell method acts as a local search procedure. This hybrid scheme is found to improve convergence and to be very robust even when applied to complex test cases. The approach was successful in reconstructing highly irregular internal objects and, by adding one extra parameter, was also able to reconstruct an image without a priori knowledge of the exact number of internal objects. The authors also determined that their approach remains robust in the presence of noise for the test cases analyzed.

Kleeman and Lamont [867] modify the General Multiobjective Parallel Genetic Algorithm (GENMOP), in order to incorporate a neighborhood search process. This approach is called the "multi-tiered memetic MOEA" and is adopted to *design quantum cascade lasers*. This approach incorporates domain knowledge to change the temporal focus of the neighborhood search based on the number of generations. The multi-tiered local search procedure is able to focus the local search on specific critical variables of the quantum cascade laser design at different stages in the optimization process. It is empirically shown that this multi-tiered memetic MOEA is able to find excellent solutions to the quantum cascade laser design problem.

7.3.4 Medicine

Table 7.12: Summary of medicine applications

Specific Applications	Reference(s)	Type of MOEA
Treatment planning	[1743]	GA with Pareto ranking
	[1271]	SPEA
Allocation in radiological facilities	[239]	GA with an aggregating function
	[940]	MOGA, the NPGA and SPEA
Prognostic models	[1078]	Diffusion genetic algorithm
Left ventricle 3D reconstruction	[15]	GA with Pareto ranking
Functional brain imaging	[908]	4D-Miner
Breast cancer diagnosis	[4]	Pareto differential evolution

Yu [1743] uses a GA with Pareto ranking for *treatment planning optimization in radiation therapy*. Two problems are considered: stereotactic radio surgery optimization and prostate implant optimization. In the first problem, four objectives are minimized: maximum dose in the target volume, average dose above the mean in the anterior critical structure, average dose above the mean in the posterior critical structure, and average dose above the mean in the normal tissue shell. In the second problem, three objectives are considered: maximize the minimum peripheral dose, minimize the dose to the urethra and minimize the number of needles used. Binary representation and binary tournament selection are used.

Petrovski and McCall [1271] use the Strength Pareto Evolutionary Algorithm (SPEA [1782]) to optimize a *chemotherapeutic treatment*. Two objectives are maximized: tumor eradication and the patient survival time. The authors really optimize treatment schedules using a progressive articulation of preferences. They use binary representation and a penalty function to incorporate the constraints of the problem.

Chen et al. [239] use a GA with a linear combination of weights to *optimize the worker allocation problem in radiological facilities*. Two criteria are minimized: (a) the unsatisfaction of the soft constraints of the problem that are violated, and (b) the number of times that a worker changes workplaces. The second criterion is really applied only to break a tie between two individuals with the same value for the first criterion. However, the approach is considered multiobjective because the first criterion is made up of a weighted combination of up to five values, corresponding to the soft constraints of the problem: (1) workplace dose constraints, (2) working time limits, (3) special skill requirements, (4) limits due to extreme environments, and (5) industrial safety requirements. Different priorities are assigned to each of them. The authors use a two-dimensional matrix representation, proportional selection, arithmetic crossover with a repair algorithm, elitism, and a mutation operator with a mechanism that minimizes the number of times that workers change workplaces (the second of the two criteria previously described). Results are compared against goal programming and the simplex method based on the numerical values of the two criteria previously mentioned. The GA outperformed the two other methods in terms of the quality of solutions found.

Lahanas et al. [940] and Milickovic et al. [1112] use MOGA [504], NPGA [709] and SPEA [1782] for *distribution problems in brachytherapy* (a treatment method for cancer). Two objectives are minimized: the variance of the dose distribution of sampling points uniformly distributed on the Planning Target Volume and the dose distribution variance inside the Planning Target Volume. The authors use real-numbers representation with several crossover operators (blend, geometric, two-point and arithmetic) and several mutation operators (uniform, non-uniform, flip, swap and Gaussian). Results are compared against an aggregating function with variable weights. This aggregating function is optimized using three global optimization methods (the Polak-Ribiere variant of the Fletcher-Reeves algorithm, the Broyden-Fletcher-Goldfarb-Shanno quasi-Newton based algorithm and the Powell method). The MOEAs adopted produce solutions that are either equivalent or better than those generated by algorithms based on phenomenological methods used in the majority of treatment planning systems. Within the MOEAs compared, SPEA produced the best results in terms of closeness to PF_{true} and uniform distribution along the Pareto front.

Marvin et al. [1078] use a diffusion genetic algorithm to *derive prognostic models*. Prognostic models are used to determine whether or not a certain patient who suffers from an uncommon type of cancer has probabilities of surviving. Three objectives are considered: maximize the correct number of survival predictions, maximize the correct number of death predictions, and minimize the number of factors used. The approach is particularly applied to predict the survival of women with a certain type of cancer (high risk gestational trophoblastic tumors). The diffusion GA [1382] is a GA whose population is spatially distributed. Normally, each individual resides at its own vertex on a square lattice. Selection is applied locally rather than globally: an individual chooses an immediate neighbor at random and mates with it to produce an offspring. The offspring competes with its parent and the fittest (in this case, the nondominated vector) wins. Since selection is local, this algorithm is more efficient than traditional Pareto ranking. Also, diversity naturally emerges from the topology adopted. Furthermore, the approach naturally follows certain parallel processor architectures where its implementation is therefore straightforward. The authors use one-point crossover, an *ad-hoc* uniform mutation operator, and integer representation.

Aguilar and Miranda [15] apply several MOEAs to the solution of the *left ventricle 3D reconstruction problem*, which is related to the diagnosis of cardiac patients. Six objectives are considered, related to slice fidelity, internal energy of the reconstructed slice, and energy of similarity between the current slice configuration and the adjacent slice previously reconstructed. Four population-based approaches and Pareto ranking are compared to simulated annealing and a GA using a linear combination of weights. The best results in terms of spread and quality of results were provided by Pareto ranking, although the population-based approaches were more efficient in terms of CPU time required.

Krmicek and Sebag [908] use an algorithm called 4D-Miner for *functional brain imaging*, where 3 objectives are maximized: length, area and alignment. Due to the nature of the problem, a solution is considered important even if it is dominated with respect to alignment and area, provided that it is located in different regions of the brain. This led the authors to introduce diversity as an additional objective, and called this restatement of the problem: multimodal multi-objective optimization (MoMOO). The 4D-Miner algorithm uses a MOEA with Pareto archive-based selection similar to PESA [301], restricted mating (only sufficiently close parents are allowed to mate), and a steady-state scheme (at each step, a single individual is selected as parent, to generate an offspring via crossover and mutation). A special procedure is adopted for the initialization of the approach, so that relevant spatio-temporal patterns are generated and the extremities of the Pareto front are excluded. *Ad-hoc* crossover and mutation operators are adopted. This approach was validated using real-world datasets collected from subjects observing a moving ball. Experiments are conducted in two stages. In the first, certain active areas of the brain are identified. In the second, such specific areas are attempted to be related to specific cognitive processes. In this case, the authors experimented with "catch" and "no-catch" activities (the subject sees a ball and decides either to catch it or to let it go). At this stage, new objectives and constraints are considered, in order to find discriminant spatio-temporal patterns. The discriminant patterns extracted were found to be satisfactory by the neuroscientists to which they were presented.

Abbass [4] use an evolutionary artificial neural network approach based on the Pareto Differential Evolution (PDE) algorithm [7] augmented with local search, for the *prediction of breast cancer*. Two objectives are minimized: error and number of hidden units. The author adopts multi-layer perceptrons, which are evolved using the PDE algorithm, such that the training and number of hidden units of the network are simultaneously determined. The author uses a memetic version of the PDE algorithm [3], because a local search procedure is coupled to the approach. The algorithm is called Memetic Pareto Artificial Neural Network (MPANN), and is validated using the Wisconsin

396   7 Applications

dataset from the UCI Machine Learning Repository [1184]. Results are compared with respect to those reported by Fogel et al. [495] and by the same author using an artificial neural network with the backpropagation algorithm [2]. The author reports that MPANN presents a slightly better performance than the two other approaches with respect to which it was compared, but with a much lower standard deviation and a much lower computational cost.

7.3.5 Ecology

Table 7.13: Summary of ecology applications

Specific Applications	Reference(s)	Type of MOEA
Assessment of ecological models	[1356, 1357]	GA with an aggregating function and a GA with Pareto ranking
Fitting of ecological process models	[894]	NSGA [1509]

Reynolds [1356, 1357] use an evolutionary multiobjective optimization technique to *assess ecological process models*. The approach consists of an evolutionary algorithm that uses a fitness function that incorporates both the number of objectives of the problem and membership in the Pareto optimal set. The initial population of the algorithm consists of a Pareto optimal set generated from a preliminary search process, augmented with some random solutions intended to preserve diversity. The approach is applied to the assessment of the canopy competition model WHORL [1505]. Binary interval error measures are selected for each objective, based on data from a permanent plot. Ten objectives are considered: mortality, stand height frequency distribution, median live tree height, number of live whorls, crown angle, crown length ratio, suppressed tree growth rate, variability in suppressed tree height increment rates, dominant tree slope and variability in dominant tree height increment rates.

Komuro [894] uses a software called Pareto_Evolve, which is based on the use of the NSGA [1509], for *fitting an ecological process model to a set of data*. Fourteen objectives are considered, two for each day of the week. Each objective corresponds to a difference between measured and simulated data. Due to the high number of objectives adopted, the author had to add a series of mechanisms to the NSGA. Namely, the author adopted an elitist strategy by which an individual is retained if it satisfies one of four possible conditions, which include different forms of weak dominance. Also, the number of nondominated solutions was restricted to one half of the population size (this was evidently necessary given the high number of objectives which rapidly increased the number of nondominated solutions in the population) and the other half of the population was generated by breeding. The author also proposed the use of dynamic crossover and mutation rates as a way to improve the efficiency of the search. An interesting outcome of this study was that the model (rather than the selection criteria adopted by the author) was found to have deficiencies. The author proposed a revised version of the model which reduces the bias of the original model, but can not decrease the error produced. The author then concludes that more biological information is required in the model in order to improve its accuracy.

7.3.6 Computer Science and Computer Engineering

Table 7.14: Summary of computer science and computer engineering applications

Specific Applications	Reference(s)	Type of MOEA
Coordination of agents	[205, 206, 529]	GA with an aggregating function
Exploration of software implementations for DSP algorithms	[1779]	SPEA
Computer-generated animation	[606]	Genetic programming with an aggregating function
	[1481]	Interactive genetic algorithm with a Pareto optimal selection strategy
Machine learning	[924, 920, 916, 921, 922]	Pareto converging genetic algorithm
	[1556, 1555]	GA with Pareto optimal selection
	[1440]	VEGA
	[1738]	Genetic programming with Pareto selection
	[156]	Genetic programming with Pareto ranking
	[124]	GA with Pareto ranking and an aggregating function
	[332]	GA with an aggregating function
Image processing	[130, 19]	GA with an aggregating function
	[899]	Hybrid of a genetic algorithm and a multilayer backpropagation neural network
	[21, 20, 22]	MOGA
Facial modeling and detection	[694]	GA with an aggregating function
	[1642]	GA with Fuzzy-Pareto-Dominance
Handwritten word recognition	[1130]	NSGA
Simulation	[134, 135]	GA with different aggregating functions
Object partition and allocation	[250]	NPGA
Games	[254]	GA with an aggregating function
Sorting networks	[1408]	GA with an aggregating function
Traveling salesperson problem	[777, 784]	GA with an aggregating function and local search
Genetic programming	[436]	Tournament selection based on Pareto dominance
	[140]	SPEA2
	[349]	Find only and complete undominated sets

Table 7.14: (continued)

Specific Applications	Reference(s)	Type of MOEA
Automatic programming	[950, 951, 952]	Genetic programming with Pareto-based tournament selection
Data mining	[678]	SPEA2
Natural language processing	[55]	aggregating function, NSGA, MOGA [504] and NSGA-II
Bioinformatics	[194]	Indicator-based Selection Method (IBEA)
	[1747]	Multi-Objective Scatter Search (MOSS), SPEA and the $(\mu + \lambda)$ MOEA

Cardon et al. [205, 206] and Galinho et al. [529] use a GA with a linear combination of weights to *coordinate micro and meta-agents adopted to optimize the Gantt diagram of a job shop scheduling problem*. They use as many objectives as jobs are being considered (up to 500 are considered in their experiments). The authors also experiment with the ϵ-constraint method coupled with a GA. They use a multi-agent system that simulates the behavior of each entity (a job) that collaborates to accomplish actions on the Gantt diagram so as to resolve a given economic function. Each agent has a fitness based on its impact on the Gantt diagram. The authors use binary representation with Gray coding, Gaussian mutation, and uniform crossover (using a toroidal chain of bits to represent strings prior to recombination).

Zitzler et al. [1779] use the Strength Pareto Evolutionary Algorithm (SPEA [1782]) to *explore trade-offs of software implementations for DSP algorithms* (SPEA is really used in a scheduling problem). Three objectives are minimized: program memory requirements (code size), data memory requirements, and execution time. Given a software library for a target processor (Programmable Digital Signal Processor, or PDSP for short), and a dataflow-based block diagram specification of a DSP application in terms of this library, the authors compute Pareto optimal solutions, trading-off the three objectives previously mentioned. The authors use a mixed representation (characters, integers and binary numbers), and a mixture of operators: order-based uniform crossover with scramble sublist mutation [339] for the characters, and one-point crossover and uniform mutation for the rest of the string. The proposed approach is tested on several commercial PDSPs for which it produced reasonable trade-offs among the three objectives considered.

Gritz and Hahn [606] use genetic programming and a linear combination of weights to derive control programs for *articulated figure motion*. Two main objectives are considered: achievement of the desired motion sequence, and style points. These style points are handled as a penalty or a reward depending on the characteristics of the motion sequence (e.g., hitting an obstacle is penalized, while performing the action quickly is rewarded). The approach is applied to the animation of a lamp and a humanoid figure. In both cases, the main goal is the distance between base center and a desired goal at the end of the time allotment. Style points are a weighted sum of four things (i.e., five objectives are considered): bonus for completing the motion early, penalty for excess movement after the goal is met, penalty for hitting obstacles, and bonus for ending with joints at neutral angles. The authors use one-point crossover and a reproduction operator, but do not use mutation.

Shibuya et al. [1481] use the Interactive Genetic Algorithm (IGA) [1553] with a Pareto Optimal Selection Strategy (POSS) [1738] to *generate animation of human-like motion* (the motion produced when passing a small object from one hand to another). Both a 16-link mechanism and a 4-link mechanism are used to model the upper half part of the human body and its two hands. Four objectives are minimized: change of the joint torques, joint torques, acceleration of the handled object, and completion time of the motion. Real-numbers representation, unimodal normal distribution crossover, roulette wheel selection and a penalty function (to incorporate the constraints of the problem) are adopted. Genotypic clustering is used to search local Pareto optimal sets in parallel. The user has to participate interactively, since a certain degree of subjectivity is involved in the decision-making process. Results compared well against the use of the simple IGA and an interactive simplex method.

Kumar and Rockett [920, 916, 921, 922, 924] use the Pareto Converging Genetic Algorithm (PCGA) [923] to *partition the pattern space into hyperspheres for subsequent mapping onto a hierarchical neural network for subspace learning*. Seven objectives are considered: minimize the number of hyperspheres, minimize the learning complexity, maximize the regularity of the decision surface, maximize the fraction of included patterns of each class, minimize the maximum fraction of included patterns in a single hypersphere, minimize the overlap of partitions, and minimize the surface area. The authors use real-numbers representation, variable-length chromosomes, one-point crossover, and Gaussian mutation. Results are compared against clustering algorithms using some benchmark problems. The PCGA was found to be superior to K-means clustering because it does not rely on a similarity measure.

Tamaki et al. [1556, 1555] use a GA with Pareto optimal selection to solve a *decision tree induction problem*. Two objectives are minimized: the accuracy (error rate of classification) and the simplicity (the number of leaves) of the decision tree. The authors use S-expressions in LISP, subtree exchange crossover and uniform mutation. Results are compared against feature subset selection [35] (FSS), ID3, and OPT [147]. The GA was found to be better than the other techniques in terms of the number of elements of the Pareto optimal set found.

Schaffer and Grefenstette [1440] use VEGA for *machine learning* in a *pattern classification domain*. The approach is applied to the classification of muscle activity patterns for several human gait classes (normal and abnormal). Schaffer extends a GA-based machine learning system called LS-1 [1499] to handle several objectives. Five objectives are considered (one for each gait class used). The use of VEGA allows to overcome the limitations of the scalar critic used to evaluate the learning task of each production system encoded by the GA. These limitations are related to complementary knowledge (i.e., conflicting objectives) from different structures that are being forced to compete. A simple GA can not be applied in this case and therefore the motivation to use a multiobjective approach.

In similar work, Yoshida et al. [1738] use GP with Pareto selection to *generate decision trees*. The approach is called Pareto Optimal and Amalgamated Induction for Decision Trees (PARADE), and it considers two objectives: accuracy (minimize error rate), and simplicity (minimize the number of leaves of the decision tree). Selection is applied to the population made of the offspring and mutated individuals after applying crossover and mutation to the original population. S-expressions in LISP are used to encode the decision trees. The authors use subtree exchange crossover and uniform mutation. Results are compared to combinational approaches such as ID3, FSS, and OPT. GP with Pareto selection was able to generate a larger number of elements of the Pareto optimal set and in less time than the other approaches.

Bot [156] uses genetic programming with Pareto ranking to *induce optimal linear classification trees*. Two objectives are minimized: the number of errors and the number of nodes in the tree. The author also experiments with a domination-based fitness sharing scheme, which is found to perform better or equally well than Pareto ranking. The author uses tournament selection, a combination of phenotypic and genotypic sharing, strong typing and elitism. The approach is compared to several decision tree classification algorithms (OC1 [1160], C5.0 [1310] and M5 [519]) on a set of classification benchmarks in machine learning. The proposed approach performed as well or better than the classification algorithms against which it was compared in two of the examples, but it performed worse in the another one. Also, genetic programming was found to be slower than the other approaches.

Bernadó i Mansilla and Garrell i Guiu [124] use several evolutionary multiobjective optimization approaches to *develop optimal set of rules for classifier systems*. Two objectives are considered: maximize accuracy and generality of each rule. The approaches used are two versions of Pareto ranking, a population-ranking method, and an aggregating function. Several types of crowding are used to maintain diversity. Several metrics are adopted to evaluate the performance of each approach: coverage, accuracy, size of the solution set, learning speed and optimal population characteristics. The approach is applied to the Wisconsin breast cancer database. Results are competitive with those obtained using state-of-the-art classifier systems.

Dasgupta and González [332] use a GA with a linear combination of weights to *extract comprehensible linguistic rules*. Three objectives are considered: maximize the sensitivity, maximize the specificity, and minimize the length of the chromosome. The authors adopt a linear representation of tree structures, restricted crossover, uniform mutation, gene elimination and gene addition. The approach produces rules with an accuracy comparable to the results of other techniques using public-domain data sets such as IRIS [758], VOTE [501] and WINE [753].

Aguirre et al. [19] use a GA with an aggregation function to solve the *image halftoning problem*. Two objectives are maximized: the gray level resolution of an image and its spatial resolution. Several weights are defined to specify search directions that are explored in parallel. The idea is to split an image into several segments and apply a GA to each of these segments. Each GA locates elements that are locally Pareto optimal. To enforce global Pareto optimality, two error functions are applied globally (i.e., to all the GAs) and those nondominated individuals with lowest values for both error functions are chosen. The authors use a GA with two genetic operators with complementary roles applied in parallel (self-reproduction with mutation and crossover and mutation), extinctive selection (i.e., individuals with lower fitness are given a zero survival probability), and an adaptive mutation probability that goes from high to low values based on the contribution of self-reproduction with mutation to the performance of the algorithm. This type of GA, called GA-SRM, has been found useful (in terms of reducing computational costs) in the same sort of application (i.e., halftoning), considered as a single-objective optimization problem. The approach is compared against different types of GAs (single- and multiobjective). The GA-SRM was able to simultaneously generate several high-quality images at a lower computational cost.

Bhanu and Lee [130] use a GA with a linear combination of weights to solve *adaptive image segmentation* problems. Five objectives are considered: edge-border coincidence, boundary consistency, pixel classification, object overlap, and object contrast. The authors use a specialized crossover operator, and binary representation. The population used by the GA is taken from a training database. An elitism scheme is used to replace the least fit elements of this training database over time. The GA in this case operates only on the parameters of the problem.

Köppen and Rudlof [899] use a hybrid between an evolutionary algorithm and a multilayer backpropagation neural network to solve a *texture filtering problem*. The evolutionary algorithm (a GA) is used to balance the weights of the neural network. The approach uses as many output neurons as objectives has the problem. Each of these neurons determines its error value for its corresponding objective independently. During each network cycle, a single objective is randomly selected by each hidden neuron (this is similar to the lexicographic method, discussed on Section 1.7.1 from Chapter 1). The update of weights and the genetic operators applied (a trasduction operator and Gaussian mutation) are performed on the basis of the objective selected by each hidden neuron. Two objectives are considered, related to minimizing the differences between an image produced by the algorithm and a goal image given by the user. Results are compared against the use of a weighted sum approach. The proposed algorithm produced solutions of a higher quality than the weighted sum approach.

Aherne et al. [21, 20, 22] use MOGA to optimize the *selection of parameters for an object recognition scheme*: the pairwise geometric histogram paradigm. Three objectives are considered: minimize repeatability, minimize the area of the histograms, and maximize histogram consistency across different examples of a given object subject to variable segmentation/fragmentation. The population used by MOGA consists of several subpopulations (which are randomly generated during the first generation) each of which is stored as an array of individuals. The authors adopt a simple selection scheme in which a newly generated subpopulation replaces the old one only if it is fitter. Different subpopulations are allowed to breed. Mutation consists of creating (randomly) a new subpopulation with a best-replacement policy (each individual of the new subpopulation is compared against the individuals of the old subpopulation, and replaces them only if they are fitter).

Ho and Huang [694] use a genetic algorithm with a linear aggregating function for *facial modeling from an uncalibrated face image*. The authors adopt a flexible generic parameterized facial model (FGPFM), which can be easily modified using the facial features as parameters to construct an accurate specific 3D facial model from only a photograph of an individual. The authors adopted the so-called "Intelligent Genetic Algorithm" IGA [695], (which they had previously proposed) to tackle this problem. The main innovation of the IGA is that it uses orthogonal arrays and factor analysis (taken from quality control) to improve the crossover operator (the new operator is called "intelligent crossover"). The problem of reconstructing a 3D facial model (e.g., from a photograph) is transformed into the problem of how to acquire 3D control points that are sufficiently accurate as to provide the desired model. This gives raise to a large parameter optimization problem that is solved using the IGA previously indicated. Four objectives are minimized: the projection function, the symmetry function, the depth value function and the model ratio function. The main aims were that: (a) the projection of the facial model from some viewpoint must coincide with the features in the given face image, and (b) the facial model must adhere to the generic knowledge of human faces accepted by the human perception. The chromosomic strings adopted consist of integer values and the search proceeds in stages, in an approach called coarse-to-fine, where "coarse" refers to global searches and "fine" refers to local searches. The proposed approach was found to be robust in the presence of different poses of a human face (although the weights of the linear aggregating function had to be tuned by hand in some cases) and outperformed other GA-based approaches.

Verschae et al. [1642] use a standard genetic algorithm with Fuzzy-Pareto-Dominance [900] to improve a *face detection* system. Rather than just being a concept, Fuzzy-Pareto-Dominance is considered a metaheuristic by its authors. As such, this approach makes single-objective optimization algorithms capable of handling multiple objectives by introducing ranking values of the fitness objective vectors. Roughly, the idea is to fuzzify the Pareto dominance relation such that we can express relationships of the form "A is being dominated by B in a degree μ". These degrees are then used to rank a set of data which in this case are the fitness values of the individuals in the population of a genetic algorithm. Three objectives are minimized: false positive rate (relative number of non-faces that the system has wrongly output a positive face detection), false negative rate (relative number of faces the system did not detect as such) and number of evaluated features (number of features that were taken into account for the face detection decision). The goal of this work was to optimize a face detection system based on a boosted cascade architecture. This system uses a grid to avoid classifying every possible window of the image in this cascade architecture. By finding appropriate thresholds for this grid, the processing time of the system can be reduced, by preserving low error rates. The approach was validated using images taken from a standard face detection benchmark dataset. The authors optimized 22 decision variables which were encoded as integers using a precision of 8 bits. The results indicated that the multi-objective evolutionary algorithm was able to improve the processing speed of the system by reducing the number of features to be considered. The approach also improved the false positive rates.

Morita et al. [1130] propose a methodology for feature selection in unsupervised learning which is applied to *handwritten word recognition tasks*. The approach is based on the Nondominated Sorting Genetic Algorithm (NSGA) [1509]. The goal of this work was to find a set of nondominated solutions that contained the more discriminant features and the more pertinent number of clusters. The two objectives to be minimized are: the number of features and a validity index that measures the quality of clusters. A standard K-Means algorithm was then applied to form the given number of clusters based on the selected features and the number of selected clusters. After performing two experiments on synthetic data sets, the authors adopted a word classifier used in some of their previous work [1129]. In their study, a word image was segmented into graphemes, each of which consisted of a correctly segmented, under-segmented, or an over-segmented character. Then, two feature sets were extracted from the sequence of graphemes to feed the classifiers. In order to allow a better assessment of their results, the authors considered only one feature set based on a mixture of segmentation primitives and concavity and contour features. The methodology of the authors consisted of applying the NSGA to obtain a set of nondominated solutions. Then, the incorporation of user's preferences was necessary as to choose only one solution from the set produced. The authors decided to train each nondominated solution obtained. Then they used each classifier in the system and chose the solution that supplied the best word recognition result on the validation set proposed. The results were compared with respect to the use of the traditional methodology previously adopted by the authors for this problem. The authors reported that the use of an evolutionary multiobjective optimization approach kept the recognition rates at the same level as the traditional strategy while reducing the time required for training the discrete Hidden Markov Models adopted. This was achieved because the approach reduced the number of features (from 34 to 29) and the number of clusters (from 80 to 36) with respect to the use of a traditional technique.

Bingul et al. [134, 135] use a GA with different aggregating functions to *determine force allocations for war simulation*. Such allocations are based on the capabilities of the threat forces, the conditions of the war and the capabilities of the friendly forces. All of these are simulated using a software system developed for the Air Force Studies and Analyses Agency called THUNDER. Four objectives are considered: minimize the territory that the friendly forces side losses, minimize the friendly forces side aircraft lost, maximize the number of enemy forces side strategic targets killed, and maximize the number of enemy forces side armor killed. Three fitness assignment schemes are adopted. One, is a conventional aggregating approach in which each of the objectives is squared and all of them are added together. The second approach raises each objective at a different power, aiming to prioritize the objectives as when using the lexicographic method. The third approach is a target vector approach in which the differences between the values achieved for the objectives and certain target values are minimized.

Choi and Wu [250] use the NPGA [709] for the *partitioning and allocation of objects in heterogeneous distributed environments*. Three objectives are minimized: network communication cost, loss of concurrency and load imbalance. The authors use binary representation, one-point crossover, fitness sharing, and tournament size of seven (for the NPGA) in their experiments.

Chow [254] uses a GA with a linear combination of weights to *search three-colored 10×10 NCR boards*. The author uses a GA with a hybrid selection scheme ($\frac{3}{4}$ of the population is selected using tournament selection and the rest using roulette wheel), two-point crossover, uniform mutation, and a two-dimensional integer representation. Four objectives are considered: the number of red, white, and blue chromatic rectangles[8] from a 10 × 10 chessboard, and a distribution factor for the three colors previously mentioned.

Ryan [1408] uses a steady state (variable-length) GA and a linear combination of weights to solve *K−sorting network problems*. Two objectives are considered: efficiency and length of the solution generated. To avoid premature convergence, the author uses a method of disassortative mating called the "Pigmy Algorithm". In this scheme, two populations are maintained, each with its corresponding fitness function. The first population is called "Civil Servants" and it uses its performance (efficiency in this case) as its fitness. The second population is called "Pygmies", and is composed of individuals that can not qualify for the first population. Individuals in this second population are evaluated using length as a second criterion in their fitness function. Recombination is applied considering that each of these two populations represents a gender (i.e., breeding is allowed only between individuals from different populations). This aims to produce individuals that have a good balance of efficiency and length in their encoded solutions. The same approach is also used with genetic programming.

---

[8] A *chromatic rectangle* is a rectangle whose four corners are of the same color.

Jaszkiewicz [777] uses a GA with local search and a linear combination of weights to solve *multiple objective symmetric traveling salesperson problems*. Two and three objectives related to costs are solved. Instances with 50 and 100 cities are considered. The author uses permutation encoding, distance preserving crossover [520], and a local search algorithm that operates in two phases (in the first phase, arcs common to both parents are not considered, and in the second, all arcs are considered). No mutation operator is used. The approach, called RD-MOGLS (Random Directions Multiple Objective Genetic Local Search) is compared to Ishibuchi and Murata's MOGLS [751], and against a version of MOGLS based on Multiple Objective Simulated Annealing (MOSA) [1617]. In related work, Jaszkiewicz [776], compares his approach to MOGA [504]. The estimation of the expected value of a weighted Tchebycheff utility function over a set of normalized weight vectors is used as a quality measure to evaluate performance of the approaches [653]. RD-MOGLS performs considerably better than the other approaches with respect to the metric adopted.

Ekárt and Németh [436] use a tournament selection scheme based on non-dominance (similar to the NPGA) to *control code growth in genetic programming*. Two objectives are minimized: standardized fitness and program size. The approach is tested on the Boolean multiplexer problem and on the symbolic regression problem. Both program size and processing time are reduced by this approach with respect to the solutions generated by a conventional GP implementation.

Bleuler et al. [140] use SPEA2 to *control code size and reduce bloating in genetic programming*. Two objectives are considered: functionality of the program and code size. Preference incorporation is required in order to apply SPEA2 to this domain. SPEA2 uses a close-grained fitness assignment strategy and an adjustable elitism scheme [1775]. SPEA2 compared well with respect to standard GP, Constant Parsimony Pressure and Adaptive Parsimony Pressure in several even-parity problems.

de Jong et al. [349] use multiobjective optimization techniques to *reduce bloat and promote diversity in genetic programming*. Three objectives are considered: maximize fitness, minimize tree size and maximize diversity. The approach, called "Find Only and Complete Undominated Sets" (FOCUS), consists of using a stricter version of Pareto dominance to select an individual for recombination.[9] Diversity is maintained by considering it as another objective (i.e., those individuals located at a certain distance from the others are preferred). Only nondominated vectors are kept at each generation, and duplicates are eliminated from the population by mutating them. The approach is tested on three instances of the n-parity problem. FOCUS was found to require a lower computational effort and to produce more compact

---

[9] The definition of Pareto dominance is modified in order to get a better spread of the solutions generated.

solutions than both basic genetic programming and another version that uses automatically defined functions.

Langdon [950, 951, 952] uses genetic programming with Pareto-based tournament selection to *evolve primitives implementing a FIFO queue*. Six objectives are used: number of tests passed by each of the operations and the number of memory cells used. The author uses an encoding in which each individual is composed of six trees, each of which implements a trial solution to one of five operations that forms the FIFO queue program. An additional tree can be called from any tree. To reduce premature convergence, the population is treated as a toroidal grid in which each grid point contains a single individual. When a new individual is created, its parents are selected from the same deme where it resides the replaced individual.

Hetland and Saetrom [678] use SPEA2 [1775] for *rule mining in time series databases*. A few aspects of SPEA2 are modified for this application. For example, only individuals having different objective values are selected in the initial archive filling procedure in order to avoid premature convergence. Thus, if two or more individuals have the same values in objective function space, only one of them is randomly selected to be added to the external archive. Since a tree-encoding is adopted, this is really a multi-objective genetic programming scheme. Three objectives were considered: the confidence of a rule, its J-measure [1501], which identifies surprising rules, and rule simplicity. The authors compare their results with respect to a single-objective version of the algorithm in which only the J-measure is optimized. The MOEA adopted was able to generate rules different from those produced by the single-objective algorithm, and the authors argued that it provided the user a choice of several different rules for further study and evaluation.

Araujo [55] uses several MOEAs for *performing simultaneously statistical parsing and tagging*. Parsing and tagging are two very important tasks in Natural Language Processing. Parsing refers to searching for the correct combination of grammatical rules from among those compatible with a given sentence. Tagging refers to labeling each word in a sentence with its lexical category and, applying disambiguation criteria to identify its appropriate lexical class. Two objectives are considered in this case: find the best match of grammar rules and the best combination of lexical categories for the words in a sentence. The author compares several approaches: (1) an aggregating function, (2) MOGA [504], (3) NSGA [1509] and (4) NSGA-II [374]. The author uses genetic programming, because the individuals considered are parses of segments of a sentence (i.e., they are trees obtained from applying a probabilistic context-free grammar to a sequence of words of a sentence). Because of the specific application, the initial population is generated in such a way that it consists of individuals that are leave trees formed only by a lexical category of the word (a different individual is generated for each lexical category of the word). A specialized crossover operator is also adopted by the author. In the comparative study, the author observes that all the MOEAs outperform a classical *best-first chart parsing* algorithm. Two interesting outcomes of the study are that MOGA outperformed the other approaches regarding precision, and that the NSGA-II was the MOEA with the largest execution time.

Calonder et al. [194] use the Indicator-Based Selection Method [1774] for *the identification of gene modules on the basis of different types of biological data* (e.g., gene expression and protein-protein interaction data). In this problem, the aim is to measure distances between genes and data sets, and such distances are the objectives, because it is normally the case that if a gene is close to one data set it is not close on the others. The authors analyze three different types of biological data: (1) gene expression, (2) protein-protein interaction and (3) metabolic pathway data. They adopt binary encoding with uniform crossover and a repair mechanism for satisfying the constraints of the problem. Additionally, they also incorporate a flip-bit mutation operator and experiment with a local search mechanism, in order to analyze if it helps to improve the performance of the algorithm. Results are compared with respect to a Tchebycheff scalarization approach with multiple independent runs. The authors also analyze the different trade-offs resulting from different data type combinations, and they compare the outcomes with those produced by a clustering algorithm (k-means). The MOEA adopted is found to be advantageous with respect to the co-clustering techniques traditionally adopted in this problem (which are based on aggregating functions that combine distance measures on the different data types into one distance measure).

Romero Zaliz et al. [1747] use their own version of multi-objective scatter search (MOSS) to *identify interesting qualitative features in biological sequences*. The authors adopt a generalized clustering methodology, where the features being sought correspond to the solutions of a multi-objective problem. In this problem, some of the objectives correspond to the degree of resemblance between features and prototypical structures considered interesting by database users. Other objectives include feature distance and, even performance criteria in some cases. The authors apply a method called Generalized Analysis of Promoter (GAP) for identifying one of the most important factors involved in the gene regulation problem in bacteria: the RNA polymerase motif. Because of the way in which the problem is modeled (using fuzzy logic), three objectives are considered, each corresponding to one of the degrees of matching to the fuzzy models adopted. The algorithm proposed was able to outperform Consensus/Patser [677], which is a typical DNA sequence analysis method, and also outperforms both SPEA [1782] and the $(\mu + \lambda)$ MOEA [1427].

7.4 Industrial Applications

After engineering, industrial applications are the most popular. Industrial applications are subdivided in four areas: design and manufacture, scheduling, management and grouping & packing. From these sub-areas, design and manufacture is the most popular, followed by scheduling. The less common sub-discipline is grouping and packing with only five application reviewed.

7.4.1 Design and Manufacture

Table 7.15: Summary of design and manufacture applications

Specific Applications	Reference(s)	Type of MOEA
Process planning	[1651, 1081, 1650]	Hybrid of VEGA and Pareto ranking
	[607]	GA with Pareto ranking
	[1613, 1125]	GA with an aggregating function
	[1468]	GA with Pareto ranking
	[1347, 1346]	Grouping genetic algorithm hybridized with PROMETHEE
	[734]	Pareto stratum-niche cubicle GA
	[236]	Generalized multi-objective evolutionary algorithm
VLSI	[409]	GA and satisfiability classes
	[971, 972]	GA with Pareto ranking
	[454, 411]	GA with an aggregating function
Cellular manufacturing	[395, 394]	MOGA, Pareto ranking and a GA with an aggregating function
	[1273]	NPGA
Machine design	[1215]	GA with an aggregating function
	[1116]	NSGA
	[503, 512]	MOGA
	[1089]	GA with Pareto ranking
	[260, 268]	GA with a weighted min-max approach
	[1438, 1146]	(μ, λ)-ES with SPEA
	[1595]	modified NSGA
Cutting problems SPEA and DPGA [929]	[1649]	
	[1587]	NSGA-II with tree encoding
Process plants design	[1109]	NSGA-II
Car design	[524]	GA with Osyczka and Kundu's approach [1225]
	[718]	NSGA-II [374]
Robust systems design	[701]	a modified version of SPEA
Enterprise planning	[1721]	a discrete version of multi-objective differential evolution

Viennet et al. [1650, 1651] use a hybrid between a population-based approach (such as VEGA [1440]) and Pareto ranking to *optimize the working conditions of a press used to make animal food*. In this approach, separate populations are used to optimize each of the objectives. Then, all these populations are combined using a selection scheme that relies on the concept of Pareto dominance. Three objectives are minimized: moisture, friability, and energetic consumption. Diploid chromosomes and multi-crossover are used. Constraints are handled through a death penalty approach (i.e., infeasible solutions are discarded and new strings are created to replace them). Massebeuf et al. [1081]

extend this approach by incorporating a decision support system to automate the decision making process.

Groppetti and Muscia [607] use a GA with Pareto ranking to *generate and select optimal assembly sequences and plans* with reference to the lifecycle design and redesign of complex mechanical products. Their approach is tested with the design of an automotive steering box. Six objectives are considered: minimize assembly cost, minimize assembly cycle-time, maximize product reliability, minimize maintenance costs, maximize product flexibility, and minimize redesign and/or modification flexibility. A remarkable characteristic of this problem is its high dimensionality (the authors mention the use of 224 decision variables). Real-numbers representation and roulette wheel selection are adopted.

Moon et al. [1125] use a GA with a linear combination of weights in conjunction with Pareto selection to solve a *flexible process sequencing problem*. Two objectives are minimized: the total processing and transportation time for producing the part mix and the workstation load variation between machines. The authors use an ordinal representation, a special exchange crossover operator and a uniform mutation operator adapted to handle a higher cardinality representation. Tzeng and Kuo [1613] use a GA with an aggregating function based on fuzzy logic to produce *new car sampling tests*. Three objectives are considered: minimize the sampling and decision making cost, minimize the bad-qualified product of cars manufactured, and ensure a high representativeness of the samples. The authors use binary representation, one-point crossover, proportional selection, and uniform mutation.

Sette et al. [1468] use a neural network and a GA with Pareto ranking to *optimize a production process in the textile industry*. Two objectives are considered: tenacity and elongation. The production process function is implemented through a backpropagation neural network. The GA is then used to optimize the backpropagation function. The authors use binary representation, roulette wheel selection, one-point crossover, uniform mutation, and fitness sharing. Results are compared to some experimental results generated by the same authors. The proposed approach was able to generate solutions that were better than those previously reported for this problem.

Rekiek et al. [1347, 1346] use the Grouping Genetic Algorithm (GGA) [463] hybridized with an outranking multicriteria decision method called PROMETHEE [168] to solve the *hybrid assembly line balancing problem*. Four objectives are considered: keep the processing time of all workstations from exceeding one cycle time, minimize the total price of the resources allocated to the workstations, maximize reliability on each workstation, and reduce congestion of workstations. Each solution is ranked using PROMETHEE (based on the preferences expressed by the user) and the GGA evaluates them using this rank as if it was a single-objective optimization problem.

Hyun et al. [734] use the Pareto stratum-niche cubicle GA (PS-NC GA) to solve *sequencing problems in mixed model assembly lines*. Three objectives are considered: total utility work, the difference between the ratio of production and that of demand to all models, and setup cost. The approach proposed (PS-NC GA) is a combination of Pareto ranking and the NPGA [709]. In this approach, two concepts are used to assign fitness to each individual: Pareto optimality (main criterion) and sparseness (secondary criterion). Niche cubicles are constructed for each individual in the population. A niche cubicle (of an individual) is defined as a rectangular region whose center is the individual. Nondominated vectors that are located at less populated cubicles, are assigned higher fitness values. The authors use string representation, immediate successor relation crossover (ISRX) [859], and inversion. Results are compared to VEGA [1440], Pareto ranking, and the NPGA [709], using as metrics the number of nondominated vectors found, and a ratio that checks for nondominance between the solutions produced by each pair of methods in turn. The PS-NC GA exhibited the best performance with respect to both metrics in all the examples presented.

Chen and Ho [236] use the generalized multi-objective evolutionary algorithm (GMOEA) [693] for *process planning in flexible manufacturing systems*. Four objectives are minimized: total flow time, deviations of machine workload, the greatest machine workload, and the tool costs. The GMOEA uses elitism, a special crossover operator (called "intelligent crossover"), integer representation, and a ranking procedure that assigns fitness to an individual based on the number of individuals it dominates and on the number of individuals by which it is dominated. A highlight of the GMOEA is that it does not use fitness sharing or niching to maintain diversity (an elite clearing mechanism is used instead). The GMOEA outperformed SPEA [1782] with respect to the metric called coverage of two set [1782], and in terms of convergence speed and accuracy (i.e., closeness to PF_{true}) within the same number of fitness function evaluations.

Drechsler et al. [409] use a GA with an approach based on Satisfiability Classes (a ranking procedure is applied based on these satisfiability classes) to *minimize Binary Decision Diagrams*, which is the state-of-the-art data structure in VLSI CAD. Two objectives are minimized: the number of nodes of the Ordered Binary Decision Diagram (OBDD) and the execution time of the newly generated heuristic. The approach consists of defining a relation "favor" whose definition is similar to "dominance" in multiobjective optimization, but not equivalent (mathematically speaking, the relation "favor" is not a partial order, because it is not a transitive relation). The model adopted by the authors has several advantages: it is efficient (computationally speaking), it does not require scaling, it supports constraint-handling and preferences from the decision maker in a natural way, and it dynamically adapts to changes during the evolutionary process. However, it also has certain disadvantages. For example, it tends to produce less solutions than the use of Pareto dominance, and it considers only goals or priorities that cannot be relative to each other. A multi-valued encoding is used to represent a sequence of basic optimization modules. Tournament selection is adopted and results are compared against the NSGA [1509]. The GA proposed was considered superior to the

NSGA because it produced results of a higher quality and performed ranking of finer granularity. The approach proposed also required less computational time than the NSGA.

In a previous version of their work, the authors use a linear combination of weights [411]. In related work, Drechsler et al. [410] compare their approach against the use of an aggregating function and Pareto ranking in the same problem of *heuristic learning* discussed in [409]. In this case, the proposed approach showed considerably faster convergence times, but not always improved the quality of the solutions produced with respect to the two other techniques considered. Up to seven objective functions are considered in this work.

Lee and Esbensen [971] use a GA with Pareto ranking for *fuzzy system design*. The GA is used to identify relevant input variables, to partition the search space and to identify the rule output parameters of a fuzzy system. The approach is illustrated with the control of a robot arm in which two objectives are minimized: the training error and the number of rules in the fuzzy system. The approach is compared to neurofuzzy learning. In an extension of the system, Lee and Esbensen [972] propose to use a fuzzy system to monitor and control the parameters of the multiobjective GA used. This approach is applied in an *integrated circuit placement problem* in which three objectives are minimized: the layout area, the deviation of the aspect ratio of the layout from a user-defined target value, and the maximum estimated time it may take for some signal in the circuit to propagate from one memory element to another. The non-generational GA introduced has an inverse Polish notation in the phenotype, uses linear ranking selection and special crossover and mutation operators that preserve feasibility of the solutions at all times. Results for the integrated circuit placement problem are compared to the use of random walk and a GA with parameters defined beforehand and kept constant throughout the search process. The GA was found to provide better offline performance and lower standard deviations than random walk.

Esbensen and Kuhn [454] use a GA with a linear combination of weights to solve a *building block placement problem*. Four objectives are considered: estimate of the maximum path delay, layout area, routing congestion, and aspect ratio deviation with respect to a certain target. Results are compared to the use of a random walk algorithm using a quality measure related to the aggregating function adopted to perform selection. Preference relations are used to discard infeasible solutions. The GA consistently outperformed random walk in the examples tried using the metric proposed by the authors.

Dimopoulos and Zalzala [395, 394] use MOGA [504], Pareto ranking and an aggregating function in a *cellular manufacturing optimization problem*. Cellular manufacturing divides a plant in a certain number of cells, each of which contains machines that process similar types of products. Two objectives are considered: maximize the total number of batches processed per year and minimize the overall cost. The approach is tested with the work load projections of an existing pilot plant facility at the factory of a pharmaceutical company.

Integer representation and special genetic operators are used. The three techniques used are compared against each other. MOGA produced competitive (but not necessarily better) results than the other approaches in terms of efficiency and number of alternative solutions found.

Pierreval and Plaquin [1273] use the NPGA [709] for *manufacturing cell formation problems*. Two objectives are considered: minimize the total intercell traffic and maximize the homogeneity of workload distribution. The approach is tested on a knife factory cell formation problem. The authors use integer representation, and *ad-hoc* recombination and mutation operators, and fitness sharing (applied on objective function space).

Osman et al. [1215] use a GA with a linear combination of weights to *design vehicle water-pumps*. Three objectives are minimized: error of exit pressure, error of exit flow, and input power. Both real-numbers and binary (with and without Gray coding) representation are tried, combined with a single population and a multipopulation scheme. The authors indicate that the type of encoding adopted does not affect the results in a significant way. However, the use of several subpopulations reduces the computational time with respect to a single population scheme.

Mitra et al. [1116] use the NSGA to *optimize an industrial nylon six semibatch reactor*. Two objectives are minimized: total reaction time and the concentration of an undesirable cyclic dimer in the product. Binary representation and stochastic remainder selection are used. Results are compared to those produced by another multiobjective optimization approach used before with a less precise model of the polymer reactor. The NSGA was found to be superior to the other approach in terms of the quality and number of Pareto optimal solutions found.

Fonseca and Fleming [503, 512] use MOGA [504] to *optimize the low-pressure spool speed governor of a Pegasus gas turbine engine*. Seven objectives are considered: minimize the maximum pole magnitude, maximize the gain margin, maximize the phase margin, maintain the rise time taken by the closed-loop system to reach and stay above 70% of the final output changed demanded by a step-input, maintain the settling time taken by the closed-loop system to stay within ±10% of the final output change demanded by the step-input, maintain the maximum value reached by the output as a consequence of a step input so that it does not exceed the final output change by more than 10%, and minimize the output error. They use binary representation with Gray coding, shuffle crossover [214], uniform mutation, and stochastic universal sampling.

Meneghetti et al. [1089] use a GA with Pareto ranking to *optimize a ductile iron casting*. Three objectives are considered: maximize the hardness of the material in a particular portion of the cast, minimize the total casting weight and minimize porosity. Integer representation and directional crossover (a technique similar to the Nelder and Mead Simplex algorithm [1182]) are used. An interesting aspect of this work is the use of a multiple criteria decision making procedure: the Local UTility Approach (LUTA) [1464], which asks the user to express preferences without justifying them. The algorithm constructs an utility function consistent with the designer's preferences.

Coello [260] and Coello Coello and Christiansen [268] use a GA with a weighted min-max approach to *formulate machining recommendations under multiple machining criteria*. Four objectives are considered: minimize surface roughness, maximize surface integrity, maximize tool life and maximize metal removal rate. Machinability tests on 390 die cast aluminum cut with VC-3 carbide cutting tools are used as a basis to test the approach. Both binary and integer representation are used, together with deterministic tournament selection. Results are compared against MOGA [504], VEGA [1440], NPGA [709], NSGA [1509], Hajela and Lin's weighted min-max technique, two Monte Carlo methods, the global criterion method, a conventional weighted min-max technique, a pure weighting method, a normalized weighting method, and a GA with a linear combination of objectives. Using as a metric the closeness to the ideal vector, the proposed approach produced better results than any of the other techniques.

Sbalzarini et al. [1438] and Müller et al. [1146] apply a (μ, λ) evolution strategy with SPEA [1782] to a *fluidic microchannel design problem*. Two objectives are minimized: the final skewness of the flow inside the channel and the total deformation of the channel contour. The results produced are qualitatively equivalent to those generated by gradient-based search methods. However, the computational costs of the evolution strategy are lower. The operators used are: intermediate recombination, elitism, and the Covariance Matrix Adaptation method (employed to adapt online the step sizes of the evolution strategy).

Toivanen et al. [1595] use a modified version of the NSGA [1509] to *design the shape of a slice channel in a paper machine headbox*. Two objectives are considered: basis weight should be even and the wood fibers of paper should mainly be oriented to the machine direction across the width of the whole paper machine. The authors use a modified version of the NSGA in which they adopt tournament selection and the so-called tournament slot sharing [1047] to maintain the diversity of the population. They also incorporate floating point encoding, heuristic crossover and a special mutation operator that promotes small mutations. From the 166 nondominated solutions that appeared in the final population of the genetic algorithm six were chosen to be presented to the decision maker. An interesting aspect of this application is its high computational cost (a single run required 24 hrs in a workstation).

Vidyakiran et al. [1649] test both the Strength Pareto Evolutionary Algorithm (SPEA) [1782] and the Distance-based Pareto Genetic Algorithm (DPGA) [929] in a *three-dimensional guillotine cutting problem*. Two objectives are minimized: the amount of scrap that results after cutting the final cuboids from the master cuboid block and the number of times the whole system needs to be turned around in order to execute a cut. From the analysis of results, the authors concluded that SPEA provided better solutions, but DPGA had a faster convergence.

Tiwari and Chakraborti [1587] use the NSGA-II [374] with a tree encoding to *solve the two-dimensional cutting problem*. The problem consists of optimizing the layout of rectangular parts placed on a rectangular sheet to cut out several parts. Two objectives are minimized: the length of the mother sheet required and the total number of cuts required to obtain all the parts from the mother sheet. The authors study two types of cutting problems: (a) those in which guillotine cutting (cutting from edge to edge) is required (i.e., when using metal sheets), and (b) those in which guillotine cutting is not essential (i.e., when using paper or rubber). An interesting aspect of this work is the use of inverse Polish notation to encode the genes of the NSGA-II. Such encoding is adopted to represent trees corresponding to the possible arrangements of parts in the mother sheet. Results are not compared with respect to any other approach.

Mierswa [1109] use the NSGA-II [374] for *3D design of process plants*. This problem is really similar to the placement problem of VLSI design, where one tries to minimize the total space needed by a set of rectangular components (this is also known as the facility layout problem). Five objectives are considered: maximize the fulfillment of the absolute position constraints,[10] maximize the fulfillment of the relative position constraints (relative constraints correlate two movable components without defining the absolute positions of the components), maximize the distribution of layer contents, minimize the connection costs, and minimize the overlap degree. The author adopts real-numbers encoding, uniform crossover, tournament selection, a special initialization procedure to produce plants whose components may overlap, and three different types of mutations to allow movable components to drift, to rotate or to change to a location in another layer. An interesting aspect of this work is that the author uses fuzzy constraints to define possible conclusions of weighted design rules. The rule weight defines the importance of such rule. The complete set of rules is stored in XML format so that the rulebase can be preprocessed and indexed for any specific application at hand. Three approaches are compared: (a) a two-membered evolution strategy (one father that generates a single offspring with respect to which it is compared) using a linear aggregating function, (b) a multi-membered evolution strategy using a linear aggregating function and (c) the NSGA-II using Pareto dominance. The experiments are performed on the data of a real chemical plant that has been already built. The best results produced by the multi-membered evolution strategy were very similar to those produced by the NSGA-II. However, the author favored the use of the NSGA-II due to the diversity of designs that it produced, which are a useful aid for the plant designer. All the objectives were normalized and maximized. Results indicated that the evolutionary approaches were able to improve the values of all the objectives of the original plant.

---

[10] Absolute constraints refer to a movable component and one of the general properties of the plant such as forbidden zones.

Fujita et al. [524] use a GA with Osyczka and Kundu's approach [1225] to *design a four-cylinder gasoline engine*. Four objectives are maximized: miles covered by a certain amount of fuel, acceleration performance, starting response, and follow-up response. Real numbers representation with direct crossover and selection based on similarity between a pair of design solutions are used. Results are compared to successive quadratic programming (SQP) with a weighted sum of objectives. SQP fell into local optimal and could not find the Pareto front, whereas the GA-based approach was able to generate it.

Hu et al. [718] use the NSGA-II [374] to *optimize the fuel economy and emissions of a hybrid electric vehicle*. Four objectives are minimized: fuel consumption, Hydro Carbons emissions, Carbon Monoxyde emissions, and Nitrous Oxides emissions. The decision variables of the model include the sizes of the energy suppliers (engine, motor and battery stack) as well as the energy control strategy parameters. The simulations are run within a software platform called ADVISOR (ADvanced VehIcle SimulatOR), which was developed by the US National Renewable Energy Laboratory. The authors report significant improvements with respect to the design of the baseline vehicle, achieving a reduction in the fuel consumption of about 31% while reducing each of the emissions over 10%.

Hollingsworth [701] adopts a methodology called *Requirements Controlled Design* in which the requirements are treated as a set of behaviors that control the system, instead of treating them as variables that define the response of the system. Then, a modified version of SPEA [1782] (called Modified Strength Pareto Evolutionary Algorithm or MSPEA) is adopted for finding the technology boundaries of a system in the requirements hyperspace. The author is interested in boundary discovery, in which the goal is to devise a method that is capable of identifying points that lie on or very near to the technology boundary. Also, it is desirable to find the widest possible variety of these boundary points. This problem has several interesting aspects when considered as a multiobjective optimization task. First, it is desirable to check dominance not only in objective function space (as normally done), but also in decision variable space. However, the definition of dominance needs a further refinement in order to give preference to points that create a closed surface in the requirements hyperspace. A second requirement is that external functions have to be operated in a reverse manner. Thus, the evolutionary algorithm must be able to record the values of those requirements that are code outputs, and input those subsystem and component properties that are code inputs. Finally, it is necessary to allow a significant number of the state variables to vary as needed. This raised the need to modify SPEA such that it could handle a set of variables that are not directly optimized by the evolutionary algorithm. The proposed approach was validated using an evaluation of the U.S. Army's Light Helicopter Experimental (LHX) program. MSPEA was found to be a good search engine, being able to find points on the technology limit induced Pareto front for single or multiple systems. MSPEA exhibited good runtime scaling properties with respect to the maximum number of generations and the sizes of both its internal and its external population. Also, MSPEA was found to be a better choice than a grid search method, except for the fact that it was harder to visualize its results.

Xue et al. [1721], use a discrete version of multi-objective differential evolution (D-MODE) for *enterprise planning*. Two objectives are minimized: cycle time and cost. The authors study the design, supplier, manufacturing planning problem using design and supplier data from a real commercial electronic circuit board product, and data from commercial manufacturing facilities. D-MODE adopts the selection and diversity mechanisms of the NSGA-II [374]. Results are compared with respect to a revised version of the original NSGA [1509], which uses an external archive to store the nondominated results obtained during the evolutionary process. However, no comparisons are provided with respect to the NSGA-II or other approaches representative of the state-of-the-art in the area.

7.4.2 Scheduling

Table 7.16: Summary of scheduling applications

Specific Applications	Reference(s)	Type of MOEA
Production	[1471, 1470, 1472, 1473]	MOGA
	[1425]	GA with Pareto ranking and the NPGA
	[1556, 1555]	GA with the Pareto reservation strategy
Flowshop	[1550, 1551]	GA with the Pareto partitioning method
	[750, 751, 1149]	GA with an aggregating function
	[173]	NSGA
	[1554]	NSGA, MOGA, VEGA, weighted average ranking and an elitist version of the NSGA
	[1506]	GA with an aggregating function
	[1151]	Cellular multi-objective genetic local search
	[312]	MOGA coupled with an artificial immune system
	[100]	Adaptive genetic algorithm and a memetic algorithm
Job shop	[76]	
	[1557]	GA with the Pareto reservation strategy
	[991]	GA with an aggregating function
Machine	[210]	Multi-population genetic algorithm and a GA with local search and an aggregating function
	[258]	Multi-population genetic algorithm
Resource	[1545]	GA with an aggregating function

Table 7.16: (continued)

Specific Applications	Reference(s)	Type of MOEA
	[304]	NSGA-II [374], SPEA2 [1775], and a GA with an aggregating function
Time-tabling	[1231]	GA with a target vector approach
Personnel	[769]	GA with Pareto ranking
	[440]	GA with lexicographic ordering
	[680, 994]	GA with Pareto ranking
	[1739]	GA with an aggregating function
Real-time	[1123]	GA with an aggregating function
Multi-component	[865]	GENMOP with variable length chromosome

Shaw and Fleming [1471, 1470] use MOGA [504] to solve a *ready meals production scheduling problem* in which 45 products have to be assigned to 13 product lines, given certain constraints and aiming to minimize three objectives: number of jobs rejected, the lateness of any order and the variation in staff shift lengths. Permutation (integer) representation and special genetic operators are used. Results are compared to a GA with a linear combination of weights and to a parallel GA in which each objective is evolved using a separate population. The last approach provided the largest variety of trade-offs and was therefore considered as the most appropriate for this application. In further work, Shaw and Fleming [1472, 1473] incorporate pre-expressed user preferences to automate the decision making process.

Santos and Pereira Correia [1425] use a GA with Pareto ranking and the NPGA [709] on a problem of *production scheduling and energy management in industrial complexes* (namely, *kraft pulp and paper mill*). Two objectives are minimized: the total consumption of electrical energy and the production rate change. They use several crossover (one-point, uniform, heuristic, and arithmetic) and mutation (uniform, boundary, non-uniform and exchange) operators. Stochastic universal sampling, fitness sharing and mating restrictions are used as well. To handle the constraints of the problem, they use a method that guarantees the generation of feasible solutions. NPGA is used with tournament sizes of 10% of the population size. A mixed floating point and integer representation is used. Different choices of genetic operators and selection techniques are tried and compared with each other. The two issues compared are: convergence time (on each objective function), and population diversity through generations. The best results were produced when fitness sharing and mating restrictions were adopted.

Tamaki et al. [1556, 1555] use the Pareto reservation strategy to solve a *scheduling problem in a hot rolling process of a steel making factory*. Three objectives are minimized: a complexity index determined by the difference of width, thickness and hardness between a pair of slabs pressed subsequently, the total quantity of fuel required for heating all slabs at furnaces, and the time required for heating all slabs. The scheduling problem is divided into two sub-problems: determine the pressing order (an "ordering problem") and assign slabs to the furnaces (an "assignment problem"). The authors use parallel selection, a local search method, multi-point crossover and a mixed encoding that incorporates the variables for both sub-problems being solved.

Tagami and Kawabe [1550, 1551] use a GA with the Pareto partitioning method to solve a *flowshop scheduling* problem. Two objectives are minimized: makespan and lateness. They test their approach with simulations involving 20, 40 and 50 jobs assigned to 10 machines. The genetic algorithm proposed uses a permutation-based representation (i.e., strings of integers representing jobs), partially mapped crossover, reverse mutation and roulette wheel selection. They compare their results against MOGA [504], a single GA with sharing, and the Iterative Improvement Method (IIM). The comparison of results is only graphical and it indicates that the Pareto partitioning method produced solutions that were better distributed than those generated by MOGA.

Ishibuchi and Murata [750, 751] and Murata [1149] use a GA with a linear combination of weights coupled with a local search strategy to solve *flowshop scheduling problems*. Three objectives are minimized: makespan, maximum tardiness and total flowtime. The approach consists of a GA with an aggregating function in which the weights are randomly generated at the time of performing recombination. These same weights are used as search directions for a local search algorithm that is applied to all the offspring that constitute the following generation, after undergoing crossover and mutation. The local search procedure is bounded by a certain number of movements to avoid an excessive computational cost. The authors use two-point crossover, shift mutation, roulette wheel selection with linear scaling and elitism. Results are compared against VEGA [1440] and the use of a single-objective GA. The proposed approach outperformed VEGA and a single-objective GA in terms of quality and spread of the solutions produced.

Brizuela et al. [173] use the NSGA to solve *flowshop scheduling problems*. Three objectives are minimized: makespan, mean flow time and mean tardiness. Several crossover (order-based, precedence-based and two-point) and mutation operators are used together with roulette wheel selection and elitism. The focus of this research is to study the influence of the operators on the generation of nondominated vectors. The result of the analysis was used to design a high performance MOEA that outperformed (in terms of closeness to PF_{true}) the ENSGA proposed by Bagchi [76].

Talbi et al. [1554] perform a comparative study of approaches used to solve *flowshop scheduling problems*. Several techniques are evaluated: an aggregating function, the NSGA [1509], MOGA [504], VEGA [1440], Weighted Average Ranking [121], and an elitist version of the NSGA. The authors also experiment with different types of fitness sharing: phenotypic, genotypic, and a combination of both. Finally, the effect of local search and parallelism is also studied. Two objectives are minimized: makespan and total tardiness. Results are compared based on the number of elements of the Pareto optimal set produced and on the quality of the solutions generated. However, no clear winner could be established, although elitism and local search were found to be very useful to improve performance.

Sridhar and Rajendran [1506] use a GA with an aggregating function to solve a *scheduling problem in flowshop and flowline-based cellular manufacturing systems*. Three objectives are minimized: makespan, total flowtime and machine idletime. The authors use permutation encoding, partially mapped crossover (PMX) [586], and a swap mutation operator that exchanges the position of two jobs randomly selected along the chromosome. Two heuristic procedures developed by other authors are used to minimize (separately) makespan and flowtime. Two subpopulations are then created: one with the best solutions in terms of makespan and another with the best in terms of flowtime (this is similar to VEGA [1440]). The parents for the next generation are selected based on an aggregating function that considers the three objectives indicated above. After applying crossover to the parents, their offspring compete against them using the same aggregating function as before. The winners become the new generation of the GA to which mutation is applied. The proposed approach performed considerably better than a heuristic proposed for this type of problem by Ho and Chang [690], against which the approach was solely compared.

Murata et al. [1151] use a linear combination of weights and a cellular multi-objective genetic local search (C-MOGLS) algorithm to solve *flowshop scheduling problems*. Two objectives are minimized: makespan and total tardiness. The authors use a cellular structure in which every individual in the population is assigned to a cell in a spatially structured space (the so-called "cellular genetic algorithm" [1699]). They also adopt a procedure by which an individual is relocated in a new cell at every generation based on its current location in objective space. The authors use two-point crossover, shift mutation, elitism, and permutation encoding. Results are compared to two previous MOEAs developed by the same authors [1150, 751] using four metrics: the number of nondominated vectors found, a set quality measure proposed by Esbensen and Kuh [454], the number of solutions that are not dominated by other solution sets, and the maximum distance between two solutions in objective space. The use of a cellular structure favored the generation of solutions closer to PF_{true}. The additional use of local search improved the quality measure proposed by Esbensen and Kuh [455]. Finally, the use of immigration favored the metric that evaluated maximum distance between two solutions in objective space.

Cui et al. [312] use MOGA [504] coupled with an approach based on the immune system to solve *flowshop scheduling problems*. Two objectives are minimized: maximum makespan and maximum lateness. The artificial immune system is used to maintain diversity in the population through the measurement of similarities among the chromosomic strings and the application of the concept of entropy. The authors use generational partially mapped crossover, reverse mutation, and binary tournament selection with replacement. The approach is compared to the use of MOGA with its original fitness sharing scheme. The proposed approach was found to be more effective at maintaining diversity and was able to produce better results than MOGA with fitness sharing.

Basseur et al. [100] experiment with an Adaptive Genetic Algorithm and a Memetic Algorithm to solve *a multi-objective flowshop scheduling problem*. Two objectives are minimized: the makespan (total completion time) and the total tardiness. The Adaptive Genetic Algorithm (AGA) adopted uses the nondominated ranking of the NSGA [1509], but incorporates elitism, two-point crossover, and an adaptive selection mechanism that allows to choose from among 4 mutation operators: insertion, reciprocal exchange, random and inversion operator. The probability of selecting a certain mutation operator is modified based on its efficiency [98, 702]. They also use a combined fitness sharing scheme in which sharing takes place in both decision variable and objective function space. The authors also experiment with a memetic algorithm which adopts two-point crossover, insertion mutation and a population-based local search mechanism. An exact method called two-phases method (TPM), which is based on the branch & bound algorithm, is also adopted for small instances of the problem. Then the authors experiment with different cooperative approaches. First, they hybridize the AGA with the memetic algorithm. After that, this hybrid approach is combined with the TPM. The idea is to limit the size of the trees obtained by the two-phases method, so that the approach can be used to solve large instances of the problem. The experimental study undertaken by the authors showed the benefits of using cooperative schemes for solving multi-objective flowshop scheduling problems.

Bagchi [76] uses a variation of the NSGA [1509] for *job shop scheduling*. The algorithm adopted is called ENSGA because it uses elitism.[11] The ENSGA uses nondominated sorting, niche formation and fitness sharing as in the NSGA. Selection, however, is done differently. The ENSGA uses a plus selection strategy (parents compete against their children). Nondominated sorting is done on the mixed population of parents and offspring. This selection strategy is called "elitism" by its author, since good parents are preserved for several generations. Three objectives are minimized: makespan, mean flow time and mean tardiness of jobs. A statistical comparison between the NSGA and the ENSGA is also provided (using the Wilcoxon signed-rank test). Both approaches are applied to a 49-job 15-machine problem. It was found that the overall quality of the solutions produced by both approaches was the same. However, the ENSGA was able to find more elements of the Pareto optimal set using the same running times than the NSGA. Also, given a pre-specified number of Pareto optimal solutions that the decision maker wished to achieve, the ENSGA was able to find them faster than the NSGA.

Tamaki et al. [1557] use the Pareto reservation strategy to solve a *job shop scheduling problem* with several jobs and several machines. Two objectives are

---

[11] It is worth mentioning that the reviewed version of the NSGA, called NSGA-II, uses elitism as well [374] (see Chapter 2).

minimized: the total flow time and the total earliness and tardiness. Within each objective function, they use combinations of weights to deal with the different machines and jobs available.

Liang and Lewis [991] use a GA with a linear combination of weights to solve a *job shop scheduling problem*. Two objectives are minimized: mean flow time and mean lateness. The authors use permutation encoding, partially mapped crossover (PMX) [586], and mating restrictions (only the fittest individuals are allowed to mate).

Carlyle et al. [210] compare the performance of two GAs used to solve *parallel machine scheduling problems*. Two objectives are minimized: makespan and total weighted tardiness. The authors use the GA proposed by Murata et al. [1153], which incorporates an aggregating function, local search and elitism. Results are compared against the Multi-Population Genetic Algorithm (MPGA) of Chochran et al. [258], which combines an aggregating function with VEGA [1440]. An interesting aspect of this work is that the authors propose a new metric called integrated convex preference measure, which is used to compare the two algorithms previously mentioned. According to this metric, the MPGA generates better sets of approximate solutions than Murata's algorithm.

Cochran et al. [258] use a two-stage Multi-Population Genetic Algorithm (MPGA) to solve *parallel machine scheduling problems*. Up to three objectives are minimized: total weighted completion times, total weight tardiness, and makespan. The approach proposed adopts a hybrid selection scheme in which an aggregating function is used in the first stage, and a subpopulation approach similar to VEGA [1440] is adopted in the second stage. The aggregating function is a multiplication of the relative measure of each objective (i.e., the result of dividing each objective function value by the best value found so far for that objective). In the second stage, the approach generates as many subpopulations as objectives has the problem plus one. This additional subpopulation uses the same aggregating function as before to combine all the objectives; the other subpopulations optimize a single objective each (as in VEGA). Each subpopulation is evolved separately and the genetic operators (crossover and mutation) are applied only within each subpopulation (unlike VEGA). They also use an elitist strategy in which the best solutions for each separate objective and for the aggregating function are stored and then used to replace the worst solutions of each subpopulation at the next generation. The authors use permutation representation, proportional selection, one-point crossover, and uniform mutation. The approach was found to be better than the multiobjective genetic algorithm with local search proposed by Murata et al. [1153], both in terms of closeness to PF_{true} and in terms of the number of elements of the Pareto optimal set found.

Syswerda and Palmucci [1545] use a GA with an aggregating function for a *resource scheduling application*. The authors use a priority value for each task to be scheduled and then add or subtract from it depending on the violations of the constraints imposed on the problem. Weights are associated to this priority value and to the constraint violations to produce a single fitness value for each solution. The authors use permutation-based representation, position-based crossover, swap mutation and domain knowledge to implement their steady state GA.

Cowling et al. [304] compare several multi-objective evolutionary algorithms on a *workforce scheduling problem*, which consists of assigning resources with the appropriate skills to geographically dispersed task locations, while satisfying certain time window constraints. Three objectives are considered: maximize schedule priority, minimize travel to and from home locations, and minimize the completion of tasks or the use of resources at inconvenient times. The authors compare the results produced by the NSGA-II [374], SPEA2 [1775], and a GA with a linear aggregating function. The authors indicate that both, the NSGA-II and SPEA2 find solutions which are within 2% of the best solution found by the aggregating approach. However, such performance is assessed using an aggregating function, as well. As expected, the NSGA-II and SPEA2 produced much more diverse populations than the GA with an aggregating function. The issue of choosing one solution from the Pareto optimal set is addressed in this paper, and therefore the interest of the authors for using aggregating functions in which the preferences are expressed *a priori*.

Paechter et al. [1231] use a GA with a weighted target vector approach for *time-tabling the classes of Napier University*. Twelve objectives are considered (e.g., room changes, time restrictions, etc.). A permutation indirect representation with a multi-point recombination and three mutation operators are adopted. Objectives are given a priority by the user. Local search is used to deal with hard constraints and genetic operators to deal with soft constraints.

Jan et al. [769] use a GA with Pareto ranking for the *nurse scheduling problem*. Two objectives are considered: maximize the average of the fitness of all nurses (defined in terms of three factors), and minimize the variance of the fitness of all nurses. There are several constraints (e.g., human related limitations, restrictions on individual nurses' monthly schedule, etc.), and also several preferences (from the nurses) have to be taken into consideration when solving the problem. The authors use a cooperative genetic algorithm (CGA), which is a population-less approach in which new individuals are created through the application of an extended two-point crossover operator. The CGA requires a special "escape" operator (consisting of an exchange of portions of the chromosome between two individuals) to keep diversity. Results are compared against a multi-agent approach based on the ant colony system. The authors indicate that their approach produced schedules that satisfied all the hard constraints unlike the multi-agent approach. Additionally, the proposed approach was able to produce better schedules than the multi-agent system and at a lower computational cost.

El Moudani et al. [440] use a GA with lexicographic ordering to solve the *nominal airline crew rostering problem*. Two objectives are considered: the airline operations cost, and the overall satisfaction degree of the staff. Cost is considered as the main objective, and satisfaction as a secondary objective. The approach consists of adopting a heuristic that aims to maximize the overall degree of satisfaction, regardless of the operations cost that these solutions have. Then, the GA is used to generate new solutions whose cost is lower. Through generations, lower crew satisfaction levels are being considered, aiming to reach a compromise between the two objectives. The authors use integer representation, roulette wheel selection, inversion and a specialized mutation operator. GAs with different combinations of operators are compared against the use of a greedy heuristic. Results indicated that the GA was able to produce competitive results in all cases, but produced better results when a combination of crossover, mutation and inversion was used.

Hilliard et al. [680] and Liepins et al. [994] use a GA with Pareto ranking in a *military airlift scheduling problem*. Two objectives are minimized: the distance traveled and the lateness of any delayed forces. Constraints are handled through a penalty function. Binary representation and proportional selection are used in a parallelized GA. Results are compared to both VEGA [1440] and a purely random search procedure. The Pareto GA found Pareto fronts better or equal than those produced by the other approaches.

Yoshimura and Nakano [1739] use a GA with an aggregation function to *solve a telephone operator scheduling problem*. Two objectives are minimized: total operator shortage and total operator surplus. The authors use a steady state GA with integer representation, uniform crossover, four types of mutation, three types of fitness scaling, and a population reinitialization process to avoid premature convergence.

Montana et al. [1123] use a GA with a linear combination of weights for *real-time scheduling of large-scale problems*. Two examples are solved in this paper. The first is a field service scheduling problem. Seven costs are minimized in this case: missed target, travel, slack, return home, parts order, unscheduled and skills mismatch cost. Ordered-paired representation is used in this case, together with a mixture of genetic operators (two types of crossover and two types of mutation). Greedy optimization algorithms are used to keep diversity in the population. The second problem concerns military land move scheduling. Two objectives are minimized: staging cost and link overuse cost. A string-based representation is adopted together with a mixture of genetic operators similar to the one employed in the first problem. This second problem is used in a further paper by the same authors [1122], with the same GA-based approach. In this other paper, the approach is applied to a real-world problem (move all the equipment of the 1st brigade of the Army's 3rd Infantry Division from its home base of Fort Stewart to the port of Savannah).

Kleeman and Lamont [865] apply the General Multiobjective Parallel Genetic Algorithm (GENMOP) to the *multi-component maintenance scheduling problem*. This problem consists of a combination of the generic flow-shop and job-shop (or open-shop) problems. Various real-world problems are discussed that can be modeled using the proposed multi-component scheduling model. With a variable length chromosome, a multi-component engine maintenance scheduling problem is addressed. A list of engines is provided listing each engine's arrival time, due time, priority (weight), and mean time to repair (MTTR) for all of its components. The two objectives are to minimize the makespan and to keep MTTR values within a predetermined range for all engines. GENMOP is a real-valued, parallel MOEA for constrained MOPs based on GENOCOP [1103]. GENMOPs real-valued crossover and mutation operators along with a repair operator generated acceptable maintenance scheduling results along PF_{known}.

7.4.3 Management

Table 7.17: Summary of management applications

Specific Applications	Reference(s)	Type of MOEA
Facility	[176]	GA with an aggregating function
	[907]	Evolution strategy with Pareto selection
	[1657]	NSGA-II, PAES and the ϵ-constraint method
Forest	[415]	MOGA with elitism and NSGA-II
Distribution system	[784]	Multiple objective genetic local search, Pareto ranking, hybrid of Pareto ranking and local search and the multiple start local search algorithm
Warehouse	[1290]	GA with a linear aggregating function
Availability allocation to repairable systems	[442]	GA with a linear aggregating function

Broekmeulen [176] uses a GA with a linear combination of weights to solve a *facility management problem* in a distribution center for vegetables and fruits. Two objectives are minimized: quality loss and capacity overflow. The author proposes a hierarchical solution strategy divided in two decision levels: cluster properties and product group assignment. In the first level, capacity is allocated to the clusters, and conditions (e.g., temperature) in the different clusters are to be determined. In the second level, a cluster is assigned to every product group with respect to quality loss and capacity utilization. The GA is used only at the first level. The second level is linearized and solved with the simplex method. The author uses binary representation, two-point crossover, uniform mutation, a selection operator based on a crowding model, and elitism.

Krause and Nissen [907] use an evolution strategy with Pareto selection to solve a *facility layout problem*. Two objectives are minimized: cost and constraint violation. The authors propose two approaches. The first is a $(\mu + \lambda)$-ES with Pareto selection (i.e., only the best μ individuals survive to form the next generation based on Pareto dominance). The second is a $(2, \lambda)$-ES with an external set containing the nondominated vectors found so far (any offspring generated is compared against this set to determine its possible inclusion within it). The authors use permutation encoding and a swap mutation operator. The first approach (Pareto selection) was found to be better than the second.

Villegas et al. [1657] use three different algorithms to *solve an uncapacitated facility location problem*. Two objectives are considered: minimize the total operating cost and maximize the sum of the demand of purchasing centers attended by depots within the maximal covering distance (this is called "coverage"). The model proposed by the authors represents the Colombian coffee supply network, and they actually solve a real-world example provided by the Colombian National Coffee-Growers Federation. The three algorithms implemented are based, respectively, on: (1) the NSGA-II [374], PAES [886] and a (3) mathematical programming technique (the ϵ-constraint method). The authors perform a first series of experiments in which the NSGA-II and PAES are compared using a randomly generated test instance. The dominated-space metric [1781] is adopted for this first empirical study, as a way of quantitatively assessing the results obtained by each approach. The results indicated that the NSGA-II was the clear winner. Then, the NSGA-II was compared to the ϵ-constraint method. In this case, the results were mixed, since the quality of the approximations obtained was practically the same (there was a slight improvement from the mathematical programming technique, but no higher than 2% in its best scenario). However, regarding CPU times, no clear winner could be determined, although the authors admitted that the ϵ-constraint method was highly variable regarding the CPU time that it required. Nevertheless, when solving a real-world instance of this problem, the authors decided to adopt only the mathematical programming technique, with which they identified unique tradeoff opportunities for the reconfiguration of the Colombian coffee supply network.

Ducheyne et al. [415] compare MOGA [504] and the NSGA-II [363] in a *forest management problem*. Two objectives are maximized: harvest volume and the benefit that people obtain from the standing forest. They use binary representation, uniform crossover, linear fitness scaling, uniform mutation and elitism. MOGA, the NSGA-II and a random search algorithm are compared using two metrics: size of the dominated space and coverage difference of two sets. MOGA with elitism performed better than the NSGA-II with respect to the two metrics adopted, and the random search strategy exhibited the lowest overall performance.

Jaszkiewicz et al. [784] use several evolutionary multiobjective optimization techniques to *design a distribution system* for a company. Three objectives are minimized: the total annual distribution cost, the worst case riding time, and the number of distribution centers. Several MOEAs are applied to this problem: multiple objective genetic local search (MOGLS) [777], Ishibuchi and Murata's MOGLS [751], Pareto ranking [581], a hybrid of Pareto ranking with local search (called Pareto GLS) and a multiple start local search algorithm (MOMSLS). Two metrics are used: coverage [1782], and an estimation of the expected value of the weighted Tchebycheff scalarizing function used to generate the ideal point that guides the search of their MOGLS [777]. Their results indicate that local search and mating restrictions play a very important role in multiobjective combinatorial optimization and that their correct use improves performance.

Poulos et al. [1290] use a GA with a linear aggregating function for *multiobjective warehouse management*. Five objectives are considered: distance from delivery point, distance from collection point, distance from picking locations, expiration date and seasonal demand. The authors adopt a permutation-based encoding (using integers in the chromosomic string), restricted mating, and a specialized crossover and mutation operators that preserve the properties of a valid permutation (i.e., that avoid the repetition of any integer in the sequence) upon their application. A fuzzy rule is adopted to define the weights to be adopted in the (normalized) aggregating function that the GA attempts to optimize. The authors investigate the dependence of the diversity of the Pareto optimal solutions obtained with respect to two things: (1) the weights adopted and (2) the mutation levels adopted. Data from a real warehouse was used to validate the proposed approach. The authors report improvements in the aggregate cost of up to 80% with respect to the previously known solution to the problem studied. The main drawback reported by the authors is the presence of (apparently unnecessary) small displacements of some products, which may be easily fixed with human intervention.

Elegbede and Adjallah [442] use a GA with a linear aggregating function to *optimize the availability and the cost of reparable parallel-series systems*. Two objectives are considered: maximize the availability and minimize the total cost of the system. The constraints of the problem are relaxed and are handled using an exterior penalty function. The authors adopt a matrix encoding, tournament selection, one-point crossover, an *ad-hoc* mutation operator, and a reinitialization process that the authors call *immigration*, by which 50% of the population is replaced by random solutions when they have not changed after a certain number of generations (the best half of the population is retained). The authors also perform an empirical study in order to determine the most appropriate parameters for their GA. The results obtained are claimed to be promising by the authors, although no comparisons with respect to other approaches are presented.

7.4.4 Grouping and Packing

Table 7.18: Summary of grouping and packing applications

Specific Applications	Reference(s)	Type of MOEA
Machine-component grouping	[1641]	Population-based GA
Truck packing	[603]	GA with an aggregating function
Object packing	[743]	GA with an aggregating function

Table 7.18: (continued)

Specific Applications	Reference(s)	Type of MOEA
Rectangular packing	[1685, 1681]	Neighborhood Cultivation Genetic Algorithm (NCGA)

Venugopal and Narendan [1641] use a GA with a technique consisting of selecting good trade-offs from two subpopulations (evolved separately and representing an objective function each) to solve a *machine-component grouping problem*. Two objectives are minimized: the volume of intercell moves and the total within cell load variation. The approach used is population-based (as VEGA [1440]), but in this case, each subpopulation is evolved separately and the genetic operators are applied over one subpopulation at a time. Two subpopulations are used, one for each objective function. A mechanism similar to Pareto ranking is used to find the best compromises from the results generated from each subpopulation. Despite the fact that the authors claim that their approach produces multiple solutions, a single solution is reported in the paper. For the implementation of the GA, the authors use integer representation, one-point and two-point crossover, and uniform crossover. Diversity is checked at each generation using a measure proposed by Grefenstette [599], and mutation is used to enforce it whenever it drops below a certain threshold. Only offspring better than their parents are included at each new generation.

Grignon et al. [603] use a GA with a linear combination of weights to solve *truck packing problems*. Two objectives are considered: minimize the differences in the centers of gravity in X and Y for each box (to be moved into a truck) with respect to some desired values, and find the optimum ordering of the boxes as to minimize wasted space. The authors use integer representation, ranking selection, uniform mutation, one-point crossover (for the first objective), and order-based crossover (for the second objective). A quadratic penalty function is used to incorporate the constraints of the problem (i.e., all the boxes have to fit within the bounds of the trailer) into the fitness function. Results are compared using some benchmark problems available in the literature. The GA proposed was found to be competitive and even able to explore regions of search space not accessible to other heuristics traditionally used to solve these packing problems.

Ikonen et al. [743] use a GA with a linear combination of weights for *packing three-dimensional non-convex objects having cavities and holes*. Three objectives are minimized: the sum of distances of parts from the global origin, the amount of intersection between parts, and the amount of intersection between parts and the built cylinder. A special representation in which a chromosome is a list of three sublists of integers is used. This requires special crossover and mutation operators (since each sublist contains permutations, they experimented with several order-based crossover operators, deciding to adopt PMX at the end).

Watanabe et al. [1685, 1681] use the Neighborhood Cultivation Genetic Algorithm (NCGA) to solve *rectangular packing problems*. Rectangular packing is a well-known discrete combinatorial optimization problem which has a number of applications in industry. In this case, the authors consider two objectives: width and height of the packing area (rather than minimizing directly the packing area, they minimize the aspect ratio of the packing area). The authors adopt an *ad-hoc* encoding which consists of the sequence-pair of each module and some orientation information. The authors adopt the Placement-based Partially Exchanging Crossover (PPEX) operator [1162] and a bit flip mutation which only acts on the bit corresponding to the orientation. The NCGA sorts the population based on a single objective at each generation, and adopts an external archive and a neighborhood crossover operator. Results indicate that the NCGA is able to outperform the NSGA-II and SPEA2. In fact, when dealing with large instances of the problem, the NSGA-II and SPEA2 tend to concentrate their solutions around the central part of the Pareto front, while the NCGA is able to spread them along the Pareto front.

7.5 Miscellaneous Applications

Finally, a group of miscellaneous applications is considered. This group is subdivided in two subgroups: finance and classification & prediction. From these two sub-disciplines, classification & prediction is the most popular.

7.5.1 Finance

Table 7.19: Summary of finance applications

Specific Applications	Reference(s)	Type of MOEA
Investment portfolio optimization	[1638]	NSGA
	[227, 1484]	GA with an aggregating function
	[433]	Customized local search, simulated annealing, Tabu search and a genetic algorithm; all of them adopt an additive global utility function
Financial time series	[1406, 1787]	NPGA
Stock ranking	[1145]	GA with an aggregating function
Bank loan management	[1144]	NSGA-II
Economic models	[1061, 1062]	GA coupled with a weighted goal programming approach

Vedarajan et al. [1638] use the NSGA [1509] for *investment portfolio optimization*. Two objectives are initially considered: maximize the expected return of the portfolio and minimize its risk. Later on, another objective is added: minimize transaction costs. Binary representation and phenotypic sharing on the parameter value space are used. Results are compared to the use of a GA with a linear combination of weights and to the use of quadratic programming. The approach is applied to a portfolio consisting of five large capital stocks from diverse industries, considering a period of five years. The NSGA and the GA with an aggregating function produced similar results. Both GA-based techniques were considered better than quadratic programming because of their ability to generate several nondominated solutions in a single run.

Chang et al. [227] use a GA with an aggregating function to solve *portfolio optimization problems*. Two objectives are considered: minimize the total variance (risk) associated with the portfolio and ensure that the portfolio has a certain expected return. Constraints related to the desired number of assets in the portfolio are also considered. An interesting aspect of this application is that the expected return of the portfolio has to be met exactly. The authors use a steady state GA with binary tournament selection, uniform crossover and a boundary mutation operator. The GA is compared against Tabu search and simulated annealing using the same aggregating function.

Shoaf and Foster [1484] use a GA with a linear combination of weights for *portfolio selection* based on the Markowitz model. Two objectives are considered: minimize portfolio variance and maximize the expected return of the portfolio. The authors use binary representation, one-point crossover and uniform mutation. Results are compared against the use of a traditional approach to select portfolios. The GA proposed outperformed the traditional method in terms of the quality of the solutions produced.

Ehrgott et al. [433] propose a model for *portfolio optimization* which extends the Markowitz mean-variance model. The authors maximize five objectives (derived from a cooperation with Standard and Poor's): 12-month performance of an asset, 3-year performance of an asset, annual dividend of a portfolio, Standard and Poor's star ranking, and volatility. The authors also allow the incorporation of the user's preferences through the construction of decision-maker specific utility functions and an additive global utility function. Using this global utility function as the objective function to be optimized, the authors perform a study in which they compare four approaches: (1) a two phase local search algorithm, (2) simulated annealing, (3) Tabu search, and (4) a genetic algorithm. The two phase local algorithm, simulated annealing and Tabu search, shared the same neighborhood structure. Results on a fund database indicated that the genetic algorithm was the best performer, followed by simulated annealing. In randomly generated instances, however, the two phase local search algorithm had a better performance, followed by the genetic algorithm.

Ruspini and Zwir [1406] & Zwir and Ruspini [1787] use the NPGA [709] for *automatic derivation of qualitative descriptions of complex objects*. In particular, they apply their methodology to the identification of significant technical-analysis patterns in financial time series. Two objectives are considered: quality of fit (measures the extent to which the time-series values correspond to a financial uptrend, downtrend, or head-and-shoulders interval) and extent (measures, through a linear function, the length of the interval being explained). The NPGA is really used to determine crisp intervals[12] corresponding to downtrends, uptrends and head-and-shoulders intervals. Niching and tournament selection are used in this application.

Mullei and Beling [1145] use a GA with a linear combination of weights to select rules for a classifier system used to *rank stocks based on profitability*. Up to nine objectives are considered, related to conjunctive attribute rule tests. The Pitt approach is used for the classifier system [581]. The authors use binary representation, roulette wheel selection, one-point crossover and uniform mutation. Results are compared against a technique related to the synthesis of polynomial networks called STATNET. Results were inconclusive since no technique was able to outperform the other in all cases.

Mukerjee et al. [1144] use the NSGA-II [374] to *determine risk-return trade-offs for a bank loan portfolio manager*. Two objectives are considered: maximize mean return on the portfolio, and minimize the variance on the return. An interesting aspect of this work is that the authors compare the performance of the NSGA-II with respect to the ϵ-constraint method (using a simple genetic algorithm for the individual single-objective optimizations performed by this method). Based on a simple graphical comparison, the authors conclude that the two methods have a high degree of overlap.

Mardle et al. [1061, 1062] use a GA with a weighted goal programming approach to *optimize a fishery bioeconomic model*. Four objectives are considered: maximize profit, maintain historic relative quota shares among countries, maintain employment in the industry and minimize discards. GENOCOP III [1100] is used for the evolutionary optimization process. Real-numbers representation and arithmetic crossover are employed. The evolutionary approach is compared to the application of traditional goal programming (developed in GAMS (General Algebraic Modeling System) [177] and solved with CONOPT[13]) in a model of the North Sea demersal fishery. The GA is considered competitive but not necessarily better than goal programming in this application.

7.5.2 Classification and Prediction

Table 7.20: Summary of classification and prediction applications

Specific Applications	Reference(s)	Type of MOEA
Prediction problems	[735]	Genetic programming with an aggregating function
	[1758]	Genetic programming with an aggregating function
	[856]	Evolutionary local selection algorithm
Feature selection	[444]	NPGA
	[1088]	Evolutionary local selection algorithm

---

[12] Fuzzy logic is used to describe the model.
[13] See http://www.conopt.com/

Table 7.20: (continued)

Specific Applications	Reference(s)	Type of MOEA
Pattern classification	[752, 753, 1155]	GA with an aggregating function
Partial classification	[350, 351]	NSGA-II
Data classification	[446, 445]	NPGA
Failure prediction	[257]	PAES
Intrusion detection	[624]	Multiobjective Artificial Immune System

Iba et al. [735] use genetic programming and a linear aggregating function to solve *pattern recognition* and *time series prediction* problems. Two objectives are considered: tree coding length and exception coding length. The authors use a minimum description length (MDL) principle [1363] to define fitness functions that allow to control the growth of trees used by genetic programming.

Zhang and Mühlenbein [1758] use genetic programming and a linear aggregating function to solve two *prediction* problems: water pollution and far-infrared laser data (a real-world time series prediction problem). Two objectives are considered: fitting error and complexity of the programs produced by the approach. The weights used by the aggregating function are adapted during the evolutionary process. The trees used in this case are neural programs. The authors use a stochastic hillclimber to adapt the weights of each neural network between generations of the evolutionary algorithm.

Kim et al. [856] use the Evolutionary Local Selection Algorithm (ELSA) [1088] to search for promising subsets of features that are used to train an artificial neural network that *predicts customers patterns*. Two objectives are considered in this case: minimize complexity and maximize the hit rate of the feature set selected. Results are compared to the use of principal component analysis (PCA) followed by logistic regression. The hybrid of ELSA and the neural network had a better performance than the PCA model.

Emmanouilidis et al. [444] use the NPGA [709] for *feature selection* with applications in *neurofuzzy modeling*. Two objectives are considered: minimization of the number of features and maximization of the modeling performance (considered as the estimated misclassification rate). An additional term (the cost function) is used to solve ties between individuals with the same misclassification rate. The authors use binary representation, polygamy (every individual in the population has the chance to mate twice on average, at each generation), elitism (i.e., they keep the nondominated solutions generated over time), variable population size, fitness sharing (on both genotypic and subset size space), and a special crossover operator (called subset size oriented common features crossover operator, or SSOCF, for short), which preserves the common features of their parents (in previous work, the authors use two-point crossover [446, 445]). The mutation rate is adapted over time according to the progress observed (in terms of the ability of the GA to produce new nondominated solutions). In order to reduce the computational cost associated with the use of the GA for feature subset evaluation, two methods are adopted: probabilistic neural networks and multilayer perceptrons. Two benchmark problems are used: a data set to classify good or bad implying evidence of some type of structure in the ionosphere, and a data set to train a network to discriminate between sonar signals bounced off a metal cylinder and those bounced off a roughly cylindrical rock. In previous work, they apply the method to classification of cancer data [445], vibration analysis data [445], fault diagnosis in rotating machinery [446], and energy consumption prediction [446]. The approach is compared against sequential feature selection. The NPGA was found to perform better (in terms of computational efficiency) than sequential feature selection.

Menczer et al. [1088] use the Evolutionary Local Selection Algorithm (ELSA) to solve a *feature selection problem in inductive learning*. Two objectives are maximized: the accuracy of the classifier and its complexity (i.e., the number of features being used). ELSA uses a local selection scheme in which fitness depends on the consumption of certain limited shared resources. Results are compared to VEGA [1440] and NPGA [709] in terms of spread and coverage of the Pareto front. ELSA was found to be superior to the other approaches in both aspects.

Ishibuchi and Murata [752] use a GA with a linear combination of weights to *minimize the number of fuzzy rules for pattern classification problems*. Two objectives are considered: minimize the number of selected fuzzy if-then rules (i.e., the fuzzy rule base), and maximize the number of correctly classified patterns (i.e., classification performance). The weights are randomly specified during selection and prior to crossover. Nondominated vectors are stored in an external file for further use. Local search is used to improve performance. The authors use binary representation, roulette wheel selection with linear scaling, uniform crossover and uniform mutation.

In related work, Ishibuchi and Nakashima [753] and Murata et al. [1155] use a GA with the same type of linear aggregating function to *extract linguistic classification knowledge from numerical data for pattern classification problems*. Three objectives are considered: maximize the number of correctly classified training patterns, minimize the number of linguistic rules, and minimize the total number of antecedent conditions. The approach is a hybrid of a Pittsburgh- and a Michigan-style genetic-based machine learning technique. The first is used to represent the linguistic rules of the problem, and the second is used as a form of mutation. The authors use variable-length strings in this case. This same approach has also been used for *linguistic function approximation* [755]. Three objectives are minimized in this case: number of linguistic rules, their total length and the total squared error between inputs and outputs. The same approach has also been used to solve *pattern classification problems* [754]. Three objectives are considered in this case: minimize number of features, minimize the number of instances, and maximize a performance measure based on the classification results of a certain number of instances. The authors use in this case binary representation and uniform mutation (with a biased probability for certain chromosomic segments). The GA proposed was able to outperform some traditional classification techniques in terms of the quality of the solutions produced.

Iglesia et al. [350, 351] use the NSGA-II [374] for *partial classification* (the so-called *nugget discovery task*). Two objectives are maximized: confidence and coverage. The authors use binary encoding (with Gray codes), binary tournament selection and the crossover and mutation operators provided within the NSGA-II. The initial population, however, is created using a special procedure that looks at the data to ensure that no rules with zero coverage are produced (as happens when a randomly generated population is adopted). In order to achieve this, a default rule is adopted (in this default rule, all limits are maximally spaced and all labels are included), and the remainder of the population are just mutations of the default rule. This was found to be effective by the authors. Results are compared with respect to another approach developed by some of the same authors. This approach is called ARAC [1359] (All Rules Algorithm Cc-optimal). The authors adopt databases taken from the UCI repository [1184] in their comparative study. The results indicate that the NSGA-II can find good sets of rules, being able to match those of ARAC in some cases, and even to improve them in others. At the end of the paper, the authors suggest the possibility of hybridizing ARAC with the NSGA-II to produce a more powerful classifier.

Cochenour et al. [257] use the Pareto Archive Evolution Strategy (PAES) [886] to *design a radial basis function neural network that predicts the failures in overhead distribution lines of power delivery systems*. Two objectives are minimized: network size and errors. The idea is to use the multi-objective approach to generate several radial basis function networks of varying sizes. Each of these networks is trained using historical data, and they are compared among themselves, so that the best trade-off designs can be selected. An interesting aspect of this work is that the authors adopt a crowding operator similar to the one incorporated in the NSGA-II [374]. Mutation is an *ad-hoc* operator which consists of invoking an orthogonal least squares procedure which changes the size of the radial basis function networks. Results were compared with respect to a fuzzy inference system and with respect to a multi-layered perceptron trained with standard backpropagation. The multiobjective evolutionary algorithm outperformed the two other approaches.

Haag [624] develops an innovative Artificial Immune System-inspired Multiobjective Evolutionary Algorithm as part of a *distributed intrusion detection system*. This extended MOEA measures the vector of tradeoff solutions among detectors with regard to two independent objectives: best classification fitness and optimal hypervolume size. The antibody detectors promiscuously monitor network traffic for exact and variant abnormal system events based on only the detector's own data structure and the application domain truth set, responding heuristically. The system structure integrates RNA transcription from the REtrovirus ALGOrithm [424] (REALGO) into the evaluation operator and the framework from the Multiobjective Immune System Algorithm (MISA) [273] with an user friendly menu structure. As applied to the MIT-DARPA

1999 insider intrusion detection data set, the software engineered algorithm in Java (jREMISA: the Java retrovirus-inspired MISA) correctly classifies normal and abnormal events at a high level which is directly attributed to a detector affinity threshold found.

7.6 Future Applications

Despite the vast number of applications reported in the literature, there are still several research areas that have been only scarcely explored or that have not been approached using multi-objective evolutionary algorithms. The following are a few examples:

Finite Element Model Tuning: Structural models that use finite element theory normally require to be adjusted or tuned, before they can accurately simulate the real structure. This tuning process is a natural consequence of the approximations used to model a real structure (through finite elements) that cannot exactly duplicate a structural member's physical characteristics and boundary conditions. The finite element model can be improved in at least two ways: by increasing the number of nodes or by adjusting the model's parameters. In the first case, the main drawback is that as the number of nodes in the model is increased, the computational cost of the process also increases. The second choice, however, is more appropriate for a multiobjective optimization approach. One can select a few parameters of the model (i.e., the decision variables) to be adjusted so that the model is forced to have a certain desired set of performances (the objectives). DeVore et al. [382] proposes such an approach and solves the resulting multiobjective optimization problem using a variant of the ϵ-constraint technique. The approach is applied to the finite element model of the stabilizer of the T-38 aircraft. Apparently, no MOEAs have been applied to this domain.

Coordination of distributed agents: The coordination of distributed agents frequently involves globally conflicting solutions to multiple (local) objectives. Therefore, the use of the concept of Pareto optimality is quite appropriate in this sort of application. An approach like this, based on tracking Pareto optimality (using classical multiobjective optimization techniques), is proposed by Petrie et al. [1269], but there are very few attempts of using evolutionary techniques in this domain (see for example [206]). Distributed agents also open other interesting possibilities for evolutionary multiobjective optimization techniques, because there are many other applications where such techniques would be quite useful. For example, negotiation mechanisms (in electronic commerce) where complex decision making strategies might be involved.

Shape design: The design of the shape of structural elements has been studied by engineers for a long time. Vitruvius, for example, devised the aesthetic ideal for the shape of a column at around 25 B.C. (the design consisted of a subtle variation on the traditional cylindrical shape, with a bulge at approximately one third of the column's height and a diminution near its top). Several other scientists and engineers have studied shape optimization problems over the years. Multiobjective optimization techniques are easily applicable in this domain, since engineering optimization problems normally have several, conflicting objectives. In fact, one of the earliest applications of multiobjective optimization concepts to engineering is related to shape optimization [1514]. Besides the applications of evolutionary multiobjective optimization techniques to airfoil design reviewed in this chapter, there are very few attempts to apply such techniques to this domain reported in the literature. Deb et al. [367, 366], for example, apply the NSGA-II [374] to several shape optimization problems (a simply supported plate, a hoister plate and a bicycle frame design). However, there are many other shape design problems that have not been dealt with using evolutionary multiobjective optimization techniques. An example is the shape optimization of arch dams, which has been of great interest for many years to structural engineers [1680, 1729].

7.7 Summary

In this chapter, a considerable number of applications of evolutionary multiobjective optimization techniques have been reviewed. Engineering applications are the most common in the current literature, followed (in a distant second place) by industrial applications (mainly scheduling). Interestingly, this seems to follow the historical roots of multiobjective optimization and clearly indicates the discipline in which most of the current interest in evolutionary multiobjective optimization lies.

However, despite the large number of applications reviewed, several areas remain to be explored. That is precisely the subject of the section on future applications. A few examples of application areas not covered yet by the current literature are briefly discussed. This intends to motivate researchers to work in those areas, so that the distribution of applications becomes more homogeneous over the following years.

Further Explorations

Class Exercises

1. Read an application paper of your choice (which has not been reviewed in this chapter) and discuss the following issues:
   a) Dimensionality of the problem (how many decision variables and objectives does the problem have?)
   b) Details of the problem (does the author provide enough details about the application as to reproduce his/her results?) If not, what is missing?
   c) Does the author justify the use of a metaheuristic in this problem? Do you think that it is justifiable to use a metaheuristic in this application?
2. What would you consider to be the most important aspects to highlight in a paper dealing with a novel application of an existing MOEA? Discuss.
3. When dealing with an application that has been previously tackled using a MOEA, what would you consider to be the most important issues to highlight? How is this different from an application that has never been solved with a MOEA? Discuss.
4. Do you consider that the choice of programming language and operating system for which a MOEA is implemented plays an important role for a practitioner? Discuss.
5. What particular MOEA design considerations have to be taken into account when dealing with problems in which the evaluation of the objective functions requires a very high computational time? Discuss.

Class Software Projects

1. Adopt a MOEA whose source code is available in the public domain (e.g., the NSGA-II [374]) to develop an application of your choice. Write a report

indicating issues such as: encoding, fitness function, parameters, statistics of the results, etc.
2. Write a library of generic functions in C/C++ that facilitates the use of different MOEAs based on Pareto ranking to a practitioner. Write a report documenting the use of the library, and include some examples of its use.
3. Repeat the previous project, but using MatLab as the programming language in which the implementation takes place.
4. Repeat the previous project, but using Java as the programming language in which the implementation takes place.
5. Perform a comparative study among the 3 implementations required in the 3 previous software projects (i.e., in C/C++, MatLab and Java). Indicate the main advantages and disadvantages of using each of these programming languages, including execution time, memory and space requirements, ease of use, and implementation time, among others.
6. Using the Java retrovirus-inspired MISA (jREMISA) [624], modify various operator parameters and evaluate intrusion detection performance for the MIT data (such data are encoded as part of jREMISA).

Discussion Questions

1. Discuss the different aspects that may be of interest when reporting an application of a MOEA (e.g., novelty of the application, difficulties associated with the application, complexity of the search space, novelty of the genetic operators adopted, etc.).
2. If a MOEA is used to solve a problem that has never been treated considering multiple objectives, how would you evaluate the performance of the algorithm used? Would you consider valid to compare your results against those produced when using a single objective function?
3. What would you consider to be the major issues when trying to apply evolutionary multiobjective optimization techniques to real-world problems (e.g., computational cost, complexity of the model used, availability of real test data, etc.)? Discuss.
4. Why do you think that most of the applications reviewed in this chapter consider just a few objective functions? What are the main problems that you anticipate if one attempts to apply MOEAs (particularly those that use Pareto ranking) to problems with hundreds (or perhaps thousands) of objectives?
5. Under what conditions would you consider appropriate to use a hybrid approach (e.g., goal programming coupled with a MOEA) in an application? Discuss.
6. Would you consider more appropriate to use a local search technique (e.g., Tabu search [572]) to deal with multiobjective combinatorial optimization problems, instead of using a MOEA (see Chapter 3)? Discuss.

7. Luke and Patnaik [1027] proposed the use of lexicographic ordering (see Chapter 1) to control bloat in genetic programming. In this proposal, fitness is treated as the main objective, and tree size as a secondary objective. Analyze this proposal and compare it to other research in which multiobjective concepts have been used to control bloat (see for example [349, 436, 140]). Do you see any limitations in using lexicographic ordering to control bloat? Do you think that multiobjective optimization concepts are properly applied by the authors? Discuss.
8. Thomson and Arslan [1583, 1584] proposed a MOEA to optimize FIR filter designs. Given the characteristics of this problem, what type of MOEA would be more appropriate to use? Discuss the importance of incorporating user's preferences (see Chapter 9) into the MOEA used to solve this problem. Compare this approach to other related proposals (see for example [659, 1707]).
9. Pullan [1300] proposed a MOEA with special genetic operators to maximize average network survivability, while minimizing its variability. Analyze the MOEA proposed and discuss possible improvements. Would you consider important the use of a parallel MOEA (see Chapter 8) in this application?
10. Consider the different types of scheduling problems that are discussed in this chapter (e.g., job shop, flowshop, timetabling, etc.). Perform a comparative study of MOEAs using a set of different scheduling problems. Discuss what MOEA operators are the most appropriate for each class of scheduling problem. Are the Pareto fronts of each class of scheduling problems different among them? Discuss. See for example [1588, 174].

Possible Research Ideas

1. Think of a possible application of multiobjective optimization to your main area of interest/expertise. Justify the use of multiobjective optimization in this application. Then, investigate if there is any previous application of evolutionary multiobjective optimization techniques reported for this problem. If none is found, discuss why do you think that this problem has not been dealt with in the specialized literature? Make sure to discuss issues such as: representation, genetic operators, ways to evaluate performance (what techniques are currently used to solve this problem?), and definition of the fitness function. Then, design and implement a MOEA to solve this problem. Compare your results against any other technique normally used to solve it. Has your MOEA improved previous results?
2. Design a new MOEA that emphasizes aspects such as computational efficiency and ease of use. What type of multiobjective technique do you consider more appropriate to speed up the generation of nondominated vectors if time is an important issue? Can you justify the use of non-Pareto approaches in real-world applications? If yes, in what cases? If not, what

alternatives are available? Investigate approaches used to improve the computational efficiency of a MOEA and propose new ideas/approaches. Justify and validate your proposal.
3. Develop a novel application of a parallel MOEA (see Chapter 8). Justify the use of parallelism in this application, as well as the type of approach taken. Compare the performance of the proposed approach against the use of a sequential MOEA.
4. Investigate possible applications of evolutionary multiobjective optimization techniques into the following areas: computer vision, pattern recognition, image processing and computer animation. Propose new applications within these areas. See for example [160, 1234, 1379, 1622, 1661].
5. Propose a hybridization of a MOEA with a deterministic technique in solving real-world nonlinear constrained multiobjective problems. What attributes of the MOEA are advantageous in this sort of application? How is the deterministic technique integrated to the MOEA? What are the main advantages of the proposed hybrid approach? See for example [943].
6. Consider the development of MOEAs that are appropriate for dynamic environments. What are the main issues to consider when dealing with dynamic environments? What type of changes do you need to perform on a MOEA to make it suitable for such environments? See for example [1725, 125, 1055].
7. Analyze possible novel applications of MOEAs in engineering. See for example [1075, 852].
8. One of the emergent application areas of MOEAs is DNA computing. Propose a novel application in this area (see for example [1483, 1482]).
9. Another interesting application area of MOEAs that has been only scarcely explored is cellular automata. Propose a novel application in this area (see for example [1210]).
10. Microelectrical Mechanical Systems (MEMS) is an emerging research field within electrical and electronics engineering, in which MOEAs have also been used (see for example [815, 816, 1235, 1762, 1761]). Propose a novel application in this area.
11. Knowledge extraction from very large databases is also an area worth exploring with MOEAs (see for example [966]). Propose a novel application in this area.
12. The application of MOEAs in virtual reality (and computer graphics, in general) is very scarce (see for example [1621]). Propose a novel application in this area.
13. Few applications of MOEAs in classifier systems exist (see for example [124, 1004, 1005, 350, 351]). Propose a novel application in this area.
14. An interesting area for using MOEAs is in software engineering (e.g., planning of software development projects [646] and in software quality estimation [848, 847]). Propose a novel application in this area.

15. Cryptography is another area in which there are very few applications of MOEAs reported in the specialized literature (see for example [1178, 1180]). Propose a novel application in this area.
16. An application domain in which MOEAs have been only scarcely applied is in game playing (see for example [1189, 1735]). Propose a novel application in this area.
17. Bioinformatics is another area in which MOEAs can find interesting applications (see for example [723, 1747, 198]). Propose a novel application in this area.

8

MOEA Parallelization[1]

> One friend in a lifetime is much, two are many, three are hardly possible. Friendship needs a certain parallelism of life, a community of thought, a rivalry of aim.
>
> Henry B. Adams

8.1 Introduction

Successfully engineering Multiobjective Evolutionary Algorithms (MOEAs) involves thoroughly addressing many different issues. However, the performance concepts of efficiency and effectiveness are paramount. MOEAs are stochastic, population-based computational procedures mimicking evolutionary concepts and operations in attempts to find satisfactory, if not optimal, solutions of problems with multiple objectives. Evolutionary Algorithms (EAs) and MOEAs are adaptive stochastic search techniques classified under the umbrella of soft computing [1577]; generic EAs such as Genetic Algorithms, Evolution Strategies, Evolutionary Programming, and Genetic Programming are all successfully used in MOEA implementations [265].

Satisfactorily solving specific and ever-larger Multiobjective Optimization Problems (MOPs) with MOEAs is an increasingly popular goal reflected by the large number of recent research efforts attacking pedagogical and real-world problems with these algorithms [266, 361]. Such real-world MOPs typically involve highly constrained design optimization tasks with high computational cost. Through these applications, MOEA structures continue to evolve into effective (i.e., "useful") search algorithms. Once convinced of a MOEA's effectiveness (how well it solves the problem), the researcher is then often interested in increasing its efficiency (how "quickly" or "cheaply" it solves the

---

[1] The authors wish to thank Jesse Zydallis for his permission to include previous collaborative effort in this discussion on MOEA parallelization.

problem). The desire to reduce execution time and/or resource expenditures naturally leads to considering the use of parallel and distributed processing techniques.

A major computational bottleneck in many contemporary MOEA applications (as well as in other numerical or real-world design/optimization problems) is the calculation of complex nonlinear MOP functions, implying algorithmic parallelization may improve computational efficiency. Just as in single-objective optimization, multiple "expensive" objective function evaluations (in terms of CPU time) are often completed in less wall clock time by decomposing the computational load across two or more processors. Evaluating more solutions in the same or reduced time may then result in a larger or higher-fidelity representation of possible outcomes. This may be especially productive in MOEA applications, due primarily to the fact that identifying a (possibly large) set of "good" objective vectors is often the primary goal driving search. A parallel MOEA (pMOEA) might then be the preferred EA implementation for solving complex real-world applications where (multiple) objective function evaluations are the computational bottleneck.

Given a MOEA's obviously inherent parallelism as well as the relative ease of gaining access to contemporary multi-processor computing platforms, interest is increasing for developing pMOEAs (see [266]). However, in publications solving engineering design and numerical optimization problems with pMOEAs, very few discuss algorithmic development issues and for the most part ignore the parallel aspects of the implementation. In general, these papers lack a thorough presentation of pMOEA employment rationale, algorithmic settings and structure, test problem selection and metrics, etc. Yet, analyzing the current corpus yields several key insights into the current pMOEA state-of-the-art and suggests areas in which further research may be focused. The most significant findings from this analysis are detailed later in this chapter.

Critical engineering practice requires a disciplined approach as embodied in the following quote: "...the essence of sound engineering [lies] in clearly stating the assumptions upon which calculations are based so that they may be checked at all times for lapses in logic and other errors. It is this imperative that engineering premises be set down clearly, and that the calculations that follow be systematically and unambiguously presented, so that they may be checked by another engineer with perhaps a different perspective on the problem [1270, p. 44]." With that thought in mind, this chapter clearly presents pMOEA symbolic formulations, describes pMOEA design and implementation issues, proposes options for satisfactory issue resolution and discusses various practical considerations. Known research approaches and various insights gained from analysis are also integrated into the presentation. Throughout this discussion a template is evolved for generating a pMOEA from either an existing (MO)EA or from first principles. The result is a generic design plan with a list of parameter considerations to be considered in designing and implementing efficient and effective pMOEAs, regardless of application problem domain. Perhaps this presentation might inspire contemplation by other pMOEA developers; their different perspectives and critical comments may well result in furthering these ideas or creating other innovative designs. The reader should note that as much readily available material describes generic parallel processing techniques and implementations, incorporation into EAs and various hardware/software configuration issues [24, 25, 28, 29, 27, 31, 72, 204, 926, 1521, 464], this chapter focuses only on exploring and analyzing possible benefits of pMOEA development and instantiation.

The remainder of this chapter is organized as follows. Fundamental background material and basic underlying pMOEA philosophy are briefly presented in Sections 8.2 through 8.4; this includes basic pMOEA background, motivation, paradigms and design issues. Key analyses derived and extended from known pMOEA implementations are discussed in Section 8.5, where additional relevant topics (e.g., hardware platforms, test functions, test suites, metrics and theory) are also presented. General pMOEA development issues, along with specific design considerations when implementing pMOEAs follow in Section 8.6. The development and advantages of a new "generic" pMOEA are detailed in Section 8.7, followed by conclusions and recommendations for future pMOEA research in Section 8.8.

8.2 pMOEA Fundamental Background

This section broadly addresses key pMOEA concepts and includes selected background material to catalyze a better understanding and appreciation of this research. It defines notational definitions used to describe pMOEA operation and also discusses the motivation for, and major issues involved with, pMOEAs.

8.2.1 pMOEA Notation

A pMOEA's complicated algorithmic structure and operation can confuse the process of identifying and manipulating sets whose members are selected based on Pareto concepts. A precise notation is thus required to explicitly identify various Pareto sets *and* the specific time at which they exist for parallel operation. Van Veldhuizen's MOEA notation [1626] is discussed in Section 1.2.4 and thus extended in this section to the pMOEA domain.

Certain pMOEAs use multiple processors and/or populations where each generates unique MOP solutions. The notation must then allow for representing the specific Pareto sets resident on a given processor at any given time. Thus, the sets existing on a given processor are represented by $PF_{current}\binom{p}{t}$ and $P_{current}\binom{p}{t}$, where p represents the processor ID and t the generation number. If so desired, the final computed sets found by each processor can be tracked and represented in a similar fashion, $PF_{known}\binom{p}{}$ and $P_{known}\binom{p}{}$.

In some pMOEAs, migration events occur in which select population members are sent to or exchanged with other processors. Thus, in these cases the current Pareto sets on any given processor are represented by $PF_{current}\left(^{\ \ p}_{t\_bm\_x}\right)$ and $P_{current}\left(^{\ \ p}_{t\_bm\_x}\right)$ prior to the migration event; the resident sets after migration occurs are $PF_{current}\left(^{\ \ p}_{t\_am\_x}\right)$ and $P_{current}\left(^{\ \ p}_{t\_am\_x}\right)$. $\left(^{\ \ p}_{t\_bm\_x}\right)$ denotes processor p before the xth migration event at generation t. Once migration occurs it is possible that some immigrants' vectors dominate some of the current members.' The notation $\left(^{\ \ p}_{t\_am\_x}\right)$ then denotes processor p *after* the xth migration event at generation t and the corresponding sets have been updated. In this scenario, $PF_{current}\binom{p}{t}$ and $P_{current}\binom{p}{t}$ may also be used to represent the sets after *all* migration events have occurred for processor p and generation t. This distinction is necessary as each processor may conduct multiple migration/replacement events at a given generation, dependent upon its neighborhood size and membership in different neighborhoods. At pMOEA termination, $PF_{known}\left(^{p}\right)$ and $P_{known}\left(^{p}\right)$ again represent the sets found on each processor with PF_{known} and P_{known} representing the sets found by all pMOEA processors combined.

The pMOEA notation described can be employed in the development of a specific computational pMOEA (algorithm and data structures) to help ensure the design is implemented with a large degree of confidence that the code is correctly written. The notation may also be used in mathematically proving various pMOEA properties such as convergence.

8.2.2 pMOEA Motivation and Issues

Although general MOPs may be solved via a variety of search techniques developed to address various deficiencies, there is a continuing need to solve real-world, high-dimensional, complex MOPs with increased effectiveness and efficiency.

Results achieved using single-objective EAs lacked efficiency and were not effective for many MOPs. For example, single-objective EA approaches using aggregated objective functions (e.g., weighted sums) generate only a single solution per run and thus require multiple executions varying weight aggregations in order to generate a set of multiple solutions. Additionally, some traditional single-objective EA approaches are known to be unable to identify the complete Pareto front if certain problem constraints exist. Although it is possible to solve MOPs with single-objective EAs, MOEAs are expressly designed to return a number of MOP solutions per run and are typically more efficient and effective in this domain. Many MOEAs are successfully applied to real-world design and constrained optimization problems, leading to increased visibility and use, but achieving better efficiency still remains a major rersearch goal. Thus, attractive pMOEA design characteristics include concurrent search for multiple solutions, ease of parallelizing serial MOEAs, reducing wall clock execution time, hybrid interfaces to other search techniques, and achieving better overall effectiveness.

Generally speaking, pMOEAs may be useful when one addresses situations in which the fitness functions are computationally expensive. One approach utilizes parallel computational function decomposition techniques; another approach spatially decomposes the population across a given set of processors. Of course, the particular pMOEA architectural selection does not have to directly map onto a given physical computational platform. If carefully planned the algorithmic model can be explicitly designed and subsequently refined for mapping to various parallel platforms. Associated testing should include consideration of the specific implementation's performance across a variety of serial, parallel, and distributed architectures. Interesting architectural characteristics may include numerical processing hardware, communication protocols, network topology, processor speed, memory access, I/O, and the like.

In order to fully understand pMOEAs one must first identify the basic EA components lending themselves to asynchronous execution and hence parallelization [204]. Parallelizing the objective functions is a simple and potentially useful idea but one inherently only decreasing execution time and not affecting effectiveness. Although some pMOEAs are potentially more effective than their serial counterpart, one typically does not gain improved effectiveness without some cost, which may be realized by increased execution time. These are tradeoffs researchers must consider. Fully understanding parallel concepts and their potential usefulness in MOEAs thus requires a brief discussion of MOEA structures and operators.

Figure 8.1 presents generic MOEA pseudo code reflecting the more common MOEAs found in the literature (see Chapter 2). As many variants of these MOEAs are continually evolving the reader is encouraged to study their details and operators as found in many recently published MOEA papers [266]. In this regard, consider possible areas of parallelization in Figure 8.1. Use of the previous section's notation is suggested to represent the populations associated with each processor, migration events, and generational updates as the developer transitions from thought to a detailed code implementation. Subsequent discussion draws upon proposed meta-level design structures in pursuing high-performance pMOEAs.

Taken as a whole, pMOEAs are *not* complex algorithms. Represented as a Directed Acyclic Graph, it is easily seen that MOEA tasks show more precedence relationships than asynchrony. In other words, a pMOEA has a large grain size with its algorithmic decomposability rapidly reaching a limit (i.e., there is only "so much" that can be parallelized). However, in broad terms, a pMOEA implementation should result in some computational speedup.

An obvious option for parallelizing MOEAs is an exact task to processor mapping but this is probably not wise.[2] Each identified task in Figure 8.2 executes for varying time periods, some extremely short. Additionally, Task 1

---

[2] Note that this discussion is based on Pareto-based MOEAs employing both sharing and niching, as this implementation class is not only the most popular, but has the most developed theory [1626].

8 MOEA Parallelization

> Perform Population Initialization (Size P)
> Compute Each Population Member's Fitness (w.r.t. k functions)
> Loop
>   Perform Clustering/Niching/Crowding
>   Execute EVOPs
>   Compute Each Population Member's Fitness (w.r.t. k functions)
>   Conduct Selection
>   Generate $PF_{current}(t)$; Update $PF_{known}(t)$
>   Conduct Local Search (If Specified)
> End Loop
> Conduct Local Search (If Specified)
> Generate PF_{known} and Present to Decision Maker

Fig. 8.1. Generic MOEA Pseudo code

executes only once. Thus, it is easy to see the proposed mapping's inefficiency. The first processor completes its task and then sits idle until MOEA termination; many of the other processors are also unable to operate asynchronously, a condition resulting in more idle than computational time.

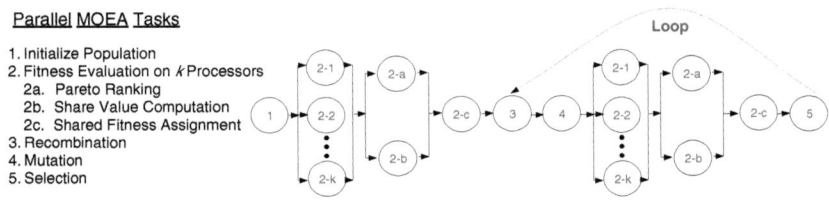

Fig. 8.2. Parallel MOEA Task Decomposition

The four steps in the execution loop (fitness evaluation, recombination, mutation, and selection) *must* occur sequentially. Mutation cannot operate until recombination finishes. Selection does not (normally) occur until all fitness values are computed. It is conceivable that the fitness evaluation task can operate on solutions immediately after mutation does or does not occur, but the resultant overhead of opening/closing a communication channel between two processors seems prohibitively expensive compared to the minimal computational gains. Additionally, since data required by some tasks is resident on separate processors additional communication costs are involved. One can thus safely conclude the proposed implementation is not very effective. "Pipelining" the MOEA's tasks is also ineffective as it is just a special case of the exact task to processor mapping.

8.2 pMOEA Fundamental Background   449

Another possible pMOEA is an implementation simultaneously executing several MOEAs on different processors and comparing, contrasting, and/or combining their results. As sequentially executing the same number of MOEAs achieves (conceptually) identical results this implementation has obvious speedup, however, it is also desirable to consider parallelizing innate MOEA tasks such as objective function evaluation.

Although some EVolutionary OPerators (EVOPs) could theoretically be parallelized, their operation is so quick (especially as compared to objective function evaluation costs) that any parallel implementation would most likely be slower than a serial one, due to the increased communication costs associated with population partitioning and distribution between processors. Selection is (generally) especially ill-suited for parallelization; several selection methods require information about the entire population that would again require excessive interprocessor communication.

pMOEA Objective Function/Data Decomposition

As was just shown, affecting the ability to effectively and efficiently parallelize a MOEA is its inherently sequential nature. But, by definition, Task 2 in Figure 8.2 computes k ($k \geq 2$) objective functions. This task can and has been parallelized and it is instructive to consider how parallelizing multiple objective function computations may be performed.

Parallelizing MOEA objective function evaluation can occur in one of three ways. One can assign each function's evaluation (for a given individual) to different processors, assign subpopulations for evaluation on different processors, or assign each individual's evaluation for a single objective function across several processors. These options are shown in Figure 8.3 and each discussed in turn.

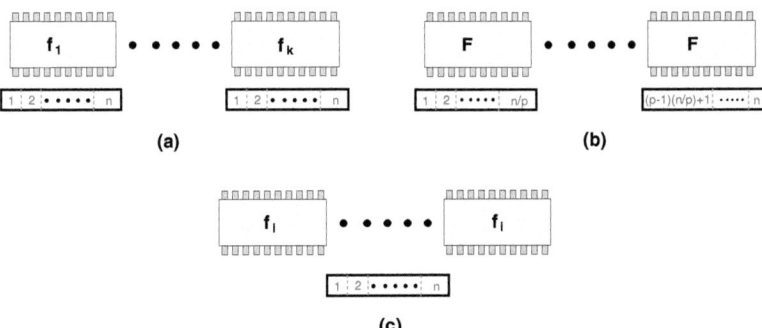

Fig. 8.3. Parallel Objective Function Evaluation Possibilities

When implementing the first option (see Figure 8.3a), one must consider that each objective function's calculation time may be radically different.

Thus, blindly assigning the entire population and each of the k functions to a different processor may then be imprudent if one objective function evaluation takes several times longer than the others. Statically or dynamically load balancing these computations may help equalize computational effort but the effort expended may not be worthwhile.

When implementing the second option (see Figure 8.3b), equal fractions of the population are assigned to different processors where they are evaluated in light of *all* objective functions. Here, identical numbers of individuals are evaluated via identical fitness functions. As long as communication time is not a significant fraction of each subpopulation's calculation time, this appears an efficient objective function evaluation parallelization method.

Finally, the third option (see Figure 8.3c) may be implemented in the case of extremely expensive objective function computations where each individual's, and possibly each function's evaluations, are split among processors. This might be the case in problem domains such as computational electromagnetics or fluid dynamics where such parallel codes already exist.

Note the preceding discussion focuses on objective function calculations only. Additional processing is sometimes required to then transform the resultant objective value *vectors* into fitness vectors or scalars. Several variants of MOEA fitness assignment and selection techniques exist (e.g., ordering, scalarization, independent sampling, and cooperative search), not all amenable to parallelization.

A MOEA's underlying data structures may also affect the ability to effectively and efficiently parallelize the algorithm. In other words, how and where necessary data is stored, its quantity, and to where (and when) it needs to be communicated may well affect how easily a MOEA is parallelized and how well the resultant implementation executes. For example, consider a generic parallel MOEA implementation evaluating k objective functions. If each slave processor evaluates only a part of one particular objective function's value perhaps just given components of the underlying data set are needed by each processor. However, if each processor computes a different objective function, each may require the entire underlying data set. In real-world design and engineering problems this data set may be quite large! As data communication significantly affects parallel programs' efficiency, reducing communication delays may well speed up overall algorithm execution.

8.3 pMOEA Paradigms

Parallel paradigms can be utilized to decompose some problem (task and/or data) and in turn decrease execution time. These paradigms may also allow for exploring more of the solution space, potentially finding "better" solutions in the same amount of time as a serial implementation. A pMOEA seeks to find as good or better MOP solutions in less time than its serial MOEA counterpart, using less resources, and/or searching more of the solution space in the same amount of execution time (i.e., increased efficiency and effectiveness). With these general performance objectives in mind generic pMOEA architectures are now addressed.

The four major pMOEA computational paradigms are considered here. They are the "Master-Slave," "Island," and "Diffusion" paradigms; the fourth includes "hierarchical" or "hybrid" paradigms that may be seen as a combination(s) of the three other forms. Island paradigms are sometimes referred to as coarse-grained paradigms and diffusion paradigms as cellular or fine-grained paradigms. Note that each paradigm may be implemented in either a synchronized or non-synchronized fashion; each has its own particular considerations. For purposes of this discussion, synchronized implementations are defined as utilizing "same-generation" populations where some sort of inter-processor communication synchronizes all processes at each generation's end. Non-synchronized implementations can greatly reduce processor idle time (assuming varying processor speeds, memory, hardware limitations, and/or data decomposition), but this implies communications occur at random times and possibly without guaranteed delivery of messages to their destinations.

Because of the relatively inexpensive cost of Commercial Off The Shelf (COTS) parallel computing platforms, numerous researchers across many academic fields are now utilizing parallel or distributed processing in their applications. However, one soon observes that their level of expertise and familiarity with engineering effective and efficient parallel or distributed codes of any sort may vary widely. The reduced cost and relative ease of setting up these systems allows many possible configurations in terms of hardware and software platforms. While it is virtually impossible to describe each of the above paradigms in relation to all possible multi-processor configurations, a generalized technique and explanation is presented such that the reader can extend the basic designs to more complex systems for use in exploring their own specialized hardware or software systems.

Researchers can utilize resources ranging from multi-million dollar homogeneous parallel supercomputer platforms to COTS heterogeneous workstation clusters costing only thousands of dollars (see Table 8.1). As these systems' costs can vary greatly so also do their respective capabilities. For example, supercomputers typically have components orders of magnitude "better" than those of PC Clusters, such as the amount of disk storage, available RAM, or the electronics required to complete an extremely quick disk/memory access. These advanced capabilities come at a cost, of course.

For discussion purposes, the design and implementation details presented in this section assume homogeneous platforms where each processor has identical available resources (e.g., CPU and RAM) and a homogeneous communications backbone (equal communication costs between each processing unit). These paradigms are easily extended to heterogeneous systems through the use of load balancing techniques, specialized system hardware/software, and/or other parallel processing concepts [926]. The following sections discuss paradigm-specific details and issues for widely available, easily obtained, and low-cost generic solutions for use in successful pMOEA development culminating in algorithms achieving desired performance levels.

Note the following discussion focuses only on objective function evaluations whereas additional processing is sometimes required to transform resultant objective value *vectors* into fitness vectors or scalars. Several variants of MOEA fitness assignment and selection techniques exist (e.g., ordering, scalarization, independent sampling, and cooperative search), not all of which may be amenable to parallelization. This fact must be accounted for in the final pMOEA implementation, i.e., the choice of a suitable paradigm.

Table 8.1. Parallel Processing System Characteristics

	System	Attributes
Homogeneous	Cluster of PCs	– Homogeneous COTS PCs – Homogeneous COTS communications backbone
	Supercomputer	– Specialized hardware/software – Homogeneous CPUs, RAMs, caches, memory access times, storage capabilities, and communications backbones
Heterogeneous	Cluster of PCs	– Heterogeneous COTS PCs – Heterogeneous CPUs, RAMs, caches, memory access times, storage capabilities, and communications backbones

8.3.1 Master-Slave pMOEA Model

The Master-Slave paradigm is quite easy to visualize from an algorithmic management perspective and is fairly simple to implement. Objective function evaluations are distributed among several slave processors while a master processor executes evolutionary operators (EVOPs) and other miscellaneous overhead functions (e.g., computing the current Pareto front, distributing/collecting subpopulations, etc.). Its search space exploration is conceptually identical to that of a MOEA executing on a serial processor. In other words, the number of processors being used is independent of the particular solutions being evaluated, but does affect execution time. This paradigm is illustrated in Figure 8.4 where the master processor distributes population members, controls when/where objective function evaluations are performed, and stores returned objective values. Although the master processor may also be used to perform objective function calculations that computational effort appears best-performed by the slaves.

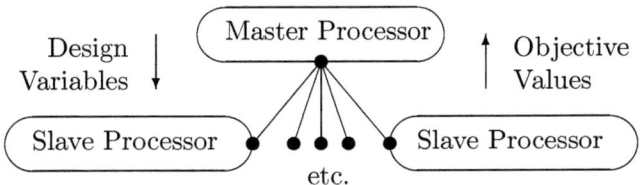

Fig. 8.4. Master-Slave pMOEA Paradigms

It is important to note that objective function calculations need to be fairly complex and time consuming in order to realize any *computational speedup*, informally defined as the ratio of the best serial MOEA run time divided by the pMOEA's run time [926]. In some cases a pMOEA may take more execution time than a serial MOEA. This is usually encountered with (relatively) simplistic objective functions; communication time overwhelms computation time and poor speedup is thus realized.

Distributing objective function evaluations over a number of slave processors can generally be implemented in three different ways (see Figure 8.3).

1. Evenly distribute population members across the slaves where each slave performs all k objective function evaluations.
2. Evenly distribute (sets of) population members across (sets of) k slaves where each slave performs one of the k objective function evaluations.
3. Evenly distribute each objective function calculation for the entire population across multiple processors.

In the first method, if the population cannot be evenly distributed the master may evaluate objective functions for a limited number of individuals. Each slave processor calculates all k objective function values for each assigned individual. Since all slaves compute identical objective functions, for (almost) identical numbers of solutions, each slave *usually* completes execution at the same time.

Master-slave pMOEA execution time is easily modeled based on a single-objective EA master-slave base case [204]. First, EVOP execution time (e.g., selection, crossover, and mutation) is ignored as researchers generally accept their cost to be much less than that of any objective function computation. Then, letting T_{c_p} be the time required to communicate between processors prior to calculating the objective function (the entire population must be transmitted), T_{c_a} the time required to communicate between processors after objective function evaluation (just objective values must be transmitted), P the number of processors used, n the total population size, $\sum_{i=1}^{k} T_{f_i}$ the time required to evaluate one individual for all k fitness functions, and G the number of generations, the running time for the algorithmic variant just described, T_{MS}, may be *estimated* as presented in equation (8.1). This equation could be used to predict performance bounds on various architectures.

$$T_{MS} = G \times \left(P(T_{c\_p} + T_{c\_a}) + \frac{n \sum_{i=1}^{k} T_{f_i}}{P} \right) \qquad (8.1)$$

In the second method, the k objective functions are uniquely mapped to the slaves (i.e., P_1 evaluates f_1, P_2 evaluates f_2, etc.); a total of P processors may be used if $P = ka$, a an integer. Each slave then evaluates its assigned function for its share of the population. That share may range from one to all individuals. Letting T_c be the time required to broadcast the entire population to the processors, T_{c_a} the time required to communicate between processors after objective function evaluation (just objective values must be transmitted), $P = ka$ the number of processors used, n the total population size, $max(T_{f_i})$, where $i = 1, \ldots, k$, the time required to evaluate the most complex objective function, and G the number of generations, the running time for the algorithmic variant just described, T'_{MS}, may be *estimated* as presented in equation (8.2). This equation may be used to predict performance bounds on various architectures.

$$T'_{MS} = G \times (T_c + PT_{c\_a} + n * (max(T_{f_i}))) \qquad (8.2)$$

This method may yield varying computational loads among the slaves as each is computing a different objective function of potentially radically different complexity and resultant wall clock execution time. Thus, blindly assigning each of the k functions to different processors may then be imprudent if one function's evaluation takes several times longer than any of the others (i.e., $T_c(f_1) >> T_c(f_i) >> \ldots >> T_c(f_k)$). Compounding the problem is the possible dissimilarity of the utilized platforms' capabilities. Static or dynamic load balancing may help equalize computational cost among employed processors, but the effort expended there could negate or exceed any gains resulting from parallelization [926].

The last method distributes the objective function calculations themselves across multiple processors due to associated high levels of complexity and execution time. The objective functions are partitioned so that each is decomposed across multiple processors. Here again, no guarantee exists that each slave experiences equal computational loads, hence, some slaves may sit idle while others work, indicating load balancing use may be fruitful. For example, real-world problem domains such as Computational ElectroMagnetics or Fluid Dynamics (CEM or CFD) often partition evaluations among slaves. Parallel CEM and/or CFD codes with load balancing already exist; they would be of great use in efficiently solving these optimization problems with a pMOEA [1045, 1046, 1047, 1266]. pMOEAs solving these MOP types are often interfaced to a specialized problem domain dynamic simulation model written by scientists or engineers from the specific discipline. Such interfacing is in general "relatively easy" and occurs through file data structures [217].

Letting T_c be the time required to broadcast the entire population to the processors, T_{com} the time required to combine the decomposed objective function values, T_{c_a} the time required to communicate between processors after objective function evaluation (just objective values must be transmitted), P the number of processors in use, n the total population size, $max(T_{f_{ij}})$ the time required to evaluate the most complex fitness function where $i = 1,\ldots,k$ and j is the number of partitions each fitness function i is decomposed into, and G the number of generations, the running time for the algorithmic variant just described, $T^{''}_{MS}$, may be *estimated* as presented in equation (8.3). This equation may be used to predict performance bounds on various architectures.

$$T^{''}_{MS} = G \times (T_c + T_{com} + PT_{c\_a} + n * (max(T_{f_{ij}}))) \qquad (8.3)$$

Efficiency is the master-slave pMOEA's main objective and hence the actual variant utilized is crucial to achieving the highest efficiency levels. One variant may be a preferred implementation due to available computational resources and the MOP currently being solved. Lastly, note that in some design problems different solutions may have widely varying objective function evaluation costs, thus negating the underlying logic for implementing a master-slave pMOEA.

8.3.2 Island pMOEA Models

"Island" paradigm pMOEAs are based on the phenomenon of natural populations evolving in relative isolation, such as might occur within some ocean island chain with limited migration between various islands. These pMOEAs are also termed "distributed" as they are sometimes implemented on distributed memory computers; they are also called multiple-population or multiple-deme. Finally, this paradigm is sometimes termed *coarse-grained* parallelism because each island (processor) contains a large number of individual solutions. Communication backbones can connect multiple processors in logical or physical geometric structures such as rings, meshes, toruses, triangles, and hypercubes. A generic island paradigm is illustrated in Figure 8.5 using a ring topology. Observe the notional communication channels for migration of selected individuals; specific paths are assigned as part of the pMOEA's design strategy and are then mapped to the physical communication backbone of the selected parallel platform upon which pMOEA implementation is realized.

The island pMOEA paradigm conceptually divides the overall pMOEA population into a number of independent, separate (sub)populations or *demes*; an alternate view observes several small, separate, simultaneously executing MOEAs (each processor often hosts a separate island). Although each island evolves in isolation for the majority of pMOEA execution, individuals occasionally migrate between an island and its neighbor(s) based on some selection

456   8 MOEA Parallelization

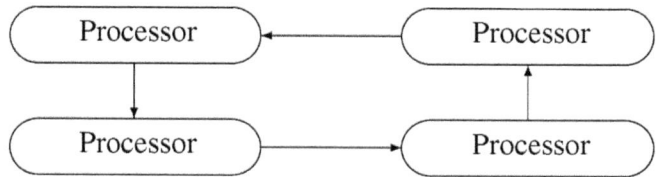

Fig. 8.5. Island pMOEA Migration Paradigm

or fitness criteria. Thus, all island pMOEA paradigms require identification of suitable migration policies defining how often migration occurs (the number of generations between events), the number of solutions to migrate, and how to select emigrating solutions and the solutions replaced by immigrants. Migration allows relatively thorough gene mixing within each deme but restricts gene flow between different demes or islands. EVOPs (may) operate differently within each island, strongly implying each population is searching many different regions of the overall search space. When using different random number generators and seeds on each island, as well as different EVOP parameter values and MOEA structures, this implication is further strengthened.

Four basic island pMOEA variants are seen to exist, each requiring appropriate migration operators. These are:

1. All islands execute identical MOEAs/parameters (homogeneous),
2. All islands execute different MOEAs/parameters (heterogeneous),
3. Each island evaluates different objective function subsets, and
4. Each island represents a different region of the genotype or phenotype domains.

Given that within each generation T_{ce} is the time for all islands to complete execution, T_{mig} the time to complete neighborhood migration, T_{coll} the time to collect/compute the overall Pareto front/Pareto optimal set, and G the number of generations, the island model's meta level running time, T_I, may be *estimated* as presented in equation (8.4). This equation may be used to predict performance bounds on various architectures. Particular island pMOEA variants would have very similar equations.

$$T_I = G \times (T_{ce} + T_{mig} + T_{coll}) \tag{8.4}$$

The first two presented variants are self-explanatory. As a specialized example of the second variant, consider the utilization of different representations (i.e., differing resolutions) on each island. Each island's population represents solutions to the same problem but are then solving the problem at different resolutions. However, note that this concept then restricts migration flow from lower to higher resolution processors.

In the third variant, islands have a (possibly) reduced problem domain, each executing either a single-objective EA or MOEA to solve j objective functions, $1 \leq j \leq k$. The employed EAs may be homogeneous or heterogeneous as described above. Due to each processor's searching a (possibly) reduced problem domain (i.e., mixing single-objective EA & MOEA results), the migration policy used requires careful thought to ensure the best possible convergence. A very important example of this approach is the distributed cooperation MOEA model which uses a parallel multi-objective genetic algorithm (MOGA), the SPEA2 and NSGA-II [1208].

The fourth variant isolates each processor to solve specific, non-overlapping regions of phenotype (or genotype) space. All possible phenotype space must be covered, which in general is a very difficult, if not impossible, *a priori* allocation for a general MOP. Moreover, when searching regions of phenotype space each island likely generates phenotype values outside its constrained phenotype region. One can force processors to generate points until a suitable number are found within its assigned region, but for certain MOPs this process may take too long, if at all possible. Of course, individuals could migrate to the processor assigned that region (causing communication overhead) or just be deleted from the population, but each of these methods requires phenotype sorting. Such an approach could also result in some islands not contributing much to the overall search. The "standard" island model, utilizing each processor to identify the complete Pareto front, appears a more efficient method. Regardless, neither case guarantees total Pareto front identification.

A specific example illustrating this variant assumes a convex Pareto front. It uses the concept of isolating each processor to search within a specific region but still lets each processor devote some effort to exploring the entire possible space. To address this obvious incongruity a guided *domination concept* is defined using a weighted function transformation of the objectives and a variant non-dominated definition [381]. equation (8.5) reflects this transformation.

$$\Omega_i(f(x)) = f_i(x) + \sum_{j=1, j \neq i}^{M} a_{ij} f_j(x), i = 1, 2, ...M \tag{8.5}$$

With this Ω transform based upon appropriate weights determined from known points on the Pareto front, a different definition of domination is defined for which the dominated region (by this new definition) is enlarged (see Figure 8.6).

This new definition permits processor state-space domain overlap. Thus, the returned Pareto front may not be completely nondominated. Therefore, each processor using this new definition finds only the "real" convex Pareto front in a region based upon the standard nondominated definition. The other points found above and below this convex Pareto front in the entire search space are dominated in the real sense, but not in this new sense. In order to successfully perform the transformation one needs to know a limited number of vectors on the known Pareto front. The number of vectors selected for allocating the overlapping regions of processor search is then the number of processors required.

458   8 MOEA Parallelization

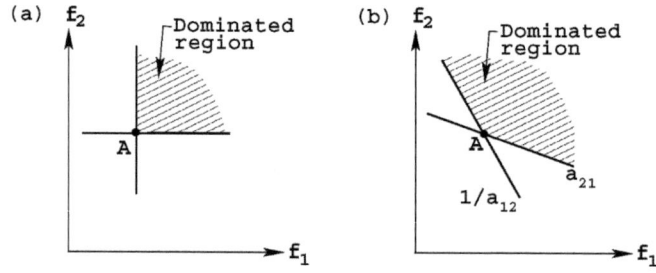

Fig. 8.6. Vector A Domination: (a) Dominated Region (Standard Definition) and (b) Dominated Region (New Definition) [381]

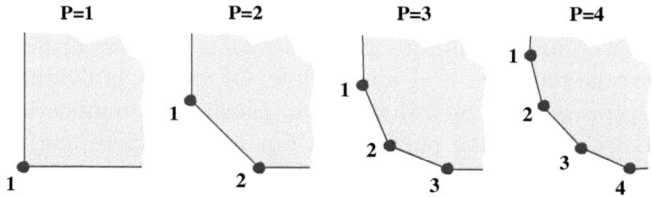

Fig. 8.7. Generic Objective Function Processor Allocation ([381])

Using the Ω transformation, all regions of the convex Pareto front are accounted for with what one hopes is an "equal slice" of the front. With known convex examples this concept deserves attention, but it appears to be somewhat simplistic. In real-world problems it may be infeasible due to Pareto front structures. Even for convex fronts, problems occur in determining how to find initial members of the front to ensure each processor generates vectors in their restricted search region, which is the intent of the revised domination definition. Figure 8.7 presents a generalized allocation for two objective functions, given 1 to 4 processors. Observe there is no guarantee that selecting two or more members of the front on a given processor and in a certain region, and subsequently performing EVOPs with them, yields another vector within that region. Therefore, general migration schemes are suggested by the authors.

8.3.3 Diffusion pMOEA Model

Like the master-slave paradigm, the "diffusion" pMOEA paradigm deals with one conceptual population, except that each processor holds only between one and a few individuals, leading some to term it *fine-grained* parallelism. The imposition of some neighborhood structure on the employed processors is this paradigm's hallmark; EVOPs occur only within these (possibly) overlapping neighborhoods. Neighborhood geometry could be a square, rectangle, cube, or other shape depending upon the number of dimensions associated with the diffusion algorithm's topological design. Each geometry reflects some associated number and arrangement of neighbors within a multi-dimensional grid. As "good" solutions arise in different areas of the local topology, the intent is for them to then spread or *diffuse* slowly throughout the entire population due to the overlapping or dynamically changing neighborhoods.

One easily sees this model involves low-level parallelization. That is, there is no migration *per se* and communication costs may then be very high within a neighborhood. This paradigm is illustrated in Figure 8.8. Observe the example shown is implemented on a logical mesh with a square neighborhood of four processors; the overall logically gridded communication structure is mapped onto some physical communication backbone. This paradigm's meta-level time equation model would be very similar to equation (8.4).

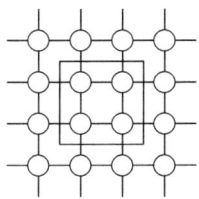

Fig. 8.8. Diffusion pMOEA Migration Paradigm

One may also vary the population density in a diffusion pMOEA. Using a percolation approach one systematically increases the population size until the diffusion lattice's carrying capacity is obtained, which occurs through a merging of smaller demes into larger ones as the pMOEA executes [983].

8.3.4 Hierarchical Hybrid pMOEA Models

An alternative to the presented paradigms is what Cantú-Paz terms the class of *hierarchical hybrids*; at a high level of abstraction these are multiple-deme algorithms with each associated island executing a particular MOEA instantiation [204, pp.126-128]. He proposes three island model hybrids (here abstracted to the pMOEA domain) where:

1. Each island contains a diffusion pMOEA,
2. Each island contains a master-slave pMOEA and
3. Each island contains an island pMOEA.

Another innovative computational pMOEA design in this class incorporates a co-evolutionary hierarchical pMOEA [216]. A tree or graph search structure is employed in an attempt to find better pMOEA algorithmic structures for a given problem domain. The bottom leaves are various MOEA instantiations with associated parameter values. As their concurrent execution completes, the next level of the tree evaluates lower leaf performance and selects new parameter values based on those results; leaves at this new level may also deterministically or stochastically add other algorithmic constructs to improve local MOEA instantiations. This development process continues up through the tree's levels to the root node. In fact, one can conceive this bottom-up, co-evolutionary approach evolving new and improved MOEAs as meta-level evolutionary search. Associated research should address the design, testing, and analysis of this hierarchical parallel-based process to evolve better (p)MOEAs [713, 1674]. Assuming a binary tree structure with n MOEAs executing at the lowest level and l levels, this design is illustrated in Figure 8.9. Note that each leaf may have any number of children, and although the overall complexity would then be much higher, any particular node may itself be a pMOEA.

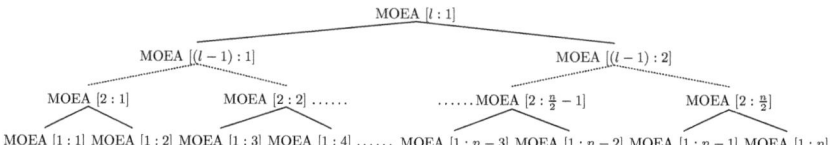

Fig. 8.9. Example Hierarchical Design

8.4 pMOEAs From the Literature

This section discusses pMOEAs from the current literature in order to provide understanding of implementation variations and their impact on results. Each pMOEA example is classified into one of the paradigm categories. As previously noted, although some implementations *may* be classified in multiple categories, the focus is on understanding pMOEA example characteristics.

8.4.1 Master-Slave pMOEAs

The following briefly describes key elements of known master-slave pMOEA implementations in the literature, highlighting key algorithmic issues and selected comments by the authors.

- In the earliest reported master-slave MOEA, a microprocessor cache memory design problem is attacked using an implementation named 'Genetic Algorithm running on the INternet' (GAIN) [1518]. This pMOEA used an Internet-connected network of workstations as slave processors (between 80 and 120 simultaneously) to distribute one chromosome per workstation for fitness evaluation; a master processor then collated results and executed the EVOPs. A major issue faced by GAIN's creators was the network's heterogeneous composition consisting of different workstation models and configurations, thus giving rise to disparate performance between machines. Additionally, their network's protocol mandates suspending remote computations when a user physically logs into a machine; the user also has the option of terminating the evaluation.

To deal with the (possibly) widely varying objective function evaluation times and externally terminated evaluations, two extensions to GAIN are developed by decomposing it into both generational and evaluation processes communicating through queues. For synchronization purposes, the generation process causes the master processor to "sleep" when some maximum number of pending evaluations was exceeded. The evaluation process controls Internet workstation communication and tracks chromosomes sent out for fitness evaluation. If some evaluation is terminated the evaluation process resubmits it, but after five resubmissions that particular chromosome's evaluation is terminated and GAIN moves to the next in line. GAIN also monitors the time between a solution's creation and its first opportunity to become a parent, as the authors expected lengthy "childhoods" might adversely affect pMOEA stability and performance. The authors also expected the implemented steady-state generational replacement policy (with a small generation gap) would stabilize this largely decoupled pMOEA; they claim this was confirmed via their limited experiments.

- A modified NSGA [1509] with Pareto ranking is used in solving a multidisciplinary design problem optimizing two-dimensional airfoils [1045]. The objective functions' high computational cost drove the authors to use a distributed parallel architecture. NSGA modifications included replacing its original roulette wheel selection by tournament selection and also adding an elitist selection mechanism.

  This pMOEA executes on an IBM SP2 computer; the MPICH library is used to control interprocessor communication. A "high-performance" switch and 8 processors (Model 390) are employed. It is noted that even when parallelized, wall-clock time for a single run is about 52 hours. To determine solution fitness, a two-dimensional Euler flow analysis solver and two-dimensional time-harmonic Maxwell field analysis solver reduced to a Helmholtz equation are used. Each solution computed using the Euler solver takes about 160 CPU seconds; using the Helmholtz solver takes about 8 CPU seconds.

- An aerodynamic and aeroacoustic airfoil optimization problem with extremely expensive objective function computations is solved via a master-slave pMOEA [805]. Because individual fitness is evaluated independently the authors feel this problem well-suited for coarse-grained parallelization. However, although the master processor performs high-level algorithmic operations and distributes individuals to slave processors, it has the additional task of dynamically load balancing computational effort across all slaves. The authors note that as airfoil design evolution begins with a random population, large differences may initially exist in the time needed to evaluate differing solutions (designs). Thus, by using dynamic load balancing, individuals are distributed for evaluation only as slave processors become available in an attempt to avoid the potentially large bottleneck associated with static load balancing.

Their pMOEA incorporates MPI to avoid limiting execution to a specific computational platform(s); it currently executes on an IBM SP2 computer. Although no formal efficiency analysis is performed, the authors report runs conducted using 8, 16, 32, and 64 nodes resulted in speedup values close to theoretical expectations.

- A modified NSGA [1509] with Pareto ranking is used in solving a Computational Fluid Dynamics (CFD) problem [1057]. As do others, the authors note their main concern with employing MOEAs in solving complex design problems is the computational effort required. To address that concern they employ a 2-level parallelization strategy: (1) Flow solver parallelization combining problem domain partitioning techniques with MPI, and (2) MOEA parallelization. Each processor appears responsible for evaluating *all* of an individual's k criteria; each must also contain the same number of individuals. Thus, with p processors (generalizing their pMOEA strategy and implementation), p individuals are evaluated in parallel, as it appears only a single individual per processor is used. The pMOEA is executed on an SGI Origin 2000 with 8 R10000/195 MHz processors.

- A master-slave pMOEA named the Parallel Multi-Objective Genetic Algorithm (PMOGA) attacks eigenstructure assignment problems [157, 325]. An identical copy of the initial randomly generated population is sent to each slave processor, which then executes a separate MOEA. Each slave uses different decision-making logic (e.g., solution evaluation may be performed with different fitness function combinations, selection may occur using Pareto principles). Once the slaves reach some user-defined stopping criterion they asynchronously communicate their final population to the master processor, which uses these populations to form a final population it then evolves. PMOGA executes on a SPARC-SUN network.

- A mobile telecommunication network design problem is tackled by another master-slave pMOEA, in particular, the positioning of base stations on potential geographic sites in order to fulfill certain objectives and constraints is investigated [1096]. Solutions identify a set of antenna sites from some pre-defined candidate set, determine the type and number of antennas used, and also the particular antenna configurations (tilt, azimuth, power, etc.).

  The authors claim a high computational cost associated with solution evaluation and constraint testing exists; a large amount of memory is also required. Computing each individual's objective function involves key feature identification (within some area), calculating radiofrequency wave fields at given points, and updating both network global fitness and handover/processed traffic for each cell.

Each slave processor is assigned a cell (piece) of the total geographical working area, computing relevant measures only for points in its cell. Allowances are made to communicate required values from neighboring cells to determine total fitness. The authors use a hardware platform composed of 24 "modern" homogeneous workstations running Linux; the pMOEA is implemented using the 'C' programming language and PVM routines for interprocessor communication.

- A master-slave pMOEA is applied to an X-ray plasma spectroscopy application [591]. This elitist implementation is based on the Niched Pareto Genetic Algorithm (NPGA) [709]. The pMOEA is developed and tested on an 8-node Beowulf cluster (PII-400 processors, 100Mbs network) using PVM 3.4 for interprocessor communication, which is kept to a minimum by occurring only when distributing solutions to evaluate and returning objective values. The authors measure each solution evaluation at about 1.6 CPU seconds in their experiments. Thus, for load balancing they suggest a total population size evenly divisible by the number of slave processors multiplied by packet size. Their experiments observe no noticeable change in pMOEA performance for different packet sizes as long as this condition is satisfied.

  The authors perform further experiments on a 20-node Beowulf cluster and reportedly achieve linear speedup. They claim a pMOEA allows use of larger populations thus resulting in improved effectiveness. As a significant reduction in computing time due to MOEA parallelization is realized, population size can be increased and thus improve the algorithm's effectiveness in finding good solutions.

- A multi-processor computer and master-slave pMOEA is used in solving a jet actuator placement problem [1372]. After testing on a single processor the code was modified to execute on four processors and ported to an SGI Origin 2000 machine. Each of three subproblems executes on a different processor with the fourth serving as master. Although theoretical speedup is computed to be 3-fold, system limitations require a manual algorithm restart each generation. However, this implementation reduces typical optimization time from 1100 hours (exhaustive search) to 22 hours.

- A master-slave pMOEA is successfully applied to finding solutions for a set of benchmark process scheduling problems [1548]. Two algorithmic versions are implemented, one with heterogeneous (termed HT) and the other with homogeneous (termed HM) populations. Both versions use k subpopulations (generalizing the authors' pMOEA strategy) sending immigrants to a separate "main" population.

  In the HT version each of the k subpopulations evolve using one of the k objective functions as their sole criterion. However, the main population is Pareto-oriented and attempts to evolve nondominated solutions. In contrast, all subpopulations in the HM version are Pareto-oriented. Both versions utilize unidirectional immigrant flow in attempts to preserve a higher level of genetic diversity in the total population. Additionally, when a population has met some user-defined convergence criteria the pMOEA is restarted (noted by the terms HTR and HMR). The authors note a restart is often necessary as subpopulations working on a single objective tend to converge much faster than a Pareto-oriented one.

The HTR and HMR variants are then directly compared, apparently finding that for problems in which an inverse relation exists between the different objective functions the HTR strategy is more effective. Conversely, for problems in which a direct relation exists between the different objective functions the HTR strategy is more efficient. Based on their experimental results the authors feel that generally speaking, the HTR strategy is consistently superior to the HMR.

- A Master-Slave with Local Cultivation (MSLC) model is proposed and implemented in the parallel domain to "gain higher diversity of the solutions [1683]." A mobile telecommunication antenna arrangement problem is used as the basis with which to compare the MSLC against three other MOEAs. Their implementation uses a few master and several slave processors. Master processors randomly select two individuals and send them to a slave, where EVOPs based on the minimal generation gap model are executed. Selected individuals are then returned by the slaves to the masters where they are ranked. The authors' experiments ran on a PC-cluster consisting of 16 Pentium II 500 MHz machines each with 128 MB memory. Comparisons were made between experiments using 2, 4, 8, and 16 slave processors. Very few MSLC operational details are presented in this paper.
- The master-slave Parallel Single Front Genetic Algorithm (PSFGA) is applied to several benchmark test problems [353]. The master processor sorts the population according to objective function values and splits it into m subpopulations; this process is periodically repeated throughout PSFGA execution. The master sorts on only one of the k objective functions at a time, yet the populations upon which sorts are performed are evolved via global Pareto dominance principles. Population diversity is obtained by filtering solutions based on a grid overlaid on objective space, using distance between discovered objective values (the plotted vectors) as the discriminating factor between which solutions to keep and which to not. The authors implement PSFGA on a cluster of eight PCs connected by Fast Ethernet. They test the algorithm using different numbers of processors and architectures; PSFGA exhibited speedup in most cases although in others it produced lower-quality solutions.
- A master-slave pMOEA is used to investigate technologies that might be applied to biologically remediate contaminated groundwater while in place (i.e., underground) [871]. The authors were driven into the parallel domain due to the intensive computational requirements of their technology model, which incorporates important factors affecting the bioremediation process. They create a pMOEA called the GENeral Multi-Objective Program (GENMOP). Chromosomes contain auxiliary genes devoted to objective function values, Pareto-ranking, and problem-specific factors, but

these auxiliary genes are not involved in either crossover or mutation. Their pMOEA executes using RedHat Linux and MPI enabling parallel computation among 32 Aspen dual 1 GHz Pentium III processors, each with 1GB memory. GENMOP was deemed useful for selecting their technology model's parameter values.

8.4.2 Island pMOEAs

The following briefly describes key elements of known island pMOEA implementations in the literature, highlighting key algorithmic issues and selected comments by the authors.

- In the earliest reported island pMOEA a vehicle scheduling problem for Venice's urban public transportation system is successfully solved [78]. Although the authors did not execute their algorithm on a parallel computer they did implement an island pMOEA construct on a single processor and their effort is thus included here. Local geographic selection, which is an EVOP based on some particular spatial structure imposed over the population, is used within each island. With local geographic selection the probability of two individuals mating is a fast-declining function based on the "geographical distance" between them. Thus, islands are defined based on the distance from their internal solutions to those within other islands. Individuals are placed on a toroidal grid with one individual per grid intersection point; selection takes place locally on this grid as each individual competes with its nearby neighbors. Specifically, a solution finds a mate during a "random walk" beginning at its current location. Once a mate is identified the solution with the best fitness value is selected.

  The authors note their adoption of local geographic selection is specifically due to its applicability within multiobjective optimization. They state this particular EVOP is a niching technique used as an alternative to fitness sharing; they believe it naturally creates niches without the difficulties of problem-dependent parameter tuning. They also employ tournament selection to complement local geographic selection but only about 15% of the time.

- An island pMOEA is used to find acceptable solutions in multiple points airfoil designs [1281]. As does the preceding implementation, local geographic selection is employed by placing individuals on a 1-D or 2-D 16×8 toroidal grid. The author suggests that although preliminary tests show no clear advantages of one grid dimensionality over the other, it does appear a correlation between the number of objectives and the grid's cardinality exists. It is observed that in this particular problem, distributing the objective evaluations on all available processors is efficient only when all individuals have an equal computational cost of determining their associated objective values. The CPU time needed for a single CFD simulation widely varies between different design geometries, thus causing possible unequal processor loading.

A distributed-memory 64-node Cray T3D computer is used. Its architecture enables shared arrays to store the population, which is immediately updated as new individuals are evaluated. This implementation thus requires no generational synchronization, allows processors to be maximally utilized, yet limits interprocessor communication to database updates. The author reports linear speedup and claims test results prove a high computational efficiency can be obtained on massively parallel computers. However, he notes this claim is valid only if the problem size is small enough to completely reside within one processor, which may well not be the case in large CFD problems.

- This same parallel MOEA is later applied to two other aerodynamic design optimization problems, again on a Cray T3D [1282]. The authors again observe that CPU loads for different CFD simulations may be quite different so synchronization problems may well result between processors at generation's end. Shared arrays store the population, limiting interprocessor communication to database updates, as each computational node holds a copy of the flow solver along with all data necessary to produce a converged solution. Their experiments used 128 nodes for 8- and 16-generation pMOEA runs, and 32 nodes for another 16-generation pMOEA run.
- An island pMOEA is used in solving a rotor-blade design problem [970]. Neural networks are employed in developing causality relations facilitating problem decomposition; optimal solutions are then obtained by solving a number of coordinated (smaller) subproblems. Once these subproblems are generated each is assigned to a different island and evolves in parallel for some fixed number of generations. Changes in one island's designs are then communicated to other islands through the migration process. However, the authors state manual intervention is necessary to coordinate solutions in different islands as decoupling is seldom complete. Execution ends when some user-defined decision point is satisfied.
- An island pMOEA is employed in optimizing aerodynamic structural designs [402]. The authors' experience in solving aerodynamic optimization problems leads them to conclude algorithmic parallelization is often required to achieve "reasonable" computation time. Additionally, they find pMOEAs generally outperform conventional serial MOEAs in many different applications. The pMOEA is implemented on nine Ethernet-connected workstations using MPI; a separate computational process (an individual MOEA) manages each deme, which may be only one of several resident on a single processor. Exchange of genetic material occurs only through migration and occurs to/from a deme's nearest neighbors (North, South, East, and West) only. A toroidal array is used to minimize communication cost.

The authors conclude that given a dedicated parallel computer with one process per processor, global instructions achieve synchronized exchange of individuals along rows and columns, thus greatly reducing communication overhead. The authors also find their pMOEA is far more efficient than

a serial version (as one would expect) even though population sizes are identical.
- Object recognition problems are attacked by an island pMOEA [20, 22].[3] This implementation stores each randomly initialized, equally-sized subpopulation as an array of individuals; each individual is assigned fitness based on Pareto rank. These fitness values are then used in deriving a fitness value *for each subpopulation*, each of which is then randomly paired for crossover and mutation operations resulting in two new subpopulations. Unique to this pMOEA is that selection is performed at the subpopulation level – the fitness of the original and newly-created subpopulations are compared and the two subpopulations with highest fitness kept.
- An island pMOEA is employed in solving constrained placement, facility layout, and Very Large Scale Integration (VLSI) macro-cell layout generation problems [1449]. Although the authors term their creation a "stepping stone" model closer examination reveals it as an island implementation. Their model's name is selected to reflect periodic migration of individuals between islands, or "stepping" between them.

  The pMOEA's islands are each composed of 10 individuals, however, each island uses differing parameters (e.g., mutation frequency, ratio of mutation to recombination) and mating strategies. The algorithm is executed on a 16-node Motorola MPC 601 processor network. The authors report results competitive with those resulting from other approaches taking longer to execute and requiring specialized tools.
- Although the authors report using a master-slave pMOEA in optimizing wing shape for transonic flow conditions [1307], their approach is classified as an island pMOEA due to the authors' explicit definition of islands and migration strategies. Their implementation is named the Virtual Subpopulation Genetic Algorithm and incorporates a finite-difference, full-potential flow solver for objective function evaluation.

  pMOEA execution begins by distributing all population members across a single toroidal grid. Islands are created by defining logical boundaries and selection occurs via a process termed a "random walk." From some initial point, two random walks (of some number of steps) are taken. The individuals located where the random walks end become parents. To implement a migration strategy the random walks are allowed to cross island boundaries based on some given probability function.

  The pMOEA is implemented using native UNIX processes and executes on an SGI Power Challenge with 16 processors, an architecture allowing data to be copied to/read from shared memory. One node is designated as the master processor; the population is then split into subsets and assigned to slaves (the islands). The master processor waits for all evaluations to

---

[3] Note that regardless of the papers' titles, of the two given cites the former only alludes to the authors' pMOEA. All details discussed here are drawn from the latter citation.

complete (one generation) before synchronizing processes; the authors note this implementation is effective only when each processor has an equal computational load, else efficiency is lost as the master stalls each generation waiting for outstanding evaluations to complete. Of note is the authors questioning the use of subpopulations in multiobjective optimization because they observe any advantages to be strongly dependent on the problem at hand. They feel it might be useful in some situations but not in their particular problem, due to limits on the maximum number of objective function evaluations allowed.

- Another island pMOEA is used in solving numeric optimization problems [139, 903].[4] It is implemented in MATLAB with interprocessor communication handled via PVM. The authors mention their algorithm can be alternatively implemented using a master-slave model but they are unclear as to whether this was ever accomplished.

  A Linux PC and 3 Sun SPARC workstations were used in their experiments. The authors report that (as expected) quicker run-times and a better Pareto optimal set representation resulted with their approach.

- An island pMOEA designs models for use in solving a fault diagnosis problem [1060] where the authors allow for the possibility of seeding the initial population. Through experimentation they determine fastest results for this problem occur when using seven islands and conclude their pMOEA converges faster than a serial version.

- An earlier described pMOEA [1282] instantiation is used in textile machine guide construction and manufacture [1089]. Although the authors report implementing only a serial version their code does allow for parallelization. This algorithm is apparently only one of several variants embedded in an overall simulation framework implemented in Java, which the authors selected because of its multithreading features.

- Some researchers target a military application with an island pMOEA, attempting to optimize a flare pattern to be effective against multiple threat types and angles of attack, as opposed to solution approaches focusing on a single flare pattern performing well against a single threat at a single angle of incidence [213, 1373]. This pMOEA incorporates very detailed infrared seeker and missile simulation models in computing objective functions. The authors use a 380-bit chromosome, giving about 2.4×10^{114} possible solutions. They report being forced into using a pMOEA implementation due to the very large search space and computationally expensive objective functions.

  Initial experimentation indicated simulation times ranging from 20 minutes to 4 hours for a single objective function evaluation, implying a single pMOEA run taking between 180 days and 6 years. The authors felt an island pMOEA model best-suited for the purpose at hand as no node syn-

---

[4] Note the former citation is most likely written in German; attempts to verify this with the authors were unsuccessful.

chronization is required and only low-bandwidth gene pool information is transferred between islands.

An existing DEC Alpha network is utilized as the computational platform; inter-node communication uses the standard Network File System. Up to 18 processors are utilized (16-500 MHz and 2-275 MHz). A "gauntlet" approach is employed to further speed objective evaluation. This approach first prioritizes all objectives by importance, thus forming the gauntlet. After the first objective is evaluated for each solution, subsequent tests (on the lower priority objectives) are not evaluated if the current test fails some user-defined criterion. As each objective may have largely disparate evaluation times (and 12 objectives were used) this novel approach appears quite effective in sparing unneeded computational expense and focusing on the most important factors.

- The Distributed Genetic Algorithm, applied to solving numeric optimization problems, is actually an island pMOEA and classified as such here [686]. Although reportedly executed on only a single processor this MOEA's extension to the parallel domain is easily accomplished. In this implementation, sharing occurs not only within each island but throughout the entire population and is performed when the number of "frontier" solutions ($|PF_{known}|$) exceeds some user-defined cardinality.

The authors state that distributing the population among islands leads to high solution accuracy and that the "sharing effect" leads to high diversity among solutions. However, although their pMOEA (with ten islands) executes more quickly than one with a single island, solution diversity is found to be less. Although not employed here the authors suggest using dynamic load balancing in appropriate situations.

- A similar pMOEA named the Divided Range Multi-Objective GA is also applied to solving numeric optimization problems [688]. Here, the overall population is sorted by objective function values (f_1 first); N/m individuals (where N is the population size) are then selected based on f_1's value. These individuals are placed in an island. This process then repeats for the $f_2, f_3, \ldots, f_k$ functions. There are thus m separate islands, each executing for a set number of generations, at which time all solutions are placed together and the process repeats.

The authors favorably compare their pMOEA's results to those derived via both a "typical" island pMOEA and a serial (one population) MOEA. Their implementation executes on a PC cluster system using five 500 MHz PII processors (each with 128 MB of memory), the Linux operating system, a Fast Ethernet TCP/IP network, and the MPICH communication library.

- An island pMOEA named "Parameter Free" GA is applied to solving numeric optimization problems [1434]. The authors use parallel processing in the aim of reaching "better" solutions "faster" than those found by sequential processing; they expect this result due to the extended search space exploration performed when several processors are simultaneously used. The authors employ and compare four pMOEA variations that differ

in island connectivity. The pMOEA executes on eight processors connected via local-area networks and uses PVM for communication; dynamic load balancing is also employed. The authors are pleased with the pMOEA's results when applied to a benchmark problem set.
- The authors apply Jones and Crossley's simple GA [805], Vicini's Distributed Genetic Algorithm (DGA) [1647], and their Divided Range Multi-Objective Genetic Algorithm (DRMOGA) to block layout problems [1682]. These problems are selected as they observe test functions used in evolutionary multiobjective optimization studies are almost always continuous problems. They claim the DGA produces "calculation waste" in that some islands might discover identical Pareto solutions, but that their DRMOGA reduces that waste. Their experiments execute on a 4-node PC cluster where each node is a Pentium II 400MHZ with 128 MB memory, each hosting a separate island. A rather superficial treatment of experimental results is presented.
- An island pMOEA is the vehicle used to investigate possible benefits of an asynchronous migration scheme [705]. The authors believe synchronous migration introduces migrants to a population before search converges, thus destroying good schemata and making it harder to generate better ones. Additionally, their algorithm does not require identification of effective migration parameters and topologies. They propose migration from and to subpopulations only when their native solutions have converged; migrants are then selected from the subpopulation with the "most different" individuals. The authors define measures determining convergence and difference, and use four test functions in judging performance when implemented on a PC cluster. They report experimental results of three test problems gave no significant differences between asynchronous and synchronous implementations, but the asynchronous pMOEA obtained a better Pareto front representation, and performed better overall, when applied to their fourth test problem.
- The Multi-Objective Genetic Algorithm with Distributed Environment Scheme (MOGADES) is proposed in [817]. Some of the authors of this approach were involved with the development of DRMOGA [1682], results of which were somewhat disappointing when compared to other MOEAs. In MOGADES, each island has a different weight parameter (since it employs a weighted-average approach), and elite and Pareto archives. MOGADES incorporates search mechanisms of both the SPEA2 [1775] and NSGA-II [374]. As it executes, solutions with top fitness values are preserved in the local elite archive; locally Pareto optimal solutions are stored in the local Pareto archive. After each migration event each island's weight is changed based upon the population's current values and the distance between the best individuals in both the current and next island in the logical topology. MOGADES is tested with two numerical MOPs and results compared to those of SPEA2 and NSGA-II.

- A specialized island pMOEA is implemented and demonstrated using a three objective MOP, but it must be noted the algorithm is only loosely based on the island paradigm [1715]. The authors' implementation divides an EA into some number of sub-EAs, where each sub-EA is specialized to solve a modified MOP by searching with respect to a subset of the original k objective functions. They liken their pMOEA to a generalized version of VEGA, however, they are silent as to whether or not VEGA's shortfalls and criticisms remained equally valid with respect to their algorithm. Seven scenarios were tested with differences in the number of sub-EAs used, migration strategies, and objective function subsets.
- The authors use a modified version of PGAPack in their pMOEA implementation used to investigate MOEA scalability in the context of *de novo* peptide identification [1049, 1050]. Their interest in using pMOEAs for this particular application arises from three major issues: Many existing algorithms used in this problem domain are very computationally expensive; algorithms may not be working with complete (or sufficient) information to satisfactorily determine possible solutions; and many algorithms have difficulty generating candidate solutions satisfying problem constraints. The authors implement a ring-based island model incorporating local search; at each migration interval each island sends one member to its left neighbor in the ring and replaces that individual by the incoming immigrant from its right. Pareto ranks are computed separately on each island. The algorithm is executed on a Terascale HP cluster composed of 1.5 GHz Itanium 64-bit dual-processor workstations, linked by a Quadrics QSNet 1 interconnect. One, two, four, and eight islands are used in testing. Although each island may have a different population size each overall test uses the same total population size.
- López Jaimes and Coello Coello [1010, 1011] propose an approach called Multiple Resolution Multi-Objective Genetic Algorithm (MRMOGA), which consists of a pMOEA based on the island paradigm, with heterogeneous nodes. The main idea of this approach is to encode the solutions using a different resolution in each island. Then, variable decision space is divided into hierarchical levels with well-defined overlaps. Evidently, migration is only allowed in one direction (from low resolution to high resolution islands). A conversion scheme is required when migrating individuals, so that the resolution is properly adjusted. MRMOGA uses an external population, and the migration strategy considers such population as well. The approach also uses a strategy to detect nominal convergence of the islands in order to increase their initial resolution. The rationale behind this approach is that the true Pareto front can be reached faster using this change of resolution in the islands, because the search space of the low resolution islands is proportionally smaller and, therefore, convergence is faster. This issue was originally identified by Parmee and Vekeria [1251] when they used an injection island strategy to solve a single-objective engineering optimization problem. MRMOGA was implemented in a cluster

with 16 nodes (with 2 processors per node) and 2Gbytes of memory per node. Each processor was an Intel Xeon, running at 2.45MHz, linked with FastEthernet and using MPICH 1.12 as its communications library (under Red Hat Linux 3.2.2.5). This approach was validated using several test functions taken from the specialized literature, and results were compared with respect to a parallel version of the NSGA-II [374] (pNSGA-II). The results indicated that MRMOGA outperforms the pNSGA-II, with a more significant difference as the number of processor increases (with one processor, the NSGA-II outperforms MRMOGA in all the test problems adopted). This performance improvement is even more remarkable in problems with very large search spaces.

- Streichert el al. [1528] suggest a divide-and-conquer approach to parallelizing MOEAs which aims to improve the speed of convergence beyond an island model MOEA with migration parameters. They limit subpopulations to specific regions and implement zone constraints based upon the dominance principle using k-means cluster centroids. With real-valued representations, one-point crossover, and self-adaptive mutation, they employ this cluster-based parallelization scheme for the NSGA-II [374] and statistically compare it to four alternative MOEA parallelization schemes on four standard multiobjective test functions. The hypervolume metric is averaged over 25 runs. Results with this performance measure over their test functions indicate that the other (less complicated) island approaches are statistically equivalent, but for complex real-world problems the proposed approach may produce some performance benefit.
- Xiong and Li [1718] developed a parallel Strength Pareto Multi-objective Evolutionary Algorithm (PSPMEA) in Java. PSPMEA is a parallel computing model designed for solving multiobjective optimization problems using both global parallelization and island parallel GA models. Each island subpopulation evolves separately with different crossover and mutation probabilities, but all use binary tournament selection. The islands exchange individuals with an elitist archive using a dynamic island migration frequency and associated number of individuals. Each island GA can be steady-state or generational. The benchmark problems adopted to validate this approach include convex, non-convex, discrete, multimodal and non-uniform test problems such as those discussed in Chapter 4. Experiments indicate that the proposed method with three islands can rapidly converge close to individual Pareto optimal fronts with regard to the hypervolume metric. This performance is due in no small measure to the global nondominated archive, size = 100. As to efficiency, no relative time data was given, but 250 generations were indicated. Both these values were used in serial SPEA2 [1775] experiments whose results are being compared. It is worth noting that Gonzalez et al. [594] proposed a parallel structure for the SPEA2 different from the one adopted in this paper.

8.4.3 Diffusion pMOEAs

The following briefly describes key elements of known diffusion pMOEA implementations in the literature, highlighting key algorithmic issues and selected comments by the authors.

- In the earliest reported diffusion pMOEA, solutions to a robot task and route planning problem are generated [768]. The approach is classified as a diffusion pMOEA because the authors assign a logical structure to their processors and allow recombination only within a given neighborhood. Although originally "solved" using a panmictic population model they felt the serial MOEA took too long to run and thus implemented a parallel version. The pMOEA's population is mapped onto a ring topology where each individual may choose a mate from some local neighborhood (in this case set to a total of eight neighbors). Thus, neighborhoods overlap and can then exchange genetic information. While varying population size from a minimum of eight to a maximum of sixty, experiments showed the pMOEA to be at a minimum 15 times faster than the original panmictic population version.

    The pMOEA is executed on a MultiCluster 2 system with thirty-two T800 processors. With this parallel architecture a neighborhood size of eight requires a maximum communication radius of two nodes, thus helping minimize communication costs. Finally, the authors believe their algorithm's speedup is nearly proportional to the number of processors employed and that their implementation is scalable.

- A diffusion pMOEA solving a sensitivity analysis problem is notable as it is one of very few pMOEA citations to explicitly discuss algorithmic operation [1382]. Selected details are noted here to allow better comprehension of this implementation's operation.

    The randomly generated initial population $(x_1, x_2, \ldots, x_n)$ is spatially distributed with each member x_i residing on a unique vertex in a square lattice. Recombination and selection are accomplished as follows. For each x_i a random neighbor x_j is chosen; these individuals are then crossed to produce an offspring y_i. Then, for each x_i, if y_i is better than x_i, $x_i = y_i$. Lastly, x_i is mutated. All individuals in the population are operated upon simultaneously. The authors hypothesize that as genetic information is only locally exchanged, allelic diversity is maintained at a higher level than in a "standard" MOEA, and thus gives rise to a niching effect as different population "areas" converge to different optima. This pMOEA utilizes Pareto dominance but does not calculate the Pareto rank of each individual in the traditional manner (i.e., by comparing each member's objective function evaluations to every other's). Here, dominance is used only in the local comparisons between two population members. The authors compare the results of two diffusion pMOEA variants (distinguished by mutation occurring before and after replacement) with a panmictic

sharing MOEA, finding both pMOEAs execute faster and require fewer generations to achieve convergence.
- A diffusion pMOEA evolves prognostic models predicting whether patients suffering from an uncommon form of cancer might survive [1078]. The authors claim an advantage of distributed populations is different local niches emerging, representing different ways of trading off the various objectives. This pMOEA is a close variant of the one just discussed [1382]. The authors state that a square lattice geometry is typically used in diffusion GAs as it is simple to program and easy to execute on parallel computers. Thus, they randomly generate an initial population and place it on a 13 by 13 square lattice connected at its opposing edges resulting in a toroidal topology. For each population member x_i, a random neighbor x_j of x_i is selected. The two solutions (x_i and x_j) are then crossed to produce y_i, which is then mutated to produce z_i. Then, for each x_i, if z_i is better than x_i $x_i = z_i$.

    In trying to improve their results by applying a simple hillclimbing heuristic, evolved solutions are somewhat less robust than those derived via their pMOEA alone. It is worthy of note that the authors report upon showing their pMOEA's results to decision makers, "too many" viable options were produced. Further heuristic application is then indicated to prune the discovered possibilities.
- A pMOEA capable of arbitrarily scaling between the island and diffusion paradigms is named the Parallel Evolutionary Multiobjective Optimization using Hypergraphs EA (PMOHYPEA) [1087]. It uses a hypergraph representation of the population. The authors recognize that unless fitness function evaluation times are constant, the target multiprocessor architecture is homogeneous in terms of computing capability, and processors are connected by a homogeneous communication network, the highest PMOHYPEA efficiency can be reached only with some effort on the researcher's part. The algorithm is based on the NSGA-II and is tested with several well-known test functions.
- Lim et al. [995] present an efficient Grid Enabling Hierarchical Parallel Genetic Algorithm framework (GE-HPGA). The grid computing framework is developed using standard grid technologies, and has two distinctive features: (1) an extended GridRPC API to conceal the high complexity of the grid environment, and (2) a scheduler for seamless resource discovery and selection. To assess the practicality of the framework, a theoretical analysis of the possible speedup offered is presented. The authors also present an empirical study focused on the GE-HPGA using a benchmark problem, and a realistic aerodynamic airfoil shape optimization problem. The computational environment involved diverse grid environments having different communication protocols, cluster sizes, processing nodes, and at geographically disparate locations. Operators and parameter values were one-point crossover with probability 0.9, uniform mutation with probability 0.01, a subpopulation of size 50, a maximum generation count of 100,

and migration interval of five. Results with 28 heterogeneous processors indicate that the GE-HPGA offers a credible framework for providing a significant speedup to evolutionary design optimization. Of course, speedup can be attained as long as appropriate bounds on fitness function cost, cluster size, and communication overheads of the grid environment are satisfied.

- Nebro et al. [1176] introduce a new cellular genetic algorithm for solving multiobjective continuous optimization problems, a multiobjective cellular genetic algorithm (MOCell). Their approach is characterized by using an external archive to store nondominated solutions and a feedback mechanism in which solutions from this archive randomly replace existing individuals in the population after each iteration. Testing was with both constrained and unconstrained problems from the ZDT problems and the WFG toolkit (see Chapter 4). Results were compared against NSGA-II [374] and SPEA2 [1775]. Preliminary experiments indicate that MOCell obtains competitive results in terms of convergence, and it clearly statistically outperforms the other two compared MOEAs concerning the diversity of solutions along the Pareto front. Metrics employed were the generational distance (GD), the spread, and the hypervolume measurement (see Chapter 5).

8.5 pMOEA Analyses and Issues

This section presents a qualitative analysis of currently known pMOEA research. Relevant meta-level topics are addressed, highlighting several issues that are treated lightly or even ignored in the literature. Detailed discussions of pMOEA suitability, hardware/software, test suite, metric, and parameter issues then follow. The section concludes by broadly discussing major factors to consider when developing, implementing, and analyzing pMOEAs, as well as identifying important issues currently unexplored by the field.

In fact, a quick review of Section 8.4 shows a total of only eight master-slave, eighteen island, and three diffusion parallel MOEAs. As over 2500 total MOEA citations are known as of this writing (end of 2006) [266], one sees less than 1.2% of the total MOEA research effort devoted to parallel implementations. This result is somewhat surprising due to the fact that so many MOEAs are used in solving engineering applications (often which have inherent computationally expensive objective functions) that obviously stand to benefit from parallelization. Additionally, as earlier stated, parallel computing capability has become more and more accessible. One might then think more researchers would be employing parallel implementations in the search for more effective and efficient MOEAs but this appears to not be the case. However, it is interesting to note that of the twenty-nine known parallel MOEAs, two are devoted to solving scheduling applications, five to numeric optimization problems, and twenty-two to solving design & engineering problems.

8.5.1 pMOEA Observations

Active research interest in pMOEAs is continuing to slowly improve [266]. However, many researchers may not be seriously pursuing pMOEA techniques because of various complexities involved in detailing parallel and distributed computation. Thus, discussion in this section and the following sections attempts to motivate a broader interest in developing pMOEAs by providing appropriate insight.

Although many approaches use a master-slave pMOEA to speed complex objective function evaluations, island pMOEAs are the most popular, just as is the case with parallel single-objective EAs [204, p. 49]. Why is this paradigm so popular? Is it because pMOEA developers love to fiddle with inputs and parameters? If so, this paradigm certainly gives many knobs and dials to turn, as beyond the usual EA parameters there exist those such as deme size, number of demes, deme interconnection topology, migration rate (how many individuals migrate and how often), identifying which individuals migrate, and those which are replaced. Another possible reason for this paradigm's popularity may be the island model's ease of implementation due to the easy integration of legacy MOEA code. Additionally, little interprocessor communication may be necessary, at least as compared to the other paradigms.

Note that current pMOEA papers generally fail to address the theoretical development of either the algorithm(s) contained within or their specific parallel aspects. Little or no discussion exists of why the selected pMOEA appears most appropriate for solving the given MOP. Additionally, few papers make any effort to explain why the MOP was even suitable for a parallel algorithmic solution. Few or no details regarding communication topology, migration, or selection are presented. A well-engineered pMOEA may likely be a "good" (in theory) integration of the problem and algorithm domains, which can be then fine-tuned for even better results. However, without adequate discussion of the utilized pMOEA and the MOP being solved one cannot evaluate the algorithm's quality or performance.

Few publications exist in which researchers adequately address pMOEA implementation concepts. Additionally, many important implementation details are generally missing. No clear explanation or justification for this situation is apparent. No theoretical or practical studies are yet known comparing the efficiency and effectiveness of major pMOEA paradigms when applied to the same problem or some test suite. These contributions are essential to support well-informed pMOEA implementation decisions. Additionally, the literature lacks sufficient background details, statistical studies comparing and contrasting results, and suitable metrics for use in judging pMOEA performance. These concepts are addressed in the following sections.

8.5.2 pMOEA Suitability Issues

pMOEAs currently give spectacular results in several disparate engineering design fields (at least according to their developers), but little or no discussion regarding the application's suitability for solution with a pMOEA is available. This begs further explanation of salient details from the problem and algorithmic domains, e.g., the factors making the problem so computationally expensive and the rationale for selected data structures. Search space discussions range from nonexistent to providing little insight. Is global knowledge of other solutions required in determining some solution's fitness? Is the value of one candidate solution dependent upon that of another? For a given problem certain solution restrictions or manipulations may be required; global knowledge of all evaluated solutions to date (i.e., clone avoidance) may be desired in attempts to improve performance. Background information such as this is vital in understanding both the applied pMOEA and its resulting performance.

Developers must closely examine the problem domain before selecting and implementing a pMOEA to ensure their expended effort has some promise of good or improved performance. One should carefully consider whether conditions are suitable for a pMOEA to (probably) find better solutions than a serial MOEA, just as one would do when pondering using a MOEA vice a single-objective EA. In some cases pMOEA implementation may be obvious, as when optimized parallelized code already exists for computationally expensive objective function evaluations. Given that code's existence, splitting the evaluations between processors using a master-slave approach seems obvious, but on the other hand, if a researcher can only access a network of workstations, then an island approach may be better as communication costs can be held to some desired level. This example illustrates that the researcher's goals and/or available computational platforms may be the deciding/limiting factors in selecting a suitable pMOEA paradigm for their particular application.

Sound engineering principles should be employed when implementing pMOEAs. A structured development approach emphasizes the careful integration of problem domain specifications with the algorithm domain solution, thus, initial efforts should be focused towards studying and understanding all relevant facets of the problem at hand before detailed algorithmic integration, execution and refinement (discussed in Section 8.6). These are the details now generally lacking in the literature yet are necessary for a better understanding of how to develop effective and efficient pMOEAs. If researchers begin to share their thoughts and decision-making processes regarding these pMOEA development issues, others can then profitably use that knowledge to improve results of their own applications.

8.5.3 pMOEA Hardware and Software Architecture Issues

Parallel and distributed EA/pMOEA references often indicate immersion in a particular problem environment or broadly discuss possible application areas. Such presentations are too superficial to be of much help in detailed implementation decisions, thus, prospective pMOEA developers should consult definitive detailed documents, texts, and experts specializing in appropriate areas.

In other words, *detailed pMOEA implementations generally require interdisciplinary team effort in order to obtain desired performance objectives.* For example, algorithmic characteristics such as data collection and interchange options, network dynamics, temporal characteristics, remote execution, and security issues lie in or cross several computational domains and require expert input to ensure effective and efficient products.

Standard parallel computer architecture models include Single Instruction, Single Data stream (SISD), Single Instruction, Multiple Data stream (SIMD), and Multiple Instruction, Multiple Data stream (MIMD) [926]. SISD architectures are standard single processor computers. Each processor in a SIMD architecture executes an identical broadcast instruction on different local data; this architecture might be useful for a diffusion pMOEA. MIMD architectures are useful for executing different EA/MOEAs with different data. Most MIMD homogeneous architectures and associated compilers reflect a distributed memory structure, this being probably the "more" generic environment for master-slave, island or hierarchical pMOEAs. Computational platform selection should also address the issue of shared memory in Symmetric Multi-Processor (SMP) architectures versus distributed memory in other architectures. Moreover, Internet implementations should be considered for pMOEA implementations using emerging computational grid concepts [810].

As described, various MIMD architectures are used in current pMOEA implementations. For example, master-slave and island model pMOEAs implementations in the literature are executed on distributed and shared memory systems for a multitude of providers. These range from large mainframes to heterogeneous computational grid of workstations, from large LINUX clusters to personal computers with multiple duo-processors. Master-slave, island, and distributed paradigms can be implemented on each. The increasing integrated chip speed and capability along with improved communication backbone performance continues to improve computational performance. This phenomenon should provide improved multiprocessor computational capabilities at the desktop, thus permitting in the future the local solution of more and more complex MOPs using MOEAs.

Common sense dictates some pMOEA paradigms are better-suited for execution on one given multiprocessor architecture than another. For example, diffusion pMOEAs can use a much greater number of SIMD processors (say greater than or equal to 128) for effective operation than other paradigms. An island pMOEA may execute quite well using as few as 4 processors. Thus, researchers' access to specific parallel computational platforms may well initially limit their choice of pMOEA paradigm(s) to employ. Consider also that the platforms' computational capabilities have tremendous bearing on pMOEA performance, both effectiveness and efficiency. These capabilities include processor speed, cache and local memories, and communication backbones. Commonly used high-speed backbones include Ethernet, Fast-Ethernet, Gigabit-Ethernet, Myrinet, Wulfkit, and the Fiber Distributed Data Interface (FDDI). These backbones connect processors in physical or logical geometric structures such as rings, meshes, toruses and hypercubes. Their associated communication software is then generally tuned or optimized for the given hardware configuration. Although users have parameter control over such variables as buffer size and placement, execution speed, protocol use, and so on, optimizing those values for a pMOEA application likely requires an expert.

From a programming standpoint, the choice of implementation language (e.g., FORTRAN-90, High-Performance FORTRAN, C++, C#, XML, etc.) is beyond this discussion's scope. Such decisions depend upon the individual programmer's knowledge, expertise, and specific computational hardware/software environment. Moreover, as previously discussed, interfacing existing physical model simulation packages to a pMOEA may require knowledge of different language communication interfaces. The novice pMOEA developer should study the many programming languages as well as available parallelizing tools that might be appropriate for the specific problem domain. Also, note the problem domain model's execution efficiency can depend upon the individual language compiler(s) and parallel communication library(ies) available in a given computational environment.

Interprocessor exchanges can be implemented via communication libraries such as the Message Passing Interface (MPI) [1230], or other specialized software such as the Parallel Virtual Machine (PVM) [549] or Open-MP [222]. All include communication routines that are readily incorporated into pMOEA implementations. Many are portable across a wide variety of homogeneous or heterogeneous parallel computer architectures or SMPs. Note the above are only some of many possible methods for controlling pMOEA execution; others normally associated with distributed systems include C++ sockets, JAVA, JAVA RMI (Remote Method Invocation), DCOM (Distributed Component Object Model), and CORBA (Common Object Request Broker Architecture), all of which use multi-threading middleware communication techniques [1042]. These protocols are constructed in a hierarchical fashion above the specific communications backbone. Associated middleware communication libraries may also open the Internet or other heterogeneous computational grids and sub-networks for pMOEA execution [517].

In order to have transparency at the level of implementation with computer languages and parallel communication libraries, a distributed MATLAB approach is suggested. This offers an easier but efficient method of using a pMOEA, although computation time could be longer than tuned parallel code. A distributed MOEA approach would employ the MATLAB distributed computing toolbox, together with the MATLAB distributed computing engine, which allows the execution of MATLAB code, either as a series of distributed tasks or as a parallel program, on a cluster of computers. A set of jobs is defined on a client machine using the distributed computing toolbox, and a job scheduler then sends tasks to a set of distributed nodes each having the MATLAB distributed computing engine installed. The primary advantage of using a MATLAB multiobjective evolutionary algorithm toolbox such as GEATbx in this distributed mode is that code can be run simultaneously on several nodes in one of the three parallel paradigms.

As a specific example, observe that the choice of heterogeneous or homogeneous systems requires careful consideration. In the reported case of a heterogeneous Sun workstation cluster, the master-slave pMOEA distributes one solution per workstation for evaluation; the master processor collates results and executes the algorithm's control components [1518]. A major issue faced by this pMOEA's designers is the network's heterogeneous composition (different workstation models and configurations) that gives rise to disparate machine performance. Additionally, the network's protocol automatically suspends remote computations when a user physically logs into that machine. The user also has the option of terminating any remote computations executing on their system. Thus, to deal with possibility of widely varying objective evaluation times and externally suspended or terminated evaluations, the authors implement unique pMOEA extensions to solve generational synchronization issues.

See Table 8.1 for a succinct listing of the major issues delimiting homogeneous and heterogeneous systems. Although considerable references are provided here, this general discussion of parallel hardware/software possibilities emphasizes the need for collaborative research when addressing high performance pMOEA design and implementation.

8.5.4 pMOEA Test Function Issues

pMOEA implementations are tested in order to evaluate, compare, classify and improve algorithm performance. Observe that these comparisons should mainly focus on the pMOEA's effectiveness and efficiency and not on which algorithm performs "best" on a particular problem. Valid tests may include pedagogical MOP test functions, a pMOEA test suite, combinatoric multiobjective problems, benchmarks, MOPs, and/or real-world problems. However, before constructing such tests one should first search the literature and evaluate historical test function use, test generators, and/or well-known real-world applications. Many researchers have already spent considerable time on constructing, evaluating and analyzing various MOP test suites [361, 1626]. MOPs may have a spectrum of characteristics that lend themselves to parallelization, which should be included in any specific pMOEA test suite.

Appropriate pMOEA tests should be developed and evaluated based upon validated assumptions, computational platform selection, available statistical tools, metric selection, and experimental design; this is most definitely an on-going process. Therefore, considerable effort must be spent not only in defining/generating proper tests, test suites, and experimental processes, but also in selecting appropriate test metrics and their associated statistical validations/comparisons. Statistical validation can be based upon max, min, mean, hypothesis testing, confidence intervals, student T-testing and Kruskal-Wallis testing [1668] (see Chapter 5). This point is stressed as many researchers identify statistical analyses of results as being extremely important, yet such analyses are still often lacking from many current EA, MOEA, and in particular, pMOEA publications [361, 581].

Many pMOEA research efforts initially select continuous numeric MOP functions as examples to show or judge algorithmic execution performance. In order to appreciate the rationale for such selections, a suitable discussion of MOP landscape issues is required along with explanations of why the selected MOPs may or may not be appropriate pMOEA test functions. MOP characteristics for discussion include objective function structures and complexity, constrained and unconstrained genotype/phenotype formulations, and the impacts of numerical approximation of continuous forms.

pMOEA Prototype Test Suite

Since general test suite "pros and cons" and guidelines for construction are extensively addressed in Chapter 5, only a brief highlight of major issues germane to pMOEA test functions is presented. To set the stage, note Holland's statement that

> using "sample" pedagogical problems is of little use in understanding EA performance when employed in solving complex real-world engineering, scientific design, and analysis problems [700].

In other words, sample problems are useful in comparing various EA performances, but those results may not provide useful insight into real-world EA applications due to problem domain characteristics and complexities. The same applies to MOP test functions used in comparing pMOEAs, thus, one should be very careful in defining and selecting MOPs for inclusion in some test suite. Moreover, the No Free-Lunch (NFL) theorems imply that if problem domain knowledge is not properly incorporated into the algorithm domain no formal assurances of the algorithm's general robust effectiveness exist [1708]. The NFL theorems additionally imply that incorporating too much problem domain knowledge into a search algorithm reduces its effectiveness on other problems outside that particular class; perhaps even within the class, robustness suffers!

Numeric test functions can be suitable representatives of real-world continuous MOPs. Currently, many modeled real-world problems are defined by a mathematical functional structure using a single optimization criterion. MOPs arguably capture more information about the modeled problem as they allow for incorporation and simultaneous consideration of several problem characteristics. Regardless, accurately modelling a real-world problem may involve several numeric or combinatorial functions, perhaps simple or perhaps complex. An MOP may contain continuous, discrete, or integer-constrained functions, or even a mixture of these types. Any acceptable test suite must then contain problems reflective of these aforementioned characteristics and possibly related to a selected complex application domain.

Since general MOEA test suite "pros and cons" and guidelines for construction are discussed in Chapter 4, the unique development of testing benchmarks for pMOEAs is addressed in this section. As discussed in Chapter 4, sample problems are useful in comparing various EA performances, but those results generally do not provide useful insight into real-world EA applications due to problem domain characteristics and complexities. The same applies to MOP test functions used in comparing pMOEAs, one should be very careful in defining and selecting MOPs for inclusion in some test suite. Moreover, the No Free-Lunch (NFL) theorems imply that if problem domain knowledge is not properly incorporated into the algorithm domain no formal assurances of the algorithm's general robust effectiveness exist [1708]. The NFL theorems additionally imply that incorporating too much problem domain knowledge into a search algorithm reduces its effectiveness on other problems outside that particular class; perhaps even within the class, robustness is sacrificed.

The MOEA research community continues to create and validate various test suites as discussed in Chapter 4; the pMOEA community is only beginning to develop such suites. It is not immediately clear that the test functions so far employed in the current MOEA literature are appropriate for inclusion in any generic pMOEA test suite. The community generally accepts that MOEAs are useful search algorithms when the problem domain has numerous decision variables and the search space is very large. But in the current literature many numerical examples do *not* explicitly reflect this multidimensional decision variable criteria. Additionally, of the many distinct pMOEA numerical test MOPs currently known, most use only two objective functions. This situation implies that unless the MOP's search space or landscape is very large, pMOEA performance claims and comparisons based on these functions may not be meaningful when considering real-world problems. The pMOEA may be operating in a problem domain not particularly well-suited to its capabilities or perhaps in one which is not challenging. For example, in his study of parallel single-objective EAs, Cantú-Paz was forced to construct artificial test functions; solving the initially selected functions in parallel did not provide enough computational loading to provide meaningful results regarding parallel speedup [204]. The same may well be true for MOEAs. Existing MOP test suites may be too simplistic to indicate likely pMOEA performance. The key is applying a pMOEA in a problem domain the algorithm is well suited for, one that may be much "harder" than some current test MOPs.

Researchers should also incorporate additional guidelines (suggested by Whitley et al. [1701]) in their generic test suite development (note the guidelines' extension into the parallel domain):

- Some test functions are *resistant* to simple search strategies; parallel approaches may yield "better" results.
- Test suites contain nonlinear, nonseparable, and nonsymmetric problems requiring the increased computational resources available in parallel processing systems.

8.5 pMOEA Analyses and Issues 483

- Test suites contain problems with increasing genotype dimensionality.
- Some test suite problems have scalable evaluation cost; this cost may also be separated and calculated on multiple CPUs.
- Test problems have a canonical or natural representation allowing easy use by a generic pMOEA.

In general, researchers should include functions both easy and hard for a pMOEA to solve. This helps identify the strengths and weaknesses of a pMOEA and possibly the particular parallel paradigm utilized. For example, one pMOEA strength might be directing an entire island (subpopulation) to focus on each promising search space area. Some problems might be solved more effectively by high inter-island communication; conversely, some problems might be hampered by that same high communication rate. Obviously, future research is needed in order to characterize function classes in which multiple populations are advantageous [204, p. 139].

Selecting appropriate test functions for testing and comparing pMOEAs is a difficult task. Numerous factors must be considered. The MOEA community has largely agreed on utilizing a variety of problems from two proposed test suites; several researchers are extending the state-of-the-art in this area and their efforts are continuing [361, 374, 1626, 1630, 720]. Many researchers agree that these test suites, composed of MOPs exhibiting various genotypical and phenotypical characteristics, are useful in making generalized statements about MOEA performance in solving MOPs. However, if the test's goal is to illustrate performance on a specific type or class of problem then the test suite should only be composed of MOPs from that class. These test guidelines and objectives are also applicable in pMOEA comparisons, thus, to better understand and validate any future pMOEA test function choices a limited historical perspective is presented.

Binh and Korn [139, 903] use three constrained and three unconstrained test functions but give no discussion regarding their suitability for pMOEA testing. All six problems have two objective functions and two decision variables. Sawai and Adachi [1434] use nine "classic" single-objective EA test functions (or close variants thereof); they note five of these were used in an international Evolutionary Computation contest. Again, no discussion as to their suitability for pMOEA testing is included. All of these problems use two objective functions and have two cases in which one case uses five decision variables and the other case ten. Hiroyasu et al. [687] implement five constrained test functions; two are three objective and all use two decision variables. Again, little discussion is offered regarding problem characteristics or function suitability for testing. Hiroyasu et al. [688] use four constrained test problems; three have two objective functions and one has three. Two problems have two decision variables; the other two have N (with the value of N not stated). They claim these functions address some range of easy to hard problems, but leave it unknown and unstated how this is true in a parallel sense. No one in the current literature has yet specified what problem characteristics might be pMOEA easy or hard.

Kirley [862] states his selected test problems represent two extremes: one relatively easy (with a convex Pareto front) and the other very difficult (with a multimodal front). Both of these unconstrained functions use two objective functions and are created using Deb's methodology [361]; one has fifty decision variables and the other one hundred. Again, it remains to be clarified as to whether a problem easy for a MOEA is also easy for a pMOEA. Finally, Zhu and Leung [1767, 1768] use three test problems, two of which are created with Deb's methodology [361]. Each function has two objectives where one uses three decision variables and the other two use thirty. Although the phenotype domain (Pareto front) is complex in these tests, observe that the genotype domain (Pareto optimal solution set) reflects a quite simple structure. Thus, although a wide variety of test functions are currently employed their use is not well-substantiated.

Based on the preceding discussion several known functions appear appropriate for initial pMOEA experimentation. Do not treat these suggestions as gospel; these functions must undergo extensive use and examination before inclusion in a permanent pMOEA test suite. Initially selected are unconstrained functions in order to clearly concentrate on pMOEA performance and not muddy the issue with questions regarding constraint-handling techniques. Subsequently, constrained functions and benchmark problems, as well as real-world MOPs, must also be selected in order to better determine overall pMOEA performance. Pedagogical, medium-, and large-dimensionality problems should be selected to stress tested algorithms.

8.5.5 pMOEA Metric/Parameter Issues

Is there any "real" difference between various pMOEA implementations? Does a given implementation generally perform better than another? These questions have no simple answer. Keeping the NFL Theorem in mind [1708], the lack of universal metrics prevents directly comparing different pMOEAs solving identical problems. Thus, appropriate measures or metrics must be selected and/or developed in order to analyze pMOEA efficiency and effectiveness when applied to a given test suite or application.

A major focus of current MOEA research efforts is identification of suitable metrics "truly" reflecting some code's efficiency/effectivess; several authors (see Chapter 5) propose and support using specific metrics to analyze and compare general MOEA performance. These metrics are not perfect, often due to the fact they are single-point mappings of multiple values. Taken in isolation they can easily mask or inflate some MOEA's true performance, although certain metrics are identified as better than others in certain situations [874]. However, as much as one might wish for mathematical certainty, as long as a derived/known Pareto front can be visualized, pMOEA effectiveness may still best be initially observed and estimated by the human eye and brain.

As the known pMOEA publications give no recommendations regarding specific metrics to use in analyzing *pMOEA performance* or in comparing various pMOEA instantiations, some MOEA metrics are then obviously useful and relevant when extended to pMOEA performance measurements [874, 883]. These include both relative and exact metrics dependent upon knowledge of PF_{true}. These metrics analyze PF_{known}'s cardinality and its associated dispersion through objective space. The reader should consult appropriate references as the selection of meaningful metrics depends upon the specific MOP being solved. Note also that additional metrics are necessary to analyze pMOEA performance in a parallel sense, so one can determine if any efficiency improvements exist as compared to a companion MOEA or single-objective EA. Taken together, proper metrics then indicate some pMOEA instantiation's utility when applied to a particular problem class.

Major parallel domain factors affecting execution time and/or resulting performance are the number of solution evaluations (problem size) and hardware/software system architecture, including memory size and the number of available processors. An additional factor affecting efficiency is the scheme employed to transfer problem data to and from the processors. Several general parallel computation metrics already exist and are commonly used by researchers to aid in measuring and judging some parallel implementation's performance. For example, the speedup metric captures the relative benefit of solving a problem in parallel [926]. T_s denotes the execution time of the fastest known serial (one-processor) MOEA implementation; of course one may not know the fastest true implementation. T_p denotes parallel run time and is the execution time for a given pMOEA implementation assuming identical processors and input sizes (search space size or number of evaluations). The speedup metric is then defined as $S = T_s/T_p$ [926, p. 118].

Scaled speedup, efficiency, cost, scalability and isoefficiency are also important metrics measuring parallel algorithm, and thus pMOEA, performance. Scaled speedup is defined as the speedup obtained when the size of the given problem is increased linearly with respect to the number of processors [926, p. 144]. Efficiency is a measurement of the fraction of time a processor is conducting work; it is calculated as the ratio of speedup over the number of processors [926, p. 120]. A pMOEA's cost for solving an MOP is defined as the product of parallel execution time and the number of processors utilized [926, p. 120]. A cost-optimal pMOEA is found when the parallel cost of solving an MOP is proportional to the execution time of the fastest-known sequential algorithm on a single processor [926, p. 120]. Scalability indicates the pMOEA's ability to increase its speedup as the number of available processors is increased [926, p. 128]. Finally, the isoefficiency metric determines how well a pMOEA maintains a constant efficiency and increases speedup as the number of employed processors increases. Such metrics can also be modeled by equations using detailed communication backbone values resulting in a symbolic complexity function useful for comparison with empirical measurements [926]. All these metrics are important when analyzing overall pMOEA performance. The pMOEA community must incorporate these metrics into their analyses to support any conclusions that a pMOEA achieves improved statistical performance versus that of a serial MOEA. Other relevant performance metrics and evaluation techniques may be found in the abundant parallel and distributed programming literature.

As for current state-of-the-art, note that few known pMOEA papers report performing formal parallel experiments and analyses, let alone their details. In fact, only two papers present a speedup graph charting their results [1769, 591]. Better understanding of the various pMOEA paradigms' effectiveness and efficiency, or that of some specific pMOEA instantiation, can only come through well-planned experimentation and extended statistical analysis of results.

Metrics specific to the parallel paradigm implemented may also be required when analyzing pMOEA results. For instance, the speedup metric applies to any of the paradigms, but if an island paradigm is utilized, one may wish to measure each deme's performance, the migration scheme's effectiveness, and/or some other measure specific to the island paradigm. Note the process of suitable pMOEA migration and replacement strategy selection is largely unaddressed and unanalyzed. Although Cantú-Paz presents a thorough analysis of several migration and replacement schemes for parallel single-objective EAs he does not address their extension to the multiobjective arena [204]. Effective single-objective EA/pMOEA schemes may radically differ and thus impact performance if unwisely employed.

The current situation does not allow for determining statistical performance differences between pMOEA implementations, not even how they perform when applied to some standard test suite. Additionally, it is readily apparent in studying the known citations that no *de facto* pMOEA parameter set (*parameters*, not their values) exists for reporting purposes. To clarify this, realize that for result reproducibility and comparison purposes most (MO)EA citations report key parameter values, e.g., crossover and mutation rates, selection methodology, etc. Few citations make any attempt to formally report these parameter values specifically describing a pMOEA implementation [688]; other papers might mention some subset of these parameters but do so only in passing.

Table 8.2 proposes key computational reporting characteristics for those researchers interested in exploring pMOEAs. Although *all* parameters are perhaps not always relevant, some subset is certainly necessary for a fuller understanding of the particular pMOEA paradigm selected, for repeatability of previous results, and for comparison purposes. Interesting comparisons might include determining the efficacy of MPI or PVM in a given pMOEA, or perhaps the performance of a master-slave and island pMOEA applied to the same MOP.

Another issue concerns reports of superlinear speedups in certain pMOEA experiments (e.g., [1769]). Claims of superlinear speedup (where parallel execution time is reduced by a factor greater than the number of employed

Table 8.2. Key Parallel Computational Characteristics

Key Characteristic	Description
Computer	Machine name (e.g., LINUX Cluster)
CPU	CPU Type(s) (e.g., Pentium 2.4 GHz)
# Nodes	Number of CPUs (e.g., 8, 16, 32, ..., 256, ...)
Memory	Memory per Machine (e.g., 256 GB)
Operating System	Name/Version (e.g., Red Hat Linux vx.x)
Communication Network	Network (e.g., Ethernet)
Communication Library	Library (e.g., MPICH vx.x)

processors) causes controversy. It is suggested that in general, fair comparisons between serial and parallel EAs may only result from each algorithm giving *identical* results; as EAs are stochastic algorithms comparisons should perhaps be based on *expected* solution quality. The above citation does not clearly state whether these considerations are taken into account. Cantú-Paz addresses the superlinear speedup issue in some detail [204, pp. 114-117]. Many researchers report superlinear speedups when parallelizing their EAs; one might expect pMOEA researchers to do the same. However, the real issue lies within comparing the serial EA's effort versus that of the parallel EA's. For example, one might attribute superlinear speedup to the fact that a pMOEA finds better solutions by examining a different number of solutions than a serial MOEA, or through the use of smaller phenotypical regions for parallel search.

To clarify the issue further, Cantú-Paz follows Punch [1302] and argues the main reason to distrust superlinear claims is that if one was to execute all of a parallel program's tasks using threads on a single processor, the total execution time cannot be less than that of a serial program performing the *same* computations [203]. The underlying assumption in these superlinear speedup claims is that the serial and parallel programs are executing the exact same tasks, which is typically not true for parallel EAs or pMOEAs. Researchers must be careful in utilizing parallel metrics to compare pMOEA performance.

A pMOEA's deterministic "work" can be argued to equal an associated serial MOEA's. The number of fitness evaluations can be equal and identical chromosomes could be created. Depending upon the objective function's complexity one may then obtain performance close to linear speedup. However, an island or diffusion paradigm's implementation requires statistically uncorrelated random number generators and the resulting landscapes are completely different than that of some serial MOEA. A pMOEA may then find better or equivalent solutions faster or possibly slower. Researchers should be very careful in claiming their pMOEAs exhibit superlinear speedup as the basic speedup metric may be ill-advised for most pMOEA paradigms. They should also clearly explain exactly *what* they are comparing in their speedup calculations.

8.6 pMOEA Development Issues

The preceding discussion lays a foundation to now address pMOEA development in some detail, whether one is building the parallel algorithm from scratch or modifying an existing (MO)EA. Obviously, the first question to ask is whether or not a MOEA is even suitable for solving the given MOP. If the answer is affirmative, one then thoroughly studies the MOP to determine whether a pMOEA truly appears the best algorithmic choice to generate candidate solutions. One must establish the particular conditions in which a pMOEA is *likely* to perform better than any algorithmic alternatives; if these conditions can't be easily determined it is then likely a pMOEA is not worth the time and trouble of implementing for solving this MOP.

Given that a pMOEA does appear appropriate the logical flow of an overall design strategy is illustrated in Figure 8.10. In a nutshell, an MOP solution process is seen as the study and specification of the problem domain (i.e., the problem's data structure(s)), study and specification of the algorithm domain (i.e., the problem's control structure(s)), integration of the two domains, and subsequent refinement to evolve an effective and efficient software design.

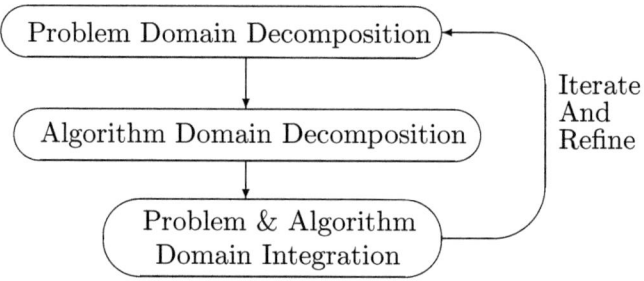

Fig. 8.10. Problem-Algorithm Domain Interaction

Serious consideration of the problem domain is the first step. Issues for examination here include data input/output, various constraints on solution states, and the set of candidate solutions. One must also consider whether candidate solutions meet constraint criteria (feasibility), how to extract one or more promising feasible candidate solutions (selection), how to define acceptable solutions (the solution set), and whether members of that set satisfactorily reflect the selected optimization criteria (the objective functions).

The next step considers the algorithm domain and is primarily concerned with input set(s) of candidates, output set(s) of solutions, and partial-solution set(s), those "working" solutions created/identified/tracked during algorithm execution. One should ensure the MOP is most suitable for attack by the EA class instead of some other (e.g., random search, branch and bound, depth-first search, breadth-first search, etc.). Any existing algorithm (with single or multiple objective functions) already being used to solve the MOP should also be mined for lessons learned. Other significant algorithm domain issues are the allowable or desirable operations upon candidate solutions; within pMOEAs these are the EVOPs.

After accomplishing these steps an intelligent decision can be made regarding the applicability and ease of implementing/employing a pMOEA. By this point one must surely know if the MOP lends itself to parallelism. Furthermore, although it appears MOPs of any complexity can be efficiently solved via the master-slave pMOEA paradigm, one must still understand the problem formulation and related details to see if additional parallel techniques or paradigms might be more fruitfully applied to produce better results.

Problem and algorithm domain integration is next. This step is where any necessary refinement occurs; the end result is a direct mapping of low-level designs to some selected computer language. But the pMOEA development process is not yet complete! Possibilities for error exist in any creative human endeavor and as a consequence, implemented code should be thoroughly checked for correct operation over the range of potential inputs. Algorithm refinement is another must. However, note that variations in algorithmic implementations may well imply different problem domain designs and structures; the reverse also holds. Thus, a series of data and control structure refinements may be required in order to reach an efficient and effective computer language implementation, which is this process' overall goal.

When considering data structures, the data required to compute objective function values, and when/where that data is needed during code execution, one must carefully examine how problem data may be assigned to and transferred between processors. The parallel system's architecture and communications backbone have significant impacts on interprocessor data transfer rates. One pMOEA paradigm may utilize them more efficiently than another can. Perhaps not obvious at first, the problem data's storage and transfer, as well as its temporal and physical requirements, also impact a pMOEA's efficiency and effectiveness. The integration step is where one should determine the best data and control structure(s) for the available parallel architecture(s). Of course, if only one parallel architecture is available the choice of structures and eventually the implemented pMOEA "flavor" may be constrained.

Instantiated pMOEAs may well benefit from applying one of the many available static or dynamic processor scheduling and load balancing techniques, e.g. [441, 926]. As pMOEAs are more often applied to real-world scientific and engineering problems where objective function calculation time is quite significant, these scheduling heuristics become more and more important [190].

One should also examine current utilization trends of the targeted parallel system. For example, many High Performance Computing (HPC) centers limit the number of processors or amount of wall-clock time some particular process may consume. These limits are often dependent on the number of users currently utilizing the system and the priority of currently running projects. These are important issues as researchers may find greater overall processing time available on a local cluster than on an HPC supercomputer. Even though an HPC facility may have more computational resources and greater raw processing power, the amount of work one may actually see their program complete in a given time at an HPC may be orders of magnitude less than the same program executed on a local cluster. This is not meant to discourage utilizing HPC resources but instead to aid researchers in making educated decisions regarding specific computational platforms to utilize.

In general, several issues arise during pMOEA design and implementation not particularly relevant to parallel single-objective EA development; they present additional challenges for researchers already familiar with parallelism and MOEAs. Some are initially discussed in Section 8.3 regarding specific pMOEA paradigms. These and other issues, and suggested solutions, are presented below in order to better the problem and algorithm domain integration process.

Researchers should note that no currently known studies (either theoretical or practical) compare or contrast the major pMOEA paradigms' efficacy in solving various test or real-world problems. Any guidelines suggested here must then be taken with a grain of salt. As with any parallel algorithm development one must carefully consider the issues raised by the proposed architecture's communication backbone and physical capabilities. Although the literature is largely silent on this point one must also examine how parallelization affects MOEA performance. For instance, are more evaluated solutions in a given length of time desired (a master-slave implementation) or is the focus on discovery and exploration (what solutions arise from several simultaneously executing MOEAs in an island implementation)?

The following discussion gives general guidelines for implementing one of the major pMOEA paradigms given some specific system configuration. Researchers may also possibly utilize these ideas to combine concepts from multiple paradigms into a new pMOEA.

8.6.1 pMOEA Creation Options

Based on the assumption a pMOEA is desired and well-suited for solving some MOP, four major development options exist.

1. **Parallelize an Existing MOEA.** Parallelizing working MOEA code is an attractive option. The process of efficiently parallelizing an existing MOEA does involve knowledge of parallel processing techniques, parallel routines and libraries, and the major pMOEA paradigms. However, by applying the concepts presented in this discussion collaborative developers can then intelligently extend existing MOEAs into the pMOEA domain.
2. **Utilize Existing pMOEA Code.** Researchers may utilize existing pMOEAs in order to reduce algorithmic development time (or cost). While in general this seems a good idea caution is required. Sound engineering

practices are recommended when modifying an algorithm to operate in different problem domains, as well as in different parallel environments. The key issue is understanding the MOP at hand and pMOEA (the problem and algorithm domains) and the physical computational platform; without that understanding adapting a pMOEA to effectively and efficiently solve a new MOP is unlikely.
3. **Design a New pMOEA.** Designing and implementing a new pMOEA is an exciting opportunity. This process allows more freedom in code implementation (e.g., ignoring pre-existing data structures) and also allows for incorporating interesting search concepts and new EVOPs. A new design could possibly provide efficient execution across a variety of parallel architectures but does involve a development time cost.
4. **Extend an Existing Parallel EA.** Extending a proven existing parallel EA to a pMOEA may appear relatively easy but in actuality is a process requiring much thought. Modifying a parallel EA to utilize additional objective functions may not be difficult, however, incorporating migration and replacement schemes as well as other EVOPs (e.g., niching) are not trivial tasks.

8.6.2 Master-Slave Implementation Issues

The master-slave pMOEA may be the simplest to understand and implement (see Section 8.3.1 for details). Generally used for splitting objective function evaluations among several processors (see Figure 8.4), this paradigm appears especially useful for computationally expensive objective function evaluations (e.g., those often found in CFD/CEM problems and where parallel codes may already exist). Sub-problems may also be solved by separate slave processors with final values being coordinated/computed by the master processor.

Implementation details are relatively minor. One should first determine that objective function calculations are suitably complex to justify pMOEA implementation. This can easily be done by comparing the wall-clock time (estimated or actual) required to compute objective function values for the entire population against the communication time spent by all processor units. The issue here is ensuring each processor receives enough work so that communication costs do not overwhelm computational costs. Using equation (8.1) this is determined by ensuring in a given generation that

$$PT_c << \frac{n \sum_{i=1}^{k} T_{f_i}}{P} . \tag{8.6}$$

In other words, additional processors are useful only to the point that each processor's idle time remains small compared to the interprocessor communication time. If a processor is starved for work or its computational time is of the same magnitude as the communication time necessary to conduct the work in parallel, a performance degradation may then result compared to the serial implementation. Figure 8.11 presents the experimental results of a two-node island implementation with each island using from zero to eight slave nodes (termed *farming nodes* in the experiments) for calculating objective function values.[5] The y-axis shows execution time and the x-axis shows the test number, where the test's required execution time increases with each test.

In the figure, one sees a significant decrease in processing time as the number of slave nodes per island goes from zero to one. This is due to the large computational effort necessary to complete objective function calculations now being shared by more processors. As the number of slave nodes per island increases to two and four, further minor improvements are seen even as interprocessor communications time becomes a larger share of the overall execution time. However, the increase from four to eight slaves is detrimental as execution time becomes worse than in the single slave case. At this point communication expenses are overwhelming computational costs due to the population's partitioning across the slaves. With more slaves in each island node, each is evaluating fewer members, and communication time (overhead) becomes a larger percentage of the total execution time. This results in each slave being under-utilized.

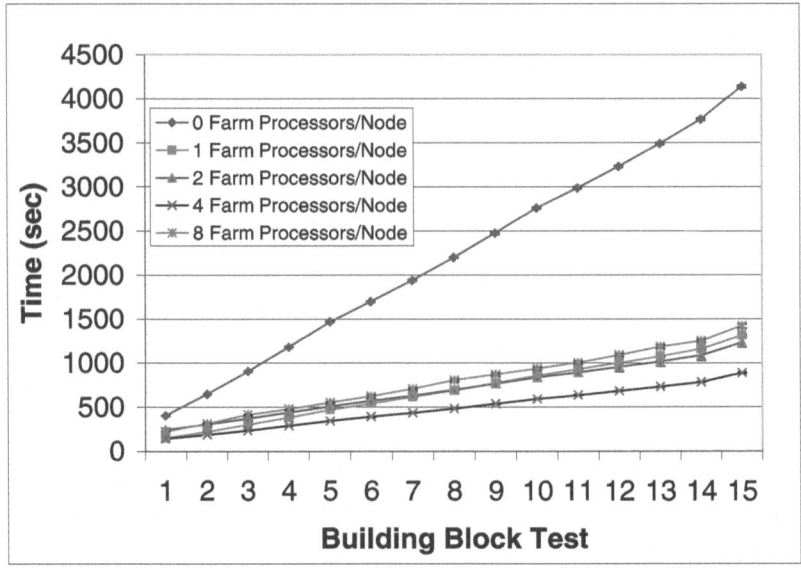

Fig. 8.11. Island Paradigm Execution Time Example

Generational synchronization should not be a large issue as the typical master-slave approach evenly distributes population members across all slaves.

---

[5] The reader is referred elsewhere for the experimental details; the results are important here because of what they illustrate [347].

Each receives a (set of) chromosome(s) along with other necessary parameters; each slave returns k objective function values for each chromosome, resulting in relatively small data sets passing between the master and slaves. However, researchers must note the worst case synchronization time in generational master-slave pMOEAs is equal to the evaluation of the most "computationally expensive" individual on the slowest processor used. If solution evaluation costs vary widely among utilized processors, idle processor times ranging from seconds to hours could result.

8.6.3 Island Implementation Issues

The island paradigm may be implemented in numerous ways but is generally used to simultaneously execute several individual MOEAs. This is perhaps most useful for concurrently examining different areas of the search space or a larger amount of the search space in the same wall-clock time as a serial MOEA, for investigating effects of varied algorithmic parameters (e.g., population size and EVOPs), and/or to determine how employed migration/replacement policies affect the MOP solution process.

When considering communication costs it is instructive to look at the island paradigm's extremes. One end of the spectrum is represented by separate MOEAs simultaneously executing on some number of processors with no communication occurring between nodes until each MOEA completes and its respective PF_{known} is transmitted to process "0;" these local sets are combined to arrive at the global front discovered. The other extreme occurs when each processor performs a one-to-all broadcast of each population member's chromosome and fitness at each generation's end. The former option makes no use of the information available to improve search; the latter is overwhelmed with the global 'state of affairs.'

A more typical implementation involves periodic migration of selected individuals among selected processors throughout algorithm execution. This can be conceptually viewed as a single pMOEA population divided among numerous processors. An identically-sized population as some typical serial MOEA can reside on each processor to keep pMOEA run time approximately the same as its serial counterpart. This is not a requirement, however, and population size can be set to whatever number of individuals is indicated. Basic migration and replacement schemes have been discussed in numerous parallel EA papers; the interested reader is referred to Cantú-Paz' thorough study for further information and background [204].

It must be noted that implementing pMOEA migration and replacement strategies is far more complex than for parallel EA instantiations, which may explain why the known literature fails to address these issues in any detail. While these topics may seem easy in concept, in actuality, a number of pMOEA-related issues arise making them more complex. A discussion of all possible single-objective EA schemes is not presented here due to the large number of possibilities and because there are MOEA-specific issues making some irrelevant.

Numerous pMOEA migration schemes and subsequent replacement strategies are available. A variety of these are structurally different from one another and are thus classified as such in this presentation. Table 8.3 presents a concise listing of potential pMOEA migration schemes and some of their attributes; analogous replacement schemes are presented in Table 8.4. Note that both migration and replacement focus on each islands' local Pareto front ($PF_{current}$), as identifying an approximation of PF^* is the *raison d'etre* of (p)MOEAs. As each $P_{current}$ contains locally known optimal solutions they must be the focus of identifying the appropriate information to share between islands. Note also that these identified strategies apply to *each* utilized processor.

Table 8.3. pMOEA Migration Schemes

	Scheme	Attributes	Selection Pressure
Non-Uniform	Elitist (random)	Migrate a random sample of individuals (equal to some percentage of the population) from $PF_{current}$ $\binom{p}{t\_bm\_x}$	Relatively High
	Elitist (niching)	Migrate a "uniform" distribution of individuals (equal to some percentage of the population) from $PF_{current}$ $\binom{p}{t\_bm\_x}$	High
	Elitist (front)	Migrate the entire local front, $PF_{current}$ $\binom{p}{t\_bm\_x}$	High
	Elitist (front+)	Migrate the entire local front, $PF_{current}$ $\binom{p}{t\_bm\_x}$, plus some number of randomly selected individuals or some number of individuals randomly selected from the ranked Pareto fronts	High
Uniform	Random	Migrate x randomly selected individuals	Low
	Elitist (random)	Migrate x randomly selected individuals from $PF_{current}$ $\binom{p}{t\_bm\_x}$, randomly selecting individuals from the ranked Pareto fronts if necessary	Relatively High
	Elitist (niching)	Migrate x "uniformly" distributed individuals from $PF_{current}$ $\binom{p}{t\_bm\_x}$, "uniformly" selecting individuals from the ranked Pareto fronts if necessary	High

In the ensuing discussion a number of parameters may be specified. Two of most interest here are the number of population members to migrate (x) and the destination processor(s) (p_{dest}). The former value is of interest as migrating too many individuals negatively impacts communications cost and may force receiving processors to converge to identical solution sets. A trade-off must be made between exploration and exploitation; in single-objective parallel EAs it is typical to migrate a low percentage of the total number of individuals in the population and there is not immediate evidence to suggest changing that in pMOEAs. The process of selecting the destination processor(s) for immigration is also of high interest. As mentioned above, a total one-to-all broadcast involves every processor migrating individuals to every other processor but this potentially (probably?) overwhelms the communications backbone as well as forces all processors to search much the same space. Conversely, not employing migration defeats the purpose of implementing an island pMOEA, but not surprisingly, just as is the case with EA parameters in general, there exists no "best" method for choice of p_{dest}.

For purposes of the following discussion let a given processor have a neighborhood of size N. At specified migration intervals each processor migrates selected individuals to its N "closest" neighbors. Note the actual definition of "closest" neighbors often depends on how processors are arranged in various logical topologies (e.g., ring, mesh, hypercube) or according to a specific physical hardware configuration. A neighborhood can contain any of the system's processors, and can be a static as well as a dynamically changing list.

The migration schemes in Table 8.3 are decomposed into two categories: non-uniform and uniform. The *uniform migration schemes* are labeled as such since a constant number x of population members migrate with each event, whereas the *non-uniform migration schemes* involve a varying number of migrating members with each event. Uniform schemes are advantageous in that the algorithm's incurred communication costs are consistent and predictable. However, non-uniform schemes have potential advantages in relation to anticipated pMOEA efficiency and effectiveness.

The simplest scheme to choose migrants is a purely *random* one, randomly selecting x population members from a given processor for migration to its N neighbors. This strategy provides low selection pressure as no guarantee exists for any members of $PF_{current}\left(^{p}_{t\_bm\_x}\right)$ to be migrated.

This is an obvious option; more thought is required to define and implement *effective* migration strategies. Parallel single-objective EA optimization allows for easy selection of the best x population members or a top percentage of the population for migration; these methods are termed elitist. Migrating the best individuals has an associated high selection pressure as this method could cause other processors to converge to (closely) identical solution sets, especially if a destination processor lacks any "better" members. But the concept of "best" takes on a different meaning in MOEAs as there is generally no single best individual solution. Identifying and selecting the top x solutions for migration is meaningless within a MOEA as all members of any given Pareto front are equally optimal (in the Pareto sense). Thus, one can randomly choose x members from $PF_{current}\left(^{p}_{t\_bm\_x}\right)$. This is an *elitist random* migration method as a portion of the best local members are utilized. However, this method has the potential to migrate members from only a concentrated region of $PF_{current}\left(_{t\_bm\_x}^{p}\right)$ and hence not provide a good representation of the current local optimum. An additional problem arises if if $|x| > |PF_{current}\left(_{t\_bm\_x}^{p}\right)|$.

In this latter case one can send fewer solutions than $|x|$, all members of $PF_{current}\left(_{t\_bm\_x}^{p}\right)$, migrate $PF_{current}\left(_{t\_bm\_x}^{p}\right)$ plus a random selection of the remaining population, or migrate all members of $PF_{current}\left(_{t\_bm\_x}^{p}\right)$ plus members existing in the successively ranked Pareto fronts until x members are identified.[6] Migrating members from multiple fronts is anticipated to achieve the quickest convergence due to this method's "strong" elitism. However, a possible disadvantage lies in the computational overhead required to determine successive Pareto fronts until x members are identified. This then adds to migration's computational complexity as determining Pareto optimality is generally of order $O(k \times n^2)$ [265]. To implement this proposed elitist random method, if all x members cannot be found in $PF_{current}\left(_{t\_bm\_x}^{p}\right)$, the rest should be selected either randomly from the remaining population or from the remaining successively ranked Pareto fronts.

Another uniform elitist migration method, termed *elitist niching*, utilizes niching in conjunction with individual selection. As opposed to randomly selecting x individuals, niching allows for (the possibility of) migrating a fuller representation of $PF_{current}\left(_{t\_bm\_x}^{p}\right)$, not just a localized one. This is accomplished by selecting a "uniform" distribution of individuals from the local Pareto front. If all x members cannot be found in $PF_{current}\left(_{t\_bm\_x}^{p}\right)$, the rest are selected from the remaining successively ranked Pareto fronts in the same manner.

Moving on to non-uniform migration schemes, a conceptually simple method termed *elitist front* migrates the entire locally known Pareto front $(PF_{current}\left(_{t\_bm\_x}^{p}\right))$ from each processor. This method is non-uniform because there is no way to a priori determine $PF_{current}\left(_{t\_bm\_x}^{p}\right)$'s size; it varies both by processor and by generation. This method has the greatest selection pressure. A modified version of this scheme, termed *elitist front+*, migrates the entire local Pareto front with an additional number of solutions either selected randomly from the remaining population or from the remaining successively ranked Pareto fronts. This method has the advantage of migrating an elitist set with some additional individuals for diversity; a disadvantage is unknown and possibly high communication costs. In an attempt to prevent premature multiobjective convergence the elitist method with niching (*elitist niching*) selects a "uniform" distribution of individuals (equal to some percentage of the population) across $PF_{current}\left(_{t\_bm\_x}^{p}\right)$. The *elitist random* method simply selects a random number of individuals (equal to some percentage of the population) from $PF_{current}\left(_{t\_bm\_x}^{p}\right)$.

Regarding these random and niching methods (whether uniform or non-uniform), note that randomly selecting individuals has the least processing

---

[6] The remaining successively ranked Pareto fronts are determined using the NSGA/NSGA-II methodology [374, 1509].

8.6 pMOEA Development Issues

overhead as niching calculations are not necessary. Niching complexity is high since some distribution density measure must be calculated for all local Pareto front members in order to select a "uniform" distribution. Although this complexity may preclude implementing a niching scheme, its potential benefits may outweigh the additional computational overhead.

Just as multiple pMOEA migration methods exist, so also do multiple methods for replacing individuals on receiving processors. Replacement strategies generally assume population size remains constant between generations. Without this requirement each processor's population size can quickly grow. Even with a partial replacement scheme, where some number y, $y < x$ of population members are replaced with migrants and the remaining immigrants added to the population, overall population size still increases. So then does computational cost, especially when dealing with multiple objectives. Thus, the assumption is made here to hold each processor's population size constant. The x individuals replaced on each receiving processor are "thrown away" after the migration event. Table 8.4 presents a concise listing of replacement schemes holding the greatest potential and some of their attributes.

Table 8.4. pMOEA Replacement Schemes

Scheme	Attributes	Selection Pressure
Random	Randomly replace x individuals	Low
None	No replacement, yields increasing population size	Medium
Elitist (random)	Maintain $PF_{current}\left(^p_{t\_bm\_x}\right)$ and randomly replace individuals $\notin PF_{current}\left(^p_{t\_bm\_x}\right)$	Relatively High
Elitist (ranking)	Rank all Pareto Fronts and replace individuals from the "worst" ranked front(s) with the immigrants	High
Elitist (100% ranking)	Combine immigrants with the current population, rank all Pareto Fronts and remove individuals from the "worst" ranked front(s)	High

As before, assume each processor has a neighborhood of size N. At specified replacement intervals each processor replaces selected individuals with those received from its N neighbors. Just as for migration, the simplest replacement scheme is a *random* one in which immigrants randomly replace individuals in the target population. This scheme has the lowest selection pressure as no guarantee exists regarding the quality of solutions being replaced, similar to that presented in [204]. For example, a solution might not be a member of the resulting Pareto front $\left(PF_{current}\left(^p_{t\_am\_x}\right)\right)$ on the receiving processor, yet it could potentially replace a solution that would have been on that front. To actually determine that fact requires combining the destination processor's population and immigrants, then ranking this newly formed population. If concerned, researchers may spend some (too much?) overhead processing to determine this situation's existence. Otherwise, researchers may lose desirable solutions and over time the population may then diverge rather than converge to optimal solutions.

Non-random replacement strategies likely lead to increased execution times. The major issue with pMOEA replacement strategies really centers on ranking population members in an efficient manner so an "elitist" replacement strategy can be implemented. Generally speaking, since all processors perform migration, $PF_{current}\left(_{t\_bm\_x}^{p}\right)$ is already calculated. However, replacement most likely changes $PF_{current}\left(_{t\_bm\_x}^{p}\right)$ on each target processor resulting in $P_{current}\left(_{t\_am\_x}^{p}\right)$. No guarantee exists that the local pre-/post-migration Pareto fronts are identical, i.e., $PF_{current}\left(_{t\_bm\_x}^{p}\right) = P_{current}\left(_{t\_am\_x}^{p}\right)$. Thus, the question is whether to compute a local front composed of the union of $PF_{current}\left(_{t\_bm\_x}^{p}\right)$ and the immigrants, or to keep all members belonging to $PF_{current}\left(_{t\_bm\_x}^{p}\right)$ and replace other individuals. The following replacement schemes are all elitist as they rely on some type of ranking or replacement to ensure the best individuals are kept for future generations.

One strategy termed *elitist random* keeps all members of $P_{current}\left(_{t\_bm\_x}^{p}\right)$, determining which of the remaining population is not Pareto optimal with respect to the immigrants, and then randomly replacing x of those individuals with the immigrants. A disadvantage of this strategy is computing which existing solutions' objective vectors are dominated by the immigrants.' Additionally, problems arise if the set of remaining members is smaller than the immigrant set, i.e., $|P_{op}S_{ize} - PF_{current}\left(_{t\_bm\_x}^{p}\right)| < |x|$. This method does not guarantee the worst individuals are removed but does ensure $PF_{current}\left(_{t\_bm\_x}^{p}\right)$'s retention.

A strategy termed *elitist ranking* has increased selection pressure and replaces individuals constituting the worst ranked Pareto front. Ranking the fronts increases computational complexity and as before, individuals must be randomly selected from some front(s) if the number of individuals to be replaced does not match the number of individuals represented by the front being replaced. For example, assume some processor contains four separate fronts with $|F_0| = 12$ (F_0 is $PF_{current}\left(_{t\_bm\_x}^{p}\right)$), $|F_1| = 8$, $|F_2| = 6$ and $|F_3| = 4$. Unless $x = 4, 10, 18$ or 30 some random replacement(s) is then required. This method provides the greatest selection pressure as the "worst" individuals are continually removed from the population, however, it likely incurs the greatest computational/time expense spent in identifying the nondominated fronts [265]. A slight modification to this method gives *elitist 100% ranking* in which immigrants are combined with the current population and the resulting population ranked as above to determine all Pareto fronts. The resulting Pareto front ($PF_{current}\left(_{t\_am\_x}^{p}\right)$) is kept and individuals in the "worst" fronts are discarded. This scheme offers a slightly higher selection pressure. Note that problems may arise here if $|PF_{current}\left(^{p}_{t\_am\_x}\right)| > |P_{op}S_{ize} - x|$, i.e., the resulting local Pareto front does not leave enough individuals in the population for discard.

pMOEAs may also employ dynamic migration and replacement schemes. Dynamic methods have a possible advantage if one has some knowledge of the fitness landscape or of how one wishes search to progress. The migration/replacement schemes presented here may also be combined in multiple ways. Decisions on which schemes to pair are likely driven by their computational cost, but one must tradeoff this cost against resulting effectiveness gains. Other EVOPs, such as mutation and crossover, can also be defined *a priori* for each island [204] or can dynamically change [18]; different islands may use static or dynamic settings.

As with many other search algorithms a local search operation such as hill-climbing may also be utilized. This may increase the quality of MOP solutions (i.e., PF_{known}). Periodic use of local search operators throughout algorithm execution may be beneficial; it remains for researchers to tune their implementations for their intended applications.

8.6.4 Diffusion Implementation Issues

The diffusion paradigm appears the least discussed paradigm in the EA and (p)MOEA literature [265]. This is possibly due to its associated complexity and potentially higher communication costs as compared to the other paradigms (see Figure 8.8). This model uses one conceptual population distributed across a number of processors, each typically holding one to a few individuals. Topological and neighborhood structures are imposed over the processors with EVOPs operating only within some specified neighborhood. Due to most EVOPs' serial nature, much communication is necessary to execute them as they may require complete knowledge of an entire neighborhood.

Few diffusion parallel EAs and pMOEAs are found in the literature [204, p.140]. Much work must yet be accomplished before achieving a good understanding of this paradigm's strengths and weaknesses. This paradigm does appear a good choice when paired with shared memory and SMP systems. Furthermore, although no proof is offered, some have suggested this paradigm allows formation of local niches where each niche represents a different way of trading off the k objectives.

A basic diffusion paradigm involves a pMOEA executing on a number of processors where communication is restricted and occurs only between processors in a specified neighborhood. The neighborhood structure may be static or dynamically change over time. In either case the neighborhoods overlap to some degree, allowing for a slow diffusion of desirable individuals throughout the entire processor pool. The only constraint is that the possibility must exist for some individual to reach any processor used within a reasonable number of generations.[7] The diffusion process continues until some specified stopping criteria is met.

Shared memory systems are ideal for diffusion pMOEA implementations as each processor has access to the entire population through a very high speed backplane, which is likely orders of magnitude faster than any current network communications structure. A disadvantage is that shared memory systems typically require using propriety communication calls in order to achieve desired performance levels; these calls cannot be easily ported to other environments. This may discourage researchers wanting efficient and portable pMOEA codes.

Implementing a diffusion pMOEA on a shared memory system is fairly straightforward. Using an SMP-based cluster is a bit more complicated and should likely use a neighborhood size equal to the number of resident processors in each SMP unit, allowing one to fully realize the superior communications backbone capability within an SMP unit. When utilizing a cluster of multiple networked SMP nodes in executing a pMOEA the communications picture becomes more complicated, mainly because inter-SMP unit communication time is much higher than that occurring intra-SMP.

One possible diffusion implementation operates by dividing the population into small groups of individuals and assigning those groups to different processors. Another implementation involves a diffusion paradigm operating as part of an hierarchical pMOEA. With some neighborhood structure imposed on the overall SMP cluster, an island paradigm treats each SMP unit as a deme, and a diffusion paradigm operates within any given SMP unit. Other variants of this general system design are certainly possible. These examples are provided only to ease understanding of how one might implement an efficient diffusion pMOEA on a non-shared memory based system.

8.6.5 Parallel Niching Issues

Numerous researchers highly suggest using niching in any (p)MOEA, but how to optimally implement parallelized niching is a currently open question (e.g., see [265, 361]). Parallel niching is addressed here as it is another area of discussion lacking in the current literature, even though the concept is acknowledged to be of high importance in obtaining high-performance (p)MOEAs.

Niching, crowding, clustering. and σ_{share} are terms widely used within the MOEA community when discussing the concept of finding a good distribution of vectors along the Pareto front or solutions among the Pareto optimal solution set. The many methods for accomplishing MOEA niching are not discussed in detail here; the interested reader is referred elsewhere [265, 361].

---

[7] In other words, all destination processors must be reachable within a small number of generations (small as regards the total number of generations during pMOEA execution).

The concept of parallel niching, however, is fairly new. The underlying question is how to parallelize niching for pMOEAs and what, if anything, is gained by doing so.

Niching in either the phenotype or genotype domain involves analyzing some Pareto front or Pareto optimal solution set, and attempting to achieve uniform knowledge of the set. In other words, one tries to avoid a clustering of solutions or vectors in any one particular region. Niching typically occurs in the phenotype domain since most MOEA researchers desire a uniform distribution of points along the (known) Pareto front [265]. Phenotype niching involves removing solutions from a given population, solutions corresponding to vectors lying within some user-defined distance of others on the Pareto front. New solutions are then generated in hopes of filling in a sparse frontal region, i.e., gaining knowledge of a previously or unsatisfactorily unexplored area of the MOP's Pareto front. Note that solutions *cannot* be generated directly by the process of analyzing some front, selecting a region with a sparse point distribution, specifying objective function values in that region and then performing a reverse mapping to find the corresponding decision variable values. One typically does not know the reverse mapping for objective functions of any real difficulty. Genotype niching involves analyzing some Pareto optimal solution set and then generating new solutions based on movements away from previously identified solutions, which is nothing more than a clustering process [265].

Niching is basically a serial process and is thus difficult to efficiently parallelize. Only a master-slave pMOEA can perform niching in the usual manner; the other paradigms require different niching processes (implementations). An island pMOEA offers several niching options. For example, a separate niching operator can execute on each island. There may be additional utility in executing a variety of niching operators across the different processors. These methods do not require extra communication, however, one may well wish to transmit locally known Pareto front information between (some set of) processors in an attempt to perform global niching.

Another parallel niching process divides the known front or solution set into a number of static regions equal to the number of available processors, so that each processor concentrates search in that specific region. At specified intervals, the currently known global front is generated by combining currently known local fronts and the global population is then redistributed according to the restricted search regions assigned to each processor. This method appears to be fairly straightforward, but in fact, is so only for those members on the local Pareto fronts. Dominated members do not map as easily to the imposed regional structure and then require utilization of some ranking method in order to determine each processor's mapping. Ranking the entire population allows for determining which region each dominated member belongs to. This method requires additional processing and communication overhead as each processor's entire population is redistributed amongst all processors. Another potential issue encountered here is the possibility for all of the combined populations to be allocated to a single processor; the remaining processors then receive zero individuals if a static regional scheme is specified *a priori*. A simple solution allocates the individuals based on where they currently are in the landscape, i.e. utilizing a dynamically changing regional structure. This method may be more useful when employing genotypic-based niching. Here, one may partition the genotype space into regions and allow the processors assigned to those regions to explore some space ($\pm \epsilon$) outside of the region. This allows the processors the possibility of finding optimal solutions outside of the currently known and defined areas.

In order to minimize parallel niching's communications costs neighborhood constraints can be imposed on each generation of new solutions. In other words, a given processor is restricted to generating solutions corresponding to vectors within its defined region, and continues generating solutions until enough acceptable ones are generated, discarding the others. This means the generation of specific frontal areas to analyze is restricted to processors within a given neighborhood, similar to the diffusion paradigm. This implementation effectively reduces overall communications costs and allows more of a parallel searching of various frontal areas, but it may increase overall execution time as each processor must continue generating solutions until it reaches the specified number of vectors in its region. Another viable method reducing communications overhead is communicating only every epoch (one epoch corresponds to a specified number of generations), thereby reducing overall communications cost yet still achieving the goal of global niching. Dynamic niching is another promising scheme involving creating variable-sized niches that (possibly) change over time. Additional constraints must be imposed so as to focus more on exploring or exploiting during certain pMOEA phases.

When utilizing parallel niching in a pMOEA one must balance the requirements for a satisfactory representation of the Pareto front and Pareto optimal solution set, and the time required to execute the algorithm. Parallel niching may improve final pMOEA performance but may also drive communications cost to unacceptable levels if not implemented well, thus, the underlying hardware and communications backbone must always be carefully considered. With some forethought the concepts presented here can also be applied in either the island or diffusion paradigms. With an island pMOEA, the best choice for parallel niching may be using a local niching operator or niching operators using different parameter values across the processors. On the other hand, especially when implemented on some type of shared memory system, the diffusion paradigm appears well-suited to handle a parallel niching operation because each processor has fast access to the entire population and can then perform global niching fairly efficiently.

8.6.6 Parallel Archiving Issues

Archiving strategies are another area of great interest. They have the potential to increase pMOEA efficiency if intelligently addressed. The concept behind archiving is maintaining a subpopulation or external database of the best members yet found by the pMOEA. Issues then arise of how to maintain the archive, how many archives to maintain, and how often to communicate information between the archives. One method of pMOEA archiving is having each processor maintain unique archives and at the conclusion of the search process have one processor combine the archives and present the final Pareto front or Pareto optimal solution set to the user. More complicated strategies exist and may increase pMOEA effectiveness, dependent upon their particular archive use. For those pMOEAs maintaining an archive and actively utilizing an archive(s) throughout the search process, communication between archives may be beneficial. However, a potential issue arises when combining the archives each generation, in that this action may lead to pre-mature convergence in some processors. In terms of actual archive implementation, synchronization is required to ensure multiple processors do not try and update the archive at the same time, which may increase algorithm execution time but in certain pMOEAs has the potential to increase their effectiveness.

8.6.7 pMOEA Theory Issues

When considering standard EAs many researchers have contributed to a body of EA theory regarding a variety of characteristics. Such theory encompasses issues such as representations, operators, algorithm convergence, equivalence, etc. Specific theories associated with diffusion and island paradigms relating to selection pressure and neighborhood size exist. For parallel EAs some extensions have been made regarding population sizing [204]. For pMOEAs little has been done except for the population sizing [1788]. The field must develop formal modelling insights as to pMOEA operator and parameter value impact as well as the impact of various computational architectures across classes of MOPs. Use of the notation in Section 8.2 is suggested.

8.7 A "Generic" pMOEA

The previous sections discuss major issues in pMOEA development and testing. A case study is now presented that applies these concepts to create a generic pMOEA (in that it can be executed in a master-slave, island, or diffusion mode). Note this generic pMOEA's evolution is an excellent example of the design strategy suggested in Section 8.6.

Pseudo code for a generic Master-Slave, Island, and Diffusion pMOEA implementation is presented to provide insight into pMOEA creation using any of these paradigms. This pseudo code is at a fairly high level as exact implementation details can vary based on one's selection of specific EVOPs, migration and replacement schemes, hardware, parallel libraries, and so on.

In these generic pMOEA formulations the following variables are of importance: p is the process ID, n is number of processes, N is the neighborhood size, P is the Master-Slave's population, and P' is the Island or Diffusion's population size (which could be P or $\frac{P}{n}$). The term process is used instead of the term processor as multiple processes can execute on a single processor. For simplicity and generality, it is assumed each process is running on a dedicated processor (each of equal computational ability) and that the communications backbone is homogeneous (in terms of bandwidth available for each process). Note that nothing prevents pMOEA implementation on a heterogeneous cluster or system whose communications backbone provides different bandwidth to different processors, as long as load balancing techniques are used to maximize performance.

Randomly Generate The Population
    Randomly Generate Population Of Size P On Processor 0
Evaluate Each Population Member's Fitness
    Send $\frac{P}{n}$ Population Members To Each Processor From Process 0
    Each Processor Conducts k Fitness Evaluations For Each Of The
        $\frac{P}{n}$ Population Members
    Each Processor Sends $k * \frac{P}{n}$ Fitness Values To Process 0
    Process 0 Determines $PF_{current}(t)$, Updates PF_{known} And
        Assigns Rank If Necessary
Perform Clustering / Niching / Crowding on Processor 0
Execute Evolutionary Operators (Crossover, Mutation) on Processor 0
Evaluate the New Populations' Fitness
    Send $\frac{P}{n}$ Population Members To Each Processor From Process 0
    Each Processor Conducts k Fitness Evaluations For Each Of The
        $\frac{P}{n}$ Population Members
    Each Processor Sends $k * \frac{P}{n}$ Fitness Values To Process 0
    Process 0 Determines $PF_{current}(t)$, Updates PF_{known} And
        Assigns Rank If Necessary
Conduct Selection on Processor 0
Repeat Until Termination Criteria is Met
Conduct Local Search on Processor 0 if Specified
Processor 0 Generates and Presents PF_{known} as the Solution

---

Fig. 8.12. Generic Master-Slave pMOEA Pseudo code

8.7.1 Engineering a pMOEA

Goldberg et al. indicated too much attention was being paid to "neat" GA genotype encodings [585]. They proposed a scheme where genotypes can exhibit redundancy, over- and under- specification, and changing structure and length, believing this algorithmic modification forms tighter and more useful building blocks than those formed by standard GAs. Their resultant messy GA (mGA) proved successful in optimizing *deceptive functions*; these functions

---

Randomly Generate The Population
    Randomly Generate Population Of Size P' On Each Processor
Evaluate Each Population Member's Fitness
    Evaluate Each Population Member's Fitness on each processor
    Each Processor Determines $PF_{current}(t)$, Updates PF_{known} And
        Assigns Rank If Necessary
Perform Clustering / Niching / Crowding on each Processor
Execute Evolutionary Operators (Crossover, Mutation) on each Processor
Evaluate Each Population Member's Fitness
    Evaluate Each Population Member's Fitness on each process
    Each Processor Determines $PF_{current}(t)$, Updates PF_{known} And
        Assigns Rank If Necessary
Conduct Selection on each Processor
Migrate Population Members Among Processes According To Desired Scheme
Repeat Until Termination Criteria is Met
Conduct Local Search on each Processor if Specified
Processor 0 Combines and Presents PF_{known} as the Solution

Fig. 8.13. Generic Island pMOEA Pseudo code

Randomly Generate The Population
    Randomly Generate Population Of Size P' On Each Processor
Evaluate Each Population Member's Fitness
    Evaluate Each Population Member's Fitness on each processor
    Each Processor Determines $PF_{current}(t)$, Updates PF_{known} And
        Assigns Rank If Necessary
Perform Clustering / Niching / Crowding on each Processor or
    Within the Neighborhood (N)
Execute Evolutionary Operators (Crossover, Mutation) Within
    Neighborhood (N) Processors
Evaluate Each Population Member's Fitness
    Evaluate Each Population Member's Fitness on each process
    Each Processor Determines $PF_{current}(t)$, Updates PF_{known} And
        Assigns Rank If Necessary
Conduct Selection Within Neighborhood (N) Processors
Diffuse Population Members Among Processes According To Desired Scheme
Repeat Until Termination Criteria is Met
Conduct Local Search on each Processor if Specified
Processor 0 Combines and Presents PF_{known} as the Solution

Fig. 8.14. Generic Diffusion pMOEA Pseudo code

misdirect GA search toward some local optimum when the global optimum is actually elsewhere [585]. The mGA initializes a population of building blocks via a deterministic process producing all possible building blocks of a specified size, thus fully deserving its reputation as a (potentially) computationally expensive algorithm.

Goldberg et al. then proposed the fast messy GA (fmGA) to reduce mGA complexity via probabilistic building block initialization schemes creating a controlled number of building block clones of specified size [584]. These clones are filtered ensuring that (in a probabilistic sense) all desired building blocks exist in the initial population. They claim this variant is as effective as the mGA but without the associated computational expense. Later, the fmGA was modified to create a parallel fmGA (pfmGA), where several issues associated with EA parallelization were considered, addressed, and tested on the real-world Protein Structure Prediction Problem [547, 1091, 1106]. These experiments obtained satisfactory speedup results; a large efficiency improvement was also noted as compared to the serial fmGA implementation.

Other original research investigated the role of building blocks in solving MOPs; the Multi-Objective mGA (MOMGA) was created for that purpose by heavily modifying the original mGA code [1630]. Good performance resulted when applying the MOMGA against a validated MOP test function suite. The MOMGA also compared favorably to or outperformed three other well-engineered MOEA variants in wide use at the time. The next evolutionary step was creating the MOMGA-II, which imported the fmGA's probabilistic building block initialization scheme into the MOMGA [1790]. The MOMGA-II differs from the fmGA in that multiple objective functions are solved simultaneously and its selection mechanisms are based on the concepts of Pareto dominance and niching.

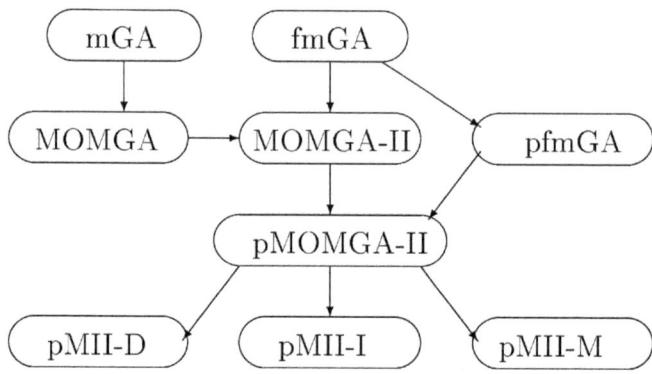

Fig. 8.15. pMOEA Evolution

8.7.2 "Genericizing" a pMOEA

The MOMGA-II performed well when compared to other MOEAs and when applied to real-world applications [1789, 1790]. These results partially motivated parallelization of the MOMGA-II (or pMOMGA-II). The pMOMGA-II also appeared to be a good test vehicle with which to thoroughly test the Master-Slave, Island, and Diffusion pMOEA paradigms as applied to a single MOEA variant. In other words, use the same MOEA as the basis with which to create three different parallel versions, yet not prevent further modifications. The pMOMGA-II is also generic in that it can execute on any parallel architecture hosting an ANSI 'C' standard compiler.

A generic pMOEA has an additional advantage in any experimental design and execution process, as algorithmic differences between variants are minimized. Of course, any serious pMOEA testing must address not only pMOEA-specific issues but must provide meaningful parallel statistics as well. This rigorous testing is still lacking in the pMOEA literature. A thorough analysis of competing parallel paradigms, while also considering parallel performance issues, should give valuable insights into pMOEA effectiveness and efficiency. Note also that the (p)MOMGA-II development process is detailed here not only to illustrate the iterative algorithmic development process depicted in Figure 8.15, but to bring home the point that engineering effective and efficient algorithms, especially algorithms applied to interesting and difficult problems, is an extensive effort that can span many years and several researchers.

8.8 Conclusions

This chapter paints a picture of pMOEA state-of-the-art with a broad brush, yet several significant contributions and innovations are clearly seen. A previous MOEA notation is extended into the pMOEA domain enabling precise description and identification of various sets of interest. A thorough discussion of four pMOEA paradigms (Master-Slave, Island, Diffusion, and Hierarchical) is included, and many succinct observations made regarding analyses of the current literature. Innovative concepts for pMOEA migration, replacement, and niching schemes are discussed and the first known generic pMOEA formulation is presented. Taken together, this chapter's analyses and original pMOEA development serve as a pedagogical framework for and example of the necessary process to design and implement efficient and effective pMOEAs. Interspersed throughout the preceding discussion are various recommendations for creating pMOEA abstractions, designs and implementations. This aids the community in achieving a fuller understanding of pMOEAs and appropriate contexts for comparing their performance.

Each pMOEA computational model must be evaluated to determine its effectiveness and efficiency across a wide spectrum of applications and implementations on various software/hardware architectures. Moreover, if possible, pMOEA theory should continue to evolve in order to provide a solid foundation for selecting appropriate computational architectures, associated operators, and parameter values. This chapter motivates the direction of such efforts.

In closing, the following quote is offered for the reader's contemplation [1270, p.51]:

> "It is of much greater importance that the engineer, on whom rests the responsibility of a work, should be competent to select the best devices proposed by his assistants, or used by others, than to be able to invent novel ones himself. The obligation to adopt the best is, however, not incompatible with a generous regard for the rights or merits of others, and the professional reputation will never suffer by giving such credit to them as may be justly due."

This chapter presents and adopts the best devices yet seen (in the authors' opinion), invents new solutions where needed, and gives credit where credit is due. Obviously, other pMOEA developers with varying levels of experience may implement different solutions. These new or similar solutions are welcome! Well-designed pMOEA studies can only aid in better understanding critical issues, assessing differing paradigms' impacts on pMOEA efficiency and effectiveness, and gaining a better grasp of the class' performance through careful experimentation and analysis.

Further Explorations

Class Exercises

1. Take the pseudo code of a state-of-the-art MOEA (e.g., the NSGA-II [374]) and indicate the key components of the approach that have to be considered when parallelizing it.
2. Discuss possible ways of incorporating external populations to a parallel MOEA (see Chapter 2).
3. Discuss possible migration policies for parallel MOEAs.
4. What are the main issues when incorporating a migrant into an island, when dealing with parallel MOEAs? Discuss.
5. Discuss possible topologies for parallel MOEAs and the potential advantages of each of them.
6. Consider the issues of using multiple threads (e.g., in Java) in a MOEA. What possible benefits do you foresee from incorporating threads into a MOEA?
7. What are the possible benefits of using different operators (i.e., selection, crossover and mutation) and different encodings in each deme of an island-based parallel MOEA? Discuss.
8. What are the main issues when parallelizing a niching technique?
9. Does it make sense to attempt to parallelize a Pareto ranking approach? Why or Why not? Do you think that there are other selection techniques used for multiobjective optimization that would benefit from being parallelized? Discuss.
10. Discuss possible parallelization strategies for clustering techniques adopted with MOEAs.
11. Assuming a homogenous set of processors and a backplane communication mesh, generate a theoretical speedup equation for an island MOEA paradigm. Develop your equation for at least three different island migration schemes. How do these mathematical models relate to MOEA scalability models considering the communication overhead?

Class Software Projects

1. Use MPI to develop a parallel implementation of the NSGA-II [374] adopting an island model. Use the test functions provided in [379] to validate this approach.
2. Repeat the previous project, but using SPEA2 [1776] instead of the NSGA-II.
3. Use PVM to develop a parallel implementation of the NSGA-II [374] adopting an island model and compare this implementation to one based on MPI (see previous project).
4. Repeat the previous project, but using SPEA2 [1776] instead of the NSGA-II.
5. Perform a comparative study of the 4 implementations developed in the 4 previous software projects adopting the DTLZ test problems [379] (see Chapter 4). Indicate the main advantages and disadvantages of each implementation language (i.e., PVM vs. MPI) in terms of ease of use, features and functionality. Analyze the scalability of each of the 2 pMOEAs compared using 2, 4, 8 and 16 processors. Can you define a clear winner between the NSGA-II and SPEA2? What performance measures do you consider appropriate to assess performance of parallel MOEAs (see Chapter 5)?
6. Develop an experimental study in which you compare parallel versions of several pMOEAs. Consider performance measures that are relevant both to evolutionary computation and to parallel computing.
7. Implement 5 of the approaches discussed in the surveys of parallel multiobjective metaheuristics by Nebro et al. [1174] and Luna et al. [1032].
8. Develop a software library for supporting the development of parallel MOEAs. See for example [26, 1693].
9. Implement a selected MOEA algorithm in Java RMI. Using several of the test functions from Chapter 4, compare MOEA results with others using some of the performance from Chapter 5. Does the performance of the Java RMI version justify its continuing development? Discuss.
10. Using the MATLAB distributed computing toolbox along with the MATLAB distributed computing engine, develop a MOEA master-slave or island paradigm software package on a cluster selecting an appropriate MOEA toolbox. Apply this software package to the solution of some of the test functions from Chapter 4. Evaluate performance in terms of not only effectiveness (using the metrics of Chapter 5), but also in terms of the standard metrics normally adopted in parallelism (speedup, efficiency and scalability).
11. Using a computational grid environment such as Globus, implement a MOEA master-slave or island model. Select a large dimensional real-world problem and evaluate grid results in terms of effectiveness and parallel efficiency. Also, compare performance results to the analytical grid model

proposed by Lim et al. [995]. Is a computational grid an effective method for solving the selected MOP? Discuss.

Discussion Questions

1. Based upon problem domain characteristics, when is a specific parallel MOEA paradigm appropriate? For example, consider the parameters optimization for various applications in Chapter 7.
2. Discuss the theoretical performance of various logical grid communication structures (e.g., ring, star, hypercube) used to implement a parallel island MOEA. Additionally, relate to various EVOPs and associated parameter values.
3. Consider a variety of parallel and distributed computational platforms with different communication backbone configurations. What MOEA forms utilize such communication backbones more efficiently than others? Why?
4. Given a parallel MOEA paradigm, and a real-world application, consider the performance of different MOEA types and/or different EVOP parameter values per processor.
5. Assuming that one has the use of a very high-performance computational platform with hundreds or thousands of high-speed processors (e.g. IBM RS/6000 SP or the ASCI (Accelerated Strategic Computing Initiative)-Red/White/Blue systems), how can a parallelized MOEA be effective and efficient in using such an architecture? Given that architecture, what are the limits of MOEA parallelization for a specific application? Do those limits also hold for a high-performance parallel platform having a very high number of processors (1024, 4096)? Also relate anticipated performance to server clusters with 64 processors as well as low-cost multiple processor desk computers (4-8 processors) having many Gigabytes of local memory.
6. Consider the various parallel MOEAs detailed in this chapter. Do they fully exploit MOEA decomposition from a task, objective function, or data perspective? If not, how could the implementations be improved?
7. When attempting to find acceptable solutions to real-world optimization problems, how can other optimization techniques be integrated into a given parallelized MOEA?
8. How might dynamic MOEA parameter variation improve performance for a given parallel implementation? Relate the performance impact to problem classes or characteristics.
9. Few comparative studies of pMOEA implementations exist (see for example [1191], where several parallel implementations of Pareto Simulated Annealing (PSA) [320] are compared). Why do you think that such comparative studies are scarce? Discuss issues that you consider important when comparing different pMOEA implementations.

10. Rudolph [1400] has raised an interesting issue related to parallelizing randomized algorithms: since a randomized algorithm must be executed more than once to get a reliable solution, we have the choice of executing the sequential version of the algorithm in parallel in an independent manner, or we can execute the parallel version of the algorithm simultaneously on the parallel hardware in a successive manner. Which of these two approaches is better? Analyze this issue and discuss the scenarios presented by Rudolph [1400].
11. Parallel niching techniques have been only scarcely studied in the specialized literature (see for example [1038]). Elaborate a report on this topic, in which you discuss its main topics in the context of evolutionary multiobjective optimization.
12. What is the MOEA performance effect of controlling explicitly distributed memory of a cluster as compared to shared-memory machine programming? Relate to ease of MOEA software development.

Possible Research Ideas

1. Propose a parallel ant colony optimization algorithm for solving multiobjective problems. See for example [771].
2. Propose an island-based parallelization of the NSGA-II [374]. Discuss issues such as migration policies and topologies. Do you consider necessary to have an external population?
3. Propose a parallel particle swarm optimization algorithm for solving multiobjective problems. Discuss issues such as selection of leaders, migration policies and topologies, variability of the cognitive and social factors. See for example [1137].
4. Propose a parallel simulated annealing algorithm for solving multi-objective problems. Analyze the possibility of adopting different cooling schedules in each subpopulation. See for example [67, 69].
5. Investigate about multi-threaded Tabu search [1552, 294]. Discuss the potential benefits of performing multiple local searches in parallel when solving multiobjective optimization problems (e.g., in the context of multiobjective memetic algorithms [879]). Develop a parallel multi-threaded version of a multiobjective Tabu search algorithm.
6. Read the survey on parallel strategies for meta-heuristics by Crainic and Toulouse [305]. Analyze the future research directions that they indicate in the paper and, if appropriate, explore one of them.
7. The cooperation of different search strategies executed in parallel that exchange information is another interesting research topic (see for example [1601]). Propose a cooperative search algorithm that exploits this concept.
8. Compare niching and iterated local search (ILS) [1020] in the context of evolutionary multiobjective optimization. See for example [1293].

9. Extend the concept of *takeover time* regarding selection methods used in evolutionary algorithms [583] to parallel MOEAs with migration.
10. Propose a reconfigurable parallel hardware architecture for MOEAs (see for example [1179, 1549]).
11. Analyze the use of surrogate methods in the context of parallel MOEAs (e.g., for solving problems with very costly objective functions evaluations). See for example [1334]. Propose a novel hybridization of a parallel MOEA with a surrogate method.
12. Propose a parallel genetic programming approach for solving multiobjective problems. See for example [481, 502].
13. Read about parallelization of exact methods for multiobjective optimization (see for example [383], which focuses on multiobjective combinatorial optimization), and devise a potential use of such techniques within evolutionary multiobjective optimization (e.g., hybridizing them with MOEAs).
14. Study the decomposition of the objective space by Deb et al. [381] and Streichert et al. [1528] for distributed MOEA computation. Generalize and combine their approaches directed towards a more realistic dynamic and adaptive decomposition of the objective space for other than convex Pareto fronts. Implement and evaluate statistical performance of this new MOEA.
15. Address the appropriate parallel algorithmic concepts for developing and using a MOEA on a very large computation grid. Formulize and extend the efforts of Lim et al. [995] as a first step. Consider MOEA operator variations, niching approaches, population archiving methods, and data storage techniques from the distributed perspective.

9
Multi-Criteria Decision Making

> Consider what you think justice requires, and decide accordingly. But never give your reasons; for your judgment will probably be right, but your reasons will certainly be wrong.
>
> Lord Mansfield

9.1 Introduction

One aspect that most of the current research on evolutionary multiobjective optimization (EMO) often disregards is the fact that the solution of a multiobjective optimization problem (MOP) really involves three stages: measurement, search, and decision making.

Being able to find P_{true} does not completely solve an MOP. The decision maker (DM) still has to choose a single solution out of this set. The process of selecting a single solution is not trivial. In fact, there is a set of methodologies regarding how and when to incorporate decisions from the DM into the search process.

Having several nondominated vectors does not provide any insight into the process of decision making itself. Nondominated vectors are really a useful generalization of a utility function under the conditions of minimum information (i.e., all attributes are considered as having equal importance; the DM does not express any preferences) [315, 316, 504, 719, 598].

Most of the current EMO research concentrates on adapting an evolutionary algorithm to generate P_{true} (i.e., search). However, the articulation of preferences has been dealt with by few researchers (see for example [1347, 1346, 315, 316, 314, 706, 504, 719, 598, 263]).

In this chapter, a brief review of the main concepts related to Multi-Criteria Decision Making (MCDM[1]) is provided. The most representative

---

[1] After the 1960s, the main emphasis of operations researchers has been to study the area known as "Multi-Criteria Decision-Aid" (MCDA) [1387, 1658]. The main

research on preference articulation found in the EMO literature is then reviewed, analyzing their contributions and weaknesses.

9.2 Multi-Criteria Decision Making

From the Operations Research (OR) perspective, there are two main lines of thought regarding MCDM [730]:

1. The French school, which is mainly based on the outranking concept [1659], and
2. The American Multi Attribute Utility Theory (MAUT) school [836].

The French school is based on an outranking relation which is built up under the form of pairwise comparisons of the objects under study (see Section 1.7.1 from Chapter 1). The main goal is to determine, on the basis of all relevant information for each pair of objects, if there exists preference, indifference, or incomparability between the two. For this purpose, preference or dominance indicators are defined and compared with certain threshold values.

The main disadvantage of this approach is that it can become very expensive (computationally speaking) when there is a large number of alternatives. Also, some authors consider the use of outranking methods as complementary to other techniques (e.g., MAUT) and are therefore intended for problems that present certain characteristics (e.g., at least one criterion is not quantitative) [1388].

MAUT is based, in contrast, on the formulation of an overall utility function, and its underlying assumption is that such a utility function is available or can be obtained through an interactive process. When this utility function is not available, the task is then to identify a set of nondominated solutions. In this case, strong preference can only be concluded if there exists enough evidence that one of the vectors is clearly dominating the vector against which it is compared. Weak preference (modeled as weak dominance[2] [997]), on the other hand, expresses a certain lack of conviction. Indifference means that both vectors are "equivalent" and that it does not matter which of them is selected. It is important to distinguish this "indifference" from the "incomparability" used with outranking methods, since the second indicates vectors with strong

---

difference between MCDM and MCDA is that MCDA assumes multiobjective optimization problems are ill-defined mathematical problems. That is, depending on the algorithm used and the preference information incorporated, different solutions to the same problem could be obtained. While MCDM focuses on finding a solution to a multiobjective optimization problem, MCDA focuses on the decision process itself. According to Roy [1387], the main aim of MCDA is "to construct or create something which is viewed as liable to help an actor taking part in a decision process either to shape, and/or to argue, and/or to transform the DM's preferences".

[2] See Section 1.2.3 from Chapter 1.

opposite merits [730]. MAUT does not work when there are intransitivities in the preferences, which is something that frequently arises when dealing with "incomparable" objects using an outranking approach [1752].

There are a few issues related to MAUT deserving some discussion. First, it is important to distinguish between global and local approaches to MAUT (see Section 1.7.1 from Chapter 1). Despite the fact that it is common practice to assume a global approach to MAUT in which an overall utility function that expresses the DM's global preferences is assumed, operations researchers tend to favor local approaches. In local approaches, the utility function is decomposed into simple utility functions (e.g., single attribute functions) that are easier to handle [1519, 556].

Second, it is important to mention that a utility function does not really reflect the DM's inner (psychological) intensity of preference. It just provides a model of the DM's behavior [1136]. This is an important distinction, since behavior should then be consistent (i.e., it should not originate intransitivities), according to MAUT's practitioners.

The main criticism towards MAUT is its inability to handle intransitivities. There are, however, reasons for not dealing with intransitivities in MAUT [1024]:

- MAUT is only concerned with behavior of the DM, and behavior is transitive.
- The transitive is often a "close" approximation to reality.
- MAUT's interest is limited to "normative" or "idealized" behavior.
- Transitive relations are far more mathematically tractable than intransitive ones

From these arguments, it is often assumed that the main reason to support MAUT is in fact the mathematical tractability of utility functions [1519].

9.2.1 Operational Attitude of the Decision Maker

The French and American schools of thought lead to three types of operational attitude of the DM [1385]:

1. Exclude incomparability and completely express preferences by a unique criterion. This leads to an aggregating approach in which all the criteria are combined using a single utility function representing the DM's global preferences. An example of this approach is the technique called "maximim programming" [422].
2. Accept incomparability and to use an outranking relation to model the DM's preferences. In this case, the DM only has to model those preferences that is capable of establishing objectively and reliably, using outranking only when such preferences cannot be established. In this case, the DM is asked to compare all criteria two by two; each objective is assigned a weight derived from the eigenvector of the pairwise comparison matrix

[730]. It is important to be aware that these pairwise comparisons can lead to intransitive or incomplete relations. One example of this approach would be ELECTRE in its different versions [1383, 1386, 1488].

3. Determine, through an interactive process, the different possible compromises based on local preferences. In this case the DM experiments with local preferences at each stage of the search process, which allows exploration of only a certain region of the search space. These local preferences can be expressed in different ways (e.g., a ranking of objectives, an adjustment of aspiration levels, or even detailed trade-off information) [461]. The main issue here is that the DM is not asked preliminary (specific) preference information. Such preferences are derived from the behavior exhibited by the DM through the search process. When further improvement is no longer necessary or is impossible then a compromise has been reached. This can be seen as a local optimum relative to an implicit criterion. An example of this approach is the STEP Method (STEM) [116].

9.2.2 When to Get the Preference Information?

A very important issue in MCDM is the moment at which the DM is required to provide preference information. There are three ways of doing this [461, 706]:

1. Prior to the search (*a priori* approaches).
2. During the search (interactive approaches).
3. After the search (*a posteriori* approaches).

There is a considerable body of work in OR involving approaches performing prior articulation of preferences (i.e., *a priori* techniques) (see for example [289, 836, 722]). The reason for its popularity is that any optimization process using this *a priori* information becomes trivial. The main difficulty (and disadvantage of the approach) is finding this preliminary global preference information.

That is the reason why despite the popularity of *a priori* schemes to articulate preferences, interactive approaches (i.e., the progressive articulation of preferences) have been normally favored by researchers [539] for several reasons [1121]:

1. Perception is influenced by the total set of elements in a situation and the environment in which the situation is embedded.
2. Individual preference functions or value structures cannot be expressed analytically, although it is assumed that the DM subscribes to a set of beliefs.
3. Value structures change over time, and preferences of the DM can change over time as well.
4. Aspirations or desires change as a result of learning and experience.

5. The DM normally looks at trade-offs satisfying a certain set of criteria, rather than at optimizing all the objectives at a time.

However, interactive approaches also have some problems, mainly related to the preference information that the DM has to provide during the search [1519]. For example, the DM can be asked to rank a set of solutions, to estimate weights or to adjust a set of aspiration levels for each objective. None of these tasks is trivial and very often DMs have problems providing answers that can guide the search in a systematic way towards a best compromise solution [461]. In fact, despite the existence of sophisticated algorithms to transform the information given by the DM to a mathematical model that can be used to guide the search, it has been shown that interactive approaches that adopt a simple trial-and-error procedure tend to be highly competitive [1672]. This indicates that the preference information provided by a DM tends to be so contradictory and inconsistent that in some cases it can be even disregarded without significantly affecting the outcome of a decision-making algorithm. That is the reason why MCDA emphasizes the decision-making process itself, as there are many factors that could contribute to this inconsistent behavior.

The use of *a posteriori* approaches is also popular in OR field [289, 732]. The main advantage of these approaches is that no utility function is required for the analysis, since they rely on the use of a "more is better" assumption [461]. The main disadvantages of *a posteriori* approaches are [461]:

1. The algorithms used with these approaches are normally very complex and tend to be difficult for the DM to understand.
2. Many real-world problems are too large and complex to be solved using this sort of approach.
3. The number of elements of the Pareto optimal set that tends to be generated is normally too large to allow an effective analysis from the DM.

It is also possible to combine two or more of these approaches. For example, one could devise an approach in which the DM is asked some preliminary information before the search, and then ask the DM to adjust those preferences during the search. This may be more efficient than using either of the two approaches independently.

Finally, it is worth mentioning that EMO researchers often disregard the importance of MCDM without taking into consideration that it normally requires a considerable amount of time (perhaps more than the search itself). It is well known in the OR community that defining a good utility function or preference structure for a real-world problem is normally a complex task that may take days or even weeks [1135]. In fact, operations researchers distinguish a trade-off between spending more time working on a good utility function (or preference structure) and spending more time searching through a larger set of solutions [461].

9.3 Incorporation of Preferences in MOEAs

The previous classification of stages at which preferences can be provided by the DM (i.e., *a priori*, interactively, *a posteriori*) is also used with respect to MOEAs as indicated in Chapter 1 [706].

The current EMO literature indicates *a priori* approaches, i.e., aggregating approaches in which weights are defined beforehand to combine all the objectives into a single objective function, are very popular.

Specific *interactive* approaches are less common in the EMO literature, although several of the approaches reviewed in this chapter can be used interactively. Additionally, several MOEAs *could* be used interactively if desired. From the current volume of research, however, one infers that most MOEA researchers assume an *a posteriori* incorporation of preferences. That is because the main research emphasis is in generating Pareto optimal solutions assuming no prior information from the DM.

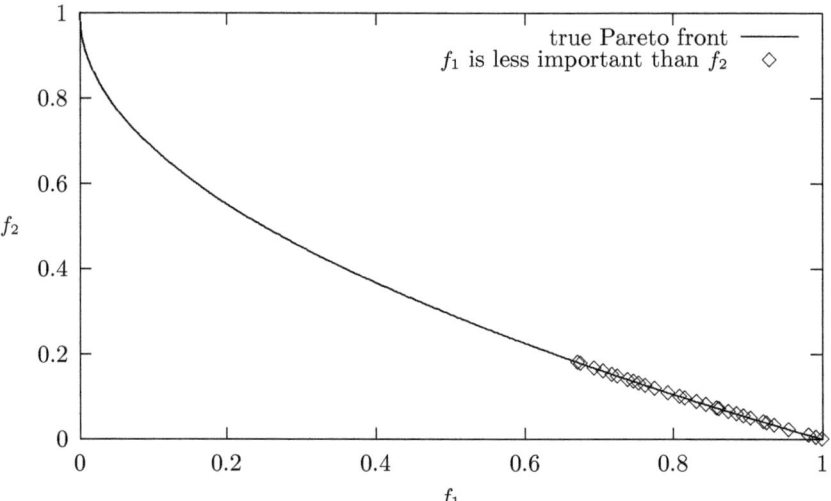

Fig. 9.1. This plot indicates with diamonds the case in which f_1 is considered less important than f_2 (only f_2 is minimized). PF_{true} for the problem (obtained by enumeration) is shown as a continuous line.

Regardless of the stage at which preferences are incorporated into a MOEA, the goals are clear: the aim is to magnify (i.e., concentrate search on) a certain portion of the Pareto front by favoring certain objectives (or trade-offs) over others. An example better illustrates this goal. Assume the following multiobjective optimization problem is to be solved [357]:

9.3 Incorporation of Preferences in MOEAs

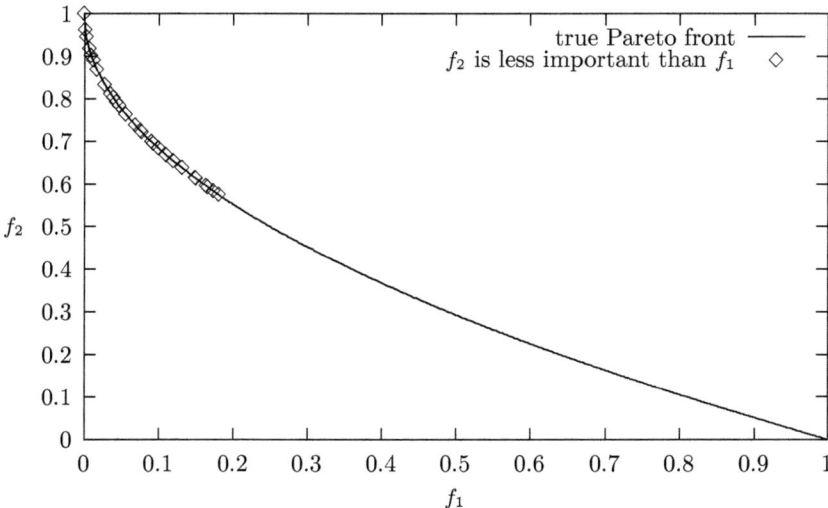

Fig. 9.2. This plot indicates with diamonds the case in which f_2 is considered less important than f_1 (only f_1 is minimized). PF_{true} for the problem (obtained by enumeration) is shown as a continuous line.

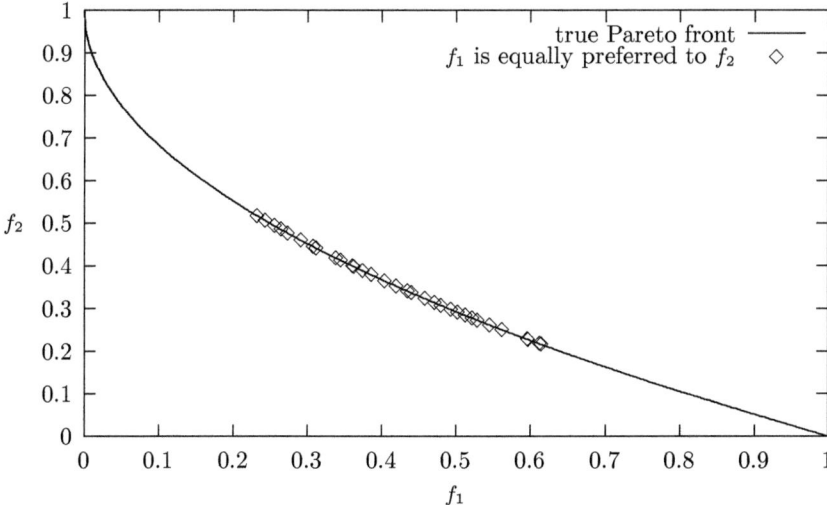

Fig. 9.3. This plot indicates with diamonds the case in which f_1 and f_2 are equally preferred (both are minimized). PF_{true} for the problem (obtained by enumeration) is shown as a continuous line.

$$\text{Minimize } f_1(x_1, x_2) = x_1 \tag{9.1}$$

$$\text{Minimize } f_2(x_1, x_2) = g(x_1, x_2) \cdot h(x_1, x_2) \tag{9.2}$$

where:

$$g(x_1, x_2) = 11 + x_2^2 - 10 \cdot \cos(2\pi x_2) \tag{9.3}$$

$$h(x_1, x_2) = \begin{cases} 1 - \sqrt{\frac{f_1(x_1, x_2)}{g(x_1, x_2)}} & \text{if } f_1(x_1, x_2) \leq g(x_1, x_2) \\ 0 & \text{otherwise} \end{cases} \tag{9.4}$$

and $0 \leq x_1 \leq 1$, $-30 \leq x_2 \leq 30$.

For this example there are three main preference situations that could be expressed by the DM:

1. f_1 is considered less important than f_2. In this case, only the lower portion of the Pareto front is generated, because only f_2 is being optimized, regardless of the values achieved by f_1. This situation is depicted in Figure 9.1.
2. f_2 is considered less important than f_1. This is the exact opposite of the previous case, and only the upper portion of the Pareto front is generated, because only f_1 is being optimized. This situation is depicted in Figure 9.2.
3. Both objectives are given equal importance. In this case, solutions around the "knee" of the Pareto curve are generated because none of the objectives is given preference over the other. This is the situation assumed by most MOEAs when the DM does not express any preferences. This situation is shown in Figure 9.3. Note that solutions could be generated so that the entire Pareto front is covered (as normally done with MOEAs), but only solutions around the "knee" of the curve are illustrated to emphasize the fact that this preference situation is the complement of the two previous.

The goal of incorporating DM's preferences is to find a mechanism allowing a certain MOEA to generate only the portion of the Pareto front corresponding to the preferences expressed by the DM (i.e., each of the three situations depicted in Figures 9.1, 9.2 and 9.3).

A comprehensive search through the current EMO literature makes evident that there is very little work in which the handling of preferences is explicitly dealt with. Most of this research is briefly described and analyzed in the remainder of this section.

9.3.1 Definition of Desired Goals

Apparently, the earliest attempt to incorporate preferences from the user in a MOEA is the multicriteria decision support system developed by Tanaka

9.3 Incorporation of Preferences in MOEAs

and Tanino [1570] and further developed by Tanino et al. [1574]. In this work, a genetic algorithm was used to generate members of the Pareto optimal set (using selection based on nondominance) and an *interactive approach* was adopted to incorporate preferences from the DM. This system allowed the DM to express preferences in three different ways:

1. By choosing satisfactory and unsatisfactory solutions from the set presented to the DM.
2. By defining aspiration levels (i.e., goals) in objective space. Solutions far from these goals are then considered unsatisfactory and vice-versa.
3. By defining the worst acceptable levels for each objective. Any value below this level is then considered unsatisfactory.

Regardless of the approach taken, satisfactory solutions are replicated whereas unsatisfactory solutions are eliminated from the population.

Fonseca and Fleming [504] made a similar proposal at about the same time as Tanaka & Tanino. The proposal consisted of extending the ranking mechanism of MOGA to accommodate goal information as an additional criterion. The goal attainment method [550] (see Section 1.7.1 in Chapter 1) was used, so that the DM could supply goals at each generation of the MOEA, reducing in consequence the size of the set under inspection and learning, at the same time, about the trade-offs between the objectives. It should be clear that this is an *interactive* approach (Fonseca and Fleming call it "progressive") since the DM must express preferences along the evolutionary process.

This work was extended in a further paper in which Fonseca & Fleming mathematically define a relational operator incorporating the preference information given by the DM [511]. The ranking performed by MOGA is then based on this operator. This operator relies on the use of priorities defined by the DM. In the absence of information all objectives are then given the same priority. These mathematical definitions provide a much more flexible scheme to incorporate preferences, since their formulation can encompass different decision making strategies (e.g., goal programming and lexicographic ordering).

Hinchliffe et al. [683] use Fonseca & Fleming's approach to incorporate DM's preferences into a multiobjective genetic programming (MOGP) system applied to chemical process systems modeling. His approach is also used *interactively*, since the goals are tightened as the search progresses.

Shaw and Fleming [1472, 1473] experiment with different approaches to incorporating preferences into a multiobjective scheduling problem (solved using MOGA). Their experiments showed that the use of preferences expressed *a priori* (in the form of attainable or desirable goals provided in a matrix) by the DM was better than using a problem-specific heuristic. They concluded that over constraining the preferences has a significant impact on the performance of the MOEA and therefore, the use of flexible schemes is encouraged. Since interactive approaches provide some extra flexibility (particularly as it is assumed that certain events may change the DM's preferences over time), an implication of these experiments is that interactive approaches may be more appropriate in real-world applications.

Another approach based on the definition of desired goals is proposed by Tan et al. [1562, 1559]. In this case the DM can define goals that are then used to modify the MOEA's ranking assignment process. The approach also allows the use of both soft and hard constraints. The approach ranks separately those individuals that satisfy the goals from those that do not satisfy them. This scheme is more flexible than the original proposal by Fonseca and Fleming [504] because it allows the use of the logical operators AND & OR (these operators were defined using concepts from fuzzy logic), and also allows the specification of "don't care" priorities and non-attainable goals. This aims to reduce the human intervention in the decision-making process. Although the authors suggest a possible interactive use of the approach they seem to have used it only as an *a priori* technique.

Sait and Youssef [1411] and Sait et al. [1410] propose a similar technique in which the DM defines a set of goals (or acceptable limits) for each objective function. Using these goals, the selection mechanism of a MOEA can be modified such that only those solutions that are "nearest" to all the individual goals established are selected. The definition of "nearest" is, in this case, also made using fuzzy logic and is similar to the proposal of Tan et al. [1562].

Deb [359] proposes a technique to transform goal programming problems into multiobjective optimization problems which are then solved using a MOEA. As discussed in Chapter 1 (see Section 1.7.1), in goal programming the DM has to assign targets or goals that wishes to achieve for each objective, and these values are incorporated into the problem as additional constraints. The objective function then attempts to minimize the absolute deviations from the targets to the objectives. Deb's approach is used only to perform the transformation from goals to objectives, but it could also be used for incorporating preferences into a MOEA. The use of weights (i.e., an utility function), deviations from ideal goals (e.g., min-max method) or the direct use of priorities (e.g., the lexicographic method) are all good candidates to incorporate preferences in an approach of this kind (in fact, these three techniques have been coupled to goal programming in the past [1111]).

Shibuya et al. [1481] propose an approach in which the DM has to provide preferences interactively during the evolutionary process. The DM expresses preferences through pairwise comparisons of images in a computer-generated animation application. The approach then sorts the solutions available based on the DM's preferences (the approach seems to use a mechanism similar to the lexicographic method).

Barbosa and Barreto [91] propose a co-evolutionary genetic algorithm with two populations: a population of solutions to the problem (i.e., individuals that encode coordinates of all vertices of a graph, since the application is a graph layout problem) and a population of weights (i.e., individuals that contain, each one, a set of weights to be applied on the different aesthetic objectives imposed on the problem). The decision maker is then presented a set of nondominated solutions and asked to rank them based on subjective preferences. This ranking is then used to determine fitness of the population of weights. The process is *interactive*, since these preferences are expressed at each iteration of the system. The subjective criteria adopted by the decision maker can be seen as goals that the system has to achieve.

Borges and Barbosa [151] propose an approach that is similar to goal attainment. The decision maker is required to provide a constant vector that contains the limit values for each objective function. These limit values represent the minimum (or maximum) values that are acceptable by the decision maker. Then, the original objective functions are penalized based on the values attained (if the values are within the requirements of the decision maker, there is no penalty and the original objective function values are adopted). The form of the penalty function proposed is similar to compromise programming (see Section 1.7.1 from Chapter 1). This is clearly an *a priori* approach, because the decision maker needs to have prior knowledge of the behavior of each objective function.

Kato et al. [824] propose an *interactive* fuzzy satisficing[3] method in which fuzzy goals are defined and a membership function is elicited from the decision maker for each objective function of the problem to be solved. The method relies on the concept of M-Pareto optimality [1414] in which an aggregating function is adopted to represent the degree of satisfaction or preference of the decision maker for all the (fuzzy) goals of the problem. Under this approach, there is one membership function for each objective function. All the membership functions are aggregated into a single expression that uses a minimax formulation, and the decision maker expresses his/her aspiration levels using the so-called reference membership levels. The approach then finds a solution which is as near as possible to the aspiration levels from the decision maker (or even better if such aspiration levels are attainable). This approach is incorporated into a genetic algorithm which is used to solve integer programming problems (namely, multidimensional integer knapsack problems). The GA adopted uses double strings [1416], a special method for generating the initial population that uses linear programming relaxation based on reference solution updating, linear scaling of the fitness function, partially matched crossover (PMX) [586], inversion, Gaussian mutation and a combination of elitist preserving selection with expected value selection. This approach outperformed a branch and bound method in several test problems.

Yun et al. [1744] propose the use of generalized data envelopment analysis (GDEA) [1745] with aspiration levels for choosing desirable solutions from the Pareto optimal set. This is an *interactive* approach in which a nonlinear aggregating function is optimized by a genetic algorithm in order to generate the Pareto optimal solutions of the multiobjective optimization problem. The

---

[3] *Satisficing* defines the DM's selection of an acceptable result but the underlying computation does not address the finding of the "optimal" solution possibly due to temporal computational constraints.

decision maker must define his/her aspiration levels for each objective, as well as the ideal values for each of them (these "ideal" values may be, but not necessarily are the components of the ideal vector). Then, the aspiration levels are adopted as constraints during the optimization, so that the Pareto optimal solutions are filtered out and those closest to the aspiration levels are given the highest fitness. The approach allows the definition of more than two aspiration levels simultaneously. However, the authors indicate that if the number of aspiration levels is increased, then the computational cost required by the approach also increases.

Criticism of Definition of Desired Goals

The main advantage of using some previously defined (or ideal) goals is that the approach is easy to implement, flexible and relatively simple. Additionally, approaches previously developed in OR (such as goal-attainment) can be used to incorporate these preferences into a MOEA. The main disadvantage of this approach is that it requires the DM to know beforehand the ranges of variation of each objective in order to establish coherent goals. This can be an expensive process (in terms of CPU time) if not impossible in many real-world applications. Also, the use of an interactive method implies the generation of a significant number of nondominated vectors so that the decision making process can be meaningful. This may also become very expensive (computationally speaking).

9.3.2 Utility Functions

Fonseca and Fleming [504] also propose the use of an expert system to automate the task of the DM. Such an expert system uses built-in knowledge obtained from the preferences expressed *a priori* by the (human) DM. In this case, a utility function continuously evaluated through the evolutionary process is used.

Tanaka et al. [1571] experiment with an *interactive* approach using a utility function. In this case the DM expresses preferences and the system uses a normalized radial basis function network to bias the selection pressure of a GA towards those regions that better approximate the preferences.

Hu et al. [719] and Greenwood et al. [598] used elements of *imprecisely specified multi-attribute value theory* (ISMAUT) to perform imprecise ranking of attributes [1698]. The idea is that the DM has to rank a set of solutions to the MOP instead of explicitly rank the attributes of the problem (this is implicitly done by the approach). Preference information is also incorporated into the survival criteria used by the MOEA in order to bias the search towards the region of main interest by the DM. Using as a basis the preferences expressed by the DM, the approach derives a set of equations that are solved to find the values of the weights of a utility function that bias the ranking procedure used by the MOEA. Assuming that no intransitivities occur, only

9.3 Incorporation of Preferences in MOEAs

linear equations have to be solved to find these weights. This approach is then a compromise between using no preference information of the problem (i.e., pure nondominance) and a utility function (the weights of each attribute are not provided by the DM). This is also an *a priori* approach, since the DM has to express preferences before the search begins.

An interesting approach called "Guided Multi-Objective Evolutionary Algorithm" also exploiting the concept of utility function is proposed by Branke et al. [163, 164]. The idea is to express the DM's preferences in terms of maximal and minimal linear weighting functions, corresponding directly to slopes of a linear utility function. The authors determine the optimal solution from a population using both of the previously mentioned weighting functions. Those individuals are given rank one and are considered the borderline solutions (since they represent extreme cases of the DM's preferences). Then all nondominated vectors are evaluated in terms of these two linear weighting functions. After that, all solutions that have a better fitness than either of the two borderline individuals (according to at least one of the two linear weighting functions) are assigned the same rank (these are the individuals preferred by the DM). These solutions are removed from the population and a similar ranking scheme is applied to the remaining individuals. It should be clear that this approach is a biased version of the NSGA [1509]. The authors use a biased version of fitness sharing, in which the maximum and minimum niche counts are incorporated into a formula assigning each individual a fitness at least as good as that of any other individual with inferior rank [163]. In more recent work [161], this approach was coupled to the NSGA-II [374]. This is also an *a priori* technique.

Meneghetti et al. [1089] use a technique called LUTA (*Local UTility Approach*) [1464]. In this approach, the DM is asked which solution prefers (out of a set of nondominated vectors previously generated). No other specific information is required (e.g., to perform pairwise comparisons or to justify a certain choice). The algorithm then proceeds in two stages. In the first stage, it checks any possible inconsistencies in the DM's preferences (e.g., intransitivities) and advises the DM to remove them. In the second stage, it proposes a composite utility function made up of the sum of a set of piecewise linear utility functions, one for each objective function under consideration. The DM's preferences are then stated as inequalities between the utility functions and the solutions available. In other words, these inequalities define a feasible search space for the algorithm, so that the definition of the composite utility function can be modified such that the search is constrained to solutions that lie within the feasible region of the problem. This is really an *a posteriori* approach, although some further (local) search may be required to refine the final solution that is presented to the DM. This search is, however, performed with a hillclimber and not with the MOEA used to generate the elements of the Pareto optimal set.

Criticism of Utility Functions

Fonseca and Fleming [504] recognized that the use of a utility function assumes that the DM can determine *a priori* what sort of trade-offs prefers (i.e., the DM can state preferences in a precise way), and this may not be always the case, particularly in real-world problems. However, utility functions have been quite popular among operations researchers (particularly in the USA) because of their mathematical tractability. In contrast, Europeans tend to prefer outranking approaches that allow intransitivities. In fact, the incapability of utility functions to handle intransitivities is their most commonly criticized feature.

The main disadvantage of Greenwood et al.'s proposal [598] is that their method assumes that all attributes are mutually, preferentially independent (i.e., the value function associated with attribute a_i is not affected by the values of some other attribute a_j, where $j \neq i$). That is not always the case, and despite the fact that the approach would still work when this assumption does not hold, it would certainly become more complicated since a nonlinear system of equations would have to be solved. In fact, Greenwood et al. [598] implemented an algorithm to check intransitivities in the preferences expressed by the DM. The algorithm is also capable of identifying minimum sets of preferences that, if removed, can produce a consistent set of preferences. However, this increases the computational cost of the approach.

The main problem with the proposal of Branke et al. [163] is that it has been used only with two objective functions. The generalization of this approach to a higher number of objectives does not seem trivial. In fact, in more recent work, Branke & Deb [161] indicate that the approach requires the specification of a higher number of trade-offs as the number of objectives increases and the dominance calculations involved also become increasingly complex.

LUTA has the same problems commonly associated with utility functions (e.g., it cannot deal with intransitivities). Also, due to the way in which it operates, the approach may have the same problems as Greenwood et al.'s methodology.

9.3.3 Preference Relations

Cvetković and co-workers [316, 318, 314, 319, 315] propose the use of binary preference relations that can be expressed qualitatively (i.e., using words such as "less important"). These preferences are translated to quantitative terms (i.e., weights) to narrow the search of a MOEA.

The weights generated can be used with a simple aggregating approach (i.e., a sum of weights) or with Pareto ranking. In the second case, the weights are used to modify the definition of nondominance used by the ranking scheme of the MOEA.[4] This approach has some resemblance with the Surrogate

---

[4] The new nondominance relationship defined is really a form of weak-dominance [1111].

Worth Trade-Off method [631], but unlike that method, binary preference relations can find concave portions of PF_{true}.

This is also an *a priori* approach since the weights are assumed constant throughout the optimization process, but nothing in the approach really precludes its use in an *interactive* way. However, there may be some practical issues to take into account if the approach is used interactively, since the DM is asked a considerably high number of questions to make it possible to translate qualitative preferences into quantitative values. This could become too expensive (computationally speaking) if done repeatedly along the evolutionary process.

The direct use of weights to estimate the importance of solutions that have been already identified as Pareto optimal has been suggested by other researchers in the evolutionary computation community in the past. For example, Bentley and Wakefield [121] define a property called "importance" which provides additional information regarding the sort of solutions that are preferred by the DM. This property is then coupled to a GA using ranking, but no Pareto ranking. The authors sort the fitness values of each objective separately and then rank solutions based on their ordering; the average ranking of each individual is then used as its fitness. This ranking scheme is very similar to the one previously proposed by Vemuri and Cedeño [1325, 1326] and does not quite correspond to the concept of Pareto ranking normally adopted by EMO researchers.

Once individuals are ranked, the "importance" that the DM assigns to each objective can be used as a weight and the fitness of an individual is now a weighted sum of its average ranking. Since the weights are assigned by the DM before actually performing the search, this is also an *a priori* technique. Obviously, it could also be used interactively, but apparently was never tested that way.

It is important to mention that the focus of the research conducted by Cvetković [314] is broader and more related to some OR work (see for example [1009]). Also, it is worth mentioning that this is the only known attempt to develop a formal decision making model explicitly for evolutionary multiobjective optimization algorithms.

Drechsler and co-workers [410, 409, 1447] propose the use of satisfiability classes to model preferences from the decision maker. The approach consists of defining a relation "favor", whose concept is similar to Pareto dominance, but not equivalent (mathematically speaking, the relation "favor" is not a partial order, because it is not a transitive relation). In this case, the search space is divided into several categories (e.g., superior, very good, good, satisfiable, and invalid). Solutions generated are then analyzed in terms of their "quality" (defined in terms of the priorities of the decision maker) and divided into several satisfiability classes (i.e., solutions of similar quality belong to the same satisfiability class). After sorting the satisfiability classes with respect to their quality, a ranking of the solutions is obtained. Preference relations are modeled in this case using a directed graph that is recomputed at each generation without any human intervention. In this approach, incomparable solutions may be placed in different satisfiability classes, even if the decision maker wants them to be within the same class. Also, goals are considered in parallel and not relative to each other.

Another interesting proposal to directly modify the ranking procedure of an evolutionary algorithm using preference information provided by the user is made by Hughes [724, 725]. In this case, the author proposes the use of expressions that incorporate information about constraint violation and priority satisfaction into the formulas used to compute domination probabilities (i.e., this information is used to alter the ranking procedure of the population). The approach is simple and elegant, and the author finds it particularly useful when dealing with noisy fitness functions.

Criticism of Preference Relations

Since Cvetković's approach relies on the use of transitive relationships, it is also incapable of handling intransitivities. Scalability is also an important disadvantage of the approach. Cvetković [314] and Cvetković and Parmee [319] empirically show that the number of questions that the DM needs to answer in their approach is, on average, much less than the theoretical upper bound,[5] but the figures still get fairly large as the number of objectives increases (e.g., for 21 objectives, "only" 62 questions must be answered although the theoretical upper bound is 210). The authors argue that a dynamic ordering of the questions could significantly reduce the number of questions needed. However, such a scheme has not been implemented.

Hughes' proposal is quite interesting and could be a way to solve both the constraint-handling and the priority handling problems at the same time. Although the idea is very promising, it is necessary to test the approach more extensively and to analyze how other MOEAs behave with this integrated scheme. Of particular interest is analyzing how strong the search bias of each of these processes (constraint-handling and preference articulation) is and how they interact. For example, problems with highly constrained search spaces may strongly bias the search towards certain specific regions from which it may not be possible to move away. Although such behavior may be desirable, it is important to be aware of it before using this sort of technique.

The model adopted by Drechsler et al. [410] has several advantages: it is efficient (computationally speaking), it does not require scaling, it supports constraint-handling and preferences from the DM in a natural way, and it dynamically adapts to changes during the evolutionary process. However, it also has certain disadvantages. For example, it tends to produce less solutions than the use of Pareto dominance, and it considers only goals or priorities that cannot be relative to each other. As a historical note on the work of

---

[5] The theoretical upper bound is $\frac{k(k-1)}{2}$ questions, where k is the number of objectives.

Drechsler, it is interesting to indicate that the *favor* relation that he proposed was originally established by baron de Condorcet in 1785 [1056] in the context of voting. Marchant et al. [1056] indicate some of the potential problems of this approach with an example:

Let $\{a, \ldots, z\}$ be the set of 26 candidates for 100 voters election. Suppose that:

- 51 voters have preferences $aPbPcP \ldots PyPz$;
- 49 voters have preferences $zPbPcP \ldots PyPa$.

It is clear that 51 voters vote for a while 49 vote for z. Thus, a will win although almost 50% of the voters consider this as the worst one. In the above example, b would be a much better compromise.

9.3.4 Outranking

Rekiek and co-workers [1347, 1346, 1348] propose the use of an outranking method called PROMETHEE II [166] (for more details on the PROMETHEE methods see section 1.7.1 from Chapter 1) combined with a MOEA. PROMETHEE II computes a net flow for each individual and this value is used to rank the population (it imposes a complete preordering of preferences). Weights assigned to each objective by the DM (called preference indexes) have an impact in the computation of the net flows and impose an ordering on the solutions found. The approach is used in an *interactive* way since the DM has to adjust the weights along the evolutionary process. An interesting aspect of this work is the use of what the authors called "branching on population" which basically consists of creating intermediate states based on the preferences expressed by the DM interactively. This allows the evolutionary algorithm to restart from one of these intermediate states rather than from the very beginning.

Guimarães Pereira [613, 614] uses ELECTRE I (see Section 1.7.1 from Chapter 1) coupled with a genetic algorithm to generate alternative routes between two geographical locations. Routes generated are compared in a pairwise fashion for each of the objectives under consideration. An index describing the outranking relation (i.e., the DM's preferences) of a pair of individuals is calculated based on these pairwise comparisons. Then, an aggregating formula is used to combine the different indexes generated from these comparisons. Thus, this aggregating formula is used in an *interactive* way to narrow the search of the GA.

In further work, Guimarães Pereira [617] and Haastrup & Guimarães Pereira [625] couple PROMETHEE to a MOEA. In this case, at each generation of the MOEA, a set of alternatives (i.e., nondominated vectors) are generated and pairwise comparisons of alternatives are performed. The different alternatives available are then ranked in terms of preferences defined over certain suitability values defined by the DM. An aggregating approach is used to measure if a certain alternative is preferred over another one.[6] This approach is applied in an *interactive* way during the evolutionary process.

Massebeuf et al. [1081] also use PROMETHEE II combined with a MOEA (in this case, a technique called "Diploid Genetic Algorithm" [1268]). The DM is asked to express preferences for each pairwise comparison of alternatives (e.g., preference or indifference). Additionally, the DM has to assign weights that express relative importance of the objectives. This information is used to compute a global concordance index and a discordance index (as in ELECTRE III [1390]). Using these indexes, the authors generate outranking degrees for every pair of alternatives. This allows to rank the alternatives so that a final recommendation can be made to the DM.

Parreiras and Vasconcelos [1253] also use PROMETHEE II combined with a MOEA. In this case, the authors adopt a Gaussian preference function and a (normalized) aggregating function for computing the global preferences. The implementation developed by the authors is called *Smart* and is applied in an *a posteriori* way, to nondominated solutions generated by the NSGA-II [374].

Criticism of Outranking

Brans et al. [169] criticize outranking methods because they require too many parameters (this is more evident in papers such as [1081]), the values of which are to be fixed by the DM and the analyst. They argue that even though some of these parameters have a real practical meaning and can therefore be fixed clearly, some others (such as concordance discrepancies and discrimination thresholds) playing an essential role in the procedure, only have a technical character, and their influence on the results is not always well understood. Moreover, in some outranking approaches, the notion of "degree of credibility" is rather difficult for the DM to assess [169].

Also, the procedure involved in the generation of the final ranking of alternatives is rather complex. Since intransitivities are allowed in outranking approaches, contradictions may occur (e.g., when dealing with group preferences) that modify the contents of a previously established global outranking relation. In fact, the problem is of such complexity that some researchers (e.g., [980, 480, 479, 981, 979]) have suggested the use of a GA to improve the quality of an outranking relation by reducing differences between the global model of preferences and the final ranking produced by the algorithm used to incorporate the DM's preferences.

Researchers who have adopted outranking methods normally omit many important details that are required to reproduce their results. For example, PROMETHEE uses six different preference types. Also, it is rather unfortunate that several authors have adopted PROMETHEE II which gives the

---

[6] Actually, the author computes differences between the intensity of preference (this is defined as a membership function, using fuzzy logic) on each objective for each pair of alternatives under consideration.

complete order of alternatives without taking advantage of the incomparability information that outranking methods usually provide.

9.3.5 Fuzzy Logic

Voget [1662] and Voget and Kolonko [1663] use a fuzzy controller that automatically regulates the selection pressure of a MOEA by using a set of pre-established goals defining the "desirable" behavior of the population. A set of fuzzy rules is used to modify the selection mechanism of the MOEA when it is deviating from the goals defined by the DM. Although the approach is used only to keep diversity in the population, it could easily be extended to incorporate preferences of the DM. The idea is similar to goal attainment, except that in this case membership functions are used to express goals in vague terms (i.e., it allows uncertainties). This is an *a priori* approach, but it could also be used interactively.

A similar fuzzy controller was proposed by Lee et al. [973], Lee & Hartani [974], Esbensen & Kuh [454], and Lee & Esbensen [972] but in this last case, on-line and off-line performance (properly generalized for MOPs) of the MOEA are used to guide the search, so that the following conditions are satisfied [972]:

- Maximize the diversity of the nondominated vectors in the population.
- Maximize the number of nondominated vectors in the population.
- Maximize the bounding volume of the set of nondominated vectors.
- Make the center of gravity of the final solution set close to the origin.

All of these conditions aim to promote diversity and bias the MOEA towards PF_{true} of the problem solved. Additionally, the DM's preferences are incorporated using a simple utility function (a weighted sum). The approach is used *a priori*, but the authors suggest its possible use as an *interactive* method.

Pirjanian [1275] uses fuzzy rules to generate weights that would narrow the search of a traditional multiobjective optimization technique. This work is extended in Pirjanian & Matarić [1276] where, instead of adopting the usual fuzzy behavior-based control, where fuzzy rules are combined using standard fuzzy inferencing, the authors use multicriteria decision making for optimizing the behavior of a robot: an action is chosen which maximizes the objective corresponding to different behaviors. The process operates in four stages:

1. Find feasible actions, based on physical and other hard constraints.
2. Find Pareto optimal solutions.
3. Find satisficing actions from the Pareto optimal solutions incorporating subjective knowledge.
4. Find the most preferred action, using other criteria.

534   9 Multi-Criteria Decision Making

In [1276], a single robot is used with the task of reaching a certain target. However, in [1277], the method is generalized to several robots (in the experiments reported, two robots are used) with the task of surrounding and capturing the target. The authors use weighted sums to generate the Pareto front (by varying the weights) and adopt lexicographic ordering, goal programming and interval criterion weights for finding satisficing actions [1275, 1276, 1278]. In all cases, the approaches seem to be applied *a posteriori*.

Jin & Sendhoff [802] propose an approach for converting fuzzy preference relations into interval-based weights which are then combined with a dynamic weighted aggregation method proposed by the same authors [801]. This approach is very similar to the proposal of Cvetković & Parmee [319], but instead of converting the fuzzy preferences into single-valued weights, they are converted to interval-based weights. So, the authors adopt preference matrices and real-valued preference relation matrices. One interesting side-effect of using intervals is that the dynamic weighted aggregation method cannot properly work with non-convex Pareto fronts in this case, because in such situations the movements of the individuals cannot be controlled based on the fuzzy preferences. However, this approach could obviously be used with an alternative multi-objective evolutionary algorithm that does not have this limitation.

Wang and Terpenny [1675] propose an approach based on the use of a fuzzy set-based aggregating function to express the preferences from the user. This aggregating function relies on two things: (1) a set of weights that express the importance of the design attributes and (2) a degree of compensation among these design attributes. By changing the weights and the compensation factors, different portions of the Pareto front can be obtained. The authors also combine their fuzzy preference aggregating function with a penalty function in order to transform a constrained problem into an unconstrained one. This approach is used in an *interactive* way within an agent-based system in which the user expresses his/her preferences and a set of agents produce subsolutions that are later refined based on the constraints and the fuzzy preference aggregating function previously indicated. An interesting aspect of this work is that the actual parameters of the fuzzy preference aggregating function are really *learnt* using a neural network that attempts to minimize the cumulative error of the network. This approach is validated using a panel meter configuration design problem, which is a combinatorial optimization problem. The authors found the approach to be very efficient and quick in terms of convergence to the Pareto optimal set.

Farina and Amato [469, 470, 471] analyze the limitations and drawbacks of the Pareto optimality definition when dealing with problems that have more than 3 objectives. The three main reasons that the authors provide for considering the definition of Pareto optimality as unsatisfactory are the following:

1. It does not consider the number of improved or equal objective function values.
2. It does not consider the (normalized) size of the improvements.
3. It does not consider preference among objectives.

Based on this analysis, they propose three alternative definitions of optimality that aim to generalize the definition of Pareto optimality: k-optimality, k_F-optimality (with fuzzy numbers) and fuzzy optimality. These definitions cope with the previous limitations of the definition of Pareto optimality. An important aspect of this work is the fact that in this case fuzzy logic is not adopted for the treatment of the user's preferences, but for modeling the size of the improvements done in each objective. Thus, in this model, all objectives are given the same importance as traditionally done with the Pareto optimality definition. The approach, however, is adopted to extract subsets of solutions from the Pareto optimal set, but instead of directly expressing preferences as membership functions, the approach allows the definition of fuzzy tolerances for the objectives.

Criticism of Fuzzy Logic

The main issue that deserves attention when extending fuzzy logic to incorporate user's preferences is (just as in the case of goal-attainment) the definition of the goals. The definition of an appropriate membership function that can manage the uncertainties implied in the multi-criteria decision making stage remains as a key issue when using fuzzy logic to incorporate user's preferences into a MOEA.

9.3.6 Compromise Programming

Deb [356] suggests a variation of compromise programming (see Section 1.7.1 from Chapter 1) to bias the sharing procedure of the NSGA [1509]. Deb uses a normalized Euclidean distance between objective vectors (as normally used to compute sharing distances in the NSGA), but introduces unequal weights such that different importance can be assigned to each objective. This allows one to bias the niche-formation procedure of a MOEA, but does not produce a single final solution as normally done with multi-criteria decision making techniques. That means that further intervention of the DM is still required. Therefore, this is an *a posteriori* approach. In more recent work, Branke & Deb [161] propose a biased crowding measure that is applied to the NSGA-II [374]. The idea is that the user provides his/her preference as a direction vector, which is really a central linearly weighted utility function. The crowding measure will be biased based on this direction vector, so that only a fraction of the Pareto front is produced. This crowding measure has the advantage of being easily scalable to any number of objectives and to work well even in the presence of non-convex portions of the Pareto front. It is worth indicating that the version of the NSGA-II adopted in this work uses a modified crowding distance that is more appropriate to deal with problems of higher dimensionality than the original proposal [161].

Sakawa et al. [1415] also suggest an approach based on compromise programming. In this case, the DM establishes goals for each objective using membership functions (i.e., fuzzy logic). A *minimum* operator is defined so that it can integrate all the preferences from the DM into a single membership function. Several expressions inspired on compromise programming are then used to guide the search of the MOEA. These expressions are used as online performance measures having a direct impact on the selection process. The approach was applied to multidimensional versions of the knapsack problem.

Criticism of Compromise Programming

One of the main problems with Deb's approach [356] is the definition of the weights determining the importance of each objective. Moreover, it is not trivial to estimate the effect that a certain weight combination produces in the search and it may be necessary to perform a considerable number of experiments to obtain the desired effect. The more recent proposal from Branke & Deb [161] for using a biased crowding distance is not as fast (in terms of convergence) as other schemes (namely, the guided dominance scheme proposed by Branke et al. [164]), but it scales better as the number of dimensions increases.

Regarding Sakawa et al.'s approach [1415], the definition of membership functions for each of the DM's goals and the further integration of preferences into a single membership function are tasks that are far from trivial.

It is worth mentioning that more sophisticated articulations of preferences are also possible with compromise programming. Some approaches such as dynamic compromise programming were suggested in OR long ago [1546], but have not been coupled with a MOEA so far, to the authors' best knowledge.

9.4 Issues Deserving Attention

Regardless of the approach used to handle DM's preferences in an MOEA, there are several issues that should be kept in mind:

- Preserving dominance.
- Transitivity.
- Scalability.
- Group decision making.

Each of these is briefly discussed in the remainder of this section.

9.4.1 Preserving Dominance

It is important to make sure that the preference relationships introduced in the MOEA preserve existing dominance relationships. Otherwise, the search would be biased towards undesired regions of the search space. Despite the fact that this property can be easily preserved in most cases (e.g., [598]), it should be nevertheless kept in mind when proposing approaches that incorporate preferences into a MOEA.

9.4.2 Transitivity

The use or lack of intransitivities has been the subject of much debate in the OR field [318, 314, 1147]. It has been argued by some researchers that human beings tend to define preferences that are not necessarily transitive, and there are several examples in the OR literature in which intransitivities of preferences easily occur (e.g., [730, 490]).

The main argument against allowing intransitivities is that their absence considerably simplifies the modeling of preferences; intransitivities can lead to contradictions that are much more difficult to handle. Also, by leaving intransitivities out, the decision model becomes mathematically tractable.

However, the issue remains open, and the French school of MCDM prefers to use outranking procedures that allow intransitivities. However, outranking procedures have been combined with MOEAs by only a few researchers (as seen in the previous section) and utility functions seem to be preferred.

9.4.3 Scalability

Some early researchers indicated that MAUT was sound only when few attributes were considered [1752]. MOEAs in general are victim of the "dimensionality curse" [115], because they tend to become cumbersome or even useless as the number of objectives is increased (see for example [1304, 1303, 727]).

Some of the approaches reviewed in this chapter, such as preference relations, are very sensitive to the number of objectives and to changes in the order of the questions asked to the DM. Therefore, they are most likely impractical in applications with a large number of objectives.

Interestingly, aggregating approaches are less sensitive to scalability. An aggregating approach can easily manipulate a large number of objective functions and preferences, because they are integrated into a single scalar value (e.g., [206] where up to 500 objectives are considered at a time). However, the effect of aggregation is diluted as the number of factors to be mixed is increased.

9.4.4 Group Decision Making

It is not trivial to get a DM to express preferences in a consistent way for an arbitrary problem. If this task is by itself difficult, incorporation of preferences from a group of DMs is even more complicated. Unfortunately, the use of group preferences is not an uncommon situation in real-world applications but its introduction raises additional questions.

If there is a group of DMs, each of them probably has their own objectives and priorities. Therefore, some form of negotiation is necessary in order to reach a consensus. Normally, a moderator intervenes to solve the many conflicts that could arise from these situations. The members of the group can express their preferences independently and leave it to the moderator to integrate them. Alternatively, they could be asked to debate and to establish a consensus regarding their priorities (even if this could take a considerable amount of time). In the latter case, the approach used to integrate preferences has to adapt to the ordering of preferences that represents the collective opinion of the group [731].

Table 9.1: The voting preferences of three rational individuals on three candidates. A > B means that A is preferred over B.

Individual	Preferences	A vs. B	B vs. C	A vs. C
1	A > B > C A > C	A	B	A
2	B > C > A B > A	B	B	C
3	C > A > B C > B	A	C	C
Group Preferences		A > B	B > C	C > A

The most common approach is the first, in which the preferences of every individual DM are aggregated into a single utility function that represents the unified preferences of the group. However, the economist Kenneth J. Arrow [59] showed that apart from some very special cases, utility functions cannot be used to aggregate individual preferences into a group utility function. The so-called *Arrow's Impossibility Theorem* has very important consequences in MCDM. To explain how it works, consider the following assumptions [836]:

- **Complete Domain**: The utility function should be able to define an ordering for the group, regardless of the individual members' ordering.
- **Positive Association of Social and Individual Values**: If the group ordering indicates that alternative x is preferred to alternative y for a certain set of individual rankings, and (1) if there are no changes on the ordering of each individual, and (2) each individual's paired comparison against x remains unchanged or is modified in x's favor, then the group ordering must imply that x is still preferred to y.
- **The Independence of Irrelevant Alternatives**: If an alternative is eliminated and the preference relations for the remaining alternatives

remain unchanged for all the individuals, then the new group ordering should remain the same as before.
- **Individual's Sovereignity**: For each pair of alternatives x and y, there is some set of individual orderings which causes x to be preferred to y.
- **Non-dictatorship**: It is impossible that the preferences of the group be always in agreement with the preferences of a single individual.

So, what Arrow's Impossibility Theorem says is that any joint decision process which is reasonably democratic and respectful of individuality (following the assumptions described before) is also irrational or unreliable. It is likely to have at least one of the following problems: a) the order of the decisions affects the final outcome, b) the independence of its elements might not be respected, and c) the unanimous will of its elements might be ignored. A classical example of Arrow's Impossibility Theorem uses three candidates and three voters with the preferences indicated on Table 9.1.

While each individual has a rational set of preferences, it is obvious that combining these to form a group utility function presents a problem (the group utility relation is cyclic, i.e., it is not transitive). In consequence, optimization for the group using this data is impossible. Some authors such as Hazelrigg [668] argue that this situation is not a rare case, but is in fact the norm and as greater detail about the preferences of individuals within a group are provided, the higher the chance of encountering this type of problem.

Some authors have shown that Arrow's conditions can be ignored in practical problems [1461, 1147], but its mere existence has triggered a considerable amount of research in economics [1389], and cannot be disregarded by EMO researchers.

9.4.5 Other important issues

Finally, it is important to establish a set of characteristics that the ideal scheme to incorporate preferences should have. In that regard, the seven prerequisites for a good MCDA approach, defined by Brans and Mareschal [166] are provided next:

1. The approach should take into account the amplitude of the deviations between the alternatives.
2. Scaling should not be required, despite the fact that the criteria of a problem could be (and normally are) expressed in different units.
3. When comparing two alternatives a and b, the MCDA technique should arrive at one of the following conclusions:
    - a is preferred to b, or b is preferred to a.
    - a and b are indifferent.
    - a and b are incomparable.
4. The method should be understandable by the DM. Therefore, black-box effects should be avoided.

5. Parameters that have no economical significance should not be included in the approach.
6. The analysis of the conflicting aspects of the criteria must be available.
7. It is important to have a clear interpretation of the weights of the criteria.

Some recent EMO research (e.g., [318]) has considered these issues, but much more work is still needed.

9.5 Summary

In this chapter, some of the main concepts relating to multi-criteria decision making that have been developed by operations researchers are reviewed. Aspects such as the operational attitude of the DM and the different stages at which preferences can be incorporated are discussed. Also, the two main schools of thought regarding MCDM (outranking and multi-attribute utility theory), are discussed and compared.

Other important issues such as scalability, transitivity and group decision making are also briefly discussed. However, the main emphasis of this chapter is to describe the most representative work regarding preference articulation into MOEAs. The review is very comprehensive and includes brief descriptions of the approaches reported in the literature as well as some discussion of their advantages and disadvantages.

Further Explorations

Class Exercises

1. Explain the importance of adopting mechanisms for the incorporation of user's preferences into a MOEA.
2. Explain the main differences between the French school and the American school of multi-criteria decision making.
3. Design a set of preferences for a certain problem that you propose (or choose from the literature) such that intransitivities occur. These preferences should look sound and logical when analyzed independently, but should "naturally" lead to contradictions when put together.
4. Real-world problems pose great challenges, since many other factors such as uncertainty, noise, and human errors, among others, must be considered in the model of preferences to be used with a MOEA [1249, 1248]. Discuss the role of preference incorporation in the context of real-world applications.
5. Design an artificial multi-criteria decision making problem in which several people play the role of decision makers. Then, simulate a situation in which the decision makers express conflicting preferences. Propose several schemes to incorporate the conflicting preferences into a single set of preferences and discuss the advantages and disadvantages of each of these schemes.

Class Software Projects

1. Implement one of the methods to incorporate user's preferences that was discussed in this chapter (e.g., Cvetković's approach [319]).
2. Implement a scheme for incorporation of user's preferences based on a utility function and incorporate it into a multi-objective evolutionary algorithm.

542   Further Explorations

3. Implement a scheme for incorporation of user's preferences based on outranking and incorporate it into a multi-objective evolutionary algorithm.
4. Design and implement a new scheme to incorporate preferences from the decision maker into a MOEA. Justify your design choices and validate your approach with a set of test functions such as those discussed in Chapter 4. There is a considerable number of MCDM approaches reported in the OR literature that have not been coupled with MOEAs. For example: PROTRADE (see Section 1.7.3 from Chapter 1), SEMOPS (see Section 1.7.3 from Chapter 1), the concept of *stochastic dominance* [489], the *conflict analysis method* [730], the *expected utility maximization* [1620], EVAMIX [1666], NAIADE [1148], QUALIFLEX [1232], and the *multiobjective statistical method* (this is an extension of the surrogate worth trade-off method discussed in Section 1.7.1 from Chapter 1) [627].
5. Read the available surveys on the incorporation of preferences in MOEAs by Coello Coello [263] and Rachmawati & Srinivasan [1313], and discuss the relevance of the research topics covered. Do you consider that the salient research issues that they discuss are still relevant nowadays? Discuss.

Discussion Questions

1. Investigate the different versions of ELECTRE in more detail (see Section 1.7.1 from Chapter 1). Then, choose an application domain in which some version of ELECTRE could be used. What advantages and disadvantages did you find when using ELECTRE to incorporate preferences into a MOEA with respect to the use of a utility function?
2. Write a report containing a comparative study of outranking vs. utility functions. Discuss issues such as ease of use, foundations, parameters required, generality, advantages and disadvantages and limitations. Can you identify certain "standard" situations in which one of these two families of approaches could be more appropriate than the other?
3. Investigate the "conflict analysis method" [730]. What are the preference articulation mechanisms that this method hybridizes? What advantages does the author argue that his hybrid method has? Do you consider this method as a good candidate to be coupled to a MOEA? Why? Why not?
4. Read:
   Bernard Roy, "A Conceptual Framework for a Prescriptive Theory of Decision-Aid", In Martin K. Starr and Milan Zeleny, editors, *Multiple Criteria Decision Making*, volume 6 of *TIMS Studies in the Management Sciences*, pages 179–210. North-Holland Publishing Company, Amsterdam, 1977.

   What is the formal definition provided by Roy for the fundamental problem of global preference modeling? What is the axiom of complete transitive comparability? From the three types of operational attitude of the

DM described by Roy, which do you think that is more appropriate in general terms (i.e., in the absence of specific information)? Why? Discuss.
5. Investigate "evolutionary game theory" [1493, 698, 1688] and discuss its potential applicability to multi-criteria decision making using MOEAs. What are the differences between classical game theory and evolutionary game theory?
6. Some authors have suggested that the so-called "transitivity of indifference" must be removed in multi-criteria decision making. Luce [1023] gives an example: since a DM cannot make a difference between coffee with n grains of sugar and $n+1$ grains of sugar, using transitivity we can infer that the DM cannot make a difference between coffee with 1 grain of sugar and a coffee with 1000 grains of sugar (2 full spoons). This is evidently a bad thing. Investigate this notion and its corresponding order, which is called *semiorder*, and write an essay where you discuss some of its possible implications when designing preference models.
7. Some researchers (see for example [1056]) have discussed the implications of "indifference" in the modeling of preferences. It is usually assumed that if the DM is indifferent between A and B, and prefers C to A, then the DM prefers C to B. However, this should not be taken for granted as the following example (taken from [1056]) shows:
A child is asked to choose between two birthday presents: a pony and a blue bicycle. Since it likes both of them very much, it cannot choose and is therefore indifferent between the pony and the bicycle. The child is further asked to choose between the blue bicycle and a red bicycle with a small bell. The child then chooses the red bicycle. However, we cannot infer that the child would prefer the red bicycle to the pony, since it can still be indifferent between them.
Investigate the notion of "indifference" in more depth and write an essay where you discuss some of its possible implications when designing preference models.
8. Read the comparative study of progressive preference articulation techniques presented by Adra et al. [13] and summarize the main issues raised by the authors when dealing with many-objective problems. Do you agree with such issues? Would you add others? Discuss.

Possible Research Ideas

1. Develop a formal framework for incorporating preferences into a MOEA. Your framework should support an interactive MCDM approach and should be coupled to a MOEA based on Pareto ranking. It is desirable that the framework can also deal with uncertainties in the preferences expressed by the DM. Refer to Cvetković & Parmee [319] for work in this area.

2. Relate the concept of ε-dominance proposed by Laumanns et al. [959] to the incorporation of preferences from the DM into an MOEA. Derive a formal framework that allows the use of this concept interactively to narrow the search space explored by a MOEA.
3. Develop a new scheme to incorporate preferences into a MOEA using as a basis an outranking method. Compare your approach to the use of a utility function with respect to: ease of use, degree of difficulty of the implementation, flexibility, and computational cost. Can you think of an application in which the use of this sort of scheme (i.e., based on an outranking approach) may be undesirable?
4. The use of MOEAs coupled with a decision-making approach could be useful in many application domains. It would be interesting, for example, to use such an approach to control a behavior-based system (e.g., a robot). This problem has a multiobjective nature since there are multiple (heterogeneous) behaviors that interact with each other and it is desirable to take an action such that all of them are simultaneously satisfied. The use of fuzzy rules to generate weights that would narrow the search of a traditional multiobjective optimization technique was proposed by Pirjanian [1275]. Propose a MOEA-based decision-making approach for this application domain.
5. Another interesting application of preference articulation schemes is in constraint-handling [1105, 265]. If preference information is incorporated into a constraint-handling approach, the search can become more efficient, because domain knowledge can be used to bias the selection mechanism of a MOEA [684]. Furthermore, the integration of a multiobjective scheme to handle constraints with some preference information extracted from the problem itself, can produce very efficient constraint-handling schemes to be used with evolutionary algorithms (see for example [1329]). Develop a novel MOEA-based constraint-handling scheme using incorporation of preferences.
6. Some researchers showed several years ago that some of the most commonly used multiobjective optimization approaches of that time could be integrated in what they called a "unified algorithm" [539, 538]. The idea is to factor out common portions of several algorithms. Then, these portions can be used as an engine to which other small modules (representing each a different technique) can be coupled. Adopt this same idea to unify the different approaches used within Multi Attribute Utility Theory (MAUT).
7. The analogy between decision making and social choice was identified several years ago [59, 58], and it is still an active area of research in MCDM [306]. The main idea of this approach is to replace criteria by voters and the choices by candidates. This transforms a multi-criteria decision making problem into one of voting [158]. Since social choice theory has a very solid foundation and the use of a voting scheme does not impose constraints in the number of voters, this transformation is appealing. It should be clear, however, that this transformation is not as simple as it may seem.

For example, the mapping from criteria to voters requires the consideration of several important differences [60]. Propose a MOEA-based voting scheme to incorporate preferences. Some interesting work regarding voting schemes in the context of behavior-based robotics is presented by Pirjanian [1275].
8. Consider the use of rough sets [1262, 1263] for multicriteria decision analysis and their possible incorporation into a MOEA. See for example the survey presented by Greco et al. [597].
9. Venkat et al. [1639] proposed a method for obtaining a set of (preferred) solutions from a larger set. This approach is called the *Greedy Reduction* algorithm and its selection is based on maximizing a scalarizing function of the vector of percentile ordinal rankings of the Pareto optimal solutions obtained. The user is allowed to set the size of the subset desired, and the quality of the solutions obtained is a function of the size of the subset required (high percentile values of the Pareto optimal solutions correspond to a small subset). Study the way in which this sort of mechanism could be coupled to a MOEA.
10. Some researchers have proposed modifications to the selection mechanism of a MOEA, such that it converges towards the "knees", which are solutions in which, a small improvement in one objective produces a large deterioration in at least one other objective (see for example [162, 1312, 1311]). Discuss the relevance of these knees in real-world applications, and propose a novel selection scheme that moves the population of a MOEA towards them.
11. Physical programming [1094] is a multi-objective optimization technique that requires that the decision maker provides a general classification of the goals and objectives using his/her knowledge of the problem. Then, this information is mapped into a utility function for which it is not necessary to define weights. Study physical programming, and propose a hybrid of this technique with a MOEA.
12. Coupling evolutionary algorithms to outranking methods for classification tasks is an interesting area that has been scarcely explored in the specialized literature (see for example [589]). Propose a novel application within this area.

10

Alternative Metaheuristics

> Nature is trying very hard to make us succeed, but nature does not depend on us. We are not the only experiment.
>
> R. Buckminster Fuller

10.1 Introduction

Evolutionary Algorithms (EAs) are not the only search techniques that have been used to deal with multiobjective optimization problems. In fact, as other search techniques (e.g., Tabu search and simulated annealing) have proved to have very good performance in many combinatorial (as well as other types of) optimization problems, it is only natural to think of extensions of such approaches to deal with multiple objectives.

The Operations Research (OR) and EA communities have shown a clear interest in pursuing these extensions. Since the multiobjective formulation of combinatorial optimization problems (e.g., the quadratic assignment problem) are known to be NP-complete [1465, 429], they present real challenges to researchers. Additionally, many real-world problems (e.g., scheduling) require efficient approaches that can at least approximate P_{true} and PF_{true} in a reasonable amount of time.

Any search technique such as those discussed in this chapter, can be extended to deal with multiple objectives in several ways, just as in the case of EAs (see Chapter 2). One could just aggregate the objective functions to form a single scalar value, or to use a target vector approach (defining ideal goals to be achieved by each objective and aggregating their differences with respect to the values obtained). However, dominance can also be checked locally (between two solutions generated by the algorithm) and then keep in an archive every nondominated solution generated over time, so that dominance can also be checked globally (i.e., with respect to this archive). Knowles and Corne [886] have argued that the use of a naive two-membered evolution strategy (with an external archive) is sufficient to generate PF_{true} for relatively complex multiobjective optimization problems.

The issues are then of a different nature. For example: how to move from a certain state to another, or how to ensure that different portions of PF_{true} are being generated rather than only a certain fraction of it. Additionally, other issues such as diversity are an important concern with the heuristics of this chapter as well as with MOEAs.

This chapter is organized as follows. Section 10.2 discusses simulated annealing and the main proposals to extend it to problems with multiple objectives. Tabu search and scatter search as well as their corresponding multiobjective extensions are discussed in Section 10.3. The ant system (including the Ant-Q algorithm) is the subject of Section 10.4. Distributed reinforcement learning is analyzed in Section 10.5. Particle swarm optimization, differential evolution and artificial immune systems are discussed in Sections 10.6, 10.7 and 10.8, respectively.

Finally, Section 10.9 covers other promising heuristics that are good candidates for solving multiobjective optimization problems (i.e., cultural algorithms and cooperative search).

10.2 Simulated Annealing

As mentioned in Chapter 1 (Section 1.4), simulated annealing is a stochastic search algorithm based on the concept called "annealing". The annealing process consists of first raising the temperature of a solid to a point where its atoms can freely (i.e., randomly) move and then to lower the temperature, forcing the atoms to rearrange themselves into a lower energy state (i.e., a crystallization process). During this process the free energy of the solid is minimized (the crystalline state is the state of minimum energy of the system). The cooling schedule is vital in this process. If the solid is cooled too quickly, or if the initial temperature of the system is too low, it is not able to become a crystal and instead the solid arrives at an amorphous state with higher energy. In this case, the system reaches a local minimum (a higher energy state) instead of the global minimum (i.e., the minimal energy state) [407, 1292].

10.2.1 Basic Concepts

Nicholas C. Metropolis et al. [1095] proposed an algorithm to simulate the evolution of a solid in a heat bath until it reached its thermal equilibrium. The Monte Carlo method was used to simulate the process, which started from a certain thermodynamic state of the system, defined by a certain energy and temperature. Then, the state was slightly perturbed. If the change in energy produced by this perturbation was negative, the new configuration was accepted. If it was positive, it was accepted with a probability given by

$e^{\frac{-\Delta E}{kT}}$, where k is the so-called Boltzmann constant, which is a constant of nature that relates temperature to energy [1292]. This process is repeated until a frozen state is achieved [407, 1410].

Thirty years after the publication of Metropolis' approach, Kirkpatrick et al. [861] and Černy [219] independently pointed out the analogy between this "annealing" process and combinatorial optimization. These researchers indicated several important analogies: a system state is analogous to a solution of the optimization problem; the free energy of the system (to be minimized) corresponds to the cost of the objective function to be optimized; the slight perturbation[1] imposed on the system to change it to another state corresponds to a movement into a neighboring position (with respect to the local search state); the cooling schedule corresponds to the control mechanism adopted by the search algorithm; and the frozen state of the system corresponds to the final solution generated by the search algorithm (using a population size of one). These important analogies led to the development of an algorithm called "Simulated Annealing".

---

1. Select an initial (feasible) solution s_0
2. Select an initial temperature $t_0 > 0$
3. Select a cooling schedule CS
4. Repeat
     Repeat
       Randomly select $s \in N(s_0)$ (N = neighborhood structure)
       $\delta = f(s) - f(s_0)$ (f = objective function)
       If $\delta < 0$ then $s_0 \leftarrow s$
       Else
         Generate random x (uniform distribution in the range $(0, 1)$)
           If $x < \exp(-\delta/t)$ then $s_0 \leftarrow s$
     Until max. number of iterations $ITER$ reached
     $t \leftarrow CS(t)$
5. Until stopping condition is met

Fig. 10.1. Simulated annealing pseudo code

The pseudo code of simulated annealing is shown in Figure 10.1 [407]. In this pseudo code, s_0 contains the solution, and minimization is assumed. This algorithm generates local movements in the neighborhood of the current state, and accepts a new state based on a function depending on the current "temperature" t. The two main parameters of the algorithm are $ITER$ (the number of iterations to apply the algorithm) and CS (the cooling schedule), since they have the most serious impact on the algorithm's performance.

---

[1] This slight perturbation is analogous to the mutation operator used in EAs.

Despite the fact that it was originally intended for combinatorial optimization, other variations of simulated annealing have been proposed to deal with continuous search spaces (e.g., [1635]).

10.2.2 Multiobjective Versions

The use of simulated annealing in multiobjective (combinatorial) optimization was initially proposed by Serafini [1466]. His proposal is to use a target-vector approach to solve a bi-objective optimization problem (several possible transition rules are proposed). A solution $\mathbf{x}'$ is generated in the neighborhood of the current solution $\mathbf{x}$. If $f(\mathbf{x}')$ is nondominated with respect to $f(\mathbf{x})$, then it is accepted as the current state, and a set of nondominated solutions is also updated. This is the basic approach used with local search procedures. The set or archive of nondominated solutions constitutes the "memory" of the approach and it allows the generation of several elements of the Pareto optimal set in a single run. Notice, however, that in this case, only local nondominance is used to fill up the archive of solutions and a further filtering procedure is required to reduce the number of nondominated solutions presented to the decision maker (DM).

The key in extending simulated annealing to handle multiple objectives lies in determining how to compute the probability of accepting an individual $\mathbf{x}'$ where $f(\mathbf{x}')$ is dominated with respect to $f(\mathbf{x})$. Serafini [1466] proposed the use of an L_∞-Tchebycheff norm:

$$P(\mathbf{x}', \mathbf{x}, T) = \min \left\{ 1, e^{\max_j \{\lambda_j (f_j(\mathbf{x}) - f_j(\mathbf{x}'))/T\}} \right\} \quad (10.1)$$

where $P(\mathbf{x}', \mathbf{x}, T)$ is the probability of accepting $\mathbf{x}'$, given $\mathbf{x}$, and the temperature T. The weights λ_j are initialized to one and modified during the search process. Serafini [1466] also proposed several other rules, including "cone ordering", which is similar to lexicographic ordering (see Section 1.7.1 in Chapter 2).

Ulungu [1614] and Ulungu et al. [1616, 1618, 1617] propose an approach very similar to Serafini's (called "Multi-Objective Simulated Annealing", or MOSA for short). In their case, however, besides experimenting with the same L_∞-Tchebycheff norm, they also use the following weighted sum in computing acceptance probability:

$$P(\mathbf{x}', \mathbf{x}, T) = \min \left\{ 1, e^{\sum_{j=1}^{k} \lambda_j (f_j(\mathbf{x}) - f_j(\mathbf{x}'))/T} \right\} \quad (10.2)$$

where $P(\mathbf{x}', \mathbf{x}, T)$ is defined as before, and k is the number of objective functions of the problem. The weights λ_j are again defined by the user based on a set of goals (or minimum satisfaction levels for each objective) defined by the DM. Results in this case are compared against PF_{true} of several instances of bi-objective knapsack problems. Such solutions are generated with an enumerative approach based on the branch and bound algorithm [1660]. Obviously, other norms are possible for computing the probability of accepting a dominated solution, and the DM is free to choose any that can bias the algorithm towards promising regions of the search space. Also, in combinatorial optimization problems it may be particularly useful to start the search not with a randomly generated solution, but with a value previously found by another heuristic (e.g., a greedy algorithm [1611]).

Ray et al. [1328] use simulated annealing to solve a multiobjective design problem in which the objectives are handled through a weighted sum. Weights are computed following the guidelines of MAUT (see Chapter 9) [836]. To handle the incommensurable units of each objective function, fuzzy membership functions are used. Because of the aggregating function adopted, the resulting global optimization problem becomes highly multimodal and the authors propose several approaches to deal with it. One of these proposals is a hybrid of simulated annealing and a nonlinear local optimization algorithm that explores the neighborhood of solutions produced by a random perturbation method, attempting to improve the quality of the comparison set used by the acceptance criterion function. The Hooke and Jeeves algorithm [703] with an external penalty function is used. Simulated annealing is found to produce better solutions (in terms of closeness to a global optimum defined by assigning equal weights to the three objective functions of the problem) than two random multi-start search methods that use traditional global optimization approaches (e.g., Hooke and Jeeves algorithm).

Ruiz-Torres et al. [1403] use simulated annealing with Pareto dominance as the selection criterion to solve bi-objective parallel machine scheduling problems. Two types of searches are employed in this case: one to minimize the number of jobs late and other to minimize the average flow-time. Then, neighborhood search is performed and nondominated solutions with respect to the two objectives are considered as the "best" moves. The approach is compared against another heuristic based on pure neighborhood search and against an enumerative approach. Simulated annealing performed well, generating more than sixty percent of the elements of P_{true}.

Czyzak and Jaszkiewicz [322, 323] propose a technique called Pareto Simulated Annealing (PSA). This approach also uses a weighted sum like MOSA. However, the technique adopts a population instead of relying on a single solution at each iteration. An external file is still used to store nondominated vectors, but quad trees are adopted to ensure efficient storage and retrieval of such vectors [626]. Also, when a solution $f(\mathbf{x}')$ is generated in the neighborhood of $f(\mathbf{x})$ (actually, $f(\mathbf{x}')$ represents the closest neighborhood solution to $f(\mathbf{x})$), the weights are increased in those objectives in which $f(\mathbf{x})$ dominates $f(\mathbf{x}')$ and are decreased in those objectives in which $f(\mathbf{x})$ is dominated by $f(\mathbf{x}')$. This is intended to increase the probability of moving as far as possible from $f(\mathbf{x})$. Another interesting aspect of this approach is that it can be easily parallelized, since the computations required for each solution can be done independently from each other. PSA was compared against Serafini's algorithm [1466]. PSA was able to generate a larger number of elements of PF_{true} and distribute them more uniformly. PSA has been applied to capital budgeting [320], the design of a cellular manufacturing system [321], software project scheduling under fuzziness [784], agricultural project scheduling [656], and to a nurse scheduling problem [773].

Hansen [648] uses a normalized aggregating function to combine several objective functions related to a decision making problem in education. Since a single aggregating function is produced, the author uses conventional simulated annealing with a geometric cooling scheme [1648]. An interesting aspect of this work, however, is the use of a fitness sharing function [368] to allow the generation of diverse solutions (something difficult to achieve when a pure aggregating approach is used). In order to make the application of the fitness sharing function possible, the author defines euclidean distances on variable decision space. The approach outperforms both a random search procedure and a hillclimbing search algorithm.

A more naive aggregating function is used by Chang et al. [226] in a portfolio optimization problem. In this case, two objectives are combined into a single value. An interesting aspect of this application, however, is that one of the constraints of the problem (expected return value) has to match exactly a certain specific value. This requires the use of an additional heuristic that can manage this constraint. The same aggregating function is used with a genetic algorithm and with Tabu search. Their results show the GA is able to generate more elements of the Pareto optimal set, but the authors favor the combination of results produced by the three techniques (i.e., Tabu search, the genetic algorithm and simulated annealing).

Lučić & Teodorović [1033] also use an aggregating function in a aircrew rostering problem with two objectives. A set of weights is used to combine the two objective functions of this application, and the result of this combination is used to compute the probability of acceptance of the new solution produced. They also propose the use of a certain threshold to determine thermal equilibrium and therefore, convergence of the algorithm. This threshold is computed using another aggregating function in which the intention is to check the improvement achieved in each objective after an epoch (some number of iterations) and stop whenever such improvement is minimum. The authors also propose an interactive algorithm in which the DM proposes a set of aspiration levels for each objective, given the ideal points for them (i.e., the set of optima considering each of the objectives separately). Weights are also defined in this case, but their values are computed using the aspiration levels defined by the DM and the ideal points. A min-max approach is used in this case to compute the probability of acceptance of the new solution produced. Results are not compared to any other approach.

Matos and Melo [1082] propose the use of simulated annealing for the multi-objective reconfiguration of radial distribution networks. Two objectives are minimized: (1) power losses and (2) number of switching actions. The authors do not provide much information about their multi-objective scheme, but they seem to use some form of lexicographic ordering, since one objective is minimized while the other is maintained at a certain level of satisfaction. Constraints are handled using penalty functions, and the approach is validated using a 52-bus distribution network. An interesting aspect of this work is that the authors introduce, in one of their case studies, a second objective which is non-conflicting, and this leads to better solutions. This is an early application of the concept of "multi-objectivity" that other researchers would later explore with great success [890, 787].

Chipperfield et al. [246] use simulated annealing with an energy function that transforms the multiobjective problem into a single objective min-max problem. The resulting problem is to minimize the maximum deviations of each objective with respect to a set of goals previously defined by the user. The approach also requires a set of weights to properly scale the different units in which the objective functions may be expressed. After performing this transformation, the authors are able to use the conventional Metropolis acceptance criterion described above. The authors suggest the use of simulated annealing for problems in which the choice of initial parameters is difficult or when hillclimbing methods can get easily trapped in local optima. From their experiments, they conclude that a MOEA (MOGA [504]) is easier to use than their multiobjective simulated annealing algorithm, mainly because the MOEA does not require weights for the objectives. However, MOGA was found to be slower than simulated annealing, and the quality of results produced by both techniques was similar.

Suppapitnarm et al. [1541, 1539] propose an approach in which Pareto dominance is used to select nondominated vectors. These nondominated vectors are then stored in an external file. Although the approach manipulates a single solution at a time (which is modified randomly at each iteration), to add it to the file, it has to be nondominated with respect to the contents of the file. Therefore, the external file takes the role of the population in this case (such as in PAES [886]). If a newly generated solution is archived, then it becomes the new search starting point. If the new solution is not archived, then its acceptance probability is given by the following expression:

$$p = \prod_{i=1}^{k} \exp\left\{-\frac{f_i(\mathbf{x}) - f_i(\mathbf{x})}{T_i}\right\} \quad (10.3)$$

where k is the number of objective functions, and T_i is the temperature associated with objective $f_i(\mathbf{x})$. Based on this probability, a potential solution (which is not added to the external file) is evaluated. If accepted, then it becomes the new starting point for the search. If rejected, then the previous solution is adopted again as the starting point. Initially, every $T_i = \infty$ so that all feasible perturbations are accepted. Periodically, T_i is reduced using $T_i' = \alpha_i T_i$, where $0.5 \leq \alpha < 1$. The authors also use a strategy called "return to base" by which at some point along the optimization process, the currently accepted solution is replaced by another one randomly selected from the external file. This aims to maintain diversity and avoid convergence to a local

Pareto front. This approach has been used to solve pressurized water reactor reload core design problems [1243] and to design bicycle frames [1540]. However, the authors of this approach admit that in their own comparative studies, this technique was not able to produce better results (or in less time) than MOGA [504], and its only obvious advantage (according to its authors) is its simplicity.

Karasakal and Köksalan [820] propose the use of simulated annealing for solving bi-objective scheduling problems on a single machine. In a first series of experiments, two objectives are minimized: (1) total flowtime and (2) maximum earliness. In this case, the objectives are optimized separately, and the goal is to minimize total flowtime for a (given) maximum earliness value. In this case, simulated annealing provided better results than those obtained with a heuristic proposed by Koksalan et al. [893]. In a second series of experiments, two different objectives are minimized: (1) flowtime and (2) number tardy. In this case, a nonlinear aggregating function (a weighted Tchebycheff function) is adopted. The results in this case are found to be superior to those generated using a descent algorithm.

Nam and Park [1164] propose a multiobjective simulated annealing algorithm which is based on Pareto dominance. The main novelty of this proposal relies on the six criteria that the authors propose for the transition probability: (1) minimum cost criterion, (2) maximum cost criterion, (3) random cost criterion, (4) self cost criterion, (5) average cost criterion, and (6) fixed cost criterion. After performing a small comparative study, the authors conclude that the criteria that work best are the random, average and fixed criterion. So, in order to compare this approach with respect to a MOEA, the authors adopt the average criterion. Results are compared with respect to the NPGA [709] using a multidimensional version of Kauffman's NK fitness landscape model [827]. The proposed approach presents a competitive performance, but has some diversity problems. The authors suggest the use of niches to deal with this problem. In further work, Nam and Park [1165] propose the Pareto-based Cost Simulated Annealing (PCSA), which estimates the cost of a state by sampling (either the neighborhood or the whole population). These two schemes are really analogous to adopting the tournament selection of the NPGA using a small tournament size in the first case and the whole population in the second. The authors compare their PCSA with respect to MOGA [504], NSGA [1509] and NPGA [709] in 18 test problems of low dimensionality (two objective functions, and between 2 and 4 decision variables). For a quantitative analysis of performance, a uniformity measure is adopted by the authors. Their approach is able to outperform the three other algorithms 67% of the time, but fails to do so in cases in which the Pareto front is disconnected.

Another aggregating approach is used by Thompson [1582] in an application where multiobjective simulated annealing is compared to a multiobjective genetic algorithm. In this case, the total probability of acceptance is the product of the single probabilities of each of the objective functions.[2] A comparative study against MOGA [504] both with and without elitism, indicated that multiobjective simulated annealing was better both in terms of the quality of the solutions produced and in terms of convergence time.

Sarker and Netwon [1428] use simulated annealing with an approach based on deviations from the best values found so far (similar to compromise programming, but linear in nature). This approach is used to solve bi-objective linear programming problems.

Baykasoğlu [105] proposes the use of simulated annealing with preemptive goal programming, for solving multi-objective optimization problems. Preemptive goal programming is a special case of goal programming in which the most important goals are optimized first, followed by the secondary goals. This is then, a form of lexicographic ordering approach in which the aim is to minimize deviations with respect to a certain set of pre-defined goals. For handling constraints, a death penalty approach is adopted by the author (i.e., infeasible solutions are discarded). Six examples are adopted to validate the proposed approach (3 are linear problems, and the other 3 are nonlinear problems). The results obtained are either comparable with those generated by other methods, or better than them.

Suman [1533, 1534] propose a multi-objective version of simulated annealing based on Pareto dominance, an external archive and a scheme that handles constraints within the expression used to determine the probability of moving to a certain state. For that sake, the author uses a weight vector for the acceptance criterion. Such weight vector considers the number of constraints satisfied by a certain solution. In further work, Suman [1535] compares five multi-objective extensions of simulated annealing in several constrained multi-objective optimization problems. The algorithms compared are: SMOSA [1539], UMOSA [1616, 1618], PSA [322, 323], WMOSA [1533, 1534], and PDMOSA, which is proposed in the paper. PDMOSA uses the selection criterion adopted by SPEA [1782], in which the solutions stored in the external archive participate in the selection process. A penalty function is adopted to handle the constraints of the problems studied. From the comparative study, the author concludes that PSA provides the best overall results with respect to quality of the solutions obtained, and PDMOSA with respect to diversity.

It is also possible to define hybrids between simulated annealing and other heuristics. For example, Dick and Jha [389] use clusters of solutions (similar to niches), and each of them is assigned a rank (the sum of ranks of all the solutions contained within it). Cluster-level operators (mutation) and solution-level operators (crossover) are adopted in this case. Reproduction is restricted to individuals within the same cluster and partial domination is considered in ranking clusters. Individuals that violate hard constraints are removed and those which violate soft constraints are handled through a penalty function. Boltzmann trials are also used to select solutions within the same cluster. A

---

[2] This sort of nonlinear aggregating function was also proposed by Serafini [1466].

global temperature-dependent criteria is used to keep diversity in the population. As the authors recognize, their approach shares similarities both with MOGA [504] and with parallel recombinative simulated annealing [1039].

Alves and Clímaco [38] propose an interactive method for solving 0-1 multiobjective linear problems using simulated annealing and Tabu search. First, the authors describe separately simulated annealing and Tabu search, and use each of them to solve multiple-constraint knapsack problems, a p-median problem and a set covering problem. Then, they propose an interactive method in which they impose bounds on the objective function values (which they call "reservation levels"). The idea is to focus the search towards regions in which there are potentially nondominated solutions (which are solutions that are currently nondominated, but could be globally dominated by other solutions that haven't been found yet). The decision maker then chooses a subregion to be explored, and a metaheuristic is run to explore such region. Simulated annealing or Tabu search are selected at each run performed. However, the same objective function is adopted, in order to avoid hurting the convergence of the approach. Pareto dominance is not used as the acceptance criterion, to avoid converging to a local Pareto front. Instead, each objective is separately optimized in turns, aiming to find a good compromise between the two objectives when changing the objective to be optimized. The authors argue that despite the convergence towards the extreme portions of the Pareto front, they can find a diversified set of nondominated solutions due to both the unsteady behavior of simulated annealing at high temperatures, and the diversification phase of Tabu search. These diverse results are shown in graphical form with several examples.

10.2.3 Advantages and Disadvantages of Simulated Annealing

One of the main reasons for the popularity of simulated annealing in single-objective combinatorial optimization has been the existence of convergence proofs for this method [1625, 1]. These convergence proofs are based on the fact that the behavior of simulated annealing can be modeled using Markov chains. Hajek [633], for example, has proved that if the cooling schedule defined by $t_k = c/\log(1+k)$ is used (where k is the number of iterations and c is at least as great as the depth of the deepest local minimum), simulated annealing is guaranteed to converge in asymptotic time. This result, although interesting, is not very useful in practice because it implies that the computational time required by simulated annealing grows exponentially with respect to the size of the problem. Therefore, under certain circumstances, simulated annealing may end up requiring more iterations than exhaustive search [407]. Nevertheless, convergence proofs provide a more solid foundation to the technique.

The first attempt to extend the convergence proofs of simulated annealing to multiobjective optimization problems seems to be the work of Serafini [1466]. Using arguments substantiated on some analysis based on Markov chains, Serafini [1466] shows that by combining two different acceptance rules, it is possible to obtain an expression that gives a higher probability of acceptance to Pareto optimal solutions. The cooling schedule $t_k = c/\log(1+k)$ (mentioned above) is used in this analysis. Villalobos et al. [1654] provide a more complete proof of convergence of multi-objective simulated annealing adopting a suitable choice of acceptance probabilities.

Approaches such as the one proposed by Suppapitnarm et al. [1541, 1539] may be an interesting future research path, because they combine the advantages of local search with Pareto ranking. However, in practice, the authors of this approach found that it had not really been able to outperform a MOEA as indicated before, and it is offered more as an alternative possibly easier to implement rather than being a heuristic designed to replace a MOEA.

A well-known disadvantage of simulated annealing is the difficulty in defining a good cooling schedule. This issue is important both in single- and in multiobjective optimization. Also, the multiobjective strategy adopted should be able to keep diversity along the Pareto front. This is difficult for any local search method.

Finally, the suitability of simulated annealing for parallel implementation (i.e., the main search algorithm) is an important advantage [1039, 323] when efficiency is emphasized. However, few applications of parallel versions of this algorithm applied to multiobjective optimization problems have been reported in the literature, and comparative studies against other heuristics (e.g., parallel MOEAs) are lacking.

10.3 Tabu Search and Scatter Search

Fred Glover proposed Tabu search in a paper that dates back to the mid-1980s [566][3], although this technique has roots that date back to the late 1960s and early 1970s [568, 572]. As credited by Glover [568], the basic ideas of Tabu search (which is nowadays a well-established optimization technique) were also sketched by Hansen in the mid-1980s [655].

The optimum in Tabu search is approached iteratively (as in simulated annealing). At each iteration, an admissible move is applied to the current solution, accepting the neighbor with the smallest cost. Tabu search acts then as a local search procedure. However, unlike conventional hillclimbing schemes, Tabu search allows movements to positions that may not seem favorable as seen from the current state. Tabu search also forbids reverse moves to avoid cycling (these forbidden movements are recorded in a data structure called *Tabu list*). Restrictions are based on the maintenance of a short term memory function which determines for how long a Tabu restriction is enforced or, alternatively, which moves are admissible at each iteration. A given movement can override its forbidden or "tabu" status when a certain (aspiration) criterion is

---

[3] As an interesting historical note, the term *metaheuristic* (or *metaheuristic*), was also proposed in this same paper from Glover [566].

satisfied (e.g., a reduction of the total cost). Due to its memory Tabu search is able to escape local optima (unlike conventional local search procedures).

10.3.1 Basic Concepts

In general terms, Tabu search has the three following components [572]:

- A short-term memory that stores the recently visited points and considers them as "tabu" (or forbidden), such that they are not revisited. This avoids cycling.
- An intermediate-term memory, which stores optimal or near-optimal points that are used as seeds to intensify the search.
- A long-term memory, which records the regions of the search space which have been explored and is used to diversify the search, since it redirects the search towards regions that have been under-explored so far.

---

1. Select $x \in \mathcal{F}$ ($\mathcal{F}$ represents feasible solutions)
2. $x^* = x$ (x^* is the best solution found so far)
3. $c = 0$ (iteration counter)
4. $T = \emptyset$ (T set of "tabu" movements)
5. If $\mathcal{N}(x) - T = \emptyset$, goto step 4
    ($\mathcal{N}(x)$ is the neighborhood function)
6. Otherwise, $c \leftarrow c + 1$
    Select $n_c \in \mathcal{N}(x) - T$ such that:
    $n_c(x) = opt(n(x) : n \in \mathcal{N}(x) - T)$
    $opt()$ is an evaluation function defined by the user
7. $x \leftarrow n_c(x)$
    If $f(x) < f(x^*)$ then $x^* \leftarrow x$
8. Check stopping conditions:
    Maximum number of iterations has been reached
    $\mathcal{N}(x) - T = \emptyset$ after reaching this
        step directly from step 2.
9. If stopping conditions are not met, update T
    and return to step 2

Fig. 10.2. Tabu search pseudo code

The general algorithm of Tabu search is shown in Figure 10.2 [567]. The basic idea of Tabu search is to create a subset T of $\mathcal{N}$, whose elements are called "tabu moves" (historical information of the search process is used to create T). Membership in T is conferred either by a historical list of moves previously detected as not productive, or by a set of tabu conditions (e.g., constraints that need to be satisfied). Therefore, the subset T constrains the search and keeps Tabu search from becoming a simple hillclimber. At each step of the algorithm, a "best" movement (defined in terms of the evaluation function *opt*()) is chosen. Note that this approach is more aggressive than the gradual descent of simulated annealing.

Scatter search was originally introduced by Fred Glover in 1977 [565], as a heuristic for integer programming. It was originally conceived as an extension of a heuristic called surrogate constraint relaxation, which was designed for the solution of integer programming problems. In his original proposal, Glover described scatter search as a method that adopts a series of different initializations to generate new solutions. Thus, a *reference set* of solutions is adopted in this approach. An interesting aspect of scatter search is that, in this case, the approach performs a deterministic search instead of a random one (as done, for example, with genetic algorithms). The idea is to start by identifying a convex combination of the points in the reference set, and then combine them with subsets of the initial reference points in order to define new subregions of the search space to be explored. After that, the central points of these new subregions are examined in a systematic way until no further changes are detected in the reference set. It was until the mid-1990s that Glover provided more implementation details of scatter search [569], which allowed its extension to nonlinear, binary and permutation optimization problems. Glover also proposes to couple scatter search with Tabu search using different types of adaptive memories and aspiration criteria to influence the selection of points from the reference set. However, the interest in scatter search really increased after the publication of a seminal paper in which Glover presents the so-called *scatter search template* [570]. This template is an algorithmic description of scatter search which simplifies the details previously available, and provides more information regarding the generation of initial solutions and their diversification [1077].

10.3.2 Multiobjective Versions

There are several proposals to extend Tabu search in order to handle multiple objectives. Gandibleux et al. [534] propose a technique (also called multiobjective Tabu search, or MOTS) based on the use of an utopian reference point.[4] A measure of each objective improvement with respect to this utopian point is recorded in the tabu memory and used later to update the search direction (this is basically an aggregating function). The utopian point used is the best objective function value for each objective from the solutions in the neighborhood of the current solutions. The weights used in the aggregating function are changed periodically to encourage diversity. Two tabu lists are used in this case: one of normal attributes considered tabu that prevents the

---

[4] It is interesting to note that Gandibleux et al. [534] used the term MOTS to denote their method shortly before Hansen's method, which was also called MOTS. That is the reason why Hansen renamed his technique MOTS* [651].

algorithm from returning to already visited solutions, and another one used to vary the weights. In a further paper, Gandibleux and Freville [533] apply a variant of MOTS to the 0-1 multiobjective knapsack problem. In this case, the authors adopt a procedure, based on the definition of bounds, to reduce decision variable space. The experiments performed validated this procedure in instances up to 500 variables. The algorithm works in two steps. In the first, initial solutions are generated using a greedy algorithm with a convex combination of the objectives. During the second step, the algorithm performs an exploration along the Pareto front based on Tabu search. This procedure allows to produce a quick first approximation of the Pareto optimal set, so that most of the search effort concentrates on this region with a less intensive exploration in other regions.

Hansen [651] proposes MOTS* (*Multi-Objective Tabu Search*). MOTS* first generates random solutions as starting points for the algorithm. A weight vector is determined for each of these solutions, based on a λ-weighted Tchebycheff metric (weak dominance is the concept adopted by Hansen[5]). The idea is to set these weights in such a way that points can be uniformly spread over the Pareto front. Alternative solutions are generated by varying the weights. Then, one chooses the best neighbor (determined by the maximum value obtained from the product of the weights and the objective functions) that does not violate the tabu list. Nondominated solutions are archived throughout the process. As mentioned before, Tabu search tends to generate moves that are in the area surrounding a candidate solution. Therefore, the main problem when extending this technique to deal with multiple objectives is how to maintain diversity so that the entire Pareto front can be generated. Hansen [651] proposes the use of a counter to keep track of the number of solutions by which a point is dominated (this is similar to MOGA's ranking procedure [504]). Diversity is then introduced whenever it is needed according to this counter. To introduce diversity, certain solutions can, for example, be replaced by others that are randomly generated. Hansen [651] also discusses possible ways to incorporate preferences from the user into his algorithm in an interactive way. MOTS* is applied to multiobjective knapsack problems in [649]. Additionally, other extensions are discussed, mainly related to operators normally associated with (single-objective) Tabu search. In a further paper, Hansen [647] presents TAMOCO (Tabu search for multiobjective combinatorial optimization), which is a variation of his MOTS*. In TAMOCO, Hansen adopts a mechanism that forces solutions to drift over the Pareto front. This can be done, for example, by simply replacing a randomly selected solution by another one, which is randomly generated. Hansen also discusses schemes to deal with situations in which most solutions are nondominated, and a "degree of domination" has to be considered, stressing the inherent limitations of pure Pareto ranking. Hansen uses quad trees [626], and finds the point in a quad tree that is closest to a given new point according to a λ-weighted

---

[5] For a definition of "weak dominance", see Section 1.2.3 in Chapter 2.

Tchebycheff metric. He also discusses other issues such as restricting the search to a certain nondominated region, and the possible use of interactive procedures. TAMOCO is validated using a multiobjective capital budgeting problem with 3 objectives: (1) maximize expected short term profit, (2) maximize expected long term profit and (3) minimize the negative environmental impact. This problem is modeled as a knapsack problem and results are compared with respect to Pareto Simulated Annealing (PSA) [320]. Results indicated that PSA produced better solutions with respect to a metric defined in terms of an achievement scalarizing function. However, Hansen indicated that PSA could not generate the entire Pareto front, whereas TAMOCO succeeded at this. There is also another paper by Hansen [652], where he discusses the use of substitute scalarizing functions based on the L_p norm as an alternative mechanism to guide a local search based heuristic. This idea is tested using a Tabu search algorithm which is applied to the multi-objective traveling salesperson problem (moTSP). The experiments performed by the author show a significant improvement when introducing substitute scalarizing functions, although the author admits that this sort of approach may not be as useful in other multi-objective combinatorial optimization problems.

Hertz et al. [676] propose three approaches to deal with multiple objectives. The first is a simple weighted sum of objectives. However, the authors realize that the approach cannot handle properly more than two objectives. In such cases, it becomes too difficult to generate proper weights for each objective so that such weights reflect the importance of each objective. This process can also become computationally expensive due to the interaction required with the DM (see Chapter 9).[6] The second approach is to define a hierarchy of objective functions. This is basically lexicographic ordering. One objective is considered at each iteration, but additional objectives can be used to break ties. The authors indicate that this approach does not work properly either, because the ordering imposed on the objectives affects the search (i.e., the most important objective in the hierarchy is given higher priority in all the trade-offs generated). The third approach, which the authors finally adopt for the cell formation problem under study, is the ϵ-constraint method (see section 1.7.2 in Chapter 2): objectives are processed sequentially, using only one at a time and treating the others as constraints. A strategic oscillation procedure [572] is used to deal with the constraints (i.e., the secondary objectives) of the problem. The main idea of the strategic oscillation is to drive the search in such a way that alternatively seeks inside and outside the feasible region, whose boundaries are previously selected. This is achieved either by directly manipulating the objective function (e.g., with a penalty function [303, 212, 482]) or by enforcing a set of moves that lead to certain specified regions.

---

[6] Other researchers have also used simple aggregating functions to handle multiple objectives with Tabu search (e.g., [226]).

Baykasoğlu [104, 106] proposes an approach, which he also calls MOTS, to solve multiobjective optimization problems using Tabu search. The proposed approach has two lists. The first is called the *Pareto list*, and stores the nondominated solutions found during the search. The second is called the *candidate list* and stores all the solutions which are not globally nondominated, but were locally nondominated at some stage during the search. Such solutions in the candidate list are used as seeds in order to diversify the search. For the diversification, variable movement strategies are applied to generate neighboring solutions around the seed solutions. The author claims that his approach can be used to problems with any type of variables (integer, zero-one, discrete or continuous) and to any type of functions (linear, nonlinear, convex and nonconvex). He also suggests the use of compromise programming to choose a single solution out of the Pareto optimal set. The approach is validates using several engineering optimization problems taken from the specialized literature. Results are indirectly compared with respect to the NSGA [1509], but the author only argues that no much detail about the solutions produced by the NSGA are provided and only indicates that his solutions are of similar quality. Results are also compared with respect to MOSES [260], but in this case, numerical values are actually compared.

Ho et al. [692] extend Tabu search in several ways, in order to solve multiobjective optimization problems. The authors adopt Pareto ranking (similar to Fonseca and Fleming's ranking scheme [504]), an external archive (which is bounded in size), fitness sharing, and a neighborhood generation based on the construction of concentric "hypercrowns", as suggested by Siarry and Berthiau [1486]. An interesting aspect of this work is that the authors adopt the contact theorem [1217] to test if a solution is Pareto optimal. The approach has two possible termination criteria. The first consists of reaching a maximum number of (pre-defined) iterations. The second is reached when the external archive is full and the point density of each of its members exceeds a certain threshold value. This approach is validated on four different cases of an engineering optimization problem, in which the aim is to determine the optimal geometry parameters of a multisectional pole arc of large hydro-generators. Two objectives are considered: (1) maximize the amplitude of the fundamental component of the flux density in the air gap, and (2) minimize the distortion factor of a sinusoidal voltage of the machine on no-load condition. The results are not compared with respect to any other approach.

Jaeggi et al. [766] propose a multi-objective parallel Tabu search approach that operates on continuous search spaces. This approach (which, was also called MOTS) uses as its search engine a Tabu search implementation proposed by Connor and Tilley [295], which uses the Hooke and Jeeves optimization method [703] coupled with short, medium and long term memories. In this case, however, the search point comparison adopts Pareto dominance and the Hooke and Jeeves movements are also generalized in order to consider problems with multiple objectives. The multiobjective version of Hooke and Jeeves allows both downhill and uphill movements, since the next point is simply any nondominated solution from those produced so far. The authors also adopt a pattern move strategy that repeats the previous move before applying Hooke and Jeeves. This accelerates the search along known downhill directions. The medium term list is replaced by an external archive that stores the nondominated solutions found so far. The intensification process adopted relies on the use of the points discarded by the Hooke and Jeeves method during the search. The authors also adopt a functional decomposition scheme that allows to execute in parallel the objective function evaluations at two stages during the search: when performing the Hooke and Jeeves' moves and during the diversification process. The parallel Tabu search algorithm is implemented using a master/slave scheme, under MPI. This approach is compared with respect to the NSGA-II [374] using some benchmark problems taken from the specialized literature, in which it's found to produce competitive results, except for one problem where it converges to a local Pareto front. Additionally, the approach is also adopted to solve an aerodynamic shape optimization problem. In this case, however, results are not compared with respect to any other technique. In a further paper, Jaeggi et al. [767] add a constraint-handling mechanism to this approach. Constraints are handled in a very simple way: a solution that violates any constraint is considered tabu and the search is not allowed to visit that point. During diversification, the algorithm is forced to loop until a feasible solution is generated, but the use of a penalty function is also considered as an alternative by the authors. In this case, the approach is validated using several constrained multiobjective optimization problems, and results are compared with respect to the NSGA-II [374].

Kulturel-Konak [915] proposes the multinomial Tabu search (MTS) algorithm for multi-objective combinatorial optimization problems. The idea is to use a multinomial probability mass function to select an objective at each iteration. Such objective is called "active" and is optimized at that iteration. The approach uses an external archive in which solutions are added based on Pareto dominance. The approach also performs neighborhood moves, and uses a diversification scheme based on restarts (i.e., if, during a certain number of moves, the external archive hasn't been updated, then one of its solutions is kept as the new current solution and all the others are deleted). The length of the tabu list is dynamically varied at every 20 iterations, and the algorithm adopts a stopping criterion based on the maximum number of iterations during which the external archive hasn't been updated. Finally, a constraint-handling technique, based on a penalty function, is also adopted. MTS is used to solve different versions of the series parallel system redundancy allocation problem, including instances with two and three objectives. Results are compared with respect to TAMOCO [647]. MTS clearly outperformed TAMOCO, since it produced more nondominated solutions and such solutions dominated the solutions produced by TAMOCO.

Xu et al. [1719] use Tabu search with an aggregating function to solve a multi-objective flight instructor scheduling problem. Three objectives are considered: (1) minimize labor cost, (2) maximize workload consistency and (3) maximize flight instructor satisfaction and their assignments. Dynamic neighborhood moves are adopted, which allows, for example, to jump into different feasible regions, and also changes the weights used in the aggregating function, so that a single objective is explored during each move. However, an interesting aspect of this algorithm is that a set of rules based on Pareto dominance are used when evaluating neighborhood moves, so that some moves during the search may be based on Pareto dominance. The authors adopt a greedy algorithm to generate a starting solution (this is called the "pre-heuristic"). The proposed approach is validated using real-world data from an US airline. The authors also indicate that this approach has been implemented in a computerized decision-making system for flight instructor scheduling at a major US airline.

Tan et al. [1560] present the Exploratory Multiobjective Evolutionary Algorithm (EMOEA), which combines features of Tabu search and evolutionary algorithms for multiobjective optimization. EMOEA uses an evolutionary algorithm with Pareto ranking as its search engine, but it incorporates several mechanisms from Tabu search in order to improve its performance. EMOEA uses two lists: an individual list and a tabu list. Both lists interact and influence each other along the evolutionary process. The approach adopts conventional genetic operators such as tournament selection, crossover and mutation. However, it also incorporates a mechanism called lateral interference, which is based on a population distribution method which can be applied either in decision variable space or in objective function space. This method is inspired on a resources competition scheme, which is based on two general rules: exploitation competition and interference competition. In exploitation competition, individuals with higher fitness are more likely to survive. Interference competition only takes place among individuals with the same fitness level, and it consists of a procedure similar to fitness sharing in which a niche radius is defined (this is called "territory" in EMOEA). Lateral interference is, however, claimed to be advantageous with respect to fitness sharing, since it does not require any user-defined parameter settings. EMOEA also incorporates an individual examination rule that incorporates a tabu restriction that avoids repetition of currently found good individuals. Given a tabu list and an individual list, every individual in the EMOEA is examined (with respect to the tabu list) for acceptance or rejection from the individual list. At the same time, the tabu list is updated when an individual dominates any member of the tabu list (the individual being dominated is replaced in that case). Individuals that are dominated by the tabu list are kept in the individual list, as long as they are not in the tabu list. If an individual is not dominated by the tabu list, and such tabu list still has room, then the individual will be added to it. EMOEA is validated using several benchmark problems reported in the specialized literature. Results are compared mainly with respect to MOGA [504], although in one test problem, the authors also refer to the NSGA [1509].

An interesting hybrid approach between genetic algorithms and Tabu search for multiobjective optimization is proposed by Kurahashi and Terano [932]. This approach keeps two tabu lists with the best individuals generated by a GA. The best individual produced at each generation (i.e., the only nondominated vector) is stored in a short-term tabu list, so that it cannot bias the selection process. The best individuals found along the evolutionary process and having different genotypes are stored in a long-term tabu list (i.e., P_{known}).

Another hybrid approach is proposed by Khor et al. [846]. In this case, an evolutionary algorithm is used to perform multiobjective optimization based on Pareto ranking. A tabu list is used to avoid the repetition of any search paths previously explored. The aspiration criteria are defined in such a way that a uniform distribution of elements is enforced within the tabu list. The authors also use lateral interference (as in [1560]), which intends to promote diversity efficiently and without the need of extra parameters (as in the case of fitness sharing). As indicated before, lateral interference is a form of dynamic niching technique in which there are two types of competition: exploitation competition (based on fitness) and interference competition (based on territory).[7] The authors argue that the use of Tabu search coupled with an evolutionary algorithm improves the search and avoids the algorithm being trapped in local optima.

Burke et al. [185] propose a population-based hybrid metaheuristic that combines hillclimbing, simulated annealing, Tabu search and a mutation operator. The algorithm initializes and improves the population using hillclimbing. Then, simulated annealing with a distributed cooling scheme for all individuals is applied as a self-adaptation process. Mutation is employed as a way to maintain diversity. Finally, cooperation among individuals is induced using principles from Tabu search (i.e., a list of forbidden and attractive moves is kept). The approach is used to solve a bi-objective space allocation problem. Results are compared to a single-objective optimization technique. The authors indicate that their proposed approach produces results with better overall quality, but at a higher computational expense than a single-objective optimization technique.

Balicki and Kitowski [82] propose the use of Tabu search to perform mutation in an adaptive evolutionary multiobjective optimization algorithm. The idea is to use Tabu search to perform local search around solutions that have been previously produced by an evolutionary algorithm. Since the authors were interested in combinatorial optimization problems, the use of local search seems an obvious choice to enhance the performance of an evolutionary algorithm. The authors use both genetic algorithms and evolution strategies in their work.

---

[7] A *territory* refers to the area within which a certain individual can interfere with others in consuming a certain resource.

Beausoleil Delgado [111] proposes an interesting approach in which Tabu search is used with an aggregating function to generate an initial set of solutions. The weights used by Tabu search are modified such that a sufficient variety of points can be generated. Instead of using Pareto ranking, the approach uses the Kramer choice function [1044]. These solutions are later used as reference points for scatter search [569] and path relinking [572, 573]. Scatter search then creates new points from linear combinations of subsets of the current reference points. Path relinking generates new solutions by exploring trajectories that "connect" high-quality solutions. The idea is to start from a certain solution and then generate a path in neighborhood space that leads towards the other solutions, which are called "guiding" solutions. As Glover and Laguna [572, 573] indicate, path relinking is really a direct extension of scatter search. The author uses path relinking in conjunction with extrapolated path relinking [572] in order to integrate intensification and diversification strategies. Also, some further refinements to reduce the number of iterations required by the algorithm are used. For example, duplicates are eliminated from the reference set of solutions and a filtering is performed to ensure that the contents of such a set consists of only nondominated vectors (this has to be done separately because the approach does not use nondominance to select solutions). The approach is used to solve multiobjective combinatorial optimization problems. In further work, Beausoleil [110], proposes a revised version of his approach, which he calls "MOSS" (multiobjective scatter search). The main extensions have to do with the capabilities to handle constrained nonlinear optimization problems (the original version was designed for unconstrained combinatorial optimization problems). MOSS adopts a Multistart Tabu Search (TS) procedure as a diversification generation method. This procedure can be seen as a sequence of Tabu searches where each of these searches has a different starting point, recency memory, and aspiration threshold. All these searches share the frequency memory to bias the search to unvisited or less visited regions. Initially, the seeds for the search are just randomly generated positions. However, over time, the current nondominated set is used as the reference set from which new solutions are generated. This allows the combination of exploration and exploitation of the search space. The approach also has a mechanism (called *critical event design*) that avoids duplicates both in decision variable space and objective function space. The Kramer Choice function is used to divide the reference solutions in two subsets. Euclidean distances are adopted as a measure of dissimilarity in order to identify diverse solutions to be combined (using linear combinations). MOSS is compared with respect to the NSGA-II [374], SPEA2 [1775] and PESA [301] in a set of benchmark problems of two and three objectives.

Zaliz et al. [1747] propose a multi-objective version of scatter search which adopts Pareto ranking. The reference set is divided in two halves: one contains nondominated solutions stored in an arbitrary order, and the other half contains the most diverse solutions[8] from the first half. If there are not enough solution to fill the second half, then dominated solutions are adopted. This approach also uses a local search technique, and combination operators which, however, are not properly described. This approach is able to outperform both SPEA [1782] and the $(\mu + \lambda)$ MOEA [1427] in the identification of interesting qualitative features in biological sequences.

Vasconcelos et al. [1636] proposes an approach in which the scatter search algorithm is modified with a nondominated sorting procedure similar to the one adopted by the NSGA [1509]. The proposed approach is called M-scatter search. In this case, the *reference set* contains the solutions obtained so far, and they are ranked based on Pareto dominance and on a niched penalty technique. This niched penalty technique penalizes each solution of a certain front (the NSGA creates several fronts or layers of nondominated individuals) based on the number of points of this front that are closer than a certain predefined niche radius. The authors adopt an external archive (called *offline set*). M-scatter search is validated using two benchmark problems and an electromagnetic problem. Results are compared with respect to the NSGA and the NPGA.

Corberán et al. [297] use scatter search to solve the problem of routing school buses in a rural area. Two objectives are minimized: (1) the number of buses used to transport students from their homes to school and back, and (2) the time that a given student spends on the bus. This approach considers each objective separately, but uses a filtering procedure in which the best trade-off solutions with respect to the two objectives are identified. Two constructive heuristics are adopted: one is based on a clustering mechanism, where each cluster corresponds to a route, and the other is based on creating sectors around locations that are sequentially selected. The approach also contains a mechanism that allows to create new solutions from the combination of two other existing solutions (a voting scheme is used for this sake), as well as other mechanisms to reduce the length of a route and to remove a location from one route and insert it into another route. In this case, the reference set contains nondominated solutions, although the objectives are really optimized separately. Thus, this approach can be seen as a variation of lexicographic ordering in which there is an external archive that considers Pareto optimality as its filtering criterion. Results are compared with respect to a Tabu search scheme previously proposed. The authors indicate that scatter search produces an average improvement of 23.4% over the approach based on Tabu search.

Gomes da Silva et al. [592] propose a scatter search approach to solve bi-criteria multi-dimensional {0,1}-knapsack problems. The approach, however, consists on using surrogate multipliers in order to transform the original bi-criteria problem into a single-objective problem. Nevertheless, in a further paper, Gomes da Silva et al. [593] develop a truly multiobjective approach based on scatter search. In this case, the initial set of solutions is obtained by

---

[8] The authors do not indicate in which space is diversity measured.

optimizing each of the two objectives separately. These solutions are used to generate the others. During the improvement stage, two heuristics are adopted in order to satisfy each of the two objectives separately, and then a combination method is adopted, but maintaining feasibility at all times. Nondominated solutions are retained in a secondary population (the reference set). The update method for the reference set is based on a property that says that given a nondominated solution, another solution may be found by changing a small number of variables. Both proximity and diversity of the nondominated solutions produced by the approach are evaluated. Additionally, results were also compared with respect to the use of exact methods for small problem instances. The authors conclude that their approach requires a relatively small amount of time to obtain a good estimate of the Pareto optimal set, even when considering fairly large problem instances. However, the method is also found to be unable to generate the complete Pareto optimal set for large problem instances. Thus, the authors recommend to adopt this approach to locate promising regions of the search space which could be further explored using exact methods.

Nebro et al. [1177] propose a scatter search algorithm for solving both constrained and unconstrained numerical multiobjective optimization problems. The proposed approach is based on a template introduced in [570], and contains five methods:

1. **Diversification generation method:** It is responsible for obtaining an initial set of diverse solutions.
2. **Improvement method:** Uses local search to improve the initial set of solutions. Pareto dominance and feasibility rules are adopted as part of this method.
3. **Reference set update method:** The reference set contains solutions which combine high quality with diversity. Such solutions are used to generate new individuals through the application of the solution combination method. The authors use in this case nondominated sorting and a crowding procedure similar to the one adopted in the NSGA-II [374].
4. **Subset generation method:** It generates subsets of individuals which are used for creating new solutions.
5. **Solution combination method:** It finds linear combinations of the reference solutions.

The proposed approach, which is called Scatter Search for Multiobjective Optimization (SSMO) is compared with respect to the NSGA-II using a wide variety of test functions. Two versions of SSMO are evaluated: the first (called SSMOv1) select the best individuals from the reference set using Pareto ranking and crowding, and the second (called SSMOv2) uses clustering to obtain the centroids that compose the reference set. Both approaches are found to be competitive with respect to the NSGA-II. SSMOv2 is found to provide better convergence than SSMv1, and the authors claim that the use of centroids to build the reference set is a promising scheme to improve the accuracy of the algorithm. In further work, Nebro et al. [1173] propose an approach called Archive-based Scatter Search (AbSS). This approach is based on the same template (and on the same five methods) than SSMO. However, AbSS introduces features from three other MOEAs. It uses the archive management scheme of PAES [886], but instead of using the adaptive grid of PAES, it adopts the crowding distance of the NSGA-II [374]. Additionally, the selection of solutions from the initial set adopted to build the reference set, relies on the density estimation mechanism of SPEA2 [1775]. AbSS is compared with respect to SSMO [1177], NSGA-II [374] and SPEA2 [1775], using a variety of test functions (both constrained and unconstrained) and three performance measures: (1) generational distance [1626], spread [374], and Hypervolume [1770]. Results indicate that AbSS outperforms the other approaches with respect to diversity, but in terms of convergence, there is no clear winner.

Molina et al. [1118] propose a Scatter Search Procedure for nonlinear Multiobjective Optimization (SSPMO). This approach includes two different phases: (1) generation of an initial set of nondominated solutions using several Tabu searches and (2) combination of solutions and updating of the reference set via scatter search. For the first phase, the authors adopt MOAMP [191], which links several tabu searches using a global criterion method (see Section 1.7.1). The ideal vector is also found during this initial phase. In the second phase, the solutions previously found are combined to produce new ones. The improvement method in this case tries to fill the gap between the extreme points of the Pareto front and the compromise point (located in the "knee" of the Pareto front). SSPMO is compared with respect to MOAMP [191], SPEA2 [1775] and MOSS [110], using four performance measures. Results indicate that SSPMO is a competitive approach, generating better solutions in terms of both convergence and distribution along the Pareto front.

Rao and Arvind [1319] propose the use of scatter search for lay-up sequence optimization of laminate composite panels. Two objectives are minimized: (1) weight and (2) cost. The problem is subject to buckling and frequency constraints. In the initial population creation method, the authors adopt two different approaches to ensure that the solutions generated are sufficiently diverse. Since this is a combinatorial optimization problem, local search schemes are used for the improvement solution method adopted. In the first set of experiments, a single-objective version of scatter search is compared with respect to other approaches previously reported in the specialized literature. In this case, buckling load is maximized. Then, a constrained version of the problem is considered. Finally, the authors consider the optimization of a composite laminate made from two materials. This problem is considered both as single-objective and was later transformed into a multiobjective one. The multiobjective approach consists of a linear aggregating function in which the weights are varied in order to generate the Pareto front. The authors discuss the advantages of the multiobjective approach, indicating that when cost is a primary consideration, the plate was made of glass-epoxy and when weight is a primary consideration, it is made of graphite-epoxy. In between, several compromises are possible.

Basseur et al. [100, 97] propose to hybridize a genetic algorithm with path relinking for solving bi-objective permutation flowshop problems. The authors adopt as their search engine an adaptive genetic algorithm, which uses the Pareto ranking scheme from the NSGA [1509], two-point crossover, a combined sharing scheme (i.e., sharing is applied both in decision variable space and objective function space), and an adaptive selection scheme that chooses among four mutation operators designed for permutation problems. The main idea of the hybrid proposed in this paper is to use path relinking to improve the solutions produced by the genetic algorithm. For that sake, the authors propose the following mechanisms:

- A neighborhood operator which generates intermediate solutions.
- A distance measure (in decision variable space).
- A selection criteria for the solutions to be linked. In this case, Pareto optimal solutions generated by the genetic algorithm are randomly selected.
- Path selection and path generation schemes.
- A Pareto local search algorithm, which refines the solutions found by the path relinking scheme.

This hybrid approach is validated using several instances of the flowshop scheduling problem, whose sizes are 50 and 100 jobs. Results are compared with respect to the use of the genetic algorithm itself. Results indicate that the introduction of path relinking significantly improves convergence.

Finally, it is important to mention that there are few comparative studies of heuristics such as multiobjective simulated annealing and multiobjective Tabu search. Viana and Pinho de Sousa [1643], for example, compare PSA [323] against MOTS* [650] in the resource constrained project scheduling problem (a generalization of the job shop scheduling problem). Three objectives are minimized in this case: project completion time (makespan), mean weighted lateness of activities, and the sum of the violation of resource availability. In their study, the authors use two metrics suggested by Czyzak and Jaszkiewicz [323] and Zeleny's ideal point [1752] to compare their results. MOTS* is found to be better than PSA both in terms of the quality of the solutions produced and in terms of their corresponding computational cost.

Gil et al. [562] perform a study in which they compare the performance of Serafini's Multi-Objective Simulated Annealing (SMOSA) [1466], Ulungu's Multi-Objective Simulated Annealing (UMOSA) [1616], Czyzak's Pareto Simulated Annealing (PSA) [320], Hansen's Multi-Objective Tabu Search (MOTS) [651] and Knowles' Pareto Archived Evolution Strategy (PAES) [886], in multi-objective network partitioning. Two objectives are considered: (1) load balancing among subnetworks, and (2) the amount of communication among nodes belonging to different sub-domains. The authors adopt two performance measures to compare their results: (1) coverage of two sets [1782] and (2) average size of the space covered [1782]. The results indicated that simulated annealing was able to outperform both Tabu search and evolutionary algorithms. From the three multi-objective versions of simulated annealing that the authors evaluated, UMOSA had the best performance, but PSA produced the highest number of nondominated solutions. When comparing results with respect to single-objective versions of the network partitioning problem, the authors found that in several cases, the results obtained by the multi-objective algorithms studied (particularly, simulated annealing) were very close to the values obtained by the single-objective optimizers.

Marett and Wright [1064] perform a study in which multiobjective versions of simulated annealing and Tabu search are compared in flow scheduling problems. Four objectives are minimized: total setup time, total setup cost, total holding time, and total late time. A linear combination of weights is used by all the heuristics used. The authors use a geometric temperature reduction scheme with feedback [426] for their simulated annealing implementation, and ten versions of Tabu search (using different acceptance criteria). The authors also implement a best improvement repeated descent (BIRD) algorithm, and a first improvement repeated descent (FIRD) algorithm. Simulated annealing and Tabu search are found to be superior to both BIRD and FIRD. However, when comparing simulated annealing against Tabu search, the results seem to indicate that the first becomes better than the second only when the complexity of the problem increases.

10.3.3 Advantages and Disadvantages of Tabu Search and Scatter Search

Tabu search has been widely and successfully used in combinatorial optimization [572]. However, its use in continuous search spaces has not been common due to the difficulties of performing neighborhood movements in continuous search spaces. In fact, the extension of multiobjective Tabu search to continuous search spaces while feasible, may become impractical because of the discretization of the search space required. Hybrids with other techniques (e.g., GAs) seem more promising to deal with continuous search spaces [569, 571].

The main issue when extending Tabu search to handle multiple objectives is how to keep diversity so that points not necessarily within the neighborhood of a candidate solution can be generated. It is important to devise clever procedures to build the tabu lists and to decide the next state to which the algorithm should move.

The use of Tabu search for exploring the neighborhood of solutions produced by another approach (e.g., an evolutionary algorithm) seems a natural choice when dealing with combinatorial optimization problems. However, the main disadvantage of this hybrid approach is the extra computational cost associated with the local search. For example, Balicki and Kitowski [82] indicate that the use of Tabu search implies an extra $O(n^3)$ complexity in their approach. It also becomes harder to design a good Tabu search method as the number of objective functions increases, as indicated by some experimental studies [1064].

10.4 Ant System

The Ant System (AS) is a metaheuristic developed by Marco Dorigo, which was inspired by colonies of real ants, that deposit a chemical substance on the ground called *pheromone* [403, 293, 405, 406]. This substance influences the behavior of the ants: they tend to take those paths where there is a larger amount of pheromone. Pheromone trails can thus be seen as an indirect communication mechanism among ants. From a computer science perspective, the AS is a multi-agent system where low level interactions between single agents[9] (i.e., artificial ants) result in a complex behavior of the entire ant colony.

10.4.1 Basic Concepts

Figure 10.3 graphically shows an example of the typical behavior of a colony of real ants. When the ants leave initially the nest, (1) they follow random patterns. (2) Over time, they start following a common path. (3,4) When faced with an obstacle, some choose to go around it through the left side of the obstacle and others avoid it going through the right. (5) Over time, the whole colony follows a common path (the shortest way) due to the pheromone trials.

There are three main ideas from colonies of real ants that have been adopted in the AS:

1. Indirect communication through pheromone trails.
2. Shortest paths tend to have a higher pheromone growth rate.
3. Ants have a higher preference (with a certain probability) for paths that have a higher amount of pheromone.

Additionally, an AS has certain capabilities nonexistent in colonies of real ants. For example:

1. Each ant is capable of estimating how far it is from a certain state.
2. Ants have information about the environment and use it to make decisions. Therefore, their "behavior" is not only adaptive, but also exhaustive.
3. Ants have memory, since this is necessary to ensure that only feasible solutions are generated at each step of the algorithm.

---

[9] An "agent" in artificial intelligence is something that perceives and acts, and is normally used as an abstraction model of human behavior [1407].

10.4 Ant System

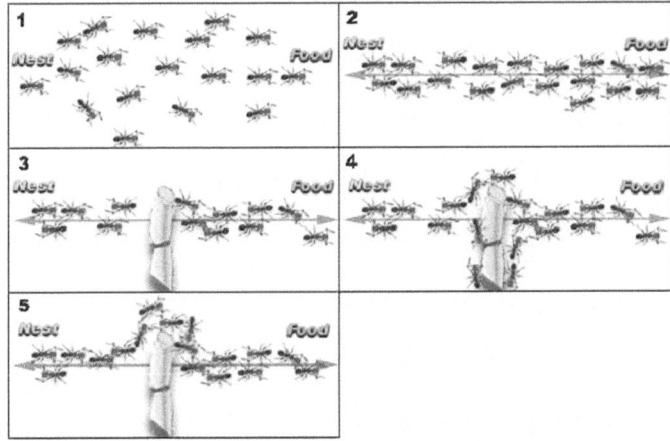

Fig. 10.3. Behavior of a colony of real ants

The AS was originally proposed for the traveling salesman problem (TSP), and most of the current applications of the algorithm require the problem to be reformulated as one in which the goal is to find the optimal path of a graph. A way to measure the distances between nodes is also required in order to apply the algorithm [404].

Gambardella and Dorigo [530] realized the AS can be interpreted as a particular kind of distributed learning technique and proposed a family of algorithms called Ant-Q. This family of algorithms is really a hybrid between Q-learning [1687] and the AS. The algorithm is basically a reinforcement learning approach [1544] with some aspects incrementing its exploratory capabilities:

1. Ant-Q uses several agents (each of which is looking for a solution to the problem) instead of only one.
2. Ant-Q uses a domain dependent heuristic function that indicates how good is an action performed by an agent going from a certain state s to another state s'.
3. Ant-Q uses an action selection rule. This selection rule considers both the heuristic function previously mentioned and the evaluation function for each pair state-action, $Q(s,a)$. This combined function can be called $C(s,s')$, and the action selection rule can be described as:

$$s' = \begin{cases} \arg\max_{s'} C(s,s') & \text{if } q \leq q_0 \\ s_{rand} & \text{otherwise} \end{cases} \quad (10.4)$$

where q is a value randomly selected with a uniform probability in the range [0,1], q_0 ($0 \leq q_0 \leq 1$) is a constant parameter. The value of q_0 must be chosen such that to greater q_0, lower probability of randomly selecting $s_{rand} \in \mathcal{S}$ ($\mathcal{S}$ is a state randomly selected according to a probability function formed by the values of the heuristic function and the value function $Q(s, a)$).

---

1. Initialize $Q(s, a)$ arbitrarily
2. For i = 1 to N (N = number of episodes)
      For i = 1 to m (m = number of agents)
        Initialize $s = s_0$ for the m agents
        Repeat for f steps in the episode
           For i = 1 to m
              Select a in s using rule (10.4)
              Apply a and observe r, s'
              $Q(s, a) \leftarrow Q(s, a) + \alpha[\gamma \max_{a'} Q(s', a') - Q(s, a)]$
              $s \leftarrow s'$
           End Loop
        Until s is a terminal state
      End Loop
3. Compare the m solutions found and select best
      For all the $Q(s, a)$ in the best solution
        $Q(s, a) \leftarrow Q(s, a) + \alpha[r + \gamma \max_{a'} Q(s', a') - Q(s, a)]$
        (α is the learning step)
        (γ is the discount factor)
4. End Loop
5. Report best solution found

Fig. 10.4. Ant-Q pseudo code

The general Ant-Q algorithm consists of four stages as shown in Figure 10.4 [530]. In the first, all the evaluation functions $Q(s, a)$ are initialized. As in traditional reinforcement learning the initialization is performed using arbitrary values.

The second stage consists of a cycle that starts by assigning to each agent a starting state s_0 from which they build their solutions. At each iteration, the agents select one of the possible actions for their current state, based on the action selection rule (10.4). At each transition, the agents update the value function for the pair state-action used in the following equation:

$$Q(s, a) \leftarrow Q(s, a) + \alpha[\gamma \max_{a'} Q(s', a') - Q(s, a)] \quad (10.5)$$

This is the update rule of Q-learning, without considering the reward term r.

The cycle terminates when all the n agents have finished building a solution (i.e., when a terminal state is reached).

In the third stage, the solutions built by the agents are evaluated and the best overall solution is rewarded. All the action value functions involved in the generation of this (best) solution are rewarded as well. Updates are performed using the following equation:

$$Q(s,a) \leftarrow Q(s,a) + \alpha[r + \gamma \max_{a'} Q(s',a') - Q(s,a)] \qquad (10.6)$$

where r is the reward.

Finally, in the fourth stage the termination condition is evaluated to see if the algorithm should stop or continue. The values of α and γ, as well as the number of agents, have to be defined by the user and their setup normally requires a trial-and-error process.

10.4.2 Multiobjective Versions

Mariano and Morales [1067, 1068] propose an extension of the Ant-Q algorithm (called MOAQ, or Multi-Objective Ant-Q) to deal with multiple objectives. This algorithm makes the following considerations [1072, 1067]:

- There is a family of agents for each objective function of the problem, (i.e., there are as many families as objectives in the problem).
- The agents within each family search for optimal solutions to their assigned objective function. This search is conducted by an Ant-Q algorithm using a cooperation mechanism (any information updated by a certain agent becomes available for all the remaining agents within the same family).
- Families are "independent" of each other, but action selection is performed using the same action value functions for all families.
- Solutions proposed by the agents of a family are transmitted to the agents of another family. The agents that receive a certain solution try to modify it in terms of their own objective function, following a certain negotiation mechanism (i.e., the agent tries to satisfy its own objective without altering in a significant way the solution previously obtained from the agents of another family).
- Objective functions are solved incrementally. This means that agents in families propose solutions following a certain predefined order imposed on the objective functions (this is similar to the approach called "lexicographic ordering", which is discussed in Section 1.7.1 from Chapter 2).
- A solution is obtained when all the families have participated in its construction.
- Solutions found after completing an iteration (for all the families) are evaluated using dominance relationships. Nondominated solutions are rewarded and stored in an external file that contains the Pareto optimal solutions found in the process. These nondominated solutions are used as starting points for the following iteration of the algorithm.

The algorithm of MOAQ is shown in Figure 10.5 [1072]. Its basic idea is to consider a family of m agents for each of the k objective functions of a problem (i.e., $m \times k$ agents cooperate to produce a solution to the problem).

1. Initialize $Q(s,a)$ arbitrarily
2. Arbitrarily sort objective functions
3. For i = 1 to N (N = number of episodes)
      For i = 1 to m (m = number of agents in $family_1$)
        Initialize $s = s_0$ for the m agents in $family_1$
        Apply step 2 of Ant-Q algorithm and find
          $solution_1$ for $objective_1$
      End Loop
      For j = 2 to k (k = objective functions)
        For i = 1 to m
          Use solution found by agent i of $family_{j-1}$
          as a basis to apply step 2 of the Ant-Q algorithm
          to the solution of $objective_j$ attempting to
          improve the value found so far
        End Loop
4. End Loop
5. Evaluate the m solutions compromised and find
      nondominated vectors (ND)
6. Compare ND to $\mathcal{P}$; Update $\mathcal{P}$
7. For the value functions in the elements of $\mathcal{P}$, apply:
      $Q(s,a) \leftarrow Q(s,a) + \alpha[r + \gamma \max_{a'} Q(s',a') - Q(s,a)]$

Fig. 10.5. MOAQ pseudo code

An iteration of MOAQ starts by establishing an arbitrary ordering of the objective functions. Agents of $family_1$ that solve $objective_1$ are required to behave according to the guidelines of the Ant-Q algorithm (see Figure 10.4). Once all the agents in $family_1$ have a solution, this is transferred to the agents in $family_2$. Agents in $family_2$ go through each of the states that made up the solution received and evaluate the possibility of applying an action that improves the performance of the solution in $objective_2$, but without harming $objective_1$. Once the agents in $family_2$ finish, their solutions are transferred to $family_3$ and the process is repeated as before. This continues until all the objective functions have been considered. Note how this algorithm has a great resemblance with the "behavioral memory" method proposed to handle constraints by Schoenauer and Xanthakis [1451].

Once all the families have finished an iteration, there are m negotiated solutions (considering m agents). These negotiated solutions are evaluated according to Pareto dominance, producing the set ND. These solutions are used to update the archive $\mathcal{P}$ that contains the globally (i.e., with respect to all the iterations) nondominated vectors found. The update implies a comparison of ND with the elements of $\mathcal{P}$. Dominated vectors are discarded from this comparison. Finally, the elements of ND that dominate elements of $\mathcal{P}$ are rewarded (with a term r as in the Ant-Q). The solutions in $\mathcal{P}$ are used to

10.4 Ant System

initiate the new iteration of MOAQ, looking to improve them (i.e., to get closer to PF_{true} of the problem).

MOAQ has been validated with several test functions defined in the specialized literature and has been applied in a real-world application (the optimization of a water distribution irrigation network where two objectives were considered: minimize network cost and maximize profits) [1067, 1068, 1072]. The results produced by the MOAQ have compared satisfactorily against those produced by the NPGA [709] and VEGA [1440].

MOAQ is not the only attempt reported in the literature to solve multiobjective optimization problems using the ant system. Gambardella et al. [531] propose a scheme based on the ant system to solve a bi-criterion vehicle routing problem (the approach was called "Multiple Ant Colony System for Vehicle Routing Problems with Time Windows," or MACS-VRPTW for short). The idea in this case is to use two ant colonies (one for each objective). One of the colonies tries to decrease the number of vehicles used and the other to optimize the feasible solutions found by the other ant colony based on total travel time (the second objective). The approach is incremental and it can be clearly seen that one objective (number of vehicles in this case) takes precedence over the other. This is also another variation of the lexicographic method.

Iredi et al. [748] propose a multi colony approach to handle the two objectives of a single machine total tardiness problem. The idea is to use heterogeneous colonies of ants, each of which weights objectives differently. Then, a global criterion is used to combine the solutions found by each of the colonies. The cooperation mechanism in this case is an exchange of solutions that belong to the other colony. Two pheromone matrices are considered in this approach (one for each objective), and each ant decides its following action using a probability defined in terms of the two objectives and a weight that estimates the relative importance of each of them. Each colony finds locally nondominated solutions, but a global selection mechanism forces the ants to obtain globally nondominated solutions. An interesting aspect of the approach is that the solutions found are sorted with respect to one objective so that the Pareto front can be divided in segments. This information is also used to force the ants to explore less densely populated regions of the search space and is intended to produce a smoother distribution of solutions. Notice the similarities of this scheme with the adaptive grid proposed by Knowles and Corne [886].

Gagné et al. [527, 528] propose an approach in which the heuristic values used to decide the movements of an ant take into consideration several objectives. However, when updating the pheromone trails, only one objective is considered. Therefore, the approach requires that the DM decides what is the most important objective to ensure that the agents converge towards solutions that primarily favor that objective and that consider trade-offs with respect to other objectives only as a secondary priority. This technique is applied to solve a single machine total tardiness problem with sequence dependent setups, where the main objective is to minimize changeover costs. The ant colony optimization algorithm incorporates look-ahead information in the transition rule, in an attempt to improve its performance. The approach performed well in a set of standard test functions when compared to a branch-and-bound algorithm, a genetic algorithm, simulated annealing and a local improvement procedure proposed by Rubin and Ragatz [1392]. In related work, Gravel et al. [595, 596] use the same approach to solve an industrial scheduling problem in an aluminum casting center. The objectives considered in this case are: minimize total tardiness for all orders, minimize unused production capacity over the planning horizon, and minimize the total number of drainings for the furnaces. Efficient transportation is treated as a constraint, incorporated through a penalty function. Results are reported using each of the objectives as the "primary objective function", and are compared to a single-objective ant colony optimization algorithm.

McMullen [1085] uses ant colony optimization to solve a just-in-time (JIT) sequencing problem. Two objectives are minimized: (1) setups and (2) usage rates. The authors transforms the problem into spatial data so that a traveling salesperson problem (TSP) approach can be used to find production sequences with the levels of setups and usage rates desirable for the user. Six different spatial approaches are then analyzed. Each of them offers a specific strategy to attempt to find desirable production sequences. Some of such strategies only focus on optimizing one objective, while others attempt to optimize both. When both objectives are considered, they are optimized separately (e.g., during the first 50% of the sequence, minimal usage rates are preferred, and during the second 50% of the sequence, minimal setups are preferred). Thus, the approach adopts lexicographic ordering. Results are compared with respect to simulated annealing, Tabu search, genetic algorithms, and artificial neural networks, using several standard test problems.

T'kindt et al. [1589] propose an approach called SACO, which uses ant colony optimization to solve a 2-machine bicriteria flowshop scheduling problem. The two objectives minimized are: (1) total completion time and (2) makespan criteria. This approach also adopts lexicographic ordering, since total completion time is optimized before makespan criteria. SACO incorporates a local search operator which is applied for a certain (fixed) number of iterations. This local search procedure is *ad-hoc* to the problem and is computationally expensive (the algorithm is $O(N^3)$). An interesting aspect of this approach is that the authors consider that diversification is preferred at the beginning of the search and intensification is preferred at the end. This is enforced using a variable selection probability which is based on simulated annealing. This makes unnecessary to initially randomly generate the pheromone trails. Results are compared with respect to other heuristics, including branch and bound when small problems are considered.

Shelokar et al. [1475, 1476] propose a multiobjective ant algorithm which is used to solve both combinatorial and continuous optimization problems. The approach can really be considered as a version of SPEA [1782] in which the search engine is an ant system, because it incorporates most of its mechanisms (strength Pareto fitness assignment, an external archive and a clustering procedure to prune the contents of the external archive when overcrowded). In fact, the approach even adopts crossover and mutation, as any regular evolutionary algorithm, although it also incorporates a local search mechanism. An interesting aspect of this work is the use of thermodynamic clustering, which is based on the thermodynamic free energy minimization technique proposed by Kita et al. [863]. Results are compared with respect to a random search technique and two simulated annealing approaches in several reliability problems.

Barán and Schaerer [90] extends the MACS-VRPTW (A Multiple Ant Colony System for Vehicle Routing Problems with Time Windows) algorithm [531] using a Pareto-based approach. In this case, all the objectives share the same pheromone trails, so that the knowledge of good solutions is equally important for every objective function. The approach maintains a list of Pareto optimal solutions, and each new generated solution is compared with respect to the contents of this list. An interesting aspect of this work is that the pheromone trail is reinitialized at the end of each generation using the average values of the Pareto optimal set. This aims to improve the exploration capabilities of the algorithm. Results are compared with respect to the original MACS-VRPTW.

Cardoso et al. [207] propose an approach called Multi-Objective Network Optimization using ACO (MONACO). This approach was designed for a dynamic problem and it is therefore atypical with respect to other MOEAs, because only the individual edge costs of the networks are used to guide the search, without waiting for the algorithm to finish building an entire network. The partial construction of a network considers a multi-pheromone trail (the number of trails corresponds to the number of objectives to be optimized) and performs a local search over a partially built solution in order to assess efficiency of the network. An interesting aspect of this work is that it apparently does not enforce a Pareto-based selection, although the authors indicate in a further paper [209] that MONACO does use an external archive to store the nondominated solutions found so far and that the solutions are evaluated using a Pareto-based selection operator. Nevertheless, reference [207] indicates the use of a random heuristic selection which is not based on Pareto dominance but on efficiency of the network built. García Martínez et al. [537] propose a static version of MONACO in which the notion of nondominance is incorporated,[10] together with the use of an external archive to retain all the nondominated solutions found so far. In [209], MONACO is applied to the multiobjective minimum spanning tree problem. The authors show that MONACO outperforms a linear aggregating function, which evidently cannot generate nondominated solutions in the concave portions of the Pareto front. In a related paper, Cardoso et al. [208] apply MONACO to the multiple

---

[10] García Martínez et al. [537] also indicate that MONACO does not produce nondominated solutions.

objective traveling salesperson problem. In this case, results are compared with respect to the Multiple Objective Genetic Local Search (MOGLS) approach proposed by Jaszkiewicz [779].

Doerner et al. [397, 398] propose an approach called Pareto Ant Colony Optimization (P-ACO) which is adopted to solve portfolio selection problems. An interesting aspect of this approach is that it uses a quadtree data structure [488] for identifying, storing and retrieving nondominated solutions. This approach considers the use of weights to assess the relative importance of each objective from the user's perspective. The approach follows the conventional structure of the ant colony optimization heuristic, except for the pheromone updates, which are done using two ants: the best and the second best values generated in the current iteration for each objective function. Results are compared with respect to Pareto Simulated Annealing (PSA) [323] and with respect to the NSGA-II [374] in 18 randomly generated problem instances and one problem with real-world data. In further work, Doerner et al. [396] use P-ACO is extended with an integer linear programming (ILP) preprocessing procedure that identifies several efficient portfolios. This information is used to initialize the pheromone trails before running P-ACO. This preprocessing is shown to clearly benefit P-ACO in a numerical example based on real world data. P-ACO has also been used for solving bi-objective flowshop scheduling problems in [1260]. In this case, besides adopting the original P-ACO algorithm, the authors also experiment with a second implementation that incorporates path relinking. The use of path relinking is found to be advantageous and very promising by the authors.

Guntsch and Middendorf [621] propose the Population-based Ant Colony Optimization (PACO) approach which is used to solve single-machine scheduling problems. The idea of this approach was originally proposed in [620] in the context of dynamic optimization. Under the population-based ACO, the first generation of ants works the same as in the standard ACO: the ants search solutions using the initial pheromone matrix and the best agent (ant) in the population adds pheromone to the pheromone matrix. However, in this case, no pheromone evaporation takes place. The best solution is placed in a population which is initially empty. This process is repeated for a (given) number of generations after which one solution in the population is removed and its corresponding amount of pheromone is subtracted from the elements of the pheromone matrix. This amount of pheromone subtracted corresponds to the amount that this solution contributed when it was added to the population. In a multiobjective context, the population is formed with a subset of the nondominated solutions found so far. First, one solution is selected at random, but then the remainder solutions are chosen so that they are closest to this initial solution with respect to some distance measure. An average-rank-weight method is adopted to construct a selection probability distribution for the ants and the new derivation of the active population to determine the pheromone matrices. The results are not compared with respect to any other approach, although the authors studied the effect of different parameter settings on the behavior of their algorithm.

Doerner et al. [399] propose COMPETants, which is inspired on the rank-based ant system originally proposed in [184]. COMPETants was specifically designed for a bi-objective optimization problem and it consists of two ant populations with different priority rules. The first of these colonies uses a priority rule that emphasizes one of the objectives (maximize utilization), and the second one emphasizes the other objective (minimize empty vehicle movements). The idea is to combine the best solutions from these two populations as to find good trade-offs. An interesting aspect of this algorithm is that the population sizes are not constant, but are adapted based on performance. The population that finds better solutions gets more ants for the next iteration. Additionally, information spillovers between the two populations occur, because ants can observe and utilize not only their own pheromone information, but also the foreign (i.e., from the other population) pheromone information. The decision of whether or not to utilize the foreign pheromone is based on the best solution found in each population. The ants who decide to utilize foreign information are called *spies*. The main motivation for using these spies is to get solutions of good overall quality (i.e., good trade-offs with respect to the two objectives).

García Martínez et al. [536, 537] perform a comprehensive comparative study in which they include eight multiobjective ant systems plus SPEA2 [1775] and the NSGA-II [374]. For the comparative study, several instances of the symmetric TSP are adopted. The assessment of results is done using the coverage of two sets performance measure [1772]. Results indicate that the ant system approaches are able to outperform the MOEAs adopted in the study. In [537] the same authors also discuss and analyze each of the eight multiobjective ant systems adopted.

10.4.3 Advantages and Disadvantages of the Ant System

There are several interesting aspects of the MOAQ. For example, it does not require an explicit mechanism to keep diversity (e.g., niching), since diversity is really produced in an emergent fashion by the action selection mechanism adopted (this mechanism performs a random action with a low probability). Another interesting aspect is that the changes proposed by the negotiation mechanism of the algorithm are done in decision variable space rather than in objective function space. This follows the suggestion of some researchers with respect to the definition of an appropriate sharing function [1509].

One of the main disadvantages of the MOAQ is that the algorithm requires the DM to impose a certain ordering in the objective functions, because despite the fact that the authors argue that an arbitrary ordering can be used, in certain cases (e.g., the application reported in [1067]), a specific ordering is required so that the algorithm can work properly. Another problem is that the MOAQ requires the fine tuning of several parameters. Also, it requires the definition of a certain heuristic function which is domain-dependent, and not necessarily easy to establish.

MACS-VRPTW also requires certain domain knowledge in order to decide which objective takes precedence and, therefore, the algorithm is also sensitive to the particular ordering imposed on the objective functions.

The multiple colony approach of Iredi et al. [748] is really a linear aggregating function in which the weights are distributed among several colonies of ants. The interval of weights can be automatically computed. However, this approach has the typical problems associated with linear aggregating functions (e.g., it cannot cover a concave portion of a Pareto front). Furthermore, the technique requires extra parameters (e.g., the number of colonies and their size) and it does not seem trivial to extend it to more than two objectives.

10.5 Distributed Reinforcement Learning

Mariano Romero [1072] and Mariano & Morales [1066, 1069, 1070, 1071] propose an extension to Q-learning [1686] to solve multiobjective optimization problems. Q-learning was originally proposed in the PhD thesis of Christopher J.C.H. Watkins [1687]. The algorithm of Mariano Romero, called multiobjective distributed Q-learning (MDQL) was motivated by some of the disadvantages of the MOAQ [1072].

10.5.1 Basic Concepts

Instead of requiring a heuristic function in the action selection rule (as does the MOAQ), MDQL uses an algorithm developed by the authors, which is called distributed Q-learning (DQL). The main idea behind DQL is allowing several agents to interact in a common environment, cooperating to establish an optimal policy over states and actions that allows achievement of a common goal. The communication mechanism among the agents is a "map" of the environment. This map is built based on the updates of the value function adopted for the problem. Such updates are performed each time an agent visits a state and selects an action (i.e., when it leaves a trace for the other agents). In order to decide its next move, an agent considers those traces previously left by other agents.

The algorithm of MDQL is shown in Figure 10.6. The main concept is the following: a family of agents is assigned to each of the objectives of the problem to be solved. The solutions obtained by the agents in one family are compared with the solutions obtained by the other families. To perform this comparison, a negotiation mechanism is adopted. The proposed mechanism aims to produce elements of P_{true} by favoring those solutions that are nondominated with respect to all objective functions. The states involved in such solutions are then rewarded (this is done dynamically, using a comparison of reinforcement values) and stored for further use. An external file $\mathcal{P}$ is used (as

---
1. Initialize $Q(s,a)$ arbitrarily
2. Arbitrarily sort objective functions
3. Assign a family to each objective function
4. For i = 1 to N (N = number of episodes)
    Copy $Q(s,a)$ in $Q_c(s,a)$ (update "map")
    For j = 1 to k (k = number of families)
        For r = 1 to m (m = agents in family$_j$)
            Initialize $s = s_0$ for the m agents
            in family$_r$ (start from initial state)
            Repeat
                Select $a \in s$ using a policy derived from Q-learning
                (i.e., use ϵ-greedy)
                Apply a; observe r and s'
                $Q_c(s,a) \leftarrow Q_c(s,a)$
                $+\alpha[\gamma\max_{a'} Q_c(s',a') - Q_c(s,a)]$
            Until s is a terminal state
        End Loop
    End Loop
5. End Loop
6. Compromise solution is obtained through
   negotiation among f solutions
7. If (Compromise solution is nondominated) add to $\mathcal{P}$
8. For the elements of $\mathcal{P}$ apply:
   $Q(s,a) \leftarrow Q(s,a) + \alpha[r + \gamma\max_{a'} Q(s',a') - Q(s,a)]$

Fig. 10.6. MDQL pseudo code

in the MOAQ of Section 10.4) to maintain the nondominated vectors generated during the search. Note that the parameters α and γ, and the reward r adopted in the MOAQ, are still used here.

MDQL has been used by its authors to solve real-world problems, mainly related to water-using systems (see [1073]).

10.5.2 Advantages and Disadvantages of Distributed Reinforcement Learning

MDQL has several advantages with respect to MOAQ. For example, it does not require a heuristic function in the action-selection rule. Also, it needs fewer parameters than the MOAQ (it requires only two, instead of the five required by the MOAQ) and it does not require that the DM imposes a certain ordering in the processing of the objective functions. More remarkable is the fact that DQL can apparently maintain the nice convergence properties of Q-learning which is an important step towards formalizing the MDQL algorithm [1072]. The algorithm can be easily parallelized. It also uses a relatively easy penalty-based approach (these penalties are applied to the rewards rather than to the values of the objective functions) to handle constraints [1066, 1072].

One of the most important disadvantages of the MDQL is that it requires a discretized decision variable space for each variable of the problem. Since the authors consider the intersections between intervals of the decision variables as states of the environment, there is an implicit exponential growth in the number of possible actions as the number of decision variables is increased [1072]. This may certainly limit the use of this approach in real-world problems.

The algorithm may also become very expensive in terms of memory requirements (this is related to the storage of nondominated vectors and value functions associated to each state) [1072].

10.6 Particle Swarm Optimization

James Kennedy and Russell Eberhart [838] proposed an approach called "particle swarm optimization" (PSO) inspired by the choreography of a bird flock. The idea of this approach is to simulate the movements of a group (or population) of birds which aim to find food. The approach can be seen as a distributed behavioral algorithm that performs (in its more general version) multidimensional search.

10.6.1 Basic Concepts

The general algorithm of PSO is shown in Figure 10.7 [1479]. In the simulation, the behavior of each individual (or particle) is affected by either the best local (i.e., within a certain neighborhood) or the best global individual. The approach uses then the concept of population and a measure of performance similar to the fitness value used with evolutionary algorithms. Also, the adjustments of individuals are analogous to the use of a crossover operator. Additionally, this approach introduces the use of flying potential solutions through hyperspace (used to accelerate convergence). Note that PSO allows individuals to benefit from their past experiences whereas in an evolutionary algorithm, normally the current population is the only "memory" used by the individuals. PSO has been successfully used for both continuous nonlinear and discrete binary optimization [838, 423, 839, 840, 450].

The analogy of particle swarm optimization with evolutionary algorithms makes evident the notion that using a Pareto ranking scheme can be the straightforward way to extend the approach to handle multiobjective optimization problems. The historical record of best solutions found by a particle (i.e., an individual) can be used to store nondominated solutions generated in the past (this would be similar to the notion of elitism used in evolutionary multiobjective optimization). The use of global attraction mechanisms combined with a historical archive of previously found nondominated vectors motivates convergence towards P_{true}.

```
1. For i = 1 to M (M = population size)
       Initialize P[i] randomly
       (P is the population of particles)
       Initialize V[i] = 0 (V = speed of each particle)
       Evaluate P[i]
       GBEST = Best particle found in P[i]
2. End For
3. For i = 1 to M
       PBESTS[i] = P[i]
       (Initialize the "memory" of each particle)
4. End For
5. Repeat
       For i = 1 to M
           V[i] = w × V[i] + C₁ × R₁ × (PBESTS[i] − P[i])
                  +C₂ × R₂ × (PBESTS[GBEST] − P[i])
           (Calculate speed of each particle)
           (W = Inertia weight, C₁ & C₂ are positive constants)
           (R₁ & R₂ are random numbers in the range [0..1])
           POP[i] = P[i] + V[i]
           If a particle gets outside the pre-defined hypercube
               then it is reintegrated to its boundaries
           Evaluate P[i]
           If new position is better then PBESTS[i] = P[i]
           GBEST = Best particle found in P[i]
       End For
6. Until stopping condition is reached
```

Fig. 10.7. Particle swarm optimization pseudo code

10.6.2 Multiobjective Versions

In the last few years, a variety of proposals for extending PSO to handle multiple objectives have been published in the specialized literature. We will review next the most representative of them.

Moore and Chapman [1126] present an algorithm based on Pareto dominance in an unpublished document. The authors emphasize the importance of performing both an individual and a group search (a cognitive component and a social component). However, the authors did not adopt any scheme to maintain diversity.

Ray and Liew [1332] propose the so-called *swarm metaphor* algorithm, which also uses Pareto dominance and combines concepts of evolutionary techniques with the particle swarm. The approach uses crowding to maintain diversity and a multilevel sieve to handle constraints (for this, the authors adopt the constraint and objective matrices proposed in some of their previous research [1329]).

Unlike the previous proposals, Parsopoulos and Vrahatis [1259] adopted an aggregating function (three types of approaches were implemented: a conventional linear aggregating function, a dynamic aggregating function and the bang bang weighted aggregation approach [801]) for their multi-objective PSO approach. In more recent work, Parsopoulos et al. [1257] studied a parallel version of the Vector Evaluated Particle Swarm (VEPSO) method for multiobjective problems. VEPSO is a multi-swarm variant of PSO, which is inspired on the Vector Evaluated Genetic Algorithm (VEGA) [1440]. In VEPSO, each swarm is evaluated using only one of the objective functions of the problem under consideration, and the information it possesses for this objective function is communicated to the other swarms through the exchange of their best experience.

Hu and Eberhart [713] propose an approach called "dynamic neighborhood", in which only one objective is optimized at a time using a scheme similar to lexicographic ordering. In further work, Hu et al. [716] adopt a secondary population (called "extended memory") and introduce some further improvements to their dynamic neighborhood PSO approach.

Fieldsend and Singh [486] propose an approach which uses an unconstrained elite archive (in which a special data structure called "dominated tree" is adopted) to store the nondominated individuals found along the search process. The archive interacts with the primary population in order to define local guides. Their approach also uses a "turbulence" operator that is basically a mutation operator that acts on the velocity value used by PSO.

Coello Coello and Salazar Lechuga [282] and Coello Coello et al. [286] propose an approach based on the idea of having a global repository in which every particle deposits its flight experiences after each flight cycle. Additionally, the updates to the repository are performed considering a geographically- based system defined in terms of the objective function values of each individual; this repository is used by the particles to identify a leader that will guide the search. The approach also uses a mutation operator that acts both on the particles of the swarm, and on the range of each design variable of the problem to be solved. In more recent work, Toscano Pulido and Coello Coello [1599] use the concept of Pareto dominance to determine the flight direction of a particle. The authors adopt clustering techniques to divide the population of particles into several swarms in order to have a better distribution of solutions in decision variable space. In each sub-swarm, a PSO algorithm is executed and, at some point, the different sub-swarms exchange information: the leaders of each swarm are migrated to a different swarm in order to vary the selection pressure. Also, this approach does not use an external population since elitism in this case is an emergent process derived from the migration of leaders. A variation of Coello's approach [286] is adopted by Baltar and Fontane [88] for solving water quality problems. The main change is that Baltar and Fontane [88] do not use the adaptive grid of the original proposal, but instead, they calculate in objective function space, the density of points around each solution stored in the repository[11] and perform a roulette wheel selection such that the probability of choosing a point is inversely proportional to its density. Thus, the repository in this case is a simple archive that stores the nondominated solutions found along the evolutionary process, but it does not work as a diversity-preserving mechanism, as in the original proposal [286]. An interesting aspect of this work is that the algorithm is implemented in a spreadsheet format using Microsoft Excel© and Visual Basic.

Tayal [1575] use PSO with a linear aggregating function to solve several engineering optimization problems, including the design of: (1) a 2 degrees-of-freedom spring mass system, (2) a coil compression spring, (3) a two-bar truss, (4) a gear train and (5) a welded beam. Constraints are handled through the use of an external penalty, and the weights for the linear aggregating function are varied from one run to the other (i.e., several independent runs are required in order to generate the Pareto front of each problem). Summarizing, this work illustrates the most straightforward way of using PSO as a single-objective optimizer to solve multiobjective optimization problems.

Mostaghim and Teich [1139] propose a sigma method in which the best local guides for each particle are adopted to improve the convergence and diversity of a PSO approach used for multiobjective optimization. They also use a "turbulence" operator, but applied on decision variable space. The idea of the sigma method is similar to compromise programming. The use of the sigma values increases the selection pressure of PSO (which was already high). This may cause premature convergence in some cases. In further work, Mostaghim and Teich [1138] study the influence of ϵ-dominance [959] on MOPSO methods. ϵ-dominance is compared with existing clustering techniques for fixing the archive size and the solutions are compared in terms of computational time, convergence and diversity. The results show that the ϵ-dominance method can find solutions much faster than the clustering technique with a comparable (and even better in some cases) convergence and diversity. The authors suggest a new diversity measure (sigma method) inspired on their previous work [1139]. Also, based on the idea that the initial archive from which the particles have to select a local guide has influence on the diversity of solutions, the authors propose the use of successive improvements adopting a previous archive of solutions. In more recent work, Mostaghim and Teich [1140] propose a new method called *covering*MOPSO (cvMOPSO). This method works in two phases. In phase 1, a MOPSO algorithm is run with a restricted archive size and the goal is to obtain a good approximation of the Pareto-front. In phase 2, the nondominated solutions obtained from phase 1 are considered as the input archive of the cvMOPSO. The particles in the population of the cvMOPSO are divided into subswarms around each nondominated solution after the first generation. The task of the subswarms is to cover the gaps between the nondominated solutions obtained from the phase 1. No restrictions on the archive are imposed during phase 2.

---

[11] This seems to be similar to niching.

Li [985] proposes an approach that incorporates the main mechanisms of the NSGA-II [374] to the PSO algorithm. This approach combines the population of particles and all the personal best positions of each particle, and selects the best particles among them to conform the next population. It also selects the leaders randomly from the leaders set among the best of them, based on two different mechanisms: a niche count and a crowding distance. In more recent work, Li [987] proposes the *maximinPSO*, which uses a fitness function derived from the maximin strategy [84] to determine Pareto domination. The author shows that one advantage of this approach is that no additional clustering or niching technique is needed, since the maximin fitness of a solution can tell us not only if a solution is dominated or not, but also if it is clustered with other solutions, i.e., the approach also provides diversity information.

Srinivasan and Hou [1510, 1511] propose an approach, called Particle Swarm Inspired Evolutionary Algorithm (PS-EA), which is a hybrid between PSO and an evolutionary algorithm. The authors argue that the traditional PSO equations are too restrictive when applied to multiconstrained search spaces. Thus, they propose to replace the PSO equations with the so-called self-updating mechanism, which emulates the workings of the equations. Such mechanism uses an inheritance probability tree to update each individual in the population. An interesting aspect of this approach is that the authors also use a dynamic inheritance probability adjuster to dynamically adjust the inheritance probabilities in the inheritance probability tree based on the status of the algorithm at a certain moment in time. The approach uses a memory to store the elite particles and does not use a recombination operator.

Zhang et al. [1759] propose an approach that attempts to improve the selection of g_{best} and p_{best} when the velocity of each particle is updated. For each objective function, there exists both a g_{best} and a p_{best} for each particle. In order to update the velocity of a particle, the algorithm defines the g_{best} of a particle as the average of the complete set of g_{best} particles. Analogously, the p_{best} is computed using either a random choice or the average from the complete set of p_{best} values. This choice depends on the dispersion degree between the g_{best} and p_{best} values of each particle.

Zhao & Cao [1763] propose a multi-objective particle swarm optimizer based on Pareto dominance. This is very similar to the proposal of Coello and Lechuga [282], since it adopts a *gbest* topology. However, this approach maintains not one but two repositories additionally to the main population: one keeps the global best individuals found so far and the other one[12] keeps a single local best for each member of the swarm. A truncated archive is adopted to store the nondominated solutions found along the evolutionary process. This truncated archive is similar to the adaptive grid of PAES [886]. An interesting aspect of this work is that the authors adopt a linear membership function to represent the goals of each objective function. The membership

---

[12] This second "repository" is actually a list.

function is adopted to modify the ranking of the nondominated solutions as to focus the search on the single solution that attains the maximum membership in the fuzzy set. This approach was adopted to solve an economic load dispatch problem.

Bartz-Beielstein et al. [95] propose an approach that starts from the idea of introducing elitism (archiving) into PSO. Different methods for selecting and deleting particles from the archive are analyzed to generate a satisfactory approximation of the Pareto front. The selection methods analyzed are based on the contribution of each particle to the diversity of the Pareto front. Deleting methods are either inversely related to the selection fitness or based on the previous success of each particle. The authors provide some statistical analysis in order to assess the impact of each of the parameters used by their approach.

Baumgartner et al. [103] propose an approach which uses weighted sums (i.e., linear aggregating functions) to solve multiobjective optimization problem. In this approach, the swarm is equally partitioned into n subswarms, each of which uses a different set of weights and evolves into the direction of its own swarm leader. The approach adopts a gradient technique to identify the Pareto optimal solutions.

Chow and Tsui [253] propose an autonomous agent response learning algorithm. The authors propose to decompose the award function into a set of local award functions and, in this way, to model the response extraction process as a multiobjective optimization problem. A modified PSO called "Multi-Species PSO" is introduced by considering each objective function as a species swarm. A communication channel is established between the neighboring swarms for transmitting the information of the best particles, in order to provide guidance for improving their objective values. Also, the authors propose to modify the equation used to update the velocity of each particle, considering also the global best particle of its neighboring species.

Mahfouf et al. [1041] present an enhancement of the original PSO algorithm which is aimed to improve the performance of this heuristic in multi-objective optimization problems. The approach is called the Adaptive Weighted PSO (AWPSO) algorithm, and its main idea is to modify the velocity by including an acceleration term which increases with the number of iterations. This aims to enhance the global search ability of the algorithm towards the end of the run thus helping the approach to escape from local optima. A weighted aggregating function is also used to guide the selection of the personal and global best leaders. The authors use dynamic weights to generate different elements of the Pareto optimal set. A nondominated sorting scheme is adopted to select the particles from one iteration to the next one. The approach was applied to the design of heat treated alloy steels based on data-driven neural-fuzzy predictive models.

Ho et al. [691] propose a new PSO-based algorithm for multiobjective optimization. As a result of three main modifications to the known formula for updating velocity and position of particles, a novel formula is described. Also, the authors introduce a "craziness" operator in order to maintain diversity into the swarm. This "craziness" operator is applied (with certain probability) to the velocity vector before updating the position of a particle. On the other hand, in order to assign a fitness value to particles, the authors adopt the mechanism proposed originally in [1778], for the SPEA algorithm. Finally, the authors introduce one repository for each particle and one global repository for the whole swarm. The repository of each particle stores the latest Pareto solutions found by the particle and the global repository stores the current Pareto set. Every time a particle updates its position, it selects its personal best from its own repository and the global best from the global repository. In both cases, the authors use a roulette selection mechanism based on the fitness values of the particles and also based on an "age" variable proposed by the authors. The proposed method is tested on two numerical examples with promising results.

Reyes and Coello [1351] propose an approach based on Pareto dominance and the use of a crowding factor for the selection of leaders (by means of a binary tournament). This proposal uses two external archives: one for storing the leaders currently being used for performing the flight and another one for storing the final solutions. The crowding factor is used to filter out the list of leaders whenever the maximum limit imposed on such list is exceeded. Only the leaders with the best crowding values are retained. On the other hand, the concept of ϵ-dominance is used to select the particles that will remain in the archive of final solutions. Additionally, the authors propose a scheme in which they subdivide the population (or swarm) into three different subsets. A different mutation operator is applied to each subset. Note however, that for all other purposes, a single swarm is considered (e.g., for selecting leaders).

Villalobos-Arias et al. [1656] propose a new mechanism to maintain diversity in multi-objective optimization problems. Although the approach is independent of the search engine adopted, they incorporate it into the multi-objective particle swarm optimizer introduced in [286]. The new approach is based on the use of stripes that are applied on the objective function space. Based on an analysis for a bi-objective problem, the main idea of the approach is that the Pareto front of the problem is "similar" to the line determined by the minimal points of the objective functions. In this way, several points (that the authors call stripe centers) are distributed uniformly along such line, and the individuals of the population are assigned to the nearest stripe center. When using this approach for solving multi-objective problems with PSO, one leader is used in each stripe. Such leader is selected minimizing a weighted sum of the minimal points of the objective functions. The authors show that their approach overcomes the drawbacks on other popular mechanisms such as ϵ-dominance [959] and the sigma method proposed in [1139].

Raquel and Naval [1327] incorporate the concept of crowding distance for selecting the global best particle and also for deleting particles from the external archive of nondominated solutions. When selecting a leader, the set of nondominated solutions is sorted in descending order with respect to the crowding factor, and a particle is randomly chosen from the top part of the list. On the other hand, when the external archive is full, it is again sorted in descendent order with respect to the crowding value and a particle is randomly chosen to be deleted, from the bottom part of the list. This approach uses the mutation operator proposed in [286] in such a way that it is applied only on a certain number of generations at the beginning of the process. The authors adopt the constraint-handling technique proposed in [374].

Alvarez-Benitez et al. [37] propose PSO-based methods based exclusively on Pareto dominance for selecting guides from an unconstrained nondominated archive. Three different techniques are presented: *Rounds* which explicitly promotes diversity, *Random* which promotes convergence and *Prob* which is a weighted probabilistic method and forms a compromise between *Random* and *Rounds*. Also, the authors propose and evaluate four mechanisms for confining particles to the feasible region, that is, constraint-handing methods. The authors showed that probabilistic selection favoring archival particles that dominate few particles provides good convergence towards and coverage of the Pareto front. Also, they concluded that allowing particles to explore regions close to the constraint boundaries is important to ensure convergence to the Pareto front.

Salazar-Lechuga and Rowe [1417] propose an approach in which PSO is used to guide the search with the help of fitness sharing (applied on objective function space) [587] to spread the particles along the Pareto front. The approach uses an external archive to store the best particles (nondominated particles) found by the algorithm. Since this repository helps to guide the search, fitness sharing is calculated for each of the particles in the repository and leaders are chosen from this set by means of an stochastic sampling method (roulette wheel). Also, fitness sharing is used as a criterion to update the repository. Each time the repository is full and a new particle wants to get in, its fitness sharing is compared with the fitness sharing of the worst solution of the repository. If the new particle is better than the worst particle, then the new particle enters into the repository and the worst particle is deleted. Fitness sharing is updated when inserting or deleting a particle from the repository.

Xiao-hua et al. [1717] propose the Intelligent Particle Swarm Optimization (IPSO) algorithm for multi-objective problems based on an Agent-Environment-Rules (AER) model to provide an appropriate selection pressure to propel the swarm population towards the Pareto optimal front. In this model, the authors modify the *global* best flight formula including the *local* best position of the neighborhood of each particle. On the other hand, each particle is taken as an agent particle with the ability of memory, communication, response, cooperation and self-learning. Each particle has its position, velocity and energy, which is related to its fitness. All particles live in a latticelike environment, which is called an agent lattice, and each particle is fixed on a lattice-point. In order to survive in the system, they compete or cooperate with their neighbors so that they can gain more resources (increase energies). Each particle has the ability of cloning itself, and the number of clones produced depends of the energy of the particle. General agent particles and latency agent particles (those who have smaller energy but contain certain features—e.g., favoring diversity—that make them good candidates to be cloned) will be cloned. The aim of the clonal operator (which is modeled in the clonal selection theory also adopted with artificial immune systems [1193]) is to increase the competition between particles, maintain diversity of the swarm and improve the convergence of the process. Finally, a clonal mutation operator is used.

Janson and Merkle [770] proposed a hybrid particle swarm optimization algorithm for multi-objective optimization, called ClustMPSO. ClustMPSO combines the PSO algorithm with clustering techniques to divide all particles into several subswarms. For this aim, the authors use the K-means algorithm. Each subswarm has its own nondominated front and the total nondominated front is obtained from the union of the fronts of all the subswarms. Each particle randomly selects its neighborhood best (*nbest*) particle from the nondominated front of the swarm to which it belongs. Also, a particle only selects a new *nbest* particle when the current is no longer a nondominated solution. On the other hand, the personal best (*pbest*) of each particle is updated based on dominance relations. Finally, the authors define that a subswarm is dominated when none of its particles belongs to the total nondominated front. In this way, when a subswarm is dominated for a certain number of consecutive generations, the subswarm is relocated. The proposed algorithm is tested on an artificial multi-objective optimization function and on a real-world problem from biochemistry, called the molecular docking problem. The authors reformulate the molecular docking problem as a multi-objective optimization problem and, in this case, the updating of the *pbest* particle is also based on the weighted sum of the objectives of the problem. ClustMPSO outperforms a well-known Lamarckian Genetic Algorithm that had been previously adopted to solve such problem.

Branke and Mostaghim [165] present a study of the influence of the personal best particles in a MOPSO. Different strategies for updating the personal guide for each particle in the population are analyzed. The authors conclude that the selection of a proper personal guide has a significant impact on the performance of MOPSOs, and they also propose to allow each particle to memorize all nondominated personal best particles it has encountered. When having such a scheme, it becomes crucial how is the best particle selected from the personal archive of each particle. Thus, several strategies for performing such a selection of the personal best are also empirically compared in this work.

Santana-Quintero et al. [1424] propose a hybrid algorithm, in which particle swarm optimization is used to generate a few solutions on the Pareto front (or very close to it), and rough sets [1262] are adopted as a local search mechanism to generate the rest of the front. The authors adopt a PSO algorithm with a very small population size (only 5 particles). Leaders are selected such that each objective is separately optimized (disregarding the others), while the rest of the particles try to converge towards the "knee" of the Pareto front. So, when solving problems with two objectives, two particles try to optimize these two objectives, while the other three use a nonlinear aggregating function (based on compromise programming) in order to converge towards the ideal vector (which is composed by the best values known so far). This approach adopts an external archive, which uses Pareto dominance as its main entrance criterion. For the secondary population, the authors adopt a variant of ϵ-dominance [959] called paϵ-dominance (see [672] for further details). Individuals rejected from the secondary population are sent to a third population, which retains individuals which have been locally nondominated, but are globally dominated. Such individuals are used in the second phase of the algorithm in order to define the search region in which rough sets operate. This approach uses both crossover (BLX-α [457]) and mutation (parameter-based mutation [374]). This approach is able to generate good approximations of the true Pareto front of problems with relatively high dimensionality (between 10 and 30 decision variables), while performing only 4,000 fitness function evaluations.

Krami et al. [906] use the MOPSO proposed in [286] to solve reactive power planning problems in which two objectives are minimized: (1) cost and (2) active power losses. This problem has constraints related to the acceptable voltage profiles at each node, which are treated using a death penalty approach (i.e., solutions not satisfying the voltage constraints are discarded).

The numerous PSO variants for multiobjective optimization that have been proposed in the last few years has motivated the first surveys and comparative studies on MOPSOs (see for example [1353, 483, 451]).

It is worth mentioning that PSO is an unconstrained search technique. Therefore, it is also necessary to develop an additional mechanism to deal with constrained multiobjective optimization problems. The design of such a mechanism is also a matter of current research even in single-objective optimization [1331, 1258, 1236, 256, 715, 714, 1598].

10.6.3 Advantages and Disadvantages of Particle Swarm Optimization

With no doubt, the greatest advantages of PSO are its simplicity (both conceptually, and at the implementation level), its ease of use and its high convergence rate. In fact, PSO is a good candidate to design an "ultra-efficient" MOEA, and some initial steps in that direction have been already reported (see for example [1596, 1600]).

The main disadvantages of PSO, when used for multiobjective optimization, are mainly related to the apparent difficulties to control diversity. The loss of diversity of multiobjective PSO approaches is normally compensated using mutation (also called "turbulence") operators. However, the role of the parameters of the PSO algorithm in its convergence and its loss of diversity

10.7 Differential Evolution

Differential Evolution (DE) is a relatively recent heuristic (it was created in the mid-1990s) proposed by Kenneth Price and Rainer Storn [1525, 1526, 1295], which was designed to optimize problems over continuous domains. This approach originated from Kenneth's Price attempts to solve the Tchebycheff Polynomial fitting Problem that had been posed to him by Rainer Storn. In one of the different attempts to solve this problem, Price came up with the idea of using vector differences for perturbing the vector population. Upon its implementation, the algorithm was further improved and refined, after many discussions between Price and Storn [1295]. DE is an evolutionary (direct-search) algorithm which has been mainly used to solve continuous optimization problems. DE shares similarities with traditional EAs. However it does not use binary encoding as a simple genetic algorithm [581] and it does not use a probability density function to self-adapt its parameters as an Evolution Strategy [1460]. Instead, DE performs mutation based on the distribution of the solutions in the current population. In this way, search directions and possible step sizes depend on the location of the individuals selected to calculate the mutation values.

There is a nomenclature scheme developed to reference the different DE variants. The most popular is called *"DE/rand/1/bin"*, where "DE" means Differential Evolution, the word "rand" indicates that individuals selected to compute the mutation values are chosen at random, "1" is the number of pairs of solutions chosen and finally "bin" means that a binomial recombination is used. The corresponding algorithm of this variant is presented in Figure 10.8.

The "CR" parameter controls the influence of the parent in the generation of the offspring. Higher values mean less influence of the parent. The "F" parameter scales the influence of the set of pairs of solutions selected to calculate the mutation value (one pair in the case of the algorithm in Figure 10.8).

It is important to note that, increasing either the population size or the number of pairs of solutions to compute the mutation values will also increase the diversity of possible movements, promoting the exploration of the search space. However, the probability to find the correct search direction decreases considerably. Then, the balance between the population size and the number of differences determines the efficiency of the algorithm [478]. Besides this balance, another important factor when using DE is the selection of the variant. Each one varies the way mutation is computed and also the type of recombination operator.

Several DE variants are possible. To exemplify this point, we took from the paper by Mezura et al. [1099] the eight DE variants adopted, each of which

```
1  Begin
2      G=0
3      Create a random initial population $\mathbf{x}_{i,G}$ $\forall i$, $i = 1, \ldots, NP$
4      Evaluate $f(\mathbf{x}_{i,G})$ $\forall i$, $i = 1, \ldots, NP$
5      For G=1 to MAX_GEN Do
6          For i=1 to NP Do
7 ⇒            Select randomly $r_1 \neq r_2 \neq r_3$ :
8 ⇒            $j_{rand} = \text{randint}(1, D)$
9 ⇒            For j=1 to D Do
10 ⇒               If $(rand_j[0,1) < CR$ or $j = j_{rand})$ Then
11 ⇒                   $u_{i,j,G+1} = x_{r_3,j,G} + F(x_{r_1,j,G} - x_{r_2,j,G})$
12 ⇒               Else
13 ⇒                   $u_{i,j,G+1} = x_{i,j,G}$
14 ⇒               End If
15 ⇒            End For
16             If $(f(\mathbf{u}_{i,G+1}) \leq f(\mathbf{x}_{i,G}))$ Then
17                 $\mathbf{x}_{i,G+1} = \mathbf{u}_{i,G+1}$
18             Else
19                 $\mathbf{x}_{i,G+1} = \mathbf{x}_{i,G}$
20             End If
21         End For
22         $G = G + 1$
23     End For
24 End
```

Fig. 10.8. "DE/rand/1/bin" algorithm. randint(min,max) is a function that returns an integer number between min and max. rand[0, 1) is a function that returns a real number between 0 and 1. Both are based on a uniform probability distribution. "NP", "MAX_GEN", "CR" and "F" are user-defined parameters. "D" is the dimensionality of the problem. Steps pointed with arrows change depending on the version adopted.

will be briefly described next. The modifications from variant to variant are in the recombination operator used (steps 9 to 15 in Figure 10.8) and also in the way individuals are selected to calculate the mutation vector (step 7 in Figure 10.8). The variants adopted by Mezura et al. [1099] are the following:

- Four variants whose recombination operator is discrete, always using two individuals: the original parent and the DE mutation vector (step 11 in Figure 10.8). Two discrete recombination operators: binomial and exponential. The main difference between them is that for binomial recombination, each variable value of the offspring is taken at every time from one of the two parents, based on the "CR" parameter value. On the other hand, in the exponential recombination, each variable value of the offspring is taken from the first parent until a random number surpasses the "CR" value. From this point, all the offspring variable values will be taken from

the second parent. These variants are: *"DE/rand/1/bin"*, *"DE/rand/1/exp"*, *"DE/best/1/bin"* and *"DE/best/1/exp"* [1294]. The "rand" variants select all the individuals to compute mutation at random and the "best" variants use the best solution in the population besides the random ones.
- Two variants with arithmetic recombination, which, unlike discrete recombination, is rotation invariant. These are *"DE/current-to-rand/1"* and variant *"DE/current-to-best/1"* [1294]. The only difference between them is that the first selects the individuals for mutation at random and the second one uses the best solution in the population besides random solutions.
- *"DE/rand/2/dir"* [478], which incorporates objective function information to the mutation and recombination operators. The aim of this approach is to guide the search to promising areas faster than traditional DE. Their authors argue that the best results are obtained when the number of pairs of solutions is two [478].
- Finally, a variant with a combined discrete-arithmetic recombination, the *"DE/current-to-rand/1/bin"* [1294].

Each variant's implementation details are summarized in Table 10.1.

10.7.1 Multiobjective Versions

Apparently, Chang et al. [224] constitutes the first reported attempt to extend differential evolution for multiobjective problems. In this paper, the authors adopt an external archive (called "Pareto optimal set" by the authors) to store the nondominated solutions obtained during the search. The approach also incorporates fitness sharing to maintain diversity. An interesting aspect of this approach is that the selection mechanism of the differential evolution algorithm is modified in order to enforce that the members of the new generation are both nondominated and at a certain minimum distance from the previously found nondominated solutions. This approach is adopted to fine-tune the fuzzy automatic train operation (ATO) for a typical mass transit system, in which three objectives are considered: (1) punctuality (least deviation from scheduled arrival time), (2) least energy consumption and (3) maximum passenger comfort. This application is discussed in further detail in [223].

As indicated by Kukkonen and Lampinen [914], Bergey [123] also reported a multi-objective evolutionary algorithm based on differential evolution at about the same time as Chang et al. [224]. This approach (which was called Pareto Differential Evolution, or PDE) is not described in detail in the paper. The author only indicates that PDE generated nondominated solutions and that it is implemented as a general-purpose spreadsheet solver designed as an add-in for Microsoft Excel. PDE is then meant to be used for multi-criteria decision making by offering a set of alternative nondominated solutions from which the user has to choose the most appropriate one(s).

Nomenclature	Variant
rand/p/bin	$u_{i,j} = \begin{cases} x_{r_3,j} + F \cdot \sum_{k=1}^{p}(x_{r_1,j}^p - x_{r_2,j}^p) & \text{if } U_j(0,1) < CR \text{ or } j = j_r \\ x_{i,j} & \text{otherwise} \end{cases}$
rand/p/exp	$u_{i,j} = \begin{cases} x_{r_3,j} + F \cdot \sum_{k=1}^{p}(x_{r_1,j}^p - x_{r_2,j}^p) & \text{from } U_j(0,1) < CR \text{ or } j = j_r \\ x_{i,j} & \text{otherwise} \end{cases}$
best/p/bin	$u_{i,j} = \begin{cases} x_{best,j} + F \cdot \sum_{k=1}^{p}(x_{r_1,j}^p - x_{r_2,j}^p) & \text{if } U_j(0,1) < CR \text{ or } j = j_r \\ x_{i,j} & \text{otherwise} \end{cases}$
best/p/exp	$u_{i,j} = \begin{cases} x_{best,j} + F \cdot \sum_{k=1}^{p}(x_{r_1,j}^p - x_{r_2,j}^p) & \text{from } U_j(0,1) < CR \text{ or } j = j_r \\ x_{i,j} & \text{otherwise} \end{cases}$
current-to-rand/p	$\mathbf{u}_i = \mathbf{x}_i + K \cdot (\mathbf{x}_{r_3} - \mathbf{x}_i) + F \cdot \sum_{k=1}^{p}(\mathbf{x}_{r_1}^p - \mathbf{x}_{r_2}^p)$
current-to-best/p	$\mathbf{u}_i = \mathbf{x}_i + K \cdot (\mathbf{x}_{best} - \mathbf{x}_i) + F \cdot \sum_{k=1}^{p}(\mathbf{x}_{r_1}^p - \mathbf{x}_{r_2}^p)$
current-to-rand/p/bin	$u_{i,j} = \begin{cases} x_{i,j} + K \cdot (x_{r_3,j} - x_{i,j}) + F \cdot \sum_{k=1}^{p}(x_{r_1,j}^p - x_{r_2,j}^p) & \text{if } U_j(0,1) < CR \text{ or } j = j_r \\ x_{i,j} & \text{otherwise} \end{cases}$
rand/2/dir	$\mathbf{v}_i = \mathbf{v}_1 + \frac{F}{2}(\mathbf{v}_1 - \mathbf{v}_2 + \mathbf{v}_3 - \mathbf{v}_4)$ where $f(\mathbf{v}_1) < f(\mathbf{v}_2)$ and $f(\mathbf{v}_3) < f(\mathbf{v}_4)$

Table 10.1. DE variants adopted by Mezura et al. [1099]. j_r is a random integer number generated between $[0, n]$, where n is the number of variables of the problem. $U_j(0, 1)$ is a real number generated at random between 0 an 1. Both numbers are generated using a uniform distribution. In their experiments, Mezura et al. [1099] use $p = 1$.

Abbass et al. [7, 6, 1426] propose the Pareto-frontier Differential Evolution (also abbreviated as PDE) approach. The algorithm works as follows. The initial population is initialized using a Gaussian distribution with mean 0.5 and standard deviation 0.15. Only the nondominated solutions are retained in the population for recombination (all dominated solutions are removed). If the number of nondominated solutions exceeds a certain threshold (50 was adopted in [7]), a distance metric is adopted to remove parents which are too close from each other (this can be seen as a niching procedure in which this distance metric is the niche radius). Three parents are randomly selected and a child is generated with them. The offspring is placed in the population only if it dominates the first selected parent; otherwise, the selection process is repeated. This process continues until the population is completed. In this approach, the step-length parameter F is generated from a Gaussian distribution $N(0, 1)$ and the boundary constraints are preserved either by reversing the sign if the variable is ≤ 0 or by repetitively subtracting 1 if it is ≥ 0, until the variable is within the allowable boundaries. PDE is compared with respect to SPEA [1778] in [7] and with respect to many other approaches (including PAES [886], the NSGA [1509] and the NPGA [709]) in [6]. In [5], a new version of PDE is introduced. This version is called Self-Adaptive Pareto Differential Evolution (SPDE) algorithm, because it self-adapts its crossover and its mutation rates. Abbass [3] proposes an approach called Memetic Pareto Artificial Neural Networks (MPANN). This approach consists of a version of Pareto Differential Evolution (PDE) [6] enhanced with local search, which is used to evolve neural networks in which an attempt is made to obtain a trade-off between the architecture and generalization ability of the network. So, two objectives are minimized: (1) error and (2) the number of hidden units. MPANN is validated using two benchmark data sets: the Australian credit card assessment problem and the diabetes problem (both were taken from the UCI Machine Learning Repository [1184]). Results are compared with respect to 23 algorithms, which include decision trees, rule-based methods, neural networks, and statistical algorithms. MPANN was able to outperform the traditional back propagation approach and obtained results competitive against the other 23 algorithms with respect to which it was compared.

The Pareto-Based Differential Evolution approach is proposed in [1037]. In this algorithm, Differential Evolution is extended to multi-objective optimization by incorporating a nondominated sorting and ranking selection procedure proposed by Deb et al. [363, 374]. Once the new candidate is obtained using DE operators, the new population is combined with the existing parents population and then the best members of the combined population (parents plus offspring) are chosen. This algorithm is not compared with respect to any other approach and is tested on 10 different unconstrained problems performing 250,000 evaluations. The authors indicate that the approach has difficulties to converge to the true Pareto front in two problems (Kursawe's test function [934] and ZDT4 [1772]).

Xue et al. [1722, 1720] propose the Multi-Objective Differential Evolution (MODE) approach. This algorithm uses a variant of the original DE, in which the best individual is adopted to create the offspring. A Pareto-based approach is introduced to implement the selection of the best individual. If a solution is dominated, a set of nondominated individuals can be identified and the "best" turns out to be any individual (randomly picked) from this set. Also, the authors adopt $(\mu+\lambda)$ selection, Pareto ranking and crowding distance in order to produce and maintain well-distributed solutions. MODE is used to solve five high dimensionality unconstrained problems with 250,000 evaluations and the results are compared only to those obtained by SPEA [1782].

Babu and Jehan [71] propose the Differential Evolution for Multi-Objective Optimization approach. This algorithm uses the single-objective Differential Evolution strategy with an aggregating function to solve bi-objective problems. A single optimal solution is obtained after N iterations using both a *Penalty Function Method* (to handle the constraints) and the *Weighting Factor Method* (to provide the importance of each objective from the user's perspective) [361] to optimize a single value. The authors present results for two bi-objective problems and compare them with respect to a simple GA. The authors indicate that the DE algorithm provides the exact optimum with a lower number of evaluations than the GA.

The Vector Evaluated Differential Evolution for Multi-Objective Optimization (VEDE) is proposed by Parsopoulos et al. [1256]. It is a parallel, multi-population Differential Evolution algorithm, which is inspired by the Vector Evaluated Genetic Algorithm (VEGA) [1440] approach. A number M of subpopulations are considered in a ring topology. Each population is evaluated using one of the objective functions of the problem, and there is an exchange of information among the populations through the migration of the best individuals. VEDE is validated using four bi-objective unconstrained problems and is compared with respect to VEGA. The authors indicate that the proposed approach outperformed VEGA in all cases.

Iorio and Li [746] propose the Nondominated Sorting Differential Evolution (NSDE). This approach is a simple modification of the NSGA-II [374]. The only difference between this approach and the NSGA-II is in the method for generating new individuals. The NSGA-II uses a real-coded crossover and mutation operator, but in the NSDE, these operators are replaced with the operators of Differential Evolution. New candidates are generated using the *DE/current-to-rand/1* strategy. NSDE is used to solve rotated problems with a certain degree of rotation on each plane. The results of the NSDE outperformed those produced by the NSGA-II. In further work, Iorio and Li [747] propose a variation of NSDE that incorporates directional information regarding both convergence and spread. For convergence, the authors modify NSDE so that offspring are generated in the direction of the previously generated solutions with better rank. For spread, the authors modify NSDE so that it favors the selection of individuals from different regions of decision variable space. The modified approach is called NSDE-DCS (DCS stands for "directional convergence and spread") and is compared with respect to the NSGA-II, the original NSDE, NSDE-DC (NSDE only with the directional convergence mechanism), and NSDE-DS (NSDE only with the directional spread mechanism). Results indicate that all the NSDE versions outperform the NSGA-II, but NSDE-DS practically provides the same results as NSDE-DCS. This is a very interesting outcome that indicates that improving spread may, in some cases, also improve convergence.

Kukkonen and Lampinen [913] propose a revised version of Generalized Differential Evolution (GDE), which was originally proposed in [946]. This approach extends the selection operation of the basic DE algorithm for constrained multi-objective optimization. The basic idea in this selection rule is that the trial vector is required to dominate the old population member used as a reference either in constraint violation space or in objective function space. If both vectors are feasible and nondominated with respect to each other, the one residing in a less crowded region is chosen to become part of the population of the next generation. GDE is validated using five bi-objective unconstrained problems. Results are compared with respect to the NSGA-II and SPEA [1782]. The authors report that the performance of GDE is similar to the NSGA-II, but they claim that their approach requires a lower CPU time. GDE is able to outperform SPEA in all the test functions adopted. In a further paper, Kukkonen and Lampinen [914] introduce GDE3, which is a new version of Generalized Differential Evolution that can handle both single- and multi-objective optimization problems (either constrained or unconstrained). Kukkonen and Lampinen [914] indicate that the main drawback of GDE2 (reported in [913]) is that its selection mechanism slows down convergence. Also, GDE2 seems to be too sensitive to its selection and control parameters. So, GDE3 extends the *DE/rand/1/bin* method to problems with any number of objectives and constraints. This approach is in fact a combination of the earlier GDE versions and the Pareto-Based Differential Evolution algorithm [1037]. The selection mechanism in GDE3 considers Pareto dominance (in objective function space) when comparing feasible solutions, and weak dominance (in constraint violation space) when comparing infeasible solutions. Feasible solutions are always preferred over infeasible ones, regardless of Pareto dominance. Nondominated sorting and crowding (as in the NSGA-II [374]) are also adopted in this approach. GDE3 is compared with respect to the NSGA-II in several test functions, including some from the DTLZ test suite [379].

Robič and Filipič [1365] propose an approach called Differential Evolution for Multi-Objective Optimization (DEMO). This algorithm combines the advantages of DE with the mechanisms of Pareto-based ranking and crowding distance sorting. DEMO only maintains one population and it is extended when newly created candidates take part immediately in the creation of the subsequent candidates. This enables a fast convergence towards the true Pareto front, while the use of nondominated sorting and crowding distance (derived from the NSGA-II [374]) of the extended population promotes the uniform spread of solutions. DEMO is compared in five high-dimensionality unconstrained problems outperforming in some problems to the NSGA-II, PDE [5], PAES [886], SPEA [1782] and MODE [1722].

Santana-Quintero and Coello Coello [1423] propose the ϵ-MyDE, whose pseudo code is shown in Algorithm 28. This approach keeps two populations: the main population (which is used to select the parents) and a secondary (external) population, in which the concept of ϵ-dominance [959] is adopted to retain the nondominated solutions found and to distribute them in an uniform way. The concept of ϵ-dominance does not allow two solutions with a difference less than ϵ_i in the i-th objective to be nondominated with respect to each other, thereby allowing a good spread of solutions. ϵ-MyDE uses real numbers representation, and incorporates a constraint-handling mechanism that allows infeasible solutions to intervene during recombination.

Algorithm 28 Proposed Algorithm: ϵ - MyDE
1: Initialize vectors of the population P
2: Evaluate the cost of each vector
3: **for** $i = 0$ to G **do**
4:    **repeat**
5:       Select three distinct vectors randomly
6:       Perform crossover using DE scheme
7:       Perform mutation
8:       Evaluate objective values
9:       **if** offspring is better than main parent **then**
10:          replace main parent in the population
11:       **end if**
12:    **until** population is completed
13:    Identify nondominated solutions in the population
14:    Add nondominated solutions into secondary population
15: **end for**

At the beginning of the evolutionary process, ϵ-MyDE randomly initializes all the individuals of the population. Each decision variable is normalized within its allowable bounds. The approach has two selection mechanisms that are activated based on the total number of generations and a parameter called $sel_2 \in (0.2 - 1)$, which regulates the selection pressure:

$$\text{Type of Selection} = \begin{cases} \text{Random,} & gen < (sel_2 * G_{max}) \\ \text{Elitist,} & \text{otherwise} \end{cases}$$

where:
gen = generation number
G_{max} = total number of generations

In both selections (random and elitist), a single parent is selected as a reference. This parent is used to compare the offspring generated by three different parents. This mechanism guarantees that all the parents of the main population will be reference parents for only one time during the generating process. Both types of selection are described next:

1. **Random Selection.-** 3 different parents are randomly selected from the primary population.
2. **Elitist Selection.-** 3 different parents are selected from the secondary population such that they maintain a close distance f_{near} among them. If no parent exist which fulfills this condition, another parent is randomly selected from the secondary population.

$$f_{near} = \frac{\sqrt{\sum_{i=0}^{FUN} (X_{i,max} - X_{i,min})^2}}{2FUN}$$

where:
FUN = number of objective functions
$X_{i,max}$ = upper bound of $i-th$ objective function of the secondary population
$X_{i,min}$ = lower bound of $i-th$ objective function of the secondary population

Recombination in ϵ-MyDE is performed using the following procedure. For each parent vector $\vec{p_i}$; $i = 0, 1, 2, \ldots, P-1$ (P = population), the offspring vector $\vec{h}$ is generated as:

$$h_j = \begin{cases} p_{r1,j} + F \cdot (p_{r2,j} - p_{r3,j}), & \text{if } x < p_{crossover}; \\ p_{ref,j}, & \text{otherwise.} \end{cases}$$

where: $j = 0, 1, 2, \ldots, var-1$ (var = number of variables for each solution vector), $x \in U(0,1)$, $p_{r1}, p_{r2}, p_{r3} \in [0, P-1]$, are integers and mutually different. $F > 0$. The integers r_1, r_2 and r_3 are the indexes of the selected parents randomly chosen from the interval [0, N-1] and ref is the index of the reference parent. F is a constant factor (a real number) which controls the amplification of the differential variation $p_{r2,j} - p_{r3,j}$.

In a further paper [671], ϵ-MyDE is hybridized with rough sets to give raise to a new approach called DEMORS (Differential Evolution for Multiobjective Optimization with Rough Sets). DEMORS operates in two phases. During the first phase, an improved version of ϵ-MyDE[13] is applied for 2000 fitness function evaluations. During the second phase, a local search procedure based on rough sets theory [1262] is applied for 1000 fitness function evaluations, in order to improve the solutions produced at the previous phase. The idea is

---
[13] The main improvement is the incorporation of the so-called Pareto-adaptive ϵ-grid [672] for the secondary population. The concept of Pareto-adaptive ϵ-dominance eliminates several of the drawbacks of ϵ-dominance [959].

to combine the high convergence rate of differential evolution with the high local search capabilities of rough sets. DEMORS is able to converge to the true Pareto front (or very close to it) in test problems with up to 30 decision variables, while only performing 3000 fitness function evaluations. Results are compared with respect to the NSGA-II.

Portilla Flores [1287] proposes a multi-objective version of differential evolution, which is used for concurrent design of pinion–rack continuously variable transmission (CVT). This mechatronic design problem is formulated as a dynamic multi-objective optimization problem in which two objectives are considered: (1) maximize the mechanical CVT efficiency, and (2) minimize the controller energy. The multi-objective algorithm adopted is based on Pareto ranking, it incorporates a secondary population to retain the nondominated solutions found during the evolutionary process, and it uses the feasibility rules from [1097] to handle the constraints of the problem. However, the approach does not include an explicit mechanism to maintain diversity (although a set of diverse solutions is actually generated). An interesting aspect of this work is that results are compared with respect to a mathematical programming technique: the goal attainment method. The comparison of results indicated that, as expected, the goal attainment method was very sensitive to its initial search point. Also, in several runs, it was not able to converge to a feasible solution. In contrast, the differential evolution algorithm was able to converge to feasible solutions in all the runs performed. However, the solutions generated by the goal attainment method were nondominated with respect to the solutions produced by differential evolution. Additionally, the CPU time required by differential evolution was about twice the time required by the goal attainment method.

Landa Becerra and Coello Coello [948] propose the use of the ϵ-constraint technique [632] hybridized with a single-objective evolutionary optimizer: the cultured differential evolution [947]. The ϵ-constraint method transforms a multi-objective optimization problem into several single-objective optimization problems (each of these optimizations leads to a single Pareto optimal point). This method has been normally disregarded in the evolutionary multi-objective optimization literature due to its high computational cost [1507, 1318]. However, the authors argue that, if care is placed in the single-objective optimizer, this sort of hybrid can generate the true Pareto front of very difficult multi-objective optimization problems at a reasonable computational cost. Such a hypothesis is validated by solving DTLZ8 and DTLZ9 from the benchmark proposed in [379] together with several other test problems from the benchmark proposed in [720, 721]. All of these test functions are considered very hard to solve by current MOEAs, and this is illustrated by showing the results obtained by the NSGA-II in them. In most cases, even when performing a very high number of fitness function evaluations, the NSGA-II is unable to reach the true Pareto front. In contrast, the hybrid algorithm proposed in this paper is able to converge to the true Pareto front (or very close to it) of all the problems.

Li and Zhang [982] propose a multi-objective differential evolution algorithm based on decomposition (MODE/D) for continuous multi-objective optimization problems with variable linkages. The authors use the weighted Tchebycheff approach to decompose a multi-objective optimization problem into several scalar optimization subproblems. The differential evolution operator is used for generating new trail solutions, and a neighborhood relationship among all the subproblems generated is defined, such that they all have similar optimal solutions. For validating their approach, the authors adopt test problems with variable linkages [1207] and propose variants of some of the ZDT test problems [1772]. Results are compared with respect to the NSGA-II [374], the Nondominated Sorting Differential Evolution (NSDE) [746] and GD3 [914]. The authors report that MODE/D clearly outperformed the other approaches with respect to which it was compared.

10.7.2 Advantages and Disadvantages of Differential Evolution

Differential evolution is with no doubt, a very powerful search engine in single-objective optimization, which has been found to be very robust by a number a researchers in a wide variety of (mainly nonlinear) optimization problems [1295, 1099]. However, its use in multiobjective optimization still raises some issues. For example, differential evolution seems to have a high convergence rate (similar to PSO, but with a higher degree of robustness), but has difficulties to reach the true Pareto front. Most multi-objective versions of differential evolution seem to converge very fast to the vicinity of the true Pareto front, but present problems to actually reach it and to spread solutions along the front (see for example [1423]). This seems to indicate that multi-objective differential evolution approaches require additional mechanisms to maintain diversity (e.g., crowded-based operators or good mutation operators, such as those adopted in multi-objective particle swarm optimizers).

Another issue that deserves attention is that differential evolution was proposed only for problems in which the decision variables are real numbers, unlike PSO, for which binary versions exist [840]. Thus, one topic of interest is to develop alternative encodings that allow the use of differential evolution in problems requiring alternative encodings (e.g., combinatorial optimization problems). The use of encodings such as the random keys [108] or other proposals that have been made mostly in the context of PSO (see for example [839, 717, 276, 1233]) may be alternatives worth exploring in such cases.

10.8 Artificial Immune Systems

Our immune system protects the organism from bacteria, viruses and other foreign pathogens. Its main task is to recognize all cells within the body and characterize them into self and foreign (or antigens). The immune system further characterizes foreign cells and develops defensive mechanisms against them (i.e., antibodies). If a foreign pathogen (i.e., an antigen) enters the body, then the immune system can launch a specific response against it. Specialized B cells must interact with Helper T cells (other specialized white blood cells) to initiate antibody production. Antibodies are specific to only one type of antigen, and they immobilize antigens, preventing them from causing infections.

Computationally speaking, the immune system is a highly parallel intelligent system that is able to learn and retrieve previous knowledge (i.e., it has "memory") to solve recognition and classification tasks. Due to these interesting features, several researchers have developed computational models of the immune system and have used it for a variety of tasks [331, 944].

Hughes Bersini and Francisco J. Varela are considered the first to apply immune algorithms to problem solving in the early 1990s [128]. Stephanie Forrest and Alan S. Perelson also developed important pioneering work on what is now known *computer immunology*. For example, they used genetic algorithms to study the pattern recognition capabilities of binary immune system models in the early 1990s [516].

10.8.1 Basic Concepts

```
Repeat
    1. Select an antigen A from PA
       (PA = Population of Antigens)
    2. Take (randomly) R antibodies from PS
       (PS = Population of Antibodies)
    3. For each antibody r ∈ R, match it against
       the selected antigen A
       Compute its match score (e.g., using Hamming distance)
    4. Find the antibody with the highest match score
       Break ties at random
    5. Add match score of winning antibody to its fitness
Until maximum number of cycles is reached
```

Fig. 10.9. Immune system model (fitness scoring) pseudo code

One of the applications in which the emulations of the immune system has been found useful is to maintain diversity in the population of a genetic algorithm used to solve multimodal optimization problems [516, 1496, 1498]. The proposal in this case has been to use binary strings to model both antibodies and antigens. Then, matching of an antibody and an antigen is determined if their bit strings are complementary (i.e., maximally different). The algorithm proposed in this case to compute fitness is shown in Figure 10.9 (this algorithm, assumes a population that includes antigens and antibodies both represented with binary strings) [1498]. The main idea of this approach is to construct a population of antigens and a population of antibodies. Antibodies are then matched against antigens and a fitness value is assigned to each antibody based on this matching (i.e., maximize matching between antigens and antibodies). This is precisely the process illustrated in the algorithm of Figure 10.9. Finally, a conventional genetic algorithm is used to replicate the antibodies that better match the antigens present.

10.8.2 Multiobjective Versions

Smith et al. [1498] show that fitness sharing emerges when their emulation of the immune system is used. Furthermore, this approach is more efficient (computationally speaking) than traditional fitness sharing [368], and it does not require additional information regarding the number of niches to be formed.

This same approach has been used to handle constraints in evolutionary optimization [634, 635], has also been hybridized with a MOEA [933, 312] (see Chapter 7 for more details of these applications), and has even been parallelized [272]. However, the first direct use of the immune system to solve multiobjective optimization problems reported in the literature is the work of Yoo and Hajela [1737]. This approach uses a linear aggregating function to combine objective function and constraint information into a scalar value that is used as the fitness function of a GA. Then, the best designs according to this value are defined as antigens and the rest of the population as a pool of antibodies. The simulation of the immune system is then done as in the previous work of the authors where the technique is used to handle constraints [635]. The algorithm is the following

1. Select randomly a single antigen from the antigens population.
2. From the population of antibodies, take a sample (randomly selected) without replacement (Yoo and Hajela [1737] suggest three times the number of antigens).
3. Each antibody in the sample is matched against the selected antigen, and a match score (based on the Hamming distance measured on the genotype) is computed.
4. The antibody with the highest score is identified, and ties are broken at random.
5. The matching score of the winning antibody is added to its fitness value (i.e., it is "rewarded").
6. The process is repeated a certain number of times (typically three times the number of antibodies).

This approach is applied to some structural optimization problems with two objectives (a two-bar truss structure, a simply supported I-beam, and a 10-bar truss structure). The use of different weights allows the authors to converge to a certain (pre-specified) number of points of the Pareto front, since they make no attempt to use any specific technique to preserve diversity. In this study, the approach is not compared to any other technique.

Anchor et al. [39] use both lexicographic ordering and Pareto-based selection in an evolutionary programming algorithm combined with an artificial immune system to detect virus and computer intrusion. In this work, however, emphasis is placed on the application rather than on the multiobjective aspects of the algorithm, since that is the main aim of this work. Therefore, the algorithm is not compared to other multiobjective optimization approaches.

Luh et al. [1026] propose the multi-objective immune algorithm (MOIA) which adopts several biologically inspired concepts. This is a fairly elaborate approach which adopts a binary encoding. In MOIA, each decision variable is divided into two parts: a heavy chain (high order bits) and a light chain (low order bits). The ratio between heavy and light portions of the string is a user-defined parameter. The antigens are the objectives and the constraint values that we wish to achieve. Thus, affinity of the antibodies is measured in such a way that the best antibodies are the feasible nondominated solutions. The approach has a germinal center where the nondominated solutions are cloned and hypermutated. Hypermutation (i.e., mutation with high mutation rates) is only applied to the light chain portion of the strings in order to avoid abrupt changes in the current solutions. The antibodies with the highest affinity (called *mature antibodies*) are divided into plasma cells (i.e., solutions to which local search has been applied) and antibodies, both of which are stored in a memory pool. Dominated solutions (called *immature antibodies*) are deleted. Dominated antibodies are removed from the memory pool, but some of them become part of the so-called germ-line DNA library (another user-defined parameter specifies how many of them become part of the library). After that, the *avidity* of the nondominated antibodies is computed as the inverse of the rank value of an antibody multiplied by its similar value with other antibodies. This aims to combine nondominance with diversity, since the expression adopted is similar to the computation of fitness sharing, and also requires a niche radius (allowable difference between antibodies). The approach adopts a tournament selection scheme to select antibodies exhibiting higher avidity values to serve in the construction of germ-line DNA libraries. These DNA libraries are made up of fragments of antibodies that come from two sources: the tournament selection process and the memory pool. A random selection of some fragments from the DNA library takes place and new antibodies are produced using gene fragments randomly selected from the DNA library. Then, a somatic point mutation is applied. This mutation changes a bit from 1 to 0 and vice versa according to a predefined mutation rate, and aims to act as a local search mechanism. The approach also uses other operators such as: somatic recombination, gene conversion, gene inversion, gene shift and nucleotide addition. At the end of the run, the memory pool contains the approximation of the Pareto optimal set produced by the algorithm. The approach was validated using six test functions taken from the specialized literature. Results were compared against SPEA [1782] adopting five performance measures: (1) generational distance [1626], (2) spacing [1452], (3) spread [363], (4) two set coverage [1782] and (5) extreme distance, which computes the Euclidean distance between the extremes of the Pareto front. The results indicated that MOIA outperformed SPEA in most cases. The authors carefully justify the (many) mechanisms of this approach using biological analogies. Nevertheless, from a purely algorithmic point of view, the approach seems complex, and it may be argued that some of its mechanisms are perhaps unnecessary or even redundant (e.g., the different mechanisms used to maintain diversity could be merged into a single operator).

In related work, Luh & Chueh [1025] present a slight variation of MOIA called constrained multi-objective immune algorithm (CMOIA). The only change is the use of a penalty function to handle constraints. CMOIA is validated using six test functions taken from [375] and two truss optimization problems. Results of the six test functions are compared with respect to the NSGA-II [374]. Results of the trusses are compared with respect to mathematical programming techniques and with respect to single-objective genetic algorithms. However, the comparisons are only graphical,[14] since no performance measures were adopted in this case.

Campelo et al. [195] propose the Multiobjective Clonal Selection Algorithm (MOCSA). This approach combines ideas from both CLONALG [1194] and opt-aiNet [1192]. MOCSA uses real-numbers encoding, nondominated sorting, cloning, maturation (i.e., Gaussian mutation) and replacement (based on nondominated sorting). The worst solutions (those not selected for cloning) are eliminated and generated anew in a random manner. MOCSA also uses an external population where the nondominated individuals found along the search process are stored. Niching is used to avoid the storage of very similar solutions (similarity is measured both in decision variable space and objective function space). In a further paper, Guimarães et al. [618] adopt this approach to solve a multi-objective problem from the area of electromagnetic design.

Balicki [80, 81] extend the Adaptive Multicriteria Evolutionary Algorithm with tabu mutation (AMEA) [82] with a constraint-handling mechanism based on an artificial immune system. Basically, he adopts the constraint-handling scheme from Coello Coello & Cruz Cortés [272] with a slight modification[15] and couples it to his AMEA. The algorithm is used to solve a task assignment problem in which the goal is to find allocations of program modules that reduce the total time of a program execution by taking advantage of the properties of some workstations or of a certain computer load within a distributed system. Results are compared with respect to the previous version of AMEA. The

---

[14] The comparisons with respect to the NSGA-II are not shown in the paper.
[15] The change is that instead of using only the Hamming distance between antigens and antibodies to determine affinity, Balicki also adds Pareto ranking. Thus, antibodies that are closest (with respect to a Hamming distance) from their antigens AND are nondominated, are selected as the winners.

results indicated that the use of the constraint-handling mechanism based on an artificial immune system produced an improvement of about 30% with respect to the previous version of AMEA.

Coello Coello and Cruz Cortés [271, 273] propose a multiobjective optimization approach called Multiobjective Immune System Algorithm (MISA), which is based on the clonal selection principle. MISA adopts a secondary population which uses the adaptive grid from PAES [886]. The antibodies in this case are the decision variables of the problem, and there is no population of antigens. Instead, only Pareto dominance and feasibility are used to identify the best solutions, which are cloned and then hypermutated. The hypermutation rate is proportional to the affinity value of each antibody. The approach also adopts a non-uniform mutation operator to the "not so good" antibodies found. This operator is initially applied with a high probability, but such probability value is linearly decreased over time. This approach is validated using several standard test functions taken from the specialized literature. Results are compared with respect to the NSGA-II [374], PAES [886] and the micro genetic algorithm for multiobjective optimization [283, 284].

Cutello et al. [313] extend PAES [886] with a different representation (*ad hoc* to the protein structure prediction problem of their interest) and with immune inspired operators.[16] The original mutation stage of PAES, which consists of two steps (mutate and evaluate) is replaced by four steps: (1) a clonal expansion phase, (2) an affinity maturation phase, (3) an evaluation phase, and (4) a selection phase (the best solution is chosen). During the affinity maturation phase, the authors adopt two mutation operators, which are also specifically designed for the problem of their interest. Two mutation rates are analyzed: (1) a static scheme and (2) a dynamic scheme in which the number of mutations decreases in a nonlinear way. The proposed approach (called I-PAES) is compared with respect to several techniques (both single-objective and multiobjective), including the NSGA-II [374].

Jiao et al. [791] propose the Immune Dominance Clonal Multiobjective Algorithm (IDCMA). This approach is based on clonal selection and adopts Pareto dominance (called "immune dominance" by the authors). In IDCMA, the antigens are the objective functions and constraints that must be satisfied. The antibodies are the candidate solutions. The affinity antibody-antigen is based on the objective function values and the feasibility of the candidate solutions. The authors also determine an antibody-antibody affinity using Hamming distances (the authors adopt the scheme proposed in [271]). This approach also adopts the so-called "immune differential degree", which is a value that denotes the relative distribution of nondominated solutions in the population, and is computed using a expression similar to the one adopted in fitness sharing [368]. The complete algorithm of IDCMA consists of 13 steps, and divides all the antibodies into three groups, which are stored in three

---

[16] It is interesting to note that Cutello et al. [313] indicate that this paper reports the first attempt to use a MOEA to fold medium size proteins (40–70 residues).

different populations. A different search strategy is applied to each of these populations, but at some point, the populations are combined in order to increase the search ability of the approach. The approach is validated using several standard test functions taken from the specialized literature. Results are compared with respect to SPEA2 [1775], the Random Weight Approach [751] and the Multiobjective Immune System Algorithm (MISA) [271].

Lu et al. [1022] propose the Immune Forgetting Multiobjective Optimization Algorithm (IFMOA), which adopts the fitness assignment scheme of SPEA [1782], a clonal selection operator, and an Immune Forgetting Operator. The clonal selection operator implements clonal proliferation, affinity maturation, and clonal selection on the antibody population (the antibodies are the possible solutions to the problem). Affinity maturation is implemented through a mutation operator. Affinity is determined using both the strength value (computed as in SPEA [1782]), and the sum of distances between one antibody and its two nearest individuals (in decision variable space). Clonal selection is used to update the antibody population. The best individuals (i.e., those with the highest affinity) are cloned, and an antibody archive is used to store the nondominated solutions obtained during the search. An interesting aspect of IFMOA is that a certain percentage of the main population is periodically "forgotten" (such solutions are sent to a clonal forgetting pool), and are replaced by solutions from the antibody archive (i.e., the external population). In [1760], IFMOA is used for unsupervised feature selection.

Freschi and Repetto [521] propose an approach called Vector Artificial Immune System (VAIS). This approach is based on the multimodal AIS optimization algorithm proposed by De Castro and Timmis [1192], called opt-aiNet. VAIS assigns fitness using the *strength* value of SPEA [1782], so that all the nondominated solutions have fitness values lower than 1, while all dominated solutions have fitness values greater than 1. After assigning fitness to an initially random population, the approach clones each solution and mutates them. Then, it applies a Pareto-based selection and the nondominated individuals are stored in an external memory. The best mutated clone for each individual replaces its parent (clonal selection). An affinity operator is applied to the external memory, acting as a diversity mechanism, since it computes Euclidean distances among solutions using thresholds in a way very similar to fitness sharing [368]. Then, the solutions that are too close together in objective function space are suppressed (i.e., deleted) from the external memory. Then, the contents of this memory is copied into the original population, and the rest of the individuals are randomly generated. If an infeasible solution is produced when creating new individuals, then it is discarded and generated again (i.e., the constraint-handling mechanism is a death penalty [265]). If a mutation produces an infeasible individual, then VAIS adopts a bisection rule to decrease the mutation rate in a progressive manner, until a feasible solution is produced. The authors indicate that this sort of approach works without making any prior assumptions about the types of constraints of the problem, but this is not true. If active constraints are considered, this sort of constraint-handling approach may not work properly or may become too expensive, computationally speaking (see for example [1097]). VAIS is compared with respect to the NSGA-II [374] using some standard test functions and performance measures reported in the specialized literature. A revised version of this approach, called Vector Immune System (VIS) is presented in [522]. VIS uses an adaptive mutation operator based on the deterministic 1/5 success rule (the self-adaptation mechanism normally adopted with evolution strategies [1460]). In this case, results are compared with respect to MISA [273] and PAES [886].

Wang and Mahfouf [1677] propose the Adaptive Clonal Selection Algorithm for Multiobjective Optimization (ACSAMO), which is a multi-objective artificial immune system based on the clonal selection principle. This approach adopts a linear aggregating function, so that the multi-objective problem is transformed into a single-objective one. The weights are dynamically varied during the search, so that several Pareto optimal solutions can be generated. However, selection is based on Pareto dominance (the authors adopt non-dominated sorting), and a crowding operator is used for eliminating extra individuals (individuals residing in the least crowded regions are eliminated). The number of clones generated is fixed and is proportional to the population size. The mutation rate, on the other hand, is proportional to the affinity (i.e., the distance from an individual to the best found so far), so that the worst individuals (i.e., those farthest from the best individual) are mutated at a higher rate than the best one. Evidently, this approach shares several similarities with the NSGA-II [374], since it uses nondominated sorting and crowding, besides adopting the main population to retain the nondominated solutions found during the search (rather than using an external population for that sake). Results are compared with respect to SPEA [1782] and the NSGA-II [374] using four of the ZDT test functions [1772]. The results shown indicate that the proposed approach outperforms the other two MOEAs in most cases.

Finally, and just to give an idea of the increasing popularity of multi-objective artificial immune systems, Campelo et al. [196] recently presented a survey of such types of approaches, as well as a proposal for a common framework for the description and analysis of multi-objective artificial immune systems.

10.8.3 Advantages and Disadvantages of Artificial Immune Systems

Artificial immune systems are relatively simple algorithms with very nice properties that make them naturally suitable for pattern recognition and classification tasks. Additionally, researchers have realized that this heuristic can also be successfully adopted for optimization tasks. However, not many researchers have explored this area, since most of the current work on artificial immune systems is focused on architectures, models and applications not related to optimization [1161]. Note however, that extending an artificial immune system so that it can deal with multiobjective optimization problem is not a trivial task. If one wants to keep the essence of the biological metaphor, recombination must not be incorporated. Although the use of clonal selection, hypermutation and local search mechanisms can compensate for the lack of recombination, in practice it may take longer to converge to the true Pareto front of certain test functions in current use if no recombination is adopted. There are, however, highly competitive multiobjective artificial immune systems, as we have seen in this section, but their main weakness is normally the additional parameters that they require (some of which may be difficult to fine tune for an arbitrary problem). In fact, some of these parameters may not be obvious at first sight (e.g., the proportion of antigens and antibodies in the population, the number of clones to be produced, etc.), but can become cumbersome when trying to use them. It is worth noting that current multiobjective artificial immune systems have mainly focused on the solution of standard test functions, rather than on application. It seems, for example, that their great potential for (multiobjective) classification and pattern recognition tasks is yet to be explored.

The clonal selection principle has been found suitable to solve multiobjective optimization problems by some researchers who justify choosing this artificial immune system model (see for example [273]), but this does not mean that other models cannot be used (e.g., immune networks [789]). It would be also interesting to explore the potential of some other concepts normally adopted in the specialized literature of artificial immune systems (e.g., gene libraries [218]) in the context of multiobjective optimization.

10.9 Other Heuristics

In this section, other heuristics that can be extended to handle multiple objectives are briefly discussed. For at least some of them, there are no implementations currently available that handle multiple objectives.

10.9.1 Cultural Algorithms

Some social researchers have suggested that culture might be symbolically encoded and transmitted within and between populations, as another inheritance mechanism [421, 1350]. Using this idea, Robert Reynolds [1358] developed a computational model in which cultural evolution is seen as an inheritance process that operates at two levels: the micro-evolutionary and the macro-evolutionary levels.

At the micro-evolutionary level, individuals are described in terms of "behavioral traits" (which can be socially acceptable or unacceptable). These behavioral traits are passed from generation to generation using several socially motivated operators. At the macro-evolutionary level, individuals are able to generate "mappa" [1350], or generalized descriptions of their experiences. Individual mappa can be merged and modified to form "group mappa" using a set of generic or problem specific operators. Both levels share a communication link.

Reynolds [1358] proposed the use of genetic algorithms to model the micro-evolutionary process, and Version Spaces [1115] to model the macro-evolutionary process of a cultural algorithm.

---

1. $t = 0$ (t = iteration counter)
2. Initialize $POP(0)$ (POP = Population)
3. Initialize $BELF(0)$ ($BELF$ = Belief Network)
4. Initialize $CHAN(0)$ ($CHAN$ = Communication Channel)
5. Evaluate $POP(0)$
6. t=1
Repeat
    Communicate $(POP(0), BELF(t))$
    Adjust $(BELF(t))$
    Communicate $(BELF(t), POP(t))$
    Modulate Fitness $(BELF(t), POP(t))$
    $t \leftarrow t + 1$
    Select $POP(t)$ from $POP(t-1)$
    Evolve $POP(t)$
    Evaluate $POP(t)$
Until Stopping Condition is Reached

Fig. 10.10. Cultural algorithm pseudo code

The pseudo code of a cultural algorithm is shown in Figure 10.10 [1358]. A population of individuals is used in this case, as with genetic algorithms. Each of these individuals are, however, described in terms of a set of traits or behaviors. An evaluation function is required to evaluate the performance of each individual in solving a problem, analogously to the fitness function of GAs. Each individual has its own set of beliefs, but these are adjusted over time using the "group mappa" or general experiences from the population. Each individual contributes to such "group mappa" at the end of each generation. When an individual mixes its individual mappa with the group mappa, there is a certain combination of beliefs. If an individual has a resulting combined mappa less than certain acceptable value, it is then pruned from the belief space. A selection process is then used to choose the parents to be evolved in the next generation (in fact, selection can be parallelized). The evolution process is done with certain operators that tend to be domain-specific. The interactions between belief space and the population depend on the communication channel used, as well as its protocols. For details refer to Reynolds [1358].

The concept is to preserve beliefs that are socially accepted and discard (or prune) unacceptable beliefs. It is possible to extend this technique to multiobjective optimization problems if nondominance is incorporated in the acceptance mechanism of the approach. The approach could work in a similar way to some proposals to extend the ant system to handle multiple objectives. In this case, an individual's cultural component could lead it to a local nondominated solution, and the global mechanism of the approach (intended for sharing group's solving experiences and behaviors) could lead the population towards global nondominated solutions. The same acceptance mechanism could incorporate additional criteria to encourage a smooth distribution of nondominated solutions (e.g., make unacceptable a nondominated solution generated in a region of the search space that is already too densely populated).

Multiobjective Versions

The only proposal known to date to extend the framework of cultural algorithms to multiobjective optimization is the one introduced by Coello & Landa [278]. In this approach, the search engine is evolutionary programming, it uses Pareto ranking and the belief space consists of two parts: the phenotypic normative part and a grid which is used to emphasize the generation of nondominated solutions that are uniformly distributed along the Pareto front. This grid is a variation of the adaptive grid proposed by Knowles and Corne [886]. The phenotypic normative part contains only the lower and upper bounds of the intervals for each objective function within which the grid is built. This grid is used to place each nondominated solution in some sort of coordinate system where the values of the objective functions are used to place each solution. The results obtained by this approach were found to be competitive with respect to those generated by the NSGA-II [374].

There is, however, plenty of room to exploit the different spaces adopted with cultural algorithms in order to produce a MOEA that can reduce its total number of objective function evaluations by exploiting domain knowledge extracted during the evolutionary search. Also, other search engines may be adopted (e.g., particle swarm optimization [840] or differential evolution [1525, 1294]).

10.9.2 Cooperative Search

Murthy et al. [1159, 1158] and Salman et al. [1418] propose an approach in which multiple agents (i.e., problem-solving methods) work together on the solution of a common problem. The approach is a sort of blackboard system [1187] in which communication and cooperation of the different agents takes place through a shared population of candidate solutions.

```
Repeat
    1. Use constructors to generate initial population P
    2. Apply improvers to generate new individuals
       that are added to P
    3. Apply destroyers to remove redundant individuals
       from P
Until maximum number of cycles is reached
```

Fig. 10.11. Cooperative search pseudo code

It is worth mentioning that the population-based nature of the approach has certain resemblance with a genetic algorithm. However, unlike genetic algorithms, the operators in this case are based on domain specific knowledge. The general algorithm of cooperative search as defined by Murthy et al. [1159, 1158] is shown in Figure 10.11.

Three types of agents are considered in this work:

1. **Constructors**: Are used to create initial solutions.
2. **Improvers**: Modify existing solutions and produce new solutions that are added to the current population. The control strategy of the improvers produces nondominated vectors.
3. **Destroyers**: Delete redundant or bad solutions from the population. The control strategy of the destroyers encourages diversity.

This approach is used to generate P_{true} for a paper mill scheduling problem. This technique allows that the DM's preferences can be added interactively.

Berro and Duthen [125] propose another approach based on a set of autonomous agents randomly created by a control system. In this case, each agent does not really know the function that it is solving, but simply tries to "colonize" an optimum. If an agent succeeds at colonizing an optimum, then it protects it by creating a zone of influence around it. Any other agent trying to penetrate this zone of influence is eliminated. An agent that has found an optimum also acquires reproduction capacity. This allows it to explore the surrounding area looking for new optima. This searching and "fighting" process encourages the dispersion of agents in the search space. The approach, that does not really use evolutionary algorithms, is proposed as an alternative to deal with multimodal and multiobjective problems. In this last case, it is argued that this system may be highly competitive with MOEAs. Additionally, it seems to be particularly appropriate to deal with dynamic environments in which traditional MOEAs are normally not applied.

There is also some work in which an agent-based model has been used as a basis to develop an evolutionary multiobjective optimization algorithm. For example, Menczer et al. [1088] propose an approach called "evolutionary local search algorithm" (ELSA), that uses a local selection scheme in which fitness depend on the consumption of certain limited shared resources. The authors show how local selection can actually act as a fitness sharing scheme in a natural way. The approach has several advantages (e.g., it is efficient and easy to be parallelized). However, they also detect some of its limitations (e.g., the potentially high communication costs when dealing with too many agents). Consequently, they advise its use mainly in distributed tasks (such as inductive learning and distributed robotics). ELSA is compared against VEGA [1440] and NPGA [709] on two problems (unitation versus pairs and feature selection in inductive learning), producing better results than the other approaches in terms of the fraction of the Pareto front covered.

A number of researchers have also used approaches (mainly based on parallel procedures) in which different evolutionary algorithms work separately and, at some point, cooperate so that globally nondominated solutions can emerge from the search (see for example the cellular multiobjective genetic algorithm of Murata et al. [1151, 1152] and the co-evolutionary multiobjective genetic algorithm of Parmee and Watson [1252]). However, the idea can still be exploited further, particularly when dealing with problems that have a high number of objectives. In that regard, the use of approaches such as cooperative game theory seem a promising alternative (see for example [92, 1189]).

10.10 Summary

In this chapter, several alternative heuristic search techniques (some of which are not evolutionary algorithms) that have been used to solve multiobjective optimization problems are discussed: simulated annealing, Tabu search (and scatter search), the ant system, distributed reinforcement learning, particle swarm optimization, artificial immune systems and differential evolution.

In the second part of this chapter, some other techniques that can be extended to deal with multiple objectives are discussed (namely, cultural algorithms and cooperative search). Each of these algorithms is briefly described, together with some discussion of how can they be extended to deal with multiple objectives. Such extensions constitute obvious paths for future research.

Further Explorations

Class Exercises

1. Compare and contrast the algorithmic components of a particle swarm optimization algorithm and a genetic algorithm. Discuss similarities and differences regarding operators, search bias and biological inspiration.
2. Describe the basic components of Tabu search and scatter search and discuss ways to adapt such components to the solution of multi-objective problems.
3. Describe the basic components of the ant system, and contrast them with the components of the ant colony system [148], and the approximated non-deterministic tree search (ANTS) [1052].
4. Discuss the role of the leader selection in particle swarm optimization and its possible impact in multi-objective optimization (see for example [165]).
5. Describe possible diversification techniques that could be adopted in a multi-objective version of a scatter search algorithm.

Class Software Projects

1. Implement three multi-objective versions of particle swarm optimization and compare them using some challenging test functions (see for example [1772, 379, 721]) and standard performance measures [1783, 874].
2. Design a generic framework for supporting different types of metaheuristics used for multi-objective optimization. See for example: A Platform and Programming Language Independent Interface for Search Algorithms (PISA) [141].
3. Implement 4 of the multi-objective versions of simulated annealing described in [1536] and compare them using the DTLZ test functions [379]. Adopt 3 of the performance measures suggested in [1783].

4. Implement 4 of the multi-objective particle swarm optimizers described in [1353] and compare them using the test functions proposed in [721]. Adopt 3 of the performance measures suggested in [1783].
5. Implement 3 different multi-objective algorithms based on Tabu search and compare them using the test functions proposed in [379]. Adopt 3 of the performance measures suggested in [1783]. Analyze the impact of using different diversification strategies on the performance of the approach.
6. Implement 4 different multi-objective algorithms based on differential evolution and compare them using the test functions proposed in [1772]. Adopt 3 of the performance measures suggested in [1783]. Analyze the impact of changing the differential evolution model on the performance of each multi-objective version.
7. Implement 3 different multi-objective algorithms based on the ant system and compare them using the test functions proposed in [379]. Adopt 3 of the performance measures suggested in [1783].
8. Implement 3 different multi-objective algorithms based on artificial immune systems and compare them using the test functions proposed in [379]. Adopt 3 of the performance measures suggested in [1783].

Discussion Questions

1. Investigate *stochastic combinatorial optimization* [523] and develop a bio-inspired heuristic for solving such problems. See for example the Stochastic Pareto-Ant Colony Optimization (SP-ACO) and Stochastic Pareto Simulated Annealing (SPSA) approaches proposed by Gutjahr [623].
2. Analyze the possibility of adding Mendel's laws of inheritance to an evolutionary multiobjective optimization algorithm. What sort of data structures would you need? What advantages do you think that this new algorithm could present? Discuss and experiment with such an algorithm. See for example the Mendelian Multi-Objective Genetic Algorithm (MMOSGA) [812] which attempts to accurately model the reproductive process of meiosis. The MMOSGA uses diploids and a dominance table that helps the algorithm to converge to P_{true}.
3. It is known that genetic algorithms are normally implemented with a haploid chromosomic structure. However, diploids (and multiploids) have been considered as a form of "historical record" that protects certain alleles (and combinations of them) from the damage that natural selection could cause them in a hostile environment. Diploids have been used by some evolutionary computation researchers to deal with dynamic fitness functions (i.e., that change over time). What role would diploids have in the context of multiobjective optimization? Do you see an advantage of using a diploid chromosomic structure in this context? Investigate about this topic (see for example [1651, 889]).

4. Suppose that you are asked to design an evolutionary multiobjective optimization algorithm that deals with a real-time application. What are the implications of such an algorithm? What processes could be parallelized? What processes should better be implemented in hardware? Discuss. The current attempts to design GA architectures for real-time response should be helpful. See for example [199].
5. Read the survey of Suman and Kumar [1536] on simulated annealing (particularly, the discussion on multi-objective optimization). Discuss some of the future areas of research discussed by the authors. Discuss the importance of the annealing schedule and the avoidance of repeated solutions in multi-objective simulated annealing.
6. Read the survey of Reyes-Sierra and Coello Coello [1353] on multi-objective particle swarm optimization. Which are the future paths of research that you consider the most promising? Discuss the role of the parameters of the particle swarm optimization algorithm in the context of multi-objective problems.
7. Analyze and criticize some of the multi-objective particle swarm optimizers not reviewed in this chapter (see for example [1090, 710, 1188, 563, 1608, 580, 1732]).
8. Analyze and criticize some of the multi-objective artificial immune systems not reviewed in this chapter (see for example [233, 259, 181]).
9. Discuss the main issues involved in designing a parameterless multi-objective particle swarm optimizer. Are these issues applicable to any other type of MOEA? Discuss.
10. Zhu and Leung [1769] proposed an enhanced annealing genetic algorithm for multiobjective optimization. Analyze the way in which simulated annealing is hybridized with a genetic algorithm and compare it to other similar hybrids previously proposed in the literature (see for example [389]). Relate the Coverage Quotient used by this method with any of the metrics discussed in Chapter 5. Criticize this approach and outline some of its possible limitations/disadvantages.

Possible Research Ideas

1. Design and implement a new MOEA with some biological inspiration (e.g., one of the algorithms discussed in this chapter that have not been extended to handle multiple objectives). Analyze the time and space complexity of your algorithm, and use the metrics discussed in Chapter 5 to validate its performance. Compare it against other MOEAs (e.g., NSGA-II [363], NPGA2 [453], PESA-II [299]). Adopt the ZDT [1772] and DTLZ [379] test functions.
2. There are several algorithms that exploit the property called *global convexity*, which is discussed by Castro Borges and Pilegaard Hansen [215]

and that have been rarely extended (or not extended at all) to multiobjective combinatorial optimization problems. For example: greedy randomized adaptive search procedure (GRASP)[17] [477], heuristic concentration [1378], and jump search [1609]. Propose a multiobjective extension of one of these algorithms and implement it. Evaluate the performance of such an algorithm using some well-known multiobjective combinatorial optimization problems and metrics such as those adopted in [1783, 874, 779, 777, 886, 1782, 1626, 1630]. What advantages (if any) presents your algorithm? What are the main drawbacks that you foresee? Discuss.
3. Consider the development of a multiobjective optimization technique that combines the ant system and particle swarm optimization (see for example [1462]). What potential benefits could you obtain from this sort of hybrid? In what domain(s) could such an algorithm be advantageous?
4. Although several researchers have considered the use of hybrids between MOEAs and artificial neural networks for diverse tasks (see for example [998, 1089]), the use of artificial neural networks themselves as a heuristic for multi-objective optimization is still scarce in the specialized literature (see for example [1048, 1537, 1538]). Explore the possible neural network models that could be suitable for approaching multi-objective optimization, and identify potential areas of opportunity within this field.
5. Investigate about *Optimal Pattern Matching* [1754] and discuss possible ways of integrating this scheme with MOEAs. Do you consider this as a viable multi-objective optimization technique when using MOEAs?
6. Design an experimental framework that allows a fair comparison of different types of metaheuristics used for multi-objective optimization. Your framework must consider issues such as memory usage, CPU time per evaluation, features of the test problems selected (e.g., some that are more suitable for methods based on local search and others that are very difficult for such methods), and number of evaluations performed. Few comparative studies of this type exist (see for example [291]), which makes this an interesting research path.
7. Propose a multi-objective version of the Max-Min Ant System (MMAS) [1529]. Attempt to extend the theoretical bounds of maximum pheromone concentration derived by Stützle and Hoos [1529] to the multi-objective case.
8. Sunil Nakrani and Craig Tovey [1163] proposed the *honey bee algorithm* for dynamically allocating servers to satisfy unpredictable request loads. This algorithm is based on the self-organization of honey bee colonies to allocate foragers among food sources. Propose a multi-objective extension of the *honey bee* algorithm and discuss some of its possible applications.

---

[17] One of the few multi-objective extensions of GRASP reported in the specialized literature is the paper of Vianna and Arroyo [1644].

9. Analyze the strategies for multi-objective simulated annealing proposed by Smith [1494] and design a new algorithm based on them. Explore the idea, discussed by Smith, of using an estimate of the Jacobian matrix to produce perturbations that properly guide the search.
10. Propose a multi-objective cultural algorithm that properly exploits knowledge extracted during the evolutionary process to speed up convergence towards PF_{true}. Any state-of-the-art MOEA can be used as the search engine of this approach (see Chapter 2). Show evidence of the efficiency gains achieved by the proposed approach.
11. Incorporate local search to a multi-objective particle swarm optimizer used to solve continuous problems. Show evidence of the improvements achieved with this memetic MOPSO. See Chapter 3 for more information on memetic MOEAs.
12. Propose a MOEA based on hybridizing differential evolution and particle swarm optimization. See for example [1211].

Epilog

As indicated in the Preface, the intent of this book is to provide a synergistic foundation for the study, development and use of multiobjective evolutionary algorithms (MOEAs). Together, the chapters provide a comprehensive framework for the examination and extension of such stochastic search algorithms; the general goal being to generate the Pareto optimal front with a uniform density of points along with the optimal values of the associated decision variables. The varieties of proposed MOEAs are classified in Chapter 2 along with their subjective advantages and disadvantages. In order to support obtainment of more accurate Pareto front solutions, local search and co-evolutionary MOEA methods are discussed in 3. Various generic test suites are given in Chapter 4 including unconstrained numerical problems, constrained numerical problems, NP-complete problems and real-world applications. A review of performance measures (metrics) that have been proposed to evaluate MOEAs is provided in Chapter 5. This chapter also includes information on statistical testing and visualization.

With the considerable number of MOEA techniques and the volume of applications of MOEAs reported in the contemporary literature as discussed in Chapters 2 and 7, it may be quite confusing for a student or practitioner new to the field to determine what type of MOEA to use for a specific application. The material exhibited attempts to demystify the selection of specific MOEA operators for a given multiobjective optimization problem (MOP) through an appropriate flow of prose, examples and historical perspective. Starting with Chapters 1 and 2, the reader is initially presented with foundational concepts and development, then simple MOPs and MOEA models leading to more sophisticated techniques and operators. Given this approach, an appropriate selection of a MOEA can be made for a given MOP. Given the current state-of-the-art in MOEA theory as reflected in Chapter 6, convergence to the Pareto optimal front in finite time is not guaranteed, but one can usually get close given enough time and assuming a stable algorithm. Considerable more theoretical MOEA research is required in order to glean insight to the convergence effect of various operators.

Although no specific guidelines for choosing a certain MOEA in a particular case can be generally provided, at least some generic suggestions can be drawn from an analysis of the current literature. Obviously, from an evolutionary computation perspective, the use of Pareto ranking and niching together with a secondary population is always advisable, regardless of the application domain. However, some researchers have reported that the use of linear aggregating functions with adaptive weights is highly competitive (and computationally efficient) when dealing with certain types of multiobjective combinatorial optimization problems (see Chapter 4). On the other hand, if the problem to be solved has a very large number of objectives, Pareto ranking tends to generate nondominated vectors easier, but in relatively limited quantity. Furthermore, the computational complexity of Pareto ranking may become significant. Therefore, some researchers have proposed as an alternative the use of criterion selection and linear aggregating functions despite their known limitations. MOEA techniques for solving these many multiobjective problems is an exciting area of study and research (see Chapter 2).

In real-world applications, the incorporation of preferences from the decision maker is normally required. Therefore, it is advisable to have at least a general understanding of the main preference incorporation techniques used with MOEAs (see Chapter 9). This discussion is paramount to providing the decision maker with a very limited number of possibilities if "quick" decisions are to be achieved.

Parallel or distributed MOEAs are an obvious recommendation for computationally intensive applications, but their use has to be justified from a performance perspective (efficiency and effectiveness). Additionally, since the design and implementation of a parallel MOEA is generally not straightforward, it is advisable to be aware of at least the basic MOEA parallelization schemes currently in use and their associated computational performance metrics (see Chapter 8).

Specific genotypic encodings such as binary, integer and real representations may be particularly useful in certain applications. Also, the use of a tree encoding (i.e., genetic programming) is commonly adopted in nonlinear regression, classification and prediction tasks [1254, 678, 54]. Multilevel encodings (such as the structured GA [333]) have been adopted in applications that deal with hierarchical systems (e.g., in telecommunications [1572, 1096]). Some authors also report the use of EAs that employ only mutation because recombination turns out to be too disruptive in those cases due to the high epistasis of the problem (e.g., [965]); additional research is needed to answer this question. Many of these general issues are discussed in Chapter 5. Additional issues such as uncertainty, noise, high-dimensionality and epistasis are also important when solving real-world MOPs. These topics are only beginning to be studied in the current MOEA literature and thus, more research is required. Nevertheless the interested reader is referred to the few sources that discuss these topics (e.g., [1586, 575, 101, 880, 1303, 484, 996, 386]). These MOEA approaches and the possible extension of other single-objective innovative approaches such as simulated annealing, Tabu search, particle swarms, and artificial ant colonies are summarized in Chapter 10.

This extensive text indicates that MOEAs are quite effective in moving the initial nondominated population close to the Pareto front and providing the decision maker with viable decision alternatives over a wide range of MOP classes. As indicated, the ultimate generation of decision variable values has been shown to be achievable across a wide range of MOPs. The material provided makes the structure and execution of MOEAs more understandable, selectable, and applicable to new and exciting real-world problems. The positive research directions and discussion questions at the end of each chapter provide more insight and areas for creative investigation. Again, the Evolutionary Multi-Objective (EMOO) Repository at:

```
http://delta.cs.cinvestav.mx/~ccoello/EMOO
```

is continually updated with new papers and theses from the MOEA research community. Additionally, a website has been devoted to this book at:

```
http://www.cs.cinvestav.mx/~emoobook
```

This website includes the Appendices of the book (not included due to space constraints), tutorial slides, and some other material to support the use of this text in a course.

References

1. E. H. Aarts and J. H. Korst. *Simulated Annealing and Botzmann Machines: A Stochastic Approach to Combinatorial Optimization and Neural Computing*. John Wiley & Sons, Chichester, UK, 1989.
2. H. Abbass, M. Towsey, and G. Finn. C-net: a method for generating non-deterministic and dynamic multivariate decision trees. *Knowledge and Information Systems*, 3:184–197, 2001.
3. H. A. Abbass. A Memetic Pareto Evolutionary Approach to Artificial Neural Networks. In *The Australian Joint Conference on Artificial Intelligence*, pages 1–12, Adelaide, Australia, December 2001. Springer. Lecture Notes in Artificial Intelligence Vol. 2256.
4. H. A. Abbass. An Evolutionary Artificial Neural Networks Approach for Breast Cancer Diagnosis. *Artificial Intelligence in Medicine*, 25(3):265–281, 2002.
5. H. A. Abbass. The Self-Adaptive Pareto Differential Evolution Algorithm. In *Congress on Evolutionary Computation (CEC'2002)*, volume 1, pages 831–836, Piscataway, New Jersey, May 2002. IEEE Service Center.
6. H. A. Abbass and R. Sarker. The Pareto Differential Evolution Algorithm. *International Journal on Artificial Intelligence Tools*, 11(4):531–552, 2002.
7. H. A. Abbass, R. Sarker, and C. Newton. PDE: A Pareto-frontier Differential Evolution Approach for Multi-objective Optimization Problems. In *Proceedings of the Congress on Evolutionary Computation 2001 (CEC'2001)*, volume 2, pages 971–978, Piscataway, New Jersey, May 2001. IEEE Service Center.
8. Y. Abdel-Magid and M. Abido. Optimal Multiobjective Design of Robust Power System Stabilizers Using Genetic Algorithms. *IEEE Transactions on Power Systems*, 18(3):1125–1132, August 2003.
9. M. Abido. Multiobjective Evolutionary Algorithms for Electric Power Dispatch Problem. *IEEE Transactions on Evolutionary Computation*, 10(3): 315–329, June 2006.
10. A. Abraham, L. Jain, and R. Goldberg, editors. *Evolutionary Multiobjective Optimization. Theoretical Advances and Applications*. Springer, USA, 2005. ISBN 1-85233-787-7.
11. S. Adra, I. Griffin, and P. Fleming. An Adaptive Memetic Algorithm for Enhanced Diversity. In I. Parmee, editor, *Adaptive Computing in Design and*

Manufacture 2006. Proceedings of the Seventh International Conference, pages 251–254, Bristol, UK, April 2006. The Institute for People-centred Computation.

12. S. F. Adra, I. Griffin, and P. J. Fleming. Hybrid Multiobjective Genetic Algorithm with a New Adaptive Local Search Process. In H.-G. B. et al., editor, *2005 Genetic and Evolutionary Computation Conference (GECCO'2005)*, volume 1, pages 1009–1010, New York, USA, June 2005. ACM Press.
13. S. F. Adra, I. Griffin, and P. J. Fleming. A Comparative Study of Progressive Preference Articulation Techniques for Multiobjective Optimisation. In S. Obayashi, K. Deb, C. Poloni, T. Hiroyasu, and T. Murata, editors, *Evolutionary Multi-Criterion Optimization, 4th International Conference, EMO 2007*, pages 908–921, Matshushima, Japan, March 2007. Springer. Lecture Notes in Computer Science Vol. 4403.
14. N. Agrawal, G. Rangaiah, A. Ray, and S. Gupta. Multi-Objective Optimization of the Operation of an Industrial Low-Density Polyethylene Tubular Reactor Using Genetic Algorithm and Its Jumping Gene Adaptations. *Industrial and Engineering Chemistry Research*, 45:3182–3199, 2006.
15. J. Aguilar and P. Miranda. Approaches Based on Genetic Algorithms for Multiobjective Optimization Problems. In W. Banzhaf, J. Daida, A. E. Eiben, M. H. Garzon, V. Honavar, M. Jakiela, and R. E. Smith, editors, *Proceedings of the Genetic and Evolutionary Computation Conference (GECCO'99)*, volume 1, pages 3–10, Orlando, Florida, USA, 1999. Morgan Kaufmann Publishers.
16. H. Aguirre and K. Tanaka. Selection, Drift, Recombination, and Mutation in Multiobjective Evolutionary Algorithms on Scalable MNK-Landscapes. In C. A. Coello Coello, A. Hernández Aguirre, and E. Zitzler, editors, *Evolutionary Multi-Criterion Optimization. Third International Conference, EMO 2005*, pages 355–369, Guanajuato, México, March 2005. Springer. Lecture Notes in Computer Science Vol. 3410.
17. H. E. Aguirre, M. Sato, and K. Tanaka. Preliminary Study on the Performance of Multi-objective Evolutionary Algorithms with MNK-Landscapes. In *Proceedings of the 2004 RISP International Workshop on Nonlinear Circuits and Signal Processing (NCSP 2004)*, pages 315–318, Hawaii, USA, March 2004. The Research Institute of Signal Processing Japan.
18. H. E. Aguirre and K. Tanaka. Parallel Varying Mutation Genetic Algorithms. In *Proceedings of the 2002 IEEE World Congress on Computational Intelligence*, pages 795–800, Piscataway, NJ, May 2002. IEEE Service Center.
19. H. E. Aguirre, K. Tanaka, T. Sugimura, and S. Oshita. Halftone Image Generation with Improved Multiobjective Genetic Algorithm. In E. Zitzler, K. Deb, L. Thiele, C. A. Coello Coello, and D. Corne, editors, *First International Conference on Evolutionary Multi-Criterion Optimization*, pages 501–515. Springer-Verlag. Lecture Notes in Computer Science No. 1993, 2001.
20. F. J. Aherne, N. A. Thacker, and P. I. Rockett. Automatic Parameter Selection for Object Recognition using a Parallel Multiobjective Genetic Algorithm. In *Proceedings of the 7th International Conference on Computer Analysis of Images and Patterns (CAIP'97)*, Lecture Notes in Computer Science 1296, pages 559–566, Kiel, Germany, September 1997. Springer Verlag.
21. F. J. Aherne, N. A. Thacker, and P. I. Rockett. Optimal Pairwise Geometric Histograms. In A. F. Clark, editor, *Electronic Proceedings of the Eighth British*

Machine Vision Conference, BMVC97, pages 480–490, University of Essex, United Kingdom, September 1997.
22. F. J. Aherne, N. A. Thacker, and P. I. Rockett. Optimising Object Recognition Parameters using a Parallel Multiobjective Genetic Algorithm. In *Proceedings of the 2nd IEE/IEEE International Conference on Genetic Algorithms in Engineering Systems: Innovations and Applications (GALESIA'97)*, pages 1–6, Glasgow, Scotland, September 1997. IEE.
23. C. W. Ahn. *Advances in Evolutionary Algorithms. Theory, Design and Practice*. Springer, 2006. ISBN 3-540-31758-9.
24. S. G. Akl. *The Design and Analysis of Parallel Algorithms*. Prentice-Hall, Englewood Cliffs, NJ, 1989.
25. J. T. Alander. An Indexed Bibliography of Genetic Algorithms: Years 1957–1993. Technical report, University of Vaasa, Department of Information Technology and Production Economics, Vaasa, Finland, 1994. Technical Report Report Series No. 94-1.
26. E. Alba, F. Almeida, M. J. Blesa, J. Cabeza, C. Cotta, M. Díaz, I. Dorta, J. Gabarró, C. León, J. Luna, L. M. Moreno, C. Pablos, J. Petit, A. Rojas, and F. Xhafa. MALLBA: A Library of Skeletons for Combinatorial Optimisation. In *Proceedings of the 8th International Euro-Par Conference on Parallel Processing*, pages 927–932. Springer-Verlag. Lecture Notes in Computer Science Vol. 2400, 2002.
27. E. Alba and M. Tomassini. Parallelism and Evolutionary Algorithms. *IEEE Transactions on Evolutionary Computation*, 6(5):443–462, 2002.
28. E. Alba and J. Troya. Analyzing Synchronous and Asynchronous Parallel Distributed Genetic Algorithms. *Future Generation Computer Systems*, 17(4):451–465, 2001.
29. E. Alba and J. Troya. Improving Flexibility and Efficiency by Adding Parallelism to Genetic Algorithms. *Statistics and Computing*, 12(2):91–114, 2002.
30. I. Alberto and P. Mateo. Representation and management of MOEA populations based on graphs. *European Journal of Operational Research*, 159(1): 52–65, November 2004.
31. V. N. Alexandrov and G. M. Megson. *Parallel Algorithms for Knapsack Type Problems*. World Scientific Publishing Company, New Jersey, 1999.
32. A. O. Allen. *Probability, Statistics, and Queuing Theory with Computer Science Applications*. Academic Press, Inc., Boston, Massachusetts, second edition, 1990.
33. R. Allenson. Genetic Algorithms with Gender for Multi-function Optimisation. Technical Report EPCC-SS92-01, Edinburgh Parallel Computing Centre, Edinburgh, Scotland, 1992.
34. K. T. Alligood. *Chaos: An Introduction to Dynamical Systems*. Springer, New York, 1996.
35. H. Almuallin and T. G. Dietterich. Learning with many irrelevant features. In *Proceedings of the Ninth National Conference on Artificial Intelligence*, pages 547–552, Menlo Park, California, 1991. AAAI Press.
36. P. Alotto, A. V. Kuntsevitch, C. Magele, G. Molinari, C. Paul, K. Preis, M. Repetto, and K. R. Richter. Multiobjective Optimization in Magnetostatics: A Proposal for Benchmark Problems. Technical report, Institut für Grundlagen und Theorie Electrotechnik, Technische Universität Graz, Graz, Austria, 1996. http://www-igte.tu-graz.ac.at/team/berl01.htm.

37. J. E. Alvarez-Benitez, R. M. Everson, and J. E. Fieldsend. A MOPSO Algorithm Based Exclusively on Pareto Dominance Concepts. In C. A. Coello Coello, A. Hernández Aguirre, and E. Zitzler, editors, *Evolutionary Multi-Criterion Optimization. Third International Conference, EMO 2005*, pages 459–473, Guanajuato, México, March 2005. Springer. Lecture Notes in Computer Science Vol. 3410.
38. M. J. Alves and J. Clímaco. An Interactive Method for 0-1 Multiobjective Problems Using Simulated Annealing and Tabu Search. *Journal of Heuristics*, 6(3):385–403, August 2000.
39. K. P. Anchor, J. B. Zydallis, G. H. Gunsch, and G. B. Lamont. Extending the Computer Defense Immune System: Network Intrusion Detection with a Multiobjective Evolutionary Programming Approach. In J. Timmis and P. J. Bentley, editors, *First International Conference on Artificial Immune Systems (ICARIS'2002)*, pages 12–21. University of Kent at Canterbury, UK, September 2002. ISBN 1-902671-32-5.
40. J. M. Anderson, T. M. Sayers, and M. G. H. Bell. Optimization of a Fuzzy Logic Traffic Signal Controller by a Multiobjective Genetic Algorithm. In *Proceedings of the Ninth International Conference on Road Transport Information and Control*, pages 186–190, London, April 1998. IEE.
41. M. B. Anderson. The Potential of Genetic Algorithms for Subsonic Wing Design. In *1st AIAA Aircraft Engineering, Technology, and Operations Congress*, Los Angeles, California, September 1995. AIAA Paper 95-3925.
42. M. B. Anderson and G. A. Gebert. Using Pareto Genetic Algorithms for Preliminary Subsonic Wing Design. Technical Report AIAA-96-4023-CP, AIAA, Washington, D.C., 1996.
43. M. B. Anderson and W. R. Lawrence. Launch Conditions and Aerodynamic Data Extraction By An Elitist Pareto Genetic Algorithm. In *AIAA Atmospheric Flight Mechanics Conference*, San Diego, California, July 1996. AIAA Paper 96-3361.
44. M. B. Anderson, W. R. Lawrence, and G. A. Gebert. Using an Elitist Pareto Genetic Algorithm for Aerodynamic Data Extraction. In *4th Aerospace Sciences Meeting and Exhibit*, Reno, Nevada, January 1996. AIAA Paper 96-0514.
45. S. Anderson, V. Kadirkamanathan, A. Chipperfield, V. Sharifi, and J. Swithenbank. Multi-objective optimization of operational variables in a waste incineration plant. *Computers & Chemical Engineering*, 29(5):1121–1130, April 2005.
46. J. Andersson and P. Krus. Metamodel Representations for Robustness Assessment in Multiobjective Optimization. In *Proceedings of the 13th International Conference on Engineering Design (ICED 01)*, Glasgow, UK, August 2001.
47. J. Andersson and P. Krus. Multiobjective Optimization of Mixed Variable Design Problems. In E. Zitzler, K. Deb, L. Thiele, C. A. C. Coello, and D. Corne, editors, *First International Conference on Evolutionary Multi-Criterion Optimization*, pages 624–638. Springer-Verlag. Lecture Notes in Computer Science No. 1993, 2001.
48. J. Andersson and D. Wallace. Pareto optimization using the struggle genetic crowding algorithm. *Engineering Optimization*, 34(6):623–643, December 2002.
49. Y. Aneja and K. Nair. Bicriteria transportation problem. *Management Science*, 25:73–78, 1978.

50. A. Angantyr, J. Andersson, and J.-O. Aidanpaa. Constrained Optimization based on a Multiobjective Evolutionary Algorithm. In *Proceedings of the 2003 Congress on Evolutionary Computation (CEC'2003)*, volume 3, pages 1560–1567, Canberra, Australia, December 2003. IEEE Press.
51. P. J. Angeline and J. B. Pollack. Competitive Environments Evolve Better Solutions for Complex Tasks. In S. Forrest, editor, *Proceedings of the Fifth International Conference on Genetic Algorithms*, pages 264–270, San Mateo, California, 1993. University of Illinois at Urbana-Champaign, Morgan Kaufmann Publishers.
52. N. E. Antoine. *Aircraft Optimization for Minimal Environmental Impact*. PhD thesis, Department of Aeronautics and Astronautics, Stanford University, Stanford, California, USA, August 2004.
53. H. Anton. *Calculus with Analytic Geometry*. John Wiley & Sons, New York, 2nd edition, 1984.
54. M. Arakawa, K. Hasegawa, and K. Funatsu. QSAR study of anti-HIV HEPT analogues based on multi-objective genetic programming and counter-propagation neural network. *Chemometrics and Intelligent Laboratory Systems*, 83(2):91–98, September 2006.
55. L. Araujo. Multiobjective Genetic Programming for Natural Language Parsing and Tagging. In T. P. Runarsson, H.-G. Beyer, E. Burke, J. J. Merelo-Guervós, L. D. Whitley, and X. Yao, editors, *Parallel Problem Solving from Nature - PPSN IX, 9th International Conference*, pages 433–442. Springer. Lecture Notes in Computer Science Vol. 4193, Reykjavik, Iceland, September 2006.
56. B. Arkov, D. Evans, P. Fleming, D. Hill, J. Norton, I. Pratt, D. Rees, and K. Rodríguez Vázquez. System Identification Strategies Applied to Aircraft Gas-Turbine Engines. In *14th IFAC World Congress*, volume 1, pages 145–152, Beijing, China, July 1999.
57. V. A. Armentano and J. E. Claudio. An Application of a Multi-Objective Tabu Search Algorithm to a Bicriteria Flowshop Problem. *Journal of Heuristics*, 10(5):463–481, September 2004.
58. K. Arrow and H. Raynaud. *Social Choice and Multicriterion Decison-Making*. The MIT Press, Cambridge, Massachusetts, 1986.
59. K. J. Arrow. *Social Choice and Individual Values*. John Wiley, New York, 1951.
60. K. J. Arrow. *Social Choice and Individual Values*. Yale University Press, second edition, 1963.
61. K. J. Arrow, E. W. Barankin, and D. Blackwell. Admissible Points of Convex Sets. In H. W. Kuhn and A. W. Tucker, editors, *Contributions to the Theory of Games*, pages 87–91. Princeton University Press, Princeton, New Jersey, 1953.
62. J. Arroyo and V. Armentano. Genetic local search for multi-objective flowshop scheduling problems. *European Journal of Operational Research*, 167(3):717–738, December 2005.
63. T. Arslan, D. H. Horrocks, and E. Ozdemir. Structural Synthesis of Cell-based VLSI Circuits using a Multi-Objective Genetic Algorithm. *IEE Electronic Letters*, 32(7):651–652, March 1996.
64. T. Arslan, E. Ozdemir, M. S. Bright, and D. H. Horrocks. Genetic Synthesis Techniques for Low-Power Digital Signal Processing Circuits. In *Proceedings*

Of The IEE Colloquium On Digital Synthesis, pages 7/1–7/5, London, UK, February 1996. IEE.
65. A. Augugliaro, L. Dusonchet, S. Favuzza, and E. R. Sanseverino. A Fuzzy-Logic based Evolutionary Multiobjective Approach for Automated Distribution Networks Management. In *2004 Congress on Evolutionary Computation (CEC'2004)*, volume 1, pages 847–854, Portland, Oregon, USA, June 2004. IEEE Service Center.
66. B. Awadh, N. Sepehri, and O. Hawaleshka. A Computer-Aided Process Planning Model Based on Genetic Algorithms. *Computers in Operations Research*, 22(8):651–652, 1995.
67. M. E. Aydin and V. Y. git. Parallel Simulated Annealing. In E. Alba, editor, *Parallel Metaheuristics*, pages 267–287. Wiley-Interscience, 2005.
68. S. Azarm, B. J. Reynolds, and S. Narayanan. Comparison of Two Multi-objective Optimization Techniques With and Within Genetic Algorithms. In *CD-ROM Proceedings of the 25th ASME Design Automation Conference*, volume Paper No. DETC99/DAC-8584, Las Vegas, Nevada, September 1999.
69. R. Azencott. *Simulated Annealing: Parallelization Techniques*. John Wiley and Sons, 1992.
70. B. Babu, P. Chakole, and J. Mubeen. Multiobjective differential evolution (MODE) for optimization of adiabatic styrene reactor. *Chemical Engineering Science*, 60(17):4822–4837, September 2005.
71. B. Babu and M. M. L. Jehan. Differential Evolution for Multi-Objective Optimization. In *Proceedings of the 2003 Congress on Evolutionary Computation (CEC'2003)*, volume 4, pages 2696–2703, Canberra, Australia, December 2003. IEEE Press.
72. T. Bäck. *Evolutionary Algorithms in Theory and Practice*. Oxford University Press, New York, 1996.
73. T. Bäck, D. Fogel, and Z. Michalewicz, editors. *Handbook of Evolutionary Computation*, volume 1. IOP Publishing Ltd. and Oxford University Press, 1997.
74. T. Bäck, D. B. Fogel, and Z. Michalewicz, editors. *Evolutionary Computation 1: Basic Algorithms and Operators*, volume 1. Institute of Physics Publishing, first edition, May 2000.
75. T. Bäck, D. B. Fogel, and Z. Michalewicz, editors. *Evolutionary Computation 2: Advances Algorithms and Operators*, volume 2. Institute of Physics Publishing, first edition, July 2000.
76. T. P. Bagchi. *Multiobjective Scheduling by Genetic Algorithms*. Kluwer Academic Publishers, Boston, Massachusetts, 1999.
77. T. P. Bagchi. Pareto-Optimal Solutions for Multi-objective Production Scheduling Problems. In E. Zitzler, K. Deb, L. Thiele, C. A. Coello Coello, and D. Corne, editors, *First International Conference on Evolutionary Multi-Criterion Optimization*, pages 458–471. Springer-Verlag. Lecture Notes in Computer Science No. 1993, 2001.
78. F. Baita, F. Mason, C. Poloni, and W. Ukovich. Genetic algorithm with redundancies for the vehicle scheduling problem. In J. Biethahn and V. Nissen, editors, *Evolutionary Algorithms in Management Applications*, pages 341–353. Springer-Verlag, Berlin, 1995.
79. J. M. Baldwin. *Development and Evolution: Including Psychophysical Evolution, Evolution by Orthoplasy and the Theory of Genetic Modes*. Macmillan, New York, 1902.

80. J. Balicki. Multi-criterion Evolutionary Algorithm with Model of the Immune System to Handle Constraints for Task Assignments. In L. Rutkowski, J. H. Siekmann, R. Tadeusiewicz, and L. A. Zadeh, editors, *Artificial Intelligence and Soft Computing - ICAISC 2004, 7th International Conference. Proceedings*, pages 394–399, Zakopane, Poland, June 2004. Springer. Lecture Notes in Computer Science. Volume 3070.
81. J. Balicki. Immune systems in multi-criterion evolutionary algorithm for task assignments in distributed computer system. In *Advances in Web Intelligence*, pages 51–56. Springer. Lecture Notes in Computer Science Vol. 3528, 2005.
82. J. Balicki and Z. Kitowski. Multicriteria Evolutionary Algorithm with Tabu Search for Task Assignment. In E. Zitzler, K. Deb, L. Thiele, C. A. Coello Coello, and D. Corne, editors, *First International Conference on Evolutionary Multi-Criterion Optimization*, pages 373–384. Springer-Verlag. Lecture Notes in Computer Science No. 1993, 2001.
83. R. Balling. Pareto sets in decision-based design. *Journal of Engineering Valuation and Cost Analysis*, 3:189–198, 2000.
84. R. Balling. The Maximin Fitness Function; Multiobjective City and Regional Planning. In C. M. Fonseca, P. J. Fleming, E. Zitzler, K. Deb, and L. Thiele, editors, *Evolutionary Multi-Criterion Optimization. Second International Conference, EMO 2003*, pages 1–15, Faro, Portugal, April 2003. Springer. Lecture Notes in Computer Science. Volume 2632.
85. R. Balling and S. Wilson. The Maximim Fitness Function for Multi-objective Evolutionary Computation: Application to City Planning. In L. Spector, E. D. Goodman, A. Wu, W. Langdon, H.-M. Voigt, M. Gen, S. Sen, M. Dorigo, S. Pezeshk, M. H. Garzon, and E. Burke, editors, *Proceedings of the Genetic and Evolutionary Computation Conference (GECCO'2001)*, pages 1079–1084, San Francisco, California, 2001. Morgan Kaufmann Publishers.
86. R. J. Balling, J. T. Taber, M. R. Brown, and K. Day. Multiobjective Urban Planning Using a Genetic Algorithm. *ASCE Journal of Urban Planning and Development*, 125(2):86–99, June 1999.
87. R. J. Balling, J. T. Taber, K. Day, and S. Wilson. City Planning with a Multi-objective Genetic Algorithm and a Pareto Set Scanner. In I. C. Parmee, editor, *Proceedings of the Fourth International Conference on Adaptive Computing in Design and Manufacture (ACDM'2000)*, pages 237–247. PEDC, University of Plymouth, UK, Springer London, 2000.
88. A. M. Baltar and D. G. Fontane. A generalized multiobjective particle swarm optimization solver for spreadsheet models: application to water quality. In *Hydrology Days 2006*, Fort Collins, Colorado, USA, March 2006.
89. W. Banzhaf, P. Nordin, R. E. Keller, and F. D. Fancone. *Genetic Programming. An Introduction*. Morgan Kaufmann Publishers, San Francisco, California, 1998.
90. B. Barán and M. Schaerer. A Multiobjective Ant Colony System for Vehicle Routing Problem with Time Windows. In *Proceedings of the 21st IASTED International Conference on Applied Informatics*, pages 97–102, Innsbruck, Austria, February 2003. IASTED.
91. H. J. Barbosa and A. M. Barreto. An interactive genetic algorithm with co-evolution of weights for multiobjective problems. In L. Spector, E. D. Goodman, A. Wu, W. Langdon, H.-M. Voigt, M. Gen, S. Sen, M. Dorigo, S. Pezeshk, M. H. Garzon, and E. Burke, editors, *Proceedings of the Genetic*

and *Evolutionary Computation Conference (GECCO'2001)*, pages 203–210, San Francisco, California, 2001. Morgan Kaufmann Publishers.
92. H. J. C. Barbosa. A coevolutionary genetic algorithm for a game approach to structural optimization. In T. Bäck, editor, *Proceedings of the Seventh International Conference on Genetic Algorithms*, pages 545–552, San Mateo, California, July 1997. Michigan State University, Morgan Kaufmann Publishers.
93. R. S. Barr, B. L. Golden, J. P. Kelly, M. G. C. Resende, and J. William R. Stewart. Designing and Reporting on Computational Experiments with Heuristic Methods. *Journal of Heuristics*, 1:9–32, 1995.
94. J. Barrow. *Theories of Everything*. Oxford University Press, 1991.
95. T. Bartz-Beielstein, P. Limbourg, K. E. Parsopoulos, M. N. Vrahatis, J. Mehnen, and K. Schmitt. Particle Swarm Optimizers for Pareto Optimization with Enhanced Archiving Techniques. In *Proceedings of the 2003 Congress on Evolutionary Computation (CEC'2003)*, volume 3, pages 1780–1787, Canberra, Australia, December 2003. IEEE Press.
96. T. Bartz-Beielstein, K. Schmitt, J. Mehnen, B. Naujoks, and D. Zibold. KEA – A software package for development, analysis, and application of multiple objective evolutionary algorithms. Interner Bericht des Sonderforschungsbereichs 531 *Computational Intelligence* CI–185/04, Universität Dortmund, November 2004.
97. M. Basseur. *Conception d'Algorithmes Coopératifs Pour L'Optimisation Multi-Objectif: Application aux Problèmes d'Ordonnancement de Type Flow-Shop*. PhD thesis, Université des Sciences et Technologies de Lille, France, 2005. (in French).
98. M. Basseur, F. Seynhaeve, and E. ghazali Talbi. Design of multi-objective evolutionary algorithms: Application to the flow-shop. In *Congress on Evolutionary Computation (CEC'2002)*, volume 2, pages 1151–1156, Piscataway, New Jersey, May 2002. IEEE Service Center.
99. M. Basseur, F. Seynhaeve, and E.-G. Talbi. A Cooperative Metaheuristic Applied to Multi-Objective Flow-Shop Scheduling Problem. In N. Nedjah and L. de Macedo Mourelle, editors, *Real-World Multi-Objective System Engineering*, pages 139–162. Nova Science Publishers, New York, 2005.
100. M. Basseur, F. Seynhaeve, and E.-G. Talbi. Path Relinking in Pareto Multi-objective Genetic Algorithms. In C. A. Coello Coello, A. Hernández Aguirre, and E. Zitzler, editors, *Evolutionary Multi-Criterion Optimization. Third International Conference, EMO 2005*, pages 120–134, Guanajuato, México, March 2005. Springer. Lecture Notes in Computer Science Vol. 3410.
101. M. Basseur and E. Zitzler. Handling Uncertainty in Indicator-Based Multiobjective Optimization. *International Journal of Computational Intelligence Research*, 2(3):255–272, 2006.
102. M. Basseur and E. Zitzler. A Preliminary Study on Handling Uncertainty in Indicator-Based Multiobjective Optimization. In F. R. et al., editor, *Applications of Evolutionary Computing. EvoWorkshops 2006: EvoBIO, EvoCOMNET, EvoHOT, EvoIASP, EvoINTERACTION, EvoMUSART, and EvoSTOC*, pages 727–739, Budapest, Hungary, April 2006. Springer, Lecture Notes in Computer Science Vol. 3907.
103. U. Baumgartner, C. Magele, and W. Renhart. Pareto Optimality and Particle Swarm Optimization. *IEEE Transactions on Magnetics*, 40(2):1172–1175, March 2004.

104. A. Baykasoğlu. Goal Programming using Multiple Objective Tabu Search. *Journal of the Operational Research Society*, 52(12):1359–1369, December 2001.
105. A. Baykasoğlu. Preemptive goal programming using simulated annealing. *Engineering Optimization*, 37(1):49–63, January 2005.
106. A. Baykasoğlu, S. Owen, and N. Gindy. A taboo search based approach to find the Pareto optimal set in multiple objective optimisation. *Engineering Optimization*, 31(6):731–748, 1999.
107. A. Baykasoğlu, L. Özbakýr, and S. A.I. A Tabu Search Based Linguistic Optimization Approach to Due Date Determination in Earliness-Tardiness Flexible Job Shop Scheduling. *International Journal of Advanced Manufacturing Systems*, 6(1):81–90, 2003.
108. J. C. Bean. Genetics and random keys for sequencing and optimization. *ORSA Journal on Computing*, 6(2):154–160, 1994.
109. J. R. Beasley. OR Library. Online, 1999. Available: http://mscmga.ms.ic.ac.uk/info.html.
110. R. P. Beausoleil. "MOSS" multiobjective scatter search applied to non-linear multiple criteria optimization. *European Journal of Operational Research*, 169(2):426–449, March 2006.
111. R. P. Beausoleil Delgado. Multiple Criteria Scatter Search. In J. P. de Sousa, editor, *Proceedings of the 4th Metaheuristics International Conference (MIC'2001)*, pages 539–543. Program Operational Ciencia, Tecnologia, Inovaçao do Quadro Comunitário de Apoio III de Fundaçao para a Ciencia e Tecnologia, Porto, Portugal, July 16–20 2001.
112. A. D. Belegundu, D. V. Murthy, R. R. Salagame, and E. W. Constants. Multiobjective Optimization of Laminated Ceramic Composites Using Genetic Algorithms. In *Fifth AIAA/USAF/NASA Symposium on Multidisciplinary Analysis and Optimization*, pages 1015–1022, Panama City, Florida, 1994. AIAA. Paper 84-4363-CP.
113. A. D. Belegundu and P. L. N. Murthy. A New Genetic Algorithm for Multiobjective Optimization. Technical Report AIAA-96-4180-CP, AIAA, Washington, D.C., 1996.
114. S. M. Belenson and K. C. Kapur. An algorithm for solving multicriterion linear programming problems with examples. *Operations Research Quarterly*, 24(1):65–77, 1973.
115. R. Bellman and S. Dreyfus. *Applied Dynamic Programming*. Princeton University Press, Princeton, New Jersey, 1962.
116. R. Benayoun, J. Montgolfier, J. Tergny, and O. Laritchev. Linear programming with multiple objective functions: Step Method (STEM). *Mathematical Programming*, 1(3):366–375, 1971.
117. R. Benayoun, B. Roy, and B. Sussman. Electre: Une méthode pour guider le choix en présence de points de vue multiple. *Direction Scientifique*, 1966. Note de Travail, No. 49.
118. E. Benini and A. Toffolo. Development of High-Performance Airfoils for Axial Flow Compressors Using Evolutionary Computation. *Journal of Propulsion and Power*, 18(3):544–554, May-June 2002.
119. H. P. Benson and S. Sayin. Towards Finding Global Representations of the Efficient Set in Multiple Objective Mathematical Programming. *Naval Research Logistics*, 44:47–67, 1997.

120. J. Bentley, H. Kung, M. Schkolnick, and C. Thomson. On the Average Number of Maxima in a Set of Vectors and Applications. *Journal of the Association for Computing Machinery*, 25(4):536–543, October 1978.
121. P. J. Bentley and J. P. Wakefield. Finding Acceptable Solutions in the Pareto-Optimal Range using Multiobjective Genetic Algorithms. In P. K. Chawdhry, R. Roy, and R. K. Pant, editors, *Soft Computing in Engineering Design and Manufacturing*, Part 5, pages 231–240, London, June 1997. Springer Verlag London Limited. (Presented at the 2nd On-line World Conference on Soft Computing in Design and Manufacturing (WSC2)).
122. J. O. Berger. *Statistical Decision Theory: Foundations, Concepts and Methods*. Springer-Verlag, New York, 1980.
123. P. Bergey. An agent enhanced intelligent spreadsheet solver for multi-criteria decision making. In *Proceedings of the Fifth Americas Conference on Information Systems (AMCIS'99)*, pages 966–968, Milwaukee, USA, August 1999.
124. E. Bernadó i Mansilla and J. M. Garrell i Guiu. MOLeCS: Using Multiobjective Evolutionary Algorithms for Learning. In E. Zitzler, K. Deb, L. Thiele, C. A. Coello Coello, and D. Corne, editors, *First International Conference on Evolutionary Multi-Criterion Optimization*, pages 696–710. Springer-Verlag. Lecture Notes in Computer Science No. 1993, 2001.
125. A. Berro and Y. Duthen. Search for optimum in dynamic environment: a efficient agent-based method. In *2001 Genetic and Evolutionary Computation Conference. Workshop Program*, pages 51–54, San Francisco, California, July 2001.
126. A. Berry and P. Vamplew. The Combative Accretion Model–Multiobjective Optimisation Without Explicit Pareto Ranking. In C. A. Coello Coello, A. Hernández Aguirre, and E. Zitzler, editors, *Evolutionary Multi-Criterion Optimization. Third International Conference, EMO 2005*, pages 77–91, Guanajuato, México, March 2005. Springer. Lecture Notes in Computer Science Vol. 3410.
127. A. Berry and P. Vamplew. An Efficient Approach to Unbounded Bi-Objective Archives—Introducing the Mak_Tree Algorithm. In M. K. et al., editor, *2006 Genetic and Evolutionary Computation Conference (GECCO'2006)*, volume 1, pages 619–626, Seattle, Washington, USA, July 2006. ACM Press. ISBN 1-59593-186-4.
128. H. Bersini and F. J. Varela. A Variant of Evolution Strategies for Vector Optimization. In H.-P. Schwefel and R. Männer, editors, *Parallel Problem Solving from Nature. 1st Workshop, PPSN I*, pages 343–354, Dortmund, Germany, October 1991. Springer-Verlag. Lecture Notes in Computer Science No. 496.
129. N. Beume, B. Naujoks, and M. Emmerich. SMS-EMOA: Multiobjective selection based on dominated hypervolume. *European Journal of Operational Research*, 181(3):1653–1669, 16 September 2007.
130. B. Bhanu and S. Lee. *Genetic Learning for Adaptive Image Segmentation*. Kluwer Academic Publishers, Boston, 1994.
131. V. Bhaskar, S. Gupta, and A. Ray. Applications of multiobjective optimization in chemical engineering. *Reviews in Chemical Engineering*, 16(1):1–54, 2000.
132. N. Bhutani, G. Rangaiah, and A. Ray. First Principles, Data Based and Hybrid Modeling and Optimization of an Industrial Hydrocracking Unit. *Industrial and Engineering Chemistry Research*, 45:7807–7816, 2006.

133. D. Billings, L. P. na, J. Schaeffer, and D. Szafron. Learning to play strong poker. In J. Furnkranz and M. Kubat, editors, *Machines that learn to play games*, pages 225–242. Nova Science Publishers, Inc., Commack, NY, USA, 2001.
134. Z. Bingul, A. Sekmen, and S. Zein-Sabatto. Adaptive Genetic Algorithms Applied to Dynamic Multi-Objective Problems. In C. H. Dagli, A. L. Buczak, J. Ghosh, M. Embrechts, O. Ersoy, and S. Kercel, editors, *Proceedings of the Artificial Neural Networks in Engineering Conference (ANNIE'2000)*, pages 273–278, New York, 2000. ASME Press.
135. Z. Bingul, A. Sekmen, and S. Zein-Sabatto. Genetic Algorithms Applied to Real Time Multi-Objective Optimization Problems. In *Proceedings of the 2000 IEEE SouteastCon Conference (SoutheastCON'00)*, pages 95–103, Nashville, Tennessee, April 2000. IEEE.
136. T. T. Binh and U. Korn. An evolution strategy for the multiobjective optimization. In *The Second International Conference on Genetic Algorithms (Mendel 96)*, pages 23–28, Brno, Czech Republic, 1996.
137. T. T. Binh and U. Korn. Multicriteria control system design using an intelligent evolution strategy. In *Proceedings of the Conference for Control of Industrial Systems (CIS'97)*, volume 2, pages 242–247, Belfort, France, 1997.
138. T. T. Binh and U. Korn. Multiobjective Evolution Strategy for Constrained Optimization Problems. In *Proc. of the 15th IMACS World Congress on Scientific Computation, Modelling and Applied Mathematics*, pages 357–362, Berlin, Germany, 1997.
139. T. T. Binh and U. Korn. A parallel multiobjective evolutionary algorithm. Technical report, Institute for Automation and Communication, Barleben, Germany, 1999.
140. S. Bleuler, M. Brack, L. Thiele, and E. Zitzler. Multiobjective Genetic Programming: Reducing Bloat Using SPEA2. In *Proceedings of the Congress on Evolutionary Computation 2001 (CEC'2001)*, volume 1, pages 536–543, Piscataway, New Jersey, May 2001. IEEE Service Center.
141. S. Bleuler, M. Laumanns, L. Thiele, and E. Zitzler. PISA—A Platform and Programming Language Independent Interface for Search Algorithms. In C. M. Fonseca, P. J. Fleming, E. Zitzler, K. Deb, and L. Thiele, editors, *Evolutionary Multi-Criterion Optimization. Second International Conference, EMO 2003*, pages 494–508, Faro, Portugal, April 2003. Springer. Lecture Notes in Computer Science. Volume 2632.
142. T. Blickle, J. Teich, and L. Thiele. System-level synthesis using evolutionary algorithms. Technical Report TIK Report-Nr. 16, Computer Engineering and Communication Networks Lab (TIK), Swiss Federal Institute of Technology (ETH), Gloriastrasse 35, 8092 Zurich, April 1996.
143. T. Blickle, J. Teich, and L. Thiele. An evolutionary approach to system-level synthesis. In *Proc. 5th International Workshop on Hardware/Software Codesign*, pages 167–172. IEEE Computer Society Press, 1997.
144. A. L. Blumel. *Robust Fuzzy Autopilot Design Using Multi-objective Optimisation for a Highly Non-linear Missile*. PhD thesis, Department of Aerospace, Power & Sensors, Cranfield University, UK, March 2001.
145. A. L. Blumel, E. J. Hughes, and B. A. White. Fuzzy Autopilot Design using a Multiobjective Evolutionary Algorithm. In *2000 Congress on Evolutionary Computation*, volume 1, pages 54–61, Piscataway, New Jersey, July 2000. IEEE Service Center.

146. K. D. Boese. *Models for iterative global optimization*. PhD thesis, Department of Computer Science, University of California at Los Angeles, Los Angeles, California, USA, 1996.
147. M. Bohanec and I. Bratko. Trading accuracy for Simplicity in Decision Trees. *Machine Learning*, 15(3):223–250, 1994.
148. E. Bonabeu, M. Dorigo, and G. Theraulaz. *Swarm Intelligence: From Natural to Artificial Systems*. Oxford University Press, 1999.
149. Y. Bonnemay, M. Sebag, and O. Teytaud. Convergence proofs for multi-modal multi-objective optimization. In *7th International Conference on Artificial Evolution (EA'05)*, University of Lille, France, October 2005.
150. C. C. Borges and H. J. Barbosa. A Non-generational Genetic Algorithm for Multiobjective Optimization. In *2000 Congress on Evolutionary Computation*, volume 1, pages 172–179, San Diego, California, July 2000. IEEE Service Center.
151. C. C. Borges and H. J. Barbosa. Obtaining a Restricted Pareto Front in Evolutionary Multiobjective Optimization. *Foundations of Computing and Decision Sciences*, 26(1):5–21, 2001.
152. C. A. Borghi, D. Casadei, M. Fabbri, and G. Serra. Reduction of the torque ripple in permanent magnet actuators by a multiobjective minimization technique. *IEEE Transactions on Magnetics*, 34(5):2869–2872, September 1998.
153. P. A. Bosman and E. D. de Jong. Exploiting Gradient Information in Numerical Multi-Objective Evolutionary Optimization. In H.-G. B. et al., editor, *2005 Genetic and Evolutionary Computation Conference (GECCO'2005)*, volume 1, pages 755–762, New York, USA, June 2005. ACM Press.
154. P. A. Bosman and E. D. de Jong. Combining Gradient Techniques for Numerical Multi-Objective Evolutionary Optimization. In M. K. et al., editor, *2006 Genetic and Evolutionary Computation Conference (GECCO'2006)*, volume 1, pages 627–634, Seattle, Washington, USA, July 2006. ACM Press. ISBN 1-59593-186-4.
155. P. A. Bosman and D. Thierens. The Naive MIDEA: A Baseline Multiobjective EA. In C. A. Coello Coello, A. Hernández Aguirre, and E. Zitzler, editors, *Evolutionary Multi-Criterion Optimization. Third International Conference, EMO 2005*, pages 428–442, Guanajuato, México, March 2005. Springer. Lecture Notes in Computer Science Vol. 3410.
156. M. C. Bot. Improving Induction of Linear Classification Trees with Genetic Programming. In D. Whitley, D. Goldberg, E. Cantú-Paz, L. Spector, I. Parmee, and H.-G. Beyer, editors, *Proceedings of the Genetic and Evolutionary Computation Conference (GECCO'2000)*, pages 403–410, San Francisco, California, 2000. Morgan Kaufmann.
157. C. P. Bottura and J. V. da Fonseca Neto. Rule-based Decision-making Unit for Eigenstructure Assignment via Parallel Genetic Algorithm and LQR Designs. In *Proceedings of the 2000 American Control Conference*, volume 1, pages 467–471, 2000.
158. D. Bouyssou, T. Marchant, P. Perny, A. Tsoukias, and P. Vincke. *Evaluation and Decision Models: A Critical Perspective*. Kluwer Academic Publishers, 2000.
159. L. M. Boychuk and V. O. Ovchinnikov. Principal Methods of Solution of Multicriterial Optimization Problems (survey). *Soviet Automatic Control*, 6:1–4, 1973.

160. H. I. Bozma and J. S. Duncan. A Game–Theoretic Approach to Integration of Modules. *IEEE Transactions on Pattern Analysis and Machine Intelligence*, 16(11):1074–1086, November 1994.
161. J. Branke and K. Deb. Integrating User Preferences into Evolutionary Multi-Objective Optimization. In Y. Jin, editor, *Knowledge Incorporation in Evolutionary Computation*, pages 461–477. Springer, Berlin Heidelberg, 2005. ISBN 3-540-22902-7.
162. J. Branke, K. Deb, H. Dierolf, and M. Osswald. Finding Knees in Multi-Objective Optimization. In *Parallel Problem Solving from Nature - PPSN VIII*, pages 722–731, Birmingham, UK, September 2004. Springer-Verlag. Lecture Notes in Computer Science Vol. 3242.
163. J. Branke, T. Kaußler, and H. Schmeck. Guiding Multi-Objective Evolutionary Algorithms Towards Interesting Regions. In I. C. Parmee, editor, *Fourth International Conference on Adaptive Computing in Design and Manufacture (ACDM 2000), Poster Proceedings*, pages 1–4. Plymouth Engineering Design Centre, University of Plymouth, April 2000.
164. J. Branke, T. Kaußler, and H. Schmeck. Guidance in Evolutionary Multi-Objective Optimization. *Advances in Engineering Software*, 32:499–507, 2001.
165. J. Branke and S. Mostaghim. About Selecting the Personal Best in Multi-Objective Particle Swarm Optimization. In T. P. Runarsson, H.-G. Beyer, E. Burke, J. J. Merelo-Guervós, L. D. Whitley, and X. Yao, editors, *Parallel Problem Solving from Nature - PPSN IX, 9th International Conference*, pages 523–532. Springer. Lecture Notes in Computer Science Vol. 4193, Reykjavik, Iceland, September 2006.
166. J. P. Brans and B. Mareschal. The PROMCALC & GAIA decision support system for multicriteria decision aid. *Decision Support Systems*, 12:297–310, 1994.
167. J.-P. Brans and B. Mareschal. PROMETHEE methods. In J. Figueira, S. Greco, and M. Ehrgott, editors, *Multiple Criteria Decision Analysis. State of the Art Surveys*, pages 163–195. Springer, New York, USA, 2005.
168. J. P. Brans and P. Vincke. A Preference Ranking Organisation Method (The PROMETHEE Method for Multiple Criteria Decision-Making). *Management Science*, 31(6):647–656, June 1985.
169. J. P. Brans, P. Vincke, and B. Mareschal. How to select and how to rank projects: the PROMETHEE method. *European Journal of Operational Research*, 24(2):228–238, February 1986.
170. G. Brassard and P. Bratley. *Algorithmics: Theory and Practice*. Prentice-Hall, Englewood Cliffs, New Jersey, 1988.
171. M. S. Bright. *Evolutionary Strategies for the High-Level Synthesis of VLSI-Based DSP Systems for Low Power*. PhD thesis, University Of Wales Cardiff, School Of Engineering, Circuits And Systems Research Group, Cardiff, Wales, UK, October 1998.
172. M. S. Bright and T. Arslan. Multi-Objective Design Strategies for High-Level Low-Power Design of DSP Systems. In *IEEE International Symposium on Circuits and Systems, ISCAS 99*, volume 1, pages 80–83, Florida, USA, May–June 1999.
173. C. Brizuela, N. Sannomiya, and Y. Zhao. Multi-Objective Flow-Shop: Preliminary Results. In E. Zitzler, K. Deb, L. Thiele, C. A. Coello Coello, and D. Corne, editors, *First International Conference on Evolutionary Multi-*

Criterion Optimization, pages 443–457. Springer-Verlag. Lecture Notes in Computer Science No. 1993, 2001.
174. C. A. Brizuela and R. Aceves. Experimental Genetic Operators Analysis for the Multi-objective Permutation Flowshop. In C. M. Fonseca, P. J. Fleming, E. Zitzler, K. Deb, and L. Thiele, editors, *Evolutionary Multi-Criterion Optimization. Second International Conference, EMO 2003*, pages 578–592, Faro, Portugal, April 2003. Springer. Lecture Notes in Computer Science. Volume 2632.
175. D. Brockhoff and E. Zitzler. Are All Objectives Necessary? On Dimensionality Reduction in Evolutionary Multiobjective Optimization. In T. P. Runarsson, H.-G. Beyer, E. Burke, J. J. Merelo-Guervós, L. D. Whitley, and X. Yao, editors, *Parallel Problem Solving from Nature - PPSN IX, 9th International Conference*, pages 533–542. Springer. Lecture Notes in Computer Science Vol. 4193, Reykjavik, Iceland, September 2006.
176. R. A. C. M. Broekmeulen. Facility Management of Distribution Centers for Vegetables and Fruits. In J. Biethahn and V. Nissen, editors, *Evolutionary Algorithms in Management Applications*, pages 199–210. Springer-Verlag, Berlin, 1995.
177. A. D. Brooke, D. Kendrick, and A. Meerhaus. *GAMS: A User's Guide*. Scientific Press, California, 1988.
178. A. J. Brown and M. Thomas. Reengineering the Naval Ship Design Process. In *Proceedings of From Research to Reality in Ship Systems Engineering Symposium*, page 277, University of Essex, United Kingdom, 1998. ASNE.
179. M. Brown and R. E. Smith. Effective Use of Directional Information in Multi-objective Evolutionary Computation. In E. C.-P. et al., editor, *Genetic and Evolutionary Computation—GECCO 2003. Proceedings, Part I*, pages 778–789. Springer. Lecture Notes in Computer Science Vol. 2723, July 2003.
180. M. Brown and R. E. Smith. Directed Multi-Objective Optimisation. *International Journal of Computers, Systems and Signals*, 6(1):3–17, 2005.
181. J. Brownlee. IIDLE: An Immunological Inspired Distributed Learning Environment for Multiple Objective and Hybrid Optimisation. In *2006 IEEE Congress on Evolutionary Computation (CEC'2006)*, pages 1614–1620, Vancouver, BC, Canada, July 2006. IEEE.
182. J. T. Buchanan. A Naïve Approach for Solving MCDM Problems: The GUESS Method. *Journal of the Operational Research Society*, 48(2):202–206, 1997.
183. L. Bull, editor. *Applications of Learning Classifier Systems*. Springer, June 2004. ISBN 3-5402-110-98.
184. B. Bullnheimer, R. F. Hartl, and C. Strauss. A New Rank Based Version of the Ant System: A Computational Study. *Central European Journal for Operations Research and Economics*, 7(1):25–38, 1999.
185. E. Burke, P. Cowling, J. Landa Silva, and S. Petrovic. Combining Hybrid Metaheuristics and Populations for the Multiobjective Optimisation of Space Allocation Problems. In L. Spector, E. D. Goodman, A. Wu, W. Langdon, H.-M. Voigt, M. Gen, S. Sen, M. Dorigo, S. Pezeshk, M. H. Garzon, and E. Burke, editors, *Proceedings of the Genetic and Evolutionary Computation Conference (GECCO'2001)*, pages 1252–1259, San Francisco, California, 2001. Morgan Kaufmann Publishers.
186. E. Burke, G. Kendall, J. Newall, E. Hart, P. Ross, and S. Schulenburg. Hyperheuristics: An emerging direction in modern search technology. In F. Glover

and G. A. Kochenberger, editors, *Handbook of Metaheuristics*, pages 457–474. Kluwer Academic Publishers, Boston/Dordrecht/London, 2003.
187. E. Burke, J. D. Landa Silva, and E. Soubeiga. Hyperheuristic Approaches for Multiobjective Optimisation. In *Proceedings of the 5th Metaheuristics International Conference (MIC 2003)*, pages 11.1–11.6, Kyoto, Japan, August 2003.
188. E. Burke and J. L. Silva. The influence of the fitness evaluation method on the performance of multiobjective search algorithms. *European Journal of Operational Research*, 169(3):875–897, March 2006.
189. E. K. Burke, J. D. Landa Silva, and E. Soubeiga. Multi-objective Hyperheuristic Approaches for Space Allocation and Timetabling. In T. Ibaraki, K. Nonobe, and M. Yagiura, editors, *Meta-heuristics: Progress as Real Problem Solvers, Selected Papers from the 5th Metaheuristics International Conference (MIC 2003)*, pages 129–158. Springer, 2005.
190. R. Buyya, editor. *High Performance Cluster Computing: Architectures and Systems*, volume 1. Prentice-Hall, NJ, 1999.
191. R. Caballero, X. Gandibleux, and J. Molina. MOAMP - A Generic Multiobjective Metaheuristic using an Adaptive Memory. Technical report, University of Valenciennes, France, 2004.
192. J. M. Cadenas and F. Jiménez. A genetic algorithm for the multiobjective solid transportation problem: a fuzzy approach. In *International Symposium on Automotive Technology and Automation, Proceedings for the dedicated conferences on Mechatronics and Supercomputing Applications in the Transportation Industries*, pages 327–334, Aachen, Germany, 1994.
193. S. Cahon, N. Melab, and E.-G. Talbi. ParadisEO: A Framework for the Reusable Design of Parallel and Distributed Metaheuristics. *Journal of Heuristics*, 10(3):357–380, 2004.
194. M. Calonder, S. Bleuler, and E. Zitzler. Module Identification from Heterogeneous Biological Data Using Multiobjective Evolutionary Algorithms. In T. P. Runarsson, H.-G. Beyer, E. Burke, J. J. Merelo-Guervós, L. D. Whitley, and X. Yao, editors, *Parallel Problem Solving from Nature - PPSN IX, 9th International Conference*, pages 573–582. Springer. Lecture Notes in Computer Science Vol. 4193, Reykjavik, Iceland, September 2006.
195. F. Campelo, F. Guimarães, R. Saldanha, H. Igarashi, S. Noguchi, D. Lowther, and J. Ramirez. A novel multiobjective immune algorithm using nondominated sorting. In *11th International IGTE Symposium on Numerical Field Calculation in Electrical Engineering*, Seggauberg, Austria, September 2004.
196. F. Campelo, F. G. Guimarães, and H. Igarashi. Overview of Artificial Immune Systems for Multi-Objective Optimization. In S. Obayashi, K. Deb, C. Poloni, T. Hiroyasu, and T. Murata, editors, *Evolutionary Multi-Criterion Optimization, 4th International Conference, EMO 2007*, pages 937–951, Matshushima, Japan, March 2007. Springer. Lecture Notes in Computer Science Vol. 4403.
197. E. Camponogara and S. N. Talukdar. A Genetic Algorithm for Constrained and Multiobjective Optimization. In J. T. Alander, editor, *3rd Nordic Workshop on Genetic Algorithms and Their Applications (3NWGA)*, pages 49–62, Vaasa, Finland, August 1997. University of Vaasa.
198. W. Cancino and A. C. B. Delbem. A Multi-objective Evolutionary Approach for Phylogenetic Inference. In S. Obayashi, K. Deb, C. Poloni, T. Hiroyasu, and T. Murata, editors, *Evolutionary Multi-Criterion Optimization, 4th*

International Conference, EMO 2007, pages 428–442, Matshushima, Japan, March 2007. Springer. Lecture Notes in Computer Science Vol. 4403.
199. B. S. Canova and J. G. Tyler. An Adaptive Distributed Architecture for Near Real-Time Genetic Algorithm Execution. In *1998 International Conference on Web-Based Modeling and Simulation*, San Diego, California, January 11–14 1998. The Society for Computer Simulation International.
200. G. Cantor. Contributions to the Foundation of Transfinite Set Theory. *Mathematische Annalen*, 46:481–512, 1895.
201. G. Cantor. Contributions to the Foundation of Transfinite Set Theory. *Mathematische Annalen*, 49:207–246, 1897.
202. E. Cantú-Paz. A Survey of Parallel Genetic Algorithms. Technical Report 97003, Illinois Genetic Algorithms Laboratory, University of Illinois at Urbana-Champaign, Urbana, Illinois, May 1997.
203. E. Cantú-Paz. Migration Policies, Selection Pressure, and Parallel Evolutionary Algorithms. Technical report, Department of Computer Science, University of Illinois, Urbana, IL, June 1999. IlliGAL Report No. 99015.
204. E. Cantú-Paz. *Efficient and Accurate Parallel Genetic Algorithms*. Kluwer Academic Publishers, Boston, Massachusetts, 2000.
205. A. Cardon, T. Galinho, and J.-P. Vacher. A Multi-Objective Genetic Algorithm in Job Shop Scheduling Problem to Refine an Agents' Architecture. In K. Miettinen, M. M. Mäkelä, P. Neittaanmäki, and J. Periaux, editors, *Proceedings of EUROGEN'99*, Jyväskyl, Finland, 1999. University of Jyváskylä.
206. A. Cardon, T. Galinho, and J.-P. Vacher. Genetic Algorithms using Multi-Objectives in a Multi-Agent System. *Robotics and Autonomous Systems*, 33(2–3):179–190, November 2000.
207. P. Cardoso, M. Jesus, and A. Márquez. MONACO - Multi-Objective Network Optimisation based on ACO. In *X Encuentros de Geometría Computacional*, Seville, Spain, June 2003.
208. P. Cardoso, M. Jesus, and A. Márquez. Multiple Objective TSP based on ACO. In *III Encuentro Andaluz de Matemáticas Discretas*, Almeria, Spain, 2003.
209. P. Cardoso, M. Jesus, and A. Márquez. Multiple Criteria Minimum Spanning Trees. In *XI Encuentros de Geometría Computacional*, Santander, Spain, 2005.
210. W. M. Carlyle, B. Kim, J. W. Fowler, and E. S. Gel. Comparison of Multiple Objective Genetic Algorithms for Parallel Machine Scheduling Problems. In E. Zitzler, K. Deb, L. Thiele, C. A. Coello Coello, and D. Corne, editors, *First International Conference on Evolutionary Multi-Criterion Optimization*, pages 472–485. Springer-Verlag. Lecture Notes in Computer Science No. 1993, 2001.
211. D. G. Carmichael. Computation of Pareto Optima in Structural Design. *International Journal for Numerical Methods in Engineering*, 15:925–952, 1980.
212. C. W. Carroll. The created response surface technique for optimizing nonlinear restrained systems. *Operations Research*, 9:169–184, 1961.
213. C. Carstensen, G. Rohling, and C. E. Hunt. Optimization of Covert Flares on the C-17 via Genetic Algorithms. In *Proceedings of Military Sensing Symposium Specialty Group Meeting on Infrared Countermeasures Conference (Classified)*, Monterey, California, May 2000.

References   643

214. R. A. Caruana, L. J. Eshelman, and J. D. Schaffer. Representation and hidden bias II: Eliminating defining length bias in genetic search via shuffle crossover. In N. S. Sridharan, editor, *Proceedings of the Eleventh International Joint Conference on Artificial Intelligence*, pages 750–755, San Mateo, California, 1989. Morgan Kaufmann.
215. P. Castro Borges and M. Pilegaard Hansen. A basis for future successes in multiobjective combinatorial optimization. Technical Report IMM-REP-1998-8, Institute of Mathematical Modelling, Technical University of Denmark, March 1998.
216. D. J. Caswell. Active Processor Scheduling Using Evolutionary Algorithms. Master's thesis, Air Force Institute of Technology, Wright-Patterson Air Force Base, Ohio, December 2002. AFIT/GCS/ENG/02-36.
217. D. J. Caswell and G. B. Lamont. Wire-Antenna Geometry Design with Multiobjective Genetic Algorithms. In *Congress on Evolutionary Computation (CEC'2002)*, volume 1, pages 103–108, Piscataway, New Jersey, May 2002. IEEE Service Center.
218. S. Cayzer, J. Smith, J. A. Marshall, and T. Kovacs. What Have Gene Libraries Done for AIS? In C. Jacob, M. L. Pilat, P. J. Bentley, and J. Timmis, editors, *Artificial Immune Systems. 4th International Conference, ICARIS 2005*, pages 86–99, Banff, Canada, August 2005. Springer. Lecture Notes in Computer Science Vol. 3627.
219. V. Cerny. A Thermodynamical Approach to the Traveling Salesman Problem: An Efficient Simulation Algorithm. *Journal of Optimization Theory and Applications*, 45(1):41–51, 1985.
220. N. Chakraborti, R. Kumar, and D. Jain. A study of the continuous casting mold using a pareto-converging genetic algorithm. *Applied Mathematical Modelling*, 25:287–297, 2001.
221. N. Chakraborti, P. Mishra, A. Aggarwal, A. Banerjee, and S. Mukherjee. The Williams and Otto Chemical Plant re-evaluated using a Pareto-optimal formulation aided by Genetic Algorithms. *Applied Soft Computing*, 6:189–197, 2006.
222. R. Chandra, R. Menon, L. Dagum, D. Kohr, D. Maydan, and J. McDonald. *Parallel Programming in OpenMP*. Morgan Kaufmann, 2000.
223. C. Chang and D. Xu. Differential Evolution Based Tuning of Fuzzy Automatic Train Operation for Mass Rapid Transit System. *IEE Proceedings of Electric Power Applications*, 147(3):206–212, May 2000.
224. C. Chang, D. Xu, and H. Quek. Pareto-optimal set based multiobjective tuning of fuzzy automatic train operation for mass transit system. *IEE Proceedings on Electric Power Applications*, 146(5):577–583, September 1999.
225. C. S. Chang, W. Wang, A. C. Liew, F. S. Wen, and D. Srinivasan. Genetic Algorithm Based Bicriterion Optimization for Traction Sustations in DC Railway System. In *Proceedings of the Second IEEE International Conference on Evolutionary Computation*, pages 11–16, Piscataway, New Jersey, 1995. IEEE Press.
226. T. J. Chang, N. Meade, and J. E. Beasley. Heuristics for Cardinality Constrained Portfolio Optimization. Technical report, The Management School, Imperial College, London SW7 2AZ, England, May 1998.
227. T. J. Chang, N. Meade, and J. E. Beasley. Heuristics for Cardinality Constrained Portfolio Optimization. *Computers and Operations Research*, 27(13):1271–1302, 2000.

228. A. Charnes and W. W. Cooper. Management Models and Industrial Applications of Linear Programming. *Management Science*, 4(1):81–87, 1957.
229. A. Charnes and W. W. Cooper. *Management Models and Industrial Applications of Linear Programming*, volume 1. John Wiley, New York, 1961.
230. A. Charnes, W. W. Cooper, R. J. Niehaus, and A. Stedry. Static and dynamic assignment models with multiple objectives and some remarks on organization design. *Management Science*, 15(8):B365–B375, 1969.
231. K. Chellapilla. Combining mutation operators in evolutionary programming. *IEEE Transactions on Evolutionary Computation*, 2(3):91–96, September 1998.
232. H. W. Chen and N.-B. Chang. Water pollution control in the river basin by fuzzy genetic algorithm-based multiobjective programming modeling. *Water Science and Technology*, 37(8):55–63, 1998.
233. J. Chen and M. Mahfouf. A population adaptive based immune algorithm for solving multi-objective optimization problems. In H. Bersini and J. Carneiro, editors, *Artificial Immune Systems, 5th International Conference, ICARIS 2006, Proceedings*, pages 280–293, Oeiras, Portugal, September 2006. Springer-Verlag, Lecture Notes in Computer Science Vol. 4163.
234. J.-H. Chen. *Theory and Applications of Efficient Multi-Objective Evolutionary Algorithms*. PhD thesis, College of Information and Electrical Engineering of the Feng Chia University, Taichung, Taiwan, R.O.C., 2004.
235. J.-H. Chen, D. E. Goldberg, S.-Y. Ho, and K. Sastry. Fitness Inheritance in Multi-Objective Optimization. In W. Langdon, E. Cantú-Paz, K. Mathias, R. Roy, D. Davis, R. Poli, K. Balakrishnan, V. Honavar, G. Rudolph, J. Wegener, L. Bull, M. Potter, A. Schultz, J. Miller, E. Burke, and N. Jonoska, editors, *Proceedings of the Genetic and Evolutionary Computation Conference (GECCO'2002)*, pages 319–326, San Francisco, California, July 2002. Morgan Kaufmann Publishers.
236. J.-H. Chen and S.-Y. Ho. Multi-Objective Optimization of Flexible Manufacturing Systems. In L. Spector, E. D. Goodman, A. Wu, W. Langdon, H.-M. Voigt, M. Gen, S. Sen, M. Dorigo, S. Pezeshk, M. H. Garzon, and E. Burke, editors, *Proceedings of the Genetic and Evolutionary Computation Conference (GECCO'2001)*, pages 1260–1267, San Francisco, California, 2001. Morgan Kaufmann Publishers.
237. Q. Chen and S.-U. Guan. Incremental Multiple Objective Genetic Algorithms. *IEEE Transactions on Systems, Man, and Cybernetics—Part B: Cybernetics*, 34(3):1325–1334, June 2004.
238. X. Chen. Pareto Tree Searching Genetic Algorithm: Approaching Pareto Optimal Front by Searching Pareto Optimal Tree. Technical Report NK-CS-2001-002, Department of Computer Science, Nankai University, Tianjin, China, 2001.
239. Y. Chen, M. Narita, M. Tsuji, and S. Sa. A Genetic Algorithm Approach to Optimization for the Radiological Worker Allocation Problem. *Health Physics*, 70(2):180–186, February 1996.
240. Y. L. Chen and C. C. Liu. Multiobjective VAR planning using the goal-attainment method. *IEE Proceedings on Generation, Transmission and Distribution*, 141(3):227–232, May 1994.
241. F. Cheng and X. Li. Generalized Center Method for Multiobjective Engineering Optimization. *Engineering Optimization*, 31:641–661, 1999.

242. F. Y. Cheng and D. Li. Multiobjective Optimization Design with Pareto Genetic Algorithm. *Journal of Structural Engineering*, 123(9):1252–1261, September 1997.
243. R. Cheng, M. Gen, and S. S. Oren. An Adaptive Hyperplane Approach for Multiple Objective Optimization Problems with Complex Constraints. In D. Whitley, D. Goldberg, E. Cantú-Paz, L. Spector, I. Parmee, and H.-G. Beyer, editors, *Proceedings of the Genetic and Evolutionary Computation Conference (GECCO'2000)*, pages 299–306, San Francisco, California, 2000. Morgan Kaufmann.
244. S. C. Chiam, C. K. Goh, and K. C. Tan. Adequacy of Empirical Performance Assessment for Multiobjective Evolutionary Optimizer. In S. Obayashi, K. Deb, C. Poloni, T. Hiroyasu, and T. Murata, editors, *Evolutionary Multi-Criterion Optimization, 4th International Conference, EMO 2007*, pages 893–907, Matshushima, Japan, March 2007. Springer. Lecture Notes in Computer Science Vol. 4403.
245. T.-C. Chiang and L.-C. Fu. Multiobjective Job Shop Scheduling using Genetic Algorithm with Cyclic Fitness Assignment. In *2006 IEEE Congress on Evolutionary Computation (CEC'2006)*, pages 11035–11042, Vancouver, BC, Canada, July 2006. IEEE.
246. A. Chipperfield, J. Whidborne, and P. Fleming. Evolutionary Algorithms and Simulated Annealing for MCDM. In T. Gal, T. Stewart, and T. Hanne, editors, *Multicriteria Decicion Making—Advances in MCDM Models, Algorithms, Theory, and Applications*, pages 16.1–16.32. Kluwer Academic Publishing, Boston, Massachusetts, 1999.
247. A. J. Chipperfield, N. V. Dakev, J. F. Whidborne, and P. J. Fleming. Multiobjective robust control using evolutionary algorithms. In *IEEE International Conference on Industrial Technology*, pages 269–274, Shanghai, China, December 1996.
248. A. J. Chipperfield and P. J. Fleming. Gas Turbine Engine Controller Design using Multiobjective Genetic Algorithms. In A. M. S. Zalzala, editor, *Proceedings of the First IEE/IEEE International Conference on Genetic Algorithms in Engineering Systems : Innovations and Applications, GALESIA'95*, pages 214–219, Halifax Hall, University of Sheffield, UK, September 1995. IEEE.
249. A. J. Chipperfield and P. J. Fleming. Multiobjective Gas Turbine Engine Controller Design Using Genetic Algorithms. *IEEE Transactions on Industrial Electronics*, 43(5):583–587, October 1996.
250. S. Choi and C. Wu. Partitioning and Allocation of Objects in Heterogeneous Distributed Environments Using the Niched Pareto Genetic-Algorithm. In K. Kelly, editor, *Proc. of 1998 Asia Pacific Software Engineering Conference (APSEC 98)*, pages 322–329. IEEE, Taipei, Taiwan, December 1998.
251. S. Y. Chong, M. K. Tan, and J. D. White. Observing the Evolution of Neural Networks Learning to Play the Game of Othello. *IEEE Transactions on Evolutionary Computation*, 9(3):240–251, June 2005.
252. S. Y. Chong and X. Yao. Behavioral Diversity, Choices and Noise in the Iterated Prisoner's Dilemma. *IEEE Transactions on Evolutionary Computation*, 9(6):540–551, December 2005.
253. C. Chow and H. Tsui. Autonomous Agent Response Learning by a Multi-Species Particle Swarm Optimization. In *2004 Congress on Evolutionary Computation (CEC'2004)*, volume 1, pages 778–785, Portland, Oregon, USA, June 2004. IEEE Service Center.

254. C. R. Chow. An Evolutionary Approach to Search for NCR-Boards. In D. B. Fogel, editor, *Proceedings of the 1998 International Conference on Evolutionary Computation*, pages 295–300, Piscataway, New Jersey, 1998. IEEE.
255. S. E. Cieniawski, J. W. Eheart, and S. Ranjithan. Using Genetic Algorithms to Solve a Multiobjective Groundwater Monitoring Problem. *Water Resources Research*, 31(2):399–409, February 1995.
256. G. Coath and S. K. Halgamuge. A Comparison of Constraint-Handling Methods for the Application of Particle Swarm Optimization to Constrained Nonlinear Optimization Problems. In *Proceedings of the Congress on Evolutionary Computation 2003 (CEC'2003)*, volume 4, pages 2419–2425, Piscataway, New Jersey, December 2003. Canberra, Australia, IEEE Service Center.
257. G. Cochenour, J. Simon, S. Das, A. Pahwa, and S. Nag. A Pareto Archive Evolutionary Strategy Based Radial Basis Function Neural Network Training Algorithm for Failure Rate Prediction in Overhead Feeders. In H.-G. B. et al., editor, *2005 Genetic and Evolutionary Computation Conference (GECCO'2005)*, volume 2, pages 2127–2132, New York, USA, June 2005. ACM Press.
258. J. K. Cochran, S.-M. Horng, and J. W. Fowler. A Multi-Population Genetic Algorithm to Solve Multi-Objective Scheduling Problems for Parallel Machines. *Computers and Operations Research*, 30(7):1087–1102, 2003.
259. G. P. Coelho and F. V. Zuben. Omni-aiNet: An immune-inspired approach for omni optimization. In H. Bersini and J. Carneiro, editors, *Artificial Immune Systems, 5th International Conference, ICARIS 2006, Proceedings*, pages 294–308, Oeiras, Portugal, September 2006. Springer-Verlag, Lecture Notes in Computer Science Vol. 4163.
260. C. A. Coello Coello. *An Empirical Study of Evolutionary Techniques for Multiobjective Optimization in Engineering Design*. PhD thesis, Department of Computer Science, Tulane University, New Orleans, LA, April 1996.
261. C. A. Coello Coello. A Comprehensive Survey of Evolutionary-Based Multiobjective Optimization Techniques. *Knowledge and Information Systems. An International Journal*, 1(3):269–308, August 1999.
262. C. A. Coello Coello. Constraint-handling using an evolutionary multiobjective optimization technique. *Civil Engineering and Environmental Systems*, 17:319–346, 2000.
263. C. A. Coello Coello. Handling Preferences in Evolutionary Multiobjective Optimization: A Survey. In *2000 Congress on Evolutionary Computation*, volume 1, pages 30–37, Piscataway, New Jersey, July 2000. IEEE Service Center.
264. C. A. Coello Coello. Treating Constraints as Objectives for Single-Objective Evolutionary Optimization. *Engineering Optimization*, 32(3):275–308, 2000.
265. C. A. Coello Coello. Theoretical and Numerical Constraint-Handling Techniques used with Evolutionary Algorithms: A Survey of the State of the Art. *Computer Methods in Applied Mechanics and Engineering*, 191(11–12):1245–1287, January 2002.
266. C. A. Coello Coello. EMOO Repository (Online), Last Download: May 28th, 2007, 2007. http://delta.cs.cinvestav.mx/~ccoello/EMOO/.
267. C. A. Coello Coello and A. D. Christiansen. Two new GA-based methods for multiobjective optimization. *Civil Engineering Systems*, 15(3):207–243, 1998.

268. C. A. Coello Coello and A. D. Christiansen. MOSES : A Multiobjective Optimization Tool for Engineering Design. *Engineering Optimization*, 31(3):337–368, 1999.
269. C. A. Coello Coello and A. D. Christiansen. Multiobjective optimization of trusses using genetic algorithms. *Computers and Structures*, 75(6):647–660, May 2000.
270. C. A. Coello Coello, A. D. Christiansen, and A. Hernández Aguirre. Using a new GA-based multiobjective optimization technique for the design of robot arms. *Robotica*, 16(4):401–414, July-August 1998.
271. C. A. Coello Coello and N. Cruz Cortés. An Approach to Solve Multiobjective Optimization Problems Based on an Artificial Immune System. In J. Timmis and P. J. Bentley, editors, *First International Conference on Artificial Immune Systems (ICARIS'2002)*, pages 212–221. University of Kent at Canterbury, UK, September 2002. ISBN 1-902671-32-5.
272. C. A. Coello Coello and N. Cruz Cortés. Hybridizing a genetic algorithm with an artificial immune system for global optimization. *Engineering Optimization*, 36(5):607–634, October 2004.
273. C. A. Coello Coello and N. Cruz Cortés. Solving Multiobjective Optimization Problems using an Artificial Immune System. *Genetic Programming and Evolvable Machines*, 6(2):163–190, June 2005.
274. C. A. Coello Coello and A. Hernández Aguirre. Design of Combinational Logic Circuits through an Evolutionary Multiobjective Optimization Approach. *Artificial Intelligence for Engineering, Design, Analysis and Manufacture*, 16(1):39–53, January 2002.
275. C. A. Coello Coello, A. Hernández Aguirre, and B. P. Buckles. Evolutionary Multiobjective Design of Combinational Logic Circuits. In J. Lohn, A. Stoica, D. Keymeulen, and S. Colombano, editors, *Proceedings of the Second NASA/DoD Workshop on Evolvable Hardware*, pages 161–170, Los Alamitos, California, July 2000. IEEE Computer Society.
276. C. A. Coello Coello, E. Hernández Luna, and A. Hernández Aguirre. Use of particle swarm optimization to design combinational logic circuits. In A. M. Tyrell, P. C. Haddow, and J. Torresen, editors, *Evolvable Systems: From Biology to Hardware. 5th International Conference, ICES 2003*, pages 398–409, Trondheim, Norway, 2003. Springer, Lecture Notes in Computer Science Vol. 2606.
277. C. A. Coello Coello and G. B. Lamont, editors. *Applications of Multi-Objective Evolutionary Algorithms*. World Scientific, Singapore, 2004. ISBN 981-256-106-4.
278. C. A. Coello Coello and R. Landa Becerra. Evolutionary Multiobjective Optimization using a Cultural Algorithm. In *2003 IEEE Swarm Intelligence Symposium Proceedings*, pages 6–13, Indianapolis, Indiana, USA, April 2003. IEEE Service Center.
279. C. A. Coello Coello and E. Mezura-Montes. Constraint-handling in genetic algorithms through the use of dominance-based tournament selection. *Advanced Engineering Informatics*, 16(3):193–203, July 2002.
280. C. A. Coello Coello and M. Reyes Sierra. A Coevolutionary Multi-Objective Evolutionary Algorithm. In *Proceedings of the 2003 Congress on Evolutionary Computation (CEC'2003)*, volume 1, pages 482–489, Canberra, Australia, December 2003. IEEE Press.

281. C. A. Coello Coello, M. Rudnick, and A. D. Christiansen. Using Genetic Algorithms for Optimal Design of Trusses. In *Proceedings of the Sixth International Conference on Tools with Artificial Intelligence*, pages 88–94, New Orleans, Louisiana, USA, November 1994. IEEE Computer Society Press.
282. C. A. Coello Coello and M. Salazar Lechuga. MOPSO: A Proposal for Multiple Objective Particle Swarm Optimization. In *Congress on Evolutionary Computation (CEC'2002)*, volume 2, pages 1051–1056, Piscataway, New Jersey, May 2002. IEEE Service Center.
283. C. A. Coello Coello and G. Toscano Pulido. A Micro-Genetic Algorithm for Multiobjective Optimization. In E. Zitzler, K. Deb, L. Thiele, C. A. Coello Coello, and D. Corne, editors, *First International Conference on Evolutionary Multi-Criterion Optimization*, pages 126–140. Springer-Verlag. Lecture Notes in Computer Science No. 1993, 2001.
284. C. A. Coello Coello and G. Toscano Pulido. Multiobjective Optimization using a Micro-Genetic Algorithm. In L. Spector, E. D. Goodman, A. Wu, W. Langdon, H.-M. Voigt, M. Gen, S. Sen, M. Dorigo, S. Pezeshk, M. H. Garzon, and E. Burke, editors, *Proceedings of the Genetic and Evolutionary Computation Conference (GECCO'2001)*, pages 274–282, San Francisco, California, 2001. Morgan Kaufmann Publishers.
285. C. A. Coello Coello and G. Toscano Pulido. Multiobjective Structural Optimization using a Micro-Genetic Algorithm. *Structural and Multidisciplinary Optimization*, 30(5):388–403, November 2005.
286. C. A. Coello Coello, G. Toscano Pulido, and M. Salazar Lechuga. Handling Multiple Objectives With Particle Swarm Optimization. *IEEE Transactions on Evolutionary Computation*, 8(3):256–279, June 2004.
287. C. A. Coello Coello, D. A. Van Veldhuizen, and G. B. Lamont. *Evolutionary Algorithms for Solving Multi-Objective Problems*. Kluwer Academic Publishers, New York, first edition, May 2002. ISBN 0-3064-6762-3.
288. J. L. Cohon. *Multiobjective Programming and Planning*. Academic Press, 1978.
289. J. L. Cohon and D. H. Marks. A Review and Evaluation of Multiobjective Programming Techniques. *Water Resources Research*, 11(2):208–220, apr 1975.
290. Y. Collette and P. Siarry. *Multiobjective Optimization. Principles and Case Studies*. Springer, August 2003.
291. Y. Collette, P. Siarry, and H.-I. Wong. A Systematic Comparison of Performance of Various Multiple Objective Metaheuristics Using a Common Set of Analytical Test Functions. *Foundations of Computing and Decision Sciences*, 25(4):249–271, 2000.
292. G. Colombo and C. Mumford. Comparing Algorithms, Representations and Operators for the Multi-Objective Knapsack Problem. In *2005 IEEE Congress on Evolutionary Computation (CEC'2005)*, volume 2, pages 1268–1275, Edinburgh, Scotland, September 2005. IEEE Service Center.
293. A. Colorni, M. Dorigo, and V. Maniezzo. Distributed optimization by ant colonies. In F. J. Varela and P. Bourgine, editors, *Proceedings of the First European Conference on Artificial Life*, pages 134–142. MIT Press, Cambridge, MA, 1992.
294. A. Connor. A multi-thread Tabu Search algorithm. *Design Optimization*, 1(3):293–304, 1999.

295. A. Connor and D. Tilley. A tabu search method for the optimisation of fluid power circuits. *Proceedings of IMechE Part I: Journal of Systems and Control Engineering*, 212(5):373–381, 1998.
296. A. Corana, M. Marchesi, C. Martini, and S. Ridella. Minimizing multimodal functions of continuous variables with the "Simulated Annealing" algorithm. *ACM Transactions on Mathematical Software*, 13(3):262–280, 1987.
297. A. Corberán, E. Fernández, M. Laguna, and R. Martí. Heuristic Solutions to the Problem of Routing School Buses with Multiple Objectives. *Journal of the Operational Research Society*, 53(4):427–435, 2002.
298. D. Corne and J. Knowles. Some Multiobjective Optimizers are Better than Others. In *Proceedings of the 2003 Congress on Evolutionary Computation (CEC'2003)*, volume 4, pages 2506–2512, Canberra, Australia, December 2003. IEEE Press.
299. D. W. Corne, N. R. Jerram, J. D. Knowles, and M. J. Oates. PESA-II: Region-based Selection in Evolutionary Multiobjective Optimization. In L. Spector, E. D. Goodman, A. Wu, W. Langdon, H.-M. Voigt, M. Gen, S. Sen, M. Dorigo, S. Pezeshk, M. H. Garzon, and E. Burke, editors, *Proceedings of the Genetic and Evolutionary Computation Conference (GECCO'2001)*, pages 283–290, San Francisco, California, 2001. Morgan Kaufmann Publishers.
300. D. W. Corne and J. D. Knowles. No Free Lunch and Free Leftovers Theorems for Multiobjective Optimisation Problems. In C. M. Fonseca, P. J. Fleming, E. Zitzler, K. Deb, and L. Thiele, editors, *Evolutionary Multi-Criterion Optimization. Second International Conference, EMO 2003*, pages 327–341, Faro, Portugal, April 2003. Springer. Lecture Notes in Computer Science. Volume 2632.
301. D. W. Corne, J. D. Knowles, and M. J. Oates. The Pareto Envelope-based Selection Algorithm for Multiobjective Optimization. In M. Schoenauer, K. Deb, G. Rudolph, X. Yao, E. Lutton, J. J. Merelo, and H.-P. Schwefel, editors, *Proceedings of the Parallel Problem Solving from Nature VI Conference*, pages 839–848, Paris, France, 2000. Springer. Lecture Notes in Computer Science No. 1917.
302. L. Costa and P. Oliveira. An Evolution Strategy for Multiobjective Optimization. In *Congress on Evolutionary Computation (CEC'2002)*, volume 1, pages 97–102, Piscataway, New Jersey, May 2002. IEEE Service Center.
303. R. Courant. Variational Methods for the Solution of Problems of Equilibrium and Vibrations. *Bulletin of the American Mathematical Society*, 49:1–23, 1943.
304. P. Cowling, N. Colledge, K. Dahal, and S. Remde. The Trade Off Between Diversity and Quality for Multi-objective Workforce Scheduling. In J. Gottlieb and G. R. Raidl, editors, *Evolutionary Computation in Combinatorial Optimization, 6th European Conference, EvoCOP 2006*, pages 13–24, Budapest, Hungary, April 2006. Springer. Lecture Notes in Computer Science Vol. 3906.
305. T. G. Crainic and M. Toulouse. Parallel Strategies for Meta-Heuristics. In F. Glover and G. A. Kochenberger, editors, *Handbook of Metaheuristics*, pages 475–513. Kluwer Academic Publishers, Boston/Dordrecht/London, 2002.
306. L. F. Cranor. *Declared Strategy Voting: An Instrument for Group Decision Making*. PhD thesis, Washington University, Department of Engineering and Policy, St. Louis, Missouri, December 1996.
307. W. A. Crossley. The Potential of Genetic Algorithms for Conceptual Design of Rotor Systems. *Engineering Optimization*, 24(3):221–238, 1995.

308. W. A. Crossley. Genetic Algorithm Approaches for Multiobjective Design of Rotor Systems. In *Proceedings of the 6th AIAA/NASA/ISSMO Symposium on Multidisciplinary Analysis and Optimization*, pages 384–394, Bellevue, Washington, September 1996. AIAA Paper 96-4025.
309. W. A. Crossley. Genetic Algorithm with the Kreisselmeier-Steinhauser Function for Multiobjective Constrained Optimization of Rotor Systems. In *AIAA 35th Aerospace Sciences Meeting and Exhibit*, Reno, Nevada, January 1997. AIAA Paper 97-0080.
310. W. A. Crossley, A. M. Cook, D. W. Fanjoy, and V. B. Venkayya. Using the Two-Branch Tournament Genetic Algorithm for Multiobjective Design. In *AIAA/ASME/ASCE/AHS/ASC Structures, Structural Dynamics, and Materials Conference*, Long Beach, California, April 1998. AIAA Paper 98-1914.
311. N. Cruz Cortés and C. A. Coello Coello. Multiobjective Optimization Using Ideas from the Clonal Selection Principle. In E. C.-P. et al., editor, *Genetic and Evolutionary Computation—GECCO 2003. Proceedings, Part I*, pages 158–170. Springer. Lecture Notes in Computer Science Vol. 2723, July 2003.
312. X. Cui, M. Li, and T. Fang. Study of Population Diversity of Multiobjective Evolutionary Algorithm Based on Immune and Entropy Principles. In *Proceedings of the Congress on Evolutionary Computation 2001 (CEC'2001)*, volume 2, pages 1316–1321, Piscataway, New Jersey, May 2001. IEEE Service Center.
313. V. Cutello, G. Narzisi, and G. Nicosia. A Class of Pareto Archived Evolution Strategy Algorithms Using Immune Inspired Operators for Ab-Initio Protein Structure Prediction. In F. R. et al., editor, *Applications of Evolutionary Computing. Evoworkshops 2005: EvoBIO, EvoCOMNET, EvoHOT, EvoIASP, EvoMUSART, and EvoSTOC*, pages 54–63. Springer. Lecture Notes in Computer Science Vol. 3449, Lausanne, Switzerland, March/April 2005.
314. D. Cvetković. *Evolutionary Multi–Objective Decision Support Systems for Conceptual Design*. PhD thesis, School of Computing, University of Plymouth, Plymouth, UK, November 2000.
315. D. Cvetković and C. A. Coello Coello. Human Preferences and Their Applications in Evolutionary Multi-Objective Optimization. In Y. Jin, editor, *Knowledge Incorporation in Evolutionary Computation*, pages 479–502. Springer, Berlin Heidelberg, 2005. ISBN 3-540-22902-7.
316. D. Cvetković and I. C. Parmee. Genetic Algorithm–based Multi–objective Optimisation and Conceptual Engineering Design. In *Congress on Evolutionary Computation – CEC99*, volume 1, pages 29–36, Washington D.C., USA, 1999. IEEE.
317. D. Cvetković and I. C. Parmee. Use of Preferences for GA–based Multi–objective Optimisation. In W. Banzhaf, J. Daida, A. E. Eiben, M. H. Garzon, V. Honavar, M. Jakiela, and R. E. Smith, editors, *Proceedings of the Genetic and Evolutionary Computation Conference (GECCO'99)*, volume 2, pages 1504–1509, Orlando, Florida, USA, 1999. Morgan Kaufmann Publishers.
318. D. Cvetković and I. C. Parmee. Designer's preferences and multi–objective preliminary design processes. In I. C. Parmee, editor, *Proceedings of the Fourth International Conference on Adaptive Computing in Design and Manufacture (ACDM'2000)*, pages 249–260. PEDC, University of Plymouth, UK, Springer London, 2000.

319. D. Cvetković and I. C. Parmee. Preferences and their Application in Evolutionary Multiobjective Optimisation. *IEEE Transactions on Evolutionary Computation*, 6(1):42–57, February 2002.
320. P. Czyzak and A. Jaszkiewicz. A multiobjective metaheuristic approach to the localization of a chain of petrol stations by the capital budgeting model. *Control and Cybernetics*, 25(1):177–187, 1996.
321. P. Czyzak and A. Jaszkiewicz. The Multiobjective Metaheuristic Approach for Optimization of Complex Manufacturing Systems. In G. Fandel and T. Gal, editors, *Multiple Criteria Decision Making. Proceedings of the XIIth International Conference*, pages 591–592, Hagen, Germany, 1997. Springer-Verlag.
322. P. Czyzak and A. Jaszkiewicz. Pareto Simulated Annealing. In G. Fandel and T. Gal, editors, *Multiple Criteria Decision Making. Proceedings of the XIIth International Conference*, pages 297–307, Hagen, Germany, 1997. Springer-Verlag.
323. P. Czyzak and A. Jaszkiewicz. Pareto simulated annealing—a metaheuristic technique for multiple-objective combinatorial optimization. *Journal of Multi-Criteria Decision Analysis*, 7:34–47, 1998.
324. N. O. Da Cunha and E. Polak. Constrained Minimization under Vector-Valued Criteria in Finite-Dimensional Spaces. Technical Report ERL–188, Electronic Research Laboratory, University of California, Berkeley, California, 1966.
325. J. da Fonseca Neto and C. P. Bottura. Parallel Genetic Algorithm Fitness Function Team for Eigenstructure Assignment via LQR. In *1999 Congress on Evolutionary Computation*, volume 21, pages 1035–1042, Washington, D.C., July 1999. IEEE Service Center.
326. N. V. Dakev, A. J. Chipperfield, J. F. Whidborne, and P. J. Fleming. An evolutionary algorithm approach for solving optimal control problems. In *Proceedings of the 13th International Federation of Automatic Control (IFAC) World Congress*, volume D, pages 321–326, San Francisco, California, 1996.
327. N. V. Dakev, J. F. Whidborne, A. J. Chipperfield, and P. J. Fleming. H_∞ design of an EMS control system for a maglev vehicle using evolutionary algorithms. *Proceedings of IMechE-I Part I: Journal of Systems and Control Engineering*, 311(4):345–355, 1997.
328. R. Dammkoehler, S. Karasek, E. Shands, and G. Marshal. Constrained Search of Conformational Hyperspace. *Journal of Computer-Aided Molecular Design*, 3:3–21, 1989.
329. I. Das and J. Dennis. A Closer Look at Drawbacks of Minimizing Weighted Sums of Objectives for Pareto Set Generation in Multicriteria Optimization Problems. *Structural Optimization*, 14(1):63–69, 1997.
330. P. Das and Y. Y. Haimes. Multiobjective optimization in water quality and land management. *Water Resources Research*, 15(6):1313–1322, December 1979.
331. D. Dasgupta, editor. *Artificial Immune Systems and Their Applications*. Springer-Verlag, Berlin, 1999.
332. D. Dasgupta and F. A. González. Evolving Complex Fuzzy Classifier Rules Using a Linear Tree Genetic Representation. In L. Spector, E. D. Goodman, A. Wu, W. Langdon, H.-M. Voigt, M. Gen, S. Sen, M. Dorigo, S. Pezeshk, M. H. Garzon, and E. Burke, editors, *Proceedings of the Genetic and Evolutionary Computation Conference (GECCO'2001)*, pages 299–305, San Francisco, California, 2001. Morgan Kaufmann Publishers.

333. D. Dasgupta and D. R. McGregor. Nonstationary Function Optimization using the Structured Genetic Algorithm. In R. Männer and B. Manderick, editors, *Proceedings of Parallel Problem Solving from Nature (PPSN 2)*, pages 145–154, Brussels, Belgium, September 1992. Elsevier Science.
334. D. Dasgupta and D. R. McGregor. A more Biologically Motivated Genetic Algorithm: The Model and some Results. *Cybernetics and Systems: An International Journal*, 25(3):447–469, May-June 1994.
335. P. Dasgupta, P. Chakrabarti, and S. DeSarkar. *Multiobjective Heuristic Search. An Introduction to intelligent Search Methods for Multicriteria Optimization*. Vieweg, Germany, 1999. ISBN 3-528-05708-4.
336. L. David and L. Duckstein. Multi-criterion ranking of alternative long-range water resource systems. *Water Resources Bulletin*, 12(4):731–745, 1976.
337. G. D'Avignon, M. Turcotte, L. Beaudry, and Y. Duperre. Degré de spécialisation des hôpitaux de Quebec. Technical report, Université Laval, Quebec, Canada, July 1983.
338. J. E. Davis and G. Kendall. An Investigation, using Co-Evolution, to Evolve an Awari Player. In *Congress on Evolutionary Computation (CEC'2002)*, volume 2, pages 1408–1413, Piscataway, New Jersey, May 2002. IEEE Service Center.
339. L. Davis. *Genetic Algorithms and Simulated Annealing*. Pitman, London, 1987.
340. R. Dawkins. *The Selfish Gene*. Oxford University Press, Oxford, UK, 1990.
341. R. O. Day. *Explicit Building Block Multiobjective Evolutionary Computation: Methods and Application*. PhD thesis, Air Force Institute of Technology, AFIT/ENG, BLDG 642, 2950 HOBSON WAY, WPAFB (Dayton) OH 45433-7765, USA, June 2005.
342. R. O. Day, M. P. Kleeman, and G. B. Lamont. Multi-Objective fast messy Genetic Algorithm Solving Deception Problems. In *2004 Congress on Evolutionary Computation (CEC'2004)*, volume 2, pages 1502–1509, Portland, Oregon, USA, June 2004. IEEE Service Center.
343. R. O. Day and G. B. Lamont. An Effective Explicit Building Block MOEA, the MOMGA-IIa. In *2005 IEEE Congress on Evolutionary Computation (CEC'2005)*, volume 1, pages 17–24, Edinburgh, Scotland, September 2005. IEEE Service Center.
344. R. O. Day and G. B. Lamont. Extended Multi-objective fast messy Genetic Algorithm Solving Deception Problems. In C. A. Coello Coello, A. Hernández Aguirre, and E. Zitzler, editors, *Evolutionary Multi-Criterion Optimization. Third International Conference, EMO 2005*, pages 296–310, Guanajuato, México, March 2005. Springer. Lecture Notes in Computer Science Vol. 3410.
345. R. O. Day and G. B. Lamont. Multiobjective Quadratic Assignment Problem Solved by an Explicit Building Block Search Algorithm - MOMGA-IIa. In G. R. Raidl and J. Gottlieb, editors, *Evolutionary Computation in Combinatorial Optimization. 5th European Conference, EvoCOP 2005*, pages 91–100, Lausanne, Switzerland, March/April 2005. Springer, Lecture Notes in Computer Science Vol. 3448.
346. R. O. Day, J. B. Zydallis, and G. B. Lamont. Solving the Protein structure Prediction Problem through a Multi-Objective Genetic Algorithm. In *Proceedings of IEEE/DARPA International Conference on Computational Nanoscience (ICCN'02)*, pages 32–35, 2002.

347. R. O. Day, J. B. Zydallis, G. B. Lamont, and R. Pachter. Fine Granularity and Building Block Sizes of the Parallel Fast Messy Genetic Algorithm. In *Proceedings of the 2002 IEEE World Congress on Computational Intelligence*, pages 127–132, Piscataway, NJ, May 2002. IEEE Service Center.
348. A. K. De Jong. *An Analysis of the Behavior of a Class of Genetic Adaptive Systems*. PhD thesis, University of Michigan, 1975.
349. E. D. de Jong, R. A. Watson, and J. B. Pollack. Reducing Bloat and Promoting Diversity using Multi-Objective Methods. In L. Spector, E. D. Goodman, A. Wu, W. Langdon, H.-M. Voigt, M. Gen, S. Sen, M. Dorigo, S. Pezeshk, M. H. Garzon, and E. Burke, editors, *Proceedings of the Genetic and Evolutionary Computation Conference (GECCO'2001)*, pages 11–18, San Francisco, California, 2001. Morgan Kaufmann Publishers.
350. B. de la Iglesia, A. Reynolds, and V. J. Rayward-Smith. Developments on a Multi-objective Metaheuristic (MOMH) Algorithm for Finding Interesting Sets of Classification Rules. In C. A. Coello Coello, A. Hernández Aguirre, and E. Zitzler, editors, *Evolutionary Multi-Criterion Optimization. Third International Conference, EMO 2005*, pages 826–840, Guanajuato, México, March 2005. Springer. Lecture Notes in Computer Science Vol. 3410.
351. B. de la Iglesia, G. Richards, M. Philpott, and V. Rayward-Smith. The application and effectiveness of a multi-objective metaheuristic algorithm for partial classification. *European Journal of Operational Research*, 169:898–917, 2006.
352. R. de Neufville. *Applied Systems Analysis: Engineering Planning and Technology Management*. McGraw-Hill, New York, New York, 1990.
353. F. de Toro, J. Ortega, J. Fernández, and A. Díaz. PSFGA: A Parallel Genetic Algorithm for Multiobjective Optimization. In F. Vajda and N. Podhorszki, editors, *10th Euromicro Workshop on Parallel, Distributed and Network-Based Processing*, pages 384–391. IEEE, 2002.
354. K. Deb. *Binary and Floating-Point Function Optimization using Messy Genetic Algorithms*. PhD thesis, University of Alabama, Tuscaloosa, Alabama, 1991.
355. K. Deb. Evolutionary Algorithms for Multi-Criterion Optimization in Engineering Design. In K. Miettinen, M. M. Mäkelä, P. Neittaanmäki, and J. Periaux, editors, *Evolutionary Algorithms in Engineering and Computer Science*, chapter 8, pages 135–161. John Wiley & Sons, Ltd, Chichester, UK, 1999.
356. K. Deb. Multi–Objective Evolutionary Algorithms: Introducing Bias Among Pareto–Optimal Solutions. KanGAL report 99002, Indian Institute of Technology, Kanpur, India, 1999.
357. K. Deb. Multi-Objective Genetic Algorithms: Problem Difficulties and Construction of Test Problems. *Evolutionary Computation*, 7(3):205–230, Fall 1999.
358. K. Deb. Non-linear goal programming using Multi-Objective Genetic Algorithms. Technical Report CI-60/98, Dortmund: Department of Computer Science/LS11, University of Dortmund, Germany, 1999.
359. K. Deb. Solving Goal Programming Problems Using Multi-Objective Genetic Algorithms. In *1999 Congress on Evolutionary Computation*, pages 77–84, Washington, D.C., July 1999. IEEE Service Center.

360. K. Deb. An Efficient Constraint Handling Method for Genetic Algorithms. *Computer Methods in Applied Mechanics and Engineering*, 186(2/4):311–338, 2000.
361. K. Deb. *Multi-Objective Optimization using Evolutionary Algorithms*. John Wiley & Sons, Chichester, UK, 2001. ISBN 0-471-87339-X.
362. K. Deb and R. B. Agrawal. Simulated Binary Crossover for Continuous Search Space. *Complex Systems*, 9:115–148, 1995.
363. K. Deb, S. Agrawal, A. Pratab, and T. Meyarivan. A Fast Elitist Non-Dominated Sorting Genetic Algorithm for Multi-Objective Optimization: NSGA-II. In M. Schoenauer, K. Deb, G. Rudolph, X. Yao, E. Lutton, J. J. Merelo, and H.-P. Schwefel, editors, *Proceedings of the Parallel Problem Solving from Nature VI Conference*, pages 849–858, Paris, France, 2000. Springer. Lecture Notes in Computer Science No. 1917.
364. K. Deb, L. Altenberg, B. Manderick, T. Bäck, Z. Michalewicz, M. Mitchell, and S. Forrest. Fitness landscapes. In T. Bäck, D. Fogel, and Z. Michalewicz, editors, *Handbook of Evolutionary Computation*, volume 1, pages B2.7:1–B2.7:25. IOP Publishing Ltd. and Oxford University Press, 1997.
365. K. Deb and T. Goel. Controlled Elitist Non-dominated Sorting Genetic Algorithms for Better Convergence. In E. Zitzler, K. Deb, L. Thiele, C. A. C. Coello, and D. Corne, editors, *First International Conference on Evolutionary Multi-Criterion Optimization*, pages 67–81. Springer-Verlag. Lecture Notes in Computer Science No. 1993, 2001.
366. K. Deb and T. Goel. A Hybrid Multi-Objective Evolutionary Approach to Engineering Shape Design. In E. Zitzler, K. Deb, L. Thiele, C. A. Coello Coello, and D. Corne, editors, *First International Conference on Evolutionary Multi-Criterion Optimization*, pages 385–399. Springer-Verlag. Lecture Notes in Computer Science No. 1993, 2001.
367. K. Deb and T. Goel. Multi-Objective Evolutionary Algorithms for Engineering Shape Design. In R. Sarker, M. Mohammadian, and X. Yao, editors, *Evolutionary Optimization*, pages 146–175. Kluwer Academic Publishers, New York, February 2002. ISBN 0-7923-7654-4.
368. K. Deb and D. E. Goldberg. An Investigation of Niche and Species Formation in Genetic Function Optimization. In J. D. Schaffer, editor, *Proceedings of the Third International Conference on Genetic Algorithms*, pages 42–50, San Mateo, California, June 1989. George Mason University, Morgan Kaufmann Publishers.
369. K. Deb and S. Jain. Running Performance Metrics for Evolutionary Multi-Objective Optimization. In L. Wang, K. C. Tan, T. Furuhashi, J.-H. Kim, and X. Yao, editors, *Proceedings of the 4th Asia-Pacific Conference on Simulated Evolution and Learning (SEAL'02)*, volume 1, pages 13–20, Orchid Country Club, Singapore, November 2002. Nanyang Technical University.
370. K. Deb and A. Kumar. Real-coded Genetic Algorithms with Simulated Binary Crossover: Studies on Multimodal and Multiobjective Problems. *Complex Systems*, 9:431–454, 1995.
371. K. Deb and T. Meyarivan. Constrained Test Problems for Multi-Objective Evolutionary Optimization. KanGAL report 200005, Indian Institute of Technology, Kanpur, India, 2000.
372. K. Deb, M. Mohan, and S. Mishra. Towards a Quick Computation of Well-Spread Pareto-Optimal Solutions. In C. M. Fonseca, P. J. Fleming, E. Zitzler,

K. Deb, and L. Thiele, editors, *Evolutionary Multi-Criterion Optimization. Second International Conference, EMO 2003*, pages 222–236, Faro, Portugal, April 2003. Springer. Lecture Notes in Computer Science. Volume 2632.
373. K. Deb, M. Mohan, and S. Mishra. Evaluating the ϵ-Domination Based Multi-Objective Evolutionary Algorithm for a Quick Computation of Pareto-Optimal Solutions. *Evolutionary Computation*, 13(4):501–525, Winter 2005.
374. K. Deb, A. Pratap, S. Agarwal, and T. Meyarivan. A Fast and Elitist Multiobjective Genetic Algorithm: NSGA–II. *IEEE Transactions on Evolutionary Computation*, 6(2):182–197, April 2002.
375. K. Deb, A. Pratap, and T. Meyarivan. Constrained Test Problems for Multiobjective Evolutionary Optimization. In E. Zitzler, K. Deb, L. Thiele, C. A. Coello Coello, and D. Corne, editors, *First International Conference on Evolutionary Multi-Criterion Optimization*, pages 284–298. Springer-Verlag. Lecture Notes in Computer Science No. 1993, 2001.
376. K. Deb and D. K. Saxena. Searching for Pareto-optimal solutions through dimensionality reduction for certain large-dimensional multi-objective optimization problems. In *2006 IEEE Congress on Evolutionary Computation (CEC'2006)*, pages 3353–3360, Vancouver, BC, Canada, July 2006. IEEE.
377. K. Deb, A. Sinha, and S. Kukkonen. Multi-Objective Test Problems, Linkages, and Evolutionary Methodologies. In M. K. et al., editor, *2006 Genetic and Evolutionary Computation Conference (GECCO'2006)*, volume 2, pages 1141–1148, Seattle, Washington, USA, July 2006. ACM Press. ISBN 1-59593-186-4.
378. K. Deb, L. Thiele, M. Laumanns, and E. Zitzler. Scalable Multi-Objective Optimization Test Problems. In *Congress on Evolutionary Computation (CEC'2002)*, volume 1, pages 825–830, Piscataway, New Jersey, May 2002. IEEE Service Center.
379. K. Deb, L. Thiele, M. Laumanns, and E. Zitzler. Scalable Test Problems for Evolutionary Multiobjective Optimization. In A. Abraham, L. Jain, and R. Goldberg, editors, *Evolutionary Multiobjective Optimization. Theoretical Advances and Applications*, pages 105–145. Springer, USA, 2005.
380. K. Deb and S. Tiwari. Multi-objective optimization of a leg mechanism using genetic algorithms. *Engineering Optimization*, 37(4):325–350, June 2005.
381. K. Deb, P. Zope, and A. Jain. Distributed Computing of Pareto-Optimal Solutions with Evolutionary Algorithms. In C. M. Fonseca, P. J. Fleming, E. Zitzler, K. Deb, and L. Thiele, editors, *Evolutionary Multi-Criterion Optimization. Second International Conference, EMO 2003*, pages 534–549, Faro, Portugal, April 2003. Springer. Lecture Notes in Computer Science. Volume 2632.
382. C. R. DeVore, H. C. Briggs, and A. R. DeWispelares. Application of Multiple Objective Optimization Techniques to Finite Element Model Tuning. *Computers and Structures*, 24(5):683–690, 1986.
383. C. Dhaenens, J. Lemesre, N. Melab, M.-S. Mezmaz, and E.-G. Talbi. Parallel Exact Methods for Multiobjective Combinatorial Optimization. In E.-G. Talbi, editor, *Parallel Combinatorial Optimization*, pages 187–210. Wiley-Interscience, 2006.
384. A. K. Dhingra and B. H. Lee. A Genetic Algorithm Approach to Single and Multiobjective Structural Optimization with Discrete-Continuous Variables. *International Journal for Numerical Methods in Engineering*, 37:4059–4080, 1994.

385. P. Di Barba, M. Farina, and A. Savini. Multiobjective Design Optimization of Real-Life Devices in Electrical Engineering: A Cost-Effective Evolutionary Approach. In E. Zitzler, K. Deb, L. Thiele, C. A. Coello Coello, and D. Corne, editors, *First International Conference on Evolutionary Multi-Criterion Optimization*, pages 560–573. Springer-Verlag. Lecture Notes in Computer Science No. 1993, 2001.
386. F. di Pierro. *Many-Objective Evolutionary Algorithms and Applications to Water Resources Engineering*. PhD thesis, School of Engineering, Computer Science and Mathematics, UK, August 2006.
387. F. di Pierro, S.-T. Khu, and D. A. Savić. An Investigation on Preference Order Ranking Scheme for Multiobjective Evolutionary Optimization. *IEEE Transactions on Evolutionary Computation*, 11(1):17–45, February 2007.
388. R. P. Dick and N. K. Jha. MOGAC: A Multiobjective Genetic Algorithm for the Co-Synthesis of Hardware-Software Embedded Systems. In *IEEE/ACM Conference on Computer Aided Design*, pages 522–529, Los Alamitos, California, 1997. IEEE Computer Society Press.
389. R. P. Dick and N. K. Jha. CORDS: Hardware-Software Co-Synthesis of Reconfigurable Real-Time Distributed Embedded Systems. In *Proceedings of the International Conference on Computer-Aided Design*, pages 62–68, November 1998.
390. R. P. Dick and N. K. Jha. MOGAC: A Multiobjective Genetic Algorithm for Hardware-Software Co-synthesis of Hierarchical Heterogeneous Distributed Embedded Systems. *IEEE Transactions on Computer-Aided Design of Integrated Circuits and Systems*, 17(10):920–935, October 1998.
391. R. P. Dick and N. K. Jha. MOCSYN: Multiobjective Core-Based Single-Chip System Synthesis. In *Proc. Design, Automation and Test in Europe*, pages 263–270, March 1999.
392. J. G. Digalakis and K. G. Margaritis. An Experimental Study of Benchmarking Functions for Genetic Algorithms. In *Proceedings of IEEE International Conference on Systems, Man, and Cybernetics (SMC'200)*, pages 3810–3815, Nashville, Tennessee, October 2000. IEEE.
393. C. Dimopoulos. A Review of Evolutionary Multiobjective Optimization Applications in the Area of Production Research. In *2004 Congress on Evolutionary Computation (CEC'2004)*, volume 2, pages 1487–1494, Portland, Oregon, USA, June 2004. IEEE Service Center.
394. C. Dimopoulos and A. M. S. Zalzala. Evolutionary Computation Approaches to Cell Optimisation. In I. Parmee, editor, *The Integration of Evolutionary and Adaptive Computing Technologies with Product/System Design and Realisation*, pages 69–83, Plymouth, United Kingdom, April 1998. Plymouth Engineering Design Centre, Springer-Verlag.
395. C. Dimopoulos and A. M. S. Zalzala. Optimization of Cell Configuration and Comparisons using Evolutionary Computation Approaches. In D. B. Fogel, editor, *Proceedings of the 1998 International Conference on Evolutionary Computation*, pages 148–153, Piscataway, New Jersey, 1998. IEEE.
396. K. Doerner, W. Gutjahr, R. Hartl, C. Strauss, and C. Stummer. Pareto ant colony optimization with ILP preprocessing in multiobjective portfolio selection. *European Journal of Operational Research*, 171(3):830–841, June 2006.
397. K. Doerner, W. J. Gutjahr, R. F. Hartl, C. Strauss, and C. Stummer. Ant Colony Optimization in Multiobjective Portfolio Selection. In *Proceedings of*

the *4th Metaheuristics International Conference (MIC'2001)*, pages 243–248, Porto, Portugal, July 2001.
398. K. Doerner, W. J. Gutjahr, R. F. Hartl, C. Strauss, and C. Stummer. Pareto Ant Colony Optimization: A Metaheuristic Approach to Multiobjective Portfolio Selection. *Annals of Operations Research*, 131(1–4):79–99, October 2004.
399. K. Doerner, R. F. Hartl, and M. Reimann. Are COMPETants more competent for problem solving? - The Case of Full Truckload Transportation. *Central European Journal of Operations Research*, 11(2):115–141, 2003.
400. D. C. Donha, D. S. Desanj, and M. R. Katebi. Genetic Algorithm for Weight Selection in h_∞ Control Design. In T. Bäck, editor, *Proceedings of the Seventh International Conference on Genetic Algorithms*, pages 599–606. Morgan Kaufmann Publishers, San Mateo, California, July 1997.
401. Y. Donoso Meisel. *Multi-Objective Optimization Scheme for Static and Dynamic Multicast Flows*. PhD thesis, Department of Electronics, Computer Science and Automatic Control, Universitat de Girona, Girona, Spain, April 2005.
402. D. J. Doorly, J. Peir, and J.-P. Oesterle. Optimisation of Aerodynamic and Coupled Aerodynamic-Structural Design Using Parallel Genetic Algorithms. In *6th AIAA/NASA/USAF Multidisciplinary Analysis & Optimization Symposium*, Monterey, California, September 1996.
403. M. Dorigo and G. D. Caro. The Ant Colony Optimization Meta-Heuristic. In D. Corne, M. Dorigo, and F. Glover, editors, *New Ideas in Optimization*. McGraw-Hill, 1999.
404. M. Dorigo, V. Maniezzo, and A. Colorni. Positive feedback as a search strategy. Technical Report 91-016, Dipartimento di Elettronica, Politecnico di Milano, Italy, 1991.
405. M. Dorigo, V. Maniezzo, and A. Colorni. The Ant System: Optimization by a colony of cooperating agents. *IEEE Transactions on Systems, Man, and Cybernetics – Part B*, 26(1):29–41, 1996.
406. M. Dorigo and T. Stützle. *Ant Colony Optimization*. The MIT Press, 2004. ISBN 0-262-04219-3.
407. K. A. Dowsland. Simulated Annealing. In C. R. Reeves, editor, *Modern Heuristic Techniques for Combinatorial Problems*, chapter 2, pages 20–69. John Wiley & Sons, 1993.
408. G. V. Dozier, S. McCullough, A. Homaifar, and L. Moore. Multiobjective Evolutionary Path Planning via Fuzzy Tournament Selection. In *IEEE International Conference on Evolutionary Computation (ICEC'98)*, pages 684–689, Piscataway, New Jersey, May 1998. IEEE Press.
409. N. Drechsler, R. Drechsler, and B. Becker. Multi-Objected Optimization in Evolutionary Algorithms Using Satisfyability Classes. In B. Reusch, editor, *International Conference on Computational Intelligence, Theory and Applications, 6th Fuzzy Days*, pages 108–117, Dortmund, Germany, 1999. Springer-Verlag. Lecture Notes in Computer Science Vol. 1625.
410. N. Drechsler, R. Drechsler, and B. Becker. Multi-objective Optimisation Based on Relation *favour*. In E. Zitzler, K. Deb, L. Thiele, C. A. Coello Coello, and D. Corne, editors, *First International Conference on Evolutionary Multi-Criterion Optimization*, pages 154–166. Springer-Verlag. Lecture Notes in Computer Science No. 1993, 2001.

411. R. Drechsler, N. Göckel, and B. Becker. Learning Heuristics for OBDD Minimization by Evolutionary Algorithms. In H.-M. Voigt, W. Ebeling, I. Rechenberger, and H.-P. Schwefel, editors, *Parallel Problem Solving from Nature (PPSN IV)*, pages 730–739, Berlin, Germany, 1996. Springer-Verlag. Lecture Notes in Computer Science No. 1141.
412. R. Drezewski and L. Siwik. Co-Evolutionary Multi-Agent System with Sexual Selection Mechanism for Multi-Objective Optimization. In *2006 IEEE Congress on Evolutionary Computation (CEC'2006)*, pages 2784–2791, Vancouver, BC, Canada, July 2006. IEEE.
413. N. Duarte, A. E. Ruano, C. Fonseca, and P. Fleming. Accelerating Multi-Objective Control System Design Using a Neuro-Genetic Approach. In *2000 Congress on Evolutionary Computation*, volume 1, pages 392–397, Piscataway, New Jersey, July 2000. IEEE Service Center.
414. E. Ducheyne, B. De Baets, and R. De Wulf. Even Flow Scheduling Problems in Forest Management. In C. A. Coello Coello and G. B. Lamont, editors, *Applications of Multi-Objective Evolutionary Algorithms*, pages 701–726. World Scientific, Singapore, 2004.
415. E. I. Ducheyne, R. R. De Wulf, and B. De Baets. Bi-objective genetic algorithm for forest management: a comparative study. In *Proceedings of the 2001 Genetic and Evolutionary Computation Conference. Late-Breaking Papers*, pages 63–66, San Francisco, California, July 2001.
416. L. Duckstein. Multiobjective Optimization in Structural Design: The Model Choice Problem. In E. Atrek, R. H. Gallagher, K. M. Ragsdell, and O. C. Zienkiewicz, editors, *New Directions in Optimum Structural Design*, pages 459–481. John Wiley and Sons, 1984.
417. L. Duckstein and M. Gershon. Multi-objective analysis of a vegetation management problem using ELECTRE II. Technical Report 81–11, Department of Systems and Industrial Engineering, University of Arizona, Tucson, Arizona, 1981.
418. L. Duckstein, M. Gershon, and R. McAniff. Development of the Santa Cruz River Basin: A Comparison of Multi-Criterion Approaches. Technical Report 51, Engineering Experimentation Station, University of Arizona, Tucson, Arizona, 1981.
419. L. Duckstein and S. Opricovic. Multiobjective optimization in river basin development. *Water Resources Research*, 16(1):14–20, feb 1980.
420. J. M. Dujardin. Une évaluation multicritère de projets de remédiation à l'échec dans l'enseignement secondaire Belge. In *XIX Meeting of the European Working Group on Multiple Criteria Decision Aid*, Liège, France, March 1984.
421. W. H. Durham. *Co-evolution: Genes, Culture, and Human Diversity*. Stanford University Press, Stanford, California, 1994.
422. R. G. Dyson. Maximim Programming, Fuzzy Linear Programming and Multi-Criteria Decision Making. *Journal of the Operational Research Society*, 31:263–267, 1980.
423. R. Eberhart and Y. Shi. Comparison between Genetic Algorithms and Particle Swarm Optimization. In V. W. Porto, N. Saravanan, D. Waagen, and A. Eibe, editors, *Proceedings of the Seventh Annual Conference on Evolutionary Programming*, pages 611–619. Springer-Verlag, March 1998.
424. K. S. Edge, G. B. Lamont, and R. A. Raines. A Retrovirus Inspired Algorithm for Virus Detection & Optimization. In M. K. et al., editor, *2006 Genetic and*

Evolutionary Computation Conference (GECCO'2006), volume 1, pages 103–110, Seattle, Washington, USA, July 2006. ACM Press. ISBN 1-59593-186-4.
425. F. Y. Edgeworth. *Mathematical Psychics*. P. Keagan, London, England, 1881.
426. R. Eglese. Heuristics in operational research. In V. Belton and R. O'Keefe, editors, *Recent Developments in Operational Research*, pages 49–67. Pergamon Press, Oxford, 1986.
427. M. Ehrgott. Approximation algorithms for combinatorial multicriteria optimization problems. *International Transactions in Operational Research*, 7:5–31, 2000.
428. M. Ehrgott. *Multicriteria Optimization*. Springer, Berlin, second edition, 2005. ISBN 3-540-21398-8.
429. M. Ehrgott and X. Gandibleux. A Survey and Annotated Bibliography of Multiobjective Combinatorial Optimization. *OR Spektrum*, 22:425–460, 2000.
430. M. Ehrgott and X. Gandibleux. Multiobjective Combinatorial Optimization—Theory, Methodology, and Applications. In M. Ehrgott and X. Gandibleux, editors, *Multiple Criteria Optimization: State of the Art Annotated Bibliographic Surveys*, pages 369–444. Kluwer Academic Publishers, Boston, 2002.
431. M. Ehrgott and X. Gandibleux. Approximative Solution Methods for Multiobjective Combinatorial Optimization. *Top*, 12(1):1–89, June 2004.
432. M. Ehrgott and K. Klamroth. Connectedness of efficient solutions in multiple criteria combinatorial optimization. *European Journal of Operational Research*, 97:159–166, 1997.
433. M. Ehrgott, K. Klamroth, and C. Schwehm. An MCDM approach to portfolio optimization. *European Journal of Operational Research*, 155(3):752–770, June 2004.
434. P. Ehrlich and P. Raven. Butterflies and Plants: A Study in Coevolution. *Evolution*, 18:586–608, 1964.
435. A. Eiben and J. Smith. *Introduction to Evolutionary Computing*. Springer, Berlin, 2003. ISBN 3-540-40184-9.
436. A. Ekárt and S. Németh. Selection Based on the Pareto Nondomination Criterion for Controlling Code Growth in Genetic Programming. *Genetic Programming and Evolvable Machines*, 2(1):61–73, March 2001.
437. N. H. Eklund and M. J. Embrechts. GA-Based Multi-Objective Optimization of Visible Spectra for Lamp Design. In C. H. Dagli, A. L. Buczak, J. Ghosh, M. J. Embrechts, and O. Ersoy, editors, *Smart Engineering System Design: Neural Networks, Fuzzy Logic, Evolutionary Programming, Data Mining and Complex Systems*, pages 451–456, New York, November 1999. ASME Press.
438. N. H. Eklund and M. J. Embrechts. Determining the Color-Efficiency Pareto Optimal Surface for Filtered Light Sources. In E. Zitzler, K. Deb, L. Thiele, C. A. Coello Coello, and D. Corne, editors, *First International Conference on Evolutionary Multi-Criterion Optimization*, pages 603–611. Springer-Verlag. Lecture Notes in Computer Science No. 1993, 2001.
439. N. H. W. Eklund. *Multiobjective Visible Spectrum Optimization: A Genetic Algorithm Approach*. PhD thesis, Rensselaer Polytechnic Institute, Troy, New York, USA, September 2002.
440. W. El Moudani, C. A. Nunes Cosenza, M. de Coligny, and F. Mora-Camino. A Bi-Criterion Approach for the Airlines Crew Rostering Problem. In E. Zitzler,

K. Deb, L. Thiele, C. A. Coello Coello, and D. Corne, editors, *First International Conference on Evolutionary Multi-Criterion Optimization*, pages 486–500. Springer-Verlag. Lecture Notes in Computer Science No. 1993, 2001.
441. H. El-Rewini, T. G. Lewis, and H. H. Ali. *Task Scheduling in Parallel and Distributed Systems*. Prentice-Hall, 1994.
442. C. Elegbede and K. Adjallah. Availability allocation to repairable systems with genetic algorithms: a multi-objective formulation. *Reliability Engineering & Systems Safety*, 82(3):319–330, December 2003.
443. T. A. Ely, W. A. Crossley, and E. A. Williams. Satellite Constellation Design for Zonal Coverage using Genetic Algorithms. In *8th AAS/AIAA Space Flight Mechanics Meeting*, Monterey, California, February 1998.
444. C. Emmanouilidis, A. Hunter, and J. MacIntyre. A Multiobjective Evolutionary Setting for Feature Selection and a Commonality-Based Crossover Operator. In *2000 Congress on Evolutionary Computation*, volume 1, pages 309–316, Piscataway, New Jersey, July 2000. IEEE Service Center.
445. C. Emmanouilidis, A. Hunter, J. MacIntyre, and C. Cox. Multiple Criteria Genetic Algorithms for Feature Selection in Neurofuzzy Modeling. In *1999 International Joint Conference on Neural Networks*, volume 6, pages 4387–4392, Washington, D.C., July 1999.
446. C. Emmanouilidis, A. Hunter, J. MacIntyre, and C. Cox. Selecting Features in Neurofuzzy Modelling by Multiobjective Genetic Algorithms. In *9th International Conference on Artificial Neural Networks*, volume 2, pages 749–754. IEE, Edinburgh, UK, September 1999.
447. M. Emmerich, N. Beume, and B. Naujoks. An EMO Algorithm Using the Hypervolume Measure as Selection Criterion. In C. A. Coello Coello, A. Hernández Aguirre, and E. Zitzler, editors, *Evolutionary Multi-Criterion Optimization. Third International Conference, EMO 2005*, pages 62–76, Guanajuato, México, March 2005. Springer. Lecture Notes in Computer Science Vol. 3410.
448. M. Emmerich and B. Naujoks. Metamodel Assisted Multiobjective Optimisation Strategies and their Application in Airfoil Design. In I. Parmee, editor, *Adaptive Computing in Design and Manufacture VI*, pages 249–260, London, 2004. Springer.
449. M. T. Emmerich and A. H. Deutz. Test Problems Based on Lamé Superspheres. In S. Obayashi, K. Deb, C. Poloni, T. Hiroyasu, and T. Murata, editors, *Evolutionary Multi-Criterion Optimization, 4th International Conference, EMO 2007*, pages 922–936, Matshushima, Japan, March 2007. Springer. Lecture Notes in Computer Science Vol. 4403.
450. A. P. Engelbrecht. *Computational Intelligence: An Introduction*. John Wiley & Sons, 2003. ISBN 0-47084-870-7.
451. A. P. Engelbrecht. *Fundamentals of Computational Swarm Intelligence*. John Wiley & Sons, Ltd, 2005. ISBN 978-0-470-09191-3.
452. A. Eremeev. A genetic algorithm with a non-binary representation for the set covering problem. In *Proceedings of Operations Research (OR'98)*, pages 175–181. Springer-Verlag, 1999.
453. M. Erickson, A. Mayer, and J. Horn. The Niched Pareto Genetic Algorithm 2 Applied to the Design of Groundwater Remediation Systems. In E. Zitzler, K. Deb, L. Thiele, C. A. Coello Coello, and D. Corne, editors, *First International Conference on Evolutionary Multi-Criterion Optimization*, pages 681–695. Springer-Verlag. Lecture Notes in Computer Science No. 1993, 2001.

454. H. Esbensen and E. S. Kuh. Design space exploration using the genetic algorithm. In *IEEE International Symposium on Circuits and Systems (ISCAS'96)*, pages 500–503, Piscataway, NJ, 1996. IEEE.
455. H. Esbensen and E. S. Kuh. EXPLORER: An Interactive Floorplaner for Design Space Exploration. In *Proceedings of the European Design Automation Conference*, pages 356–361, 1996.
456. L. J. Eshelman and J. D. Schaffer. Preventing Premature Convergence in Genetic Algorithms by Preventing Incest. In R. K. Belew and L. B. Booker, editors, *Proceedings of the Fourth International Conference on Genetic Algorithms*, pages 115–122, San Mateo, California, July 1991. Morgan Kaufmann Publishers.
457. L. J. Eshelman and J. D. Schaffer. Real-coded Genetic Algorithms and Interval-Schemata. In L. D. Whitley, editor, *Foundations of Genetic Algorithms 2*, pages 187–202. Morgan Kaufmann Publishers, San Mateo, California, 1993.
458. S. Esquivel, S. Ferrero, and R. Gallard. Parameter settings and representations in pareto-based optimization for job shop scheduling. *Cybernetics and Systems*, 33(6):559–578, September 2002.
459. S. Esquivel, S. Ferrero, R. Gallard, C. Salto, H. Alfonso, and M. Schütz. Enhanced evolutionary algorithms for single and multiobjective optimization in the job scheduling problem. *Knowledge-Based Systems*, 15(1–2):13–25, January 2002.
460. S. C. Esquivel, H. A. Leiva, and R. H. Gallard. Multiplicity in Genetic Algorithms to face Multicriteria Optimization. In *1999 Congress on Evolutionary Computation*, pages 85–90, Washington, D.C., July 1999. IEEE Service Center.
461. G. W. Evans. An Overview of Techniques for Solving Multiobjective Mathematical Programs. *Management Science*, 30(11):1268–1282, November 1984.
462. R. M. Everson, J. E. Fieldsend, and S. Singh. Full Elite Sets for Multi-Objective Optimisation. In I. Parmee, editor, *Proceedings of the Fifth International Conference on Adaptive Computing Design and Manufacture (ACDM 2002)*, volume 5, pages 343–354, University of Exeter, Devon, UK, April 2002. Springer-Verlag.
463. E. Falkenauer. *Genetic Algorithms and Grouping Problems*. John Wiley & Sons, New York, 1998.
464. A. F. Farahani, M. Kamal, and M. Salmani-Jelodar. Parallel-Genetic-Algorithm-Based HW/SW Partitioning. In *International Symposium on Parallel Computing in Electrical Engineering (PARELEC)*, pages 337–342. IEEE Computer Society, September 2006.
465. A. Farhang-Mehr and S. Azarm. Multi-Objective Genetic Algorithms With Concepts from Statistical Thermodynamics. In L. Spector, E. D. Goodman, A. Wu, W. B. Langdon, H.-M. Voigt, M. Gen, S. Sen, M. Dorigo, S. Pezeshk, M. H. Garzon, and E. Burke, editors, *Proceedings of the Genetic and Evolutionary Computation Conference (GECCO'2001)*, page 1075. Morgan Kaufmann Publishers, San Francisco, California, July 2001.
466. A. Farhang-Mehr and S. Azarm. Diversity Assessment of Pareto Optimal Solution Sets: An Entropy Approach. In *Congress on Evolutionary Computation (CEC'2002)*, volume 1, pages 723–728, Piscataway, New Jersey, May 2002. IEEE Service Center.

467. A. Farhang-Mehr and S. Azarm. Entropy-based multi-objective genetic algorithm for design optimization. *Structural and Multidisciplinary Optimization*, 24(5):351–361, November 2002.
468. M. Farina. A Neural Network Based Generalized Response Surface Multiobjective Evolutionary Algorithm. In *Congress on Evolutionary Computation (CEC'2002)*, volume 1, pages 956–961, Piscataway, New Jersey, May 2002. IEEE Service Center.
469. M. Farina and P. Amato. On the Optimal Solution Definition for Many-criteria Optimization Problems. In *Proceedings of the NAFIPS-FLINT International Conference'2002*, pages 233–238, Piscataway, New Jersey, June 2002. IEEE Service Center.
470. M. Farina and P. Amato. Fuzzy Optimality and Evolutionary Multiobjective Optimization. In C. M. Fonseca, P. J. Fleming, E. Zitzler, K. Deb, and L. Thiele, editors, *Evolutionary Multi-Criterion Optimization. Second International Conference, EMO 2003*, pages 58–72, Faro, Portugal, April 2003. Springer. Lecture Notes in Computer Science. Volume 2632.
471. M. Farina and P. Amato. A fuzzy definition of "optimality" for many-criteria optimization problems. *IEEE Transactions on Systems, Man, and Cybernetics Part A—Systems and Humans*, 34(3):315–326, May 2004.
472. M. Farina, A. Bramanti, and P. Di Barba. A GRS Method for Pareto-Optimal Front Identification in Electromagnetic Synthesis. *IEE Proceedings—Science, Measurement and Technology*, 149(5):207–213, September 2002.
473. M. Farina, K. Deb, and P. Amato. Dynamic Multiobjective Optimization Problems: Test Cases, Approximation, and Applications. In C. M. Fonseca, P. J. Fleming, E. Zitzler, K. Deb, and L. Thiele, editors, *Evolutionary Multi-Criterion Optimization. Second International Conference, EMO 2003*, pages 311–326, Faro, Portugal, April 2003. Springer. Lecture Notes in Computer Science. Volume 2632.
474. M. Farina, K. Deb, and P. Amato. Dynamic Multiobjective Optimization Problems: Test Cases, Approximations, and Applications. *IEEE Transactions on Evolutionary Computation*, 8(5):425–442, October 2004.
475. R. Farmani and J. A. Wright. Self-Adaptive Fitness Formulation for Constrained Optimization. *IEEE Transactions on Evolutionary Computation*, 7(5):445–455, October 2003.
476. C.-W. Feng, L. Liu, and S. A. Burns. Using Genetic Algorithms to Solve Construction Time-Cost Trade-Off Problems. *Journal of Computing in Civil Engineering*, 10(3):184–189, 1999.
477. T. A. Feo and M. G. Resende. Greedy Randomized Adaptive Search Procedures. *Journal of Global Optimization*, 6:109–133, 1995.
478. V. Feoktistov and S. Janaqi. Generalization of the Strategies in Differential Evolution. In *Proceedings of the 18th International Parallel and Distributed Processing Symposium (IPDPS 2004), 2004, Santa Fe, New Mexico, USA*, page 165a, New Mexico, USA, April 2004. IEEE Computer Society.
479. E. Fernández and J. C. Leyva. A method based on multiobjective optimization for deriving a ranking from a fuzzy preference relation. *European Journal of Operational Research*, 154(1):110–124, April 2004.
480. E. Fernández and R. Olmedo. An improved method for deriving final ranking from a fuzzy preference relation via multiobjective optimization. *Foundations of Computing and Decision Sciences*, 28(3):143–157, 2003.

481. F. Fernández, G. Spezzano, M. Tomassini, and L. Vanneschi. Parallel genetic programming. In E. Alba, editor, *Parallel Metaheuristics*, pages 127–153. Wiley-Interscience, 2005.
482. A. V. Fiacco and G. P. McCormick. Extensions of SUMT for nonlinear programming: equality constraints and extrapolation. *Management Science*, 12(11):816–828, 1968.
483. J. Fieldsend. Multi-Objective Particle Swarm Optimisation Methods. Technical Report 419, Department of Computer Science, University of Exeter, Exeter, UK, March 2004.
484. J. E. Fieldsend and R. M. Everson. Multi-objective Optimisation in the Presence of Uncertainty. In *2005 IEEE Congress on Evolutionary Computation (CEC'2005)*, volume 1, pages 243–250, Edinburgh, Scotland, September 2005. IEEE Service Center.
485. J. E. Fieldsend, R. M. Everson, and S. Singh. Using Unconstrained Elite Archives for Multiobjective Optimization. *IEEE Transactions on Evolutionary Computation*, 7(3):305–323, June 2003.
486. J. E. Fieldsend and S. Singh. A Multi-Objective Algorithm based upon Particle Swarm Optimisation, an Efficient Data Structure and Turbulence. In *Proceedings of the 2002 U.K. Workshop on Computational Intelligence*, pages 37–44, Birmingham, UK, September 2002.
487. J. Figueira, V. Mousseau, and B. Roy. ELECTRE methods. In J. Figueira, S. Greco, and M. Ehrgott, editors, *Multiple Criteria Decision Analysis. State of the Art Surveys*, pages 133–162. Springer, New York, USA, 2005.
488. R. A. Finkel and J. L. Bentley. Quad-Trees: A Data Structure for Retrieval on Composite Keys. *Acta Informatica*, 4:1–9, 1974.
489. P. C. Fishburn. A Survey of Multiattribute/Multicriterion Evaluation Theories. In S. Zionts, editor, *Multiple Criteria Problem Solving*, pages 181–224, Berlin, 1978. Springer-Verlag.
490. P. C. Fishburn. Nontransitive Preferences in Decision Theory. *Journal of Risk and Uncertainty*, 4:113–134, 1991.
491. M. Fleischer. The Measure of Pareto Optima. Applications to Multi-objective Metaheuristics. In C. M. Fonseca, P. J. Fleming, E. Zitzler, K. Deb, and L. Thiele, editors, *Evolutionary Multi-Criterion Optimization. Second International Conference, EMO 2003*, pages 519–533, Faro, Portugal, April 2003. Springer. Lecture Notes in Computer Science. Volume 2632.
492. J. Fliege and B. Fux Svaiter. Steepest descent methods for multicriteria optimization. *Mathematical Methods of Operations Research*, 51(3):479–494, 2000.
493. R. Flynn and P. D. Sherman. Multicriteria Optimization of Aircraft Panels: Determining Viable Genetic Algorithm Configurations. *International Journal of Intelligent Systems*, 10:987–999, 1995.
494. T. C. Fogarty. An Incremental Genetic Algorithm for Real-Time Optimisation. In *Proceedings of the 1989 IEEE International Conference on Systems, Man, and Cybernetics*, volume 1, pages 321–326. IEEE, 1989.
495. D. Fogel, E. Wasson, and E. Boughton. Evolving neural networks for detecting breast cancer. *Cancer Letters*, 96(1):49–53, 1995.
496. D. B. Fogel. *Evolutionary Computation. Toward a New Philosophy of Machine Intelligence*. The Institute of Electrical and Electronic Engineers, New York, 1995.

497. D. B. Fogel, editor. *Evolutionary Computation. The Fossil Record. Selected Readings on the History of Evolutionary Algorithms*. The Institute of Electrical and Electronic Engineers, New York, 1998.
498. D. B. Fogel and A. Ghozeil. A Note on Representations and Variation Operators. *IEEE Transactions on Evolutionary Computation*, 1(2):159–161, July 1997.
499. L. J. Fogel. *Artificial Intelligence through Simulated Evolution*. John Wiley, New York, 1966.
500. L. J. Fogel. *Artificial Intelligence through Simulated Evolution. Forty Years of Evolutionary Programming*. John Wiley & Sons, Inc., New York, 1999.
501. G. Folino, C. Pizzuti, and F. Spezzano. A Cellular Genetic Programming Approach to Classification. In W. Banzhaf, J. Daida, A. E. Eiben, M. H. Garzon, V. Honavar, M. Jakiela, and R. E. Smith, editors, *Proceedings of the Genetic and Evolutionary Computation Conference (GECCO'99)*, volume 2, pages 1015–1020, San Francisco, California, July 1999. Morgan Kaufmann.
502. G. Folino, C. Pizzuti, and G. Spezzano. CAGE: A Tool for Parallel Genetic Programming Applications. In J. Miller, M. Tomassini, P. L. Lanzi, C. Ryan, A. G. Tettamanzi, and W. B. Langdon, editors, *Genetic Programming. 4th European Conference, EuroGP 2001*, pages 64–73. Springer. Lecture Notes in Computer Science Vol. 2038, Lake Como, Italy, April 2001.
503. C. M. Fonseca. *Multiobjective Genetic Algorithms with Applications to Control Engineering Problems*. PhD thesis, Department of Automatic Control and Systems Engineering, University of Sheffield, Sheffield, UK, 1995.
504. C. M. Fonseca and P. J. Fleming. Genetic Algorithms for Multiobjective Optimization: Formulation, Discussion and Generalization. In S. Forrest, editor, *Proceedings of the Fifth International Conference on Genetic Algorithms*, pages 416–423, San Mateo, California, 1993. University of Illinois at Urbana-Champaign, Morgan Kaufmann Publishers.
505. C. M. Fonseca and P. J. Fleming. An overview of evolutionary algorithms in multiobjective optimization. Technical report, Department of Automatic Control and Systems Engineering, University of Sheffield, Sheffield, U. K., 1994.
506. C. M. Fonseca and P. J. Fleming. Multiobjective Genetic Algorithms Made Easy: Selection, Sharing, and Mating Restriction. In *Proceedings of the First International Conference on Genetic Algorithms in Engineering Systems: Innovations and Applications*, pages 42–52, Sheffield, UK, September 1995. IEE.
507. C. M. Fonseca and P. J. Fleming. An Overview of Evolutionary Algorithms in Multiobjective Optimization. *Evolutionary Computation*, 3(1):1–16, Spring 1995.
508. C. M. Fonseca and P. J. Fleming. Nonlinear System Identification with Multiobjective Genetic Algorithms. In *Proceedings of the 13th World Congress of the International Federation of Automatic Control*, pages 187–192, San Francisco, California, 1996. Pergamon Press.
509. C. M. Fonseca and P. J. Fleming. On the Performance Assessment and Comparison of Stochastic Multiobjective Optimizers. In H.-M. Voigt, W. Ebeling, I. Rechenberg, and H.-P. Schwefel, editors, *Parallel Problem Solving from Nature—PPSN IV*, pages 584–593. Springer-Verlag. Lecture Notes in Computer Science No. 1141, Berlin, Germany, September 1996.

510. C. M. Fonseca and P. J. Fleming. Multiobjective Optimization. In T. Bäck, D. B. Fogel, and Z. Michalewicz, editors, *Handbook of Evolutionary Computation*, volume 1, pages C4.5:1–C4.5:9. Institute of Physics Publishing and Oxford University Press, 1997.
511. C. M. Fonseca and P. J. Fleming. Multiobjective Optimization and Multiple Constraint Handling with Evolutionary Algorithms—Part I: A Unified Formulation. *IEEE Transactions on Systems, Man, and Cybernetics, Part A: Systems and Humans*, 28(1):26–37, 1998.
512. C. M. Fonseca and P. J. Fleming. Multiobjective Optimization and Multiple Constraint Handling with Evolutionary Algorithms—Part II: Application Example. *IEEE Transactions on Systems, Man, and Cybernetics, Part A: Systems and Humans*, 28(1):38–47, 1998.
513. C. M. Fonseca, V. Grunert da Fonseca, and L. Paquete. Exploring the Performance of Stochastic Multiobjective Optimisers with the Second-Order Attainment Function. In C. A. Coello Coello, A. Hernández Aguirre, and E. Zitzler, editors, *Evolutionary Multi-Criterion Optimization. Third International Conference, EMO 2005*, pages 250–264, Guanajuato, México, March 2005. Springer. Lecture Notes in Computer Science Vol. 3410.
514. C. M. Fonseca, L. Paquete, and M. López-Ibáñez. An Improved Dimension-Sweep Algorithm for the Hypervolume Indicator. In *2006 IEEE Congress on Evolutionary Computation (CEC'2006)*, pages 3973–3979, Vancouver, BC, Canada, July 2006. IEEE.
515. K. T. Formiga, F. H. Chaufhry, P. B. Cheung, and L. F. Reis. Optimal Design of Water Distribution System by Multiobjective Evolutionary Methods. In C. M. Fonseca, P. J. Fleming, E. Zitzler, K. Deb, and L. Thiele, editors, *Evolutionary Multi-Criterion Optimization. Second International Conference, EMO 2003*, pages 677–691, Faro, Portugal, April 2003. Springer. Lecture Notes in Computer Science. Volume 2632.
516. S. Forrest and A. S. Perelson. Genetic algorithms and the immune system. In H.-P. Schwefel and R. Männer, editors, *Parallel Problem Solving from Nature*, Lecture Notes in Computer Science, pages 320–325. Springer-Verlag, Berlin, Germany, 1991.
517. I. Foster, C. Kesselman, and S. Tuecke. The Anatomy of the Grid: Enabling Scalable Virtual Organization. *International Journal of Supercomputer Applications*, 15(3):200–222, 2001.
518. M. P. Fourman. Compaction of Symbolic Layout using Genetic Algorithms. In J. J. Grefenstette, editor, *Genetic Algorithms and their Applications: Proceedings of the First International Conference on Genetic Algorithms*, pages 141–153. Lawrence Erlbaum, Hillsdale, New Jersey, 1985.
519. E. Frank, Y. Wang, S. Inglis, G. Holmes, and I. H. Witten. Using model trees for classification. *Machine Learning*, 32(1):63–76, 1998.
520. B. Freisleben and P. Merz. New Genetic Local Search Algorithm for the Traveling Salesman Problem. In H.-M. Voigt, W. Ebeling, I. Rechenberg, and H.-P. Schwefel, editors, *Parallel Problem Solving from Nature—PPSN IV*, pages 890–899. Springer-Verlag. Lecture Notes in Computer Science No. 1141, Berlin, Germany, September 1996.
521. F. Freschi and M. Repetto. Multiobjective Optimization by a Modified Artificial Immune System Algorithm. In C. Jacob, M. L. Pilat, P. J. Bentley, and J. Timmis, editors, *Artificial Immune Systems. 4th International Conference,*

ICARIS 2005, pages 248–261, Banff, Canada, August 2005. Springer. Lecture Notes in Computer Science Vol. 3627.
522. F. Freschi and M. Repetto. VIS: an artificial immune network for multiobjective optimization. *Engineering Optimization*, 38(8):975–996, December 2006.
523. M. C. Fu. Optimization for Simulation: Theory vs. Practice. *INFORMS Journal on Computing*, 14:192–215, 2002.
524. K. Fujita, N. Hirokawa, S. Akagi, S. Kitamura, and H. Yokohata. Multiobjective optimal design of automotive engine using genetic algorithm. In *Proceedings of DETC'98 – ASME Design Engineering Technical Conferences*, page 11, 1998.
525. L. Gacôgne. Research of Pareto Set by Genetic Algorithm, Application to Multicriteria Optimization of Fuzzy Controller. In *5th European Congress on Intelligent Techniques and Soft Computing EUFIT'97*, pages 837–845, Aachen, Germany, September 1997.
526. L. Gacôgne. Multiple Objective Optimization of Fuzzy Rules for Obstacles Avoiding by an Evolution Algorithm with Adaptative Operators. In *Proceedings of the Fifth International Mendel Conference on Soft Computing (Mendel'99)*, pages 236–242, Brno, Czech Republic, June 1999.
527. C. Gagné, M. Gravel, and W. L. Price. Scheduling a single machine where setup times are sequence dependent using an ant-colony heuristic. In *Abstract Proceedings of ANTS'2000*, pages 157–160, Brussels, Belgium, September 2000.
528. C. Gagné, W. L. Price, and M. Gravel. Scheduling a Single Machine with Sequence Dependent Setup Time Using Ant Colony Optimization. Technical Report 2001-003, Faculté des Sciences de L'Administration, Université Laval, Québec, Canada, April 2001. Available at http://www.fsa.ulaval.ca/rd.
529. T. Galinho, A. Cardon, and J.-P. Vacher. Genetic Integration in a Multiagent System for Job-Shop Scheduling. In H. Coelho, editor, *Progress in Artificial Intelligence—IBERAMIA'98*, pages 76–87, Lisbon, Portugal, October 1998. Springer-Verlag.
530. L. M. Gambardella and M. Dorigo. Ant-Q: A Reinforcement Learning approach to the traveling salesman problem. In A. Prieditis and S. Russell, editors, *Proceedings of the 12th International Conference on Machine Learning*, pages 252–260. Morgan Kaufmann, 1995.
531. L. M. Gambardella, Éric Taillard, and G. Agazzi. MACS-VRPTW: A Multiple Ant Colony System for Vehicle Routing Problems with Time Windows. In D. Corne, M. Dorigo, and F. Glover, editors, *New Ideas in Optimization*, pages 63–76. McGraw-Hill, 1999.
532. X. Gandibleux and M. Ehrgott. 1984-2004 – 20 Years of Multiobjective Metaheuristics. But What About the Solution of Combinatorial Problems with Multiple Objectives? In C. A. Coello Coello, A. Hernández Aguirre, and E. Zitzler, editors, *Evolutionary Multi-Criterion Optimization. Third International Conference, EMO 2005*, pages 33–46, Guanajuato, México, March 2005. Springer. Lecture Notes in Computer Science Vol. 3410.
533. X. Gandibleux and A. Freville. Tabu Search Based Procedure for Solving the 0-1 Multi-Objective Knapsack Problem: The Two Objectives Case. *Journal of Heuristics*, 6(3):361–383, August 2000.
534. X. Gandibleux, N. Mezdaoui, and A. Fréville. A Tabu Search Procedure to Solve Combinatorial Optimisation Problems. In R. Caballero, F. Ruiz, and

R. E. Steuer, editors, *Advances in Multiple Objective and Goal Programming*, volume 455 of *Lecture Notes in Economics and Mathematical Systems*, pages 291–300. Springer-Verlag, 1997.
535. X. Gandibleux, H. Morita, and N. Katoh. The Supported Solutions Used as a Genetic Information in a Population Heuristic. In E. Zitzler, K. Deb, L. Thiele, C. A. Coello Coello, and D. Corne, editors, *First International Conference on Evolutionary Multi-Criterion Optimization*, pages 429–442. Springer-Verlag. Lecture Notes in Computer Science No. 1993, 2001.
536. C. García-Martínez, O. Cordón, and F. Herrera. An Empirical Analysis of Multiple Objective Ant Colony Optimization Algorithms for the Bi-criteria TSP. In M. Dorigo, M. Birattari, C. Blum, L. M. Gambardella, F. Mondada, and T. Stützle, editors, *Proceedings of the 4th International Workshop on Ant Colony Optimization and Swarm Intelligence*, pages 61–72. Springer. Lecture Notes in Computer Science Vol. 3172, 2004.
537. C. García-Martínez, O. Cordón, and F. Herrera. A taxonomy and an empirical analysis of multiple objective ant colony optimization algorithms for the bi-criteria TSP. *European Journal of Operational Research*, 180:116–148, 2007.
538. L. R. Gardiner and R. E. Steuer. Unified Interactive Multiple Objective Programming. *European Journal of Operational Research*, 74:391–406, 1994.
539. L. R. Gardiner and R. E. Steuer. Unified Interactive Multiple Objective Programming: An Open Architecture For Accommodating New Procedures. *Journal of the Operational Research Society*, 45(12):1456–1466, 1994.
540. J. Garen. A Genetic Algorithm for Tackling Multiobjective Job-Shop Scheduling Problems. In X. Gandibleux, M. Sevaux, K. Sörensen, and V. T'kindt, editors, *Metaheuristics for Multiobjective Optimisation*, pages 201–219, Berlin, 2004. Springer. Lecture Notes in Economics and Mathematical Systems Vol. 535.
541. M. Garey and D. Johnson. *Computers and Intractability: A Guide to the Theory of NP-Completeness*. Freeman, 1979.
542. C. A. Garrett, J. Huang, M. N. Goltz, and G. B. Lamont. Parallel Real-Valued Genetic Algorithms for Bioremediation Optimization of TCE-Contaminated Groundwater. In *1999 Congress on Evolutionary Computation*, pages 2183–2189, Washington, D.C., July 1999. IEEE Service Center.
543. A. Gaspar-Cunha and J. Covas. A Real-World Test Problem for EMO Algorithms. In C. M. Fonseca, P. J. Fleming, E. Zitzler, K. Deb, and L. Thiele, editors, *Evolutionary Multi-Criterion Optimization. Second International Conference, EMO 2003*, pages 752–766, Faro, Portugal, April 2003. Springer. Lecture Notes in Computer Science. Volume 2632.
544. A. Gaspar Cunha, P. Oliveira, and J. A. Covas. Use of Genetic Algorithms in Multicriteria Optimization to Solve Industrial Problems. In T. Bäck, editor, *Proceedings of the Seventh International Conference on Genetic Algorithms*, pages 682–688, San Mateo, California, July 1997. Michigan State University, Morgan Kaufmann Publishers.
545. A. Gaspar Cunha, P. Oliveira, and J. A. Covas. Genetic Algorithms in Multiobjective Optimization Problems: An Application to Polymer Extrusion. In A. S. Wu, editor, *Proceedings of the 1999 Genetic and Evolutionary Computation Conference. Workshop Program*, pages 129–130, Orlando, Florida, July 1999.
546. S. Gass and T. L. Saaty. The computational algorithm for the parametric objective function. *Naval Research Logistics Quarterly*, 2:39–45, 1955.

547. G. H. Gates. Predicting Protein Structure Using Parallel Genetic Algorithms. Master's thesis, Air Force Institute of Technology, Wright Patterson AFB, March 1994. AFIT/GCS/ENG/94D-03.
548. R. Ge and Y. Qin. A class of filled function for finding global minimizers of a function of several variables. *Journal of Optimization Theory and Applications*, 54:241–251, 1987.
549. A. Geist, A. Beguelin, J. Dongarra, W. Jiang, R. Manchek, and V. Sunderam. *PVM: Parallel Virtual Machine: A Users' Guide and Tutorial for Networked Parallel Computing*. MIT Press, Cambridge, Massachusetts, 1994.
550. F. W. Gembicki. *Vector Optimization for Control with Performance and Parameter Sensitivity Indices*. PhD thesis, Case Western Reserve University, Cleveland, Ohio, 1974.
551. F. W. Gembicki and Y. Y. Haimes. Approach to performance and sensitivity multiobjective optimization: the goal attainment method. *IEEE Transactions on Automatic Control*, AC-15:591–593, 1975.
552. M. Gen and R. Cheng. *Genetic Algorithms and Engineering Optimization*. Wiley Series in Engineering Design and Automation. John Wiley & Sons, New York, 2000.
553. M. Gen, K. Ida, and J. Kim. A Spanning Tree-Based Genetic Algorithm for Bicriteria Topological Network Design. In *Proceedings of the 5th IEEE Conference on Evolutionary Computation*, pages 15–20, Piscataway, New Jersey, 1998. IEEE Press.
554. M. Gen, K. Ida, and Y. Li. Solving bicriteria solid transportation problem with fuzzy numbers by genetic algorithm. *International Journal of Computers and Industrial Engineering*, 29:537–543, 1995.
555. M. Gen and Y.-Z. Li. Solving Multi-Objective Transportation Problems by Spanning Tree-based Genetic Algorithm. In I. Parmee, editor, *The Integration of Evolutionary and Adaptive Computing Technologies with Product/System Design and Realisation*, pages 95–108, Plymouth, United Kingdom, April 1998. Plymouth Engineering Design Centre, Springer-Verlag.
556. A. M. Geoffrion, J. S. Dyer, and A. Feinberg. An interactive approach for multi-criterion optimization, with an application to the operation of an academic department. *Management Science*, 19(4):357–368, 1972.
557. J. S. Gero and S. J. Louis. Improving Pareto Optimal Designs Using Genetic Algorithms. *Microcomputers in Civil Engineering*, 10(4):241–249, 1995.
558. J. S. Gero, S. J. Louis, and S. Kundu. Evolutionary learning of novel grammars for design improvement. *Artificial Intelligence for Engineering Design, Analysis and Manufacturing*, 8:83–94, 1994.
559. M. Gershon, L. Duckstein, and A. Bardossy. Differential Dynamic Programming Application to Multi-objective decision making. In *Proceedings of the CORS/ORSA/TIMS Joint Meeting*, Toronto, Canada, 1981.
560. O. Giel. Expected Runtimes of a Simple Multi-objective Evolutionary Algorithm. In *Proceedings of the 2003 Congress on Evolutionary Computation (CEC'2003)*, volume 3, pages 1918–1925, Canberra, Australia, December 2003. IEEE Press.
561. O. Giel and P. K. Lehre. On the Effect of Populations in evolutionary multi-objective optimization. In M. K. et al., editor, *2006 Genetic and Evolutionary Computation Conference (GECCO'2006)*, volume 1, pages 651–658, Seattle, Washington, USA, July 2006. ACM Press. ISBN 1-59593-186-4.

562. C. Gil, R. B. nos, M. Montoya, and J. Gómez. Performance of Simulated Annealing, Tabu Search, and Evolutionary Algorithms for Multi-objective Network Partitioning. *Algorithmic Operations Research*, 1(1):55–64, 2006.
563. M. Gill, Y. Kaheil, A. Khalil, M. Mckee, and L. Bastidas. Multiobjective particle swarm optimization for parameter estimation in hydrology. *Water Resources Research*, 42(7, Art. No. W07417), July 22 2006.
564. A. P. Giotis and K. C. Giannakoglou. Single- and Multi-Objective Airfoil Design Using Genetic Algorithms and Artificial Intelligence. In K. Miettinen, M. M. Mäkelä, P. Neittaanmäki, and J. Periaux, editors, *Proceedings of EU-ROGEN'99*, Jyväskyl, Finland, 30 May-6 June 1999. University of Jyváskylä.
565. F. Glover. Heuristics for integer programming using surrogate constraints. *Decision Sciences*, 8:156–166, 1977.
566. F. Glover. Future Paths for Integer Programming and Links to Artificial Intelligence. *Computers and Operations Research*, 13(5):533–549, 1986.
567. F. Glover. Tabu Search—Part I. *ORSA Journal on Computing*, 1(3):190–206, Summer 1989.
568. F. Glover. A user's guide to tabu search. *Annals of Operations Research*, 41:3–28, 1993.
569. F. Glover. Tabu search for nonlinear and parametric optimization (with links to genetic algorithms). *Discrete Applied Mathematics*, 49:231–255, 1994.
570. F. Glover. A Template for Scatter Search and Path Relinking. In J.-K. Hao, E. Lutton, E. Ronald, M. Schoenauer, and D. Snyers, editors, *Artificial Evolution. Third European Conference, AE'97*, pages 3–51, Nîmes, France, October 1997. Springer-Verlag. Lecture Notes in Computer Science Vol. 1363.
571. F. Glover, J. Kelly, and M. Laguna. Genetic Algorithms and Tabu Seach: Hybrid for Optimization. *Computers and Operations Research*, 22(1):111–134, 1995.
572. F. Glover and M. Laguna. *Tabu Search*. Kluwer Academic Publishers, Boston, Massachusetts, 1997.
573. F. Glover and M. Laguna. Fundamentals of scatter search and path relinking. *Control and Cybernetics*, 29:653–684, 1999.
574. T. Goel and K. Deb. Hybrid Methods for Multi-Objective Evolutionary Algorithms. In L. Wang, K. C. Tan, T. Furuhashi, J.-H. Kim, and X. Yao, editors, *Proceedings of the 4th Asia-Pacific Conference on Simulated Evolution and Learning (SEAL'02)*, volume 1, pages 188–192, Orchid Country Club, Singapore, November 2002. Nanyang Technical University.
575. C. K. Goh and K. C. Tan. Noise Handling in Evolutionary Multi-Objective Optimization. In *2006 IEEE Congress on Evolutionary Computation (CEC'2006)*, pages 4497–4504, Vancouver, BC, Canada, July 2006. IEEE.
576. A. Goicoechea. *A multi-objective stochastic programming model in watershed management*. Dept. of systems and industrial engineering, University of Arizona, Tucson, Arizona, 1977. (Unpublished).
577. A. Goicoechea, L. Duckstein, and M. Fogel. Multi-objective programming in watershed management: A study of the Charleston watershed. *Water Resources Research*, 12(6):1085–1092, December 1976.
578. A. Goicoechea, L. Duckstein, and M. Fogel. Multiple objectives under uncertainty: An illustrative application of PROTRADE. *Water Resources Research*, 15(2):203–210, April 1979.

579. A. Goicoechea, D. R. Hansen, and L. Duckstein. *Multiobjective Analysis with Engineering and Business Applications*. John Wiley and Sons, New York, 1982.
580. E. F. Goldbarg, G. R. de Souza, and M. C. Goldbarg. Particle Swarm Optimization for the Bi-objective Degree-constrained Minimum Spanning Tree. In *2006 IEEE Congress on Evolutionary Computation (CEC'2006)*, pages 1527–1534, Vancouver, BC, Canada, July 2006. IEEE.
581. D. E. Goldberg. *Genetic Algorithms in Search, Optimization and Machine Learning*. Addison-Wesley Publishing Company, Reading, Massachusetts, 1989.
582. D. E. Goldberg. From Genetic and Evolutionary Optimization to the Design of Conceptual Machines. Technical Report IlliGAl Report 98008, Department of General Engineering, University of Illinois at Urbana-Champaign, Urbana, Illinois, USA, 1998.
583. D. E. Goldberg and K. Deb. A Comparison of Selection Schemes Used in Genetic Algorithms. In G. J. E. Rawlins, editor, *Foundations of Genetic Algorithms*, pages 69–93. Morgan Kaufmann, San Mateo, California, 1991.
584. D. E. Goldberg, K. Deb, H. Kargupta, and G. Harik. Rapid, Accurate Optimization of Difficult Problems Using Fast Messy Genetic Algorithms. In S. Forrest, editor, *Proceedings of the Fifth International Conference on Genetic Algorithms*, pages 56–64, San Mateo, CA, July 1993. Morgan Kaufmann Publishers.
585. D. E. Goldberg, B. Korb, and K. Deb. Messy Genetic Algorithms: Motivation, Analysis, and First Results. *Complex Systems*, 3:493–530, 1989.
586. D. E. Goldberg and R. Lingle. Alleles, Loci, and the Traveling Salesman Problem. In J. J. Grefenstette, editor, *Proceedings of the First International Conference on Genetic Algorithms and Their Applications*, pages 154–159. Lawrence Erlbaum Associates, Hillsdale, New Jersey, 1985.
587. D. E. Goldberg and J. Richardson. Genetic algorithm with sharing for multimodal function optimization. In J. J. Grefenstette, editor, *Genetic Algorithms and Their Applications: Proceedings of the Second International Conference on Genetic Algorithms*, pages 41–49, Hillsdale, New Jersey, 1987. Lawrence Erlbaum.
588. D. E. Goldberg and L. Wang. Adaptive niching via coevolutionary sharing. In D. Quagliarella, J. Périaux, C. Poloni, and G. Winter, editors, *Genetic Algorithms and Evolution Strategies in Engineering and Computer Science. Recent Advances and Industrial Applications*, pages 21–38. John Wiley and Sons, West Sussex, England, 1998.
589. Y. Goletsis, C. Papaloukas, D. I. Fotiadis, A. Likas, and L. K. Michalis. Automated Ischemic Beat Classification Using Genetic Algorithms and Multicriteria Decision Analysis. *IEEE Transactions on Biomedical Engineering*, 51(10):1717–1725, October 2004.
590. I. Golovkin, R. Mancini, S. Louis, Y. Ochi, K. Fujita, H. Nishimura, H. Shirga, N. Miyanaga, H. Azechi, R. Butzbach, I. Uschmann, E. Förster, J. Delettrez, J. Koch, R. Lee, and L. Klein. Spectroscopic Determination of Dynamic Plasma Gradients in Implosion Cores. *Physical Review Letters*, 88(4), January 2002.
591. I. E. Golovkin, R. C. Mancini, S. J. Louis, R. W. Lee, and L. Klein. Multicriteria Search and Optimization: an Application to X-ray Plasma Spec-

troscopy. In *2000 Congress on Evolutionary Computation*, volume 2, pages 1521–1527, Piscataway, New Jersey, July 2000. IEEE Service Center.
592. C. Gomes da Silva, J. Clímaco, and J. Figueira. A scatter search method for the bi-criteria multi-dimensional {0,1}-knapsack problem using surrogate relaxation. *Journal of Mathematical Modelling and Algorithms*, 3(3):183–208, 2004.
593. C. Gomes da Silva, J. Clímaco, and J. Figueira. A scatter search method for bi-criteria {0,1}-knapsack problems. *European Journal of Operational Research*, 169(2):373–391, March 2006.
594. O. Gonzalez, C. Leon, G. Miranda, C. Rodriguez, and C. Segura. A Parallel Skeleton for the Strength Pareto Multiobjective Evolutionary Algorithm 2. In *Proceedings of 15th EUROMICRO International Conference on Parallel, Distributed and Network-Based Processing (PDP'07)*, pages 434–441. IEEE Computer Society, February 2007.
595. M. Gravel, W. L. Price, and C. Gagné. Scheduling Continuous Casting of Aluminum Using a Multiple-Objective Ant Colony Optimization Metaheuristic. Technical Report 2001–004, Faculté des Sciences de L'Administration, Université Laval, Québec, Canada, April 2001. Available at http://www.fsa.ulaval.ca/rd.
596. M. Gravel, W. L. Price, and C. Gagné. Scheduling continuous casting of aluminum using a multiple objective ant colony optimization metaheuristic. *European Journal of Operational Research*, 143(1):218–229, November 2002.
597. S. Greco, B. Matarazzo, and R. Slowinski. Rough sets theory for multicriteria decision analysis. *European Journal of Operational Research*, 129:1–47, 2001.
598. G. W. Greenwood, X. S. Hu, and J. G. D'Ambrosio. Fitness Functions for Multiple Objective Optimization Problems: Combining Preferences with Pareto Rankings. In R. K. Belew and M. D. Vose, editors, *Foundations of Genetic Algorithms 4*, pages 437–455. Morgan Kaufmann, San Mateo, California, 1997.
599. J. J. Grefenstette. Incorporating problem specific knowledge into genetic algorithms. In L. Davis, editor, *Genetic Algorithms and Simulated Annealing*, pages 49–67. Pitman Press, New York, 1987.
600. D. Greiner, G. Winter, J. M. Emperador, and B. Galván. Gray Coding in Evolutionary Multicriteria Optimization: Application in Frame Structural Optimum Design. In C. A. Coello Coello, A. Hernández Aguirre, and E. Zitzler, editors, *Evolutionary Multi-Criterion Optimization. Third International Conference, EMO 2005*, pages 576–591, Guanajuato, México, March 2005. Springer. Lecture Notes in Computer Science Vol. 3410.
601. P. Grignon and G. M. Fadel. Configuration design optimization method. In *Proceedings of DETC'99 – ASME Design Engineering Technical Conferences*, Las Vegas, Nevada, September 1999.
602. P. Grignon and G. M. Fadel. Multiobjective optimization by iterative genetic algorithm. In *Proceedings of DETC'99 – ASME Design Engineering Technical Conferences*, Las Vegas, Nevada, September 1999.
603. P. Grignon, J. Wodziack, and G. M. Fadel. Bi-Objective optimization of components packing using a genetic algorithm. In *NASA/AIAA/ISSMO Multidisciplinary Design and Optimization Conference*, pages 352–362, Seattle, Washington, September 1996. AIAA-96-4022-CP.

604. C. Grimme and J. Lepping. Designing Multi-objective Variation Operators Using a Predator-Prey Approach. In S. Obayashi, K. Deb, C. Poloni, T. Hiroyasu, and T. Murata, editors, *Evolutionary Multi-Criterion Optimization, 4th International Conference, EMO 2007*, pages 21–35, Matshushima, Japan, March 2007. Springer. Lecture Notes in Computer Science Vol. 4403.
605. C. Grimme and K. Schmitt. Inside a Predator-Prey Model for Multi-Objective Optimization: A Second Study. In M. K. et al., editor, *2006 Genetic and Evolutionary Computation Conference (GECCO'2006)*, volume 1, pages 707–714, Seattle, Washington, USA, July 2006. ACM Press. ISBN 1-59593-186-4.
606. L. Gritz and J. K. Hahn. Genetic Programming for Articulated Figure Motion. *Journal of Visualization and Computer Animation*, 6:129–142, 1995.
607. R. Groppetti and R. Muscia. On a Genetic Multiobjective Approach for the Integration and Optimization of Assembly Product Design and Process Planning. In P. Chedmail, J. C. Bocquet, and D. Dornfeld, editors, *Integrated Design and Manufacturing in Mechanical Engineering*, pages 61–70. Kluwer Academic Publishers, The Netherlands, 1997.
608. T. Grueninger and D. Wallace. Multi-modal optimization using genetic algorithms. Technical Report 96.02, CADlab, Massachusetts Institute of Technology, Cambridge, Massachusetts, USA, 1996.
609. V. Grunert da Fonseca, C. M. Fonseca, and A. O. Hall. Inferential Performance Assessment of Stochastic Optimisers and the Attainment Function. In E. Zitzler, K. Deb, L. Thiele, C. A. Coello Coello, and D. Corne, editors, *First International Conference on Evolutionary Multi-Criterion Optimization*, pages 213–225. Springer-Verlag. Lecture Notes in Computer Science No. 1993, 2001.
610. S.-U. Guan and S. Li. Incremental Learning with Respect to New Incoming Input Attributes. *Neural Processing Letters*, 14(3):241–260, December 2001.
611. S.-U. Guan and J. Liu. Incremental Ordered Neural Network Training. *Journal of Intelligent Systems*, 12(3):137–172, 2002.
612. M. Guillen. *Bridges to Infinity*. Jeremy P. Tarcher, Inc., Los Angeles, 1983.
613. A. Guimarães Pereira. Generating Alternative Routes using Genetic Algorithms and Multi-Criteria Analysis Techniques. In R. Wyatt and H. Hossain, editors, *Fourth International Conference on Computers in Urban Planning and Urban Management)*, pages 547–560, Melbourne, Australia, July 11–14 1995.
614. A. Guimarães Pereira. Generating alternative routes by multicriteria evaluation and a genetic algorithms. *Environment and Planning B: Planning and Design*, 23:711–720, 1996.
615. A. Guimarães Pereira, G. Munda, and M. Pariccini. Generating alternatives for siting retail and service facilities using genetic algorithms and multiple criteria devision techniques. *Journal of Retailing and Consumer Services*, 1(2):40–47, 1994.
616. A. Guimarães Pereira, R. J. Peckham, and M. P. Antunus. GENET: A Method to Generate Alternatives for Facilities Siting using Genetic Algorithms. In J. Harts, H. F. L. Ottens, and H. J. Scholten, editors, *Fourth European Conference and Exhibition on Geographical Information Systems (EGIS'93)*, pages 973–981, Genoa, Italy, March 29–April 1 1993.
617. A. C. M. Guimarães Pereira. *Extending Environmental Impact Assessment Processes: Generation of Alternatives for Siting and Routing Infrastructural*

Facilities by Multi-Criteria Evaluation and Genetic Algorithms. PhD thesis, New University of Lisbon, Lisbon, Portugal, 1997.
618. F. G. Guimarães, F. Campelo, R. R. Saldanha, H. Igarashi, R. H. Takahashi, and J. A. Ramírez. A Multiobjective Proposal for the TEAM Benchmark Problem 22. *IEEE Transactions on Magnetics*, 42(4):1471–1474, April 2006.
619. S. Gunawan and S. Azarm. Multi-objective robust optimization using a sensitivity region concept. *Structural and Multidisciplinary Optimization*, 29(1):50–60, January 2005.
620. M. Guntsch and M. Middendorf. A Population Based Approach for ACO. In *Applications of Evolutionary Computing. EvoWorkshops 2002: EvoCOP, EvoIASP, EvoSTIM/EvoPLAN*, pages 72–81, Kinsale, Ireland, April 2002. Springer. Lecture Notes in Computer Science Vol. 2279.
621. M. Guntsch and M. Middendorf. Solving Multi-criteria Optimization Problems with Population-Based ACO. In C. M. Fonseca, P. J. Fleming, E. Zitzler, K. Deb, and L. Thiele, editors, *Evolutionary Multi-Criterion Optimization. Second International Conference, EMO 2003*, pages 464–478, Faro, Portugal, April 2003. Springer. Lecture Notes in Computer Science. Volume 2632.
622. H. V. Gupta, S. Sorroshian, and P. O. Yapo. Towards Improved Calibration of Hydrologic Models: Multiple and Non-Commensurable Measures of Information. *Water Resources Research*, 34(4):751–763, 1998.
623. W. J. Gutjahr. Two Metaheuristics for Multiobjective Stochastic Combinatorial Optimization. In O. Lupanov, O. Kasim-Zade, A. Chaskin, and K. Steinhoefl, editors, *Stochastic Algorithms: Foundations and Applications, SAGA 2005, Proceedings*, pages 116–125, Moscow, Russia, October 2005. Springer. Lecture Notes in Computer Science Vol. 3777.
624. C. R. Haag. An Artificial Immune System-inspired Multiobjective Evolutionary Algorithm with Application to the Detection of Distributed Computer Network Intrusions. Master's thesis, Department of Electrical and Computer Engineering, Graduate School of Engineering and Management, Air Force Institute of Technology (AFIT), WPAFB, Dayton, Ohio, USA, March 2007.
625. P. Haastrup and A. Guimarães Pereira. Exploring the Use of Multi-Objective Genetic Algorithms for Reducing Traffic Generated Urban Air and Noise Pollution. In *Proceedings of the 5th European Congress on Intelligent and Soft Computing*, pages 819–825, Aachen, Germany, September 1997.
626. W. Habenicht. Quad trees: A data structure for discrete vector optimization problems. In *Lecture Notes in Economics and Mathematical Systems No. 209*, pages 136–145, 1982.
627. Y. Haimes, K. Loparo, S. C. Olenik, and S. Nanda. Multi-objective statistical method for interior drainage systems. *Water Resources Research*, 16(3):465–475, 1980.
628. Y. Y. Haimes. *Hierarchical Analysis of Water Resource Systems: Modeling and Optimization of Large-Scale Systems*. McGraw-Hill International Book Co., New York, 1977.
629. Y. Y. Haimes, P. Das, and K. Sung. Multi-objective Analysis in the Maumee River Basin: A Case Study on Level B Planning. Technical Report SED-WRG-77-1, Case Western Reserve University, Cleveland, Ohio, 1977.
630. Y. Y. Haimes, W. Hall, and H. Freedman. *Multi-Objective Optimization in Water Resources Systems: The Surrogate Trade-Off Method*. Elsevier, Amsterdam, 1975.

674   References

631. Y. Y. Haimes and W. A. Hall. Multiobjectives in water resources systems analysis: The surrogate trade-off method. *Water Resources Research*, 10(4):615–624, aug 1974.
632. Y. Y. Haimes, L. S. Lasdon, and D. A. Wismer. On a Bicriterion Formulation of the Problems of Integrated System Identification and System Optimization. *IEEE Transactions on Systems, Man, and Cybernetics*, 1(3):296–297, July 1971.
633. B. Hajek. Cooling schedules for optimal annealing. *Mathematics of Operations Research*, 13(2):311–329, 1988.
634. P. Hajela and J. Lee. Constrained Genetic Search via Schema Adaptation. An Immune Network Solution. In N. Olhoff and G. I. N. Rozvany, editors, *Proceedings of the First World Congress of Stuctural and Multidisciplinary Optimization*, pages 915–920, Goslar, Germany, 1995. Pergamon.
635. P. Hajela and J. Lee. Constrained Genetic Search via Schema Adaptation. An Immune Network Solution. *Structural Optimization*, 12:11–15, 1996.
636. P. Hajela and C. Y. Lin. Genetic search strategies in multicriterion optimal design. *Structural Optimization*, 4:99–107, 1992.
637. D. Halhal, G. A. Walters, D. Ouazar, and D. A. Savic. Multi-objective improvement of water distribution systems using a structured messy genetic algorithm approach. *Journal of Water Resources Planning and Management ASCE*, 123(3):137–146, 1997.
638. N. Hallam, P. Blanchfield, and G. Kendall. Handling Diversity in Evolutionary Multiobjective Optimisation. In *2005 IEEE Congress on Evolutionary Computation (CEC'2005)*, volume 3, pages 2233–2240, Edinburgh, Scotland, September 2005. IEEE Service Center.
639. N. Hallan, G. Kendall, and P. Blanchfield. Solving Multi-objective Optimisation Problems Using the Potential Pareto Regions Evolutionary Algorithm. In T. P. Runarsson, H.-G. Beyer, E. Burke, J. J. Merelo-Guervós, L. D. Whitley, and X. Yao, editors, *Parallel Problem Solving from Nature - PPSN IX, 9th International Conference*, pages 503–512. Springer. Lecture Notes in Computer Science Vol. 4193, Reykjavik, Iceland, September 2006.
640. M. Hampsey. *Multiobjective Evolutionary Optimisation of Small Wind Turbine Blades*. PhD thesis, Department of Mechanical Engineering, University of Newcastle, Australia, August 2002.
641. Z. X. Han, L. Xu, R. Wei, B. P. Wang, and T. Reinikainen. Reliability-Based Design Optimization for Land Grid Array Solder Joints Under Thermo-Mechanical Load. In *Proceedings of the 5th International Conference on Thermal and Mechanical Simulation and Experiments in Microelectronics and Microsystems (EuroSimE 2004)*, pages 219–224. IEEE, May 2004.
642. T. Hanne. Concepts of a learning objected-oriented problem solver LOOPS. In *Proceedings of the 12th International Conference on Multiple Criteria Decision Making*, pages 330–339. Springer-Verlag, 1995.
643. T. Hanne. On the convergence of multiobjective evolutionary algorithms. *European Journal of Operational Research*, 117(3):553–564, 1999.
644. T. Hanne. Global Multiobjective Optimization with Evolutionary Algorithms: Selection Mechanisms and Mutation Control. In E. Zitzler, K. Deb, L. Thiele, C. A. Coello Coello, and D. Corne, editors, *First International Conference on Evolutionary Multi-Criterion Optimization*, pages 197–212. Springer-Verlag. Lecture Notes in Computer Science No. 1993, 2001.

645. T. Hanne. *Intelligent Strategies for Meta Multiple Criteria Decision Making*. Kluwer Academic Publishers, Boston, 2001.
646. T. Hanne and S. Nickel. A multiobjective evolutionary algorithm for scheduling and inspection planning in software development projects. *European Journal of Operational Research*, 167(3):663–678, December 2005.
647. M. Hansen. Tabu search for multiobjective combinatorial optimization: TAMOCO. *Control and Cybernetics*, 29(3):799–818, 2000.
648. M. P. Hansen. Generating a Diversity of Good Solutions to a Practical Combinatorial Problem using Vectorized Simulated Annealing. Technical report, Institute of Mathematical Modelling, Technical University of Denmark, August 1997. Working Paper.
649. M. P. Hansen. Solving multiobjective knapsack problems using MOTS. In *2nd Metaheuristics International Conference (MIC'97)*, Sophia Antopolis, France, July 1997.
650. M. P. Hansen. Tabu Search in Multiobjective Optimisation : MOTS. In *Proceedings of the 13th International Conference on Multiple Criteria Decision Making (MCDM'97)*, Cape Town, South Africa, January 1997.
651. M. P. Hansen. *Metaheuristics for multiple objective combinatorial optimization*. PhD thesis, Institute of Mathematical Modelling, Technical University of Denmark, Lyngby, Denmark, March 1998.
652. M. P. Hansen. Use of Substitute Scalarizing Functions to Guide a Local Search Based Heuristic: The Case of moTSP. *Journal of Heuristics*, 6:419–431, 2000.
653. M. P. Hansen and A. Jaszkiewicz. Evaluating the quality of approximations to the non-dominated set. Technical Report IMM-REP-1998-7, Technical University of Denmark, March 1998.
654. N. Hansen and A. Ostermeier. Completely Derandomized Self-adaptation in Evolution Strategies. *Evolutionary Computation*, 9(2):159–195, Summer 2001.
655. P. Hansen. The steepest ascent mildest descent heuristic for combinatorial programming. In *Congress on Numerical Methods in Combinatorial Optimization*, Capri, Italy, 1986.
656. M. Hapke, A. Jaszkiewicz, and R. Slowinski. Pareto Simulated Annealing for Fuzzy Multi-Objective Combinatorial Optimization. *Journal of Heuristics*, 6(3):329–345, August 2000.
657. K. Harada, K. Ikeda, and S. Kobayashi. Hybridizing of Genetic Algorithm and Local Search in Multiobjective Function Optimization: Recommendation of GA then LS. In M. K. et al., editor, *2006 Genetic and Evolutionary Computation Conference (GECCO'2006)*, volume 1, pages 667–674, Seattle, Washington, USA, July 2006. ACM Press. ISBN 1-59593-186-4.
658. K. Harada, J. Sakuma, and S. Kobayashi. Local Search for Multiobjective Function Optimization: Pareto Descent Method. In M. K. et al., editor, *2006 Genetic and Evolutionary Computation Conference (GECCO'2006)*, volume 1, pages 659–666, Seattle, Washington, USA, July 2006. ACM Press. ISBN 1-59593-186-4.
659. S. P. Harris and E. C. Ifeachor. Nonlinear FIR Filter Design by Genetic Algorithm. In *1st Online Conference on Soft Computing*, August 1996.
660. W. E. Hart. *Adaptive Global Optimization with Local Search*. PhD thesis, The University of California, San Diego, San Diego, California, USA, 1994.

661. W. E. Hart, N. Krasnogor, and J. Smith, editors. *Recent Advances in Memetic Algorithms*. Studies in Fuzziness and Soft Computing. Springer, Germany, 2005. ISBN 3-540-22904-3.
662. J. W. Hartmann. Low-thrust Trajectory Optimization Using Stochastic Optimization Methods. Master's thesis, Department of Aeronautical and Astronautical Engineering, University of Illinois at Urbana-Champaign, January 1999.
663. J. W. Hartmann, V. L. Coverstone-Carroll, and S. N. Williams. Optimal Interplanetary Spacecraft Trajectories via a Pareto Genetic Algorithm. *The Journal of the Astronautical Sciences*, 46(3):267–282, July–September 1998.
664. J. W. Hartmann, V. L. Coverstone-Carroll, and S. N. Williams. Optimal Interplanetary Spacecraft Trajectories Via A Pareto Genetic Algorithm. In *AAS/AIAA Space Flight Mechanics Meeting*, Monterey, California, February 1998. Paper No. AAS-98-202.
665. M. Hasenjäger and B. Sendhoff. Crawling Along the Pareto Front: Tales From the Practice. In *2005 IEEE Congress on Evolutionary Computation (CEC'2005)*, volume 1, pages 174–181, Edinburgh, Scotland, September 2005. IEEE Service Center.
666. M. Hasenjäger, B. Sendhoff, T. Sonoda, and T. Arima. Three Dimensional Evolutionary Aerodynamic Design Optimization with CMA-ES. In H.-G. B. et al., editor, *2005 Genetic and Evolutionary Computation Conference (GECCO'2005)*, volume 2, pages 2173–2180, New York, USA, June 2005. ACM Press.
667. F. Hausdorff. Investigations Concerning Order Types. In *Berichte über die Verhandlungen der Königlich Sächsischen Gesellschaft der Wissenschaften zu Leipzig, Mathematisch-Physische Klasse*, volume 58, pages 106–169, 1906.
668. G. A. Hazelrigg. The implications of Arrow's impossibility theorem on approaches to optimal engineering design. *Journal of Mechanical Design*, 118:161–164, June 1996.
669. A. Hernández Aguirre and S. Botello Rionda. Evolutionary Multi-Objective Optimization of Trusses. In C. A. Coello Coello and G. B. Lamont, editors, *Applications of Multi-Objective Evolutionary Algorithms*, pages 201–226. World Scientific, Singapore, 2004.
670. A. Hernández Aguirre, S. Botello Rionda, C. A. Coello Coello, G. Lizárraga Lizárraga, and E. Mezura Montes. Handling Constraints using Multiobjective Optimization Concepts. *International Journal for Numerical Methods in Engineering*, 59(15):1989–2017, April 2004.
671. A. G. Hernández-Díaz, L. V. Santana-Quintero, C. Coello Coello, R. Caballero, and J. Molina. A New Proposal for Multi-Objective Optimization using Differential Evolution and Rough Sets Theory. In M. K. et al., editor, *2006 Genetic and Evolutionary Computation Conference (GECCO'2006)*, volume 1, pages 675–682, Seattle, Washington, USA, July 2006. ACM Press. ISBN 1-59593-186-4.
672. A. G. Hernández-Díaz, L. V. Santana-Quintero, C. A. Coello Coello, and J. Molina. Pareto adaptive - ϵ-dominance. Technical Report EVOCINV-02-2006, Evolutionary Computation Group at CINVESTAV, México, March 2006.
673. A. Herreros, E. Baeyens, and J. R. Perán. Design of Multiobjective Robust Controllers Using Genetic Algorithms. In A. S. Wu, editor, *Proceedings of the*

1999 Genetic and Evolutionary Computation Conference. Workshop Program, pages 131–132, Orlando, Florida, July 1999.
674. A. Herreros, E. Baeyens, and J. R. Perán. MRCD: A Genetic Algorithm for Multiobjective Robust Control Design. *Engineering Applications of Artificial Intelligence*, 15(3–4):285–301, June-August 2002.
675. A. Herreros López. *Diseño de Controladores Robustos Multiobjetivo por Medio de Algoritmos Genéticos*. PhD thesis, Departamento de Ingeniería de Sistemas y Automática, Universidad de Valladolid, Valladolid, España, Septiembre 2000. (In Spanish).
676. A. Hertz, B. Jaumard, C. Ribeiro, and W. F. Filho. A multi-criteria tabu search approach to cell formation problems in group technology with multiple objectives. *RAIRO/Operations Research*, 28(3):303–328, 1994.
677. G. Hertz and G. Stormo. Identifying DNA and protein patterns with statistically significant alignments of multiple sequences. *Bioinformatics*, 15(7/8):563–577, 1999.
678. M. L. Hetland and P. Saetrom. Evolutionary Rules Mining in Time Series Databases. *Machine Learning*, 58(2–3):107–125, February–March 2005.
679. T. Higashishara and M. Atsumi. Evolutionary Acquisition of Sensory—Action Network of Mobile Robot using Multiobjective Genetic Algorithm. *IPSJ SIG-ICS*, 98-ICS-111:1–6, 1998. (In Japanese).
680. M. R. Hilliard, G. E. Liepins, M. Palmer, and G. Rangarajen. The computer as a partner in algorithmic design: Automated discovery of parameters for a multiobjective scheduling heuristic. In R. Sharda, B. L. Golden, E. Wasil, O. Balci, and W. Stewart, editors, *Impacts of Recent Computer Advances on Operations Research*, pages 321–331. North-Holland Publishing Company, New York, 1989.
681. F. S. Hillier and G. J. Lieberman. *Introduction to Operations Research*. Holden-Day, Inc., San Francisco, 1967.
682. W. D. Hillis. Co-evolving pasasites improve simulated evolution as an optimization procedure. *Physica D*, 42(1–3):228–234, 1990.
683. M. Hinchliffe, M. Willis, and M. Tham. Chemical Process Systems Modelling using Multi-Objective Genetic Programming. In J. R. Koza, W. Banzhaf, K. Chellapilla, K. Deb, M. Dorigo, D. B. Fogel, M. H. Garzon, D. E. Goldberg, H. Iba, and R. L. Riolo, editors, *Proceedings of the Third Annual Conference on Genetic Programming*, pages 134–139, San Mateo, California, July 1998. University of Wisconsin at Madison, Morgan Kaufmann Publishers.
684. R. Hinterding and Z. Michalewicz. Your Brains and My Beauty: Parent Matching for Constrained Optimisation. In *Proceedings of the 5th International Conference on Evolutionary Computation*, pages 810–815, Anchorage, Alaska, May 1998.
685. K. Hirasawa, Y. Ishikawa, J. Hu, J. Murata, and J. Mao. Genetic Symbiosis Algorithm. In *2000 Congress on Evolutionary Computation (CEC'2000)*, volume 2, pages 1377–1384, Piscataway, New Jersey, July 2000. IEEE Service Center.
686. T. Hiroyasu, M. Miki, and S. Watanabe. Distributed Genetic Algorithms with a New Sharing Approach in Multiobjective Optimization Problems. In *1999 Congress on Evolutionary Computation*, pages 69–76, Washington, D.C., July 1999. IEEE Service Center.

687. T. Hiroyasu, M. Miki, and S. Watanabe. Divided Range Genetic Algorithms in Multiobjective Optimization Problems. In *Proceedings of International Workshop on Emergent Synthesis*, pages 57–66, Kobe, Japan, December 1999.
688. T. Hiroyasu, M. Miki, and S. Watanabe. The New Model of Parallel Genetic Algorithm in Multi-Objective Optimization Problems—Divided Range Multi-Objective Genetic Algorithm—. In *2000 Congress on Evolutionary Computation*, volume 1, pages 333–340, Piscataway, New Jersey, July 2000. IEEE Service Center.
689. C. J. Hitch. Sub-Optimization in Operations Research. *Operations Research*, 1(3):87–99, 1953.
690. J. Ho and Y.-L. Chang. A new heuristic for the n-job, m-machine flowshop problem. *European Journal of Operational Research*, 52:194–202, 1991.
691. S. Ho, S. Yang, G. Ni, E. W. Lo, and H. Wong. A Particle Swarm Optimization-Based Method for Multiobjective Design Optimizations. *IEEE Transactions on Magnetics*, 41(5):1756–1759, May 2005.
692. S. Ho, S. Yang, G. Ni, and H. Wong. A Tabu Method to Find the Pareto Solutions of Multiobjective Optimal Design Problems in Electromagnetics. *IEEE Transactions on Magnetics*, 38(2):1013–1016, March 2002. Part 1.
693. S.-Y. Ho and X.-I. Chang. An Efficient Generalized Multiobjective Evolutionary Algorithm. In W. Banzhaf, J. Daida, A. E. Eiben, M. H. Garzon, V. Honavar, M. Jakiela, and R. E. Smith, editors, *Proceedings of the Genetic and Evolutionary Computation Conference (GECCO'99)*, volume 1, pages 871–878, Orlando, Florida, USA, 1999. Morgan Kaufmann Publishers.
694. S.-Y. Ho and H.-L. Huang. Facial modeling from an uncalibrated face image using a coarse-to-fine genetic algorithm. *Pattern Recognition*, 34:1015–1031, 2001.
695. S.-Y. Ho, L.-S. Shu, and H.-M. Chen. Intelligent Genetic Algorithm with a New Intelligent Crossover Using Orthogonal Arrays. In *GECCO-99: Proceedings of the Genetic and Evolutionary Computation Conference*, volume 1, pages 289–296, San Francisco, California, USA, July 1999. Morgan Kaufmann Publishers.
696. W. M. Hoag. The Relevance of Cost in Operations Research. *Operations Research*, 4:448–459, 1956.
697. B.-M. Hodge, F. Pettersson, and N. Chakraborti. Re-evaluation of the optimal operating conditions for the primary end of an integrated steel plant using multi-objective genetic algorithms and nash equilibrium. *Steel Research International*, 77(7):459–461, July 2006.
698. J. Hofbauer and K. Sigmund. *Evolutionary Games and Population Dynamics*. Cambridge University Press, Cambridge, UK, 1998.
699. J. H. Holland. *Adaptation in Natural and Artificial Systems. An Introductory Analysis with Applications to Biology, Control and Artificial Intelligence*. University of Michigan Press, Ann Arbor, Michigan, USA, 1975.
700. J. H. Holland. Building Blocks, Cohort Genetic Algorithms, and Hyperplane-Defined Functions. *Evolutionary Computation*, 8(4):373–391, 2000.
701. P. M. Hollingsworth. *Requirements Controlled Design: A Method for Discovery of Discontinuous System Boundaries in the Requirements Hyperspace*. PhD thesis, School of Aerospace Engineering, Georgia Institute of Technology, USA, March 2004.

702. T.-P. Hong, H.-S. Wang, and W.-C. Chen. Simultaneous applying multiple mutation operators in genetic algorithm. *Journal of Heuristics*, 6(4):439–455, September 2000.
703. R. Hooke and T. Jeeves. Direct Search Solution of Numerical and Statistical Problems. *Journal of the ACM*, 8(2):212–229, 1961.
704. H. H. Hoos and T. Stützle. *Stochastic Local Search. Foundations and Applications*. Morgan Kaufmann Publishers, 2005. ISBN 1-55860-872-9.
705. H. Horii, M. Miki, T. Koizumi, and N. Tsujiuchi. Asynchronous Migration of Island Parallel GA for Multi-Objective Optimization Problem. In L. Wang, K. C. Tan, T. Furuhashi, J.-H. Kim, and X. Yao, editors, *Proceedings of the 4th Asia-Pacific Conference on Simulated Evolution and Learning (SEAL'02)*, volume 1, pages 86–90, Orchid Country Club, Singapore, November 2002. Nanyang Technical University.
706. J. Horn. Multicriterion Decision making. In T. Bäck, D. Fogel, and Z. Michalewicz, editors, *Handbook of Evolutionary Computation*, volume 1, pages F1.9:1–F1.9:15. IOP Publishing Ltd. and Oxford University Press, 1997.
707. J. Horn. Shape Nesting by Coevolving Species. In H.-G. B. et al., editor, *2005 Genetic and Evolutionary Computation Conference (GECCO'2005)*, volume 1, pages 557–558, New York, USA, June 2005. ACM Press.
708. J. Horn and N. Nafpliotis. Multiobjective Optimization using the Niched Pareto Genetic Algorithm. Technical Report IlliGAl Report 93005, University of Illinois at Urbana-Champaign, Urbana, Illinois, USA, 1993.
709. J. Horn, N. Nafpliotis, and D. E. Goldberg. A Niched Pareto Genetic Algorithm for Multiobjective Optimization. In *Proceedings of the First IEEE Conference on Evolutionary Computation, IEEE World Congress on Computational Intelligence*, volume 1, pages 82–87, Piscataway, New Jersey, June 1994. IEEE Service Center.
710. X. Hou, L. Shen, and H. Zhu. A smart particle swarm optimization algorithm for multi-objective problems. In *Computational Intelligence and Bioinformatics, Part 3*, pages 72–80. Springer-Verlag. Lecture Notes in Computer Science Vol. 4115, 2006.
711. C.-T. Hsiao, G. Chahine, and N. Gumerov. Application of a hybrid genetic/powell algorithm and a boundary element to electrical impedance tomography. *Journal of Computational Physics*, 173(2):433–454, November 2001.
712. J. Hu, K. Hirasawa, J. Mutata, M. Ohbayashi, and Y. Eki. A new random search method for neural network learning—RasID. In *Proceedings of the 1998 IEEE International Joint Conference on Neural Networks*, volume 3, pages 2346–2351. IEEE Press, 1998.
713. X. Hu and R. Eberhart. Multiobjective Optimization Using Dynamic Neighborhood Particle Swarm Optimization. In *Congress on Evolutionary Computation (CEC'2002)*, volume 2, pages 1677–1681, Piscataway, New Jersey, May 2002. IEEE Service Center.
714. X. Hu and R. Eberhart. Solving Constrained Nonlinear Optimization Problems with Particle Swarm Optimization. In *Proceedings of the 6th World Multiconference on Systemics, Cybernetics and Informatics (SCI 2002)*, volume 5. Orlando, USA, IIIS, July 2002.
715. X. Hu, R. C. Eberhart, and Y. Shi. Engineering Optimization with Particle Swarm. In *Proceedings of the 2003 IEEE Swarm Intelligence Symposium*, pages 53–57. Indianapolis, Indiana, USA, IEEE Service Center, April 2003.

716. X. Hu, R. C. Eberhart, and Y. Shi. Particle Swarm with Extended Memory for Multiobjective Optimization. In *2003 IEEE Swarm Intelligence Symposium Proceedings*, pages 193–197, Indianapolis, Indiana, USA, April 2003. IEEE Service Center.
717. X. Hu, R. C. Eberhart, and Y. Shi. Swarm intelligence for permutation optimization: A case study on N-queens problem. In *Proceedings of the IEEE Swarm Intelligence Symposium 2003 (SIS 2003)*, pages 243–246, Indianapolis, Indiana, USA, 2003. IEEE Press.
718. X. Hu, Z. Wang, and L. Liao. Multi-Objective Optimization of HEV Fuel Economy and Emissions using Evolutionary Computation. In *Proceedings of the Society of Automotive Engineering World Congress 2004, Electronics Simulation and Optimization (SP-1856)*, pages 117–128, Detroit, USA, March 2004. Society of Automotive Engineers.
719. X. S. Hu, G. Greenwood, and J. G. D'Ambrosio. An Evolutionary Approach to Hardware/Software Partitioning. In H.-M. Voigt, W. Ebeling, I. Rechenberg, and H.-P. Schwefel, editors, *Parallel Problem Solving from Nature—PPSN IV*, pages 900–909. Springer-Verlag. Lecture Notes in Computer Science No. 1141, September 1996.
720. S. Huband, L. Barone, L. While, and P. Hingston. A Scalable Multi-objective Test Problem Toolkit. In C. A. Coello Coello, A. Hernández Aguirre, and E. Zitzler, editors, *Evolutionary Multi-Criterion Optimization. Third International Conference, EMO 2005*, pages 280–295, Guanajuato, México, March 2005. Springer. Lecture Notes in Computer Science Vol. 3410.
721. S. Huband, P. Hingston, L. Barone, and L. While. A Review of Multiobjective Test Problems and a Scalable Test Problem Toolkit. *IEEE Transactions on Evolutionary Computation*, 10(5):477–506, October 2006.
722. G. P. Huber. Multi-Attribute Utility Models: A Review of Field and Field-like Structures. *Management Science*, 20:1393–1402, 1974.
723. R. Hubley, E. Zitzler, and J. Roach. Evolutionary algorithms for the selection of single nucleotide polymorphisms. *BMC Bioinformatics*, 4(30), July 2003.
724. E. J. Hughes. Multi-Objective Probabilistic Selection Evolutionary Algorithm. Technical Report DAPS/EJH/56/2000, Department of Aerospace, Power, & Sensors, Cranfield University, RMCS, Shrivenham, UK, SN6 8LA, September 2000.
725. E. J. Hughes. Constraint Handling With Uncertain and Noisy Multi-Objective Evolution. In *Proceedings of the Congress on Evolutionary Computation 2001 (CEC'2001)*, volume 2, pages 963–970, Piscataway, New Jersey, May 2001. IEEE Service Center.
726. E. J. Hughes. Evolutionary Multi-objective Ranking with Uncertainty and Noise. In E. Zitzler, K. Deb, L. Thiele, C. A. Coello Coello, and D. Corne, editors, *First International Conference on Evolutionary Multi-Criterion Optimization*, pages 329–343. Springer-Verlag. Lecture Notes in Computer Science No. 1993, 2001.
727. E. J. Hughes. Evolutionary Many-Objective Optimisation: Many Once or One Many? In *2005 IEEE Congress on Evolutionary Computation (CEC'2005)*, volume 1, pages 222–227, Edinburgh, Scotland, September 2005. IEEE Service Center.
728. L. Hurwicz. Programming in Linear Spaces. In K. J. Arrow, L. Hurwicz, and H. Uzawa, editors, *Studies in Linear and Nonlinear Programming*, pages 38–102. Oxford University Press, London, England, 1964.

729. P. Husbands. Genetic Algorithms in Optimization and Adaptation. In L. Kronsjö and D. Shumsherudin, editors, *Advances in Parallel Algorithms*, chapter 8, pages 227–276. Halsted Press, New York, 1992.
730. G. V. Huylenbroeck. The Conflict Analysis Method: bridging the gap between ELECTRE, PROMETHEE and ORESTE. *European Journal of Operational Research*, 82(3):490–502, May 1995.
731. C.-L. Hwang and M.-J. Lin. *Group Decision Making under Multiple Criteria. Methods and Applications*. Springer-Verlag. Lecture Notes in Economics and Mathematical Systems, Vol. 281, 1988.
732. C. L. Hwang and A. S. M. Masud. Multiple Objective Decision-Making Methods and Applications. In *Lecture Notes in Economics and Mathematical Systems*, volume 164. Springer-Verlag, New York, 1979.
733. C. L. Hwang, S. R. Paidy, and K. Yoon. Mathematical Programming with Multiple Objectives: A Tutorial. *Computing and Operational Research*, 7:5–31, 1980.
734. C. J. Hyun, Y. Kim, and Y. K. Kim. A Genetic Algorithm for Multiple Objective Sequencing Problems in Mixed Model Assembly Lines. *Computers & Operations Research*, 25(7/8):675–690, 1998.
735. H. Iba, H. de Garis, and T. Sato. Genetic Programming Using a Minimum Description Length Principle. In J. Kenneth E. Kinnear, editor, *Advances in Genetic Programming*, pages 265–284. MIT Press, 1994.
736. K. Ida, M. Gen, and Y.-Z. Li. Solving Multiobjective Chance-constrained Solid Transportation Problem by Evolutionary Computation. In *5th European Congress on Intelligent Techniques and Soft Computing EUFIT'97*, pages 743–747, Aachen, Germany, September 1997.
737. C. Igel, N. Hansen, and S. Roth. Covariance Matrix Adaptation for Multi-objective Optimization. *Evolutionary Computation*, 15(1):1–28, Spring 2007.
738. J. P. Ignizio. *Goal Programming and Extensions*. Heath, Lexington, Massachusetts, 1976.
739. J. P. Ignizio. The determination of a subset of efficient solutions via goal programming. *Computers and Operations Research*, 3:9–16, 1981.
740. J. P. Ignizio. Integrating Cost, Effectiveness, and Stability. *Acquisition Review Quarterly*, pages 51–60, Winter 1998.
741. H. Iima. Proposition of Selection Operation in a Genetic Algorithm for a Job Shop Rescheduling Problem. In C. A. Coello Coello, A. Hernández Aguirre, and E. Zitzler, editors, *Evolutionary Multi-Criterion Optimization. Third International Conference, EMO 2005*, pages 721–735, Guanajuato, México, March 2005. Springer. Lecture Notes in Computer Science Vol. 3410.
742. Y. Ijiri. *Management Goals and Accounting for Control*. North-Holland, Amsterdam, 1965.
743. I. Ikonen, W. E. Biles, A. Kumar, J. C. Wissel, and R. K. Ragade. Genetic Algorithm for Packing Three-Dimensional Non-Convex Objects Having Cavities and Holes. In *Proceedings of the 7th International Conference on Genetic Algortithms*, pages 591–598, East Lansing, Michigan, July 1997. Morgan Kaufmann Publishers.
744. IlliGAL. Illigal website. Online, 2007. Available: http://wwww-illigal.ge.uiuc.edu/.
745. A. W. Iorio and X. Li. A Cooperative Coevolutionary Multiobjective Algorithm Using Non-dominated Sorting. In K. D. et al., editor, *Genetic and*

Evolutionary Computation–GECCO 2004. Proceedings of the Genetic and Evolutionary Computation Conference. Part I, pages 537–548, Seattle, Washington, USA, June 2004. Springer-Verlag, Lecture Notes in Computer Science Vol. 3102.

746. A. W. Iorio and X. Li. Solving rotated multi-objective optimization problems using differential evolution. In *AI 2004: Advances in Artificial Intelligence, Proceedings*, pages 861–872. Springer-Verlag, Lecture Notes in Artificial Intelligence Vol. 3339, 2004.

747. A. W. Iorio and X. Li. Incorporating Directional Information within a Differential Evolution Algorithm for Multi-objective Optimization. In M. K. et al., editor, *2006 Genetic and Evolutionary Computation Conference (GECCO'2006)*, volume 1, pages 691–697, Seattle, Washington, USA, July 2006. ACM Press. ISBN 1-59593-186-4.

748. S. Iredi, D. Merkle, and M. Middendorf. Bi-Criterion Optimization with Multi Colony Ant Algorithms. In E. Zitzler, K. Deb, L. Thiele, C. A. Coello Coello, and D. Corne, editors, *First International Conference on Evolutionary Multi-Criterion Optimization*, pages 358–372. Springer-Verlag. Lecture Notes in Computer Science No. 1993, 2001.

749. H. Ishibuchi and S. Kaige. Comparison of Multiobjective Memetic Algorithms on 0/1 Knapsack Problems. In A. Barry, editor, *2003 Genetic and Evolutionary Computation Conference. Workshop Program*, pages 222–227, Chicago, Illinois, USA, July 2003. AAAI.

750. H. Ishibuchi and T. Murata. Multi-Objective Genetic Local Search Algorithm. In T. Fukuda and T. Furuhashi, editors, *Proceedings of the 1996 International Conference on Evolutionary Computation*, pages 119–124, Nagoya, Japan, 1996. IEEE.

751. H. Ishibuchi and T. Murata. Multi-Objective Genetic Local Search Algorithm and Its Application to Flowshop Scheduling. *IEEE Transactions on Systems, Man and Cybernetics*, 28(3):392–403, August 1998.

752. H. Ishibuchi and T. Murata. Multi-Objective Genetic Local Search for Minimizing the Number of Fuzzy Rules for Pattern Classification Problems. In D. B. Fogel, editor, *Proceedings of the 1998 IEEE International Conference on Evolutionary Computation*, pages 1100–1105, Piscataway, New Jersey, May 1998. IEEE.

753. H. Ishibuchi and T. Nakashima. Linguistic Rule Extraction by Genetics-Based Machine Learning. In D. Whitley, D. Goldberg, E. Cantú-Paz, L. Spector, I. Parmee, and H.-G. Beyer, editors, *Proceedings of the Genetic and Evolutionary Computation Conference (GECCO'2000)*, pages 195–202, San Francisco, California, 2000. Morgan Kaufmann.

754. H. Ishibuchi and T. Nakashima. Multi-Objective Pattern and Feature Selection by a Genetic Algorithm. In D. Whitley, D. Goldberg, E. Cantú-Paz, L. Spector, I. Parmee, and H.-G. Beyer, editors, *Proceedings of the Genetic and Evolutionary Computation Conference (GECCO'2000)*, pages 1069–1076, San Francisco, California, 2000. Morgan Kaufmann.

755. H. Ishibuchi and T. Nakashima. Three-Objective Optimization in Linguistic Function Approximation. In *Proceedings of the Congress on Evolutionary Computation 2001 (CEC'2001)*, volume 1, pages 340–347, Piscataway, New Jersey, May 2001. IEEE Service Center.

756. H. Ishibuchi and K. Narukawa. Some Issues on the Implementation of Local Search in Evolutionary Multiobjective Optimization. In K. D. et al., editor,

Genetic and Evolutionary Computation–GECCO 2004. Proceedings of the Genetic and Evolutionary Computation Conference. Part I, pages 1246–1258, Seattle, Washington, USA, June 2004. Springer-Verlag, Lecture Notes in Computer Science Vol. 3102.
757. H. Ishibuchi, Y. Nojima, and T. Doi. Comparison between Single-Objective and Multi-Objective Genetic Algorithms: Performance Comparison and Performance Measures. In *2006 IEEE Congress on Evolutionary Computation (CEC'2006)*, pages 3959–3966, Vancouver, BC, Canada, July 2006. IEEE.
758. H. Ishibuchi, K. Nozaki, N. Yamamoto, and H. Tanaka. Selecting Fuzzy If-Then Rules for Classification Problems using Genetic Algorithms. *IEEE Transactions on Fuzzy Systems*, 3(3):260–270, 1995.
759. H. Ishibuchi and Y. Shibata. An Empirical Study on the Effect of Mating Restriction on the Search Ability of EMO Algorithms. In C. M. Fonseca, P. J. Fleming, E. Zitzler, K. Deb, and L. Thiele, editors, *Evolutionary Multi-Criterion Optimization. Second International Conference, EMO 2003*, pages 433–477, Faro, Portugal, April 2003. Springer. Lecture Notes in Computer Science. Volume 2632.
760. H. Ishibuchi and Y. Shibata. Mating Scheme for Controlling the Diversity-Convergence Balance for Multiobjective Optimization. In K. D. et al., editor, *Genetic and Evolutionary Computation–GECCO 2004. Proceedings of the Genetic and Evolutionary Computation Conference. Part I*, pages 1259–1271, Seattle, Washington, USA, June 2004. Springer-Verlag, Lecture Notes in Computer Science Vol. 3102.
761. H. Ishibuchi and Y. Shibata. Single-Objective and Multi-Objective Evolutionary Flowshop Scheduling. In C. A. Coello Coello and G. B. Lamont, editors, *Applications of Multi-Objective Evolutionary Algorithms*, pages 529–554. World Scientific, Singapore, 2004.
762. H. Ishibuchi, T. Yoshida, and T. Murata. Balance Between Genetic Search and Local Search in Memetic Algorithms for Multiobjective Permutation Flowshop Scheduling. *IEEE Transactions on Evolutionary Computation*, 7(2):204–223, April 2003.
763. R. S. H. Istepanian and J. F. Whidborne. Multi-objective design of finite word-length controller structures. In *1999 Congress on Evolutionary Computation*, pages 61–68, Washington, D.C., July 1999. IEEE Service Center.
764. K. Ito, S. Akagi, and M. Nishikawa. A Multiobjective Optimization Approach to a Design Problem of Heat Insulation for Thermal Distribution Piping Network Systems. *Journal of Mechanisms, Transmissions and Automation in Design (Transactions of the ASME)*, 105:206–213, jun 1983.
765. R. H. F. Jackson, P. T. Boggs, S. G. Nash, and S. Powell. Guidelines for Reporting Results of Computational Experiments – Report of the Ad Hoc Committee. *Mathematical Programming*, 49:413–425, 1991.
766. D. Jaeggi, C. Asselin-Miller, G. Parks, T. Kipouros, T. Bell, and J. Clarkson. Multi-objective Parallel Tabu Search. In *Parallel Problem Solving from Nature - PPSN VIII*, pages 732–741, Birmingham, UK, September 2004. Springer-Verlag. Lecture Notes in Computer Science Vol. 3242.
767. D. Jaeggi, G. Parks, T. Kipouros, and J. Clarkson. A Multi-objective Tabu Search Algorithm for Constrained Optimisation Problems. In C. A. Coello Coello, A. Hernández Aguirre, and E. Zitzler, editors, *Evolutionary Multi-Criterion Optimization. Third International Conference, EMO 2005*, pages

490–504, Guanajuato, México, March 2005. Springer. Lecture Notes in Computer Science Vol. 3410.
768. W. Jakob, M. Gorges-Schleuter, and C. Blume. Application of Genetic Algorithms to Task Planning and Learning. In R. Männer and B. Manderick, editors, *Parallel Problem Solving from Nature, 2nd Workshop*, Lecture Notes in Computer Science, pages 291–300, Amsterdam, 1992. North-Holland Publishing Company.
769. A. Jan, M. Yamamoto, and A. Ohuchi. Evolutionary Algorithms for Nurse Scheduling Problem. In *2000 Congress on Evolutionary Computation*, volume 1, pages 196–203, Piscataway, New Jersey, July 2000. IEEE Service Center.
770. S. Janson and D. Merkle. A New Multi-objective Particle Swarm Optimization Algorithm Using Clustering Applied to Automated Docking. In M. J. Blesa, C. Blum, A. Roli, and M. Sampels, editors, *Hybrid Metaheuristics, Second International Workshop, HM 2005*, pages 128–142, Barcelona, Spain, August 2005. Springer. Lecture Notes in Computer Science Vol. 3636.
771. S. Janson, D. Merkle, and M. Middendorf. Parallel ant colony algorithms. In E. Alba, editor, *Parallel Metaheuristics*, pages 171–201. Wiley-Interscience, 2005.
772. C. H. Jarvis, N. Stuart, K. Kelsey, and R. H. A. Baker. Towards a Methodology for Selecting a "characteristic" Sample from an Existing Database: An Evolutionary Approach. In *Third International Conference/Workshop on Integrating GIS and Environmental Modeling*, Santa Fe, New Mexico, January 21–26 1996. National Center for Geographic Information and Analysis.
773. A. Jaszkiewicz. A metaheuristic approach to multiple objective nurse scheduling. *Foundations of Computing and Decision Sciences*, 22(3):169–184, 1997.
774. A. Jaszkiewicz. On the computational effectiveness of multiple objective metaheuristics. In *Proceedings of the Fourth International Conference on Multi-Objective Programming and Goal Programming MOPGP'00. Theory & Applications*, pages 201–214, Ustron, Poland, May 29–June 1 2000. University of Economics in Katowice.
775. A. Jaszkiewicz. Comparison of Local Search-Based Metaheuristics on the Multiple Objective Knapsack Problem. *Foundations of Computing and Decision Sciences*, 26(1):99–120, 2001.
776. A. Jaszkiewicz. *Multiple objective metaheuristic algorithms for combinatorial optimization*. Poznan University of Technology, Poznan, Poland, 2001. Habilitation thesis.
777. A. Jaszkiewicz. Genetic local search for multiple objective combinatorial optimization. *European Journal of Operational Research*, 137(1):50–71, 2002.
778. A. Jaszkiewicz. On the Computational Effectiveness of Multiple Objective Metaheuristics. In T. Traskalik and J. Michnik, editors, *Multiple Objective and Goal Programming. Recent Developments*, pages 86–100. Physica-Verlag, Heidelberg, 2002.
779. A. Jaszkiewicz. On the Performance of Multiple-Objective Genetic Local Search on the 0/1 Knapsack Problem—A Comparative Experiment. *IEEE Transactions on Evolutionary Computation*, 6(4):402–412, August 2002.
780. A. Jaszkiewicz. Do Multiple-Objective Metaheuristics Deliver on Their Promises? A Computational Experiment on the Set-Covering Problem. *IEEE Transactions on Evolutionary Computation*, 7(2):133–143, April 2003.

781. A. Jaszkiewicz. A Comparative Study of Multiple-Objective Metaheuristics on the Bi-Objective Set Covering Problem and the Pareto Memetic Algorithm. *Annals of Operations Research*, 131(1–4):135–158, October 2004.
782. A. Jaszkiewicz. Evaluation of Multiple Objective Metaheuristics. In X. Gandibleux, M. Sevaux, K. Sörensen, and V. T'kindt, editors, *Metaheuristics for Multiobjective Optimisation*, pages 65–89, Berlin, 2004. Springer. Lecture Notes in Economics and Mathematical Systems Vol. 535.
783. A. Jaszkiewicz. On the Computational Efficiency of Multiple Objective Metaheuristics. The Knapsack Problem Case Study. *European Journal of Operational Research*, 158(2):418–433, October 2004.
784. A. Jaszkiewicz, M. Hapke, and P. Kominek. Performance of Multiple Objective Evolutionary Algorithms on a Distribution System Design Problem—Computational Experiment. In E. Zitzler, K. Deb, L. Thiele, C. A. Coello Coello, and D. Corne, editors, *First International Conference on Evolutionary Multi-Criterion Optimization*, pages 241–255. Springer-Verlag. Lecture Notes in Computer Science No. 1993, 2001.
785. A. Jaszkiewicz and R. Slowinski. The Light Beam Search—Outranking Based Interactive Procedure for Multiple-Objective Mathematical Programming. In P. Pardalos, Y. Siskos, and C. Zopounidis, editors, *Advances in Multicriteria Analysis*, pages 129–146. Kluwer Academic Publishers, Dordrecht, 1995.
786. A. Jaszkiewicz and R. Slowinski. The 'Light Beam Search' approach – an overview of methodology and applications. *European Journal of Operational Research*, 113:300–314, 1999.
787. M. T. Jensen. Guiding Single-Objective Optimization Using Multi-objective Methods. In G. R. et al., editor, *Applications of Evolutionary Computing. Evoworkshops 2003: EvoBIO, EvoCOP, EvoIASP, EvoMUSART, EvoROB, and EvoSTIM*, pages 199–210, Essex, UK, April 2003. Springer. Lecture Notes in Computer Science Vol. 2611.
788. M. T. Jensen. Reducing the Run-Time Complexity of Multiobjective EAs: The NSGA-II and Other Algorithms. *IEEE Transactions on Evolutionary Computation*, 7(5):503–515, October 2003.
789. N. K. Jerne. The immune system. *Scientific American*, 229(1):52–60, 1973.
790. A. Jestel, D. Pena, and R. Zamar. A Multivariate Kolmogorov-Smirnov Test of Goodness of Fit. *Statistics and Probability Letters*, 35:251–259, 1997.
791. L. Jiao, M. Gong, R. Shang, H. Du, and B. Lu. Clonal Selection with Immune Dominance and Anergy Based Multiobjective Optimization. In C. A. Coello Coello, A. Hernández Aguirre, and E. Zitzler, editors, *Evolutionary Multi-Criterion Optimization. Third International Conference, EMO 2005*, pages 474–489, Guanajuato, México, March 2005. Springer. Lecture Notes in Computer Science Vol. 3410.
792. F. Jiménez and J. M. Cadenas. An evolutionary program for the multiobjective solid transportation problem with fuzzy goals. *Operations Research and Decisions*, 2:5–20, 1995.
793. F. Jiménez, A. Gómez-Skarmeta, and G. Sánchez. How Evolutionary Multiobjective Optimization can be used for Goals and Priorities based Optimization. In *Primer Congreso Español de Algoritmos Evolutivos y Bioinspirados (AEB'02)*, pages 460–465. Mérida, Spain, 2002.
794. F. Jiménez, A. F. Gómez-Skarmeta, H. Roubos, and R. Babuška. Accurate, transparent, and compact fuzzy models for function approximation and

dynamic modeling through multi-objective evolutionary optimization. In E. Zitzler, K. Deb, L. Thiele, C. A. Coello Coello, and D. Corne, editors, *First International Conference on Evolutionary Multi-Criterion Optimization*, pages 653–667. Springer-Verlag. Lecture Notes in Computer Science No. 1993, 2001.

795. F. Jiménez and J. L. Verdegay. Interval multiobjective solid transportation problem via genetic algorithms (IPMU'96). In *Proceedings of Information Processing and Management of Uncertainty in Knowledge-Based Systems*, pages 787–792, Granada, Spain, 1996.

796. F. Jiménez and J. L. Verdegay. Constrained multiobjective optimization by evolutionary algorithms. In *Proceedings of the International ICSC Symposium on Engineering of Intelligent Systems (EIS'98)*, pages 266–271, University of La Laguna, Tenerife, Spain, 1998.

797. F. Jiménez, J. L. Verdegay, and A. F. Gómez-Skarmeta. Evolutionary Techniques for Constrained Multiobjective Optimization Problems. In A. S. Wu, editor, *Proceedings of the 1999 Genetic and Evolutionary Computation Conference. Workshop Program*, pages 115–116, Orlando, Florida, July 1999.

798. H. Jin and M.-L. Wong. Adaptive Diversity Maintenance and Convergence Guarantee in Multiobjective Evolutionary Algorithms. In *Proceedings of the 2003 Congress on Evolutionary Computation (CEC'2003)*, volume 4, pages 2498–2505, Canberra, Australia, December 2003. IEEE Press.

799. Y. Jin, editor. *Multi-Objective Machine Learning*. Springer, Berlin, 2006. ISBN 3-540-30676-6.

800. Y. Jin, T. Okabe, and B. Sendhoff. Adapting Weighted Aggregation for Multiobjective Evolution Strategies. In E. Zitzler, K. Deb, L. Thiele, C. A. Coello Coello, and D. Corne, editors, *First International Conference on Evolutionary Multi-Criterion Optimization*, pages 96–110. Springer-Verlag. Lecture Notes in Computer Science No. 1993, 2001.

801. Y. Jin, T. Okabe, and B. Sendhoff. Dynamic Weighted Aggregation for Evolutionary Multi-Objective Optimization: Why Does It Work and How? In L. Spector, E. D. Goodman, A. Wu, W. Langdon, H.-M. Voigt, M. Gen, S. Sen, M. Dorigo, S. Pezeshk, M. H. Garzon, and E. Burke, editors, *Proceedings of the Genetic and Evolutionary Computation Conference (GECCO'2001)*, pages 1042–1049, San Francisco, California, 2001. Morgan Kaufmann Publishers.

802. Y. Jin and B. Sendhoff. Incorporation of Fuzzy Preferences into Evolutionary Multiobjective Optimization. In W. Langdon, E. Cantú-Paz, K. Mathias, R. Roy, D. Davis, R. Poli, K. Balakrishnan, V. Honavar, G. Rudolph, J. Wegener, L. Bull, M. Potter, A. Schultz, J. Miller, E. Burke, and N. Jonoska, editors, *Proceedings of the Genetic and Evolutionary Computation Conference (GECCO'2002)*, page 683, San Francisco, California, July 2002. Morgan Kaufmann Publishers.

803. Y. Jin and B. Sendhoff. Connectedness, Regularity and the Success of Local Search in Evolutionary Multi-objective Optimization. In *Proceedings of the 2003 Congress on Evolutionary Computation (CEC'2003)*, volume 3, pages 1910–1917, Canberra, Australia, December 2003. IEEE Press.

804. Y. Jin, W. von Seelen, and B. Sendhoff. On generating flexible, complete, consistent and compact(FC^3) fuzzy rule systems from data using evolution strategies. *IEEE Transactions on Systems, Man, and Cybernetics*, 29(4):829–845, 1999.

805. B. R. Jones, W. A. Crossley, and A. S. Lyrintzis. Aerodynamic and Aeroacoustic Optimization of Airfoils via a Parallel Genetic Algorithm. In *Proceed-*

ings of the 7th AIAA/USAF/NASA/ISSMO Symposium on Multidisciplinary Analysis and Optimization, AIAA-98-4811. AIAA, 1998.
806. D. Jones, S. Mirrazavi, and M. Tamiz. Multi-objective metaheuristics: An overview of the current state-of-the-art. *European Journal of Operational Research*, 137(1):1–9, February 2002.
807. D. F. Jones and M. Tamiz. Goal Programming in the Period 1990–2000. In M. Ehrgott and X. Gandibleux, editors, *Multiple Criteria Optimization. State of the Art. Annotated Bibliographic Surveys*, pages 129–170. Kluwer Academic Publishers, Boston/Dordrecht/London, 2002.
808. G. Jones, R. D. Brown, D. E. Clark, P. Willett, and R. C. Glen. Searching Databases of Two-Dimensional and Three-Dimensional Chemical Structures using Genetic Algorithms. In S. Forrest, editor, *Proceedings of the Fifth International Conference on Genetic Algorithms*, pages 597–602, San Mateo, California, 1993. Morgan Kaufmann Publishers.
809. N. Jozefowiez, F. Semet, and E.-G. Talbi. Parallel and Hybrid Models for Multi-objective Optimization: Application to the Vehicle Routing Problem. In J. J. Merelo Guervós, P. Adamidis, H.-G. Beyer, J.-L. F.-V. nas, and H.-P. Schwefel, editors, *Parallel Problem Solving from Nature—PPSN VII*, pages 271–280, Granada, Spain, September 2002. Springer-Verlag. Lecture Notes in Computer Science No. 2439.
810. Z. Juhasz, P. Kacsuk, and D. Kranzlmuller, editors. *Distributed and Parallel Systems : Cluster and Grid Computing*. Springer, 2004. ISBN 0-38723-094-7.
811. H. Jutler. Liniejnaja modiel z nieskolkimi celevymi funkcjami (Linear Model with Several Objective Functions). *Ekonomika i matematiceckije Metody*, 3:397–406, 1967. (In Polish).
812. B. A. Kadrovach, S. R. Michaud, J. B. Zydallis, G. B. Lamont, B. Secrest, and D. Strong. Extending the Simple Genetic Algorithm into Multi-Objective Problems via Mendelian Pressure. In *2001 Genetic and Evolutionary Computation Conference. Workshop Program*, pages 181–188, San Francisco, California, July 2001.
813. S. Kaige, T. Murata, and H. Ishibuchi. Performance evaluation of memetic EMO algorithms using dominance relation-based replacement rules on MOO test problems. In *Proceedings of the 2003 IEEE International Conference on Systems, Man, and Cybernetics*, volume 1, pages 14–19. IEEE Press, 2003.
814. L. Kallel, B. Naudts, and A. Rogers, editors. *Theoretical Aspects of Evolutionary Computing*. Springer, Berlin, 2001. ISBN 3-540-67396-2.
815. R. Kamalian, H. Takagi, and A. M. Agogino. Optimized Design of MEMS by Evolutionary Multi-objective Optimization with Interactive Evolutionary Computation. In K. D. et al., editor, *Genetic and Evolutionary Computation—GECCO 2004. Proceedings of the Genetic and Evolutionary Computation Conference. Part II*, pages 1030–1041, Seattle, Washington, USA, June 2004. Springer-Verlag, Lecture Notes in Computer Science Vol. 3103.
816. R. R. Kamalian. *Evolutionary Synthesis of MEMS*. PhD thesis, Mechanical Engineering, University of California, Berkeley, USA, 2004.
817. J. Kamiura, T. Hiroyasu, M. Miki, and S. Watanabe. MOGADES: Multi-objective genetic algorithm with distributed environment scheme. In *Computational Intelligence and Applications (Proceedings of the Second International Workshop on Intelligent Systems Design and Applications: ISDA'02)*, pages 143–148, 2002.

818. M. K. Karakasis and K. C. Giannakoglou. Metamodel-Assisted Multi-Objective Evolutionary Optimization. In R. Schilling, W. Haase, J. Periaux, H. Baier, and G. Bugeda, editors, *EUROGEN 2005. Evolutionary Methods for Design, Optimization and Control with Applications to Industrial Problems*, Munich, Germany, 2005.
819. M. K. Karakasis and K. C. Giannakoglou. On the use of metamodel-assisted, multi-objective evolutionary algorithms. *Engineering Optimization*, 38(8):941–957, December 2006.
820. E. K. Karasakal and M. Köksalan. A Simulated Annealing Approach to Bicriteria Scheduling Problems on a Single Machine. *Journal of Heuristics*, 6(3):311–327, August 2000.
821. S. Karlin. Mathematical Methods and Theory in Games. In *Programming and Economics*, volume 1, pages 216–217. Addison-Wesley, Reading, Massachusetts, 1959.
822. M. Karnaugh. A Map Method for Synthesis of Combinational Logic Circuits. *Transactions of the AIEE, Communications and Electronics*, 72 (I):593–599, November 1953.
823. R. Kasat, D. Kunzru, D. Saraf, and S. Gupta. Multiobjective optimization of industrial FCC units using elitist nondominated sorting genetic algorithm. *Industrial & Engineering Chemistry Research*, 41(19):4765–4776, September 2002.
824. K. Kato, C. Perkgoz, and M. Sakawa. An Interactive Fuzzy Satisficing Method for Multiobjective Integer Programming Problems through Genetic Algorithms. In Y. Jin, editor, *Knowledge Incorporation in Evolutionary Computation*, pages 503–523. Springer, Berlin Heidelberg, 2005. ISBN 3-540-22902-7.
825. K. Kato, M. Sakawa, and T. Ikegame. Interactive Decision Making for Multiobjective Block Angular 0-1 Programming Problems with Fuzzy Parameters Through Genetic Algorithms. *Japanese Journal of Fuzzy Theory and Systems*, 9(1):49–59, 1997.
826. Y. Katsumata and T. Terano. Bayesian Optimization Algorithm for Multi-Objective Solutions: Application to Electric Equipment Configuration Problems in a Power Plant. In *Proceedings of the 2003 Congress on Evolutionary Computation (CEC'2003)*, volume 2, pages 1101–1107, Canberra, Australia, December 2003. IEEE Press.
827. S. Kauffman. Adaptation on rugged fitness landscapes. In D. Sein, editor, *Lectures in the Sciences of Complexity*, pages 527–618. Addison-Wesley, Redwood City, USA, 1989.
828. S. Kauffman and S. Levin. Towards a general theory of adaptive walks on rugged landscapes. *Journal of Theoretical Biology*, 128:11–45, 1987.
829. S. A. Kauffman. *The Origins of Order: Self-Organization and Selection in Evolution*. Oxford University Press, New York, 1993.
830. T. Kawabe and T. Tagami. A New Genetic Algorithm using Pareto Partitioning Method for Robust Partial Model Matching PID Design with Two Degrees of Freedom. In *Proceedings of the Third International ICSC (International Computer Science Conventions) Symposia on Intelligent Industrial Automation (IIA'99) and Soft Computing (SOCO'99)*, pages 562–567, Genova, 1999.
831. E. Keedwell and S.-T. Khu. A hybrid genetic algorithm for the design of water distribution networks. *Engineering Applications of Artificial Intelligence*, 18(4):461–472, 2005.

832. E. Keedwell and S.-T. Khu. A novel cellular automata approach to optimal water distribution network design. *Journal of Computers in Civil Engineering*, 20(1):1–8, 2006.
833. E. Keedwell and S.-T. Khu. A novel evolutionary meta-heuristic for the multi-objective optimization of real-world water distribution network. *Engineering Optimization*, 38(3):319–336, April 2006.
834. T. W. Keelin. *A protocol and procedure for assessing multi-attribute preference functions*. PhD thesis, Department of Engineering Economic Systems, Stanford University, Stanford, California, 1976.
835. R. L. Keeney. Multi-dimensional Utility Functions: Theory, Assessment and Applications. Technical Report 43, Massachusetts Institute of Technology, Operations Research Center, Cambridge, Massachusetts, USA, 1969.
836. R. L. Keeney and H. Raiffa. *Decisions with Multiple Objectives. Preferences and Value Tradeoffs*. Cambridge University Press, Cambridge, UK, 1993.
837. N. Keerativuttiumrong, N. Chaiyaratana, and V. Varavithya. Multi-objective Co-operative Co-evolutionary Genetic Algorithm. In J. J. Merelo Guervós, P. Adamidis, H.-G. Beyer, J.-L. F.-V. nas, and H.-P. Schwefel, editors, *Parallel Problem Solving from Nature—PPSN VII*, pages 288–297, Granada, Spain, September 2002. Springer-Verlag. Lecture Notes in Computer Science No. 2439.
838. J. Kennedy and R. C. Eberhart. Particle Swarm Optimization. In *Proceedings of the 1995 IEEE International Conference on Neural Networks*, pages 1942–1948, Piscataway, New Jersey, 1995. IEEE Service Center.
839. J. Kennedy and R. C. Eberhart. A Discrete Binary Version of the Particle Swarm Algorithm. In *Proceedings of the 1997 IEEE Conference on Systems, Man, and Cybernetics*, pages 4104–4109, Piscataway, New Jersey, 1997. IEEE Service Center.
840. J. Kennedy and R. C. Eberhart. *Swarm Intelligence*. Morgan Kaufmann Publishers, San Francisco, California, 2001.
841. Z. Khairullah and S. Zionts. An experiment with some approaches for solving problems with multiple criteria. In *Third International Conference on Multiple Criteria Decision-Making*, Konigswinger, Germany, 1979.
842. S. Khajehpour. *Optimal Conceptual Design of High-Rise Office Buldings*. PhD thesis, Civil Engineering Department, University of Waterloo, Ontario, Canada, 2001.
843. S. Khajehpour and D. Grierson. Conceptual Design using Adaptive Computing. In *2001 Genetic and Evolutionary Computation Conference. Workshop Program*, pages 62–67, San Francisco, California, July 2001.
844. N. Khan. Bayesian Optimization Algorithms for Multiobjective and Hierarchically Difficult Problems. Master's thesis, Graduate College of the University of Illinois at Urbana-Champaign, Urbana, Illinois, USA, 2003.
845. N. Khan, D. E. Goldberg, and M. Pelikan. Multi-Objective Bayesian Optimization Algorithm. Technical Report 2002009, Illinois Genetic Algorithms Laboratory, University of Illinois at Urbana-Champaign, Urbana, Illinois, March 2002.
846. E. Khor, K. Tan, and T. Lee. Tabu-Based Exploratory Evolutionary Algorithm for Effective Multi-objective Optimization. In E. Zitzler, K. Deb, L. Thiele, C. A. Coello Coello, and D. Corne, editors, *First International Conference on Evolutionary Multi-Criterion Optimization*, pages 344–358. Springer-Verlag. Lecture Notes in Computer Science No. 1993, 2001.

847. T. Khoshgoftaar, Y. Liu, and N. Seliya. A Multiobjective Module-Order Model for Software Quality Enhancement. *IEEE Transactions on Evolutionary Computation*, 8(6):593–608, 2004.
848. T. M. Khoshgoftaar, Y. Liu, and N. Seliya. Genetic Programming-Based Decision Trees for Software Quality Classification. In *Proceedings of the Fifteenth International Conference on Tools with Artificial Intelligence (ICTAI 03)*, pages 374–383, Los Alamitos, California, November 2003. IEEE Computer Society.
849. S.-T. Khu. Automatic Calibration of NAM Model with Multi-Objectives Consideration. Technical Report 1298-1, National University of Singapore/Danish Hydraulic Institute, December 1998.
850. S. Khuri, T. Bäck, and J. Heitkötter. The zero/one multiple knapsack problem and genetic algorithms. In *Proceedings of the 1994 ACM Symposium on Applied Computing*, pages 188–193, New York, 1994. ASME Press.
851. A. A. Khwaja, M. O. Rahman, and M. Wagner. Inverse Kinematics of Arbitrary Robotic Manipulators using Genetic Algorithms. In J. Lenarcic and M. L. Justy, editors, *Advances in Robot Kinematics: Analysis and Control*, pages 375–382. Kluwer Academic Publishers, 1998.
852. R. Kicinger, T. Arciszewski, and K. De Jong. Evolutionary computation and structural design: A survey of the state-of-the-art. *Computers & Structures*, 83:1943–1978, 2005.
853. D. Kim. Structural Risk Minimization on Decision Trees Using an Evolutionary Multiobjective Optimization. In M. Keijzer, U.-M. O'Reilly, S. M. Lucas, E. Costa, and T. Soule, editors, *Genetic Programming. 7th European Conference, EuroGP 2004*, pages 338–348, Coimbra, Portugal, April 2004. Springer. Lecture Notes in Computer Science, Vol. 3003.
854. K. Kim and R. Smith. Systematic procedure for designing processes with multiple environmental objectives. *Environmental Science & Technology*, 39(7):2394–2405, April 2005.
855. Y. Kim, J.-H. Kim, and K.-H. Han. Quantum-inspired Multiobjective Evolutionary Algorithm for Multiobjective 0/1 Knapsack Problems. In *2006 IEEE Congress on Evolutionary Computation (CEC'2006)*, pages 9151–9156, Vancouver, BC, Canada, July 2006. IEEE.
856. Y. Kim, W. N. Street, and F. Menczer. An Evolutionary Multi-Objective Local Selection Algorithm for Customer Targeting. In *Proceedings of the Congress on Evolutionary Computation 2001 (CEC'2001)*, volume 2, pages 759–766, Piscataway, New Jersey, May 2001. IEEE Service Center.
857. Y.-J. Kim and J. Ghaboussi. A New Genetic Algorithm Based Control Method Using State Space Reconstruction. In *Proceedings of the Second World Conference on Struc. Control*, pages 2007–2014, Kyoto, Japan, 1998.
858. Y.-J. Kim and J. Ghaboussi. A New Method of Reduced Order Feedback Control Using Genetic Algorithms. *Earthquake Engineering and Structural Dynamics*, 28(2):193–212, 1999.
859. Y. K. Kim, C. J. Hyun, and Y. Kim. Sequencing in mixed model assembly lines: a genetic algorithm approach. *Computers and Operations Research*, 23:1131–1145, 1996.
860. K. E. Kinnear, R. E. Smith, and Z. Michalewicz. Derivative Methods. In T. Bäck, D. Fogel, and Z. Michalewicz, editors, *Handbook of Evolutionary Computation*, volume 1, pages B1.5:1–B1.5:15. IOP Publishing Ltd. and Oxford University Press, 1997.

861. S. Kirkpatrick, C. Gellatt, and M. Vecchi. Optimization by Simulated Annealing. *Science*, 220(4598):671–680, 1983.
862. M. Kirley. MEA: A metapopulation evolutionary algorithm for multi-objective optimisation problems. In *Proceedings of the Congress on Evolutionary Computation 2001 (CEC'2001)*, volume 1, pages 949–956, Piscataway, New Jersey, May 2001. IEEE Service Center.
863. H. Kita, Y. Yabumoto, N. Mori, and Y. Nishikawa. Multi-Objective Optimization by Means of the Thermodynamical Genetic Algorithm. In H.-M. Voigt, W. Ebeling, I. Rechenberg, and H.-P. Schwefel, editors, *Parallel Problem Solving from Nature—PPSN IV*, pages 504–512. Springer-Verlag. Lecture Notes in Computer Science No. 1141, Berlin, Germany, September 1996.
864. C. N. Klahr. Multiple Objectives in Mathematical Programming. *Operations Research*, 6(6):849–855, 1958.
865. M. P. Kleeman and G. B. Lamont. Solving the Aircraft Engine Maintenance Scheduling Problem Using a Multi-objective Evolutionary Algorithm. In C. A. Coello Coello, A. Hernández Aguirre, and E. Zitzler, editors, *Evolutionary Multi-Criterion Optimization. Third International Conference, EMO 2005*, pages 782–796, Guanajuato, México, March 2005. Springer. Lecture Notes in Computer Science Vol. 3410.
866. M. P. Kleeman and G. B. Lamont. Coevolutionary Multi-Objective EAs: The Next Frontier? In *2006 IEEE Congress on Evolutionary Computation (CEC'2006)*, pages 6190–6199, Vancouver, BC, Canada, July 2006. IEEE.
867. M. P. Kleeman, G. B. Lamont, A. Cooney, and T. R. Nelson. A Multi-tiered Memetic Multiobjective Evolutionary Algorithm for the Design of Quantum Cascade Lasers. In S. Obayashi, K. Deb, C. Poloni, T. Hiroyasu, and T. Murata, editors, *Evolutionary Multi-Criterion Optimization, 4th International Conference, EMO 2007*, pages 186–200, Matshushima, Japan, March 2007. Springer. Lecture Notes in Computer Science Vol. 4403.
868. M. P. Kleeman, G. B. Lamont, K. M. Hopkinson, and S. R. Graham. Solving Multicommodity Capacitated Network Design Problems using a Multiobjective Evolutionary Algorithm. In *IEEE Symposium on Computational Intelligence in Security and Defense Applications (CISDA 2007)*, pages 33–41. IEEE Press, April 2007.
869. R. M. Kling. *Optimization by Simulated Evolution and its Application to Cell Placement*. PhD thesis, University of Illinois at Urbana-Champaign, Urbana, Illinois, 1990.
870. A. Klinger. Vector-Valued Performance Criteria. *IEEE Transactions on Automatic Control*, AC-9(1):117–118, 1964.
871. M. R. Knarr, M. N. Goltz, G. B. Lamont, and J. Huang. *In Situ* Bioremediation of Perchlorate-Contaminated Groundwater using a Multi-Objective Parallel Evolutionary Algorithm. In *Proceedings of the 2003 Congress on Evolutionary Computation (CEC'2003)*, volume 3, pages 1604–1611, Canberra, Australia, December 2003. IEEE Press.
872. J. Knowles. ParEGO: A Hybrid Algorithm With On-Line Landscape Approximation for Expensive Multiobjective Optimization Problems. *IEEE Transactions on Evolutionary Computation*, 10(1):50–66, February 2006.
873. J. Knowles and D. Corne. M-PAES: A Memetic Algorithm for Multiobjective Optimization. In *2000 Congress on Evolutionary Computation*, volume 1, pages 325–332, Piscataway, New Jersey, July 2000. IEEE Service Center.

874. J. Knowles and D. Corne. On Metrics for Comparing Nondominated Sets. In *Congress on Evolutionary Computation (CEC'2002)*, volume 1, pages 711–716, Piscataway, New Jersey, May 2002. IEEE Service Center.
875. J. Knowles and D. Corne. Towards Landscape Analyses to Inform the Design of Hybrid Local Search for the Multiobjective Quadratic Assignment Problem. In A. Abraham, J. R. del Solar, and M. Köppen, editors, *Soft Computing Systems: Design, Management and Applications*, pages 271–279, Amsterdam, 2002. IOS Press. ISBN 1-58603-297-6.
876. J. Knowles and D. Corne. Instance Generators and Test Suites for the Multiobjective Quadratic Assignment Problem. In C. M. Fonseca, P. J. Fleming, E. Zitzler, K. Deb, and L. Thiele, editors, *Evolutionary Multi-Criterion Optimization. Second International Conference, EMO 2003*, pages 295–310, Faro, Portugal, April 2003. Springer. Lecture Notes in Computer Science. Volume 2632.
877. J. Knowles and D. Corne. Properties of an Adaptive Archiving Algorithm for Storing Nondominated Vectors. *IEEE Transactions on Evolutionary Computation*, 7(2):100–116, April 2003.
878. J. Knowles and D. Corne. Bounded Pareto Archiving: Theory and Practice. In X. Gandibleux, M. Sevaux, K. Sörensen, and V. T'kindt, editors, *Metaheuristics for Multiobjective Optimisation*, pages 39–64, Berlin, 2004. Springer. Lecture Notes in Economics and Mathematical Systems Vol. 535.
879. J. Knowles and D. Corne. Memetic Algorithms for Multiobjective Optimization: Issues, Methods and Prospects. In W. E. Hart, N. Krasnogor, and J. Smith, editors, *Recent Advances in Memetic Algorithms*, pages 313–352. Springer. Studies in Fuzziness and Soft Computing, Vol. 166, 2005.
880. J. Knowles and D. Corne. Quantifying the Effects of Objective Space Dimension in Evolutionary Multiobjective Optimization. In S. Obayashi, K. Deb, C. Poloni, T. Hiroyasu, and T. Murata, editors, *Evolutionary Multi-Criterion Optimization, 4th International Conference, EMO 2007*, pages 757–771, Matshushima, Japan, March 2007. Springer. Lecture Notes in Computer Science Vol. 4403.
881. J. Knowles and E. J. Hughes. Multiobjective Optimization on a Budget of 250 Evaluations. In C. A. Coello Coello, A. Hernández Aguirre, and E. Zitzler, editors, *Evolutionary Multi-Criterion Optimization. Third International Conference, EMO 2005*, pages 176–190, Guanajuato, México, March 2005. Springer. Lecture Notes in Computer Science Vol. 3410.
882. J. Knowles, L. Thiele, and E. Zitzler. A Tutorial on the Performance Assessment of Stochastic Multiobjective Optimizers. 214, Computer Engineering and Networks Laboratory (TIK), ETH Zurich, Switzerland, Feb. 2006. revised version.
883. J. D. Knowles. *Local-Search and Hybrid Evolutionary Algorithms for Pareto Optimization*. PhD thesis, The University of Reading, Department of Computer Science, Reading, UK, January 2002.
884. J. D. Knowles and D. W. Corne. Local Search, Multiobjective Optimization and the Pareto Archived Evolution Strategy. In B. e. a. McKay, editor, *Proceedings of Third Australia-Japan Joint Workshop on Intelligent and Evolutionary Systems*, pages 209–216, Ashikaga, Japan, November 1999. Ashikaga Institute of Technology.
885. J. D. Knowles and D. W. Corne. The Pareto Archived Evolution Strategy: A New Baseline Algorithm for Multiobjective Optimisation. In *1999 Congress*

on Evolutionary Computation, pages 98–105, Washington, D.C., July 1999. IEEE Service Center.
886. J. D. Knowles and D. W. Corne. Approximating the Nondominated Front Using the Pareto Archived Evolution Strategy. *Evolutionary Computation*, 8(2):149–172, 2000.
887. J. D. Knowles and D. W. Corne. Benchmark Problem Generators and Results for the Multiobjective Degree-Constrained Minimum Spanning Tree Problem. In L. Spector, E. D. Goodman, A. Wu, W. Langdon, H.-M. Voigt, M. Gen, S. Sen, M. Dorigo, S. Pezeshk, M. H. Garzon, and E. Burke, editors, *Proceedings of the Genetic and Evolutionary Computation Conference (GECCO'2001)*, pages 424–431, San Francisco, California, 2001. Morgan Kaufmann Publishers.
888. J. D. Knowles, D. W. Corne, and M. Fleischer. Bounded Archiving using the Lebesgue Measure. In *Proceedings of the 2003 Congress on Evolutionary Computation (CEC'2003)*, volume 4, pages 2490–2497, Canberra, Australia, December 2003. IEEE Press.
889. J. D. Knowles, M. J. Oates, and D. W. Corne. Multiobjective Evolutionary Algorithms Applied to two Problems in Telecommunications. *BT Technology Journal*, 18(4):51–64, October 2000.
890. J. D. Knowles, R. A. Watson, and D. W. Corne. Reducing Local Optima in Single-Objective Problems by Multi-objectivization. In E. Zitzler, K. Deb, L. Thiele, C. A. Coello Coello, and D. Corne, editors, *First International Conference on Evolutionary Multi-Criterion Optimization*, pages 268–282. Springer-Verlag. Lecture Notes in Computer Science No. 1993, 2001.
891. T. E. Koch and A. Zell. MOCS: Multi-Objective Clustering Selection Evolutionary Algorithm. In W. Langdon, E. Cantú-Paz, K. Mathias, R. Roy, D. Davis, R. Poli, K. Balakrishnan, V. Honavar, G. Rudolph, J. Wegener, L. Bull, M. Potter, A. Schultz, J. Miller, E. Burke, and N. Jonoska, editors, *Proceedings of the Genetic and Evolutionary Computation Conference (GECCO'2002)*, pages 423–430, San Francisco, California, July 2002. Morgan Kaufmann Publishers.
892. I. Kokolo, K. Hajime, and K. Shigenobu. Failure of Pareto-based MOEAs: Does Non-dominated Really Mean Near to Optimal? In *Proceedings of the Congress on Evolutionary Computation 2001 (CEC'2001)*, volume 2, pages 957–962, Piscataway, New Jersey, May 2001. IEEE Service Center.
893. M. Koksalan, M. Azizoglu, and S. Kondakci. Minimizing flowtime and maximum earliness in a single machine. *IIE Transactions*, 30:192–200, 1998.
894. R. Komuro. *Multi-Objective Evolutionary Algorithms for Ecological Process Models*. PhD thesis, University of Washington, Seattle, Washington, USA, December 2005.
895. B. O. Koopman. The Optimum Distribution of Effort. *Operations Research*, 1(2):52–63, 1953.
896. B. O. Koopman. Fallacies in Operations Research. *Operations Research*, 4(4):422–426, 1956.
897. T. C. Koopmans. Analysis of Production as an efficient combination of activities. In T. C. Koopmans, editor, *Activity Analysis of Production and Allocation, Cowles Commision Monograph No. 13*, pages 33–97. John Wiley and Sons, New York, New York, 1951.
898. M. Köppen. On the Benchmarking of Multiobjective Optimization Algorithm. In V. Palade, R. J. Howlett, and L. C. Jain, editors, *Proceedings of*

the 7th International Conference on Knowledge-Based Intelligent Information and Engineering Systems (KES 2003). Part I, pages 379–385, Oxford, UK, September 2003. Springer. Lecture Notes on Computer Science Vol. 2773.
899. M. Köppen and S. Rudlof. Multiobjective Optimization by Nessy Algorithm. In R. Roy, T. Furuhashi, and P. Chawdhry, editors, *Advances in Soft Computing*, pages 357–368, London, 1998. Springer.
900. M. Köppen, R. Vicente-Garcia, and B. Nickolay. Fuzzy-Pareto-Dominance and Its Application in Evolutionary Multi-objective Optimization. In C. A. Coello Coello, A. Hernández Aguirre, and E. Zitzler, editors, *Evolutionary Multi-Criterion Optimization. Third International Conference, EMO 2005*, pages 399–412, Guanajuato, México, March 2005. Springer. Lecture Notes in Computer Science Vol. 3410.
901. M. Köppen and K. Yoshida. Substitute Distance Assignments in NSGA-II for Handling Many-Objective Optimization Problems. In S. Obayashi, K. Deb, C. Poloni, T. Hiroyasu, and T. Murata, editors, *Evolutionary Multi-Criterion Optimization, 4th International Conference, EMO 2007*, pages 727–741, Matshushima, Japan, March 2007. Springer. Lecture Notes in Computer Science Vol. 4403.
902. P. Korhonen. Reference direction approach to multiple objective linear programming: Historical overview. In M. H. Karwan, J. Spronk, and J. Wallenius, editors, *Essays in Decision Making: A Volume in Honour of Stanley Zionts*, pages 72–94. Springer-Verlag, Berlin, 1997.
903. U. Korn and T. T. Binh. The parallel evolution strategy toolbox. *Automatisierungstechnik*, 4:207–208, 1998.
904. K. Koski. Multicriterion Optimization in Structural Design. In E. Atrek, R. H. Gallagher, K. M. Ragsdell, and O. C. Zienkiewicz, editors, *New Directions in Optimum Structural Design*, pages 483–503. John Wiley and Sons, 1984.
905. J. R. Koza. *Genetic Programming. On the Programming of Computers by Means of Natural Selection*. The MIT Press, Cambridge, Massachusetts, 1992.
906. N. Krami, M. A. El-Sharkawi, and M. Akherraz. Multi Objective Particle Swarm Optimization Technique for Reactive Power Planning. In *2006 Swarm Intelligence Symposium (SIS'06)*, pages 170–174, Indianapolis, Indiana, USA, May 2006. IEEE Press.
907. M. Krause and V. Nissen. On Using Penalty Functions and Multicriteria Optimisation Techniques in Facility Layout. In J. Biethahn and V. Nissen, editors, *Evolutionary Algorithms in Management Applications*, pages 153–166. Springer-Verlag, Berlin, 1995.
908. V. Krmicek and M. Sebag. Functional Brain Imaging with Multi-objective Multi-modal Evolutionary Optimization. In T. P. Runarsson, H.-G. Beyer, E. Burke, J. J. Merelo-Guervós, L. D. Whitley, and X. Yao, editors, *Parallel Problem Solving from Nature - PPSN IX, 9th International Conference*, pages 382–391. Springer. Lecture Notes in Computer Science Vol. 4193, Reykjavik, Iceland, September 2006.
909. R. Krzysztofowicz and L. Duckstein. Preference criterion for flood control under uncertainty. *Water Resources Research*, 15(3):513–520, 1979.
910. H. W. Kuhn and A. W. Tucker. Nonlinear Programming. In J. Neyman, editor, *Proceedings of the Second Berkeley Symposium on Mathematical Statistics and Probability*, pages 481–492, Berkeley, California, 1951. University of California Press.

911. T. Kuhn. *The Structure of Scientific Revolutions*. University of Chicago Press, 1962.
912. S. Kukkonen and K. Deb. A Fast and Effective Method for Pruning of Nondominated Solutions in Many-Objective Problems. In T. P. Runarsson, H.-G. Beyer, E. Burke, J. J. Merelo-Guervós, L. D. Whitley, and X. Yao, editors, *Parallel Problem Solving from Nature - PPSN IX, 9th International Conference*, pages 553–562. Springer. Lecture Notes in Computer Science Vol. 4193, Reykjavik, Iceland, September 2006.
913. S. Kukkonen and J. Lampinen. An Extension of Generalized Differential Evolution for Multi-objective Optimization with Constraints. In *Parallel Problem Solving from Nature - PPSN VIII*, pages 752–761, Birmingham, UK, September 2004. Springer-Verlag. Lecture Notes in Computer Science Vol. 3242.
914. S. Kukkonen and J. Lampinen. GDE3: The third Evolution Step of Generalized Differential Evolution. In *2005 IEEE Congress on Evolutionary Computation (CEC'2005)*, volume 1, pages 443–450, Edinburgh, Scotland, September 2005. IEEE Service Center.
915. S. Kulturel-Konak, A. E. Smith, and B. A. Norman. Multi-objective tabu search using a multinomial probability mass function. *European Journal of Operational Research*, 169:918–931, 2006.
916. R. Kumar. *Feature Selection, Representation and Classification*. PhD thesis, University of Sheffield, Sheffield, UK, 1997.
917. R. Kumar and N. Banerjee. Running time analysis of a multiobjective evolutionary algorithm on simple and hard problems. In A. H. Wright, M. D. Vose, K. A. D. Jong, and L. M. Schmitt, editors, *Foundations of Genetic Algorithms. 8th International Workshop, FOGA 2005*, pages 112–131, Aizu-Wakamatsu City, Japan, January 2005. Springer. Lecture Notes in Computer Science Vol. 3469.
918. R. Kumar and N. Banerjee. Analysis of a multiobjective evolutionary algorithm on the 0-1 knapsack problem. *Theoretical Computer Science*, 358(1):104–120, July 2006.
919. R. Kumar, S. Prasanth, and M. Sudarshan. Topological Design of Mesh Communication Networks using Multiobjecitve Genetic Optimisation. In *PPSN/SAB Workshop on Multiobjective Problem Solving from Nature (MPSN)*, Paris, France, September 2000.
920. R. Kumar and P. Rockett. Decomposition of High Dimensional Pattern Spaces for Hierarchical Classification. In *Proceedings of the Workshop on Statistical Techniques in Pattern Recognition*, Prague, Czech Republic, June 1997.
921. R. Kumar and P. Rockett. Decomposition of High Dimensional Pattern Spaces for Hierarchical Classification. *Kybernetika*, 34(4):435–442, 1998.
922. R. Kumar and P. Rockett. Multiobjective Genetic Algorithm Partitioning for Hierarchical Learning of High-Dimensional Pattern Spaces: A Learning-Follows-Decomposition Strategy. *IEEE Transactions on Neural Networks*, 9(5):822–830, 1998.
923. R. Kumar and P. Rockett. Improved Sampling of the Pareto-Front in Multiobjective Genetic Optimizations by Steady-State Evolution: A Pareto Converging Genetic Algorithm. *Evolutionary Computation*, 10(3):283–314, Fall 2002.
924. R. Kumar and P. I. Rockett. Assessing the Convergence of Rank-Based Multiobjective Genetic Algorithms. In *Proceedings of the 2nd IEE/IEEE International Conference on Genetic Algorithms in Engineering Systems: Innovations*

and Applications (GALESIA'97), pages 19–23, Glasgow, Scotland, September 1997. IEE.
925. S. V. Kumar and S. R. Ranjithan. Evaluation of the Constraint Method-Based Evolutionary Algorithm (CMEA) for a Tree-Objective Optimization Problem. In W. Langdon, E. Cantú-Paz, K. Mathias, R. Roy, D. Davis, R. Poli, K. Balakrishnan, V. Honavar, G. Rudolph, J. Wegener, L. Bull, M. Potter, A. Schultz, J. Miller, E. Burke, and N. Jonoska, editors, *Proceedings of the Genetic and Evolutionary Computation Conference (GECCO'2002)*, pages 431–438, San Francisco, California, July 2002. Morgan Kaufmann Publishers.
926. V. Kumar, A. Grama, A. Gupta, and G. Karypis. *Introduction to Parallel Computing: Design and Analysis of Algorithms*. The Benjamin/Cummings Publishing Company, Inc., Redwood City, CA, 1994.
927. S. Kundu. A multicriteria genetic algorithm to solve optimization problems in structural engineering design. In B. Kumar, editor, *Information Processing in Civil and Structural Engineering Design*, pages 225–233, Glasgow, Scotland, August 1996. Civil-Comp Press Ltd.
928. S. Kundu and S. Kawata. AI in Control System Design Using a New Paradigm for Design Representation. In J. S. Gero and F. Sudweeks, editors, *Artificial Intelligence in Design*, pages 135–150. Kluwer Academic Publishers, The Netherlands, 1996.
929. S. Kundu and A. Osyczka. The effect of genetic algorithm selection mechanisms on multicriteria optimization using the distance method. In *Proceedings of the Fifth International Conference on Intelligent Systems*, pages 164–168, Reno, Nevada, 1996. International Society for Computers and Their Applications (ISCA).
930. H. Kung, F. Luccio, and F. Preparata. On finding the maxima of a set of vectors. *Journal of the Association for Computing Machinery*, 22(4):469–476, 1975.
931. S. Künzli, S. Bleuler, L. Thiele, and E. Zitzler. A Computer Engineering Benchmark Application for Multiobjective Optimizers. In C. A. Coello Coello and G. B. Lamont, editors, *Applications of Multi-Objective Evolutionary Algorithms*, pages 269–294. World Scientific, Singapore, 2004.
932. S. Kurahashi and T. Terano. A Genetic Algorithm with Tabu Search for Multimodal and Multiobjective Function Optimization. In D. Whitley, D. Goldberg, E. Cantú-Paz, L. Spector, I. Parmee, and H.-G. Beyer, editors, *Proceedings of the Genetic and Evolutionary Computation Conference (GECCO'2000)*, pages 291–298, San Francisco, California, 2000. Morgan Kaufmann.
933. A. Kurapati and S. Azarm. Immune Network Simulation with Multiobjective Genetic Algorithms for Multidisciplinary Design Optimization. *Engineering Optimization*, 33:245–260, 2000.
934. F. Kursawe. A Variant of Evolution Strategies for Vector Optimization. In H.-P. Schwefel and R. Männer, editors, *Parallel Problem Solving from Nature. 1st Workshop, PPSN I*, pages 193–197, Dortmund, Germany, October 1991. Springer-Verlag. Lecture Notes in Computer Science No. 496.
935. F. Kursawe. Evolution strategies for vector optimization. In *Preliminary Procedings of the Tenth International Conference on Multiple Criteria Decision Making*, pages 187–193, Taipei, China, July 1992. National Chiao Tung University.

936. A. Kurup, K. Hidajat, and A. Ray. Comparative study of modified simulated moving bed systems at optimal conditions for the separation of ternary mixtures of xylene isomers. *Industrial & Engineering Chemistry Research*, 45(18):6251–6265, August 30 2006.
937. O. Labé, I. Abi-Zeid, and J.-M. Martel. Comparaison de deux méthodes multicritères traitant de l'information mixte : NAIADE et PAMSSEM. In *Journées de l'Optimisation 1999*, Montreal, Canada, May 1999.
938. M. Laguna and R. Martí. *Scatter Search : Methodology and Implementations in C*. Kluwer Academic Publishers, 2003. ISBN 1-402-07376-3.
939. J. R. F. Lagunas Jiménez. *Sintonización de controladores PID mediante un algoritmo genético multiobjetivo (NSGA-II)*. PhD thesis, Departamento de Control Automático, CINVESTAV-IPN, México, D.F., April 2004. (in Spanish).
940. M. Lahanas, N. Milickovic, D. Baltas, and N. Zamboglou. Application of Multiobjective Evolutionary Algorithms for Dose Optimization Problems in Brachytherapy. In E. Zitzler, K. Deb, L. Thiele, C. A. Coello Coello, and D. Corne, editors, *First International Conference on Evolutionary Multi-Criterion Optimization*, pages 574–587. Springer-Verlag. Lecture Notes in Computer Science No. 1993, 2001.
941. Y.-J. Lai and C.-L. Hwang. *Fuzzy Multiple Objective Decision Making: Methods and Applications*. Springer-Verlag, 1994.
942. G. B. Lamont, editor. *Compendium of Parallel Programs for the Intel iPSC Computers*. Department of Electrical and Computer Engineering, Graduate School of Engineering, Air Force Institute of Technology, Wright-Patterson AFB, OH 45433, 1993.
943. G. B. Lamont, S. M. Brown, and G. H. G. Jr. Evolutionary Algorithms Combined with Deterministic Search. In V. Porto, N. Saravanan, D. Waagen, and A. Eiben, editors, *Evolutionary Programming VII, Proceedings of the 7th Annual Conference on Evolutionary Programming*, pages 517–526, Berlin, 1998. Springer.
944. G. B. Lamont, R. Marmelstein, and D. A. Van Veldhuizen. A Distributed Architecture for a Self-Adaptive computer Virus Immune System. In D. Corne, M. Dorigo, and F. Glover, editors, *New Ideas in Optimization*, pages 167–183. McGraw-Hill, 2000.
945. G. B. Lamont, J. N. Slear, and K. Melendez. UAV Swarm Mission Planning and Routing using Multi-Objective Evolutionary Algorithms. In *IEEE Symposium on Computational Intelligence in Multicriteria Decision Making (MCDM 2007)*, pages 10–20. IEEE Press, April 2007.
946. J. Lampinen. DE's selection rule for multiobjective optimization. Technical report, Lappeenranta University of Technology, Department of Information Technology, 2001.
947. R. Landa Becerra and C. A. Coello Coello. Cultured differential evolution for constrained optimization. *Computer Methods in Applied Mechanics and Engineering*, 195(33–36):4303–4322, July 1 2006.
948. R. Landa Becerra and C. A. Coello Coello. Solving Hard Multiobjective Optimization Problems Using ε-Constraint with Cultured Differential Evolution. In T. P. Runarsson, H.-G. Beyer, E. Burke, J. J. Merelo-Guervós, L. D. Whitley, and X. Yao, editors, *Parallel Problem Solving from Nature - PPSN IX, 9th International Conference*, pages 543–552. Springer. Lecture Notes in Computer Science Vol. 4193, Reykjavik, Iceland, September 2006.

949. J. D. Landa Silva. *Metaheuristic and Multiobjective Approaches for Space Allocation*. PhD thesis, School of Computer Science and Information Technology, University of Nottingham, UK, November 2003.
950. W. B. Langdon. Evolving data structures using genetic programming. In L. Eshelman, editor, *Proceedings of the Sixth International Conference on Genetic Algorithms (ICGA'95)*, pages 295–302, San Mateo, California, July 1995. Morgan Kaufmann Publishers.
951. W. B. Langdon. Data Structures and Genetic Programming. In P. J. Angeline and K. E. Kinnear, Jr., editors, *Advances in Genetic Programming 2*, chapter 20, pages 395–414. MIT Press, Cambridge, MA, USA, 1996.
952. W. B. Langdon. Using Data Structures within Genetic Programming. In J. R. Koza, D. E. Goldberg, D. B. Fogel, and R. L. Riolo, editors, *Genetic Programming 1996: Proceedings of the First Annual Conference*, pages 141–148, Stanford University, CA, USA, 28–31 July 1996. MIT Press.
953. P. L. Lanzi, W. Stolzmann, and S. W. Wilson, editors. *Advances in Learning Classifier Systems*. Springer. Lecture Notes in Computer Science Vol. 2321, San Francisco, California, USA, August 2002.
954. P. Larrañaga and J. A. Lozano, editors. *Estimation of Distribution Algorithms. A New Tool for Evolutionary Computation*. Kluwer Academic Publishers, Boston/Dordrecht/London, 2002.
955. M. Laumanns. *Analysis and Applications of Evolutionary Multiobjective Optimization Algorithms*. PhD thesis, Swiss Federal Institute of Technology, Zürich, Switzerland, 2003.
956. M. Laumanns and J. Ocenasek. Bayesian Optimization Algorithms for Multi-objective Optimization. In J. J. Merelo Guervós, P. Adamidis, H.-G. Beyer, J.-L. F.-V. nas, and H.-P. Schwefel, editors, *Parallel Problem Solving from Nature—PPSN VII*, pages 298–307, Granada, Spain, September 2002. Springer-Verlag. Lecture Notes in Computer Science No. 2439.
957. M. Laumanns, G. Rudolph, and H.-P. Schwefel. A Spatial Predator-Prey Approach to Multi-Objective Optimization: A Preliminary Study. In A. E. Eiben, T. Bäck, M. Schoenauer, and H.-P. Schwefel, editors, *Parallel Problem Solving From Nature — PPSN V*, pages 241–249, Amsterdam, Holland, 1998. Springer-Verlag. Lecture Notes in Computer Science No. 1498.
958. M. Laumanns, G. Rudolph, and H.-P. Schwefel. Mutation Control and Convergence in Evolutionary Multi-Objective Optimization. In *Proceedings of the 7th International Mendel Conference on Soft Computing (MENDEL 2001)*, pages 24–29, Brno, Czech Republic, June 2001. Brno University of Technology.
959. M. Laumanns, L. Thiele, K. Deb, and E. Zitzler. Combining Convergence and Diversity in Evolutionary Multi-objective Optimization. *Evolutionary Computation*, 10(3):263–282, Fall 2002.
960. M. Laumanns, L. Thiele, and E. Zitzler. Running Time Analysis of Multiobjective Evolutionary Algorithms on Pseudo-Boolean Functions. *IEEE Transactions on Evolutionary Computation*, 8(2):170–182, April 2004.
961. M. Laumanns, L. Thiele, and E. Zitzler. An efficient, adaptive parameter variation scheme for metaheuristics based on the epsilon-constraint method. *European Journal of Operational Research*, 169:932–942, 2006.
962. M. Laumanns, L. Thiele, E. Zitzler, E. Welzl, and K. Deb. Running Time Analysis of Multi-objective Evolutionary Algorithms on a Simple Discrete

Optimization Problem. In J. J. Merelo Guervós, P. Adamidis, H.-G. Beyer, J.-L. F.-V. nas, and H.-P. Schwefel, editors, *Parallel Problem Solving from Nature—PPSN VII*, pages 44–53, Granada, Spain, September 2002. Springer-Verlag. Lecture Notes in Computer Science No. 2439.
963. M. Laumanns, E. Zitzler, and L. Thiele. A Unified Model for Multi-Objective Evolutionary Algorithms with Elitism. In *2000 Congress on Evolutionary Computation*, volume 1, pages 46–53, Piscataway, New Jersey, July 2000. IEEE Service Center.
964. M. Laumanns, E. Zitzler, and L. Thiele. On the Effects of Archiving, Elitism, and Density Based Selection in Evolutionary Multi-objective Optimization. In E. Zitzler, K. Deb, L. Thiele, C. A. Coello Coello, and D. Corne, editors, *First International Conference on Evolutionary Multi-Criterion Optimization*, pages 181–196. Springer-Verlag. Lecture Notes in Computer Science No. 1993, 2001.
965. N. Laumanns, M. Laumanns, and D. Neunzig. Multi-objective Design Space Exploration of Road Trains with Evolutionary Algorithms. In E. Zitzler, K. Deb, L. Thiele, C. A. Coello Coello, and D. Corne, editors, *First International Conference on Evolutionary Multi-Criterion Optimization*, pages 612–623. Springer-Verlag. Lecture Notes in Computer Science No. 1993, 2001.
966. M. Lawrence. Multiobjective Genetic Algorithms for Materialized View Selection in OLAP Data Warehouses. In M. K. et al., editor, *2006 Genetic and Evolutionary Computation Conference (GECCO'2006)*, volume 1, pages 699–706, Seattle, Washington, USA, July 2006. ACM Press. ISBN 1-59593-186-4.
967. A. Lazzaretto and A. Toffolo. Energy, economy and environment as objectives in multi-criterion optimization of thermal systems design. *Energy*, 29(8):1139–1157, June 2004.
968. R. G. Le Riche and R. T. Haftka. Optimization of Laminate Stacking Sequence for Buckling Load Maximization by Genetic Algorithm. *AIAA Journal*, 31(5):951–970, 1993.
969. D. Lee. Multiobjective Design of a Marine Vehicle with Aid of Design Knowledge. *International Journal for Numerical Methods in Engineering*, 40:2665–2677, 1997.
970. J. Lee and P. Hajela. Parallel Genetic Algorithm Implementation in Multidisciplinary Rotor Blade Design. *Journal of Aircraft*, 33(5):962–969, September-October 1996.
971. M. A. Lee and H. Esbensen. Multiobjective Optimization using Fuzzy/Evolutionary Algorithms. In *Proceedings of the International Society for Computers and Their Applications (ISCA'96)*, pages 67–70, San Francisco, California, 1996.
972. M. A. Lee and H. Esbensen. Fuzzy/Multiobjective Genetic Systems for Intelligent Systems Design Tools and Components. In W. Pedrycz, editor, *Fuzzy Evolutionary Computation*, pages 57–80. Kluwer Academic Publishers, Boston, Massachusetts, 1997.
973. M. A. Lee, H. Esbensen, and L. Lemaitre. The Design of Hybrid Fuzzy/Evolutionary Multiobjective Optimization Algorithms. In *Proceedings of the 1995 IEEE/Nagoya University World Wiseperson Workshop*, pages 118–125, Nagoya, Japan, 1995.
974. M. A. Lee and R. Hartani. A multiobjective evolutionary algorithms approach to fuzzy modeling. In *Proceedings of the Second Annual Conference on*

Information Science (JCIS'95), pages 460–463, Shell Island, North Carolina, 1995.
975. S. Lee. *Goal Programming for Decision Analysis*. Auerbach, Philadelphia, 1972.
976. S. Lee and V. Jaaskelainen. Goal Programming: Management's math model. *Industrial Engineering*, pages 30–35, February 1971.
977. H. A. Leiva, S. C. Esquivel, and R. H. Gallard. Multiplicity and Local Search in Evolutionary Algorithms to Build the Pareto Front. In *Proceedings of the XX International Conference of the Chilean Computer Science Society*, pages 7–13, Piscataway, New Jersey, 2000. IEEE Computer Society Press.
978. A. Levy, A. Montalvo, S. Gomez, and A. Calderon. Topics in Global Optimization. In J. Hennart, editor, *Numerical Analysis*, pages 18–33. Springer-Verlag. Lecture Notes in Mathematics Vol. 909, New York, 1981.
979. J. C. Leyva-Lopez and M. A. Aguilera-Contreras. A Multiobjective Evolutionary Algorithm for Deriving Final Ranking from a Fuzzy Outranking Relation. In C. A. Coello Coello, A. Hernández Aguirre, and E. Zitzler, editors, *Evolutionary Multi-Criterion Optimization. Third International Conference, EMO 2005*, pages 235–249, Guanajuato, México, March 2005. Springer. Lecture Notes in Computer Science Vol. 3410.
980. J. C. Leyva López and E. Fernández González. A Genetic Algorithm for Deriving Final Ranking from a Fuzzy Outranking Relation. *Foundations of Computing and Decision Sciences*, 24(1):33–47, 1999.
981. J. C. Leyva-López and E. Fernández-González. A new method for group decision support based on ELECTRE III methodology. *European Journal of Operational Research*, 148(1):14–27, July 2003.
982. H. Li and Q. Zhang. A Multiobjective Differential Evolution Based on Decomposition for Multiobjective Optimization with Variable Linkages. In T. P. Runarsson, H.-G. Beyer, E. Burke, J. J. Merelo-Guervós, L. D. Whitley, and X. Yao, editors, *Parallel Problem Solving from Nature - PPSN IX, 9th International Conference*, pages 583–592. Springer. Lecture Notes in Computer Science Vol. 4193, Reykjavik, Iceland, September 2006.
983. J. Li and N. Satofuka. Optimization design of a compressor cascade airfoil using a Navier-stokes solver and genetic algorithms. *Proceedings of the Institution of Mechanical Engineering Part A—Journal of Power and Energy*, 216(A2):195–202, 2002.
984. J.-P. Li, M. E. Balazs, G. T. Parks, and P. J. Clarkson. A Species Conserving Genetic Algorithm for Multimodal Function Optimization. *Evolutionary Computation*, 10(3):207–234, Fall 2002.
985. X. Li. A Non-dominated Sorting Particle Swarm Optimizer for Multiobjective Optimization. In E. C.-P. et al., editor, *Genetic and Evolutionary Computation—GECCO 2003. Proceedings, Part I*, pages 37–48. Springer. Lecture Notes in Computer Science Vol. 2723, July 2003.
986. X. Li. A Real-Coded Predator-Prey Genetic Algorithm for Multiobjective Optimization. In C. M. Fonseca, P. J. Fleming, E. Zitzler, K. Deb, and L. Thiele, editors, *Evolutionary Multi-Criterion Optimization. Second International Conference, EMO 2003*, pages 207–221, Faro, Portugal, April 2003. Springer. Lecture Notes in Computer Science. Volume 2632.
987. X. Li. Better Spread and Convergence: Particle Swarm Multiobjective Optimization Using the Maximin Fitness Function. In K. D. et al., editor, *Genetic*

and *Evolutionary Computation–GECCO 2004. Proceedings of the Genetic and Evolutionary Computation Conference. Part I*, pages 117–128, Seattle, Washington, USA, June 2004. Springer-Verlag, Lecture Notes in Computer Science Vol. 3102.

988. Y. Li, M. Gen, and K. Ida. Evolutionary Computation for Multicriteria Solid Transportation Problem with Fuzzy Numbers. In T. Bäck, Z. Michalewicz, and H. Kitano, editors, *Proceedings of the Third IEEE Conference on Evolutionary Computation*, pages 596–601, Piscataway, New Jersey, 1996. IEEE Service Center.

989. Y. Li, K. Ida, and M. Gen. Improved Genetic Algorithm for Solving Multi-objective Solid Transportation Problem with Fuzzy Numbers. *Computers in Industrial Engineering*, 33(3-4):589–592, 1997.

990. J. Liang and P. Suganthan. Dynamic Multi-Swarm Particle Swarm Optimizer with a Novel Constraint-Handling Mechanism. In *2006 IEEE Congress on Evolutionary Computation (CEC'2006)*, pages 316–323, Vancouver, BC, Canada, July 2006. IEEE.

991. S. J. Liang and J. M. Lewis. Job Shop Scheduling Using Multiple Criteria. In *Proceedings of the Joint Hungarian-British Mechatronic Conference*, pages 77–82. Computational Mechanics, September 1994.

992. E. R. Lieberman. Soviet multi-objective mathematical programming methods: An Overview. *Management Science*, 37(9):1147–1165, September 1991.

993. A. Liefooghe, M. Basseur, L. Jourdan, and E.-G. Talbi. ParadisEO-MOEO: A Framework for Evolutionary Multi-objective Optimization. In S. Obayashi, K. Deb, C. Poloni, T. Hiroyasu, and T. Murata, editors, *Evolutionary Multi-Criterion Optimization, 4th International Conference, EMO 2007*, pages 386–400, Matshushima, Japan, March 2007. Springer. Lecture Notes in Computer Science Vol. 4403.

994. G. E. Liepins, M. R. Hilliard, J. Richardson, and M. Palmer. Genetic algorithms application to set covering and travelling salesman problems. In D. E. Brown and C. C. White, editors, *Operations research and Artificial Intelligence: The integration of problem-solving strategies*, pages 29–57. Kluwer Academic, Norwell, Massachusetts, 1990.

995. D. Lim, Y.-S. Ong, Y. Jin, B. Sendhoff, and B.-S. Lee. Efficient Hierarchical Parallel Genetic Algorithms using Grid Computing. *Future Generation Computer Systems*, 23(4):658–670, May 2007.

996. P. Limbourg. Multi-objective Optimization of Problems with Epistemic Uncertainty. In C. A. Coello Coello, A. Hernández Aguirre, and E. Zitzler, editors, *Evolutionary Multi-Criterion Optimization. Third International Conference, EMO 2005*, pages 413–427, Guanajuato, México, March 2005. Springer. Lecture Notes in Computer Science Vol. 3410.

997. J. G. Lin. Maximal Vectors and Multi-Objective Optimization. *Journal of Optimization Theory and Applications*, 18(1):41–64, January 1976.

998. S.-Y. Liong, S.-T. Khu, and W. T. Chan. Novel Application of Genetic Algorithm and Neural Network in Water Resources: Development of Pareto Front. In *Eleventh Congress of the International Association for Hydraulic Research—Asia and Pacific Division*, pages 185–194, Yogyakarta, Indonesia, 1998.

999. S.-Y. Liong, S. T. Khu, and W. T. Chang. Derivation of Pareto Front with Accelerated Convergence Genetic Algorithm, ACGA. In V. Babovic and

L. Larsen, editors, *Proceedings of the Third Hydroinformatics Conference*, 1998.
1000. J. Lis and A. E. Eiben. A Multi-Sexual Genetic Algorithm for Multiobjective Optimization. In T. Fukuda and T. Furuhashi, editors, *Proceedings of the 1996 International Conference on Evolutionary Computation*, pages 59–64, Nagoya, Japan, 1996. IEEE.
1001. T. R. Liszkai. *Modern Heuristics in Structural Damage Detection using Frequency Response Functions*. PhD thesis, Civil Engineering Department, Texas A&M University, USA, August 2003.
1002. T. R. Liszkai and A. M. Raich. Solving Inverse Problems in Structural Damage Identification Using Advanced Genetic Algorithm Representations. In *6th World Congress of Structural and Multidisciplinary Optimization*, Rio de Janeiro, Brazil, June 2005.
1003. X. Liu, D. W. Begg, and R. J. Fishwick. Genetic approach to optimal topology/controller design of adaptive structures. *International Journal for Numerical Methods in Engineering*, 41:815–830, 1998.
1004. X. Llorà, D. Goldberg, I. Traus, and E. Bernadó. Accuracy, parsimony, and generality in evolutionary learning systems via multiobjective selection. In *Learning Classifier Systems*, pages 118–142. Springer. Lecture Notes in Artificial Intelligence Vol. 2661, 2002.
1005. X. Llorà and D. E. Goldberg. Bounding the Effect of Noise in Multiobjective Learning Classifier Systems. *Evolutionary Computation*, 11(3):279–298, Fall 2003.
1006. J. D. Lohn, G. L. Haith, S. P. Colombano, and D. Stassinopoulos. A Comparison of Dynamic Fitness Schedules for Evolutionary Design of Amplifiers. In A. Stoica, D. Keymeulen, and J. Lohn, editors, *Proceedings of The First NASA/DoD Workshop on Evolvable Hardware*, pages 87–92, Los Alamitos, California, USA, July 1999. IEEE Computer Society.
1007. J. D. Lohn, W. F. Kraus, and G. L. Haith. Comparing a Coevolutionary Genetic Algorithm for Multiobjective Optimization. In *Congress on Evolutionary Computation (CEC'2002)*, volume 2, pages 1157–1162, Piscataway, New Jersey, May 2002. IEEE Service Center.
1008. J. D. Lohn, W. F. Kraus, D. S. Linden, and S. P. Colombano. Evolutionary Optimization of Yagi-Uda Antennas. In Y. Liu, K. Tanaka, M. Iwata, T. Higuchi, and M. Yasunaga, editors, *Evolvable Systems: From Biology to Hardware, 4th International Conference, ICES 2001*, pages 236–243, Tokyo, Japan, October 2001. Springer. Lecture Notes in Computer Science Vol. 2210.
1009. F. A. Lootsma. *Fuzzy Logic for Planning and Decision Making*. Delft University of Technology, The Netherlands, 1997.
1010. A. López-Jaimes and C. C. Coello. MRMOGA: Parallel Evolutionary Multiobjective Optimization using Multiple Resolutions. In *2005 IEEE Congress on Evolutionary Computation (CEC'2005)*, volume 3, pages 2294–2301, Edinburgh, Scotland, September 2005. IEEE Service Center.
1011. A. López Jaimes and C. A. Coello Coello. MRMOGA: A New Parallel Multi-Objective Evolutionary Algorithm Based on the Use of Multiple Resolutions. *Concurrency and Computation: Practice and Experience*, 19(4):397–441, March 2007.
1012. A. Loraschi, A. Tettamanzi, M. Tomassini, and P. Verda. Distributed genetic algorithms with an application to portfolio selection problems. In N. Steele

and R. Albrecht, editors, *Artificial Neural Networks and Genetic Algorithms (ICANNGA'95)*, pages 384–387, Wien, 1995. Springer.
1013. P. Loridan and J. Morgan. A Theoretical Approximation Scheme for Stackelberg Games. *Optimization Theory and Applications*, 61(1):95–110, 1989.
1014. A. V. Lotov, V. A. Bushenkov, and G. K. Kamenev. *Interactive Decision Maps. Approximation and Visualization of Pareto Frontier*. Kluwer Academic Publishers, Boston, Massachusetts, February 2004. ISBN 1-4020-7631-2.
1015. D. P. Loucks. Conflict and choice: Planning for multiple objectives. In C. Blitzer, P. Clark, and L. Taylor, editors, *Economy wide Models and Development Planning*, New York, New York, 1975. Oxford University Press.
1016. D. H. Loughlin. *Genetic Algorithm-Based Optimization in the Development of Tropospheric Ozone Control Strategies: Least Cost, Multiobjective, Alternative Generation, and Chance-Constrained Applications (Air Quality Management)*. PhD thesis, North Carolina State University, February 1998.
1017. D. H. Loughlin and S. Ranjithan. The Neighborhood constraint method: A Genetic Algorithm-Based Multiobjective Optimization Technique. In T. Bäck, editor, *Proceedings of the Seventh International Conference on Genetic Algorithms*, pages 666–673, San Mateo, California, July 1997. Michigan State University, Morgan Kaufmann Publishers.
1018. S. J. Louis and G. J. E. Rawlins. Pareto Optimality, GA-easiness and Deception. In S. Forrest, editor, *Proceedings of the Fifth International Conference on Genetic Algorithms*, pages 118–123, San Mateo, California, 1993. Morgan Kaufmann Publishers.
1019. Z. Lounis and M. Z. Cohn. Multiobjective Optimization of Prestressed Concrete Structures. *Journal of Structural Engineering*, 119(3):794–808, mar 1993.
1020. H. R. Lourenço, O. C. Martin, and T. Stützle. Iterated Local Search. In F. Glover and G. A. Kochenberger, editors, *Handbook of Metaheuristics*, pages 321–353. Kluwer Academic Publishers, Boston/Dordrecht/London, 2002.
1021. H. Lu and G. G. Yen. Rank-Density Based Multiobjective Genetic Algorithm. In *Congress on Evolutionary Computation (CEC'2002)*, volume 1, pages 944–949, Piscataway, New Jersey, May 2002. IEEE Service Center.
1022. N. Lu, L. Jiao, H. Du, and M. Gong. IFMOA: Immune Forgetting Multiobjective Optimization Algorithm. In *Proceedings of the First International Conference on Advances in Natural Computation, ICNC 2005, Part III*, pages 399–408, Changsha, China, August 2005. Springer. Lecture Notes in Computer Science Vol. 3612.
1023. R. Luce. Semiorders and a theory of utility discrimination. *Econometrica*, 24:178–191, 1956.
1024. R. Luce and H. Raiffa. *Games and Decisions: Introduction and Critical Survey*. John Wiley, New York, 1957.
1025. G.-C. Luh and C.-H. Chueh. Multi-objective optimal design of truss structure with immune algorithm. *Computers and Structures*, 82:829–844, 2004.
1026. G.-C. Luh, C.-H. Chueh, and W.-W. Liu. MOIA: Multi-Objective Immune Algorithm. *Engineering Optimization*, 35(2):143–164, April 2003.
1027. S. Luke and L. Panait. Lexicographic Parsimony Pressure. In W. Langdon, E. Cantú-Paz, K. Mathias, R. Roy, D. Davis, R. Poli, K. Balakrishnan, V. Honavar, G. Rudolph, J. Wegener, L. Bull, M. Potter, A. Schultz, J. Miller,

E. Burke, and N. Jonoska, editors, *Proceedings of the Genetic and Evolutionary Computation Conference (GECCO'2002)*, pages 829–836, San Francisco, California, July 2002. Morgan Kaufmann Publishers.
1028. K.-Y. Lum, P.-M. Jacquart, and M. Sefrioui. Constrained Optimization of Multilayered Anti-Reflection Coatings using Genetic Algorithms. In L. Wang, K. C. Tan, T. Furuhashi, J.-H. Kim, and X. Yao, editors, *Proceedings of the 4th Asia-Pacific Conference on Simulated Evolution and Learning (SEAL'02)*, volume 1, pages 172–177, Orchid Country Club, Singapore, November 2002. Nanyang Technical University.
1029. E. H. Luna and C. A. Coello Coello. Using a Particle Swarm Optimizer with a Multi-Objective Selection Scheme to Design Combinational Logic Circuits. In C. A. Coello Coello and G. B. Lamont, editors, *Applications of Multi-Objective Evolutionary Algorithms*, pages 101–124. World Scientific, Singapore, 2004.
1030. E. H. Luna, C. A. Coello Coello, and A. H. Aguirre. On the Use of a Population-Based Particle Swarm Optimizer to Design Combinational Logic Circuits. In R. S. Zebulum, D. Gwaltney, G. Hornby, D. Keymeulen, J. Lohn, and A. Stoica, editors, *Proceedings of the 2004 NASA/DoD Conference on Evolvable Hardware*, pages 183–190, Los Alamitos, California, USA, June 2004. IEEE Computer Society.
1031. F. Luna, A. Nebro, B. Dorronsoro, E. Alba, P. Bouvry, and L. Hogie. Optimal Broadcasting in Metropolitan MANETs Using Multiobjective Scatter Search. In F. Rothlauf, J. Branke, S. Cagnoni, E. Costa, C. Cotta, R. Drechsler, E. Lutton, P. Machado, J. Moore, J. Romero, G. Smith, G. Squillero, and H. Takagi, editors, *Applications of Evolutionary Computing, EvoWorkshops 2006: EvoBIO, EvoCOMNET, EvoHOT, EvoIASP, EvoINTERACTION, EvoMUSART*, pages 255–266. Springer. Lecture Notes in Computer Science Vol. 3907, 2006.
1032. F. Luna, A. J. Nebro, and E. Alba. Parallel Evolutionary Multiobjective Optimization. In N. Nedjah, E. Alba, and L. de Macedo Mourelle, editors, *Parallel Evolutionary Computations*, pages 33–56. Springer, Berlin Heidelberg, 2006.
1033. P. Lučić and D. Teodorović. Simulated annealing for the multi-objective aircrew rostering problem. *Transportation Research Part A*, 33:19–45, 1999.
1034. N. Lyu and K. Saitou. Decomposition-based assembly synthesis of a three-dimensional body-in-white model for structural stiffness. *Journal of Mechanical Design*, 127(1):34–48, January 2005.
1035. N. Lyu and K. Saitou. Topology optimization of multicomponent beam structure via decomposition-based assembly synthesis. *Journal of Mechanical Design*, 127(2):170–183, March 2005.
1036. K. R. MacCrimmon. An overview of multiple objective decision making. In J. L. Cochrane and M. Zeleny, editors, *Multiple Criteria Decision Making*, pages 18–44. University of South Carolina Press, 1973.
1037. N. K. Madavan. Multiobjective Optimization Using a Pareto Differential Evolution Approach. In *Congress on Evolutionary Computation (CEC'2002)*, volume 2, pages 1145–1150, Piscataway, New Jersey, May 2002. IEEE Service Center.
1038. S. W. Mahfoud. A Comparison of Parallel and Sequential Niching Methods. In L. J. Eshelman, editor, *Proceedings of the Sixth International Confnerence on Genetic Algorithms*, pages 136–143, San Francisco, California, July 1995. Morgan Kaufmann Publishers.

1039. S. W. Mahfoud and D. E. Goldberg. Parallel recombinative simulated annealing: A genetic algorithm. *Parallel Computing*, 21:45–52, January 1995.
1040. M. Mahfouf, M. F. Abbod, and D. A. Linkens. Multi-Objective Genetic Optimization of the Performance Index of Self-Organizing Fuzzy Logic Control Algorithm Using a Fuzzy Ranking Approach. In H. J. Zimmerman, editor, *Proceedings of the Sixth European Congress on Intelligent Techniques and Soft Computing*, pages 1799–1808, Aachen, 1998. Verlag Mainz.
1041. M. Mahfouf, M.-Y. Chen, and D. A. Linkens. Adaptive Weighted Particle Swarm Optimisation for Multi-objective Optimal Design of Alloy Steels. In *Parallel Problem Solving from Nature - PPSN VIII*, pages 762–771, Birmingham, UK, September 2004. Springer-Verlag. Lecture Notes in Computer Science Vol. 3242.
1042. Q. H. Mahmoud. *Distributed Programming with Java*. Manning Publications Company, Greenwich, CT, 2000.
1043. C. Maier-Rothe and J. M. F. Stankard. A linear programming approach to choosing between multi-objective alternatives. In *Proceedings of the 7th Mathematical Programming Symposium*, The Hague, 1970.
1044. I. Makarov, T. Vinigradskaia, A. Rubinski, and V. Sokolov. *Choice Theory and Decision Making*. Nauka, 1982. (In Russian).
1045. R. Mäkinen, P. Neittaanmäki, J. Periaux, M. Sefrioui, and J. Toivanen. Parallel Genetic Solution for Multiobjective MDO. In A. Schiano, A. Ecer, J. Périaux, and N. Satofuka, editors, *Parallel CFD'96 Conference*, pages 352–359, Capri, 1996. Elsevier.
1046. R. Mäkinen, P. Neittaanmäki, J. Périaux, and J. Toivanen. A genetic Algorithm for Multiobjective Design Optimization in Aerodynamics and Electromagnetics. In K. Papailiou, editor, *Computational Fluid Dynamics '98, Proceedings of the ECCOMAS 98 Conference*, volume 2, pages 418–422, Athens, Greece, September 1998. Wiley.
1047. R. A. Mäkinen, J. Periaux, and J. Toivanen. Multidisciplinary shape optimization in aerodynamics and electromagnetics using genetic algorithms. *International Journal for Numerical Methods in Fluids*, 30(2):149–159, May 1999.
1048. B. Malakooti, J. Wang, and E. Tandler. A sensor-based accelerated approach for multi-attribute machinability and tool life evaluation. *International Journal of Production Research*, 28:23–73, 1990.
1049. J. Malard, A. Heredia-Langner, D. Baxter, K. Jarman, and W. Cannon. Constrained De Novo Peptide Identification via Multi-objective Optimization. In *Online Proceedings of the Third IEEE International Workshop on High Performance Computational Biology (HiCOMB 2004)*, Santa Fe, New Mexico, April 2004.
1050. J. Malard, A. Heredia-Langner, W. Cannon, R. Mooney, and D. Baxter. Peptide identification via constrained multi-objective optimization: Pareto-based genetic algorithms. *Computation & Concurrency: Practice and Experience*, 17(14):1687–1704, December 2005.
1051. L. Mandow and J. Pérez de la Cruz. Multicriteria heuristic search. *European Journal of Operational Research*, 150:253–280, 2003.
1052. V. Maniezzo and A. Carbonaro. An ants heuristic for the frequency assignment problem. *Future Generation Computer Systems*, 16(9):927–935, 2000.

1053. J. Mao, K. Hirasawa, J. Hu, and J. Murata. Genetic Symbiosis Algorithm for Multiobjective Optimization Problem. In *Proceedings of the 9th IEEE International Workshop on Robot and Human Interactive Communication (RO-MAN 2000)*, pages 137–142. IEEE, 2000.
1054. J. Mao, K. Hirasawa, J. Hu, and J. Murata. Genetic Symbiosis Algorithm for Multiobjective Optimization Problems. In *Proceedings of the 2001 Genetic and Evolutionary Computation Conference. Late-Breaking Papers*, pages 267–274, San Francisco, California, July 2001.
1055. D. Maravall and J. de Lope. Multi-objective dynamic optimization with genetic algorithms for automatic parking. *Soft Computing*, 11(3):249–257, February 2007.
1056. T. Marchant, D. Bouyssou, P. Perny, M. Pirlot, A. Tsoukias, and P. Vincke. Choosing on the basis of several viewpoints: the example of voting. *Service de Mathématiques de la Gestion, U.L.B.*, 99(3), 1999.
1057. N. Marco, S. Lanteri, J.-A. Desideri, and J. Périaux. A Parallel Genetic Algorithm for Multi-Objective Optimization in Computational Fluid Dynamics. In K. Miettinen, M. M. Mäkelä, P. Neittaanmäki, and J. Périaux, editors, *Evolutionary Algorithms in Engineering and Computer Science*, chapter 22, pages 445–456. John Wiley & Sons, Ltd, Chichester, UK, 1999.
1058. T. Marcu. A multiobjective evolutionary approach to pattern recognition for robust diagnosis of process faults. In R. J. Patton and J. Chen, editors, *IFAC Symposium on Fault Detection, Supervision and Safety for Technical Processes: SAFEPROCESS'97*, pages 1183–1188, Kington Upon Hull, United Kingdom, August 1997.
1059. T. Marcu, L. Ferariu, and P. M. Frank. Genetic Evolving of Dynamic Neural Networks with Application to Process Fault Diagnosis. In *Procedings of the EUCA/IFAC/IEEE European Control Conference ECC'99*, Karlsruhe, Germany, 1999. CD-ROM, F-1046,1.
1060. T. Marcu and P. M. Frank. Parallel Evolutionary Approach to System Identification for Process Fault Diagnosis. In P. S. Dhurjati and S. Cauvin, editors, *Procedings of the IFAC Workshop on 'On-line Fault Detection and Supervision in the Chemical Process Industries'*, pages 113–118, Solaize (Lyon), France, 1998.
1061. S. Mardle, S. Pascoe, and M. Tamiz. An Investigation of Genetic Algorithms for the Optimization of Multiobjective Fisheries Bioeconomic Models. In *Proceedings of the Third International Conference on Multi-Objective Programming and Goal Programming: Theory and Applications (MOPGP'98)*, Quebec City, Canada, 1998.
1062. S. Mardle, S. Pascoe, and M. Tamiz. An investigation of genetic algorithms for the optimisation of multi-objective fisheries bioeconomic models. *International Transactions of Operations Research*, 7(1):33–49, 2000.
1063. B. Mareschal and J.-P. Brans. Geometrical Representations for MCDA. *European Journal of Operational Research*, 34(1):69–77, February 1988.
1064. R. Marett and M. Wright. A Comparison of Neighborhood Search Techniques for Multi-Objective Combinatorial Problems. *Computers and Operations Research*, 23(5):465–483, 1996.
1065. S. Marglin. *Public Investment Criteria*. MIT Press, Cambridge, Massachusetts, 1967.
1066. C. Mariano and E. Morales. A New Approach for the Solution of Multiple Objective Optimization Problems Based on Reinforcement Learning. In

O. Cairo, L. E. Sucar, and F. J. Cantu, editors, *MICAI 2000: Advances in Artificial Intelligence. Mexican International Conference on Artificial Intelligence*, pages 212–223, Acapulco, Mexico, April 2000. Springer-Verlag.

1067. C. E. Mariano and E. Morales. MOAQ an Ant-Q Algorithm for Multiple Objective Optimization Problems. In W. Banzhaf, J. Daida, A. E. Eiben, M. H. Garzon, V. Honavar, M. Jakiela, and R. E. Smith, editors, *Genetic and Evolutionary Computing COnference (GECCO 99)*, volume 1, pages 894–901, San Francisco, California, July 1999. Morgan Kaufmann.

1068. C. E. Mariano and E. Morales. A Multiple Objective Ant-Q Algorithm for the Design of Water Distribution Irrigation Networks. Technical Report HC-9904, Instituto Mexicano de Tecnología del Agua, June 1999.

1069. C. E. Mariano and E. Morales. A New Distributed Reinforcement Learning Algorithm for Multiple Objective Optimization Problems. Technical Report HC-200001, Instituto Mexicano de Tecnología del Agua, January 2000.

1070. C. E. Mariano and E. F. Morales. Distributed Reinforcement Learning for Multiple Objective Optimization Problems. In *2000 Congress on Evolutionary Computation*, volume 1, pages 188–195, Piscataway, New Jersey, July 2000. IEEE Service Center.

1071. C. E. Mariano and E. F. Morales. MDQL: A Reinforcement Learning Approach for the Solution of Multiple Objective Optimization Problems. In *PPSN/SAB Workshop on Multiobjective Problem Solving from Nature (MPSN)*, Paris, France, September 2000.

1072. C. E. Mariano Romero. *Aprendizaje por Refuerzo en Optimización Multiobjetivo*. PhD thesis, Departamento de Ciencias Computacionales, Instituto Tecnológico y de Estudios Superiores de Monterrey, Cuernavaca, Morelos, México, Marzo 2001. (In Spanish).

1073. C. E. Mariano-Romero and V. H. Alcocer-Yamanaka. Multiobjective Optimization of Water-Using Systems. In N. Nedjah and L. de Macedo Mourelle, editors, *Real-World Multi-Objective System Engineering*, pages 163–192. Nova Science Publishers, New York, 2005.

1074. C. E. Mariano-Romero, V. H. Alcocer-Yamanaka, and E. F. Morales. Multi-objective optimization of water-using systems. *European Journal of Operational Research*, 181(3):1691–1707, 16 September 2007.

1075. R. Marler and J. Arora. Survey of multi-objective optimization methods for engineering. *Structural and Multidisciplinary Optimization*, 26:369–395, 2004.

1076. J.-M. Martel, L. N. Kiss, and M. A. Rousseau. PAMSSEM: Une Procédure d'Agrégation Multicritère de type Surclassement de Synthèse pour Évaluations Mixtes. Unpublished Working Document, 1996.

1077. R. Martí. Scatter Search–Wellsprings and Challenges. *European Journal of Operational Research*, 169:351–358, 2006.

1078. N. Marvin, M. Bower, and J. E. Rowe. An evolutionary approach to constructing prognostic models. *Artificial Intelligence in Medicine*, 15(2):155–165, February 1999.

1079. W. Mason, V. Coverstone-Carroll, and J. Hartmann. Optimal Earth Orbiting Satellite Constellations via a Pareto Genetic Algorithm. In *1998 AIAA/AAS Astrodynamics Specialist Conference and Exhibit*, pages 169–177, Boston, Massachusetts, August 1998. Paper No. AIAA 98-4381.

1080. W. J. Mason. Satellite Constellation Design Via Evolutionary Computation. Master's thesis, Department of Aeronautical and Astronautical Engineering, University of Illinois at Urbana Champaign, December 2001. (In process).

1081. S. Massebeuf, C. Fonteix, L. N. Kiss, I. Marc, F. Pla, and K. Zaras. Multicriteria Optimization and Decision Engineering of an Extrusion Process Aided by a Diploid Genetic Algorithm. In *1999 Congress on Evolutionary Computation*, pages 14–21, Washington, D.C., July 1999. IEEE Service Center.
1082. M. Matos and P. Melo. Multiobjective Reconfiguration for Loss Reduction and Service Restorating Using Simulated Annealing. In *International Conference on Electric Power Engineering, 1999. PowerTech Budapest 99*, pages 213–218, Budapest, Hungary, 1999. IEEE.
1083. K. B. Matthews, S. Craw, S. Elder, A. R. Sibbald, and I. MacKenzie. Applying Genetic Algorithms to Multi-Objective Land Use Planning. In D. Whitley, editor, *Genetic and Evolutionary Computation Conference*, pages 613–620, Las Vegas, Nevada, July 2000. Morgan Kaufmann Publishers.
1084. E. J. McCluskey. Minimization of Boolean Functions. *Bell Systems Technical Journal*, 35 (5):1417–1444, November 1956.
1085. P. R. McMullen. An ant colony optimization approach to addressing a JIT sequencing problem with multiple objectives. *Artificial Intelligence in Engineering*, 15:309–317, 2001.
1086. A. L. Medaglia, E. Gutiérrez, and J. G. Villegas. Solving Facility Location Problems with a Tool for Rapid Development of Multi-Objective Evolutionary Algorithms (MOEAs). In J.-P. Rennard, editor, *Handbook of Research on Nature Inspired Computing for Economy and Management*, volume 2, pages 642–660, Hershey, UK, 2006. Idea Group Reference. ISBN 1-59140-984-5.
1087. J. Mehnen, T. Michelitsch, K. Schmitt, and T. Kohlen. pMOHypEA: Parallel Evolutionary Multiobjective Optimization using Hypergraphs. Technical Report Reihe CI-189/04, SFB 531, University of Dortmund, Dortmund, Germany, ISSN 1433-3325, 2004.
1088. F. Menczer, M. Degeratu, and W. N. Street. Efficient and Scalable Pareto Optimization by Evolutionary Local Selection Algorithms. *Evolutionary Computation*, 8(2):223–247, Summer 2000.
1089. G. Meneghetti, V. Pediroda, and C. Poloni. Application of a Multi Objective Genetic Algorithm and a Neural Network to the Optimisation of Foundry Processes. In K. Miettinen, M. M. Mäkelä, P. Neittaanmäki, and J. Périaux, editors, *Evolutionary Algorithms in Engineering and Computer Science*, chapter 23, pages 457–470. John Wiley & Sons, Ltd, Chichester, UK, 1999.
1090. H. Meng, X. Zhang, and S. Liu. A co-evolutionary particle swarm optimization-based method for multiobjective optimization. In S. Zhang and R. Jarvis, editors, *AI 2005: Advances in Artificial Intelligence*, pages 349–359. Springer-Verlag. Lecture Notes in Artificial Intelligence Vol. 3809, 2005.
1091. L. D. Merkle, G. H. Gates, Jr., G. B. Lamont, and R. Pachter. Application of the Parallel Fast Messy Genetic Algorithm to the Protein Structure Prediction Problem. *Proceedings of the Intel Supercomputer Users' Group Users Conference*, pages 189–195, 1994.
1092. L. D. Merkle and G. B. Lamont. A Random Function Based Framework for Evolutionary Algorithms. In T. Bäck, editor, *Proceedings of the Seventh International Conference on Genetic Algorithms*, pages 105–112. Morgan Kaufmann Publishers, San Mateo, California, July 1997.
1093. P. Merz and B. Freisleben. Genetic Local Search for the TSP: New Results. In *Proceedings of the 1997 IEEE International Conference on Evolutionary Computation*, pages 159–164. IEEE Press, 1997.

1094. A. Messac. Physical programming: effective optimization for computational design. *AIAA Journal*, 34(1):149–158, January 1996.
1095. N. Metropolis, A. Rosenbluth, M. Rosenbluth, A. Teller, and E. Teller. Equation of State Calculations by Fast Computing Machines. *Journal of Chemical Physics*, 21(6):1087–1092, 1953.
1096. H. Meunier, E.-G. Talbi, and P. Reininger. A Multiobjective Genetic Algorithm for Radio Network Optimization. In *2000 Congress on Evolutionary Computation*, volume 1, pages 317–324, Piscataway, New Jersey, July 2000. IEEE Service Center.
1097. E. Mezura-Montes and C. A. Coello Coello. A Simple Multimembered Evolution Strategy to Solve Constrained Optimization Problems. *IEEE Transactions on Evolutionary Computation*, 9(1):1–17, February 2005.
1098. E. Mezura-Montes and C. A. Coello Coello. A Survey of Constraint-Handling Techniques Based on Evolutionary Multiobjective Optimization. Technical Report EVOCINV-04-2006, Evolutionary Computation Group at CINVESTAV, Departamento de Computación, CINVESTAV-IPN, México, October 2006.
1099. E. Mezura-Montes, J. Velázquez-Reyes, and C. A. Coello Coello. Comparing Differential Evolution Models for Global Optimization. In M. K. et al., editor, *2006 Genetic and Evolutionary Computation Conference (GECCO'2006)*, volume 1, pages 485–492, Seattle, Washington, USA, July 2006. ACM Press.
1100. Z. Michalewicz. *Genetic Algorithms + Data Structures = Evolution Programs*. Springer-Verlag, third edition, 1996.
1101. Z. Michalewicz and D. B. Fogel. *How to Solve It: Modern Heuristics*. Springer. Second, Revised and Extended Edition, Berlin, 2004. ISBN 3-540-22494-7.
1102. Z. Michalewicz and D. B. Fogel. *How to Solve It: Modern Heuristics*. Springer, Berlin, Germany, second edition, 2004. ISBN 3-540-22494-7.
1103. Z. Michalewicz and C. Z. Janikow. GENOCOP: a genetic algorithm for numerical optimization problems with linear constraints. *Communications of the ACM*, 39(12):223–240, December 1996.
1104. Z. Michalewicz and G. Nazhiyath. Genocop III: A Co-Evolutionary Algorithm for Numerical Optimization Problems with Nonlinear Constraints. In D. B. Fogel, editor, *Proceedings of the Second IEEE Conference on Evolutionary Computation*, pages 647–651, Piscataway, New Jersey, 1995. IEEE Service Center.
1105. Z. Michalewicz and M. Schoenauer. Evolutionary Algorithms for Constrained Parameter Optimization Problems. *Evolutionary Computation*, 4(1):1–32, 1996.
1106. S. R. Michaud. Solving the Protein Structure Prediction Problem with Parallel Messy Genetic Algorithms. Master's thesis, Air Force Institute of Technology, Wright-Patterson AFB, March 2001. AFIT/GCS/ENG/01M-06.
1107. E. Michielssen and D. S. Weile. Electromagnetic System Design using Genetic Algorithms. In *Genetic Algorithms and Evolution Strategies in Engineering and Computer Science*, pages 267–288. John Wiley and Sons, England, 1995.
1108. T. Miconi. When Evolving Populations is Better than Coevolving Individuals: The Blind Mice Problem. In G. Gottlob and T. Walsh, editors, *IJCAI-03, Proceedings of the Eighteenth International Joint Conference on Artificial Intelligence*, pages 647–652. Morgan Kaufmann, August 2003.
1109. I. Mierswa. Incorporating Fuzzy Knowledge Into Fitness: Multiobjective Evolutionary 3D Design of Process Plants. In H.-G. B. et al., editor, *2005 Genetic*

and *Evolutionary Computation Conference (GECCO'2005)*, volume 2, pages 1985–1992, New York, USA, June 2005. ACM Press.
1110. K. Miettinen. Interactive Nonlinear Multiobjective Procedures. In M. Ehrgott and X. Gandibleux, editors, *Multiple Criteria Optimization: State of the Art Annotated Bibliographic Surveys*, pages 227–276. Kluwer Academic Publishers, Boston, 2002.
1111. K. M. Miettinen. *Nonlinear Multiobjective Optimization*. Kluwer Academic Publishers, Boston, Massachusetts, 1999.
1112. N. Milickovic, M. Lahanas, D. Baltas, and N. Zamboglou. Comparison of Evolutionary and Deterministic Multiobjective Algorithms for Dose Optimization in Branchytherapy. In E. Zitzler, K. Deb, L. Thiele, C. A. Coello Coello, and D. Corne, editors, *First International Conference on Evolutionary Multi-Criterion Optimization*, pages 167–180. Springer-Verlag. Lecture Notes in Computer Science No. 1993, 2001.
1113. G. A. Miller. The Magical Number Seven, Plus or Minus Two: Some Limits on Our Capacity for Processing Information. *The Psychological Review*, 63(2):81–97, 1956.
1114. M. Mitchell. *An Introduction to Genetic Algorithms*. The MIT Press, Cambridge, Massachusetts, 1996.
1115. T. Mitchell. *Version Spaces: An Approach to Concept Learning*. PhD thesis, Computer Science Department, Stanford University, Stanford, California, 1978.
1116. K. Mitra, K. Deb, and S. K. Gupta. Multiobjective Dynamic Optimization of an Industrial Nylon 6 Semibatch Reactor Using Genetic Algorithm. *Journal of Applied Polymer Science*, 69(1):69–87, 1998.
1117. O. A. Mohammed and G. F. Üler. Genetic Algorithms for the Optimal Design of Electromagnetic Devices. In *Conference on the Annual Review of Progress in Applied Computational Electromagnetics*, volume 11, pages 386–393, 1995.
1118. J. Molina, M. Laguna, R. Martí, and R. Caballero. SSPMO: A Scatter Search Procedure for Non-Linear Multiobjective Optimization. Technical Report TR11-2004, Departamento de Estadística e Investigación Operativa, Universidad de Valencia, Valencia, Spain, 2004.
1119. A. Molyneaux, G. Leyland, and D.Favrat. A New, Clustering Evolutionary Multi-Objective Optimisation Technique. In *Proceedings of the Third International Symposium on Adaptive Systems—Evolutionary Computation and Probabilistic Graphical Models*, pages 41–47, Havana, Cuba, March 19–23 2001. Institute of Cybernetics, Mathematics and Physics.
1120. D. E. Monarchi. Interactive Algorithm for Multiple Objective Decision Making. Technical Report 6, Hydrology and Water Resources Department, The University of Arizona, Tucson, Arizona, 1972.
1121. D. E. Monarchi, C. C. Kisiel, and L. Duckstein. Interactive multiobjective programming in water resources: a case study. *Water Resources Research*, 9(4):837–850, August 1973.
1122. D. Montana, G. Bidwell, G. Vidaver, and J. Herrero. Scheduling and Route Selection for Military Land Moves Using Genetic Algorithms. In *1999 Congress on Evolutionary Computation*, volume 2, pages 1118–1123, Washington, D.C., July 1999. IEEE Service Center.
1123. D. Montana, M. Brinn, S. Moore, and G. Bidwell. Genetic Algorithms for Complex, Real-Time Scheduling. In *Proceedings of the 1998 IEEE Interna-*

tional Conference on Systems, Man, and Cybernetics, pages 2213–2218, La Jolla, California, October 1998. IEEE.
1124. D. Montana and J. Radi. Optimizing Parameters of a Mobile Ad Hoc Network Protocol with a Genetic Algorithm. In H.-G. B. et al., editor, *2005 Genetic and Evolutionary Computation Conference (GECCO'2005)*, volume 2, pages 1993–1998, New York, USA, June 2005. ACM Press.
1125. C. Moon, Y.-Z. Lin, and M. Gen. Evolutionary Algorithm for Flexible Process Sequencing with Multiple Objectives. In D. B. Fogel, editor, *Proceedings of the 1998 International Conference on Evolutionary Computation*, pages 27–32, Piscataway, New Jersey, 1998. IEEE.
1126. J. Moore and R. Chapman. Application of Particle Swarm to Multiobjective Optimization. Department of Computer Science and Software Engineering, Auburn University. (Unpublished manuscript), 1999.
1127. J. J. Moré, B. S. Garbow, and K. E. Hillstrom. Testing Unconstrained Optimization Software. *ACM Transactions on Mathematical Software*, 7(1):17–41, 1981.
1128. D. R. Morgan, J. W. Eheart, and A. J. Valocchi. Aquifer remediation design under uncertainty using a new chance constrained programming technique. *Water Resources Research*, 29(3):551–561, 1993.
1129. M. Morita, R. Sabourin, F. Bortolozzi, and C. Suen. Segmentation and recognition of handwritten dates. In *Proceedings of the Eighth International Workshop on Frontiers of Handwriting Recognition (IWFHR'02)*, pages 105–110, Ontario, Canada, August 2002. IEEE Computer Society.
1130. M. Morita, R. Sabourin, F. Bortolozzi, and C. Suen. Unsupervised Feature Selection Using Multi-Objective Genetic Algorithm for Handwritten Word Recognition. In *Proceedings of the 7th International Conference on Document Analysis and Recognition (ICDAR'2003)*, pages 666–670, Edinburgh, Scotland, August 2003.
1131. R. W. Morrison and K. A. De Jong. A Test Problem Generator for Non-Stationary Environments. In *1999 Congress on Evolutionary Computation*, pages 2047–2053, Washington, D.C., July 1999. IEEE Service Center.
1132. J. Morse. Reducing the size of the nondominated set: Pruning by clustering. *Computers and Operations Research*, 7(1–2):55–66, 1980.
1133. P. Moscato. On Evolution, Search, Optimization, Genetic Algorithms and Martial Arts. Towards Memetic Algorithms. Technical Report 158–79, Caltech Concurrent Computation Program, California Institute of Technology, Pasadena, California, September 1989.
1134. P. Moscato. Memetic Algorithms: A Short Introduction. In D. Corne, F. Glover, and M. Dorigo, editors, *New Ideas in Optimization*, pages 219–234. McGraw-Hill, 1999.
1135. H. Moskowitz, G. W. Evans, and I. Jiménez-Lerma. Development of A Multiattribute Value Function for Long Range Electrical Generation Expansion. *IEEE Transactions on Engineering Management*, EM–25:78–87, 1978.
1136. J. Mossin. *Theory of Financial Markets*. Prentice-Hall, Englewood Cliffs, New Jersey, 1973.
1137. S. Mostaghim, J. Branke, and H. Schmeck. Multi-Objective Particle Swarm Optimization on Computer Grids. Technical Report 502, AIFB Institute, December 2006. available at: http://www.aifb.uni-karlsruhe.de/EffAlg/smo/paper-12-06.pdf.

1138. S. Mostaghim and J. Teich. The role of ε-dominance in multi objective particle swarm optimization methods. In *Proceedings of the 2003 Congress on Evolutionary Computation (CEC'2003)*, volume 3, pages 1764–1771, Canberra, Australia, December 2003. IEEE Press.
1139. S. Mostaghim and J. Teich. Strategies for Finding Good Local Guides in Multi-objective Particle Swarm Optimization (MOPSO). In *2003 IEEE Swarm Intelligence Symposium Proceedings*, pages 26–33, Indianapolis, Indiana, USA, April 2003. IEEE Service Center.
1140. S. Mostaghim and J. Teich. Covering Pareto-optimal Fronts by Subswarms in Multi-objective Particle Swarm Optimization. In *2004 Congress on Evolutionary Computation (CEC'2004)*, volume 2, pages 1404–1411, Portland, Oregon, USA, June 2004. IEEE Service Center.
1141. S. Mostaghim, J. Teich, and A. Tyagi. Comparison of Data Structures for Storing Pareto-sets in MOEAs. In *Congress on Evolutionary Computation (CEC'2002)*, volume 1, pages 843–848, Piscataway, New Jersey, May 2002. IEEE Service Center.
1142. V. Mousseau and R. Slowinski. Inferring an ELECTRE TRI Model from Assignment Examples. *Journal of Global Optimization*, 12:157–174, 1998.
1143. V. Mousseau, R. Slowinski, and P. Zielniewicz. A user-oriented implementation of the ELECTRE-TRI method integrating preference elicitation support. *Computers & Operations Research*, 27:757–777, 2000.
1144. A. Mukerjee, R. Biswas, K. Deb, and A. P. Mathur. Multi-objective evolutionary algorithms for the risk-return trade-off in bank-load management. *International Transactions in Operational Research*, 9(5):583–597, September 2002.
1145. S. Mullei and P. Beling. Hybrid Evolutionary Algorithms for a Multiobjective Financial Problem. In *Proceedings of the 1998 IEEE International Conference on Systems, Man, and Cybernetics*, pages 3925–3930. IEEE, October 1998.
1146. S. D. Müller, I. F. Sbalzarini, J. H. Walther, and P. D. Koumoutsakos. Evolution Strategies for the Optimization of Microdevices. In *Proceedings of the Congress on Evolutionary Computation 2001 (CEC'2001)*, volume 1, pages 302–309, Piscataway, New Jersey, May 2001. IEEE Service Center.
1147. G. Munda. Multiple-Criteria Decision-Aid: Some Epistemological Considerations. *Journal of Multi-Criteria Decision Analysis*, 2:41–55, 1993.
1148. G. Munda. *Multicriteria Evaluation in a Fuzzy Environment*. Springer-Verlag, New York, 1995.
1149. T. Murata. *Genetic Algortithms for Multi-Objective Optimization*. PhD thesis, Osaka Prefecture University, Japan, 1997.
1150. T. Murata and H. Ishibuchi. MOGA: Multi-Objective Genetic Algorithms. In *Proceedings of the 2nd IEEE International Conference on Evolutionary Computing*, pages 289–294, Perth, Australia, November 1995.
1151. T. Murata, H. Ishibuchi, and M. Gen. Cellular Genetic Local Search for Multi-Objective Optimization. In D. Whitley, D. Goldberg, E. Cantú-Paz, L. Spector, I. Parmee, and H.-G. Beyer, editors, *Proceedings of the Genetic and Evolutionary Computation Conference (GECCO'2000)*, pages 307–314, San Francisco, California, 2000. Morgan Kaufmann.
1152. T. Murata, H. Ishibuchi, and M. Gen. Specification of Genetic Search Directions in Cellular Multi-objective Genetic Algorithms. In E. Zitzler, K. Deb,

L. Thiele, C. A. Coello Coello, and D. Corne, editors, *First International Conference on Evolutionary Multi-Criterion Optimization*, pages 82–95. Springer-Verlag. Lecture Notes in Computer Science No. 1993, 2001.

1153. T. Murata, H. Ishibuchi, and H. Tanaka. Multi-Objective Genetic Algorithm and Its Application to Flowshop Scheduling. *Computers and Industrial Engineering Journal*, 30(4):957–968, September 1996.

1154. T. Murata and R. Itai. Local Search in Two-Fold EMO Algorithm to Enhance Solution Similarity for Multi-objective Vehicle Routing Problems. In S. Obayashi, K. Deb, C. Poloni, T. Hiroyasu, and T. Murata, editors, *Evolutionary Multi-Criterion Optimization, 4th International Conference, EMO 2007*, pages 201–215, Matshushima, Japan, March 2007. Springer. Lecture Notes in Computer Science Vol. 4403.

1155. T. Murata, S. Kawakami, H. Nozawa, M. Gen, and H. Ishibuchi. Three-Objective Genetic Algorithms for Designing Compact Fuzzy Rule-Based Systems for Pattern Classification Problems. In L. Spector, E. D. Goodman, A. Wu, W. Langdon, H.-M. Voigt, M. Gen, S. Sen, M. Dorigo, S. Pezeshk, M. H. Garzon, and E. Burke, editors, *Proceedings of the Genetic and Evolutionary Computation Conference (GECCO'2001)*, pages 485–492, San Francisco, California, 2001. Morgan Kaufmann Publishers.

1156. T. Murata, H. Nozawa, H. Ishibuchi, and M. Gen. Modifications of Local Search Directions for Non-dominated Solutions in Cellular Multiobjective Genetic Algorithms for Pattern Classification Problems. In C. M. Fonseca, P. J. Fleming, E. Zitzler, K. Deb, and L. Thiele, editors, *Evolutionary Multi-Criterion Optimization. Second International Conference, EMO 2003*, pages 593–607, Faro, Portugal, April 2003. Springer. Lecture Notes in Computer Science. Volume 2632.

1157. T. Murata, H. Nozawa, Y. Tsujimura, M. Gen, and H. Ishibuchi. Effect of Local Search on the Performance of Cellular Multi-Objective Genetic Algorithms for Designing Fuzzy Rule-based Classification Systems. In *Congress on Evolutionary Computation (CEC'2002)*, volume 1, pages 663–668, Piscataway, New Jersey, May 2002. IEEE Service Center.

1158. S. Murthy, R. Akkiraju, R. Goodwin, P. Keskinocak, J. Rachlin, F. Wu, S. Kumaran, and R. Daigle. Enhancing the Decision-Making Process for Paper Mill Schedulers. *Tappi Journal*, 82(7):42–47, July 1999.

1159. S. Murthy, R. Akkiraju, J. Raclin, and F. Wu. Agent-Based Cooperative Scheduling. In *Proceedings of the AAAI'97 Workshop on Constraints and Agents*, pages 112–117, 1997.

1160. S. Murthy, S. Kasif, and S. Salzberg. OC1: Randomized induction of oblique decision trees. In *Proceedings of the Eleventh National Conference on Artificial Intelligence*, pages 322–327. AAAI, MIT Press, 1993.

1161. L. N. de Castro and J. Timmis. *An Introduction to Artificial Immune Systems: A New Computational Intelligence Paradigm*. Springer, London, 2002. ISBN 1-85233-594-7.

1162. S. Nakaya, S. Wakabayashi, and T. Koide. An adaptive genetic algorithm for VLSI floorplanning based on sequence-pair. In *2000 IEEE International Symposium on Circuits and Systems (ISCAS'2000)*, volume 3, pages 65–68. IEEE Press, 2000.

1163. S. Nakrani and C. Tovey. On honey bees and dynamic allocation in an internet server colony. In C. Anderson and T. Balch, editors, *Proceedings of the Second*

International Workshop on the Mathematics and Algorithms of Social Insects, pages 115–122, Atlanta, Georgia, USA, December 2003. Georgia Institute of Technology.

1164. D. Nam and C. H. Park. Multiobjective Simulated Annealing: A Comparative Study to Evolutionary Algorithms. *International Journal of Fuzzy Systems*, 2(2):87–97, 2000.

1165. D. Nam and C. H. Park. Pareto-Based Cost Simulated Annealing for Multiobjective Optimization. In L. Wang, K. C. Tan, T. Furuhashi, J.-H. Kim, and X. Yao, editors, *Proceedings of the 4th Asia-Pacific Conference on Simulated Evolution and Learning (SEAL'02)*, volume 2, pages 522–526, Orchid Country Club, Singapore, November 2002. Nanyang Technical University.

1166. D. Nam, Y. D. Seo, L.-J. Park, C. H. Park, and B. Kim. Parameter Optimization of a Voltage Reference Circuit using EP. In D. B. Fogel, editor, *Proceedings of the 1998 International Conference on Evolutionary Computation*, pages 245–266, Piscataway, New Jersey, 1998. IEEE.

1167. S. Narayanan and S. Azarm. On Improving Multiobjective Genetic Algorithms for Design Optimization. *Structural Optimization*, 18:146–155, 1999.

1168. N. Nariman-Zadeh, K. Atashkari, A. Jamali, A. Pilechi, and X. Yao. Inverse modelling of multi-objective thermodynamically optimized turbojet engines using GMDH-type neural networks and evolutionary algorithms. *Engineering Optimization*, 37(5):437–462, July 2005.

1169. J. Nash. Two-person cooperative games. *Econometrica*, 21:128–140, 1953.

1170. B. Naujoks, N. Beume, and M. Emmerich. Multi-objective Optimization using S-metric Selection: Application to three-dimensional Solution Spaces. In *2005 IEEE Congress on Evolutionary Computation (CEC'2005)*, volume 2, pages 1282–1289, Edinburgh, Scotland, September 2005. IEEE Service Center.

1171. B. Naujoks, L. Willmes, T. Bäck, and W. Haase. Evaluating Multi-criteria Evolutionary Algorithms for Airfoil Optimization. In J. J. Merelo Guervós, P. Adamidis, H.-G. Beyer, J.-L. F.-V. nas, and H.-P. Schwefel, editors, *Parallel Problem Solving from Nature—PPSN VII*, pages 841–850, Granada, Spain, September 2002. Springer-Verlag. Lecture Notes in Computer Science No. 2439.

1172. R. Neapolitan and K. Naimipour. *Foundations of Algorithms*. D. C. Heath and Company, Lexington, Massachusetts, 1996.

1173. A. Nebro, F. Luna, E. Alba, A. Beham, and B. Dorronsoro. AbYSS: Adapting Scatter Search for Multiobjective Optimization. Technical Report ITI-2006-2, Dept. Lenguajes y Ciencias de la Computación, University of Málaga, Malaga, Spain, 2006.

1174. A. Nebro, F. Luna, E.-G. Talbi, and E. Alba. Parallel Multiobjective Optimization. In E. Alba, editor, *Parallel Metaheuristics*, pages 371–394. Wiley-Interscience, New Jersey, USA, 2005. ISBN 13-978-0-471-67806-9.

1175. A. J. Nebro, J. J. Durillo, F. Luna, B. Dorronsoro, and E. Alba. A Cellular Genetic Algorithm for Multiobjective Optimization. In D. A. Pelta and N. Krasnogor, editors, *Proceedings of the Workshop on Nature Inspired Cooperative Strategies for Optimization (NICSO 2006)*, pages 25–36, Granada, Spain, 2006.

1176. A. J. Nebro, J. J. Durillo, F. Luna, B. Dorronsoro, and E. Alba. Design Issues in a Multiobjective Cellular Genetic Algorithm. In S. Obayashi, K. Deb, C. Poloni, T. Hiroyasu, and T. Murata, editors, *Evolutionary Multi-Criterion*

Optimization, 4th International Conference, EMO 2007, pages 126–140, Matshushima, Japan, March 2007. Springer. Lecture Notes in Computer Science Vol. 4403.
1177. A. J. Nebro, F. Luna, and E. Alba. New Ideas in Applying Scatter Search to Multiobjective Optimization. In C. A. Coello Coello, A. Hernández Aguirre, and E. Zitzler, editors, *Evolutionary Multi-Criterion Optimization. Third International Conference, EMO 2005*, pages 443–458, Guanajuato, México, March 2005. Springer. Lecture Notes in Computer Science Vol. 3410.
1178. N. Nedjah and L. de Macedo Mourelle. Multi-Objective Evolutionary Hardware for RSA-Based Cryptosystems. In *Proceedings of the International Conference on Information Technology: Coding and Computing (ITCC'04)*, volume 2, pages 503–507, Las Vegas, Nevada, April 2004. IEEE.
1179. N. Nedjah and L. de Macedo Mourelle. A Reconfigurable Parallel Hardware for Genetic Algorithms. In N. Nedjah, E. Alba, and L. de Macedo Mourelle, editors, *Parallel Evolutionary Computations*, pages 59–69. Springer, Berlin Heidelberg, 2006.
1180. N. Nedjah and L. M. Mourelle. Secure Evolutionary Hardware for Public-Key Cryptosystems. In *2004 Congress on Evolutionary Computation (CEC'2004)*, volume 2, pages 2130–2137, Portland, Oregon, USA, June 2004. IEEE Service Center.
1181. M. Neef, D. Thierens, and H. Arciszewski. A Case Study of a Multiobjective Recombinative Genetic Algorithm with Coevolutionary Sharing. In *1999 Congress on Evolutionary Computation*, pages 796–803, Washington, D.C., July 1999. IEEE Service Center.
1182. J. Nelder and R. Mead. A simplex method for function minimization. *Computer Journal*, 7(4):308–313, 1965.
1183. F. Neumann. Expected runtimes of a simple evolutionary algorithm for the multi-objective minimum spanning tree problem. *European Journal of Operational Research*, 181(3):1620–1629, 16 September 2007.
1184. D. Newman, S. Hettich, C. Blake, and C. Merz. UCI Repository of machine learning databases, 1998. http://www.ics.uci.edu/~mlearn/MLRepository.html.
1185. H. B. Nielsen. Methods for Analyzing Pipe Networks. *Journal of Hydraulic Engineering*, 125(2):139–157, February 1989.
1186. J. Nievergelt, R. Gasser, F. Mäser, and C. Wirth. All the Needles in a Haystack: Can Exhaustive Search Overcome Combinatorial Chaos? In J. van Leeuwen, editor, *Computer Science Today: Lecture Notes in Computer Science 1000*, pages 254–274. Springer, Berlin, 1995.
1187. P. Nii. The Blackboard Model of Problem Solving. *AI Magazine*, 7(2):38–53, Summer 1986.
1188. Y. Niu and L. Shen. Multi-resolution image fusion using AMOPSO-II. In *Intelligent Computing in Signal Processing and Pattern Recognition*, pages 343–352. Springer-Verlag. Lecture Notes in Control and Information Sciences Vol. 345, 2006.
1189. J. Noble and R. A. Watson. Pareto coevolution: Using performance against coevolved opponents in a game as dimensions for Pareto selection. In L. Spector, E. D. Goodman, A. Wu, W. Langdon, H.-M. Voigt, M. Gen, S. Sen, M. Dorigo, S. Pezeshk, M. H. Garzon, and E. Burke, editors, *Proceedings of the Genetic and Evolutionary Computation Conference (GECCO'2001)*, pages 493–500, San Francisco, California, 2001. Morgan Kaufmann Publishers.

1190. S. R. Norris and W. A. Crossley. Pareto-Optimal Controller Gains Generated by a Genetic Algorithm. In *AIAA 36th Aerospace Sciences Meeting and Exhibit*, Reno, Nevada, January 1998. AIAA Paper 98-0010.
1191. R. B. nos, C. Gil, B. Paechter, and J. Ortega. Parallelization of Population-based Multi-objective Metaheuristics: An Empirical Study. *Applied Mathematical Modelling*, 30(7):578–592, 2006.
1192. L. Nunes de Castro and J. Timmis. An Artificial Immune Network for Multimodal Function Optimization. In *Proceedings of the 2002 IEEE World Congress on Computational Intelligence*, volume 1, pages 699–704, Honolulu, Hawaii, May 2002. IEEE Service Center.
1193. L. Nunes de Castro and J. Timmis. *An Introduction to Artificial Immune Systems: A New Computational Intelligence Paradigm*. Springer-Verlag, 2002.
1194. L. Nunes de Castro and F. J. Von Zuben. Learning and Optimization Using the Clonal Selection Principle. *IEEE Transactions on Evolutionary Computation*, 6(3):239–251, 2002.
1195. M. Oates and D. Corne. QoS based GA Parameter Selection for Autonomously Managed Distributed Information Systems. In H. Prade, editor, *Proceedings of the Thirteenth European Conference on Artificial Intelligence*, pages 670–674, Chichester, England, 1998. John Wiley & Sons.
1196. S. Obayashi. Pareto Genetic Algorithm for Aerodynamic Design using the Navier-Stokes Equations. In D. Quagliarella, J. Périaux, C. Poloni, and G. Winter, editors, *Genetic Algorithms and Evolution Strategies in Engineering and Computer Science. Recent Advances and Industrial Applications*, chapter 12, pages 245–266. John Wiley and Sons, West Sussex, England, 1997.
1197. S. Obayashi. Multidisciplinary Design Optimization of Aircraft Wing Planform Based on Evolutionary Algorithms. In *Proceedings of the 1998 IEEE International Conference on Systems, Man, and Cybernetics*, La Jolla, California, October 1998. IEEE.
1198. S. Obayashi, K. Nakahashi, A. Oyama, and N. Yoshino. Design Optimization of Supersonic Wings Using Evolutionary Algorithms. In *Proceedings of the Fourth ECCOMAS Computational Fluid Dynamics Conference*, Athens, Greece, September 1998.
1199. S. Obayashi and D. Sasaki. Visualization and Data Mining of Pareto Solutions Using Self-Organizing Map. In C. M. Fonseca, P. J. Fleming, E. Zitzler, K. Deb, and L. Thiele, editors, *Evolutionary Multi-Criterion Optimization. Second International Conference, EMO 2003*, pages 796–809, Faro, Portugal, April 2003. Springer. Lecture Notes in Computer Science. Volume 2632.
1200. S. Obayashi, S. Takahashi, and I. Fejtek. Transonic Wing Design by Inverse Optimization using MOGA. In *Sixth Annual Conference of the Computational Fluid Dynamics Society of Canada*, Quebec, Canada, June 1998.
1201. S. Obayashi, S. Takahashi, and Y. Takeguchi. Niching and Elitist Models for MOGAs. In A. E. Eiben, T. Bäck, M. Schoenauer, and H.-P. Schwefel, editors, *Parallel Problem Solving From Nature — PPSN V*, pages 260–269, Amsterdam, Holland, 1998. Springer-Verlag. Lecture Notes in Computer Science No. 1498.
1202. S. Obayashi, T. Tsukahara, and T. Nakamura. Cascade Airfoil Design by Multiobjective Genetic Algorithms. In *Second International Conference on Genetic Algorithms in Engineering Systems: Innovations and Applications*, pages 24–29. IEEE Conference Publication No. 446, September 1997.

1203. S. Obayashi, T. Tsukahara, and T. Nakamura. Multiobjective Evolutionary Computation for Supersonic Wing-Shape Optimization. *IEEE Transactions on Evolutionary Computation*, 4(2):182–187, July 2000.

1204. S. Obayashi, Y. Yamaguchi, and T. Nakamura. Multiobjective Genetic Algorithm for Multidisciplinary Design of Transonic Wing Planform. *Journal of Aircraft*, 34(5):690–693, September-October 1997.

1205. C. K. Oei, D. E. Goldberg, and S.-J. Chang. Tournament Selection, Niching, and the Preservation of Diversity. Technical Report 91011, Illinois Genetic Algorithms Laboratory, University of Illinois at Urbana-Champaign, Urbana, Illinois, December 1991.

1206. I. Oesterreichter, A. Mitschele, F. Schlottmann, and D. Seese. Comparison of Multi-Objective Evolutionary Algorithms in Optimizing Combinations of Reinsurance Contracts. In M. K. et al., editor, *2006 Genetic and Evolutionary Computation Conference (GECCO'2006)*, volume 1, pages 747–748, Seattle, Washington, USA, July 2006. ACM Press. ISBN 1-59593-186-4.

1207. T. Okabe, Y. Jin, M. Olhofer, and B. Sendhoff. On Test Functions for Evolutionary Multi-objective Optimization. In X. Y. et al., editor, *Parallel Problem Solving from Nature - PPSN VIII*, pages 792–802, Birmingham, UK, September 2004. Springer-Verlag. Lecture Notes in Computer Science Vol. 3242.

1208. T. Okuda, T. Hiroyasu, M. Miki, and S. Watanabe. DCMOGA: Distributed Cooperation Model of Multi-Objective Genetic Algorithm. In *PPSN/SAB Workshop on Multiobjective Problem Solving from Nature II (MPSN-II)*, Granada, Spain, September 2002.

1209. S. C. Olenik and Y. Y. Haimes. A hierarchical multi-objective method for water resources planning. *IEEE Transactions on Systems, Man and Cybernetics*, SMC-9(9):534–544, 1979.

1210. G. M. B. Oliveira, J. C. Bortot, and P. P. de Oliveira. Multiobjective evolutionary search for one-dimensional cellular automata in the density classification task. In R. Standish, M. Bedau, and H. Abbass, editors, *Artificial Life VIII: The 8th International Conference on Artificial Life*, pages 202–206, Cambridge, Massachusetts, 2002. MIT Press.

1211. M. G. Omran, A. P. Engelbrecht, and A. Salman. Differential Evolution Based Particle Swarm Optimization. In *Proceedings of the 2007 IEEE Swarm Intelligence Symposium (SIS'2007)*, pages 112–119, Honolulu, Hawaii, USA, April 2007.

1212. C. Ong, H. Huang, and G. Tzeng. A novel hybrid model for portfolio selection. *Applied Mathematics and Computation*, 169(2):1195–1210, October 2005.

1213. K. R. Oppenheimer. A proxy approach to multi-attribute decision making. *Management Science*, 24(6):675–689, February 1978.

1214. M. Ortmann and W. Weber. Multi-Criterion Optimization of Robot Trajectories with Evolutionary Strategies. In *Proceedings of the 2001 Genetic and Evolutionary Computation Conference. Late-Breaking Papers*, pages 310–316, San Francisco, California, July 2001.

1215. K. A. Osman, A. M. Higginson, and J. Moore. Improving the efficiency of vehicle water-pump designs using genetic algorithms. In C. Dagli, M. Akay, A. Buczak, O. Ersoy, and B. Fernandez, editors, *Smart Engineering Systems: Proceedings of the Artificial Neural Networks in Engineering Conference (ANNIE '98)*, volume 8, pages 291–296, New York, 1998. ASME, ASME Press.

1216. A. Osyczka. An Approach to Multicriterion Optimization Problems for Engineering Design. *Computer Methods in Applied Mechanics and Engineering*, 15:309–333, 1978.
1217. A. Osyczka. *Multicriterion Optimization in Engineering with FORTRAN programs*. Ellis Horwood Limited, 1984.
1218. A. Osyczka. Multicriteria optimization for engineering design. In J. S. Gero, editor, *Design Optimization*, pages 193–227. Academic Press, 1985.
1219. A. Osyczka. *Evolutionary Algorithms for Single and Multicriteria Design Optimization*. Physica Verlag, Germany, 2002. ISBN 3-7908-1418-0.
1220. A. Osyczka and J. Koski. Selected Works related to Multicriterion Optimization Methods for Engineering Design. In *Proceedings of Euromech Colloquium*, University of Siegen, 1982.
1221. A. Osyczka and S. Krenich. A New Constraint Tournament Selection Method for Multicriteria Optimization using Genetic Algorithm. In *2000 Congress on Evolutionary Computation*, volume 1, pages 501–507, Piscataway, New Jersey, July 2000. IEEE Service Center.
1222. A. Osyczka and S. Krenich. Evolutionary Algorithms for Multicriteria Optimization with Selecting a Representative Subset of Pareto Optimal Solutions. In E. Zitzler, K. Deb, L. Thiele, C. A. Coello Coello, and D. Corne, editors, *First International Conference on Evolutionary Multi-Criterion Optimization*, pages 141–153. Springer-Verlag. Lecture Notes in Computer Science No. 1993, 2001.
1223. A. Osyczka, S. Krenich, and K. Karaś. Optimum Design of Robot Grippers using Genetic Algorithms. In *Proceedings of the Third World Congress of Structural and Multidisciplinary Optimization (WCSMO)*, Buffalo, New York, May 1999.
1224. A. Osyczka and S. Kundu. A Genetic Algorithm-Based Multicriteria Optimization Method. In *Proceedings of First World Congress of Structural and Multidisciplinary Optimization*, pages 909–914, Goslar, Germany, May 1995. Elsevier Science.
1225. A. Osyczka and S. Kundu. A new method to solve generalized multicriteria optimization problems using the simple genetic algorithm. *Structural Optimization*, 10:94–99, 1995.
1226. A. Osyczka and H. Tamura. Pareto set distribution method for multicriteria optimization using genetic algorithm. In *Proceedings of the Second International Conference on Genetic Algorithms (Mendel'96)*, pages 97–102, Brno, Czech Republic, June 1996.
1227. A. Oyama and M.-S. Liou. Multiobjective Optimization of Rocket Engine Pumps using Evolutionary Algorithm. In *Proceedings of the 15th AIAA Computational Fluid Dynamics Conference, Paper A01-31074*, Anaheim, California, June 2001.
1228. A. Oyama, S. Obayashi, K. Nakahashi, and N. Hirose. Coding by Taguchi Method for Evolutionary Algorithms Applied to Aerodynamic Optimization. In *Proceedings of the Fourth ECCOMAS Computational Fluid Dynamics Conference*, pages 196–203, Athens, Greece, September 1998. John Wiley & Sons.
1229. A. Oyama, K. Shimoyama, and K. Fujii. New Constraint-Handling Method for Multi-Objective Multi-Constraint Evolutionary Optimization and Its Application to Space Plane Design. In R. Schilling, W. Haase, J. Periaux, H. Baier, and G. Bugeda, editors, *Evolutionary and Deterministic Methods for Design,*

Optimization and Control with Applications to Industrial and Societal Problems (EUROGEN 2005), Munich, Germany, 2005.
1230. P. Pacheco. *Parallel Programming with MPI*. Morgan Kaufmann, 1996.
1231. B. Paechter, R. Rankin, A. Cumming, and T. C. Fogarty. Timetabling the Classes of an Entire University with an Evolutionary Algorithm. In A. E. Eiben, T. Bäck, M. Schoenauer, and H.-P. Schwefel, editors, *Parallel Problem Solving From Nature — PPSN V*, Amsterdam, Holland, 1998. Springer-Verlag. Lecture Notes in Computer Science No. 1498.
1232. J. Paelinck. Qualiflex, a Flexible Multiple Criteria Method. *Ecnomic Letters*, 3:193–197, 1978.
1233. G. Pampara, N. Franken, and A. Engelbrecht. Combining Particle Swarm Optimisation with angle modulation to solve binary problems. In *2005 IEEE Congress on Evolutionary Computation (CEC'2005)*, volume 1, pages 89–96, Edinburgh, Scotland, September 2005. IEEE Service Center.
1234. S. Pankanti and A. K. Jain. Integrating Vision Modules: Stereo, Shading, Grouping, and Line Labeling. *IEEE Transactions on Pattern Analysis and Machine Intelligence*, 17(8):831–842, September 1995.
1235. M. Papila, R. T. Haftka, T. Nishida, and M. Sheplak. Piezoresistive Microphone Design Pareto Optimization: Tradeoff Between Sensitivity and Noise Floor. *Journal of Microelectromechanical Systems*, 15(6):1632–1643, December 2006.
1236. U. Paquet and A. P. Engelbrecht. A New Particle Swarm Optimiser for Linearly Constrained Optimization. In *Proceedings of the Congress on Evolutionary Computation 2003 (CEC'2003)*, volume 1, pages 227–233, Piscataway, New Jersey, December 2003. Canberra, Australia, IEEE Service Center.
1237. L. Paquete, M. Chiarandini, and T. Stützle. Pareto Local Optimum Sets in the Biobjective Traveling Salesman Problem: An Experimental Study. In X. Gandibleux, M. Sevaux, K. Sörensen, and V. T'kindt, editors, *Metaheuristics for Multiobjective Optimisation*, pages 177–199, Berlin, 2004. Springer. Lecture Notes in Economics and Mathematical Systems Vol. 535.
1238. L. Paquete and T. Stützle. A Two-Phase Local Search for the Biobjective Traveling Salesman Problem. In C. M. Fonseca, P. J. Fleming, E. Zitzler, K. Deb, and L. Thiele, editors, *Evolutionary Multi-Criterion Optimization. Second International Conference, EMO 2003*, pages 479–493, Faro, Portugal, April 2003. Springer. Lecture Notes in Computer Science. Volume 2632.
1239. L. Paquete and T. Stützle. A study of stochastic local search algorithms for the biobjective QAP with correlated flow matrices. *European Journal of Operational Research*, 169:943–959, 2006.
1240. J. Paredis. Coevolutionary computation. *Artificial Life*, 2(4):355–375, 1995.
1241. J. Paredis. Coevolutionary algorithms. In T. Bäck, D. B. Fogel, and Z. Michalewicz, editors, *The Handbook of Evolutionary Computation, 1st Supplement*, pages 225–238. Institute of Physics Publishing and Oxford University Press, 1998.
1242. V. Pareto. *Cours D'Economie Politique*, volume I and II. F. Rouge, Lausanne, 1896.
1243. G. Parks and A. Suppapitnarm. Multiobjective optimization of PWR reload core designs using simulated annealing. In *Mathematics & Computation, Reactor Physics and Environmental Analysis in Nuclear Applications*, volume 2, pages 1435–1444, Madrid, Spain, 1999.

1244. G. T. Parks. Multiobjective PWR Reload Core Optimization Using Genetic Algorithms. In *Proceedings of the International Conference on Mathematics and Computations, Reactor Physics, and Environmental Analyses*, pages 615–624, La Grange Park, Illinois, 1995. American Nuclear Society.
1245. G. T. Parks. Multiobjective Pressurized Water Reactor Reload Core Design by Nondominated Genetic Algorithm Search. *Nuclear Science and Engineering*, 124(1):178–187, 1996.
1246. G. T. Parks. Multiobjective Pressurised Water Reactor Reload Core Design using a Genetic Algorithm. In G. D. Smith, N. C. Steele, and R. F. Albrecht, editors, *Artificial Neural Nets and Genetic Algorithms*, pages 53–57, Norwich, UK, 1997. Springer-Verlag.
1247. G. T. Parks and I. Miller. Selective Breeding in a Multiobjective Genetic Algorithm. In A. E. Eiben, T. Bäck, M. Schoenauer, and H.-P. Schwefel, editors, *Parallel Problem Solving From Nature — PPSN V*, pages 250–259, Amsterdam, Holland, 1998. Springer-Verlag. Lecture Notes in Computer Science No. 1498.
1248. I. Parmee. Poor-Definition, Uncertainty, and Human Factors—Satisfying Multiple Objectives in Real-World Decision-Making Environments. In E. Zitzler, K. Deb, L. Thiele, C. A. Coello Coello, and D. Corne, editors, *First International Conference on Evolutionary Multi-Criterion Optimization*, pages 67–81. Springer-Verlag. Lecture Notes in Computer Science No. 1993, 2001.
1249. I. C. Parmee, D. Cvetković, A. H. Watson, and C. R. Bonham. Multiobjective Satisfaction within an Interactive Evolutionary Design Environment. *Evolutionary Computation*, 8(2):197–222, Summer 2000.
1250. I. C. Parmee and G. Purchase. The development of a directed genetic search technique for heavily constrained design spaces. In I. C. Parmee, editor, *Adaptive Computing in Engineering Design and Control-'94*, pages 97–102, Plymouth, UK, 1994. University of Plymouth, University of Plymouth.
1251. I. C. Parmee and H. D. Vekeria. Co-operative Evolutionary Strategies for Single Component Design. In T. Bäck, editor, *Proceedings of the Seventh International Conference on Genetic Algorithms*, pages 529–536, San Francisco, California, USA, 1997. Morgan Kaufmann Publishers.
1252. I. C. Parmee and A. H. Watson. Preliminary Airframe Design Using Co-Evolutionary Multiobjective Genetic Algorithms. In W. Banzhaf, J. Daida, A. E. Eiben, M. H. Garzon, V. Honavar, M. Jakiela, and R. E. Smith, editors, *Proceedings of the Genetic and Evolutionary Computation Conference (GECCO'99)*, volume 2, pages 1657–1665, San Francisco, California, July 1999. Morgan Kaufmann.
1253. R. O. Parreiras and J. A. Vasconcelos. Decision Making in Multiobjective Optimization Problems. In N. Nedjah and L. de Macedo Mourelle, editors, *Real-World Multi-Objective System Engineering*, pages 29–52. Nova Science Publishers, New York, 2005.
1254. R. Parsons and S. Canfield. Developing genetic programming techniques for the design of compliant mechanisms. *Structural and Multidisciplinary Optimization*, 24(1):78–86, August 2002.
1255. K. Parsopoulos, V. Plagianakos, G. Magoulas, and M. Vrahatis. Stretch technique for obtaining global minimizers through Particle Swarm Optimization. In *Proceedings of Particle Swarm Optimization Workshop, Indiana University Purdue University at Indianapolis*, pages 22–38, Indianapolis, Indiana, April 2001.

1256. K. Parsopoulos, D. Taoulis, N. Pavlidis, V. Plagianakos, and M. Vrahatis. Vector Evaluated Differential Evolution for Multiobjective Optimization. In *2004 Congress on Evolutionary Computation (CEC'2004)*, volume 1, pages 204–211, Portland, Oregon, USA, June 2004. IEEE Service Center.

1257. K. Parsopoulos, D. Tasoulis, and M. Vrahatis. Multiobjective Optimization Using Parallel Vector Evaluated Particle Swarm Optimization. In *Proceedings of the IASTED International Conference on Artificial Intelligence and Applications (AIA 2004)*, volume 2, pages 823–828, Innsbruck, Austria, February 2004. ACTA Press.

1258. K. Parsopoulos and M. Vrahatis. Particle Swarm Optimization Method for Constrained Optimization Problems. In P. Sincak, J.Vascak, V. Kvasnicka, and J. Pospicha, editors, *Intelligent Technologies - Theory and Applications: New Trends in Intelligent Technologies*, pages 214–220. IOS Press, 2002. Frontiers in Artificial Intelligence and Applications series, Vol. 76 ISBN: 1-58603-256-9.

1259. K. Parsopoulos and M. Vrahatis. Particle Swarm Optimization Method in Multiobjective Problems. In *Proceedings of the 2002 ACM Symposium on Applied Computing (SAC'2002)*, pages 603–607, Madrid, Spain, 2002. ACM Press.

1260. J. M. Pasia, R. F. Hartl, and K. F. Doerner. Solving a Bi-objective Flowshop Scheduling Problem by Pareto-Ant Colony Optimization. In M. Dorigo, L. M. Gambardella, M. Birattari, A. Martinoli, R. Poli, and T. Stützle, editors, *Ant Colony Optimization and Swarm Intelligence. 5th International Workshop, ANTS 2006*, pages 294–305. Springer. Lecture Notes in Computer Science Vol. 4150, Brussels, Belgium, September 2006.

1261. H. Pastijn and J. Leysen. Constructing an outranking relation with ORESTE. *Mathematical and Computer Modelling*, 12(10–11):1255–1268, 1989.

1262. Z. Pawlak. Rough sets. *International Journal of Computer and Information Sciences*, 11(1):341–356, Summer 1982.

1263. Z. Pawlak. *Rough Sets: Theoretical Aspects of Reasoning about Data*. Kluwer Academic Publishers, Dordrecht, The Netherlands, 1991. ISBN 0-471-87339-X.

1264. J. Pearl. *Heuristics*. Addison–Wesley Publishing Company, Reading, Massachusetts, 1989.

1265. M. Pelikan, K. Sastry, and D. E. Goldberg. Multiobjective hBOA, Clustering, and Scalability. In H.-G. B. et al., editor, *2005 Genetic and Evolutionary Computation Conference (GECCO'2005)*, volume 1, pages 663–670, New York, USA, June 2005. ACM Press.

1266. J. Périaux, M. Sefrioui, and B. Mantel. RCS multi-objective optimization of scattered waves by active control elements using GAs. In *Proceedings of the Fourth International Conference on Control, Automation, Robotics and Vision (ICARCV'96)*, Singapore, 1996.

1267. J. Périaux, M. Sefrioui, and B. Mantel. GA Multiple Objective Optimization Strategies for Electromagnetic Backscattering. In D. Quagliarella, J. Périaux, C. Poloni, and G. Winter, editors, *Genetic Algorithms and Evolution Strategies in Engineering and Computer Science. Recent Advances and Industrial Applications*, chapter 11, pages 225–243. John Wiley and Sons, West Sussex, England, 1997.

1268. E. Perrin, A. Mandrille, M. Oumoun, C. Fonteix, and I. Marc. Optimisation Globale par Stratégie d'Evolution. *RAIRO-Recherche Opérationelle*, 32(2):161–201, 1997.
1269. C. J. Petrie, T. A. Webster, and M. R. Cutkosky. Using Pareto Optimality to Coordinate Distributed Agents. *Artificial Intelligence for Engineering Design, Analysis and Manufacturing*, 9:269–281, 1995.
1270. H. Petroski. *Engineers of Dreams: Great Bridge Builders and the Spanning of American*. Morgan Kaufmann Publishers, 1996.
1271. A. Petrovski and J. McCall. Multi-objective Optimisation of Cancer Chemotherapy Using Evolutionary Algorithms. In E. Zitzler, K. Deb, L. Thiele, C. A. Coello Coello, and D. Corne, editors, *First International Conference on Evolutionary Multi-Criterion Optimization*, pages 531–545. Springer-Verlag. Lecture Notes in Computer Science No. 1993, 2001.
1272. F. Pettersson, N. Chakraborti, and H. Saxén. A genetic algorithms based multi-objective neural net applied to noisy blast furnace data. *Applied Soft Computing*, 7:387–397, 2007.
1273. H. Pierreval and M.-F. Plaquin. An Evolutionary Approach of Multicriteria Manufacturing Cell Formation. *International Transactions in Operational Research*, 5(1):13–25, January 1998.
1274. C. Pimpawat and N. Chaiyaratana. Using a co-operative co-evolutionary genetic algorithm to solve a three-dimensional container loading problem. In *Proceedings of the Congress on Evolutionary Computation 2001 (CEC'2001)*, volume 2, pages 1197–1204, Piscataway, New Jersey, May 2001. IEEE Service Center.
1275. P. Pirjanian. *Multiple Objective Action Selection & Behavior Fusion using Voting*. PhD thesis, Department of Medical Informatics and Image Analysis, Institute of Electronic Systems, Aalborg University, Aalborg, Denmark, August 1998.
1276. P. Pirjanian and M. Matarić. A decision-theoretic approach to fuzzy behavior coordination. In *IEEE International Symposium on Computational Intelligence in Robotics & Automation (CIRA'99)*, pages 101–106, Monterey, California, USA, November 1999.
1277. P. Pirjanian and M. Matarić. Multi-robot target acquisition using multiple objective behavior coordination. In *Proceedings of the International Conference on Robotics and Automation (ICRA'2000)*, pages 2696–2702, San Francisco, California, USA, 2000.
1278. P. Pirjanian and M. Matarić. Multiple objective vs. fuzzy behavior coordination. In D. Drainkov and A. Saffiotti, editors, *Fuzzy Logic Techniques for Autonomous Vehicle Navigation*, pages 235–253. Springer. Studies on Fuzziness and Soft Computing, 2000.
1279. H. Pohlheim. Genetic and Evolutionary Algorithm Toolbox for use with MATLAB. Technical report, Technical University Ilmenau, 1998.
1280. J. B. Pollack, A. D. Blair, and M. Land. Coevolution of a Backgammon Player. In C. Langton, editor, *Artificial Life V*, pages 92–98. The MIT Press, Cambridge, Massachusetts, USA, 1996.
1281. C. Poloni. Hybrid GA for Multi-Objective Aerodynamic Shape Optimization. In G. Winter, J. Périaux, M. Galan, and P. Cuesta, editors, *Genetic Algorithms in Engineering and Computer Science*, pages 397–416. Wiley & Sons, Chichester, 1995.

1282. C. Poloni, M. Fearon, and D. Ng. Parallelisation of Genetic Algorithms for Aerodynamic Design Optimisation. In I. C. Parmee and M. J. Denham, editors, *Proceedings of the Second International Conference on Adaptive Computing in Engineering Design and Control*, pages 59–64. University of Plymouth, Plymouth, UK, 1996.
1283. C. Poloni, G. Mosetti, and S. Contessi. Multiobjective Optimization by GAs: Application to System and Component Design. In *Computational Methods in Applied Sciences '96: Invited Lectures and Special Technological Sessions of the Third ECCOMAS Computational Fluid Dynamics Conference and the Second ECCOMAS Conference on Numerical Methods in Engineering*, pages 258–264, Chichester, 1996. Wiley.
1284. C. Poloni and V. Pediroda. GA coupled with computationally expensive simulations: tools to improve efficiency. In D. Quagliarella, J. Périaux, C. Poloni, and G. Winter, editors, *Genetic Algorithms and Evolution Strategies in Engineering and Computer Science. Recent Advances and Industrial Applications*, chapter 13, pages 267–288. John Wiley and Sons, West Sussex, England, 1997.
1285. W. Ponweiser and M. Vincze. The Multiple Multi Objective Problem—Definition, Solution and Evaluations. In S. Obayashi, K. Deb, C. Poloni, T. Hiroyasu, and T. Murata, editors, *Evolutionary Multi-Criterion Optimization, 4th International Conference, EMO 2007*, pages 877–892, Matshushima, Japan, March 2007. Springer. Lecture Notes in Computer Science Vol. 4403.
1286. P. W. Poon and G. T. Parks. Application of genetic algorithms to in-core nuclear fuel management optimization. In H. Küsters, E. Stein, and W. Werner, editors, *Proceedings of the Joint Conference on Mathematical Methods and Supercomputing in Nuclear Applications*, pages 777–786, Karlsruhe, Germany, 1993. Kernforschungszentrum Karlsruhe GmbH.
1287. E. A. Portilla Flores. *Integración Simultánea de Aspectos Estructurales y Dinámicos para el Diseño Óptimo de un Sistema de Transmisión de Variación Continua*. PhD thesis, Departamento de Ingeniería Eléctrica, Sección de Mecatrónica, CINVESTAV-IPN, México, D.F., México, June 2006. (In Spanish).
1288. M. A. Potter and K. de Jong. A Cooperative Coevolutionary Approach to Function Optimization. In Y. Davidor, H.-P. Schwefel, and R. Männer, editors, *Parallel Problem Solving from Nature—PPSN III*, pages 249–257, Jerusalem, Israel, October 1994. Springer-Verlag. Lecture Notes in Computer Science Vol. 866.
1289. J.-Y. Potvin and S. Bengio. The Vehicle Routing Problem with Time Windows–Part II: Genetic Search. *INFORMS Journal on Computing*, 8:165–172, 1996.
1290. P. Poulos, G. Rigatos, S. Tzafestas, and A. Koukos. A Pareto-optimal genetic algorithm for warehouse multi-objective optimization. *Engineering Applications of Artificial Intelligence*, 14:737–749, 2001.
1291. D. Powell and M. M. Skolnick. Using genetic algorithms in engineering design optimization with non-linear constraints. In S. Forrest, editor, *Proceedings of the Fifth International Conference on Genetic Algorithms (ICGA-93)*, pages 424–431, San Mateo, California, July 1993. Morgan Kaufmann Publishers.
1292. W. H. Press, B. P. Flannery, S. A. Teukolsky, and W. T. Vetterling. *Numerical Recipes in Pascal*. Cambridge University Press, Cambridge, UK, 1989.

1293. M. Preuss. Niching Prospects. In B. Filipič and J. Šilc, editors, *Bioinspired Optimization Methods and their Applications*, pages 25–34. Jožef Stefan Institute, October 2006.
1294. K. V. Price. An Introduction to Differential Evolution. In D. Corne, M. Dorigo, and F. Glover, editors, *New Ideas in Optimization*, pages 79–108. McGraw-Hill, London, UK, 1999.
1295. K. V. Price, R. M. Storn, and J. A. Lampinen. *Differential Evolution. A Practical Approach to Global Optimization*. Springer, Berlin, 2005. ISBN 3-540-20950-6.
1296. C. Prins. A simple and effective evolutionary algorithm for the vehicle routing problem. In J. P. de Sousa, editor, *Proceedings of the 4th Metaheuristics International Conference (MIC'2001)*, pages 143–148. Program Operational Ciencia, Tecnologia, Inovaçao do Quadro Comunitário de Apoio III de Fundaçao para a Ciencia e Tecnologia, Porto, Portugal, July 16–20 2001.
1297. C. Prins. A simple and effective evolutionary algorithm for the vehicle routing problem. *Computers & Operations Research*, 31(12):1985–2002, 2004.
1298. H. Prüfer. Neuer beweis eines satzes über permutationen. *Archiv fue Mathematische und Physik*, 27:742–744, 1918.
1299. A. Pryke, S. Mostaghim, and A. Nazemi. Heatmap Visualization of Population Based Multi Objective Algorithms. In S. Obayashi, K. Deb, C. Poloni, T. Hiroyasu, and T. Murata, editors, *Evolutionary Multi-Criterion Optimization, 4th International Conference, EMO 2007*, pages 361–375, Matshushima, Japan, March 2007. Springer. Lecture Notes in Computer Science Vol. 4403.
1300. W. Pullan. Optimising Multiple Aspects of Network Survivability. In *Congress on Evolutionary Computation (CEC'2002)*, volume 1, pages 115–120, Piscataway, New Jersey, May 2002. IEEE Service Center.
1301. T. Pulliam, M. Nemec, T. Hoslt, and D. Zingg. Comparison of Evolutionary (Genetic) Algorithm and Adjoint Methods for Multi-Objective Viscous Airfoil Optimizations. In *41st Aerospace Sciences Meeting. Paper AIAA 2003-0298*, Reno, Nevada, January 2003.
1302. W. F. Punch. How Effective are Multiple Poplulations in Genetic Programming. In J. R. Koza, W. Banzhaf, K. Chellapilla, K. Deb, M. Dorigo, D. B. Fogel, M. H. Garzon, D. E. Goldberg, H. Iba, and R. L. Riolo, editors, *Proceedings of the Third Annual Conference on Genetic Programming*, pages 308–313, San Mateo, California, July 1998. University of Wisconsin at Madison, Morgan Kaufmann Publishers.
1303. R. C. Purshouse. *On the Evolutionary Optimisation of Many Objectives*. PhD thesis, Department of Automatic Control and Systems Engineering, The University of Sheffield, Sheffield, UK, September 2003.
1304. R. C. Purshouse and P. J. Fleming. Conflict, Harmony, and Independence: Relationships in Evolutionary Multi-criterion Optimisation. In C. M. Fonseca, P. J. Fleming, E. Zitzler, K. Deb, and L. Thiele, editors, *Evolutionary Multi-Criterion Optimization. Second International Conference, EMO 2003*, pages 16–30, Faro, Portugal, April 2003. Springer. Lecture Notes in Computer Science. Volume 2632.
1305. M. Qiu. Prioritizing and Scheduling Road Projects by Genetic Algorithm. *Mathematics and Computers in Simulation*, 43:569–574, 1997.
1306. D. Quagliarella and A. Vicini. Coupling Genetic Algorithms and Gradient Based Optimization Techniques. In D. Quagliarella, J. Périaux, C. Poloni,

and G. Winter, editors, *Genetic Algorithms and Evolution Strategies in Engineering and Computer Science. Recent Advances and Industrial Applications*, chapter 14, pages 289–309. John Wiley and Sons, West Sussex, England, 1997.
1307. D. Quagliarella and A. Vicini. Sub-population Policies for a Parallel Multiobjective Genetic Algorithm with Applications to Wing Design. In *1998 IEEE International Conference On Systems, Man, And Cybernetics*, pages 3142–3147, San Diego, California, October 1998. Institute of Electrical and Electronic Engineers (IEEE).
1308. D. Quagliarella and A. Vicini. Designing High-Lift Airfoils Using Genetic Algorithms. In K. Miettinen, M. M. Mäkelä, and J. Toivanen, editors, *Proceedings of EUROGEN'99 — Short Course on Evolutionary Algorithms in Engineering and Computer Science*, pages 143–149, Jyväskyl, Finland, May 1999. University of Jyväskylä. (Reports of the Department of Mathematical Information Technology, Series A. Collections, No. 2/1999, ISBN 951-39-0473-3).
1309. W. V. Quine. A Way to Simplify Truth Functions. *American Mathematical Monthly*, 62 (9):627–631, 1955.
1310. J. R. Quinlan. *C4.5 Programs for Machine Learning*. Morgan Kaufmann Publishers, 1993.
1311. L. Rachmawati and D. Srinivasan. A Multi-objective Evolutionary Algorithm with Weighted-Sum Niching for Convergence on Knee Regions. In M. K. et al., editor, *2006 Genetic and Evolutionary Computation Conference (GECCO'2006)*, volume 1, pages 749–750, Seattle, Washington, USA, July 2006. ACM Press. ISBN 1-59593-186-4.
1312. L. Rachmawati and D. Srinivasan. A Multi-Objective Genetic Algorithm with Controllable Convergence on Knee Regions. In *2006 IEEE Congress on Evolutionary Computation (CEC'2006)*, pages 6807–6814, Vancouver, BC, Canada, July 2006. IEEE.
1313. L. Rachmawati and D. Srinivasan. Preference Incorporation in Multi-objective Evolutionary Algorithms: A Survey. In *2006 IEEE Congress on Evolutionary Computation (CEC'2006)*, pages 3385–3391, Vancouver, BC, Canada, July 2006. IEEE.
1314. A. M. Raich and T. R. Liszkai. Multi-Objective Genetic Algorithms for Sensor Layout Optimization in Structural Damage Detection. In C. H. Dagli, A. L. Buczak, J. Ghosh, M. J. Embrechts, and O. Ersoy, editors, *Smart Engineering System Design: Neural Networks, Fuzzy Logic, Evolutionary Programming, Complex Systems, and Artificial Life (ANNIE'2003)*, pages 889–894. ASME Press, November 2003.
1315. H. Raiffa. Preferences for Multi-Attributed Alternatives. Technical Report RM-5868-DOT/RC, Rand Corporation, Santa Monica, California, 1969.
1316. I. J. Ramírez Rosado and J. L. Bernal Agustín. Reliability and Cost Optimization for Distribution Networks Expansion Using an Evolutionary Algorithm. *IEEE Transactions on Power Systems*, 16(1):111–118, February 2001.
1317. I. J. Ramírez Rosado, J. L. Bernal Agustín, L. M. Barbosa Proença, and V. Miranda. Multiobjective Planning of Power Distribution Systems Using Evolutionary Algorithms. In M. H. Hamza, editor, *8th IASTED International Conference on Modelling, Identification and Control—MIC'99*, pages 185–188, Innsbruck, Austria, February 1999.
1318. S. R. Ranjithan, S. K. Chetan, and H. K. Dakshima. Constraint Method-Based Evolutionary Algorithm (CMEA) for Multiobjective Optimization. In

E. Zitzler, K. Deb, L. Thiele, C. A. Coello Coello, and D. Corne, editors, *First International Conference on Evolutionary Multi-Criterion Optimization*, pages 299–313. Springer-Verlag. Lecture Notes in Computer Science No. 1993, 2001.
1319. A. R. M. Rao and N. Arvind. A scatter search algorithm for stacking sequence optimisation of laminate composites. *Composite Structures*, 70(4):383–402, October 2005.
1320. S. Rao. Game Theory Approach for Multiobjective Structural Optimization. *Computers and Structures*, 25(1):119–127, 1986.
1321. S. S. Rao. Multiobjective Optimization in Structural Design with Uncertain Parameters and Stochastic Processes. *AIAA Journal*, 22(11):1670–1678, November 1984.
1322. S. S. Rao. Game Theory Approach for Multiobjective Structural Optimization. *Computers and Structures*, 25(1):119–127, 1987.
1323. S. S. Rao. Genetic Algorithmic Approach for Multiobjective Optimization of Structures. In *ASME Annual Winter Meeting, Structures and Controls Optimization*, volume AD-Vol. 38, pages 29–38, New Orleans, Louisiana, November 1993. ASME.
1324. S. S. Rao. *Engineering Optimization*. John Wiley & Sons, third edition, 1996.
1325. V. Rao Vemuri and W. Cedeño. A New Genetic Algorithm for Multi Objective Optimization in Water Resource Management. In *Proceedings of the Second IEEE International Conference on Evolutionary Computation*, pages 495–500, Piscataway, New Jersey, 1995. IEEE Press.
1326. V. Rao Vemuri and W. Cedeño. A New Genetic Algorithm for Multi-Objective Optimization in Water Resource Management. In *1996 Knowledge-based Computer Systems*, Bombay, India, December 1996. KBIS Proceedings.
1327. C. R. Raquel and P. C. Naval, Jr. An Effective Use of Crowding Distance in Multiobjective Particle Swarm Optimization. In H.-G. B. et al., editor, *2005 Genetic and Evolutionary Computation Conference (GECCO'2005)*, volume 1, pages 257–264, New York, USA, June 2005. ACM Press.
1328. T. Ray, R. Gokarn, and O. Sha. A global optimization model for ship design. *Computers in Industry*, 26:175–192, 1995.
1329. T. Ray, T. Kang, and S. K. Chye. An Evolutionary Algorithm for Constrained Optimization. In D. Whitley, D. Goldberg, E. Cantú-Paz, L. Spector, I. Parmee, and H.-G. Beyer, editors, *Proceedings of the Genetic and Evolutionary Computation Conference (GECCO'2000)*, pages 771–777, San Francisco, California, 2000. Morgan Kaufmann.
1330. T. Ray, T. Kang, and S. K. Chye. Multiobjective Design Optimization by an Evolutionary Algorithm. *Engineering Optimization*, 33(3):399–424, 2001.
1331. T. Ray and K. Liew. A Swarm with and Effective Information Sharing Mechanism for Unconstrained and Constrained Single Objective Optimization. In *Proceedings of the Congress on Evolutionary Computation 2001 (CEC'2001)*, volume 1, pages 75–80, Piscataway, New Jersey, May 2001. IEEE Service Center.
1332. T. Ray and K. Liew. A Swarm Metaphor for Multiobjective Design Optimization. *Engineering Optimization*, 34(2):141–153, March 2002.
1333. T. Ray and K. Liew. Society and Civilization: An Optimization Algorithm Based on the Simulation of Social Behavior. *IEEE Transactions on Evolutionary Computation*, 7(4):386–396, August 2003.

1334. T. Ray and W. Smith. A surrogate assisted parallel multiobjective evolutionary algorithm for robust engineering design. *Engineering Optimization*, 38(8):997–1011, December 2006.
1335. T. Ray and K. Tai. An Evolutionary Algorithm with a Multilevel Pairing Strategy for Single and Multiobjective Optimization. *Foundations of Computing and Decision Sciences*, 26:75–98, 2001.
1336. B. J. Reardon. Optimization of Densification Modeling Parameters of Beryllium Powder using a Fuzzy Logic Based Multiobjective Genetic Algorithm. Technical Report LA-UR-98-1036, Los Alamos National Laboratory, Los Alamos, New Mexico, March 1998.
1337. B. J. Reardon. Optimization of Micromechanical Densification Modeling Parameters For Copper Powder using a Fuzzy Logic Based Multiobjective Genetic Algorithm. Technical Report LA-UR-98-0419, Los Alamos National Laboratory, Los Alamos, New Mexico, January 1998.
1338. P. Reed and V. Devireddy. Groundwater Monitoring Design: A Case Study Combining Epsilon Dominance Archiving and Automatic Parameterization for the NSGA-II. In C. A. Coello Coello and G. B. Lamont, editors, *Applications of Multi-Objective Evolutionary Algorithms*, pages 79–100. World Scientific, Singapore, 2004.
1339. P. Reed, B. S. Minsker, and D. E. Goldberg. Simplifying multiobjective optimization: An automated design methodology for the nondominated sorted genetic algorithm-II. *Water Resources Research*, 39(7):TNN 2.1–2.5, July 2003.
1340. P. M. Reed, B. S. Minsker, and D. E. Goldberg. Designing a New Elitist Nondominated Sorted Genetic Algorithm for a Multiobjective Long Term Groundwater Monitoring Application. In *Proceedings of the 2001 Genetic and Evolutionary Computation Conference. Late-Breaking Papers*, pages 352–358, San Francisco, California, July 2001.
1341. P. M. Reed, B. S. Minsker, and D. E. Goldberg. A multiobjective approach to cost effective long-term groundwater monitoring using an elitist nondominated sorted genetic algorithm with historical data. *Journal of Hydroinformatics*, 3(2):71–89, April 2001.
1342. P. M. Reed, B. S. Minsker, and D. E. Goldberg. Why Optimize Long Term Groundwater Monitoring Design? A Multiobjective Case Study of Hill Air Force Base. In D. Phelps and G. Sehlke, editors, *Bridging the Gap: Meeting the World's Water and Environmental Resources Challenges. Proceedings of the World Water and Environmental Resources Congress*, Washington, DC, 2001. American Society of Civil Engineers. ISBN 0-7844-0569-7.
1343. C. R. Reeves and J. E. Rowe. *Genetic Algorithms—Principles and Perspectives. A Guide to GA Theory*. Kluwer Academic Publishers, Boston/Dordrecht/London, 2003. ISBN 1-4020-7240-6.
1344. J. Régnier, B. Sareni, and X. Roboam. System optimization by multiobjective genetic algorithms and analysis of the coupling between variables, constraints and objectives. *COMPEL-The International Journal for Computation and Mathematics in Electrical and Electronic Engineering*, 24(3):805–820, 2005.
1345. G. Reinelt. Traveling Salesman Problem Library. Online, 1995. Available: http://softlib.rice.edu/softlib
1346. B. Rekiek. *Assembly Line Design (multiple objective grouping genetic algorithm and the balancing of mixed-model hybrid assembly line)*. PhD thesis,

Free University of Brussels, CAD/CAM Department, Brussels, Belgium, December 2000.
1347. B. Rekiek, P. D. Lit, F. Pellichero, T. L'Eglise, E. Falkenauer, and A. Delchambre. Dealing With User's Preferences in Hybrid Assembly Lines Design. In Z. Binder, editor, *Proceedings of the Management and Control of Production and Logistics 2000 (MCPL'2000) Conference*, volume 3, pages 989–994. Pergamon, Grenoble, France, July 2000.
1348. B. Rekiek, F. Pellichero, P. D. Lit, E. Falkenauer, and A. Delchambre. A Resource Planner for Hybrid Assembly Lines. In *Proceedings of the 15th International Conference on CAD/CAM Robotics & Factories of the Future CAR & FOF'99*, volume 1, pages MW6-18–MW6-23, August 1999.
1349. X. Ren and B. Minsker. Which Groundwater Remediation Objective is Better, a Realistic One or a Simple One? In *American Society of Civil Engineers (ASCE) Environmental & Water Resources Institute (EWRI) World Water & Environmental Resources Congress 2003 & Related Symposia*, Philadelphia, PA, 2003.
1350. A. C. Renfrew. Dynamic Modeling in Archaeology: What, When, and Where? In S. E. van der Leeuw, editor, *Dynamical Modeling and the Study of Change in Archaelogy*. Edinburgh University Press, Edinburgh, Scotland, 1994.
1351. M. Reyes Sierra and C. A. Coello Coello. Improving PSO-Based Multi-objective Optimization Using Crowding, Mutation and ϵ-Dominance. In C. A. Coello Coello, A. Hernández Aguirre, and E. Zitzler, editors, *Evolutionary Multi-Criterion Optimization. Third International Conference, EMO 2005*, pages 505–519, Guanajuato, México, March 2005. Springer. Lecture Notes in Computer Science Vol. 3410.
1352. M. Reyes Sierra and C. A. Coello Coello. A Study of Fitness Inheritance and Approximation Techniques for Multi-Objective Particle Swarm Optimization. In *2005 IEEE Congress on Evolutionary Computation (CEC'2005)*, volume 1, pages 65–72, Edinburgh, Scotland, September 2005. IEEE Service Center.
1353. M. Reyes-Sierra and C. A. Coello Coello. Multi-Objective Particle Swarm Optimizers: A Survey of the State-of-the-Art. *International Journal of Computational Intelligence Research*, 2(3):287–308, 2006.
1354. M. M. Reyes Sierra. *Use of Coevolution and Fitness Inheritance for Multiobjective Particle Swarm Optimization*. PhD thesis, Computer Science Section, Department of Electrical Engineering, CINVESTAV-IPN, Mexico, August 2006.
1355. C. Reynolds. Competition, coevolution and the game of tag. In R. Brooks and P. Maes, editors, *Proceedings of the Fourth International Workshop on the Synthesis and Simulation of Living Systems*, pages 59–69. The MIT Press, Cambridge, Massachusetts, USA, 1994.
1356. J. Reynolds. *Multi-Criteria Assessment of Ecological Process Models using Pareto Optimization*. PhD thesis, University of Washington, Seattle, Washington, USA, 1997.
1357. J. H. Reynolds and E. D. Ford. Multi-criteria assessment of ecological process models. *Ecology*, 80(5):538–553, may 1999.
1358. R. G. Reynolds. An Introduction to Cultural Algorithms. In A. V. Sebald and L. J. Fogel, editors, *Proceedings of the Third Annual Conference on Evolutionary Programming*, pages 131–139. World Scientific, River Edge, New Jersey, USA, 1994.

1359. G. Richards and V. Rayward-Smith. The Discovery of Association Rules from Tabular Databases Comprising Nominal and Ordinal Attributes. *Intelligent Data Analysis*, 9(3):289–307, 2005.
1360. J. T. Richardson, M. R. Palmer, G. Liepins, and M. Hilliard. Some Guidelines for Genetic Algorithms with Penalty Functions. In J. D. Schaffer, editor, *Proceedings of the Third International Conference on Genetic Algorithms*, pages 191–197, San Mateo, California, 1989. Morgan Kaufmann Publishers.
1361. R. G. L. Riche, C. Knopf-Lenoir, and R. T. Haftka. A Segregated Genetic Algorithm for Constrained Structural Optimization. In L. J. Eshelman, editor, *Proceedings of the Sixth International Conference on Genetic Algorithms*, pages 558–565, San Mateo, California, July 1995. Morgan Kaufmann Publishers.
1362. P. Rietveld. *Multiple Objective Decision Methods and Regional Planning*. North-Holland, New York, 1980.
1363. J. Rissanen. Modeling by Shortest Data Description. *Automatica*, 14:465–471, 1978.
1364. B. J. Ritzel, J. W. Eheart, and S. Ranjithan. Using genetic algorithms to solve a multiple objective groundwater pollution containment problem. *Water Resources Research*, 30(5):1589–1603, may 1994.
1365. T. Robič and B. Filipič. DEMO: Differential Evolution for Multiobjective Optimization. In C. A. Coello Coello, A. Hernández Aguirre, and E. Zitzler, editors, *Evolutionary Multi-Criterion Optimization. Third International Conference, EMO 2005*, pages 520–533, Guanajuato, México, March 2005. Springer. Lecture Notes in Computer Science Vol. 3410.
1366. J. E. Rodríguez, A. L. Medaglia, and J. P. Casas. Approximation to the Optimum Design of a Motorcycle Frame using Finite Element Analysis and Evolutionary Algorithms. In E. J. Bass, editor, *Proceedings of the 2005 IEEE Systems and Information Engineering Design Symposium*, pages 277–285. IEEE Press, 2005.
1367. K. Rodríguez Vázquez. *Multiobjective Evolutionary Algorithms in Non-Linear System Identification*. PhD thesis, Department of Automatic Control and Systems Engineering, The University of Sheffield, Sheffield, UK, 1999.
1368. K. Rodriguez-Vazquez and P. Fleming. Evolution of mathematical models of chaotic systems based on multiobjective genetic programming. *Knowledge and Information Systems*, 8(2):235–256, August 2005.
1369. K. Rodríguez Vázquez and P. J. Fleming. Functionality and Optimality in Circuit Design: A Genetic Programming Approach. In *Proceedings of the Third International Symposium on Adaptive Systems—Evolutionary Computation and Probabilistic Graphical Models*, pages 23–28, Havana, Cuba, March 19–23 2001. Institute of Cybernetics, Mathematics and Physics.
1370. K. Rodríguez Vázquez, C. M. Fonseca, and P. J. Fleming. Multiobjective Genetic Programming : A Nonlinear System Identification Application. In J. R. Koza, editor, *Late Breaking Papers at the Genetic Programming 1997 Conference*, pages 207–212, Stanford University, California, July 1997. Stanford Bookstore.
1371. J. L. Rogers. Optimum Actuator Placement with a Genetic Algorithm for Aircraft Control. In C. H. Dagli, A. L. Buczak, J. Ghosh, M. J. Embrechts, and O. Ersoy, editors, *Smart Engineering System Design: Neural Networks, Fuzzy Logic, Evolutionary Programming, Data Mining, and Complex Systems (ANNIE'99)*, pages 355–360, New York, November 1999. ASME Press.

1372. J. L. Rogers. A Parallel Approach to Optimum Actuator Selection With A Genetic Algorithm. In *AIAA Paper No. 2000-4484, AIAA Guidance, Navigation, and Control Conference*, Denver, Colorado, August 14–17 2000.

1373. G. Rohling, D. Lamm, C. Carstensen, and C. E. Hunt. Flare Pattern Design for Imaging and Reticle Based Missile Seekers Using Genetic Algorithms. In *Proceedings of Threats Countermeasures and Situational Awareness Conference*, Virginia Beach, Virginia, June 2000.

1374. P. Roosen, S. Uhlenbruck, and K. Lucas. Pareto optimization of a combined cycle power system as a decision support tool for trading off investment vs. operating costs. *International Journal of Thermal Sciences*, 42(6):553–560, June 2003.

1375. R. S. Rosenberg. *Simulation of genetic populations with biochemical properties*. PhD thesis, University of Michigan, Ann Arbor, Michigan, USA, 1967.

1376. H. Rosenbrock. An automatic method for finding the greatest or least value of a function. *Computer Journal*, 3(3):175–184, 1960.

1377. C. Rosin and R. Belew. New methods for competitive coevolution. *Evolutionary Computation*, 5(1):1–29, 1996.

1378. K. Rosing and C. S. ReVelle. Heuristic concentration: two stage solution construction. *European Journal of Operational Research*, 97(19):75–86, 1997.

1379. B. J. Ross, W. Ralph, and H. Zong. Evolutionary Image Synthesis Using a Model of Aesthetics. In *2006 IEEE Congress on Evolutionary Computation (CEC'2006)*, pages 3832–3839, Vancouver, BC, Canada, July 2006. IEEE.

1380. P. J. Ross. *Taguchi Methods for Quality Engineering: Loss Function, Orthogonal Experiments, Parameter and Tolerance Design*. McGraw-Hill, New York, second edition, 1995.

1381. M. Roubens. Preference Relations on actions and criteria in multicriteria decision making. *European Journal of Operational Research*, 10:51–55, 1982.

1382. J. Rowe, K. Vinsen, and N. Marvin. Parallel GAs for Multiobjective Functions. In J. T. Alander, editor, *Proceedings of the Second Nordic Workshop on Genetic Algorithms and Their Applications (2NWGA)*, pages 61–70, Vaasa, Finland, August 1996. University of Vaasa.

1383. B. Roy. Classement et choix en présence de points de vue multiples: La méthode Electre. *Revue Francaise d'Informatique et de Recherche Opérationalle*, 8:207–218, 1968.

1384. B. Roy. Problems and methods with multiple objective functions. *Mathematical programming*, 1(2):239–266, 1971.

1385. B. Roy. A Conceptual Framework for a Prescriptive Theory of "Decision-Aid". In M. K. Starr and M. Zeleny, editors, *Multiple Criteria Decision Making*, volume 6 of *TIMS Studies in the Management Sciences*, pages 179–210. North-Holland Publishing Company, Amsterdam, 1977.

1386. B. Roy. Electre III: Algorithme de classement basé sur une représentation floue des préférences en présence de critères multiples. *Cahiers du C.E.R.O.*, 20(1):3–24, 1978.

1387. B. Roy. Decision-aid and Decision-making. In C. Bana e Costa, editor, *Readings in Multiple Criteria Decision Aid*, pages 17–35. Springer-Verlag, Berlin, 1990.

1388. B. Roy. *Multicriteria Methodology for Decision Aiding*. Kluwer Academic Publishers, 1996.

1389. B. Roy and P. Bertier. La méthode Electre: Une application du média planning. In M. Ross, editor, *Operational Research*, pages 291–302. North-Holland, Amsterdam, 1973.
1390. B. Roy and D. Bouyssou. *Aide Multicritère à la décision: Méthodes et Cas.* Economica, Paris, 1993.
1391. B. Roy and P. Vincke. Multicriteria analysis: survey and new directions. *European Journal of Operational Research*, 11:207–218, 1981.
1392. P. A. Rubin and G. Ragatz. Scheduling in a Sequence Dependent Setup Environment with Genetic Search. *Computers and Operations Research*, 22(2):85–99, 1995.
1393. G. Rudolph. Convergence Analysis of Canonical Genetic Algorithms. *IEEE Transactions on Neural Networks*, 5:96–101, January 1994.
1394. G. Rudolph. *Convergence Properties of Evolutionary Algorithms*. Verlag Dr. Kovač, Hamburg, 1997.
1395. G. Rudolph. Evolutionary Search for Minimal Elements in Partially Ordered Finite Sets. In V. Porto, N. Saravanan, D. Waagen, and A. Eiben, editors, *Evolutionary Programming VII, Proceedings of the 7th Annual Conference on Evolutionary Programming*, pages 345–353, Berlin, 1998. Springer.
1396. G. Rudolph. On a Multi-Objective Evolutionary Algorithm and Its Convergence to the Pareto Set. In *Proceedings of the 5th IEEE Conference on Evolutionary Computation*, pages 511–516, Piscataway, New Jersey, 1998. IEEE Press.
1397. G. Rudolph. Evolutionary Search under Partially Ordered Fitness Sets. In *Proceedings of the International NAISO Congress on Information Science Innovations (ISI 2001)*, pages 818–822. ICSC Academic Press: Millet/Sliedrecht, 2001.
1398. G. Rudolph. A Partial Order Approach to Noisy Fitness Functions. In *Proceedings of the Congress on Evolutionary Computation 2001 (CEC'2001)*, volume 1, pages 318–325, Piscataway, New Jersey, May 2001. IEEE Service Center.
1399. G. Rudolph. Some Theoretical Properties of Evolutionary Algorithms under Partially Ordered Fitness Values. In C. Fabian and I. Intorsureanu, editors, *Proceedings of the Evolutionary Algorithms Workshop (EAW-2001)*, pages 9–22, Bucharest, Romania, January 2001.
1400. G. Rudolph. Deployment Scenarios of Parallelized Code in Stochastic Optimization. In B. Filipič and J. Šilc, editors, *Bioinspired Optimization Methods and their Applications*, pages 3–11. Jožef Stefan Institute, October 2006.
1401. G. Rudolph and A. Agapie. Convergence Properties of Some Multi-Objective Evolutionary Algorithms. In *Proceedings of the 2000 Conference on Evolutionary Computation*, volume 2, pages 1010–1016, Piscataway, New Jersey, July 2000. IEEE Press.
1402. G. Rudolph, B. Naujoks, and M. Preuss. Capabilities of EMOA to Detect and Preserve Equivalent Pareto Subsets. In S. Obayashi, K. Deb, C. Poloni, T. Hiroyasu, and T. Murata, editors, *Evolutionary Multi-Criterion Optimization, 4th International Conference, EMO 2007*, pages 36–50, Matshushima, Japan, March 2007. Springer. Lecture Notes in Computer Science Vol. 4403.
1403. A. J. Ruiz-Torres, E. E. Enscore, and R. R. Barton. Simulated Annealing Heuristics for the Average Flow-Time and the Number of Tardy Jobs Bi-Criteria Identical Parallel Machine Problem. *Computers and Industrial Engineering*, 33(1–2):257–260, 1997.

1404. T. P. Runarsson and X. Yao. Stochastic Ranking for Constrained Evolutionary Optimization. *IEEE Transactions on Evolutionary Computation*, 4(3):284–294, September 2000.
1405. T. P. Runarsson and X. Yao. Search biases in constrained evolutionary optimization. *IEEE Transactions on Systems, Man, and Cybernetics Part C— Applications and Reviews*, 35(2):233–243, May 2005.
1406. E. H. Ruspini and I. S. Zwir. Automated Qualitative Description of Measurements. In *Proceedings of the 16th IEEE Instrumentation and Measurement Technology Conference*, volume 2, pages 1086–1091, Venice, Italy, 1999. IEEE Press.
1407. S. Russell and P. Norvig. *Artificial Intelligence: A Modern Approach*. Prentice-Hall, Upper Saddle River, New Jersey, 1995.
1408. C. Ryan. Pygmies and Servants. In J. Kenneth E. Kinnear, editor, *Advances in Genetic Programming*, pages 243–263. The MIT Press, Cambridge, Massachussets, 1994.
1409. J. Ryoo. *Adaptation of Evolutionary Search in Topology and Decomposition Based Design Optimization*. PhD thesis, Mechanical Engineering Department, Rensselaer Polytechnic Institute, Troy, New York, USA, August 2002.
1410. S. M. Sait, H. Youseff, and H. Ali. Fuzzy Simulated Evolution Algorithm for Multi-objective Optimization of VLSI Placement. In *1999 Congress on Evolutionary Computation*, pages 91–97, Washington, D.C., July 1999. IEEE Service Center.
1411. S. M. Sait and H. Youssef. *Iterative Computer Algorithms with Applications in Engineering. Solving Combinatorial Optimization Problems*. IEEE Computer Society, Los Alamitos, California, 1999.
1412. S. M. Sait, H. Youssef, and J. A. Khan. Fuzzy Evolutionary Algorithm for VLSI Placement. In L. Spector, E. D. Goodman, A. Wu, W. Langdon, H.-M. Voigt, M. Gen, S. Sen, M. Dorigo, S. Pezeshk, M. H. Garzon, and E. Burke, editors, *Proceedings of the Genetic and Evolutionary Computation Conference (GECCO'2001)*, pages 1056–1063, San Francisco, California, 2001. Morgan Kaufmann Publishers.
1413. M. Sakawa. *Genetic Algorithms and Fuzzy Multiobjective Optimization*. Kluwer Academic Publishers, Boston, 2002. ISBN 0-7923-7452-5.
1414. M. Sakawa. *Fuzzy Sets and Interactive Multiobjective Optimization*. Springer, 2003. ISBN 0-306-44337-6.
1415. M. Sakawa, M. Inuiguchi, H. Sunada, and K. Sawada. Fuzzy Multiobjective Combinatorial Optimization Through Revised Genetic Algorithms. *Japanese Journal of Fuzzy Theory and Systems*, 6(1):77–88, 1994.
1416. M. Sakawa, K. Kato, and T. Shibano. Fuzzy Programming For Multiobjective 0-1 Programming Problems Through Revised Genetic Algorithms. *European Journal of Operational Research*, 97(1):149–158, 1997.
1417. M. Salazar-Lechuga and J. E. Rowe. Particle Swarm Optimization and Fitness Sharing to solve Multi-Objective Optimization Problems. In *2005 IEEE Congress on Evolutionary Computation (CEC'2005)*, volume 2, pages 1204–1211, Edinburgh, Scotland, September 2005. IEEE Service Center.
1418. F. S. Salman, J. Kalagnanam, and S. Murthy. Cooperative Strategies for Solving the Bicriteria Sparse Multiple Knapsack Problem. In *1999 Congress on Evolutionary Computation*, pages 53–60, Washington, D.C., July 1999. IEEE Service Center.

1419. L. Saludjian, J. L. Coulomb, and A. Izabelle. Genetic Algorithm and Taylor Development of the Finite Element Solution for Shape Optimization of Electromagnetic Devices. *IEEE Transactions on Magnetics*, 34(5):2841–2844, September 1998.
1420. M. E. Salukvadze. On the Existence of Solution in Problems of Optimization under Vector Valued Criteria. *Journal of Optimization Theory and Applications*, 12(2):203–217, 1974.
1421. J. W. Sammon. A nonlinear mapping for data structure analysis. *IEEE Transactions on Computers*, C-18(5):401–408, 1969.
1422. E. Sandgren. Multicriteria design optimization by goal programming. In H. Adeli, editor, *Advances in Design Optimization*, pages 225–265. Chapman & Hall, London, 1994.
1423. L. V. Santana-Quintero and C. A. Coello Coello. An Algorithm Based on Differential Evolution for Multi-Objective Problems. *International Journal of Computational Intelligence Research*, 1(2):151–169, 2005.
1424. L. V. Santana-Quintero, N. Ramírez-Santiago, C. A. Coello Coello, J. Molina Luque, and A. G. Hernández-Díaz. A New Proposal for Multiobjective Optimization Using Particle Swarm Optimization and Rough Sets Theory. In T. P. Runarsson, H.-G. Beyer, E. Burke, J. J. Merelo-Guervós, L. D. Whitley, and X. Yao, editors, *Parallel Problem Solving from Nature - PPSN IX, 9th International Conference*, pages 483–492. Springer. Lecture Notes in Computer Science Vol. 4193, Reykjavik, Iceland, September 2006.
1425. A. Santos and A. D. Pereira Correia. Constrained GA Applied to Production and Energy Mangement of a Pulp and Paper Mill. In J. Carroll, H. Haddad, D. Oppenheim, B. Bryant, and G. B. Lamont, editors, *Proceedings of the 1999 ACM Symposium on Applied Computing*, pages 324–332, San Antonio, Texas, 1999. ACM.
1426. R. Sarker, H. A. Abbass, and C. S. Newton. Solving Two Multi-objective Optimization Problems using Evolutionary Algorithm. In M. Mohammadian, R. A. Sarker, and X. Yao, editors, *Computational Intelligence in Control*, pages 218–232. Idea Group Publishing, USA, 2003.
1427. R. Sarker, K. Liang, and C. Newton. A New Evolutionary Algorithm for Multiobjective Optimization. *European Journal of Operational Research*, 140(1):12–23, 2002.
1428. R. Sarker and C. Netwon. Solving a Multiple Objective Linear Program using Simulated Annealing. *Asia-Pacific Journal of Operational Research*, 18:109–120, 2001.
1429. G. V. Sarma, L. Sellami, and K. D. Houam. Application of Lexicographic Goal Programming in Production Planning—Two case studies. *Opsearch*, 30(2):141–162, 1993.
1430. K. Sastry, M. Pelikan, and D. E. Goldberg. Limits of Scalability of Multiobjective Estimation of Distribution Algorithms. In *2005 IEEE Congress on Evolutionary Computation (CEC'2005)*, volume 3, pages 2217–2224, Edinburgh, Scotland, September 2005. IEEE Service Center.
1431. H. Sato, H. E. Aguirre, and K. Tanaka. Local Dominance Using Polar Coordinates to Enhance Multiobjective Evolutionary Algorithms. In *2004 Congress on Evolutionary Computation (CEC'2004)*, volume 1, pages 188–195, Portland, Oregon, USA, June 2004. IEEE Service Center.

1432. D. A. Savic, G. A. Walters, and M. Schwab. Multiobjective Genetic Algorithms for Pump Scheduling in Water Supply. In *AISB International Workshop on Evolutionary Computing. Lecture Notes in Computer Science 1305*, pages 227–236, Berlin, April 1997. Springer-Verlag.
1433. D. Savir. Multi-objective linear programming. Technical Report ORC 66-21, Operations Research Center, University of California, Berkeley, California, 1966.
1434. H. Sawai and S. Adachi. Effects of Hierarchical Migration in a Parallel Distributed Parameter-Free GA. In *2000 Congress on Evolutionary Computation*, volume 2, pages 1117–1124, Piscataway, New Jersey, July 2000. IEEE Service Center.
1435. D. K. Saxena and K. Deb. Non-linear Dimensionality Reduction Procedures for Certain Large-Dimensional Multi-objective Optimization Problems: Employing Correntropy and a Novel Maximum Variance Unfolding. In S. Obayashi, K. Deb, C. Poloni, T. Hiroyasu, and T. Murata, editors, *Evolutionary Multi-Criterion Optimization, 4th International Conference, EMO 2007*, pages 772–787, Matshushima, Japan, March 2007. Springer. Lecture Notes in Computer Science Vol. 4403.
1436. T. M. Sayers and J. M. Anderson. The multi-objective optimisation of a traffic control system. In A. Ceder, editor, *Proceedings of 14th International Symposium on Transportation and Traffic Theory*, pages 153–176. Transportation Research Institute, Haifa, Israel, July 1999.
1437. S. Sayin. Measuring the Quality of Discrete Representations of Efficient Sets in Multiple Objective Mathematical Programming. *Mathematical Programming*, 87(3):543–560, 2000.
1438. I. F. Sbalzarini, S. Müller, and P. Koumoutsakos. Microchannel Optimization Using Multiobjective Evolution Strategies. In E. Zitzler, K. Deb, L. Thiele, C. A. Coello Coello, and D. Corne, editors, *First International Conference on Evolutionary Multi-Criterion Optimization*, pages 516–530. Springer-Verlag. Lecture Notes in Computer Science No. 1993, 2001.
1439. J. D. Schaffer. *Multiple Objective Optimization with Vector Evaluated Genetic Algorithms*. PhD thesis, Vanderbilt University, Nashville, Tennessee, 1984.
1440. J. D. Schaffer. Multiple Objective Optimization with Vector Evaluated Genetic Algorithms. In *Genetic Algorithms and their Applications: Proceedings of the First International Conference on Genetic Algorithms*, pages 93–100, Hillsdale, New Jersey, 1985. Lawrence Erlbaum.
1441. J. D. Schaffer and J. J. Grefenstette. Multiobjective Learning via Genetic Algorithms. In *Proceedings of the 9th International Joint Conference on Artificial Intelligence (IJCAI-85)*, pages 593–595, Los Angeles, California, 1985. AAAI.
1442. S. Schäffler, R. Schultz, and K. Weinzierl. Stochastic method for the solution of unconstrained vector optimization problems. *Journal of Optimization Theory and Applications*, 114(1):209–222, 2002.
1443. F. Schlottmann, A. Mitschele, and D. Seese. A Multi-objective Approach to Integrated Risk Management. In C. A. Coello Coello, A. Hernández Aguirre, and E. Zitzler, editors, *Evolutionary Multi-Criterion Optimization. Third International Conference, EMO 2005*, pages 692–706, Guanajuato, México, March 2005. Springer. Lecture Notes in Computer Science Vol. 3410.
1444. F. Schlottmann and D. Seese. Hybrid multi-objective evolutionary computation of constrained downside risk-return efficient sets for credit portfolio. In

Proceedings of the 8th International Conference of the Society for Computational Economics. Computing in Economics and Finance, Aix-en-Provence, France, June 2002.

1445. F. Schlottmann and D. Seese. Financial Applications of Multi-Objective Evolutionary Algorithms: Recent Developments and Future Research Directions. In C. A. Coello Coello and G. B. Lamont, editors, *Applications of Multi-Objective Evolutionary Algorithms*, pages 627–652. World Scientific, Singapore, 2004.

1446. F. Schlottmann and D. Seese. A Hybrid Heuristic Approach to Discrete Multi-Objective Optimization of Credit Portfolios. *Computational Statistics & Data Analysis*, 47(2):373–399, September 2004.

1447. F. Schmiedle, N. Drechsler, D. Große, and R. Drechsler. Priorities in Multi-Objective Optimization for Genetic Programming. In L. Spector, E. D. Goodman, A. Wu, W. Langdon, H.-M. Voigt, M. Gen, S. Sen, M. Dorigo, S. Pezeshk, M. H. Garzon, and E. Burke, editors, *Proceedings of the Genetic and Evolutionary Computation Conference (GECCO'2001)*, pages 129–136, San Francisco, California, 2001. Morgan Kaufmann Publishers.

1448. K. Schmitt, J. Mehnen, and T. Michelitsch. Using Predators and Preys in Evolution Strategies. In H.-G. B. et al., editor, *2005 Genetic and Evolutionary Computation Conference (GECCO'2005)*, volume 1, pages 827–828, New York, USA, June 2005. ACM Press.

1449. V. Schnecke and O. Vornberger. Hybrid Genetic Algorithms for Constrained Placement Problems. *IEEE Transactions on Evolutionary Computation*, 1(4):266–277, November 1997.

1450. T. Schnier, X. Yao, and P. Liu. Digital Filter Design Using Multiple Pareto Fronts. In D. Keymeulen, A. Stoica, J. Lohn, and R. Salem Zebulum, editors, *Proceedings of the Third NASA/DoD Workshop on Evolvable Hardware*, pages 136–145, Long Beach, California, July 2001. IEEE Computer Society Press.

1451. M. Schoenauer and S. Xanthakis. Constrained GA Optimization. In S. Forrest, editor, *Proceedings of the Fifth International Conference on Genetic Algorithms*, pages 573–580. Morgan Kaufmann Publishers, San Mateo, California, July 1993.

1452. J. R. Schott. Fault Tolerant Design Using Single and Multicriteria Genetic Algorithm Optimization. Master's thesis, Department of Aeronautics and Astronautics, Massachusetts Institute of Technology, Cambridge, Massachusetts, May 1995.

1453. P. Schroder, A. J. Chipperfield, P. J. Fleming, and N. Grum. Multi-Objective Optimization of Distributed Active Magnetic Bearing Controllers. In *Genetic Algorithms in Engineering Systems: Innovations and Applications*, pages 13–18. IEE, September 1997.

1454. O. Schütze. A New Data Structure for the Nondominance problem in Multi-objective Optimization. In C. M. Fonseca, P. J. Fleming, E. Zitzler, K. Deb, and L. Thiele, editors, *Evolutionary Multi-Criterion Optimization. Second International Conference, EMO 2003*, pages 509–518, Faro, Portugal, April 2003. Springer. Lecture Notes in Computer Science. Volume 2632.

1455. O. Schütze. *Set Oriented Methods for Global Optimization*. PhD thesis, Fakultät für Elektrotechnik, Informatik und Mathematik, Universität Paderborn, Paderborn, Germany, December 2004.

1456. O. Schütze, S. Mostaghim, M. Dellnitz, and J. Teich. Covering Pareto Sets by Multilevel Evolutionary Subdivision Techniques. In C. M. Fonseca, P. J. Fleming, E. Zitzler, K. Deb, and L. Thiele, editors, *Evolutionary Multi-Criterion Optimization. Second International Conference, EMO 2003*, pages 118–132, Faro, Portugal, April 2003. Springer. Lecture Notes in Computer Science. Volume 2632.
1457. M. Schwab, D. A. Savic, and G. A. Walters. Multi-Objective Genetic Algorithm for Pump Scheduling in Water Supply Systems. Technical Report 96/02, Centre For Systems And Control Engineering, School of Engineering, University of Exeter, Exeter, United Kingdom, 1996.
1458. J. Schwarz and J. Ocenasek. Evolutionary Multiobjective Bayesian Optimization Algorithm: Experimental Study. In *Proceedings of the 35th Spring International Conference: Modelling and Simulation of Systems (MOSIS'01)*, pages 101–108, Czech Republic, 2001. MARQ, Hradec and Moravici.
1459. J. Schwarz and J. Ocenasek. Multiobjective Bayesian Optimization Algorithm for Combinatorial Problems: Theory and Practice. *Neural Network World*, 11(5):423–441, 2001.
1460. H.-P. Schwefel. *Evolution and Optimum Seeking*. John Wiley & Sons, New York, 1995.
1461. M. Scott and E. Antonsson. Arrow's theorem and engineering design decision making. *Research in Engineering Design*, 11(4):218–228, 1999.
1462. B. R. Secrest and G. B. Lamont. Communication in Particle Swarm Optimization Illustrated by the Traveling Salesman Problem. In *Proceedings of Particle Swarm Optimization Workshop, Indiana University Purdue University at Indianapolis*, pages 14–21, Indianapolis, Indiana, April 2001.
1463. M. Sefrioui and J. Periaux. Nash Genetic Algorithms: examples and applications. In *2000 Congress on Evolutionary Computation*, volume 1, pages 509–516, Piscataway, NJ, July 2000. IEEE Service Center.
1464. P. Sen and J. B. Yang. Multiple-Criteria Decision-Making in Design Selection and Synthesis. *Journal of Engineering Design*, 6(3):207–230, 1995.
1465. P. Serafini. Some considerations about computational complexity for multiobjective combinatorial problems. In J. Jahn and W. Krabs, editors, *Recent Advances and Historical Developments of Vector Optimization*, pages 222–232. Springer. Lecture Notes in Economics and Mathematical Systems Vol. 249, 1987.
1466. P. Serafini. Simulated Annealing for Multiple Objective Optimization Problems. In G. Tzeng, H. Wang, U. Wen, and P. Yu, editors, *Proceedings of the Tenth International Conference on Multiple Criteria Decision Making: Expand and Enrich the Domains of Thinking and Application*, volume 1, pages 283–292, Berlin, 1994. Springer-Verlag.
1467. P. Serafini. Simulated Annealing for Multiple Objective Optimization Problems. In G. Tzeng, H. Wang, U. Wen, and P. Yu, editors, *Proceedings of the Tenth International Conference on Multiple Criteria Decision Making: Expand and Enrich the Domains of Thinking and Application*, volume 1, pages 283–294, Berlin, 1994. Springer-Verlag.
1468. S. Sette, L. Boullart, and L. V. Langenhove. Optimizing a Production Process by a Neural Network/Genetic Algorithm Approach. *Engineering Applications in Artificial Intelligence*, 9(6):681–689, 1996.

1469. J. L. Shapiro. Statistical Mechanics Theory of Genetic Algorithms. In L. Kallel, B. Naudts, and A. Rogers, editors, *Theoretical Aspects of Evolutionary Computing*, pages 87–108. Springer, Berlin, Germany, 2001.
1470. K. J. Shaw and P. J. Fleming. Initial Study of Practical Multi-Objective Genetic Algorithms for Scheduling the Production of Chilled Ready Meals. In *Proceedings of Mendel'96, the 2nd International Mendel Conference on Genetic Algorithms*, Brno, Czech Republic, September 1996.
1471. K. J. Shaw and P. J. Fleming. An Initial Study of Practical Multi-Objective Production Scheduling using Genetic Algorithms. In *Proceedings of the UKACC International Conference on Control'96*, volume 1, pages 479–484, University of Exeter, UK, September 1996.
1472. K. J. Shaw and P. J. Fleming. Including Real-Life Preferences in Genetic Algorithms to Improve Optimisation of Production Schedules. In *GALESIA '97: Proceedings of the Conference on Genetic Algorithms in Engineering Systems: Innovations and Applications*, pages 239–244, Glasgow, Scotland, September 1997. IEE.
1473. K. J. Shaw and P. J. Fleming. Use of Rules and Preferences for Schedule Builders in Genetic Algorithms Production Scheduling. In D. Corne and J. L. Shapiro, editors, *Selected Papers from AISB Workshop on Evolutionary Computing*, pages 237–250, Manchester, UK, 1997. Springer-Verlag. Lecture Notes in Computer Science Vol. 1305.
1474. K. J. Shaw, A. L. Nortcliffe, M. Thompson, J. Love, C. M. Fonseca, and P. J. Fleming. Assessing the Performance of Multiobjective Genetic Algorithms for Optimization of a Batch Process Scheduling Problem. In *1999 Congress on Evolutionary Computation*, pages 37–45, Washington, D.C., July 1999. IEEE Service Center.
1475. P. Shelokar, V. Jayaraman, and B. Kulkarni. Ant algorithm for single and multiobjective reliability optimization problems. *Quality and Reliability Engineering International*, 18(6):497–514, November-December 2002.
1476. P. S. Shelokar, S. Adhikari, R. Vakil, V. Jayaraman, and B. Kulkarni. Multi-objective ant algorithm for continuous function optimization: Combination of strength Pareto fitness assignment and thermodynamic clustering. *Foundations of Computing and Decision Sciences*, 25(4):213–229, 2000.
1477. D. J. Sheskin. *Handbook of Parametric and Nonparametric Statistical Procedures*. Chapman & Hall/CRC, third edition, 2003. ISBN 1-5848-8440-1.
1478. R. Shi. *Studies on Multi-objective Evolutionary Algorithms with Applications to Production Scheduling*. PhD thesis, School of Economics and Management, Beihang University, Beijing, China, 2006.
1479. Y. Shi and R. C. Eberhart. Parameter Selection in Particle Swarm Optimization. In V. W. Porto, N. Saravanan, D. Waagen, and A. Eibe, editors, *Proceedings of the Seventh Annual Conference on Evolutionary Programming*, pages 591–600. Springer-Verlag, March 1998.
1480. T. Shibano and M. Sakawa. Interactive Decision Making for Fuzzy Multiobjective 0-1 Programs Through Genetic Algorithms with Double Strings. In *Proceedings of the Sixth IEEE Conference on Fuzzy Systems*, pages 1639–1644, 1997.
1481. M. Shibuya, H. Kita, and S. Kobayashi. Integration of multi-objective and interactive genetic algorithms and its application to animation design. In *Proceedings of IEEE Systems, Man, and Cybernetics*, volume III, pages 646–651, 1999.

1482. S.-Y. Shin. *Multi-Objective Evolutionary Optimization of DNA Sequences for Molecular Computing*. PhD thesis, School of Computer Science and Engineering, Seoul, South Korea, August 2005.
1483. S.-Y. Shin, I.-H. Lee, D. Kim, and B.-T. Zhang. Multiobjective Evolutionary Optimization of DNA Sequences for Reliable DNA Computing. *IEEE Transactions on Evolutionary Computation*, 9(2):143–158, April 2005.
1484. J. S. Shoaf and J. A. Foster. A Genetic Algorithm Solution to the Efficient Set Problem: A Technique for Portfolio Selection Based on the Markowitz Model. In *Proceedings of the Decision Sciences Institute Annual Meeting*, pages 571–573, Orlando, Florida, 1996.
1485. P. K. Shukla. On Gradient Based Local Search Methods in Unconstrained Evolutionary Multi-objective Optimization. In S. Obayashi, K. Deb, C. Poloni, T. Hiroyasu, and T. Murata, editors, *Evolutionary Multi-Criterion Optimization, 4th International Conference, EMO 2007*, pages 96–110, Matshushima, Japan, March 2007. Springer. Lecture Notes in Computer Science Vol. 4403.
1486. P. Siarry and G. Berthiau. Fitting of tabu search to optimize functions of continuous variables. *International Journal for Numerical Methods in Engineering*, 40(13):2449–2457, 1997.
1487. W. Siedlecki and J. Sklanski. Constrained Genetic Optimization via Dynamic Reward-Penalty Balancing and Its Use in Pattern Recognition. In J. D. Schaffer, editor, *Proceedings of the Third International Conference on Genetic Algorithms*, pages 141–150, San Mateo, California, June 1989. Morgan Kaufmann Publishers.
1488. J. M. Skalka. Electre III et IV. Aspects méthodologiques et guide d'utilisation. *Document 25*, 1984. Lamsade, Paris.
1489. A. E. Smith and D. W. Coit. Constraint Handling Techniques—Penalty Functions. In T. Bäck, D. B. Fogel, and Z. Michalewicz, editors, *Handbook of Evolutionary Computation*, chapter C 5.2. Oxford University Press and Institute of Physics Publishing, 1997.
1490. A. E. Smith and D. M. Tate. Genetic Optimization Using a Penalty Function. In S. Forrest, editor, *Proceedings of the Fifth International Conference on Genetic Algorithms*, pages 499–503, San Mateo, California, July 1993. Morgan Kaufmann Publishers.
1491. J. Smith. Co-evolving Memetic Algorithms: Initial Investigations. In J. J. Merelo Guervós, P. Adamidis, H.-G. Beyer, J.-L. F.-V. nas, and H.-P. Schwefel, editors, *Parallel Problem Solving from Nature—PPSN VII*, pages 537–546, Granada, Spain, September 2002. Springer-Verlag. Lecture Notes in Computer Science No. 2439.
1492. J. Smith. The Co-Evolution of Memetic Algorithms for Protein Structure Prediction. In W. E. Hart, N. Krasnogor, and J. Smith, editors, *Recent Advances in Memetic Algorithms*, pages 105–128. Springer, 2005.
1493. J. M. Smith. *Evolution and the Theory of Games*. Cambridge University Press, Cambridge, UK, 1982.
1494. K. I. Smith. *A Study of Simulated Annealing Techniques for Multi-Objective Optimisation*. PhD thesis, University of Exeter, UK, October 2006.
1495. R. E. Smith, B. A. Dike, and S. A. Stegmann. Fitness Inheritance in Genetic Algorithms. In *Proceedings of the 1995 ACM Symposium on Applied Computing*, pages 345–350, Nashville, Tennessee, USA, 1995. ACM Press.

1496. R. E. Smith, S. Forrest, and A. S. Perelson. Searching for diverse, cooperative populations with genetic algorithms. Technical Report TCGA No. 92002, University of Alabama, Tuscaloosa, Alabama, 1992.
1497. R. E. Smith, S. Forrest, and A. S. Perelson. Population diversity in an immune system model: Implications for genetic search. In L. D. Whitley, editor, *Foundations of Genetic Algorithms 2*, pages 153–165. Morgan Kaufmann Publishers, San Mateo, CA, 1993.
1498. R. E. Smith, S. Forrest, and A. S. Perelson. Population Diversity in an Immune System Model: Implications for Genetic Search. In L. D. Whitley, editor, *Foundations of Genetic Algorithms 2*, pages 153–165. Morgan Kaufmann Publishers, San Mateo, California, 1993.
1499. S. F. Smith. Flexible learning of problem solving heuristics through adaptive search. In A. Bundy, editor, *Proceedings of the 8th International Joint Conference on Artificial Intelligence*, pages 422–425, Karlsruhe, Germany, August 1983.
1500. T. Smith, P. Husbands, P. Layzell, and M. O'Shea. Fitness Landscapes and Evolvability. *Evolutionary Computation*, 10(1):1–34, Spring 2002.
1501. P. Smyth and R. Goodman. Rule induction using information theory. In G. Piatetsky-Shapiro and W. Frawley, editors, *Knowledge Discovery in Databases*, pages 159–176. The MIT Press, Cambridge, Massachusetts, USA, 1991.
1502. K. Socha and M. Kisiel-Dorohinicki. Agent-based Evolutionary Multiobjective Optimisation. In *Congress on Evolutionary Computation (CEC'2002)*, volume 1, pages 109–114, Piscataway, New Jersey, May 2002. IEEE Service Center.
1503. R. Solich. Zadanie programowania liniowego z wieloma funkcjami celu (Linear Programming Problem with Several Objective Functions). *Przeglad Statystyczny*, 16:24–30, 1969. (In Polish).
1504. G. A. Soremekun. Genetic Algorithms for Composite Laminate Design and Optimization. Master's thesis, Department of Mechanical Engineering, Virginia Polytechnic Institute, Blacksburgh, Virginia, February 5 1997.
1505. K. Sorrensen-Cothern, E. Ford, and D. Sprugel. A process based model of competition for light incorporating plasticity through modular foliage and crown development. *Ecological Monographs*, 63:277–304, 1993.
1506. J. Sridhar and C. Rajendran. Scheduling in Flowshop and Cellular Manufacturing Systems with Multiple Objectives – A Genetic Algorithmic Approach. *Production Planning & Control*, 7(4):374–382, 1996.
1507. K. C. Srigiriraju. Noninferior Surface Tracing Evolutionary Algorithm (NSTEA) for Multi Objective Optimization. Master's thesis, North Carolina State University, Raleigh, North Carolina, August 2000.
1508. N. Srinivas and K. Deb. Multiobjective optimization using nondominated sorting in genetic algorithms. Technical report, Department of Mechanical Engineering, Indian Institute of Technology, Kanput, India, 1993.
1509. N. Srinivas and K. Deb. Multiobjective Optimization Using Nondominated Sorting in Genetic Algorithms. *Evolutionary Computation*, 2(3):221–248, fall 1994.
1510. D. Srinivasan and T. H. Seow. Particle Swarm Inspired Evolutionary Algorithm (PS-EA) for Multiobjective Optimization Problem. In *Proceedings of the 2003 Congress on Evolutionary Computation (CEC'2003)*, volume 4, pages 2292–2297, Canberra, Australia, December 2003. IEEE Press.

1511. D. Srinivasan and T. H. Seow. Particle Swarm Inspired Evolutionary Algorithm (PS-EA) for Multi-Criteria Optimization Problems. In A. Abraham, L. Jain, and R. Goldberg, editors, *Evolutionary Multiobjective Optimization: Theoretical Advances And Applications*, pages 147–165. Springer-Verlag, London, 2005. ISBN 1-85233-787-7.

1512. P. F. Stadler and C. Flamm. Barrier Trees on Poset-Valued Landscapes. *Genetic Programming and Evolvable Machines*, 4(1):7–20, March 2003.

1513. W. Stadler. Preference optimality and applications to Pareto optimality. In G. Leitmann and A. Marzollo, editors, *Multi-Criteria Decision Making*, volume 211. Springer-Verlag, New York, 1975.

1514. W. Stadler. Natural Structural Shapes (The Static Case). *The Quarterly Journal of Mechanics and Applied Mathematics*, XXXI, Pt. 2:169–217, 1978.

1515. W. Stadler. A Survey of Multicriteria Optimization or the Vector Maximum Problem, Part I : 1776-1960. *Journal of Optimization Theory and Applications*, 29(1):1–52, sep 1984.

1516. W. Stadler. Initiators of Multicriteria Optimization. In J. Jahn and W. Krabs, editors, *Recent Advances and Historical Development of Vector Optimization*, pages 3–47. Springer-Verlag, Berlin, 1986.

1517. W. Stadler. Fundamentals of multicriteria optimization. In W. Stadler, editor, *Multicriteria Optimization in Engineering and the Sciences*, pages 1–25. Plenum Press, New York, 1988.

1518. T. J. Stanley and T. Mudge. A Parallel Genetic Algorithm for Multiobjective Microprocessor Design. In L. J. Eshelman, editor, *Proceedings of the Sixth International Conference on Genetic Algorithms*, pages 597–604, San Mateo, California, July 1995. Morgan Kaufmann Publishers.

1519. M. K. Starr and M. Zeleny. MCDM-state and future of the arts. In M. K. Starr and M. Zeleny, editors, *Multiple Criteria Decision Making*, volume 6 of *TIMS Studies in the Management Sciences*, pages 5–29. North-Holland Publishing Company, Amsterdam, 1977.

1520. E. J. Steele, R. A. Lindley, and R. V. Blanden. *Lamarck's Signature. How Retrogenes are Changing Darwin's Natural Selection Paradigm*. Perseus Books, Reading, Massachusetts, 1998.

1521. J. Stender, editor. *Parallel Genetic Algorithms, Theory and Applications*, volume 14 of *Frontiers in Artificial Intelligence and Applications*. IOS Press, The Netherlands, 1993.

1522. R. E. Steuer. *Multiple Criteria Optimization: Theory, Computation, and Application*. John Wiley, New York, 1986.

1523. B. S. Stewart and C. C. White. Multiobjective A*. *Journal of the ACM*, 38(4):775–814, October 1991.

1524. T. J. Stewart, R. Janssen, and M. van Herwijnen. A genetic algorithm approach to multiobjective land use planning. *Computers & Operations Research*, 31:2293–2313, 2004.

1525. R. Storn and K. Price. Differential Evolution: A Simple and Efficient Adaptive Scheme for Global Optimization over Continuous Spaces. Technical Report TR-95-012, International Computer Science Institute, Berkeley, California, March 1995.

1526. R. Storn and K. Price. Differential Evolution - A Fast and Efficient Heuristic for Global Optimization over Continuous Spaces. *Journal of Global Optimization*, 11:341–359, 1997.

1527. D. Stoyan. Random sets: Models and Statistics. *International Statistical Review*, 66:1–27, 1998.
1528. F. Streichert, H. Ulmer, and A. Zell. Parallelization of Multi-objective Evolutionary Algorithms Using Clustering Algorithms. In C. A. Coello Coello, A. Hernández Aguirre, and E. Zitzler, editors, *Evolutionary Multi-Criterion Optimization. Third International Conference, EMO 2005*, pages 92–107, Guanajuato, México, March 2005. Springer. Lecture Notes in Computer Science Vol. 3410.
1529. T. Stützle and H. Hoos. Max-min ant system. *Future Generation Computer Systems*, 16(8):889–914, 2000.
1530. R. Subbu, P. P. Bonissone, N. Eklund, S. Bollapragada, and K. Chalermkraivuth. Multiobjective Financial Portfolio Design: A Hybrid Evolutionary Approach. In *2005 IEEE Congress on Evolutionary Computation (CEC'2005)*, volume 2, pages 1722–1729, Edinburgh, Scotland, September 2005. IEEE Service Center.
1531. A. Sülflow, N. Drechsler, and R. Drechsler. Robust Multi-objective Optimization in High Dimensional Spaces. In S. Obayashi, K. Deb, C. Poloni, T. Hiroyasu, and T. Murata, editors, *Evolutionary Multi-Criterion Optimization, 4th International Conference, EMO 2007*, pages 715–726, Matshushima, Japan, March 2007. Springer. Lecture Notes in Computer Science Vol. 4403.
1532. A. M. Sultan and A. M. Templeman. Generation of Pareto Solutions by Entropy-Based Methods. In M. Tamiz, editor, *Multiobjecitve and Goal Programming: Theories and Application*, pages 164–195. Springer-Verlag, Berlin, 1996.
1533. B. Suman. Multiobjective simulated annealing–A metaheuristic technique for multiobjective optimization of a constrained problem. *Foundations of Computing and Decision Sciences*, 27(3):171–191, 2002.
1534. B. Suman. Simulated Annealing-Based Multiobjective Algorithms and Their Application for System Reliability. *Engineering Optimization*, 35(4):391–416, August 2003.
1535. B. Suman. Study of simulated annealing based algorithms for multiobjective optimization of a constrained problem. *Computers & Chemical Engineering*, 28:1849–1871, 2004.
1536. B. Suman and P. Kumar. A survey of simulated annealing as a tool for single and multiobjective optimization. *Journal of the Operational Research Society*, 57(10):1143–1160, October 2006.
1537. M. Sun, A. Stam, and R. Steuer. Solving multiple objective programming problems using feed-forward artificial neural networks: The interactive FFANN procedure. *Management Science*, 42:835–849, 1996.
1538. M. Sun, A. Stam, and R. Steuer. Interactive multiple objective programming using tchebycheff programs and artificial neural networks. *Computers and Operations Research*, 27:601–620, 2000.
1539. A. Suppapitnarm, K. Seffen, G. Parks, and P. Clarkson. A simulated annealing algorithm for multiobjective optimization. *Engineering Optimization*, 33(1):59–85, 2000.
1540. A. Suppapitnarm, K. Seffen, G. Parks, and A. Connor. Multiobjective optimisation of bicycle frames using simulated annealing. In *Proceedings of the First ASMO/ISSMO Conference on Engineering Design Optimization*, volume 1, pages 357–364, Ilkley, West Yorkshire, 1999.

1541. A. Suppapitnarm, K. Seffen, G. Parks, and J.-S. Liu. Design by multiobjective optimisation using simulated annealing. In *Proceedings of the 12th International Conference in Engineering Design (ICED'99)*, volume 3, pages 1395–1400, Munich, Germany, 1999.

1542. P. D. Surry and N. J. Radcliffe. The COMOGA Method: Constrained Optimisation by Multiobjective Genetic Algorithms. *Control and Cybernetics*, 26(3):391–412, 1997.

1543. P. D. Surry, N. J. Radcliffe, and I. D. Boyd. A Multi-Objective Approach to Constrained Optimisation of Gas Supply Networks : The COMOGA Method. In T. C. Fogarty, editor, *Evolutionary Computing. AISB Workshop. Selected Papers*, pages 166–180. Springer-Verlag. Lecture Notes in Computer Science No. 993, Sheffield, U.K., 1995.

1544. R. S. Sutton and A. G. Barto. *Reinforcement Learning. An Introduction*. The MIT Press, Cambridge, Massachusetts, 1999.

1545. G. Syswerda and J. Palmucci. The Application of Genetic Algorithms to Resource Scheduling. In R. K. Belew and L. B. Booker, editors, *Proceedings of the Fourth International Conference on Genetic Algorithms*, pages 502–508, San Mateo, California, 1991. Morgan Kaufmann Publishers.

1546. F. Szidarovszky. Notes on multi-objective dynamic programming. Technical Report 79-1, Department of Systems and Industrial Engineering, University of Arizona, Tucson, Arizona, 1979.

1547. F. Szidarovszky and L. Duckstein. Basic Properties of MODM problems. In *Classnotes 82-1*. Department of Systems and Industrial Engineering, University of Arizona, Tucson, Arizona, 1982.

1548. R. Szmit and A. Barak. Evolution Strategies for a Parallel Multi-Objective Genetic Algorithm. In D. Whitley, D. Goldberg, E. Cantú-Paz, L. Spector, I. Parmee, and H.-G. Beyer, editors, *Proceedings of the Genetic and Evolutionary Computation Conference (GECCO'2000)*, pages 227–234, San Francisco, California, 2000. Morgan Kaufmann.

1549. T. Tachibana, Y. Murata, N. Shibata, K. Yasumoto, and M. Ito. A Hardware Implementation Method of Multi-Objective Genetic Algorithms. In *2006 IEEE Congress on Evolutionary Computation (CEC'2006)*, pages 10922–10929, Vancouver, BC, Canada, July 2006. IEEE.

1550. T. Tagami and T. Kawabe. Genetic Algorithm with a Pareto Partitioning Method for Multi-objective Flowshop Scheduling. In *Proceedings of the 1998 International Symposium of Nonlinear Theory and its Applications (NOLTA'98)*, pages 1069–1072, Crans-Montana, 1998.

1551. T. Tagami and T. Kawabe. Genetic Algorithm based on a Pareto Neighborhood Search for Multiobjective Optimization. In *Proceedings of the 1999 International Symposium of Nonlinear Theory and its Applications (NOLTA'99)*, pages 331–334, Hawaii, 1999.

1552. E. Taillard. Parallel taboo search techniques for the job shop scheduling problem. *ORSA Journal on Computing*, 6(2):108–117, 1994.

1553. H. Takagi. Interactive Evolutionary Computation—Cooperation of Computational Intelligence and Human KANSEI—. In *Proceedings of the 5th International Conference on Soft Computing and Information/Intelligent Systems (IIZUKA'98)*, pages 41–50, Iizuka, Fukuoka, Japan, October 1998. World Scientific.

1554. E.-G. Talbi, M. Rahoual, M. H. Mabed, and C. Dhaenens. A Hybrid Evolutionary Approach for Multicriteria Optimization Problems: Application to the

Flow Shop. In E. Zitzler, K. Deb, L. Thiele, C. A. Coello Coello, and D. Corne, editors, *First International Conference on Evolutionary Multi-Criterion Optimization*, pages 416–428. Springer-Verlag. Lecture Notes in Computer Science No. 1993, 2001.

1555. H. Tamaki, H. Kita, and S. Kobayashi. Multi-Objective Optimization by Genetic Algorithms : A Review. In T. Fukuda and T. Furuhashi, editors, *Proceedings of the 1996 International Conference on Evolutionary Computation (ICEC'96)*, pages 517–522, Nagoya, Japan, 1996. IEEE.

1556. H. Tamaki, M. Mori, M. Araki, Y. Mishima, and H. Ogai. Multi-Criteria Optimization by Genetic Algorithms : A Case of Scheduling in Hot Rolling Process. In *Proceedings of the 3rd Conference of the Association of Asian-Pacific Operational Research Societies within IFORS (APORS'94)*, pages 374–381. World Scientific, 1995.

1557. H. Tamaki, E. Nishino, and S. Abe. A Genetic Algorithm Approach to Multi-Objective Scheduling Problems with Earliness and Tardiness Penalties. In *1999 Congress on Evolutionary Computation*, pages 46–52, Washington, D.C., July 1999. IEEE Service Center.

1558. K. C. Tan, E. F. Khor, and T. H. Lee. *Multiobjective Evolutionary Algorithms and Applications*. Springer-Verlag, London, 2005. ISBN 1-85233-836-9.

1559. K. C. Tan, E. F. Khor, T. H. Lee, and R. Sathikannan. An Evolutionary Algorithm with Advanced Goal and Priority Specification for Multi-objective Optimization. *Journal of Artificial Intelligence Research*, 18:183–215, 2003.

1560. K. C. Tan, E. F. Khor, T. H. Lee, and Y. J. Yang. A tabu-based exploratory evolutionary algorithm for multiobjective optimization. *Artificial Intelligence Review*, 19(3):231–260, May 2003.

1561. K. C. Tan, T. H. Lee, Y. H. Chew, and L. H. Lee. A Hybrid Multiobjective Evolutionary Algorithm For Solving Truck and Trailer Vehicle Routing Problems. In *Proceedings of the 2003 Congress on Evolutionary Computation (CEC'2003)*, volume 3, pages 2134–2141, Canberra, Australia, December 2003. IEEE Press.

1562. K. C. Tan, T. H. Lee, and E. F. Khor. Evolutionary Algorithms with Goal and Priority Information for Multi-Objective Optimization. In *1999 Congress on Evolutionary Computation*, pages 106–113, Washington, D.C., July 1999. IEEE Service Center.

1563. K. C. Tan, T. H. Lee, and E. F. Khor. Evolutionary Algorithms with Dynamic Population Size and Local Exploration for Multiobjective Optimization. *IEEE Transactions on Evolutionary Computation*, 5(6):565–588, December 2001.

1564. K. C. Tan, T. H. Lee, and E. F. Khor. Incrementing Multi-Objective Evolutionary Algorithms: Performance Studies and Comparisons. In E. Zitzler, K. Deb, L. Thiele, C. A. Coello Coello, and D. Corne, editors, *First International Conference on Evolutionary Multi-Criterion Optimization*, pages 111–125. Springer-Verlag. Lecture Notes in Computer Science No. 1993, 2001.

1565. K. C. Tan, T. H. Lee, E. F. Khor, and K. Ou. Control system design unification and automation using an incremented multi-objective evolutionary algorithm. In M. H. Hamza, editor, *Proceedings of the 19th IASTED International Conference on Modeling, Identification and Control*. IASTED, Innsbruck, Austria, 2000.

1566. K. C. Tan, T. H. Lee, E. F. Khor, and R. Sathikannan. Incremented multi-objective evolutionary design automation of robust tracking thumbprint performances in QFT. In *Proceedings of the International Conference on Evolutionary Computation for Computer, Communication, Control and Power*, pages 137–142, Chennai, India, 2000.

1567. K. C. Tan and Y. Li. Multi-objective genetic algorithm based time and frequency domain design unification of linear control systems. In *IFAC International Symposium on Artificial Intelligence and Real-Time Control*, pages 61–66, Kuala Lumpar, Malaysia, September 1997.

1568. K. C. Tan, Y. J. Yang, and C. K. Goh. A Distributed Cooperative Coevolutionary Algorithm for Multiobjective Optimization. *IEEE Transactions on Evolutionary Computation*, 10(5):527–549, October 2006.

1569. K. C. Tan, Y. J. Yang, and T. H. Lee. A Distributed Cooperative Coevolutionary Algorithm for Multiobjective Optimization. In *Proceedings of the 2003 Congress on Evolutionary Computation (CEC'2003)*, volume 4, pages 2513–2520, Canberra, Australia, December 2003. IEEE Press.

1570. M. Tanaka and T. Tanino. Global optimization by the genetic algorithm in a multiobjective decision support system. In *Proceedings of the 10th International Conference on Multiple Criteria Decision Making*, volume 2, pages 261–270, 1992.

1571. M. Tanaka, H. Watanabe, Y. Furukawa, and T. Tanino. GA-Based Decision Support System for Multicriteria Optimization. In *Proceedings of the 1995 IEEE International Conference on Systems, Man, and Cybernetics*, volume 2, pages 1556–1561, Piscataway, NJ, 1995. IEEE.

1572. K.-S. Tang, K.-F. Man, and K.-T. Ko. Wireless LAN Design using Hierarchical Genetic Algorithm. In T. Bäck, editor, *Proceedings of the Seventh International Conference on Genetic Algorithms*, pages 629–635. Morgan Kaufmann Publishers, San Mateo, California, July 1997.

1573. T. Tanino and H. Kuk. Nonlinear Multiobjective Programming. In R. Sarker, M. Mohammadian, and X. Yao, editors, *Evolutionary Optimization*, pages 71–128. Kluwer Academic Publishers, New York, February 2002. ISBN 0-7923-7654-4.

1574. T. Tanino, M. Tanaka, and C. Hojo. An interactive multicriteria decision making method by using a genetic algorithm. In *Proceedings of 2nd International Conference on Systems Science and Systems Engineering*, pages 381–386, 1993.

1575. M. Tayal. Particle Swarm Optimization for Mechanical Design. Master's thesis, The University of Texas at Arlington, Arlington, Texas, USA, December 2003.

1576. J. Teich. Pareto-Front Exploration with Uncertain Objectives. In E. Zitzler, K. Deb, L. Thiele, C. A. C. Coello, and D. Corne, editors, *First International Conference on Evolutionary Multi-Criterion Optimization*, pages 314–328. Springer-Verlag. Lecture Notes in Computer Science No. 1993, 2001.

1577. A. Tettamanzi and M. Tomassini. *Soft Computing: Integrating Evolutionary, Neural and Fuzzy Systems*. Springer, New York, 2001.

1578. D. Thierens and P. A. Bosman. Multi-Objective Mixture-based Iterated Density Estimation Evolutionary Algorithms. In L. Spector, E. D. Goodman, A. Wu, W. Langdon, H.-M. Voigt, M. Gen, S. Sen, M. Dorigo, S. Pezeshk, M. H. Garzon, and E. Burke, editors, *Proceedings of the Genetic and Evolutionary Computation Conference (GECCO'2001)*, pages 663–670, San Francisco, California, 2001. Morgan Kaufmann Publishers.

1579. D. Thierens and P. A. Bosman. Multi-Objective Optimization with Iterated Density Estimation Evolutionary Algorithms using Mixture Models. In *Proceedings of the Third International Symposium on Adaptive Systems—Evolutionary Computation and Probabilistic Graphical Models*, pages 129–136, Havana, Cuba, March 19–23 2001. Institute of Cybernetics, Mathematics and Physics.
1580. M. W. Thomas. *A Pareto Frontier for Full Stern Submarines via Genetic Algorithm*. PhD thesis, Ocean Engineering Department, Massachusetts Institute of Technology, Cambridge, Massachusetts, USA, June 1998.
1581. M. W. Thomas. Multi-Species Pareto Frontiers in Preliminary Submarine Design. *Foundations of Computing and Decision Sciences*, 25(4):273–289, 2000.
1582. M. Thompson. Application of Multi Objective Evolutionary Algorithms to Analogue Filter Tuning. In E. Zitzler, K. Deb, L. Thiele, C. A. Coello Coello, and D. Corne, editors, *First International Conference on Evolutionary Multi-Criterion Optimization*, pages 546–559. Springer-Verlag. Lecture Notes in Computer Science No. 1993, 2001.
1583. R. Thomson and T. Arslan. An Evolutionary Algorithm for the Multi-Objective Optimisation of VLSI Primitive Operator Filters. In *Congress on Evolutionary Computation (CEC'2002)*, volume 1, pages 37–42, Piscataway, New Jersey, May 2002. IEEE Service Center.
1584. R. Thomson and T. Arslan. The Evolutionary Design and Synthesis of Non-Linear Digital VLSI Systems. In J. Lohn, R. Zebulum, J. Steincamp, D. Keymeulen, A. Stoica, and M. I. Ferguson, editors, *Proceedings of the 2003 NASA/DoD Conference on Evolvable Hardware*, pages 125–134, Los Alamitos, California, July 2003. IEEE Computer Society Press.
1585. G. Timmel. Ein stochastisches suchverrahren zur bestimmung der optimalen kompromilsungen bei statischen polzkriteriellen optimierungsaufgaben. *Wiss. Z. TH Ilmenau*, 26(5):159–174, 1980.
1586. A. Tiwari, R. Roy, G. Jared, and O. Munaux. Challenges in Real-Life Engineering Design Optimization: An Analysis. In *2001 Genetic and Evolutionary Computation Conference. Workshop Program*, pages 289–294, San Francisco, California, July 2001.
1587. S. Tiwari and N. Chakraborti. Multi-objective optimization of a two-dimensional cutting problem using genetic algorithms. *Journal of Materials Processing Technology*, 173:384–393, 2006.
1588. V. T'kindt and J.-C. Billaut. *Multicriteria Scheduling. Theory, Models and Algorithms*. Springer, Berlin, 2002. ISBN 3-540-43617-0.
1589. V. T'kindt, N. Monmarché, F. Tercinet, and D. Laügt. An Ant Colony Optimization algorithm to solve a 2-machine bicriteria flowshop scheduling problem. *European Journal of Operational Research*, 142(2):250–257, October 2002.
1590. D. S. Todd. *Multiple Criteria Genetic Algorithms in Engineering Design and Operation*. PhD thesis, University of Newcastle, Newcastle-upon-Tyne, UK, October 1997.
1591. D. S. Todd and P. Sen. A Multiple Criteria Genetic Algorithm for Containership Loading. In T. Bäck, editor, *Proceedings of the Seventh International Conference on Genetic Algorithms*, pages 674–681, San Mateo, California, July 1997. Morgan Kaufmann Publishers.

1592. D. S. Todd and P. Sen. Multiple Criteria Scheduling using Genetic Algorithms in a Shipyard Environment. In *Proceedings of the 9th International Conference on Computer Applications in Shipbuilding*, Yokohama, Japan, October 1997.
1593. D. S. Todd and P. Sen. Tackling Complex Job Shop Problems Using Operation Based Scheduling. In I. Parmee, editor, *The Integration of Evolutionary and Adaptive Computing Technologies with Product/System Design and Realisation*, pages 45–58, Plymouth, United Kingdom, April 1998. Plymouth Engineering Design Centre, Springer-Verlag.
1594. A. Toffolo and E. Benini. Genetic Diversity as an Objective in Multi-Objective Evolutionary Algorithms. *Evolutionary Computation*, 11(2):151–167, Summer 2003.
1595. J. Toivanen, J. P. Hämäläinen, K. Miettinen, and P. Tarvainen. Designing Paper Machine Headbox using GA. *Materials and Manufacturing Processes*, 18(3):533–541, 2003.
1596. G. Toscano Pulido. *On the Use of Self-Adaptation and Elitism for Multiobjective Particle Swarm Optimization*. PhD thesis, Computer Science Section, Department of Electrical Engineering, CINVESTAV-IPN, Mexico, September 2005.
1597. G. Toscano Pulido and C. A. Coello Coello. The Micro Genetic Algorithm 2: Towards Online Adaptation in Evolutionary Multiobjective Optimization. In C. M. Fonseca, P. J. Fleming, E. Zitzler, K. Deb, and L. Thiele, editors, *Evolutionary Multi-Criterion Optimization. Second International Conference, EMO 2003*, pages 252–266, Faro, Portugal, April 2003. Springer. Lecture Notes in Computer Science. Volume 2632.
1598. G. Toscano Pulido and C. A. Coello Coello. A Constraint-Handling Mechanism for Particle Swarm Optimization. In *2004 Congress on Evolutionary Computation (CEC'2004)*, volume 2, pages 1396–1403, Portland, Oregon, USA, June 2004. IEEE.
1599. G. Toscano Pulido and C. A. Coello Coello. Using Clustering Techniques to Improve the Performance of a Particle Swarm Optimizer. In K. D. et al., editor, *Genetic and Evolutionary Computation–GECCO 2004. Proceedings of the Genetic and Evolutionary Computation Conference. Part I*, pages 225–237, Seattle, Washington, USA, June 2004. Springer-Verlag, Lecture Notes in Computer Science Vol. 3102.
1600. G. Toscano-Pulido, C. A. Coello Coello, and L. V. Santana-Quintero. EMOPSO: A Multi-Objective Particle Swarm Optimizer with Emphasis on Efficiency. In S. Obayashi, K. Deb, C. Poloni, T. Hiroyasu, and T. Murata, editors, *Evolutionary Multi-Criterion Optimization, 4th International Conference, EMO 2007*, pages 272–285, Matshushima, Japan, March 2007. Springer. Lecture Notes in Computer Science Vol. 4403.
1601. M. Toulouse, T. G. Crainic, and B. Sansó. Systemic behavior of cooperative search algorithms. *Parallel Computing*, 30:57–79, 2004.
1602. A. Trebi-Ollennu and B. A. White. Multiobjective Fuzzy Genetic Algorithm Optimisation Approach to Nonlinear Control System Design. In *IFAC World Congress*, pages 1–10, San Francisco, California, September 1996.
1603. A. Trebi-Ollennu and B. A. White. Multiobjective Fuzzy Genetic Algorithm Optimization Approach to Nonlinear Control System Design. *IEE Proceedings, Control Theory and Applications*, 144(2):137–142, March 1997.
1604. M. Trefzer, J. Langeheine, K. Meier, and J. Schemmel. Operational amplifiers: An example for multi-objective optimization on an analog evolvable

hardware platform. In J. M. Moreno, J. Madrenas, and J. Cosp, editors, *Evolvable Systems: From Biology to Hardware, 6th International Conference, ICES 2005*, pages 86–97, Sitges, Spain, September 2005. Springer. Lecture Notes in Computer Science Vol. 3637.

1605. M. Trefzer, J. Langeheine, J. Schemmel, and K. Meier. New Genetic Operators to Facilitate Understanding of Evolved Transistor Circuits. In R. S. Zebulum, D. Gwaltney, G. Hornby, D. Keymeulen, J. Lohn, and A. Stoica, editors, *Proceedings of the 2004 NASA/DoD Conference on Evolvable Hardware*, pages 217–224, Los Alamitos, California, USA, June 2004. IEEE Computer Society.

1606. C. H. Tseng and T. W. Lu. Minimax Multiobjective Optimization in Structural Design. *International Journal for Numerical Methods in Engineering*, 30:1213–1228, 1990.

1607. E. Tsoi, K. P. Wong, and C. C. Fung. Hybrid GA/SA Algorithms for Evaluating Trade-off Between Economic Cost and Environmental Impact in Generation Dispatch. In D. B. Fogel, editor, *Proceedings of the Second IEEE Conference on Evolutionary Computation (ICEC'95)*, pages 132–137, Piscataway, New Jersey, 1995. IEEE Press.

1608. C.-S. Tsou, H.-H. Fang, H.-H. Chang, and C.-H. Kao. An Improved Particle Swarm Pareto Optimizer with Local Search and Clustering. In T.-D. Wang, X. Li, S.-H. Chen, X. Wang, H. A. Abbass, H. Iba, G. Chen, and X. Yao, editors, *Simulated Evolution and Learning, 6th International Conference, SEAL 2006, Proceedings*, pages 400–407, Hefei, China, October 2006. Springer. Lecture Notes in Computer Science Vol. 4247.

1609. S. Tsumakitani and J. Evans. An empirical study of a new metaheuristic for the traveling salesman problem. *European Journal of Operational Research*, 104:113–128, 1998.

1610. S. Tsutsui and Y. Fujimoto. Forking Genetic Algorithm with Blocking and Shrinking Modes (fGA). In S. Forrest, editor, *Proceedings of the Fifth International Conference on Genetic Algorithms*, pages 206–213, San Mateo, California, 1993. Morgan Kaufmann Publishers.

1611. D. Tuyttens, J. Teghem, P. Fortemps, and K. V. Nieuwenhuyze. Performance of the MOSA Method for the Bicriteria Assignment Problem. *Journal of Heuristics*, 6(3):295–310, August 2000.

1612. T. Tyni and J. Ylinen. Evolutionary bi-objective optimisation in the elevator car routing problem. *European Journal of Operational Research*, 169(3):960–977, March 2006.

1613. G.-H. Tzeng and J.-S. Kuo. Fuzzy Multiobjective Double Samplig Plans with Genetic Algorithms Based on Bayesian Model. In W. Chiang and J. Lee, editors, *Proceedings of the International Joint Conference of CFSA/IFIS/SOFT95 on Fuzzy Theory and Applications*, pages 59–64, Singapore, 1995. World Scientific.

1614. E. Ulungu. *Optimisation Combinatoire multicritere: Determination de l'ensemble des solutions efficaces et methodes interactives*. PhD thesis, Faculté des Sciences, Université de Mons-Hainaut, Mons, Belgium, 1993.

1615. E. Ulungu and J. Teghem. Multi-objective Combinatorial Optimization Problems: A Survey. *Journal of Multi-Criteria Decision Analysis*, 3:83–104, 1994.

1616. E. Ulungu, J. Teghem, and P. Fortemps. Heuristics for multi-objective combinatorial optimization by simulated annealing. In J. Gu, G. Chen, Q. Wei, and S. Wang, editors, *Multiple Criteria Decision Making: Theory and*

Applications. Proceedings of the 6th National Conference on Multiple Criteria Decision Making, pages 228–238, Windsor, UK, 1995. Sci-Tech.
1617. E. Ulungu, J. Teghem, P. Fortemps, and D. Tuyttens. MOSA Method: A Tool for Solving Multiobjective Combinatorial Optimization Problems. *Journal of Multi-Criteria Decision Analysis*, 8(4):221–236, 1999.
1618. E. Ulungu, J. Teghem, and C. Ost. Efficiency of interactive multi-objective simulated annealing through a case study. *Journal of the Operational Research Society*, 49:1044–1050, 1998.
1619. R. K. Ursem. *Models for Evolutionary Algorithms and Their Applications in System Identification and Control Optimization*. PhD thesis, Department of Computer Science, University of Aarhus, Denmark, April 2003.
1620. S. Vajda. *Probabilistic Programming*. Academic Press, New York, 1972.
1621. J. J. Valdés and A. J. Barton. Multi-objective Evolutionary Optimization for Visual Data Mining with Virtual Reality Spaces: Application to Alzheimer Gene Expressions. In M. K. et al., editor, *2006 Genetic and Evolutionary Computation Conference (GECCO'2006)*, volume 1, pages 723–730, Seattle, Washington, USA, July 2006. ACM Press. ISBN 1-59593-186-4.
1622. J. J. Valdés and A. J. Barton. Virtual Reality Spaces for Visual Data Mining with Multiobjective Evolutionary Optimization: Implicit and Explicit Function Representations Mixing Unsupervised and Supervised Properties. In *2006 IEEE Congress on Evolutionary Computation (CEC'2006)*, pages 5591–5598, Vancouver, BC, Canada, July 2006. IEEE.
1623. C. L. Valenzuela. A Simple Evolutionary Algorithm for Multi-Objective Optimization (SEAMO). In *Congress on Evolutionary Computation (CEC'2002)*, volume 1, pages 717–722, Piscataway, New Jersey, May 2002. IEEE Service Center.
1624. M. Valenzuela Rendón and E. Uresti Charre. A Non-Generational Genetic Algorithm for Multiobjective Optimization. In T. Bäck, editor, *Proceedings of the Seventh International Conference on Genetic Algorithms*, pages 658–665, San Mateo, California, July 1997. Michigan State University, Morgan Kaufmann Publishers.
1625. P. van Laarhoven and E. Aarts. *Simulated Annealing: Theory and Applications*. Kluwer Academic Publishers, Dordrecht, 1987.
1626. D. A. Van Veldhuizen. *Multiobjective Evolutionary Algorithms: Classifications, Analyses, and New Innovations*. PhD thesis, Department of Electrical and Computer Engineering. Graduate School of Engineering. Air Force Institute of Technology, Wright-Patterson AFB, Ohio, May 1999.
1627. D. A. Van Veldhuizen and G. B. Lamont. Evolutionary Computation and Convergence to a Pareto Front. In J. R. Koza, editor, *Late Breaking Papers at the Genetic Programming 1998 Conference*, pages 221–228, Stanford University, California, July 1998. Stanford University Bookstore.
1628. D. A. Van Veldhuizen and G. B. Lamont. Multiobjective Evolutionary Algorithm Research: A History and Analysis. Technical Report TR-98-03, Department of Electrical and Computer Engineering, Graduate School of Engineering, Air Force Institute of Technology, Wright-Patterson AFB, Ohio, 1998.
1629. D. A. Van Veldhuizen and G. B. Lamont. Genetic Algorithms, Building Blocks, and Multiobjective Optimization. In A. S. Wu, editor, *Proceedings of the 1999 Genetic and Evolutionary Computation Conference. Workshop Program*, pages 125–126, Orlando, Florida, July 1999.

1630. D. A. Van Veldhuizen and G. B. Lamont. Multiobjective Evolutionary Algorithm Test Suites. In J. Carroll, H. Haddad, D. Oppenheim, B. Bryant, and G. B. Lamont, editors, *Proceedings of the 1999 ACM Symposium on Applied Computing*, pages 351–357, San Antonio, Texas, 1999. ACM.
1631. D. A. Van Veldhuizen and G. B. Lamont. Multiobjective Evolutionary Algorithms: Analyzing the State-of-the-Art. *Evolutionary Computation*, 7(3):1–26, 2000.
1632. D. A. Van Veldhuizen and G. B. Lamont. Multiobjective Optimization with Messy Genetic Algorithms. In *Proceedings of the 2000 ACM Symposium on Applied Computing*, pages 470–476, Villa Olmo, Como, Italy, 2000. ACM.
1633. D. A. Van Veldhuizen and G. B. Lamont. On Measuring Multiobjective Evolutionary Algorithm Performance. In *2000 Congress on Evolutionary Computation*, volume 1, pages 204–211, Piscataway, New Jersey, July 2000. IEEE Service Center.
1634. D. A. Van Veldhuizen, B. S. Sandlin, R. M. Marmelstein, and G. B. Lamont. Finding Improved Wire-Antenna Geometries with Genetic Algorithms. In D. B. Fogel, editor, *Proceedings of the 1998 International Conference on Evolutionary Computation*, pages 102–107, Piscataway, New Jersey, 1998. IEEE.
1635. D. Vanderbilt and S. Louie. A Monte Carlo Simulated Annealing Approach to Optimization over Continuous Variables. *Journal of Computational Physics*, 56:259–271, 1984.
1636. J. Vasconcelos, J. Maciel, and R. O. Parreiras. Scatter Search Techniques Applied to Electromagnetic Problems. *IEEE Transactions on Magnetics*, 41(5):1804–1807, May 2005.
1637. V. K. Vassilev, T. C. Fogarty, and J. F. Miller. Information Characteristics and the Structure of Landscapes. *Evolutionary Computation*, 8(1):31–60, Spring 2000.
1638. G. Vedarajan, L. C. Chan, and D. E. Goldberg. Investment Portfolio Optimization using Genetic Algorithms. In J. R. Koza, editor, *Late Breaking Papers at the Genetic Programming 1997 Conference*, pages 255–263, Stanford University, California, July 1997. Stanford Bookstore.
1639. V. Venkat, S. H. Jacobson, and J. A. Stori. A Post-Optimality Analysis Algorithm for Multi-Objective Optimization. *Computational Optimization and Applications*, 28:357–372, 2004.
1640. S. Venkatraman and G. G. Yen. A Generic Framework for Constrained Optimization Using Genetic Algorithms. *IEEE Transactions on Evolutionary Computation*, 9(4), August 2005.
1641. V. Venugopal and T. T. Narendran. A Genetic Algorithm Approach to the Machine-Component Grouping Problem with Multiple Objectives. *Computers and Industrial Engineering*, 22(4):469–480, 1992.
1642. R. Verschae, J. Ruiz del Solar, M. Köppen, and R. V. Garcia. Improvement of a Face Detection System by Evolutionary Multi-Objective Optimization. In N. Nedjah, L. M. Mourelle, M. M. Vellasco, A. Abraham, and M. Köppen, editors, *Fifth International Conference on Hybrid Intelligent Systems (HIS'05)*, pages 361–366, Los Alamitos, California, USA, November 2005. IEEE Computer Society.
1643. A. Viana and J. Pinho de Sousa. Using metaheuristics in multiobjective resource constrained project scheduling. *European Journal of Operational Research*, 120:359–374, 2000.

1644. D. S. Vianna and J. E. C. Arroyo. A GRASP algorithm for the multi-objective knapsack problem. In *XXIV International Conference of the Chilean Computer Science Society (SCCC'04)*, pages 69–75, Arica, Chile, November 2004. IEEE Computer Society.

1645. A. Vicini and D. Quagliarella. Inverse and Direct Airfoil Design Using a Multiobjective Genetic Algorithm. *AIAA Journal*, 35(9):1499–1505, September 1997.

1646. A. Vicini and D. Quagliarella. Multipoint transonic airfoil design by means of a multiobjective genetic algorithm. In *35th AIAA Aerospace Sciences Meeting and Exhibit*, Reno, Nevada, January 1997. American Institute of Aeronautics and Astronautics (AIAA). AIAA Paper 97-0082.

1647. A. Vicini and D. Quagliarella. Airfoil and Wing Design Through Hybrid Optimization Strategies. In *16th Applied Aerodynamics Conference*, pages 536–546, Albuquerque, New Mexico, June 1998. American Institute of Aeronautics and Astronautics (AIAA). AIAA Paper 98-2729.

1648. R. V. Vidal, editor. *Applied Simulated Annealing*. Springer-Verlag. Lecture Notes in Economics and Mathematical Systems Vol. 396, 1993.

1649. Y. Vidyakiran, B. Mahanty, and N. Chakraborti. A genetic-algorithms-based multiobjective approach for a three-dimensional guillotine cutting problem. *Materials and Manufacturing Processes*, 20(4):697–715, 2005.

1650. R. Viennet, C. Fontiex, and I. Marc. New Multicriteria Optimization Method Based on the Use of a Diploid Genetic Algorithm: Example of an Industrial Problem. In J. M. Alliot, E. Lutton, E. Ronald, M. Schoenauer, and D. Snyers, editors, *Proceedings of Artificial Evolution (European Conference, selected papers)*, pages 120–127. Springer-Verlag, Brest, France, September 1995.

1651. R. Viennet, C. Fontiex, and I. Marc. Multicriteria Optimization Using a Genetic Algorithm for Determining a Pareto Set. *Journal of Systems Science*, 27(2):255–260, 1996.

1652. G. Vignaux and Z. Michalewicz. A Nonstandard Genetic Algorithm for the Linear Transportation Problem. *IEEE Transactions on Systems, Man, and Cybernetics*, 21(2):445–452, 1991.

1653. M. Villalobos-Arias, C. A. Coello Coello, and O. Hernández-Lerma. Asymptotic Convergence of some Metaheuristics used for Multiobjetive Optimization. In A. W. et al., editor, *Foundations of Genetic Algorithms (FOGA 2005)*, pages 95–111, Aizu, Japan, 2005. Springer-Verlag. Lecture Notes in Computer Science Vol. 3469.

1654. M. Villalobos-Arias, C. A. Coello Coello, and O. Hernández-Lerma. Asymptotic convergence of a simulated annealing algorithm for multiobjective optimization problems. *Mathematical Methods of Operations Research*, 64(2):353–362, October 2006.

1655. M. A. Villalobos Arias. *Analysis of Optimization Heuristics for Multiobjective Problems*. PhD thesis, Department of Mathematics, CINVESTAV-IPN, Mexico, D.F., Mexico, August 2005.

1656. M. A. Villalobos-Arias, G. Toscano Pulido, and C. A. Coello Coello. A Proposal to Use Stripes to Maintain Diversity in a Multi-Objective Particle Swarm Optimizer. In *2005 IEEE Swarm Intelligence Symposium (SIS'05)*, pages 22–29, Pasadena, California, USA, June 2005. IEEE Press.

1657. J. Villegas, F. Palacios, and A. Medaglia. Solution methods for the biobjective (cost-coverage) unconstrained facility location problem with an illustrative example. *Annals of Operations Research*, 147(1):109–141, October 2006.

1658. P. Vincke. *Multicriteria Decision-aid*. John Wiley & Sons, Chichester, UK, 1992. ISBN 0-471-93184-5.

1659. P. Vincke. Analysis of MCDA in Europe. *European Journal of Operational Research*, 25:160–168, 1995.

1660. M. Visée, J. Teghem, M. Pirlot, and E. Ulungu. Two-phases method and branch and bound to solve bi-objective knapsack problem. *Journal of Global Optimization*, 12:139–155, 1998.

1661. I. Vite-Silva, N. Cruz-Cortés, G. Toscano-Pulido, and L. G. de la Fraga. Optimal Triangulation in 3D Computer Vision Using a Multi-objective Evolutionary Algorithm. In M. G. et al., editor, *Applications of Evolutionary Computing. EvoWorkshops 2007: EvoCOMNET, EvoFIN, EvoIASP, EvoINTERACTION, EvoMUSART, EvoSTOC and EvoTRANSLOG*, pages 330–339, Valencia, Spain, April 2007. Springer. Lecture Notes in Computer Science Vol. 4448.

1662. S. Voget. Multiobjective optimization with genetic algorithm and fuzzy control. In *Proceedings of the Fourth European Conference on Intelligent Techniques and Soft Computing (EUFIT'96)*, pages 391–394, Aachen, Germany, 1996.

1663. S. Voget and M. Kolonko. Multidimensional Optimization with a Fuzzy Genetic Algorithm. *Journal of Heuristics*, 4(3):221–244, September 1998.

1664. J. von Neumann and O. Morgenstern. *Theory of Games and Economic Behavior*. Princeton University Press, Princeton, New Jersey, 1944.

1665. J. von Neumann and O. Morgenstern. *Theory of Game and Economic Behavior*. Princeton University Press, Princeton, New Jersey, second edition, 1947.

1666. H. Voogd. *Multicriteria evaluation for urban and regional planning*. Pion Ltd., London, 1983.

1667. I. Voutchkov and A. Keane. Multiobjective Optimization using Surrogates. In I. Parmee, editor, *Adaptive Computing in Design and Manufacture 2006. Proceedings of the Seventh International Conference*, pages 167–175, Bristol, UK, April 2006. The Institute for People-centred Computation.

1668. D. D. Wackerly, William Mendenhall III, and R. L. Scheaffer. *Mathematical Statistics with Applications*. Duxbury Press, New York, 5th edition, 1996.

1669. T. Wagner, N. Beume, and B. Naujoks. Pareto-, Aggregation-, and Indicator-Based Methods in Many-Objective Optimization. In S. Obayashi, K. Deb, C. Poloni, T. Hiroyasu, and T. Murata, editors, *Evolutionary Multi-Criterion Optimization, 4th International Conference, EMO 2007*, pages 742–756, Matshushima, Japan, March 2007. Springer. Lecture Notes in Computer Science Vol. 4403.

1670. D. Wakefield. Multi-Objective Mission-Resource Value Assessment Development using Evolutionary Algoritms. Master's thesis, Air Force Institute of Technology, Graduate School of Engineering and Management, Wright-Patterson AFB, Dayton, Ohio, 2001.

1671. D. R. Wallace, M. J. Jakiela, and W. Flowers. Design Search Under Probabilistic Specifications Using Genetic Algorithms. *Computer-Aided Design*, 28(5):405–421, 1996.

1672. J. Wallenius. Comparative Evaluation of Some Interactive Approaches to Multicriterion Optimization. *Management Science*, 21:1387–1396, 1975.

1673. D. Wallin, C. Ryan, and R. Azad. Symbiogenetic coevolution. In *2005 IEEE Congress on Evolutionary Computation (CEC'2005)*, volume 1, Edinburgh, Scotland, September 2005. IEEE Service Center.
1674. G. Wang, E. Goodman, and W. Punch III. Simultaneous Multi-Level Evolution. Technical Report 96-03-01, Department of Computer Science, Michigan State University, East Lansing, Michigan, 1996.
1675. J. Wang and J. P. Terpenny. Interactive Preference Incorporation in Evolutionary Engineering Design. In Y. Jin, editor, *Knowledge Incorporation in Evolutionary Computation*, pages 525–543. Springer, Berlin Heidelberg, 2005. ISBN 3-540-22902-7.
1676. J. F. Wang and J. Periaux. Multi-Point Optimization using GAs and Nash/Stackelberg Games for High Lift Multi-airfoil Design in Aerodynamics. In *Proceedings of the Congress on Evolutionary Computation 2001 (CEC'2001)*, volume 1, pages 552–559, Piscataway, New Jersey, May 2001. IEEE Service Center.
1677. X. Wang and M. Mahfouf. ACSAMO: An Adaptive Multiobjective Optimization Algorithm using the Clonal Selection Principle. In *2nd European Symposium on Nature-Inspired Smart Information Systems*, Puerto de la Cruz, Tenerife, Spain, November 29–December 1 2006.
1678. Y. Wang, D. Liu, and Y.-M. Cheung. Preference bi-objective evolutionary algorithm for constrained optimization. In Y. H. et al., editor, *Computational Intelligence and Security. International Conference, CIS 2005*, pages 184–191, Xi'an, China, December 2005. Springer, Lecture Notes in Artificial Intelligence Vol. 3801.
1679. E. F. Wanner, F. G. Guimaraes, R. H. Takahashi, and P. J. Fleming. A Quadratic Approximation-Based Local Search Procedure for Multiobjective Genetic Algorithms. In *2006 IEEE Congress on Evolutionary Computation (CEC'2006)*, pages 3361–3368, Vancouver, BC, Canada, July 2006. IEEE.
1680. K. Wassermann. Three-Dimensional Shape Optimization of Arch Dams with Prescribed Shape Functions. *Journal of Structural Mechanics*, 11(4):465–489, 1983.
1681. S. Watanabe and T. Hiroyasu. Multi-Objective Rectangular Packing Problem. In C. A. Coello Coello and G. B. Lamont, editors, *Applications of Multi-Objective Evolutionary Algorithms*, pages 581–602. World Scientific, Singapore, 2004.
1682. S. Watanabe, T. Hiroyasu, and M. Miki. Parallel Evolutionary Multi-Criterion Optimization for Block Layout Problems. In *2000 International Conference on Parallel and Distributed Processing Techniques and Applications (PDPTA'2000)*, pages 667–673, 2000.
1683. S. Watanabe, T. Hiroyasu, and M. Miki. Parallel Evolutionary Multi-Criterion Optimization for Mobile Telecommunication Networks Optimization. In K. Giannakoglou, D. Tsahalis, J. Periaux, K. Papailiou, and T. Fogarty, editors, *Evolutionary Methods for Design, Optimization and Control with Applications to Industrial Problems. Proceedings of the EUROGEN'2001. Athens. Greece, September 19-21*, pages 167–172, Barcelona, Spain, 2001. International Center for Numerical Methods in Engineering(CIMNE).
1684. S. Watanabe, T. Hiroyasu, and M. Miki. NCGA: Neighborhood Cultivation Genetic Algorithm for Multi-Objective Optimization Problems. In E. Cantú-Paz, editor, *2002 Genetic and Evolutionary Computation Conference. Late-Breaking Papers*, pages 458–465, New York, July 2002.

1685. S. Watanabe, T. Hiroyasu, and M. Miki. Multi-objective Rectangular Packing Problem and Its Applications. In C. M. Fonseca, P. J. Fleming, E. Zitzler, K. Deb, and L. Thiele, editors, *Evolutionary Multi-Criterion Optimization. Second International Conference, EMO 2003*, pages 565–577, Faro, Portugal, April 2003. Springer. Lecture Notes in Computer Science. Volume 2632.
1686. C. J. Watkins and P. Dayan. Q-learning. *Machine Learning*, 8(3):279–292, 1992.
1687. C. J. C. H. Watkins. *Learning from Delayed Rewards*. PhD thesis, Department of Psychology, King's College, Cambridge University, Cambridge, UK, 1989.
1688. J. W. Weibull. *Evolutionary Game Theory*. The MIT Press, Cambridge, Massachusetts, 1997.
1689. D. S. Weile and E. Michielssen. Integer coded Pareto genetic algorithm design of constrained antenna arrays. Technical Report CCEM-13-96, Electrical and Computer Engineering Department, Center for Computational Electromagnetics, University of Illinois at Urbana-Champaign, November 1996.
1690. D. S. Weile and E. Michielssen. Integer coded Pareto genetic algorithm design of constrained antenna arrays. *Electronics Letters*, 32(19):1744–1745, September 1996.
1691. D. S. Weile, E. Michielssen, and D. E. Goldberg. Genetic algorithm design of Pareto optimal broadband microwave absorbers. *IEEE Transactions on Electromagnetic Compatibility*, 38(3):518–525, August 1996.
1692. D. S. Weile, E. Michielssen, and D. E. Goldberg. Multiobjective synthesis of electromagnetic devices using nondominated sorting genetic algorithms. In *1996 IEEE Antennas and Propagation Society International Symposium Digest*, volume 1, pages 592–595, Baltimore, Maryland, July 1996.
1693. T. Weise and K. Geihs. DGPF–An Adaptable Framework for Distributed Multi-Objective Search Algorithms Applied to the Genetic Programming of Sensor Networks. In B. Filipič and J. Šilc, editors, *Bioinspired Optimization Methods and their Applications*, pages 157–166. Jožef Stefan Institute, October 2006.
1694. J. F. Whidborne, D.-W. Gu, and I. Postlethwaite. Algorithms for the Method of Inequalities — A Comparative Study. In *Procedings of the 1995 American Control Conference*, pages 3393–3397, Seattle, Washington, 1995.
1695. L. While. A New Analysis of the LebMeasure Algorithm for Calculating Hypervolume. In C. A. Coello Coello, A. Hernández Aguirre, and E. Zitzler, editors, *Evolutionary Multi-Criterion Optimization. Third International Conference, EMO 2005*, pages 326–340, Guanajuato, México, March 2005. Springer. Lecture Notes in Computer Science Vol. 3410.
1696. L. While, L. Bradstreet, L. Barone, and P. Hingston. Heuristics for Optimising the Calculation of Hypervolume for Multi-Objective Optimization Problems. In *2005 IEEE Congress on Evolutionary Computation (CEC'2005)*, volume 3, pages 2225–2232, Edinburgh, Scotland, September 2005. IEEE Service Center.
1697. L. While, P. Hingston, L. Barone, and S. Huband. A Faster Algorithm for Calculating Hypervolume. *IEEE Transactions on Evolutionary Computation*, 10(1):29–38, February 2006.
1698. C. White, A. Sage, and S. Dozono. A model of multiattribute decisionmaking and tradeoff weight determination under uncertainty. *IEEE Transactions on Systems, Man, and Cybernetics*, SMC-14:223–229, 1984.
1699. D. Whitley. Cellular Genetic Algorithms. In S. Forrest, editor, *Proceedings of the Fifth International Conference on Genetic Algorithms*, page 658, San

Mateo, California, 1993. University of Illinois at Urbana-Champaign, Morgan Kaufmann Publishers.
1700. D. Whitley, S. gordon, and K. Mathias. Larmarckian Evolution, The Baldwin Effect and Function Optimization. In H.-P. S. Y. Davidor and R. Männer, editors, *Parallel Problem Solving from Nature III*, pages 6–15. Springer Verlag, 1994.
1701. D. Whitley, K. Mathias, S. Rana, and J. Dzubera. Evaluating Evolutionary Algorithms. *Artificial Intelligence*, 85:245–276, 1996.
1702. P. B. Wienke, C. Lucasius, and G. Kateman. Multicriteria target optimization of analytical procedures using a genetic algorithm. *Analytical Chimica Acta*, 265(2):211–225, 1992.
1703. A. P. Wierzbicki. On the use of Penalty functions in Multiobjective optimization. In *Proceedings of the International Symposium on Operations Research*, Mannheim, Germany, 1978.
1704. A. P. Wierzbicki. A Methodological Guide to Multiobjective Optimization. In *IIASA Working Paper, WP-79-122*, Laxenburg, Austria, 1979. International Institute for Applied System Analysis.
1705. A. P. Wierzbicki. The Use of Reference Objectives in Multiobjective Optimization. In G. Fandel and T. Gal, editors, *Multiple Criteria Decision Making Theory and Application*, pages 469–486. Springer-Verlag, New York, 1980.
1706. E. B. Wilson. *An Introduction to Scientific Research*. Courier Dover, 1991.
1707. P. B. Wilson and M. D. Macleod. Low implementation cost IIR digital filter design using genetic algorithms. In *IEE/IEEE Workshop on Natural Algorithms in Signal Processing*, pages 4/1–4/8, Chelmsford, U.K., 1993.
1708. D. H. Wolpert and W. G. Macready. No Free Lunch Theorems for Optimization. *IEEE Transactions on Evolutionary Computation*, 1(1):67–82, April 1997.
1709. J. Wright and H. Loosemore. An Infeasibility Objective for Use in Constrained Pareto Optimization. In E. Zitzler, K. Deb, L. Thiele, C. A. Coello Coello, and D. Corne, editors, *First International Conference on Evolutionary Multi-Criterion Optimization*, pages 256–268. Springer-Verlag. Lecture Notes in Computer Science No. 1993, 2001.
1710. J. Wright and H. Loosemore. The Multi-Criterion Optimization of Building Thermal Design and Control. In *7th IBPSA Conference: Building Simulation*, volume 2, pages 873–880, Rio de Janeiro, Brazil, 2001. ISBN 85-901939-3-4.
1711. J. Wright, H. Loosemore, and R. Farmani. Optimization of building thermal design and control by multi-criterion genetic algorithm. *Energy and Buildings*, 34(9):959–972, October 2002.
1712. S. Wright. The roles of mutation, inbreeding, crossbreeding, and selection in evolution. In D. Jones, editor, *Proceedings of the 6th International Congress on Genetics*, volume 1, pages 356–366, Ithaca, New York, 1932. Brooklyn Botanical Gardens.
1713. J. Wu and S. Azarm. Metrics for Quality Assessment of a Multiobjective Design Optimization Solution Set. *Transactions of the ASME, Journal of Mechanical Design*, 123:18–25, 2001.
1714. J. Wu and S. Azarm. On a New Constraint Handling Technique for Multi-Objective Genetic Algorithms. In L. Spector, E. D. Goodman, A. Wu, W. Langdon, H.-M. Voigt, M. Gen, S. Sen, M. Dorigo, S. Pezeshk, M. H. Garzon, and E. Burke, editors, *Proceedings of the Genetic and Evolutionary*

Computation Conference (GECCO'2001), pages 741–748, San Francisco, California, 2001. Morgan Kaufmann Publishers.
1715. N. Xiao and M. P. Armstrong. A Specialized Island Model and Its Application in Multiobjective Optimization. In E. C.-P. et al., editor, *Genetic and Evolutionary Computation—GECCO 2003. Proceedings, Part II*, pages 1530–1540. Springer. Lecture Notes in Computer Science Vol. 2724, July 2003.
1716. N. Xiao, D. A. Bennet, and M. P. Armstrong. Using evolutionary algorithms to generate alternatives for multiobjective site-search problems. *Environment and Planning A*, 34(4):639–656, April 2002.
1717. Z. Xiao-hua, M. Hong-yun, and J. Li-cheng. Intelligent Particle Swarm Optimization in Multiobjective Optimization. In *2005 Congress on Evolutionary Computation*, pages 714–719, Edinburgh, Scotland, UK, September 2005. IEEE Press.
1718. S. Xiong and F. Li. Parallel Strength Pareto Multi-objective Evolutionary Algorithm. In *Proceedings of the Fourth International Conference on Applications and Technologies (PDCAT'2003)*, pages 681–683. IEEE, August 2003.
1719. J. Xu, M. Sohoni, M. McCleery, and T. G. Bailey. A dynamic neighborhood based tabu search algorithm for real-world flight instructor scheduling problems. *European Journal of Operational Research*, 169:978–993, 2006.
1720. F. Xue. *Multi-Objective Differential Evolution: Theory and Applications*. PhD thesis, Rensselaer Polytechnic Institute, Troy, New York, September 2004.
1721. F. Xue, A. C. Sanderson, and R. J. Graves. Multi-Objective Differential Evolution and Its Application to Enterprise Planning. In *Proceedings of the 2003 IEEE International Conference on Robotics and Automation (ICRA'03)*, volume 3, pages 3535–3541, Taipei, Taiwan, September 2003. IEEE.
1722. F. Xue, A. C. Sanderson, and R. J. Graves. Pareto-based Multi-Objective Differential Evolution. In *Proceedings of the 2003 Congress on Evolutionary Computation (CEC'2003)*, volume 2, pages 862–869, Canberra, Australia, December 2003. IEEE Press.
1723. F. Xue, A. C. Sanderson, and R. J. Graves. Modeling and convergence analysis of a continuous multi-objective differential evolution algorithm. In *2005 IEEE Congress on Evolutionary Computation (CEC'2005)*, volume 1, pages 228–235, Edinburgh, Scotland, September 2005. IEEE Service Center.
1724. F. Xue, A. C. Sanderson, and R. J. Graves. Multi-objective differential evolution - algorithm, convergence analysis, and applications. In *2005 IEEE Congress on Evolutionary Computation (CEC'2005)*, volume 1, pages 743–750, Edinburgh, Scotland, September 2005. IEEE Service Center.
1725. K. Yamasaki. Dynamic Pareto Optimum GA against the changing environments. In *2001 Genetic and Evolutionary Computation Conference. Workshop Program*, pages 47–50, San Francisco, California, July 2001.
1726. D. Yamashiro, T. Yoshikawa, and T. Furuhashi. Visualization of Search Process and Improvement of Search Performance in Multi-Objective Genetic Algorithm. In *2006 IEEE Congress on Evolutionary Computation (CEC'2006)*, pages 3967–3972, Vancouver, BC, Canada, July 2006. IEEE.
1727. Z. Yan, L. Zhang, L. Kang, and G. Lin. A New MOEA for Multi-objective TSP and Its Convergence Property Analysis. In C. M. Fonseca, P. J. Fleming, E. Zitzler, K. Deb, and L. Thiele, editors, *Evolutionary Multi-Criterion Optimization. Second International Conference, EMO 2003*, pages 342–354, Faro, Portugal, April 2003. Springer. Lecture Notes in Computer Science. Volume 2632.

1728. X. Yang and M. Gen. Evolution Program for Bicriteria Transportation Problem. *Computers and Industrial Engineering*, 27(1–4):481–484, 1994.
1729. T.-M. Yao and K. Choi. Shape Optimal Design of an Arch Dam. *Journal of Structural Engineering*, 115(9):2401–2405, September 1989.
1730. X. Yao and Y. Liu. Fast Evolutionary Programming. In L. J. Fogel, P. J. Angeline, and T. Bäck, editors, *Evolutionary Programming V:Proceedings of the Fifth Annual Conference on Evolutionary Programming (EP '96)*, pages 451–460. MIT Press, Cambridge, Massachusetts, 1996.
1731. X. Yao and Y. Liu. Fast Evolution Strategies. In P. J. Angeline, R. G. Reynolds, J. R. McDonnell, and R. Eberhart, editors, *Evolutionary Programming VI: Proceedings of the Sixth Annual Conference on Evolutionary Programming (EP '97)*, pages 151–161. Springer-Verlag. Lecture Notes in Computer Science No. 1213, 1997.
1732. H. Yapicioglu, G. Dozier, and A. E. Smith. Neural Network Enhancement of Multiobjective Evolutionary Search. In *2006 IEEE Congress on Evolutionary Computation (CEC'2006)*, pages 6800–6806, Vancouver, BC, Canada, July 2006. IEEE.
1733. P. O. Yapo. *A multiobjective global optimization algorithm with application to the calibration of hydrologic models*. PhD thesis, Department of Systems and Industrial Engineering, The University of Arizona, Tucson, Arizona, 1996.
1734. P. O. Yapo, H. V. Gupta, and S. Sorooshian. Multi-Objective Global Optimization for Hydrologic Models. *Journal of Hydrology*, 204:83–97, 1998.
1735. Y. J. Yau, J. Teo, and P. Anthony. Pareto Evolution and Co-evolution in Cognitive Game AI Synthesis. In S. Obayashi, K. Deb, C. Poloni, T. Hiroyasu, and T. Murata, editors, *Evolutionary Multi-Criterion Optimization, 4th International Conference, EMO 2007*, pages 227–241, Matshushima, Japan, March 2007. Springer. Lecture Notes in Computer Science Vol. 4403.
1736. W. Yong and C. Zixing. A Constrained Optimization Evolutionary Algorithm Based on Multiobjective Optimization Techniques. In *2005 IEEE Congress on Evolutionary Computation (CEC'2005)*, volume 2, pages 1081–1087, Edinburgh, Scotland, September 2005. IEEE Service Center.
1737. J. Yoo and P. Hajela. Immune network simulations in multicriterion design. *Structural Optimization*, 18:85–94, 1999.
1738. K. Yoshida, M. Yamamura, and S. Kobayashi. Generating Pareto Optimal Decision Trees by GAs. In *Proceedings of the 4th International Conference on Soft Computing*, pages 854–859, 1996.
1739. K. Yoshimura and R. Nakano. Genetic Algorithms for Information Operator Scheduling. In D. B. Fogel, editor, *Proceedings of the 1998 International Conference on Evolutionary Computation*, pages 277–282, Piscataway, New Jersey, 1998. IEEE.
1740. H. Youssef, S. M. Sait, and S. A. Khan. Fuzzy Simulated Evolution Algorithm for Topology Design on Campus Networks. In *2000 Congress on Evolutionary Computation*, volume 1, pages 180–187, Piscataway, New Jersey, July 2000. IEEE Service Center.
1741. H. Youssef, S. M. Sait, and S. A. Khan. Fuzzy Evolutionary Hybrid Metaheuristic for Network Topology Design. In E. Zitzler, K. Deb, L. Thiele, C. A. Coello Coello, and D. Corne, editors, *First International Conference on Evolutionary Multi-Criterion Optimization*, pages 400–415. Springer-Verlag. Lecture Notes in Computer Science No. 1993, 2001.

1742. P. L. Yu. Decision dynamics with an application to persuasion and negotiation. In M. K. Starr and M. Zeleny, editors, *Multiple Criteria Decision Making*, pages 159–177. North-Holland Publish. Co., New York, 1977.
1743. Y. Yu. Multi-objective decision theory for computational optimization in radiation therapy. *Medical Physics*, 24:1445–1454, 1997.
1744. Y. Yun, H. Nakayama, and M. Arakawa. Multiple criteria decision making with generalized DEA and an aspiration level method. *European Journal of Operational Research*, 158(3):697–706, November 2004.
1745. Y. Yun, H. Nakayama, and T. Tanino. A generalization of DEA model. *Journal of the Society of Instrument and Control Engineers (SICE)*, 35(8):1813–1818, 1999.
1746. L. A. Zadeh. Optimality and Nonscalar-Valued Performance Criteria. *IEEE Transactions on Automatic Control*, AC-8(1):59–60, 1963.
1747. R. R. Zaliz, I. Zwir, and E. Ruspini. Generalized Analysis of Promoters: A Method for DNA Sequence Description. In C. A. Coello Coello and G. B. Lamont, editors, *Applications of Multi-Objective Evolutionary Algorithms*, pages 427–449. World Scientific, Singapore, 2004.
1748. R. S. Zebulum, M. A. Pacheco, and M. Vellasco. A multi-objective optimisation methodology applied to the synthesis of low-power operational amplifiers. In I. J. Cheuri and C. A. dos Reis Filho, editors, *Proceedings of the XIII International Conference in Microelectronics and Packaging*, volume 1, pages 264–271, Curitiba, Brazil, August 1998.
1749. R. S. Zebulum, M. A. Pacheco, and M. Vellasco. Synthesis of CMOS operational amplifiers through Genetic Algorithms. In *Proceedings of the Brazilian Symposium on Integrated Circuits, SBCCI'98*, pages 125–128, Rio de Janeiro, Brazil, September 1998. IEEE.
1750. R. S. Zebulum, M. A. Pacheco, and M. Vellasco. Artificial Evolution of Active Filters: A Case Study. In *Proceedings of the First NASA/DoD Workshop on Evolvable Hardware*, pages 66–75, Los Alamitos, California, July 1999. IEEE Computer Society.
1751. M. Zeleny. Compromise Programming. In J. Cochrane and M. Zeleny, editors, *Multiple Criteria Decision Making*, pages 262–301. University of South Carolina Press, Columbia, South Carolina, 1973.
1752. M. Zeleny. Adaptive displacement of preferences in decision making. In M. K. Starr and M. Zeleny, editors, *Multiple Criteria Decision Making*, volume 6 of *TIMS Studies in the Management Sciences*, pages 147–157. North-Holland Publishing Company, Amsterdam, 1977.
1753. M. Zeleny. *Multiple Criteria Decision Making*. McGraw-Hill Book Company, New York, 1982.
1754. M. Zeleny. Multiple criteria decision making: Eight concepts of optimality. *Human Systems Management*, 17:97–107, 1998.
1755. S. Zeng, L. Ding, Y. Chen, and L. Kang. A New Multiobjective Evolutionary Algorithm: OMOEA. In *Proceedings of the 2003 Congress on Evolutionary Computation (CEC'2003)*, volume 2, pages 898–905, Canberra, Australia, December 2003. IEEE Press.
1756. S. Zeng, S. Yao, L. Kang, and Y. Liu. An Efficient Multi-objective Evolutionary Algorithm: OMOEA-II. In C. A. Coello Coello, A. Hernández Aguirre, and E. Zitzler, editors, *Evolutionary Multi-Criterion Optimization. Third International Conference, EMO 2005*, pages 108–119, Guanajuato, México, March 2005. Springer. Lecture Notes in Computer Science Vol. 3410.

1757. S. Y. Zeng, L. S. Kang, and L. X. Ding. An Orthogonal Multi-objective Evolutionary Algorithm for Multi-objective Optimization Problems with Constraints. *Evolutionary Computation*, 12(1):77–98, Spring 2004.
1758. B.-T. Zhang and H. Mühlenbein. Adaptive Fitness Functions for Dynamic Growing/Pruning of Program Trees. In P. J. Angeline and J. Kenneth E. Kinnear, editors, *Advances in Genetic Programming 2*, pages 241–256. MIT Press, 1996.
1759. L. Zhang, C. Zhou, X. Liu, Z. Ma, and Y. Liang. Solving Multi Objective Optimization Problems Using Particle Swarm Optimization. In *Proceedings of the 2003 Congress on Evolutionary Computation (CEC'2003)*, volume 4, pages 2400–2405, Canberra, Australia, December 2003. IEEE Press.
1760. X. Zhang, B. Lu, S. Gou, and L. Jiao. Immune Multiobjective Optimization Algorithm Using Unsupervised Feature Selection. In F. R. et al., editor, *Applications of Evolutionary Computing. EvoWorkshops 2006: EvoBIO, EvoCOMNET, EvoHOT, EvoIASP, EvoINTERACTION, EvoMUSART, and EvoSTOC*, pages 484–494, Budapest, Hungary, April 2006. Springer, Lecture Notes in Computer Science Vol. 3907.
1761. Y. Zhang. *MEMS Design Synthesis Based on Hybrid Evolutionary Computation*. PhD thesis, Civil and Environmental Engineering, University of California, Berkeley, USA, 2006.
1762. Y. Zhang, R. Kamalian, A. M. Agogino, and C. H. Séquin. Design Synthesis of Microelectromechanical Systems Using Genetic Algorithms with Component-Based Genotype Representation. In M. K. et al., editor, *2006 Genetic and Evolutionary Computation Conference (GECCO'2006)*, volume 1, pages 731–738, Seattle, Washington, USA, July 2006. ACM Press. ISBN 1-59593-186-4.
1763. B. Zhao and Y. j. Cao. Multiple objective particle swarm optimization technique for economic load dispatch. *Journal of Zhejiang University SCIENCE*, 6A(5):420–427, 2005.
1764. G. Zhou and M. Gen. Evolutionary Computation on Multicriteria Production Process Planning Problem. In W. Porto, editor, *Proceedings of the 1997 IEEE International Conference on Evolutionary Computation*, pages 419–424, Piscataway, New Jersey, April 1997. IEEE Press.
1765. G. Zhou and M. Gen. Genetic Algorithm Approach on Multi-Criteria Minimum Spanning Tree Problem. *European Journal of Operational Research*, 114(1), April 1999.
1766. Y. Zhou, Y. Li, J. He, and L. Kang. Multi-objective and MGG Evolutionary Algorithm for Constrained Optimization. In *Proceedings of the 2003 Congress on Evolutionary Computation (CEC'2003)*, volume 1, pages 1–5, Canberra, Australia, December 2003. IEEE Press.
1767. Z.-Y. Zhu. *An Evolutionary Approach to Multi-Objective Optimization Problems*. PhD thesis, The Chinese University of Hong Kong, August 2002.
1768. Z.-Y. Zhu and K.-S. Leung. Asynchronous Self-Adjustable Island Genetic Algorithm for Multi-Objective Optimization Problems. In *Congress on Evolutionary Computation (CEC'2002)*, volume 1, pages 837–842, Piscataway, New Jersey, May 2002. IEEE Service Center.
1769. Z.-Y. Zhu and K.-S. Leung. An Enhanced Annealing Genetic Algorithm for Multi-Objective Optimization Problems. In W. Langdon, E. Cantú-Paz, K. Mathias, R. Roy, D. Davis, R. Poli, K. Balakrishnan, V. Honavar, G. Rudolph, J. Wegener, L. Bull, M. Potter, A. Schultz, J. Miller, E. Burke,

and N. Jonoska, editors, *Proceedings of the Genetic and Evolutionary Computation Conference (GECCO'2002)*, pages 658–665, San Francisco, California, July 2002. Morgan Kaufmann Publishers.

1770. E. Zitzler. *Evolutionary Algorithms for Multiobjective Optimization: Methods and Applications*. PhD thesis, Swiss Federal Institute of Technology (ETH), Zurich, Switzerland, November 1999.

1771. E. Zitzler, D. Brockhoff, and L. Thiele. The Hypervolume Indicator Revisited: On the Design of Pareto-compliant Indicator Via Weighted Integration. In S. Obayashi, K. Deb, C. Poloni, T. Hiroyasu, and T. Murata, editors, *Evolutionary Multi-Criterion Optimization, 4th International Conference, EMO 2007*, pages 862–876, Matshushima, Japan, March 2007. Springer. Lecture Notes in Computer Science Vol. 4403.

1772. E. Zitzler, K. Deb, and L. Thiele. Comparison of Multiobjective Evolutionary Algorithms: Empirical Results. *Evolutionary Computation*, 8(2):173–195, Summer 2000.

1773. E. Zitzler, K. Deb, L. Thiele, C. A. Coello Coello, and D. Corne, editors. *Evolutionary Multi-Criterion Optimization. First International Conference (EMO'01)*. Lecture Notes in Computer Science 1993. Springer, Berlin, 2001.

1774. E. Zitzler and S. Künzli. Indicator-based Selection in Multiobjective Search. In X. Y. et al., editor, *Parallel Problem Solving from Nature - PPSN VIII*, pages 832–842, Birmingham, UK, September 2004. Springer-Verlag. Lecture Notes in Computer Science Vol. 3242.

1775. E. Zitzler, M. Laumanns, and L. Thiele. SPEA2: Improving the Strength Pareto Evolutionary Algorithm. In K. Giannakoglou, D. Tsahalis, J. Periaux, P. Papailou, and T. Fogarty, editors, *EUROGEN 2001. Evolutionary Methods for Design, Optimization and Control with Applications to Industrial Problems*, pages 95–100, Athens, Greece, 2001.

1776. E. Zitzler, M. Laumanns, and L. Thiele. SPEA2: Improving the Strength Pareto Evolutionary Algorithm. Technical Report 103, Computer Engineering and Networks Laboratory (TIK), Swiss Federal Institute of Technology (ETH) Zurich, Gloriastrasse 35, CH-8092 Zurich, Switzerland, May 2001.

1777. E. Zitzler, M. Laumanns, L. Thiele, C. M. Fonseca, and V. Grunert da Fonseca. Why Quality Assessment of Multiobjective Optimizers Is Difficult. In W. Langdon, E. Cantú-Paz, K. Mathias, R. Roy, D. Davis, R. Poli, K. Balakrishnan, V. Honavar, G. Rudolph, J. Wegener, L. Bull, M. Potter, A. Schultz, J. Miller, E. Burke, and N. Jonoska, editors, *Proceedings of the Genetic and Evolutionary Computation Conference (GECCO'2002)*, pages 666–673, San Francisco, California, July 2002. Morgan Kaufmann Publishers.

1778. E. Zitzler, J. Teich, and S. S. Bhattacharyya. Evolutionary Algorithm Based Exploration of Software Schedules for Digital Signal Processors. In W. Banzhaf, J. Daida, A. E. Eiben, M. H. Garzon, V. Honavar, M. Jakiela, and R. E. Smith, editors, *Proceedings of the Genetic and Evolutionary Computation Conference (GECCO'99)*, volume 2, pages 1762–1769, San Francisco, California, July 1999. Morgan Kaufmann.

1779. E. Zitzler, J. Teich, and S. S. Bhattacharyya. Multidimensional Exploration of Software Implementations for DSP Algorithms. *Journal of VLSI Signal Processing*, 24(1):83–98, February 2000.

1780. E. Zitzler and L. Thiele. An Evolutionary Algorithm for Multiobjective Optimization: The Strength Pareto Approach. Technical Report 43, Computer

Engineering and Communication Networks Lab (TIK), Swiss Federal Institute of Technology (ETH), Zurich, Switzerland, May 1998.
1781. E. Zitzler and L. Thiele. Multiobjective Optimization Using Evolutionary Algorithms—A Comparative Study. In A. E. Eiben, editor, *Parallel Problem Solving from Nature V*, pages 292–301, Amsterdam, The Netherlands, September 1998. Springer-Verlag. Lecture Notes in Computer Science No. 1498.
1782. E. Zitzler and L. Thiele. Multiobjective Evolutionary Algorithms: A Comparative Case Study and the Strength Pareto Approach. *IEEE Transactions on Evolutionary Computation*, 3(4):257–271, November 1999.
1783. E. Zitzler, L. Thiele, M. Laumanns, C. M. Fonseca, and V. G. da Fonseca. Performance Assessment of Multiobjective Optimizers: An Analysis and Review. *IEEE Transactions on Evolutionary Computation*, 7(2):117–132, April 2003.
1784. X. Zou, M. Liu, L. Kang, and J. He. A High Performance Multi-objective Evolutionary Algorithm Based on the Principles of Thermodynamics. In X. Y. et al., editor, *Parallel Problem Solving from Nature - PPSN VIII*, pages 922–931, Birmingham, UK, September 2004. Springer-Verlag. Lecture Notes in Computer Science Vol. 3242.
1785. Z. Zou, Q. Jiang, P. Zhang, and Y. Cao. Application of Multi-objective Evolutionary Algorithm in Coordinated Design of PSS and SVC Controllers. In Y. H. et al., editor, *Computational Intelligence and Security. International Conference, CIS 2005*, pages 1106–1111, Xi'an, China, December 2005. Springer, Lecture Notes in Artificial Intelligence Vol. 3801.
1786. K. Zuse. Mathematical Programming Testdata. Online, 1997. Available: ftp://ftp.zib.de/Packages/mp-testdata/index.html.
1787. I. S. Zwir and E. H. Ruspini. Qualitative Object Description: Initial Reports of the Exploration of the Frontier. In *Procedings of the Joint EUROFUSE—SIC'99 International Conference*, Budapest, Hungary, 1999.
1788. J. B. Zydallis. *Explicit Building-Block Multiobjective Genetic Algorithms: Theory, Analysis, and Development*. PhD thesis, Air Force Institute of Technology, Department of the Air Force, Air University, Wright-Patterson, Airforce Base, Ohio, USA, March 2003.
1789. J. B. Zydallis and G. B. Lamont. Solving of Discrete Multiobjective Problems Using an Evolutionary Algorithm with a Repair Mechanism. In *Proceedings of the IEEE 2001 Midwest Symposium on Circuits and Systems*, volume 1, pages 470–473. IEEE, 2001.
1790. J. B. Zydallis, D. A. V. Veldhuizen, and G. B. Lamont. A Statistical Comparison of Multiobjective Evolutionary Algorithms Including the MOMGA–II. In E. Zitzler, K. Deb, L. Thiele, C. A. C. Coello, and D. Corne, editors, *First International Conference on Evolutionary Multi-Criterion Optimization*, pages 226–240. Springer-Verlag. Lecture Notes in Computer Science No. 1993, 2001.

Index

$(\mu + \lambda)$ selection, 26
$(\mu + \lambda)$-ES, 359
(μ, λ) selection, 26
(μ, λ)-ES, 368, 413
H_∞ controller design, 363
NP-Complete problems, 22, 220
ϵ-MyDE, 601
ϵ-Pareto Set, 295
ϵ-approximate Pareto Set, 295
ϵ-box, 296
ϵ-constraint method, 46, 76, 352, 430, 561
   applications, 344, 346, 398, 434
   hybridized with differential evolution, 603
ϵ-dominance, 84, 295, 544, 587, 590, 593
ϵ-indicator, 262
μGA, *see* micro genetic algorithm, 609
μGA^2, *see* micro genetic algorithm 2
k-optimality, 535
k_F-optimality, 535
$m - k$ deceptive trap problems, 320
n-parity problem, 406
$r(n)$-approximate algorithms, 129
$(\mu + \lambda)$ MOEA, 407, 567
PF_{true}
   definition, 11
0/1 knapsack problem, 139, 221, 344
1/5 success rule, 611
2-opt local search, 381
3D Navier-Stokes flow solver, 387
3D aerodynamic design design

evolutionary multiobjective optimization in, 387
3D bin packing, 427
   evolutionary multiobjective optimization in, 427
3D integer representation, 380
3D molecular structures
   evolutionary multiobjective optimization in, 391
4D-Miner, 395
7-point average distance measure, 257

A Platform and Programming Language Interface for Search Algorithms, 275, 617
a posteriori preference articulation
   advantages, 519
a priori preference articulation, 518
A*, 22
AbSS, *see* Archive-based Scatter Search
accelerated convergence genetic algorithm, 345
ACGA, *see* accelerated convergence genetic algorithm
   applications, 345
Ackley's function, 177
ACSAMO, *see* Adaptive Clonal Selection Algorithm for Multiobjective Optimization
active magnetic bearing controllers, 363
   design, 363
active mass driver, 374
actuator placement

evolutionary multiobjective optimization in, 370, 371, 385
ad-hoc networks, 357
Adaptive Clonal Selection Algorithm for Multiobjective Optimization, 611
adaptive evaluation function
    applications, 359
adaptive genetic algorithm, 420, 570
adaptive grid, 614
adaptive image segmentation, 401
adaptive multicriteria evolutionary algorithm with tabu mutation, 608
adaptive mutation, 432
adaptive parsimony pressure, 405
adaptive population size, 365
Adaptive Weighted PSO, 589
admissible solutions, 11
ADvanced VehIcle SimulatOR (ADVISOR), 415
AER, see agent-environment-rules
aerodynamic design, 466
    evolutionary multiobjective optimization in, 376, 383, 387
aerodynamic shape optimization
    using Tabu search, 563
aeronautical engineering
    applications, 382
    evolutionary multiobjective optimization in, 382–386
affinity, 607, 611
affinity maturation, 610
affinity operator, 610
agent-based model, 616
agent-based system, 534
agent-environment-rules model, 591
agents, 398, 434
agents coordination
    evolutionary multiobjective optimization in, 398
agents for multiobjective optimization, 614
aggregating function, 385, 566
    applications, 343, 345, 347–354, 359, 362, 363, 366, 368, 371, 373–375, 378–380, 383, 386–388, 391, 394, 395, 398–401, 404, 405, 409, 411, 412, 417–419, 421–424, 426, 427, 429–432, 571

aggregation selection, 75
agricultural project scheduling, 552
    Pareto Simulated Annealing in, 552
air quality management, 345
aircrew rostering, 552
airfoil design, 383, 385, 386, 465
    evolutionary multiobjective optimization in, 383, 385
airfoil shape optimization, 385
airframe
    preliminary design, 386
airline crew rostering
    evolutionary multiobjective optimization in, 423
algorithmic parameters, 239
All Rules Algorithm Cc-optimal, 433
allele, 24
alternating-objective repeated line-search, 331
AMEA, see adaptive multicriteria evolutionary algorithm with tabu mutation
analog active filters
    synthesis, 353
analog filter tuning
    evolutionary multiobjective optimization in, 352
analysis of variance, 271
analytical solutions
    of MOPs, 181
annealing, 548
ANOVA, see analysis of variance
ant colony optimization, 575, 578
    cooperation mechanism, 577
    multiobjective, 577
ant colony system, 617
ant system
    applications, 578
    multiobjective, 572
    multiobjective optimization, 575
ant-Q, 573
    pseudo code, 574
antenna arrays
    evolutionary multiobjective optimization in, 351
antenna design
    evolutionary multiobjective optimization in, 358
antenna tuning unit design, 352

Index 763

anti-reflection coatings, 392
antibody, 374, 605, 606
  archive, 610
antigen, 374, 605, 606
applications
  3D aerodynamic design, 387
  3D bin packing, 427
  actuator placement, 370, 385
  adaptive distributed database management, 359
  aerodynamic optimization, 387
  aeronautical engineering, 223, 382–386
  agents, 398
  agricultural project scheduling, 552
  air quality management, 345
  aircraft control, 385
  aircraft gas turbine engine design, 362
  airfoil design, 383, 385, 386
  airfoil shape optimization, 385
  airline crew rostering, 423
  analog filter tuning, 352
  antenna design, 358
  articulated figure motion, 399
  assembly line balancing, 409
  beam design, 372, 373, 377, 607
  bin packing, 427
  bioinformatics, 407
  bioremediation of contaminated groundwater, 343
  bloat reduction, 406
  brachytherapy, 394
  breast cancer prediction, 395
  building block placement problem, 411
  cancer treatment, 393, 394
  capital budgeting, 552
  car sampling tests, 409
  cascade compensator, 365
  cell formation, 412
  cellular manufacturing, 412, 552
  chance-constrained solid transportation problem, 380
  chemical engineering, 223
  chemistry, 390, 391
  chemotherapeutic treatment, 393
  chromatic rectangles, 404
  circuit design, 355

circuit optimization, 355
civil and construction engineering, 377
classification and prediction, 399, 400, 432
clustering, 399
code growth in GP, 405
coloring problems, 404
composite plate optimization, 373
computer aided design, 411
computer generated animation, 399
computer graphics, 399
computer science, 223, 398, 401, 404–407, 432
computer vision, 402
configuration problems, 374
conformational search, 391
constellation design, 382
constraint-handling, 116
construction, 377
control, 364, 365, 368, 369
  of a civil structure, 374
control system design, 364–366
controller design, 364, 365
core-based single-chip system synthesis, 350
data extraction, 385
data sampling, 388
data structures evolution, 406
decision tree induction, 399
design and manufacture, 371, 372, 408, 409, 411–415, 417
design of control systems, 365
design of controllers, 363
design of electromagnetic devices, 350
design of fuzzy controllers, 361
design of multiplierless IIR filters, 353
design of permanent magnet actuators, 351
design of robust controllers, 364, 366
design of turbomachinery airfoils, 372
design unification of linear control systems, 365
digital filter design, 354
distance controller, 368
distributed database management, 359
distributed object systems, 404
distributed systems, 404

Index

distribution system design, 426
ecological model assessment, 396
ecological process model fitting, 397
ecology, 396, 397
economic load dispatch, 349
economic model optimization, 430
economics, 430
electrical and electronics engineering, 223, 348–353
electrical impedance tomography problem, 392
electromagnetic devices design, 351
elevators control, 363
environmental modeling, 388
environmental, naval and hydraulic engineering, 341–347, 383, 577
extraction of munition aerodynamic characteristics, 385
face detection, 403
face modeling, 402
facility layout, 425
facility location problem, 425
facility management, 424
fault diagnosis, 362
fault tolerant system design, 352
feature selection, 403, 432
finance, 223, 429, 430
finite word-length feedback controller design, 364
fishery bioeconomic model optimization, 430
flexible manufacturing, 410
flexible process sequencing, 409
flowshop scheduling, 418–420
food industry, 408, 417
forest management, 425
fuzzy system design, 368, 369, 411
gas turbine design, 412
gear-box design, 375
generation of optimal rules, 400
genetic programming, 406
geography, 388, 389
groundwater monitoring networks, 343
grouping problem, 427
handwritten word recognition, 403
hardware-software embedded systems co-synthesis, 350
helicopter flat panels design, 383

helicopter rotor system design, 383
high-lift airfoil design, 386
hydraulic and environmental engineering, 223
image processing, 392, 401
image segmentation, 401
induction motor design, 352
inductive learning, 432
industrial, 359, 408, 409, 412, 413, 415, 417–422, 427
integrated circuit placement, 411
inverse kinematics of a robotic manipulator, 367
investment portfolio optimization, 429
job shop scheduling, 346, 398, 420, 421
knapsack problem, 142, 344
knowledge extraction, 432
lamp design, 354
land use, 377, 389
  planning, 389
launch conditions, 385
line balancing, 409
linguistic function approximation, 432
machine learning, 400, 432
machine-component grouping problem, 427
machining recommendations, 413
magnetically levitated vehicle design, 366
management, 426
manufacturing, 409
mechanical and structural engineering, 223, 374, 375
medicine, 393–395
metallurgy, 375
microchannel design, 413
microprocessor design, 354
microwave absorbers design, 351
military, 404, 423
  airlift scheduling, 423
  land moves scheduling, 423
minimum spanning tree, 359
mobile robots, 361
motor design, 352
motorway routes planning, 378

multipoint aerodynamic optimization, 383
multipoint airfoil design, 383
multivariable control system design, 364
natural language processing, 406
network design, 356, 357
network topology design, 357
neural networks architecture design, 399
nonlinear FIR filters design, 353
nonlinear system identification, 362
nuclear engineering, 392
nurse scheduling, 422, 552
object recognition, 402
obstacle avoidance, 361
office design, 377
offline routing problem, 359
operating of electrified railway systems, 378
optical filter design, 354
optimal control of a space vehicle, 367
optimization of laminated composites, 372
optimization of rocket engine pumps, 384
optimization of working conditions of a press, 408
packing problems, 374
paper mill scheduling, 616
parallel machine scheduling, 421
parameter optimization of a mobile network, 358
partial classification, 433
path planning, 361
pattern classification, 400, 433
pattern recognition, 431
pattern space partitioning, 399
permutation scheduling, 142
physics, 391, 392
plane truss design, 372
plant modeling, 369
portfolio optimization, 429
power distribution system planning, 349
prediction, 431
pressurized water reactor design, 392
process planning, 359, 410
process plants design, 414

process sequencing, 409
production process optimization, 409
production process planning, 359
production scheduling, 417
prostate implant optimization, 393, 394
radiation therapy, 393
radiology, 394
reactor design, 354, 392
ready meals production scheduling, 417
real-time scheduling, 423
rectangular packing, 427
road train design, 378
robot arm optimization, 368
robot grippers design, 367
robotics and control engineering, 361–369
robust PID controller design, 364
robust trajectory tracking problem, 367
routing, 359
satellite constellation design, 382
scheduling, 223, 393, 398, 418–423, 570
self-organizing fuzzy logic controller, 364
sensory-action network acquisition, 361
sequencing problems, 410
shape design of electromagnetic devices, 354
shipyard plane cutting shop problem, 346
site-search, 388
software implementations for DSP algorithms, 398
software project scheduling, 552
sorting networks, 404
spacecraft trajectories, 382
specification optimization problem, 371
spectroscopic analysis, 392
stock ranking, 430
structural control system design, 374
structural engineering, 370–374
structural synthesis of cell-based VLSI circuits, 349
surge tank system, 362

suspension system design, 366
symbolic layout compaction, 348
synthesis of analog filters, 353
synthesis of low-power operational amplifiers, 353
synthesis of operational amplifiers, 352
system-level synthesis, 350
task planning, 368
telecommunications, 356–359
telephone operator scheduling, 423
textile industry, 409
texture filtering, 401
thermodynamic cycle of ideal turbojet engines, 387
thinned antenna arrays, 351
three-colored NCR boards, 404
time series prediction, 431
time-optimal following paths, 367
time-optimal trajectories, 367
time-series prediction, 430
time-tabling of classes, 422
topology design, 357
transonic airfoil design, 384
transonic wing design, 384
transport engineering, 378, 379
transportation plans, 377
traveling salesperson problem, 405
tree size control, 406
truck packing, 427
truss design, 371, 607
truss optimization, 370–372
turbine design, 412
vehicle routing, 380
vehicle scheduling, 379
vehicle water-pumps design, 412
ventricle 3D reconstruction problem, 395
vibrating platform design, 372, 373
VLSI, *see* very large scale integration
VLSI CAD, 411
VLSI cell placement, 348
VLSI design, 411
voltage reference circuit design, 349
war simulation, 404
warehouse management, 426
water quality control, 343
water quality management, 344
Wiener process, 362

wing design, 386
wing planform design, 384
wing shape optimization, 385
wing subsonic design, 385
word recognition, 403
approximated non-deterministic tree search, 617
aquifer remediation, 342
arch dam design
    evolutionary multiobjective optimization in, 435
Archive-based Scatter Search, 359, 569
areto memetic algorithm, 141
arithmetic crossover, 358, 394, 417
Arrow, Kenneth J., 30
    impossibility theorem, 538
articulated figure motion
    evolutionary multiobjective optimization in, 399
artificial immune system, 592
artificial neural networks, 395, 578
    in multi-objective optimization, 620
aspiration levels, 343, 525
assembly line balancing
    evolutionary multiobjective optimization in, 409
assembly sequences and plans
    generate and select optimal, 409
attainable criteria set, 372
attainable goals
    advantages, 526
    disadvantages, 526
attained set, 246
attainment function, 243, 249
attribute, 6
    definition, 6
aussian mutation, 399
automatically defined functions, 406
automotive industry
    evolutionary multiobjective optimization in, 409, 415
automotive steering box design, 409
autonomous aerial vehicles
    mission planning and routing, 381
autonomous agent
    for multiobjective optimization, 615
autonomous vehicle guidance, 361
availability allocation, 426
average crossover, 384

average linkage method, 98, 381
average Pareto front error, 265
average size of the space covered, 571
awari, 147
AWPSO, *see* Adaptive Weighted PSO

B cells, 605
B-splines, 376
backgammon, 147
backpropagation function, 409
backscattered waves, 385
backscattering of a reflector
    minimization of, 391
backtracking, 22
Baldwin effect, 133, 382
Baldwin, James Mark, 133
Baldwinian fitness assignment, 133
bank loan management, 430
bargaining model, 370
beam design, 373, 377
    evolutionary multiobjective optimization in, 372, 373
behavior-based system, 544
behavioral trait, 613
beryllium powder modeling, 375
best improvement repeated descent, 571
best-first chart parsing algorithm, 407
best-first search, 22
best-N selection, 384
bi-citeria knapsack problem, 567
biased fitness sharing, 527
bicycle design
    evolutionary multiobjective optimization in, 435
bicycle frame design, 435, 554
bin packing
    evolutionary multiobjective optimization in, 427
binary encoding, 237
binary preference relations, 528
binary representation, 350, 364, 366, 368, 371, 375, 377, 383, 385, 386, 409, 429
binary tournament selection, 391
bioinformatics, 407
bioremediation, 343
    evolutionary multiobjective optimization in, 343

BIRD, *see* best improvement repeated descent
blackboard architecture, 614
blade design, 375
blend crossover, 384, 394
blind search, 22
bloat
    in genetic programming, 439
bloat reduction
    evolutionary multiobjective optimization in, 406
BLX-α crossover, 344, 593
Boltzmann constant, 549
Borel, Félix Édouard Émile, 29
boundary discovery, 415
boundary mutation, 354, 417, 429
box counting dimension, 287
box packing
    evolutionary multiobjective optimization in, 427
brachytherapy
    evolutionary multiobjective optimization in, 394
branch and bound, 22, 420, 578
breadth-first search, 22
breast cancer database, 400
breast cancer prediction, 395
Broyden-Fletcher-Goldfarb-Shanno quasi-Newton algorithm, 394
Buchanan, John, 59
building block filtering, 101
building block placement problem
    evolutionary multiobjective optimization in, 411

C-MOGLS, *see* cellular multiple objective genetic local search applications, 419
C5.0, 400
CAD, *see* computer aided design
calculus-based search, 22
CAMOGA, *see* cellular automaton and genetic approach to multi-objective optimization
cancer classification, 432
cancer treatment, 393
    evolutionary multiobjective optimization in, 393, 394

768   Index

CANDA, *see* cellular automaton for network design algorithm
candidate list, 562
Cantor, Georg, 30
capital budgeting, 552
  Pareto Simulated Annealing in, 552
capital budgeting problem, 561
car design, 415
car engine design, 415
car sampling tests, 409
  evolutionary multiobjective optimization in, 409
cardiac diseases
  diagnosis of, 395
cardinality, 285
CAS, *see* collision avoidance action sequence
cascade compensator
  evolutionary multiobjective optimization in, 365
Cauchy mutation, 354
cell formation, 412
cellular automata, 344, 440
cellular automaton and genetic approach to multi-objective optimization, 344
cellular automaton for network design algorithm, 344
cellular genetic algorithm, 130, 419
cellular manufacturing, 412, 419, 552
  evolutionary multiobjective optimization in, 412
  Pareto Simulated Annealing in, 552
cellular multiobjective genetic algorithm, 616
cellular multiple objective genetic local search, 419
  applications, 419
chance-constrained solid transportation problem, 380
Charnes, Abraham, 30
chemical process modeling, 523
chemical process system
  modeling of, 390
chemistry
  evolutionary multiobjective optimization in, 390, 391
chemotherapeutic treatment, 393
chi-square distribution, 59

chromatic rectangle, 404
chromosome, 24
circuit design, 355
  evolutionary multiobjective optimization in, 355
city planning, 377
civil and construction engineering
  evolutionary multiobjective optimization in, 377
classification
  of MOEA techniques, 54
classification and prediction
  applications, 399
  evolutionary multiobjective optimization in, 399, 431–433
classification trees, 400
classifier systems, 24, 130, 400, 430, 440
clonal mutation operator, 592
clonal selection, 609, 610
  algorithm, 608
  principle, 609
  theory, 592
CLONALG, *see* clonal selection algorithm
clone identification, 267
clones, 352
clustering, 399, 407, 568, 586, 592
  applications, 399
  evolutionary multiobjective optimization in, 399
clusters, 350, 556
ClustMPSO, 592
CMA-ES, *see* covariance matrix adaptation evolution strategy
CMOIA, *see* constrained multi-objective immune algorithm
co-evolutionary genetic algorithm, 525
co-evolutionary hierarchical pMOEA, 459
co-evolutionary multiobjective genetic algorithm, 616
  applications, 386
co-evolutionary shared niching, 384
coarse-grained parallelism, 455
code growth in GP
  evolutionary multiobjective optimization in, 405
code size control

evolutionary multiobjective optimization in, 405
coevolution, 146
 competitive, 147
 cooperative, 147
coevolutionary genetic algorithm, 148, 159
coevolutionary multi-objective evolutionary algorithms, 147
coevolutionary multi-objective genetic algorithm, 163
coevolutionary shared niching, 153
coevolutionary sharing, 153
collision avoidance
 evolutionary multiobjective optimization in, 368
collision avoidance action sequence, 368
coloring problems
 evolutionary multiobjective optimization in, 404
combinational circuit design, 306, 355
combinatorial MOEA test functions, 220
combinatorial optimization, 346, 388, 534
 problem
 definition, 220
COMPETants, *see* competing ant colonies
competing ant colonies, 581
competitive fitness, 147
 comprehensive, 147
 minimal, 147
competitive species, 345
completeness, 285
complex objects
 qualitative descriptions of, 430
composite plate optimization
 applications, 373
 evolutionary multiobjective optimization in, 373
composite utility function, 527
compromise programming, 33, 374, 535, 536, 555
compromise solution, 380
computable difference, 237
computational fluid dynamics, 383, 462
computational grid, 237
computer aided design, 411

evolutionary multiobjective optimization in, 411
computer generated animation
 applications, 399
 evolutionary multiobjective optimization in, 399
computer graphics
 evolutionary multiobjective optimization in, 399
computer immunology, 605
computer science
 evolutionary multiobjective optimization in, 388, 398–407, 432
computer vision
 evolutionary multiobjective optimization in, 401, 402
concavity, 9
concordance, 42
concordance index, 532
cone, 292
configuration problems, 374
conflict analysis model, 45, 542
conformational search, 391
 evolutionary multiobjective optimization in, 391
connectedness, 137
CONOPT, 430
constant parsimony pressure, 405
constellation design, 382
constrained multi-objective immune algorithm, 608
constrained test functions, 189
constraint-handling, 116, 544
 in MOPs, 113
 in multiobjective Tabu search, 563
 using nondominance, 188
constraints
 definition, 5
 equality, 6
 explicit, 6
 implicit, 6
 inequality, 6
 overconstrained problem, 6
construction time-cost trade-off problems, 377
contact theorem, 562
containership layouts
 preplanning of, 346

contaminated groundwater aquifers remediation, 342
continuously updated fitness sharing, 95, 342
control
   applications, 364–367
   of a civil structure
      applications, 374
      evolutionary multiobjective optimization in, 374
control of elevators, 362
control system design, 365
   applications, 365
   evolutionary multiobjective optimization in, 364–366
controller
   design, 364, 366, 368, 379
      applications, 364
      evolutionary multiobjective optimization in, 363, 364, 379
   design of a finite word-length feedback, 364
convergence
   evolutionary algorithm, 288
convex hull, 67
convex set
   definition, 9
   example, 9
convexity, 8
Cooper, William Wager, 30
cooperative coevolutionary genetic algorithms, 149
cooperative genetic algorithm, 422
cooperative search, 614
coordination of distributed agents, 434
   applications, 434
copper powder modeling, 375
Corana functions, 177
core-based single-chip system synthesis, 350
   evolutionary multiobjective optimization in, 350
covariance matrix adaptation evolution strategy, 130, 387
covariance matrix adaptation method, 413
coverage difference of two sets, 425
coverage of two sets, 410, 571, 581
coveringMOPSO, 587

craziness, 590
creep mutation, 391
criterion, 6
critical event design, 566
critical path method (CPM), 377
crossover, 25
   intelligent, 410
   one-point, 370
   order-based uniform, 398
   partially mapped, 418
   unimodal normal distribution, 399
crowding, 311, 351, 568, 611
   applications, 400
crowding factor, 590
crowding model, 424
cryogenic rocket engines, 384
cryptography, 441
CSN, *see* coevolutionary shared niching
cultural algorithm, 603, 613
   in multiobjective optimization, 614
cultured differential evolution, 603
cumulative distribution function, 234
customer patterns prediction
   evolutionary multiobjective optimization in, 431
cutting problem
   evolutionary multiobjective optimization in, 413, 414
cvMOPSO, *see* coveringMOPSO

D-MODE, *see* discrete multi-objective differential evolution
data extraction, 385
data mining, 406
data sampling
   evolutionary multiobjective optimization in, 388
data structures evolution, 406
   evolutionary multiobjective optimization in, 406
databases
   knowledge extraction, 440
Davidon-Fletcher-Powell method, 392
DC railway system, 378
death penalty, 358, 377, 593
Deb-Thiele-Laumanns-Zitzler (DTLZ) test problems, 200
deceptive functions, 177
decision maker, 5, 7, 515

operational attitude, 517
decision making, 409, 515
    social choice, 545
    system, 564
decision tree induction, 399
    evolutionary multiobjective optimization in, 399
decision variables
    definition, 5
degree of domination, 560
degrees of freedom, 6
Delta II 7925, 382
deme, 406
DEMO, *see* differential evolution for multi-objective optimization
DEMORS, *see* differential evolution for multiobjective optimization with rough sets
depth-first search, 22
descent algorithm, 554
design and manufacture
    evolutionary multiobjective optimization in, 371, 376, 408–417, 427
design of H_∞ controllers
    evolutionary multiobjective optimization in, 363
design of active magnetic bearing controllers
    evolutionary multiobjective optimization in, 363
design of fuzzy controllers
    evolutionary multiobjective optimization in, 361
design of multiplierless IIR filters
    evolutionary multiobjective optimization in, 353
design unification of linear control systems, 365
    evolutionary multiobjective optimization in, 365
deterministic search, 22
differential evolution, 376
    applications, 375
    multi-objective, 356, 416
    multiobjective, 594
    parallel, 599
differential evolution for multi-objective optimization, 599, 600

differential evolution for multiobjective optimization with rough sets, 602
diffusion, 451
diffusion genetic algorithm
    applications, 395
diffusion implementation issues, 499
diffusion pMOEA model, 458
diffusion pMOEAs, 473
digital filter design
    evolutionary multiobjective optimization in, 354
digital phase shifters, 351
dimensionality curse, 537
diploid genetic algorithm, 532
diploidy, 119, 408, 618
directional convergence and spread, 599
disassortative mating, 404
discordance, 42
discordance index, 43, 532
discrete multi-objective differential evolution, 416
displaced ideal, 33
distance preserving crossover, 141, 405
distance-based Pareto genetic algorithm, 63, 413
distributed cooperative coevolutionary algorithm, 161
distributed database management
    evolutionary multiobjective optimization in, 359
distributed genetic algorithm, 469
distributed object systems
    applications, 404
distributed Q-learning, 582
distributed spacing, 265
distributed systems
    applications, 404
    evolutionary multiobjective optimization in, 404
distribution system, 344
    design, 426
        evolutionary multiobjective optimization in, 426
diversity, 178, 310, 427
    as an objective in a MOEA, 129
    using entropy, 337
    using the immune system, 606
diversity preservation, 81
diversity techniques

shaking, 352
divided range multi-objective genetic
    algorithm, 469
DM, see decision maker
DNA
    computing, 440
    library, 607
    sequence analysis, 407
dominance, 119
    weak, 517
dominance compliant, 245
dominance count, 79
dominance depth, 79
dominance rank, 79
dominated tree, 586
dominated-space metric, 425
domination graph, 318
    irreducible, 319
domination probability, 530
domination-based fitness sharing, 400
domination-free quadtrees, 125
downhill simplex search, 345
DPGA, see distance-based Pareto
    genetic algorithm
DQL, see distributed Q-learning
DSP algorithms
    evolutionary multiobjective optimization in, 398
ductile iron casting
    optimization, 412, 413
duplicates removal, 389
dynamic compromise programming, 536
dynamic environment, 231, 615
    MOEAs in, 440
dynamic neighborhood, 586
dynamic neural networks, 362
dynamic sharing, 365
dynamic weight aggregation
    applications, 387
dynamic weighted aggregation, 534

EA, see evolutionary algorithm
earthquake engineering
    evolutionary multiobjective optimization in, 374
easy MOPs, 186
EC, see evolutionary computation
ecological model assessment, 396
ecological process model fitting, 397

ecology
    evolutionary multiobjective optimization in, 396, 397
economic equilibrium, 29
economic load dispatch, 349, 589
economics
    evolutionary multiobjective optimization in, 430
Edgeworth, Francis Ysidro, 10, 29
effectiveness, 234
efficiency, 234
efficiency preservation, 292
efficiency preserving MOEA, 293
efficient set, 292
efficient solutions, 11
efficient vector, 30
ELECTRE, 42, 45, 518
ELECTRE I, 42, 531
    applications, 378
ELECTRE II, 44
ELECTRE III, 44, 532
ELECTRE IS, 44
ELECTRE IV, 44
ELECTRE TRI, 44
electrical and electronics engineering
    evolutionary multiobjective optimization in, 348–356
electrical impedance tomography
    problem, 392
electrified railway systems, 378
electromagnetic design, 608
electromagnetic devices design, 350, 351, 354
    evolutionary multiobjective optimization in, 351
electromagnetic superconducting device, 351
electromagnetic suspension system
    design, 366
electromechanical systems, 355
    evolutionary multiobjective optimization in, 355
elevator car routing problem, 362
elimination and choice translating
    algorithm, see ELECTRE
elite clearing mechanism, 410
elitism, 103, 343, 355, 371, 379, 380, 383, 386, 400, 405, 413, 418–421, 424, 425, 432

elitist NSGA, 420
elitist preservation selection, 525
elitist recombinative multiobjective genetic algorithm with coevolutionary sharing, 153
elitist selection pressure, 343
ELSA, 431, *see* evolutionary local selection algorithm, *see* evolutionary local search algorithm
endosymbiosis, 144, 150
energy consumption prediction, 432
engineering configuration problems, 374
entreprise planning, 416
entropy, 337, 371
enumerative search, 21
environmental engineering
　evolutionary multiobjective optimization in, 341–345, 347, 378
environmental modeling
　evolutionary multiobjective optimization in, 388
EP, *see* evolutionary programming
epistasis, 230, 301
equivalence, 285
equivalence class sharing, 94, 242, 313, 352, 392
error ratio, 256
ES, *see* evolution strategy
escape operator, 422
estimation of distribution algorithms, 130
estuary, 344
euclidean space, 7
EVAMIX, 542
evenness coverage, 389
evolution program, 380
evolution strategy, 23, 239, 359, 414
　applications, 351, 352, 354, 368, 378, 413, 425
　convergence, 291
　multiobjective, 364, 368
evolutionary algorithm, 1, 23
evolutionary computation, 23
　terminology, 24
evolutionary dynamic weighted aggregation algorithm, 137
evolutionary game theory, 543
evolutionary local search algorithm, 616

evolutionary local selection algorithm, 432
　applications, 432
evolutionary operators, 25, 449
evolutionary programming, 349, 614
Evolutionary Standardized-Objective Weighted Aggregation Method, 362
Evolving Objects, 276
EVOP, *see* evolutionary operators
exchange mutation, 417
expected return, 429
expected utility maximization, 542
expected value selection, 525
experimental goals, 234
expert system, 526
exploration vs. exploitation, 27
exploratory multiobjective evolutionary algorithm, 564
exponential selection, 357
extent of coverage, 389
external file of nondominated vectors, 425
extinction, 380
extinctive selection, 401
extraction of munition aerodynamic characteristics, 385
extrapolated path relinking, 566

face detection system, 403
facial modeling, 402
facility layout, 414, 467
facility layout problem
　evolutionary multiobjective optimization in, 425
facility location problem, 425
　evolutionary multiobjective optimization in, 425
facility management
　evolutionary multiobjective optimization in, 424
fair population-based evolutionary multiobjective optimizer, 321
fast messy genetic algorithm, 101, 506
fault diagnosis, 362, 432
　evolutionary multiobjective optimization in, 362
fault diagnosis problems, 468
fault tolerant system design, 352

evolutionary multiobjective optimization in, 352
favor relationship, 530
    potential problems, 531
feasible descent direction, 60
feature selection, 403, 616
    evolutionary multiobjective optimization in, 432
    in biological sequences, 407
feature subset selection, 399
feedforward neural network, 376
field programmable gate array, 350
field programmable transistor array, 352
FIFO queue, 406
filled function acceleration technique, 351
finance
    evolutionary multiobjective optimization in, 429, 430
find only and complete undominated sets, 406
fine-grained parallelism, 458
finite element model tuning, 434
Finite Impulse Response (FIR) filters, 353
finite word-length feedback controller
    design of a, 364
FIR filter design, 439
FIRD, see first improvement repeated descent
first improvement repeated descent, 571
Fisher permutation test, 271
fishery bioeconomic model optimization, 430
    evolutionary multiobjective optimization in, 430
fitness, 25
fitness function, 23
fitness functions used with MOEAs, 306
fitness landscape, 231, 300
fitness landscape analysis, 139
fitness proportional, 25
fitness scaling, 241, 423
fitness sharing, 64, 242, 357, 359, 364, 367, 372, 382, 384, 389, 404, 409, 417, 420, 552, 591, 607, 610
    continuously updated, 95, 342
    domination-based, 400
    implicit, 354

phenotypic, 375
flat crossover, 361
Fletcher-Powell function, 179
Fletcher-Reeves algorithm, 394
flexible generic parameterized facial model, 402
flexible manufacturing
    evolutionary multiobjective optimization in, 410
flexible process sequencing
    evolutionary multiobjective optimization in, 409
flight instructor scheduling problem, 563
flip mutation, 394
floating point representation, 349, 384
flow crossover, 358
flowshop scheduling, 418–420, 570, 578, 580
    evolutionary multiobjective optimization in, 418–420
fluidic microchannel design, 413
fmGA, see fast messy genetic algorithm
FOCUS, see find only and complete undominated sets
food industry, 408
    evolutionary multiobjective optimization in, 408, 417
force allocations for war simulation, 404
forest management
    evolutionary multiobjective optimization in, 425
foundry processes, 413
    evolutionary multiobjective optimization in, 413
four-cylinder gasoline engine
    design, 415
FPGA, see field programmable gate array
free-form packing problems, 374
freedom
    degrees of, 6
Freudenstein-Roth functions, 177
FTS, see fuzzy tournament selection
fuel consumption, 415
fuel economy
    evolutionary multiobjective optimization in, 415
full stern submarines

feasibility, 346
functional brain imaging, 395
future applications
  evolutionary multiobjective optimization, 434, 435
future applications of evolutionary multiobjective optimization, 434
fuzzy automatic train operation, 596
fuzzy boundary local perturbation, 365
fuzzy controller
  design, 361
fuzzy decision table, 364
fuzzy logic, 220, 343, 347, 348, 357, 361, 364, 368, 369, 375, 379, 380, 430, 536
  applications, 366
fuzzy logic traffic signal controller, 379
fuzzy modeling, 369
fuzzy ranking, 364, 380
fuzzy rule, 426
fuzzy satisficing method, 525
fuzzy set-based aggregating function, 534
fuzzy system design, 368, 369, 411
  evolutionary multiobjective optimization in, 368, 369, 411
fuzzy tournament selection, 361
  applications, 361
Fuzzy-Pareto-Dominance, 402

GA, *see* genetic algorithm
GA with a target vector approach
  applications, 353, 377, 422
GA with compromise programming
  applications, 374
GA with lexicographic ordering
  applications, 361
GA with linear aggregating function, 358, 426
GA with Pareto ranking, 392
GAIN, *see* genetic algorithm running on the internet
game of tag, 147
game playing, 441
game theory, 29, 30, 386, 543, 616
  applications, 370, 383, 391
GAMS, *see* general algebraic modeling system
Gantt diagram optimization, 398

GAP, *see* generalized analysis of promoter
gas supply network optimization, 345
gas turbine blades design
  evolutionary multiobjective optimization in, 387
gas turbine design
  evolutionary multiobjective optimization in, 412
gas turbine engine
  design of its control system, 364
gas turbine stator blades
  optimization, 387
gaussian mutation, 361, 394, 398, 401, 525, 608
GD3, 604
GDE, *see* generalized differential evolution
GDEA, *see* generalized data envelopment analysis
GE-HPGA, *see* grid enabling hierarchical parallel genetic algorithm
gear-box design, 375
gene, 24
  conversion, 607
  expression, 407
  inversion, 607
  shift, 607
general algebraic modeling system, 430
general MOP
  formal definition, 8
general multi-objective program, 464
general multiobjective evolutionary algorithm, 108
generalized analysis of promoter, 407
generalized data envelopment analysis, 525
generalized differential evolution, 600
generalized multi-objective evolutionary algorithm, 410
generation gap, 311
generational distance, 256, 569
generational nondominated vector generation, 266
generational population knowledge, 234
generic attainment function, 245
generic MOEA, 86
genetic algorithm, 23, 344, 429, 578
  aggregating function, 552

with fuzzy logic, 375
with goal programming, 389
genetic algorithm running on the internet, 460
genetic diversity evolutionary algorithm, 129
genetic drift, 153, 311, 343
genetic pMOEA, 503
genetic programming, 24, 362, 390, 399, 406, 439
  applications, 399, 404, 406, 431
  evolutionary multiobjective optimization in, 405, 406
genetic symbiosis algorithm, 155
genMOGA pseudo code, 88
GENMOP, 108, see general multi-objective program
GENOCOP III, 430
genotype, 24
  in evolutionary multiobjective optimization, 54
genotypic clustering, 399
genotypic sharing, 312, 400, 419
geographic information system, 347
geography
  applications, 389
  evolutionary multiobjective optimization in, 388, 389
geometric crossover, 394
geometric temperature reduction, 571
germinal center, 607
GGA, see grouping genetic algorithm
  applications, 409
Gibbs entropy, 130
GIS, see geographic information system
global convexity, 134, 620
global criterion method, 32, 368, 569
global minimum
  definition, 4
global optimization problem
  definition, 4
global preferences, 532
global repository, 590
GMDH, see group method of data handling
GMDH-type neural networks, 387
GMOEA, see generalized multi-objective evolutionary algorithm
goal attainment, 35, 362, 523

applications, 353, 362
versus multi-objective differential evolution, 603
goal definition, 524
goal programming, 30, 34, 57, 379, 389, 523, 524, 534
  applications, 371, 390
  lexicographic, 35
Goldstein-Price functions, 177
goundwater monitoring wells
  optimal location, 341
GP, see genetic programming
gradient information
  coupled to a MOEA, 142
graph-based encoding, 389
graphical user interface for multi-objective optimization, 275
GRASP, see greedy randomized adaptive search procedure
  multi-objective, 620
Gray coding, 350, 366, 370, 383, 398, 412
greedy algorithm, 22, 564
greedy heuristic, 423
greedy randomized adaptive search procedure, 620
greedy reduction algorithm, 545
grid enabling hierarchical parallel genetic algorithm, 474
Griewank quartic, 177
groundwater monitoring networks
  evolutionary multiobjective optimization in, 343
groundwater monitoring wells
  optimal location, 342
groundwater pollution containment, 341
groundwater remediation, 342
  evolutionary multiobjective optimization in, 342
group decision making, 54, 538
group method of data handling, 387
group technology, 412
group utility, 33
grouping genetic algorithm, 409
grouping problem
  evolutionary multiobjective optimization in, 427
GUESS method, 59

guided multi-objective evolutionary algorithm, 527
guillotine cutting problem
  evolutionary multiobjective optimization in, 413
Guimoo, *see* graphical user interface for multi-objective optimization

H_∞, 366
halftoning problem, 401
Hamming cliffs, 237
Hamming distance, 312
handwritten word recognition, 403
hard selection, 27
hardware-software embedded systems co-synthesis, 350
  evolutionary multiobjective optimization in, 350
Hausdorff, Felix, 30
heating, ventilating and air conditioning system, 377
heavy chain, 607
helicopeter flat panels design, 383
helicopter design, 415
heuristic concentration, 620
heuristic crossover, 354, 413, 417
heuristic knowledge, 354
heuristic learning, 411
heuristic tie-breaking crossover, 392
HGA, *see* hierarchical genetic algorithm
  applications, 356
Hidden Markov Models, 403
hierarchical decision system, 424
hierarchical genetic algorithm, 356
hierarchical hybrid pMOEA models, 459
hierarchical systems, 375
hierarchical tree representation, 362
high performance multi-objective evolutionary algorithm, 130
high-dimensional problems, 22
high-lift airfoil design, 386
hillclimbing, 22, 351, 527, 565
honey bee algorithm, 620
Hooke and Jeeves, 562
hot rolling process, 421
HPMOEA, *see* high performance multi-objective evolutionary algorithm

HTBX, *see* heuristic tie-breaking crossover
human-like motion, 399
humanoid figure
  animation of, 399
Hurwitz, Leonid, 30
HVAC system, *see* heating, ventilating and air conditioning system
hybrid approaches, 564, 570
  genetic algorithm and simulated annealing, 349
hybrid electric vehicle
  design, 415
hybrid metaheuristic
  applications, 565
hybrid of neural network and genetic algorithm
  applications, 401
hybrid representation, 367, 374
hybrid selection, 404
  applications, 142
hydraulic engineering
  evolutionary multiobjective optimization in, 341–345, 383, 577
hydro-generators design, 562
hydrologic model calibration, 345
  evolutionary multiobjective optimization in, 345
hydrology
  evolutionary multiobjective optimization in, 345
hyperarea ratio, 258
hypercrowns, 562
hyperheuristic, 173
hypermutation, 607, 609
hypervolume, 257, 359, 569

ID3, 399
IDCMA, *see* immune dominance clonal multiobjective algorithm
ideal vector, 526
  definition, 8
identification of gene modules, 407
IFMOA, *see* immune forgetting multiobjective optimization algorithm
IGA, *see* interactive genetic algorithm
IIM, *see* iterative improvement method

778   Index

IIR filters
   design, 353
Illinois Genetic Algorithms Laboratory, 273
ILP, *see* integer linear programming
image compression, 350
image halftoning, 401
image processing
   evolutionary multiobjective optimization in, 392, 401
image segmentation
   evolutionary multiobjective optimization in, 401
immediate successor relation crossover, 410
immigration, 380, 426
immune differential degree, 609
immune dominance, 609
immune dominance clonal multiobjective algorithm, 609
immune forgetting multiobjective optimization algorithm, 610
immune forgetting operator, 610
immune system, 420, 605, 606
   applications, 375, 607
   multiobjective, 607
IMOEA, *see* incremented multi-objective evolutionary algorithm
Implanting Block of Cells crossover, 352
implicit fitness sharing, 354
impossibility theorem, 539
imprecisely specified multi-attribute value theory, 527
incest prevention, 385
incomparability, 517
incremental genetic algorithm
   applications, 349
incremental learning, 126
incremental multiple objective genetic algorithm, 126
incremented multi-objective evolutionary algorithm, 365
independent sampling techniques, 72
indicator-based evolutionary algorithm, 130, 275, 407
indifference, 543
indirect representation, 422
indiscernibility interval method, 367, 373

individual, 23, 24
induction motor design, 352
   evolutionary multiobjective optimization in, 352
inductive learning, 432
industrial applications
   evolutionary multiobjective optimization in, 422
industrial nylon 6 semibatch reactor optimization, 412
industrial scheduling, 578
industry
   evolutionary multiobjective optimization in, 427
inferential estimator
   development, 390
inferiority index, 373
initialization phase, 101
insertion mutation, 420
integer linear programming, 580
integer programming, 525, 559
integer representation, 353, 368, 373, 377, 380, 423
integrated circuit placement, 411
   evolutionary multiobjective optimization in, 411
integrated convex preference, 421
intelligent crossover, 402, 410
intelligent genetic algorithm, 402
intelligent particle swarm optimization, 591
intensities of emission lines of trace elements
   optimization of, 390
interactive articulation of preferences, 523
interactive genetic algorithm, 399
   with co-evolving weighting factors, 157
interactive preference articulation
   advantages, 518
   disadvantages, 519
intermediate crossover, 367
intermediate recombination, 364, 413
interval criterion weights, 534
interval programming problem
   evolutionary multiobjective optimization in, 380
intransitivities, 537

intrusion detection, 434
inverse kinematics of a robotic manipulator
    solution of the, 367
inverse Polish notation
    phenotype using, 411
inversion, 410, 420, 423, 525
inverted generational distance, 257
inverter-permanent magnet motor-reducer-load association
    design, 355
investment portfolio optimization, 429
    evolutionary multiobjective optimization in, 429
IPSO, see intelligent particle swarm optimization
iron blast furnace, 376
irregular problems, 22
iSIGHT, 275
island, 451
    implementation issues, 493
    model, 316
    pMOEAs, 465
isoefficiency, 485
iterated prisoner's dilemma, 147
iterative genetic algorithm
    applications, 374
iterative improvement method, 418

J-measure, 406
Java retrovirus-inspired MISA, 434
Jevons, William Stanley, 29
job shop scheduling, 346, 420, 421, 570
    evolutionary multiobjective optimization in, 398, 420, 421
jREMISA, see Java retrovirus-inspired MISA
jump mutation, 391
jump search, 620
just-in-time sequencing problem, 578
juxtapositional phase, 101, 243

K-means algorithm, 403, 592
Karnaugh maps, 355
KEA, see kit for evolutionary algorithms
kit for evolutionary algorithms, 274
knapsack problem, 142, 536, 556
    bicriteria, 567

evolutionary multiobjective optimization in, 344
    multiobjective, 139, 560
knees
    in MOEAs, 545
knowledge extraction, 432
    evolutionary multiobjective optimization in, 432
Koopmans, Tjalling C., 30
Kramer choice function, 566
Kreisselmeir-Steinhauser function, 383
Kruskal Wallis Test, 271
Kuhn, Harold W., 30
Kuhn-Tucker conditions, 30

Lamarck, Jean Baptiste Pierre Antoine de Monet, 133
Lamarckian fitness assignment, 133
Lamarckian Genetic Algorithm, 592
Lamarckism, 133
laminate composite panels, 569
laminated composites
    optimization, 372
lamp animation, 399
lamp design, 354
    evolutionary multiobjective optimization in, 354
land use, 377
    applications, 389
    planning, 389
        applications, 389
lateral interference, 564, 565
launch conditions, 385
layout problems, 222
leading ones-trailing zeroes, 320
learning
    applications, 368
    evolutionary multiobjective optimization in, 368
learning object-oriented problem solver, 292
learning speed, 400
leg mechanism, 375
Levy functions, 177
lexicographic GA
    applications, 348, 423
lexicographic goal programming, 59

lexicographic ordering, 36, 63, 67, 361, 401, 439, 523, 524, 534, 553, 578, 586
  advantages, 66
  applications, 361
  disadvantages, 66
  scatter search, 567
light chain, 607
linear aggregating function, 69, 358, 362, 414, 569
  advantages, 69
  disadvantages, 69
linear control systems, 365
linear fitness scaling, 350, 425
linear matrix inequalities, 364
linear programming, 24, 344, 555
linear programming relaxation, 525
linear ranking, 27, 351, 366, 373, 411
linear representation, 400
linear scaling, 370, 418, 432, 525
linear transportation problem, 380
  evolutionary multiobjective optimization in, 380
linear weighting function, 527
linguistic function approximation
  evolutionary multiobjective optimization in, 432
linguistic rule extraction
  evolutionary multiobjective optimization in, 400
LMI, see linear matrix inequalities
local geographic selection genetic algorithm, 379, 465
local search, 132, 142, 330, 395, 418, 419, 421, 527, 569, 578
  importance, 426
  tiered, 140
local selection, 383, 432
  role in multiobjective optimization, 616
local utility approach, 413, 527
locus, 24
logistic regression, 431
LOOPS, see learning object-oriented problem solver
low-power operational amplifiers
  synthesis, 353
LS-1, 400

LSGA, see local geographic selection genetic algorithm
  applications, 379
LUTA, see local utility approach, 527

M-PAES, 97, 124, 136, see memetic Pareto archived evolution strategy, 238
M-Pareto optimality, 525
M-scatter search, 567
M5, 400
machine design, 413
machine job scheduling, 578
machine learning, 432
  evolutionary multiobjective optimization in, 400
MACS-VRPTW, 577, see multiple ant colony system for vehicle routing problems with time windows
maglev vehicle, see magnetically levitated vehicle
magnetically levitated vehicle, 366
  design, 366
  electromagnetic suspension system for a, 366
management
  evolutionary multiobjective optimization in, 424, 426
Mann-Whitney rank-sum test, 270
manufacturing
  evolutionary multiobjective optimization in, 412
mappa, 613
marine vehicle design, 346
  evolutionary multiobjective optimization in, 346
Markov chain, 289
Markowitz mean-variance model, 429
master-slave, 451
  pMOEA, 460, 463
  pMOEA implementation issues, 491
  with local cultivation, 464
matching
  of individuals, 373
mathematical complexity
  of MOEAs, 310
mathematical programming, 24
mating groups, 316

mating restrictions, 221, 241, 313, 315, 345, 350, 354, 364, 372, 386, 404, 421
  importance, 426
MATLAB, 273
matrix encoding, 380, 426
mature antibodies, 607
MAUT, *see* multi-attribute utility theory
  criticism, 517
  intransitivities, 517
max-min ant system, 620
maximal symmetric excursion, 353
maximim programming, 517
maximin fitness function, 377
maximinPSO, 588
maximum independent set (clique) problem, 222
maximum Pareto front error, 264
mBOA, *see* multi-objective Bayesian optimization algorithm
MCDM, *see* multi-criteria decision making
MDQL, *see* multiobjective distributed Q-learning
  applications, 583
measurement, 515
mechanical and structural engineering
  evolutionary multiobjective optimization in, 374–376
mechatronic design problem, 603
medicine
  evolutionary multiobjective optimization in, 393–395
meiosis, 128
membership function, 361, 525, 589
  multiobjective, 366
memetic algorithm, 97, 382, 395, 420
  applications, 142
  definition, 132
memetic MOEAs, 330
memetic Pareto archived evolution strategy, 139
memetic Pareto artificial neural network, 396, 598
memory pool, 607
MEMS, *see* microelectrical mechanical systems
Mendel, Gregor, 618

laws, 618
mendelian multi-Objective genetic algorithm, 618
Menger, Carl, 29
mesh communication networks, 357
Message Passing Interface, 238, 387
messy genetic algorithm, 238, 344
metabolic pathway data, 407
metaheuristic, 63
metallurgy
  evolutionary multiobjective optimization in, 375
metrics, 234, 236, 267
  coverage, 426
  integrated convex preference, 421
metrics for pMOEAs, 486
metropolitan mobile ad-hoc network, 359
MICCP, *see* mixed integer chance constrained programming
Michigan-style machine learning, 432
micro genetic algorithm
  for multiobjective optimization, 63, 102, 125, 609
micro genetic algorithm 2, 63
microchannel design, 413
  evolutionary multiobjective optimization in, 413
microelectrical mechanical systems, 440
microprocessor design, 354
  evolutionary multiobjective optimization in, 354
microwave absorbers design, 351
  evolutionary multiobjective optimization in, 351
middling, 73
migration, 456
military airlift scheduling
  evolutionary multiobjective optimization in, 423
military applications, 468
  evolutionary multiobjective optimization in, 404, 423
military land moves scheduling
  evolutionary multiobjective optimization in, 423
min-max GA
  applications, 364
min-max optimization, 38, 524

782 Index

minimal element, 285
minimum description length, 431
minimum penalty rule, 114
minimum spanning tree, 222
   evolutionary multiobjective optimization in, 359
MINSGA, *see* modified Illinois NSGA
   applications, 382
MISA, 433, *see* multiobjective immune system algorithm, 610, 611
missile trajectory tracking
   evolutionary multiobjective optimization in, 367
mission ready resource, 223
mitosis, 128
mixed integer chance constrained programming, 341
mixed model assembly lines, 410
MMOSGA, *see* mendelian multi-objective genetic algorithm
MNC GA, *see* multi-niche crowding genetic algorithm
   applications, 342
MO-Turtle GA, 352
MOAMP, 569
MOAQ, *see* multi-objective ant-Q
   applications, 577
mobile networks, 357
mobile robots, 544
   applications, 361
mobile telecommunication network, 357
MOCell, *see* multiobjective cellular genetic algorithm
MOCOM-UA, *see* multi-objective complex evolution
   applications, 345
MOCSA, *see* multiobjective clonal selection algorithm
MODE, *see* multi-objective differential evolution, 601
MODE/D, 604
model optimization, 430
modified Illinois NSGA, 382
modified strength Pareto evolutionary algorithm, 415
MODM, *see* multiple objective decision making
MOEA, *see* multi-objective evolutionary algorithm

challenging functions, 179
comparison, 236
complexity, 318
computational cost, 326
convergence, 288
   with probability one, 290
generic algorithm, 78
goals, 78
metrics, 264
Primary goals, 3
storage requirements, 318
test function suite issues, 176
test suites, 175
theoretical issues, 300
used with noisy functions, 336
MOEDAs, *see* multi-objective estimation of distribution algorithms
MOEP, *see* multi-objective evolutionary programming
   applications, 349
MOGA, *see* multiple objective genetic algorithm, 158, 238, 239, 345, 364, 554, 556
   applications, 345, 346, 350, 352, 355, 361–365, 367, 368, 372, 375, 377, 379, 384, 389, 390, 394, 402, 405, 406, 412, 413, 417–419, 425
   compared to multiobjective simulated annealing, 555
   definition, 88
   parallel implementation, 362
   preferences, 523
   pseudo code, 89
   vs. EMOEA, 564
   with elitism, 352
MOGA/NPGA hybrid, 365
MOGADES, *see* Multi-Objective Genetic Algorithm with Distributed Environment Scheme
MOGLS, *see* multiple objective genetic local search, 426, 580
   applications, 405
MOGP, *see* multi-objective genetic programming, *see* multi-objective genetic programming
   applications, 355, 362, 390
MOIA, *see* multi-objective immune algorithm
molecular docking problem, 592

MOMGA, see multi-objective messy genetic algorithm, 99, 238, 364
  pseudo code, 98
MOMGA-II, 63, 101, 238
  pseudo code, 101
MOMGA-III, 63, 101, 238
MOMSLS, see multiple start local search algorithm
MONACO, see multi-objective network optimization using ant colony optimization
  static version, 579
monopolistic competition, 153
monotonicity, 289
Monte Carlo method, 23, 368, 549
  applications, 372, 413
MOP, see multiobjective optimization problem
  characteristics, 195
  domain characteristics, 180
  example, 19
    plane truss, 19
  global minimum, 13
  global minimum solution set, 13
  global optimization problem, 13
  landscape
    stretching, 231
  Pareto front determination, 237
MOPSO, 349, see multi-objective particle swarm optimizer
Morgenstern, Oskar, 30
MOSA, 140, see multi-objective simulated annealing, 550
  applications, 351, 608
MOSGA, see multi-objective struggle genetic algorithm
MOSS, see multi-objective scatter search, 566, 569
motorcycle frame design, 374
motorway routes planning
  evolutionary multiobjective optimization in, 378
MOTS, 560, see multi-objective Tabu search, 570
MOTS*, 560
  compared to PSA, 570
moving boundaries process, 365
MP-Testdata, 222

MPANN, see memetic Pareto artificial neural network, 598
MPGA, 421, see multi-population genetic algorithm
MPI, see Message Passing Interface, 387
MPICH library, 461
MRCD, see multiobjective robust control design
MRCD GA
  applications, 364
MRMOGA, see multiple resolution multiobjective genetic algorithm
MSGA, see multisexual genetic algorithm
MSPC-GA, see multi-sexual-parents-crossover genetic algorithm
MSPEA, see modified strength Pareto evolutionary algorithm
multi ant colony, 577
multi-agent system, 398
multi-attribute utility theory, 39, 516
multi-component maintenance scheduling, 424
multi-criteria decision making, 52, 66, 409, 516
multi-criteria decision-aid, 516
multi-dimensional surface, 181
multi-layered perceptron, 433
multi-membered evolution strategy, 58
multi-modal multi-objective optimization, 395
multi-niche crowding genetic algorithm, 342
multi-objective ant-Q, 575
multi-objective artificial immune systems, 608
  common framework, 611
multi-objective Bayesian optimization algorithm, 63, 94
multi-objective covariance matrix adaptation evolution strategy, 130
multi-objective differential evolution, 599, 604
  applications, 356, 603
  discrete, 416
multi-objective estimation of distribution algorithms, 319

multi-objective evolutionary algorithm, 2
multi-objective evolutionary programming, 349
Multi-Objective Evolving Objects, 276
Multi-Objective Genetic Algorithm with Distributed Environment Scheme, 470
multi-objective genetic programming, 362, 390, 405, 406
multi-objective immune algorithm, 607
multi-objective Java Genetic Algorithm, 279
multi-objective linear programming, 556
multi-objective messy genetic algorithm, 63, 99
multi-objective network optimization using ant colony optimization, 579
multi-objective optimizers, 245
multi-objective particle swarm optimization
 applications, 587, 593
 survey, 619
multi-objective particle swarm optimizer, 337, 349, 584
multi-objective scatter search, 407, 566
multi-objective simulated annealing, 550, 552, 554–556
 comparative study, 555
 compared to MOGA, 555
 survey, 619
multi-objective struggle genetic algorithm, 105
multi-objective Tabu search, 556, 560, 561, 563, 565
 applications, 562, 563, 570
 constraint-handling, 563
multi-objective traveling salesperson problem, 561
multi-objectivity, 553
multi-objectivizing, 126
multi-pheromone trail, 579
multi-point crossover, 352, 367
multi-point recombination, 422
multi-population genetic algorithm
 applications, 421
multi-sexual-parents-crossover genetic algorithm, 138

multi-species PSO, 589
multicast flows
 optimization, 358
multicommodity capacitated network design problem, 358
multicriteria optimization, *see* multiobjective optimization
multicriteria scheduling, 60
multidimensional integer knapsack problems, 525
multidisciplinary design optimization, 375
multilayer backpropagation neural network, 401
multilayer perceptron, 384, 432
multilevel sieve, 585
multimodal functions, 177
multimodal optimization, 606
multimodal problems, 22
multinomial Tabu search, 563
multiobjective 0-1 programming problems, 220
multiobjective ant algorithm, 578
multiobjective ant systems, 581
multiobjective cellular genetic algorithm, 475
multiobjective clonal selection algorithm, 608
multiobjective co-operative co-evolutionary genetic algorithm, 158
multiobjective combinatorial optimization, 81, 388, 394, 418, 419, 421, 423, 560, 566, 569
multiobjective distributed Q-learning, 582
multiobjective evolution strategy, 352
 applications, 354, 364
multiobjective flowshop scheduling, 221
multiobjective heuristic search, 22
multiobjective immune system algorithm, 609, 610
multiobjective job shop scheduling, 221
multiobjective knapsack problems, 221
multiobjective minimum spanning tree problem, 579
multiobjective optimization
 definition, 5
 formal definition, 7

metrics, 352
origins of research in, 30
using evolutionary programming, 614
multiobjective optimization problem, 5
  definition, 5
  stages, 515
multiobjective quadratic assignment, 139
multiobjective solid transportation problems, 220
multiperformance optimization, *see* multiobjective optimization
multiple ant colony system for vehicle routing problems with time windows, 579
multiple instruction, multiple data stream, 478
multiple objective decision making, 54
multiple objective genetic algorithm, 63
multiple objective genetic local search, 141, 405, 580
  applications, 426
multiple objective multiple start local search with random weight vectors, 141
multiple objective simulated annealing, 141, 405
multiple objective traveling salesperson problem, 580
multiple resolution multiobjective genetic algorithm, 471
multiple start local search algorithm, 426
  applications, 426
multiplexer, 355, 405
multiplicative aggregation genetic algorithm
  applications, 374
multipoint airfoil design, 383
multipopulation scheme, 412
multisexual genetic algorithm, 138
multistage liquid oxygen pump, 384
multistart Tabu search, 566
multivariable control system design, 364
multivariable controller design
  applications, 364
mutation, 25, 565
  neighborhood search, 359
  reverse, 418
  scramble sublist, 398
  uniform, 370

NACA64A410 airfoil optimization, 385
nadir objective vector, 59
nadir values, 389
NAIADE, *see* novel approach to imprecise assessment and decision environments, 542
NAM
  calibration, 345
  rainfall-runoff model, 345
Nash equilibrium, 383
  point, 376
Nash genetic algorithm, 383
  applications, 376, 391
natural language processing, 406
naval engineering
  evolutionary multiobjective optimization in, 346
naval ship concept design, 346
NCGA, *see* neighborhood cultivation genetic algorithm
NCM, *see* neighborhood constraint method
  applications, 345
NCR boards, 404
neighborhood constraint method, 345
neighborhood cultivation genetic algorithm, 427
neighborhood migration, 362
neighborhood operator, 570
neighborhood search, 359
Nelder-Mead method, 365
Nelder-Mead minimax, 365
NESSY, *see* neural evolutionary strategy system
network design, 356, 357
  evolutionary multiobjective optimization in, 356, 357
network optimization
  evolutionary multiobjective optimization in, 359
network survivability, 439
network topology design, 357
  evolutionary multiobjective optimization in, 357
neural evolutional strategy system, 401

neural networks, 384, 401, 431, 433, 534, 598
  applications, 365, 409
  architecture design
    applications, 399
    evolutionary multiobjective optimization in, 399
    design, 362
  hybrids with evolutionary algorithms, 365
  optimization
    evolutionary multiobjective optimization in, 361
neural programs, 431
neural-fuzzy predictive models, 589
neuroscience, 395
niche count, 313
niche cubicles, 410
niched Pareto genetic algorithm, 63, 94
niched Pareto genetic algorithm 2, 63
niching, 372, 554
  bias, 242
  techniques, 64
NK fitness landscape, 301
  model, 554
No Free-Lunch Theorems, 178, 317, 327, 481, 482
noise, 530
noisy fitness function, 336, 530
non-generational genetic algorithm, 411
non-inferior solutions, 11
non-supported points, 67
non-uniform mutation, 351, 354, 384, 394, 417
nonconvex set
  definition, 9
  example, 9
nondominated solutions, 11, 53
nondominated sorting, 388
nondominated sorting cooperative coevolutionary genetic algorithm, 165
nondominated sorting differential evolution, 599
nondominated sorting evolution strategy
  applications, 354
nondominated sorting genetic algorithm, 63, 91

nondominated sorting genetic algorithm-II, 63
nondominated sorting memetic algorithm
  applications, 382
nondominated vector addition, 266
noninferior surface tracing evolutionary algorithm, 344
nonlinear aggregating function
  applications, 371
  criticism, 70
nonlinear digital circuits
  synthesis, 140
nonlinear FIR filters
  design, 353
  evolutionary multiobjective optimization in, 353
nonlinear programming, 24
nonlinear system identification, 362
normative part, 614
novel approach to imprecise assessment and decision environments, 45
NPGA, see niched Pareto genetic algorithm, 94, 238, 273, 364, 554, 567
  applications, 142, 341, 344–346, 351, 352, 354, 359, 364, 365, 372, 378, 383, 391, 392, 394, 404, 405, 410, 412, 413, 417, 430, 432, 577
NPGA 2, see niched Pareto genetic algorithm 2, 95, 238, see niched Pareto genetic algorithm 2
  applications, 342
NPGA/MOGA hybrid, 365
NSDE, see nondominated sorting differential evolution, 604
NSDE-DCS, 599
NSGA, see nondominated sorting genetic algorithm, 91, 141, 238, 364, 383, 420, 554, 567, 570
  applications, 343, 344, 351, 359, 367, 368, 372, 380, 382, 383, 385, 391, 396, 403, 406, 411–413, 416, 418–420, 429, 462
  compromise programming, 536
  hybridized with Tabu search, 380
  parallel, 380
  pseudo code, 91
  with elitism, 419

NSGA-II, *see* nondominated sorting
    genetic algorithm-II, 93, 126, 137,
    238, 275, 319, 416, 420, 428, 532,
    566, 568, 569, 580, 581, 599–601,
    603, 604, 608, 609, 611, 614
  applications, 344, 352, 355, 366, 375,
    387, 406, 414, 415, 422, 425, 430,
    433, 435, 563
  hybridized, 344
  incorporation of preferences, 536
  parallel version, 472
  pseudo code, 93
  with ϵ elimination, 387
  with tree encoding, 414
NSMA, *see* nondominated sorting
    memetic algorithm
  applications, 382
NSTEA, *see* noninferior surface tracing
    evolutionary algorithm
  applications, 344
nuclear engineering, 554
  evolutionary multiobjective optimization in, 392
nucleotide addition, 607
nugget discovery, 433
  evolutionary multiobjective optimization in, 433
null hypothesis, 249
nurse scheduling, 422, 552
  evolutionary multiobjective optimization in, 422
  Pareto Simulated Annealing in, 552
nylon 6, 412

OBDD, *see* ordered binary decision
    diagram
object recognition, 467
  evolutionary multiobjective optimization in, 402
objective function, 6, 25
  commensurable, 6
  non-commensurable, 6
  space, 6
obstacle avoidance
  evolutionary multiobjective optimization in, 361
OC1, 400
office design
  evolutionary multiobjective optimization in, 377
offline performance, 378
offline routing, 359
offline set, 567
OMOEA, *see* orthogonal multi-objective
    evolutionary algorithm
OMOEA-II, 107
one-point crossover, 351, 354, 364, 366,
    370–373, 375, 377, 383–386, 391,
    399, 404, 409, 417, 421, 426, 429
online performance, 378
operational amplifiers
  synthesis, 352
operations research, 1
  influence on MOEAs, 54
operators, 23
opt-aiNet, 608, 610
optical filter design, 354
  evolutionary multiobjective optimization in, 354
optics, 392
optimal control, 367
optimal pattern matching, 620
optimal rules
  generation of, 400
optimal set, 292
optimization of land grid array solder
    joints, 356
optimization of rocket engine pumps
  evolutionary multiobjective optimization in, 384
optimization techniques
  classification, 21
optimum placement of pumping wells,
    342
optimum pumping schedules, 342
OR, *see* operations research
OR Library, 222
Or-opt mutation, 381
order-based crossover, 418
order-based uniform crossover, 398
ordered binary decision diagram, 411
organic selection, 133
orthogonal least squares, 433
orthogonal multi-objective evolutionary
    algorithm, 106
orthogonal regression, 362
Osyczka and Kundu's approach

applications, 415
othello, 147
outranking methods, 45, 516, 531
overall nondominated vector generation, 358
overall nondominated vector generation ratio, 358
overconstrained problem, 6

P-ACO, *see* Pareto Ant Colony Optimization
p-median problem, 556
paϵ-dominance, 593
packet switched networks, 357
packing
   evolutionary multiobjective optimization in, 427
PACO, *see* population-based ant colony optimization
PAES, *see* Pareto archived evolution strategy, 95, 126, 238, 569, 570, 588, 601, 609, 611
   adaptive grid, 96
   applications, 359, 425, 433
   hybridized with immune inspired principles (I-PAES), 609
pairwise geometric histogram, 402
PAMSSEM, *see* procédure d'aggrégation multicritère de type surclassement de syntèse pour évaluations mixtes
panel meter configuration design, 534
panmictic reproduction, 345
paper industry
   evolutionary multiobjective optimization in, 417
paper mill scheduling, 616
PARADE, *see* Pareto optimal and amalgamated induction for decision trees
ParadisEO, *see* PARAllel and DIStributed Evolving Objects
PARAllel and DIStributed Evolving Objects, 275
parallel archiving issues, 502
parallel evolutionary multiobjective optimization using hypergraphs evolutionary algorithm, 474
parallel genetic algorithm, 383, 385, 386, 391, 395, 417

parallel machine scheduling, 421
parallel MOEAs, 368, 439
   applications, 343, 354, 362, 383, 386
   examples, 460
   objective function decomposition, 449
   task decomposition, 447
parallel MOGA, 362
parallel multi-objective genetic algorithm, 462
parallel multiobjective evolutionary algorithm, 444
parallel niching issues, 500
parallel niching techniques, 512
parallel Pareto Tabu search, 381
parallel recombinative simulated annealing, 350, 556
Parallel Single Front Genetic Algorithm, 464
parallel strength Pareto multi-objective evolutionary algorithm, 472
parallel virtual machine, 387
   library, 387
parallel-series systems, 426
parallelism, 419
parameter free genetic algorithm, 469
parameter-based mutation, 593
parameters selection
   evolutionary multiobjective optimization in, 402
Pareto Ant Colony Optimization, 580
Pareto archived evolution strategy, 63, 95
Pareto compliant, 245, 253
Pareto converging genetic algorithm, 357, 399
   applications, 399
Pareto deme-based selection, 110
Pareto descent method, 330
Pareto Differential Evolution, 395, 596, 598
Pareto dominance, 11, 79, 395
   applications, 406
Pareto elitist-based selection, 111
   applications, 372
Pareto envelope-based selection algorithm, 63
Pareto envelope-based selection algorithm-II, 63
Pareto epsilon Dominance, 17

Pareto epsilon Front, 17
Pareto epsilon model, 17
Pareto epsilon Optimal Set, 17
Pareto epsilon Optimality, 17
Pareto front, 11
 approximation set, 236
 cardinality, 284, 287
 characteristics, 195
 structure, 287
 width distribution, 18
Pareto genetic algorithm, 341
 applications, 392
Pareto list, 562
Pareto noncompliant, 254
Pareto notation, 53
Pareto optimal and amalgamated induction for decision trees, 400
Pareto optimal selection strategy, 399
 applications, 399
Pareto optimal set, 11, 396
 cardinality, 286
 minimal cardinality, 285
 reducing its size, 367
Pareto optimality, 10
Pareto optimum, 10
Pareto partitioning
 applications, 418
Pareto ranking, 63, 79, 241, 308, 392
 applications, 142, 341, 342, 344–346, 349–351, 353, 354, 357, 364, 369, 370, 372, 377, 383–386, 392, 393, 395, 400, 409–413, 417, 422, 423, 426
Pareto reservation strategy
 applications, 421
Pareto selection, 425
 applications, 409
Pareto set distribution method
 applications, 367, 373
Pareto Simulated Annealing, 141, 511, 551, 561, 580
 applications, 552, 570
Pareto stratum-niche cubicle GA
 applications, 410
Pareto Tree Searching Genetic Algorithm, 126
Pareto, Vilfredo, 10, 29
Pareto-based Cost Simulated Annealing, 554

Pareto-based Differential Evolution, 598
Pareto-frontier Differential Evolution, 598
Pareto-rank histograms, 280
Pareto-tournament selection, 406
parsimony, 405
parsing and tagging, 406
partial classification, 433
 evolutionary multiobjective optimization in, 433
partial order, 52
partial order relation, 285
partially mapped crossover, 379, 418–421
partially matched crossover, 525
partially ordered sets, 284, 336
particle swarm inspired evolutionary algorithm, 588
particle swarm optimization, 584
 multiobjective, 355
 with VEGA, 355
path planning, 361
 evolutionary multiobjective optimization in, 361
path relinking, 566, 570, 580
path selection, 570
pattern classification
 evolutionary multiobjective optimization in, 399, 400, 432, 433
pattern move strategy, 563
pattern recognition
 evolutionary multiobjective optimization in, 431
pattern space partition
 applications, 399
 evolutionary multiobjective optimization in, 399
PCGA, 357, see Pareto converging genetic algorithm
 applications, 357, 391
PCSA, see Pareto-based Cost Simulated Annealing
PDE, see Pareto-frontier Differential Evolution, 601
PDMOSA, 555
PDSP, see programmable digital signal processor

penalty function, 114, 343, 357, 371, 373, 393, 399, 561, 599, 608
   exterior, 113
   interior, 114
performance measures, 234
permanent magnet actuators, 351
permutation-based encoding, 359, 419, 421, 422, 425, 426
personal best
   in MOPSOs, 592
perturbation mutation, 380
PESA, see Pareto envelope-based selection algorithm, 395, 566
PESA II, see Pareto envelope-based selection algorithm-II
phenotype, 24
   in evolutionary multiobjective optimization, 54
phenotypic sharing, 242, 312, 343, 345, 351, 378, 384, 385, 400, 419
physical programming, 545
physics
   evolutionary multiobjective optimization in, 391, 392
PID controller design, 364, 365
pinion-rack continuously variable transmission, 603
PISA, see A Platform and Programming Language Interface for Search Algorithms, 617
Pitt approach, 430
Pittsburgh-style machine learning, 432
placement of pumping wells
   optimum placement, 342
placement-based partially exchanging crossover, 428
planar mechanism
   optimization, 371
plane truss
   design, 372
planning
   evolutionary multiobjective optimization in, 377
plant cost optimization, 376
plate design
   evolutionary multiobjective optimization in, 435
pMOEA, see parallel multiobjective evolutionary algorithm

creation options, 490
development issues, 488
hardware and software, 477
implementations, 479
notation, 446
paradigms, 450
   cellular, 458
   coarse-grained, 455
   diffusion, 451, 458
   fine-grained, 458
   hierarchical, 451
   hybrid, 451, 459
   island, 451, 455
   master-slave, 451, 452
scalability, 476
task decomposition, 448
test functions, 480
theory issues, 503
PMOGA, see parallel multi-objective genetic algorithm
PMOHYPEA, see parallel evolutionary multiobjective optimization using hypergraphs evolutionary algorithm
PMX, 419, see partially mapped crossover, 525
Polak-Ribiere algorithm, 394
pollution reduction
   evolutionary multiobjective optimization in, 378
polygamy, 432
polymer extrusion optimization, 391
   evolutionary multiobjective optimization in, 391
polymer reactor optimization, 412
   evolutionary multiobjective optimization in, 412
population, 23, 24
   heuristic, 142
   reinitialization, 423
population size
   adaptive, 365
   setting up, 343
population-based ant colony optimization, 580
population-based multiobjective optimization, 385
portfolio optimization, 429, 552
portfolio selection, 580

poset, 285
position-based crossover, 422
positive variation kernel, 286
POSS, see Pareto optimal selection strategy
pot core problem, 350
potential Pareto regions evolutionary algorithm, 128
power delivery systems, 433
power dispatch problem
    evolutionary multiobjective optimization in, 349
power distribution system planning
    evolutionary multiobjective optimization in, 349
power system stabilizers, 355
PPEX, see placement-based partially exchanging crossover
PPREA, see potential Pareto regions evolutionary algorithm
Prüfer number, 119, 359
    encoding, 380
pre-heuristic, 564
precedence-based crossover, 418
predator fitness
    applications, 367
predator-prey genetic algorithm
    applications, 376
predator-prey model, 147, 314
prediction, 431, 433
    evolutionary multiobjective optimization in, 431
preemptive goal programming, 555
preference articulation, 518
    a posteriori, 518
    a priori, 518, 524, 526, 527
    automation, 526
    compromise programming, 536
    dynamic compromise programming, 536
    ELECTRE I, 531
    fuzzy logic, 524
    goals, 524, 525
    group, 538
    interactive, 518, 519, 523–526
    outranking, 531, 532
    PROMETHEE, 532
    social choice, 545
    unified model, 544

utility function, 527
voting, 545
preference by similarity to ideal solution, 380
preference incorporation, 405, 534
    example, 522
preference management, 417
preference ranking organization method for enrichment evaluations, 45
preference relations, 411, 528
preferences
    a posteriori, 536
    articulation of, 515
    expressed in a qualitative way, 528
    incorporation of, 362
    progressive articulation, 393
preliminary wing subsonic design, 385
pressurized water reactor design, 392, 554
primary-secondary fitness, 373
principal component analysis, 431
priority articulation
    novel applications, 544
probabilistic complete initialization, 101
probabilistic neural network, 432
probabilistic trade-off development method, 47
procédure d'aggrégation multicritère de type surclassement de syntèse pour évaluations mixtes, 45
process planning
    evolutionary multiobjective optimization in, 410
process plants design, 414
production process optimization
    evolutionary multiobjective optimization in, 409
production process planning
    evolutionary multiobjective optimization in, 359
production scheduling, 417
    evolutionary multiobjective optimization in, 417
prognostic models, 395, 474
programmable digital signal processor, 398
progress measure, 265
progressive MOEAs, 71
    criticism, 71

PROMETHEE, *see* preference ranking organization method for enrichment evaluations, 531, 532
applications, 378, 409
PROMETHEE I, 45
PROMETHEE II, 45, 532
proportional selection, 351, 366, 409, 421
prostate implant optimization, 393
protein structure prediction, 609
protein-protein interaction, 407
PROTRADE, *see* probabilistic trade-off development method
PSA, *see* Pareto Simulated Annealing, 570, 580
    applications, 552
    compared to MOTS*, 570
pseudo-Boolean problems, 320
PSFGA, *see* Parallel Single Front Genetic Algorithm
PSO, *see* particle swarm optimization with a linear aggregating function, 587
PSPMEA, *see* parallel strength Pareto multi-objective evolutionary algorithm
pump scheduling, 344
PVM, *see* parallel virtual machine, 468
pygmies and servants, 404

Q-learning, 573, 575, 582
quad trees, 125, 551, 561
    data structure, 580
    in multiobjective optimization, 140
QUALIFLEX, 542
quality indicator, 243
quantitative MOEA performance, 236
quantum cascade laser design, 393
quantum computing, 130
Quine-McCluskey method, 355

radial basis function, 365
    network, 433, 526
radiation therapy
    evolutionary multiobjective optimization in, 393
radiology
    evolutionary multiobjective optimization in, 394

RAE-2822
    optimization, 384
rainfall-runoff model calibration
    evolutionary multiobjective optimization in, 345
random directions multiple objective genetic local search, 140, 405
    applications, 405
random mutation, 344, 420
random nondominated point set, 246
random search, 23
random search with intensification and diversification, 157
random set theory, 249
random walk, 23, 383, 386, 411, 465
random weight approach, 610
random wires mutation, 352
random-objective conjugate gradients, 331
rank-based ant system, 581
rank-histograms, 357
ranking selection, 26, 364, 427
rapid prototyping, 427
RasID, *see* random search with intensification and diversification
Rastrigin's function, 179
Rastrigin steps, 177
RD-MOGLS, *see* random directions multiple objective genetic local search
    applications, 405
reachability, 243
reactive power planning, 593
reactor design, 354, 392
ready meals production scheduling, 417
real-coded genetic algorithm, 343, 384
real-numbers representation, 354
real-time scheduling, 423
    evolutionary multiobjective optimization in, 423
real-world MOEA test functions, 223
REALGO, *see* retrovirus algorithm
recessive genetic information, 119
reciprocal exchange mutation, 420
recombination, 25
reconfiguration of radial distribution networks, 552
rectangular packing

evolutionary multiobjective optimization in, 427
reduced Pareto set algorithm, 391
    applications, 391
reference set, 567, 568
regular MOP, 292
relation, 285
    antireflexive, 285
    antisymmetric, 285
    reflexive, 285
    symmetric, 285
    transitive, 285
relative coverage comparison of two sets, 264
relaxed forms of dominance, 84
reliability-based design, 356
repair algorithm, 114
repair procedure, 394
replacement policy, 349
reproduction, 399
requirements controlled design, 415
resolvability of conflict, 232
resource allocation, 223
resource constrained project scheduling, 570
resource scheduling
    evolutionary multiobjective optimization in, 422
restricted crossover, 400
restricted mating, 85, 316, 395, 426
restricted sharing, 313
retrovirus algorithm, 433
return to base, 553
reverse mutation, 418, 420
ring topology, 362
risk-return trade-offs, 430
RNA polymerase motif, 407
road projects
    scheduling, 379
road train design, 378
    evolutionary multiobjective optimization in, 378
robot arm optimization, 368
    evolutionary multiobjective optimization in, 368
robot behavior, 533
robot grippers design
    applications, 367

evolutionary multiobjective optimization in, 367
robot task and route planning, 473
robotics and control engineering, 544
    evolutionary multiobjective optimization in, 361–369
robust control, 366
robust systems design, 415
Rosenbrock ridge, 177
Rosenbrock's hillclimbing, 365
rotor system design, 383
rotor-blade design, 466
rough sets, 172, 545
    for multicriteria decision analysis, 545
    hybridized with particle swarm optimization, 592
rough sets theory, 602
roulette wheel selection, 25, 349, 359, 373, 377, 391, 399, 404, 409, 418, 423, 432
route-based crossover, 381
routing, 359
    evolutionary multiobjective optimization in, 359
routing school buses, 567
RPS, *see* reduced Pareto set algorithm
rule-based optimization, 383
running time analysis, 320

S metric, 344
S metric selection multi-objective evolutionary algorithm, 130
S-expressions, 399
S-MOGLS, *see* simple multiple objective genetic local search
SA, *see* simulated annealing
satisfiability classes, 411, 530
satisficing
    definition, 525
scalability, 234, 485, 537
scalable MOPs, 186
scaled speedup, 485
scatter search, 559
    applications, 567, 569
    methods, 568
    multiobjective, 359, 566, 567, 569
    reference set, 568
    template, 559, 568

scatter search for multiobjective
    optimization, 568
scatter search procedure for nonlinear
    multiobjective optimization, 569
Schaffer, David, 51
scheduling, 393, 420–422, 524
    evolutionary multiobjective optimization in, 142, 393, 398, 418–423, 570
    road projects, 379
scheduling problems, 222
scientific applications
    evolutionary multiobjective optimization in, 389
scientific method, 234
scramble sublist mutation, 398
search, 515
search space enumeration, 238
search techniques
    classification, 21
second-order attainment function, 249
secondary population, 85
seeding strategy, 142
selection, 25
selective breeding, 364
selective laser sintering, 427
self-adaptation, 354, 565
Self-adaptive Pareto Differential
    Evolution, 598
semiorder, 543
SEMOPS, *see* sequential multiobjective
    problem solving method
sensitivity analysis, 386
sensory action network acquisition
    evolutionary multiobjective optimization in, 361
SEPTOP, *see* solar electric propulsion
    trajectory optimization program
sequence-based crossover, 381
sequencing problems, 410
    evolutionary multiobjective optimization in, 410
sequential multiobjective problem
    solving method, 49
series parallel system redundancy
    allocation problems, 563
set coverage, 359
set covering problem, 556
set/vertex covering problem, 222

SGA-C, 273
Shannon-Wiener index, 389
shape design, 137, 413
    evolutionary multiobjective optimization in, 435
shape design of electromagnetic devices
    evolutionary multiobjective optimization in, 354
shape design of single-phase reactor, 354
shape optimization
    evolutionary multiobjective optimization in, 413
sharing
    equivalence class, 242, 352
    triangular, 351
sharing function, 312
shift mutation, 418, 419
ship design
    evolutionary multiobjective optimization in, 346
shipyard plane cutting shop problem, 346
shuffle crossover, 412
side-constrained MOEA test functions, 187
sigma method, 587
simple multiple objective genetic local
    search, 135
simplex method, 399, 424
simulated annealing, 23, 138, 350, 395, 429, 548, 565, 571, 578, 579
    aggregating function, 552
    and goal programming, 555
    applications, 349, 352, 554
    compared to Tabu search, 571
    coupled to an evolution strategy, 352
    hybridized with Tabu search, 556
    min-max, 552
    multi-objective, 556
    multiobjective, 351, 405, 550, 552
    parallel recombinative, 556
    Pareto dominance, 551
simulated binary crossover, 165
simulated evolution
    applications, 348, 357
single convex Pareto curve, 182
single DM/group methods, 54

single instruction, multiple data stream, 478
single instruction, single data stream, 478
single-machine scheduling problems, 580
single-objective evolutionary algorithms, 446
single-objective optimization
  definition, 4
single-point crossover, 25
site location, 347
  evolutionary multiobjective optimization in, 347
size of the dominated space, 425
Smart, 532
Smith, Adam, 29
SMS-MOEA, see S metric selection multi-objective evolutionary algorithm
social choice, 545
soft selection, 27
software engineering, 440
  evolutionary multiobjective optimization in, 404
software project scheduling, 552
  Pareto Simulated Annealing in, 552
solar electric propulsion trajectory optimization program, 382
somatic point mutation, 607
somatic recombination, 607
sonar signal discrimination, 432
sorting networks, 404
SP-ACO, see Stochastic Pareto-Ant Colony Optimization
space allocation problem
  evolutionary multiobjective optimization in, 565
space vehicle
  optimal control, 367
spacecraft trajectories
  applications, 382
  optimization, 382
spacing, 344
spanning tree problem
  evolutionary multiobjective optimization in, 380
  runtime analysis, 336
spatio-temporal patterns, 395

SPDE, see Self-adaptive Pareto Differential Evolution
SPEA, see strength Pareto evolutionary algorithm, 86, 97, 137, 140, 141, 358, 381, 407, 410, 567, 578, 590, 600, 601, 610, 611
  applications, 344, 393, 394, 398, 413, 415
SPEA2, see strength Pareto evolutionary algorithm 2, 98, 126, 238, 275, 428, 566, 569, 581, 610
  applications, 405, 406, 422
  parallel, 472
speciation, 73
species, 316
specification optimization
  evolutionary multiobjective optimization in, 371
spectroscopic analysis
  evolutionary multiobjective optimization in, 392
spies, 581
split crossover, 381
spread, 569
SPSA, see Stochastic Pareto Simulated Annealing
SSMO, see scatter search for multiobjective optimization, 569
SSPMO, see scatter search procedure for nonlinear multiobjective optimization
Stackelberg equilibrium, 383
Stackelberg genetic algorithm, 383
standard test suite, 186
standardized computational resolution, 237
starch gelatinization
  model for the degree of, 390
static var compensators, 355
statistical tests, 269
STATNET, 430
steady state chemical process system, 390
steady state generational replacement, 354, 357
steady state genetic algorithm, 357, 404, 411, 423, 429
steady state selection, 361
steel plant cost optimization, 376

steel rolling mill, 364
steepest descent methods
    for multiobjective optimization, 60
STEM, see STEP method, 518
STEP method, 48, 518
stepwise regression, 362
stereotactic radio surgery optimization, 393
stochastic combinatorial optimization, 618
stochastic dominance, 542
stochastic hillclimbing, 431
stochastic local search, 172
stochastic matrix, 290
Stochastic Pareto Simulated Annealing, 618
Stochastic Pareto-Ant Colony Optimization, 618
stochastic programming, 24
stochastic remainder selection, 350, 371
stochastic search techniques, 23
stochastic universal selection, 362, 372, 382, 384, 412, 417
stock ranking, 430
    evolutionary multiobjective optimization in, 430
strategic oscillation, 561
strength, 610
    in SPEA, 97
strength Pareto evolutionary algorithm, 63, 97, 98, 139, 344, 398
strength Pareto evolutionary algorithm 2, 63
strict Pareto optimality, 12
strict partial order, 79
string representation, 410
stripes
    to maintain diversity, 590
strong typing, 400
structural control system design
    evolutionary multiobjective optimization in, 374
structural engineering
    evolutionary multiobjective optimization in, 370–374
structural optimization
    evolutionary multiobjective optimization in, 607
structural synthesis of cell-based VLSI circuits
    evolutionary multiobjective optimization in, 349
structure, 24
structured genetic algorithm, 356, 357, 362, 364
struggling crowding genetic algorithm, 105
style points, 399
submarine design
    evolutionary multiobjective optimization in, 346
subpopulations, 419
subset size oriented common features crossover operator, 432
substitute scalarizing functions, 561
subtree exchange crossover, 399
superconducting device, 351
superlinear speedups, 486
surge tank system, 362
surrogate constraint relaxation, 559
surrogate methods, 513
surrogate multipliers, 567
surrogate objective function, 48
surrogate worth trade-off method, 41, 528, 542
swap mutation, 394, 422, 425
swarm metaphor, 585
SYBYL CSEARCH, 391
symbiosis, 144, 149
symbiotic coevolutionary algorithm, 151
symbolic layout compaction, 348
    evolutionary multiobjective optimization in, 348
symbolic regression, 405
symmetric traveling salesperson problem, 581
synthesis of amplifiers
    evolutionary multiobjective optimization in, 352, 353
synthesis of analog filters
    evolutionary multiobjective optimization in, 353
system-level synthesis, 350
    evolutionary multiobjective optimization in, 350

T cells, 605

Index   797

tabu list, 564, 565
Tabu search, 23, 357, 429, 560, 565, 567, 571, 578
   aggregating function, 552
   compared to simulated annealing, 571
   dynamic neighborhood, 564
   hybrid with genetic algorithm, 565
   multiobjective, 560, 561, 566
   multiobjective optimization, 560
   used for mutation, 565
Tabu search for multiobjective combinatorial optimization, 560
takeover time, 513
TAMOCO, *see* Tabu search for multiobjective combinatorial optimization
   applications, 563
tardiness problem, 578
target vector approaches
   applications, 404
task assignment problem, 608
task decomposition, 52
task planning
   applications, 368
   evolutionary multiobjective optimization in, 368
Tchebycheff goal programming, 59
Tchebycheff scalarization, 407
Tchebycheff scalarizing function, 141
Tchebycheff weighting GA
   applications, 342, 351
Tchebycheff weights, 140
TDGA, *see* thermodynamical genetic algorithm, 130
technique for order preference by similarity to ideal solution, 380
telecommunications, 356–359
   evolutionary multiobjective optimization in, 356–359
test functions, 176
   controlling difficulty, 192
   side-constrained, 187
   unconstrained, 182
test suite
   design, 178
   functions, 182
   guidelines, 178
testing, 234
textile industry, 409

textile machine guide construction and manufacture, 468
texture filtering
   evolutionary multiobjective optimization in, 401
The Wealth of Nations, 29
thermal systems design, 377
thermodynamic clustering, 579
thermodynamical genetic algorithm, 63, 130
thinned antenna arrays
   design, 351
threads, 509
three-colored NCR-boards
   evolutionary multiobjective optimization in, 404
three-dimensional bin packing, 427
three-dimensional cutting problem, 413
three-phase induction motor
   design, 352
thumbprint specification, 365
THUNDER, 404
tic tac toe, 147
time and frequency domain design unification, 365
time-optimal control, 367
time-optimal trajectories
   applications, 367
time-series prediction, 431
   evolutionary multiobjective optimization in, 430
time-tabling of classes, 422
   evolutionary multiobjective optimization in, 422
topology optimization
   evolutionary multiobjective optimization in, 371, 372
   of a network, 357
TOPSIS, *see* technique for order preference by similarity to ideal solution
toroidal grid, 383, 406
toroidal topology, 386
tournament selection, 26, 241, 342, 355, 358, 368, 375, 377, 379, 400, 404, 414, 420, 426, 429, 430
   based on nondominance, 405
tournament slot sharing, 413
trade-off method, 41, 47

traffic route planning
  evolutionary multiobjective optimization in, 378
traffic signal controller design
  evolutionary multiobjective optimization in, 379
trajectory tracking
  evolutionary multiobjective optimization in, 367
transition probability criteria, 554
transitivity of indifference, 543
transonic airfoil design, 384
transonic flow conditions
  wing shape optimization, 385
transonic wing design, 384
transport engineering
  evolutionary multiobjective optimization in, 378–380
transportation plans, 377
transportation problem
  evolutionary multiobjective optimization in, 380
transportation system, 465
trasduction operator, 401
traveling salesperson problem, 140, 222, 578
  evolutionary multiobjective optimization in, 405
  symmetric, 405
treatment schedules
  optimization, 393
tree crossover, 358
tree encoding
  in the NSGA-II, 414
triangular probability distribution, 345
truck packing
  evolutionary multiobjective optimization in, 427
truncated archive, 588
truss design
  evolutionary multiobjective optimization in, 371
truss optimization
  evolutionary multiobjective optimization in, 370–372
TSP, see traveling salesperson problem
TSP library of sample instances, 222
TSPLIB, see TSP library of sample instances

Tucker, Albert W., 30
turbojet engines, 387
turbomachinery airfoils
  design, 372
turbulence operator, 586, 587
two and a half dimensional packing
  evolutionary multiobjective optimization in, 427
two dimensional packing
  evolutionary multiobjective optimization in, 427
two phase local search algorithm, 429
two set coverage, 264, 344
two-branch tournament genetic algorithm
  applications, 370, 382, 383
two-dimensional cutting problem, 414
  evolutionary multiobjective optimization in, 414
two-dimensional integer representation, 404
two-link manipulator
  path control, 367
  trajectory control, 367
two-phases method, 420
two-point crossover, 349, 351, 352, 354, 355, 365, 377, 383, 394, 404, 418, 419, 424

UCI machine learning repository, 396, 433, 598
UIFDO, see unknown input fault detection observer
ULTIC controller design, 365
UMMEA, see unified model for multi-objective evolutionary algorithms
unary quality indicators, 236
unconstrained nondominated archive, 591
unified model for multi-objective evolutionary algorithms, 378
uniform crossover, 344, 350, 351, 354, 357, 359, 380, 389, 398, 407, 414, 417, 423, 425, 427, 429, 432
uniform mutation, 351, 352, 354, 361, 372, 383, 391, 394, 417
uniformity
  along the Pareto front, 195

unimodal functions, 177
unimodal normal distribution crossover, 399
uninformed search, 22
unitation function, 220
unknown input fault detection observer, 362
unrestricted migration, 362
unsupervised feature selection, 610
unsupervised learning, 403
urban planning, 377
    evolutionary multiobjective optimization in, 377
utility function, 40, 41, 370, 429, 517, 524, 526, 527
    advantages, 528
    disadvantages, 528
utopian point, 560

VAIS, *see* vector artificial immune system
variable length encoding, 361
variable length genetic algorithm, 389, 432
variable linkage, 604
vector artificial immune system, 610
vector evaluated differential evolution for multi-objective optimization, 599
vector evaluated genetic algorithm, 63, 66, 72, 74
vector evaluated particle swarm, 586
vector immune system, 611
vector maximum problem, 30
vector optimization, *see* multiobjective optimization
vector optimized evolution strategy, 63, 75
VEDE, *see* vector evaluated differential evolution for multi-objective optimization
VEGA, 51, *see* vector evaluated genetic algorithm, 66, 74, 471, 586, 599
    applications, 142, 341, 342, 344, 345, 353, 355, 368, 385, 400, 408, 410, 418, 421, 427, 432, 577
vehicle routing problem, 222, 577
    evolutionary multiobjective optimization in, 380

vehicle scheduling
    evolutionary multiobjective optimization in, 379
    optimize the planning of a, 379
vehicle water-pumps design, 412
    evolutionary multiobjective optimization in, 412
ventricle 3D reconstruction problem
    evolutionary multiobjective optimization in, 395
VEPSO, *see* vector evaluated particle swarm
version space, 613
very large scale integration, 411
vibrating platform design, 372, 373
    evolutionary multiobjective optimization in, 373
vibration analysis, 432
virtual subpopulation genetic algorithm, 386, 467
VIS, *see* vector immune system
VLSI
    circuits
        structural synthesis, 349
    design, 414
        evolutionary multiobjective optimization in, 411
    layout, 348
        evolutionary multiobjective optimization in, 348
        macro-cell layout generation problems, 467
VOES, *see* vector optimized evolution strategy
voltage reference circuit design, 349
    evolutionary multiobjective optimization in, 349
von Neumann, John, 29, 30
voting schemes, 545
VSGA, *see* virtual subpopulation genetic algorithm
    applications, 386

Walras, Léon, 29
war simulation
    evolutionary multiobjective optimization in, 404
warehouse management, 426

evolutionary multiobjective optimization in, 426
water distribution irrigation network optimization, 577
water distribution networks, 344
  improvements, 344
water quality control
  evolutionary multiobjective optimization in, 343
water quality management problem
  evolutionary multiobjective optimization in, 344
water-using systems, 583
WBGA, *see* weight-based genetic algorithm, *see* weight-based genetic algorithm
weak dominance, 517
weak Pareto optimality, 12
Weierstrass function, 177
weight-based genetic algorithm, 63, 75
weighted average ranking
  applications, 419
weighted goal programming, 59
weighted goal programming genetic algorithm
  applications, 430
weighted min-max genetic algorithm
  applications, 368, 372, 373, 413
weighted Tchebycheff utility function, 405, 426
weighting factor method, 599

weighting method, 46
WHORL, 396
Wilcoxon Rank-Sum Test, 270
Wilcoxon signed-rank test, 420
Williams and Otto chemical plant, 391
wind turbine blades, 375
wing design, 384, 386
  evolutionary multiobjective optimization in, 386
wing planform design, 384
wing shape optimization for transonic flow conditions, 385
wing-box structure
  optimization, 372
wireless local area network design
  evolutionary multiobjective optimization in, 356
WLAN, *see* wireless local area network design, 356
word classifier, 403
worker allocation in radiological facilities, 394
workforce scheduling, 422
working conditions of a press optimization, 408

X-ray plasma spectroscopy, 463

Zadeh, Lofti, 31
ZDT test problems, 611
Zeleny's ideal point, 570

If you have any concerns about our products,
you can contact us on
ProductSafety@springernature.com

In case Publisher is established outside the EU,
the EU authorized representative is:
**Springer Nature Customer Service Center GmbH
Europaplatz 3, 69115 Heidelberg, Germany**

Printed by Libri Plureos GmbH
in Hamburg, Germany